Macrophomina Phaseolina

Macrophomina Phaseolina

Ecobiology, Pathology and Management

Edited by

Pankaj Kumar

Associate Professor, Department of Microbiology, Dolphin (PG) Institute of Biomedical and Natural Science, Dehradun, Uttarakhand, India

Ramesh Chandra Dubey

Professor, Department of Botany and Microbiology, Gurukul Kangri Vishwavidyalaya, Haridwar, Uttarakhand, India

ELSEVIER

ACADEMIC PRESS

An imprint of Elsevier

Academic Press is an imprint of Elsevier
125 London Wall, London EC2Y 5AS, United Kingdom
525 B Street, Suite 1650, San Diego, CA 92101, United States
50 Hampshire Street, 5th Floor, Cambridge, MA 02139, United States
The Boulevard, Langford Lane, Kidlington, Oxford OX5 1GB, United Kingdom

ISBN: 978-0-443-15443-0

For information on all Academic Press publications visit our website at
https://www.elsevier.com/books-and-journals

Publisher: Nikki P. Levy
Acquisitions Editor: Nancy J. Maragioglio
Editorial Project Manager: Kathrine Esten
Production Project Manager: Selvaraj Raviraj
Cover Designer: Miles Hitchen

Typeset by TNQ Technologies

Contents

Section II
Resistance and pathogenicity

Contributors

Nisha Agrawal, Division of Crop Improvement, ICAR-Indian Institute of Soybean Research (IISR), Indore, Madhya Pradesh, India

Mohd. Mogees Ansari, Microbial Technology Division, Council of Scientific and Industrial Research-National Botanical Research Institute (CSIR-NBRI), Lucknow, Uttar Pradesh, India

Y.M.A.Y. Bandara, Department of Plant Pathology and Environmental Microbiology, Penn State University, University Park, PA, United States

Mohano Behara, Department of Microbiology, PVKN Government Degree College (A) Chittoor, Andhra Pradesh, India

Shiuly Bhowmick, Microbial Technology Division, Council of Scientific and Industrial Research-National Botanical Research Institute (CSIR-NBRI), Lucknow, Uttar Pradesh, India

Md. Khurshed Alam Bhuiyan, Department of Plant Pathology, Bangabandhu Sheikh Mujibur Rahman Agricultural University, Gazipur, Bangladesh

Sarita Bhutada, Department of Microbiology, Sanjivani Arts, Commerce and Science College Kopargaon, Maharashtra, India

Györgyi Biró, University of Debrecen, Faculty of Agricultural and Food Sciences, Environmental Management Plant Protection Institute, Debrecen, Hungary

Ankita Chatterjee, Department of Biotechnology, School of Applied Sciences, REVA University, Bangalore, Karnataka, India

Harshvardhan Chaudhary, Department of Botany and Microbiology, Gurukula Kangri (Deemed to be University), Haridwar, Uttarakhand, India

Puneet Singh Chauhan, Microbial Technology Division, Council of Scientific and Industrial Research-National Botanical Research Institute (CSIR-NBRI), Lucknow, Uttar Pradesh, India

Anam Choudhary, Department of Plant Protection, Faculty of Agricultural Sciences, A.M.U., Aligarh, Uttar Pradesh, India

Ndiaga Cisse, Institut Sénégalais de Recherches Agricoles/ Centre National de Recherches Agronomiques BP 53, Bambey, Senegal

Kitti Csüllög, University of Debrecen, Faculty of Agricultural and Food Sciences, Environmental Management Plant Protection Institute, Debrecen, Hungary

András Csótó, University of Debrecen, Faculty of Agricultural and Food Sciences, Environmental Management Plant Protection Institute, Debrecen, Hungary

Samadhan Bhagwan Dahikar, Department of Microbiology, Sanjivani Arts, Commerce and Science College, Kopargaon, Maharashtra, India

Ramesh Chandra Dubey, Department of Botany and Microbiology, Gurukula Kangri (Deemed to be University), Haridwar, Uttarakhand, India

Himani Gautam, Department of Entomology, Dr. Yashwant Singh Parmar University of Horticulture and Forestry, Solan, Himachal Pradesh, India

Ragini Gothalwal, Department of Biotechnology and Bioinformatics Centre Barkatullah Vishwavidhyalaya, Bhopal, Madhya Pradesh, India

Ramesh Nath Gupta, Department of Plant Pathology, Bihar Agricultural University, Bhagalpur, Bihar, India

N. Hari Prasad, Department of Microbiology, PVKN Government Degree College (A) Chittoor, Andhra Pradesh, India

S.I. Harlapur, ICAR-All India Coordinated Maize Improvement Project, University of Agricultural Sciences, Dharwad, India

K.S. Hooda, ICAR-National Bureau of Plant Genetic Resources, Pusa Campus, New Delhi, India

Sapna Jaiswal, Division of Crop Protection, ICAR-Indian Institute of Soybean Research (IISR), Indore, Madhya Pradesh, India; Division of Plant Pathology, Jawaharlal Nehru Krishi Vishwa Vidyalaya, Jabalpur, Madhya Pradesh, India

Arshad Javaid, Department of Plant Pathology, Faculty of Agricultural Sciences, University of the Punjab, Lahore, Pakistan

K. Jaya Chandra, Department of Microbiology, PVKN Government Degree College (A) Chittoor, Andhra Pradesh, India

Kriti Joshi, Department of Biotechnology and Microbiology SJHS Gujarati Innovative College of Commerce and Science, Indore, Madhya Pradesh, India

Naveen Joshi, Department of Biotechnology and Microbiology SJHS Gujarati Innovative College of Commerce and Science, Indore, Madhya Pradesh, India

Srishti Kar, Microbial Technology Division, Council of Scientific and Industrial Research-National Botanical Research Institute (CSIR-NBRI), Lucknow, Uttar Pradesh, India

Iqra Haider Khan, Department of Plant Pathology, Faculty of Agricultural Sciences, University of the Punjab, Lahore, Pakistan

Manzoor Raiees Khan, Section of Plant Pathology and Nematology, Department of Botany, Aligarh Muslim University, Aligarh, Uttar Pradesh, India

M.K. Khokhar, ICAR-National Research Centre for Integrated Pest Management, Pusa Campus, New Delhi, India

Sandeep Kumar, Department of Botany and Microbiology, Gurukula Kangri (Deemed to be University), Haridwar, Uttarakhand, India

Narendra Kumar, Department of Agriculture, Doon (PG) College of Agriculture Science and Technology, Dehradun, Uttarakhand, India

Pankaj Kumar, Department of Microbiology, Dolphin (PG) Institute of Biomedical and Natural Sciences, Dehradun, Uttarakhand, India

Sanjeev Kumar, Division of Crop Protection, ICAR-Indian Institute of Soybean Research (IISR), Indore, Madhya Pradesh, India

Garima Kumari, Forest Ecology and Climate Change Division, ForestResearch Institute, Dehradun, Uttarakhand, India

Giriraj Kumawat, Division of Crop Improvement, ICAR-Indian Institute of Soybean Research (IISR), Indore, Madhya Pradesh, India

Christopher R. Little, Department of Plant Pathology, Kansas State University, Manhattan, KS, United States

Hemant Singh Maheshwari, Division of Crop Improvement, ICAR-Indian Institute of Soybean Research (IISR), Indore, Madhya Pradesh, India

Basudev Majhi, Microbial Technology Division, Council of Scientific and Industrial Research-National Botanical Research Institute (CSIR-NBRI), Lucknow, Uttar Pradesh, India

Saloni Mandloi, Division of Crop Protection, ICAR-Indian Institute of Soybean Research (IISR), Indore, Madhya Pradesh, India; Division of Plant Pathology, Jawaharlal Nehru Krishi Vishwa Vidyalaya, Jabalpur, Madhya Pradesh, India

Manish Mathur, ICAR- Central Arid Zone Research Institute, Jodhpur, Rajasthan, India

Preet Mathur, Computer Science Department, Jodhpur Institute of Engineering and Technology, Jodhpur, Rajasthan, India

Sankalp Misra, Faculty of Biosciences, Institue of Bioscience and Technology, Shri Ramswaroop Memorial University, Lucknow-Deva Road, Barabanki, Uttar Pradesh, India

Nasreen Musheer, School of Agriculture, Glocal University, Saharanpur, Uttar Pradesh, India

Rachana Dattatray Nagarkar, Department of Microbiology, Sanjivani Arts, Commerce and Science College, Kopargaon, Maharashtra, India

Antal Nagy, University of Debrecen, Faculty of Agricultural and Food Sciences, Environmental Management Plant Protection Institute, Debrecen, Hungary

V. Nataraj, Division of Crop Improvement, ICAR-Indian Institute of Soybean Research (IISR), Indore, Madhya Pradesh, India

Mbaye Ndiaye, Centre Regional Aghrymet/ Expert in Plant Pathology, Saly, Senegal

Ram Niwas, Department of Plant Pathology, Bihar Agricultural University, Bhagalpur, Bihar, India

Afsana Noor, Department of Plant Pathology, Kansas State University, Manhattan, KS, United States

Vibha Pandey, Division of Plant Pathology, Jawaharlal Nehru Krishi Vishwa Vidyalaya, Jabalpur, Madhya Pradesh, India

Kriti Pathak, Division of Crop Protection, ICAR-Indian Institute of Soybean Research (IISR), Indore, Madhya Pradesh, India

Ramasamy Perumal, Agricultural Research Center - Hays, Kansas State University, Hays, KS, United States

Alexandra Piti, University of Debrecen, Faculty of Agricultural and Food Sciences, Environmental Management Plant Protection Institute, Debrecen, Hungary

Siva Raagini Polakala, Department of Microbiology, PVKN Government Degree College (A) Chittoor, Andhra Pradesh, India

Barsha Poudel, Centre for Crop Health, Institute of Life Sciences and the Environment, University of Southern Queensland, Darling Heights, QLD, Australia

Hema Prashad, Department of Entomology, Dr. Yashwant Singh Parmar University of Horticulture and Forestry, Solan, Himachal Pradesh, India

Gauri Rai, Microbial Technology Division, Council of Scientific and Industrial Research-National Botanical Research Institute (CSIR-NBRI), Lucknow, Uttar Pradesh, India

Laxman Singh Rajput, Division of Crop Protection, ICAR-Indian Institute of Soybean Research (IISR), Indore, Madhya Pradesh, India; Division of Plant Improvement and Pest Management, ICAR-Central Arid Zone Research Institute, Jodhpur, Rajasthan, India

M.B. Ratnaparkhe, Division of Crop Improvement, ICAR-Indian Institute of Soybean Research (IISR), Indore, Madhya Pradesh, India

Sunita Rawat, Plant Physiology Discipline, Genetics and Tree Improvement Division, Forest Research Institute, Dehradun, Uttarakhand, India

K. Shiva Raghava Rayalu, Department of Microbiology, PVKN Government Degree College (A) Chittoor, Andhra Pradesh, India

Péter Riczu, University of Debrecen, Faculty of Agricultural and Food Sciences, Environmental Management Plant Protection Institute, Debrecen, Hungary

Md. Tanbir Rubayet, Department of Plant Pathology, Bangabandhu Sheikh Mujibur Rahman Agricultural University, Gazipur, Bangladesh

Sabiha Saeed, Department of Plant Protection, Faculty of Agricultural Sciences, A.M.U., Aligarh, Uttar Pradesh, India

Mame P. Sarr, Institut Sénégalais de Recherches Agricoles/Centre National de Recherches Agronomiques BP 53, Bambey, Senegal

A. Sasi Kala, Department of Microbiology, PVKN Government Degree College (A) Chittoor, Andhra Pradesh, India

Adikshita Sharma, Department of Plant Pathology, Yashwant Singh Parmar University of Horticulture and Forestry, Solan, Himachal Pradesh, India

Monica Sharma, Department of Plant Pathology, Yashwant Singh Parmar University of Horticulture and Forestry, Hamirpur, Himachal Pradesh, India

Kishor Sharma, Department of Plant Pathology, Yashwant Singh Parmar University of Horticulture and Forestry, Mandi, Himachal Pradesh, India

Mukta Shinde, Department of Zoology, Sanjivani Arts, Commerce and Science College Kopargaon, Maharashtra, India

M. Shivakumar, Division of Crop Improvement, ICAR-Indian Institute of Soybean Research (IISR), Indore, Madhya Pradesh, India

Zaki Anwar Siddiqui, Section of Plant Pathology and Nematology, Department of Botany, Aligarh Muslim University, Aligarh, Uttar Pradesh, India

Shubhra Singh, Department of Entomology, Dr. Yashwant Singh Parmar University of Horticulture and Forestry, Solan, Himachal Pradesh, India

Tanya Singh, Microbial Technology Division, Council of Scientific and Industrial Research-National Botanical Research Institute (CSIR-NBRI), Lucknow, Uttar Pradesh, India

Ravita Tadiya, Plant Physiology Discipline, Genetics and Tree Improvement Division, Forest Research Institute, Dehradun, Uttarakhand, India

Gábor Tarcali, University of Debrecen, Faculty of Agricultural and Food Sciences, Environmental Management Plant Protection Institute, Debrecen, Hungary

Tesfaye Tesso, Department of Agronomy, Kansas State University, Manhattan, KS, United States

Gyula Tóth, University of Debrecen, Faculty of Agricultural and Food Sciences, Environmental Management Plant Protection Institute, Debrecen, Hungary

Niloofar Vaghefi, Faculty of Science, University of Melbourne, Parkville, VIC, Australia

Cintia Vartek, University of Debrecen, Faculty of Agricultural and Food Sciences, Environmental Management Plant Protection Institute, Debrecen, Hungary

Shulbhi Verma, Department of Plant Breeding and Genetics, C.P College of Agriculture, S.D Agricultural University, S.K Nagar, Gujarat, India

Anshika Verma, Microbial Technology Division, Council of Scientific and Industrial Research-National Botanical Research Institute (CSIR-NBRI), Lucknow, Uttar Pradesh, India

Vineet Kumar Vishnoi, Department of Botany and Microbiology, Gurukula Kangri (Deemed to be University), Haridwar, Uttarakhand, India

About the authors

Dr. Pankaj Kumar completed his master's and doctorate degrees at the Gurukula Kangri Vishwavidyalaya, Haridwar, Uttarakhand (India). He has more than 10 years of teaching and research experience. His research interest and expertise include the areas of rhizosphere microbiology, bacterial characterization, and biofertilizers. He has published several research papers in national and international journals. He is also a coeditor of the book *Rhizosphere Engineering* (Elsever, 2022) with Prof. R. C. Dubey. He is also the coauthor of *An Objective Compendium on Food Science* (Brillion publishing, 2022). He is a life member of the Association of Microbiology of India (AMI), New Delhi, and Indian Science Congress, Kolkata, West Bengal, and serves as a reviewer and editorial board member of national and international journals. Currently, Dr. Pankaj is affiliated with the Department of Microbiology at the Dolphin (PG) Institute of Biomedical and Natural Science, Dehradun, Uttarakhand, India.

Dr. Ramesh Chandra Dubey is a Professor, Dean-Research and Dean-Faculty of Medical Science and Health, and also has been the former Head of the Department of Botany and Microbiology at Gurukula Kangri Vishwavidyalaya, Haridwar, Uttarakhand, India. He obtained his M.Sc. and Ph.D. degrees from the internationally renowned Banaras Hindu University, Varanasi, India. He has more than 35 years of teaching and research experiences in the field of agricultural microbiology. He has published over 242 research papers in national and international scientific journals. His Google Citation Index is above 5138 till date. He has authored 7 books and coedited 10 books on varied aspects, such as soil and agriculture, microbiology, industrial microbiology, and biotechnology. In 2022, Dr. Dubey published a unique book entitled *Vedic Microbiology* with Motilal Banarsidass International, New Delhi (India).

Dr. Dubey is a life member and Fellow of the Indian Botanical Society, Indian Phytopathological Society, and the International Society for Conservation of Natural Resources. He has also previously served as the Councilor of the Indian Botanical Society for 3 years (2011–2013) and Institutional Animal Ethical Committee (IAEC) for 6 years (2015–2021).

Preface

The idea for publishing this edited book (*Macrophomina phaseolina*: Ecobiology, Pathology and Management) was given by Prof Ramesh Chandra Dubey, who earned Ph.D. for conducting research on this phytopathogen. *M. phaseolina* infects over 700 plant species. Various diseases of crop plants are a serious threat to global food production. We hope this book will be a very pragmatic resource for information about the fungus and its control measures. This book includes articles contributed by the researchers working on *Macrophomina* across the globe.

Macrophomina species are necrotrophic, seed-borne, and soil-borne in nature that were first introduced in 1923. Until recently, *Macrophomina* was believed to be monotypic with *M. phaseolina.* However, more recent multilocus phylogenetic analyses revealed the presence of several cryptic species, namely, *M.euphorbiicola*, *M. pseudophaseolina*, *M. tecta*, and *M. vaccinii.* Charcoal rot disease caused by *M. phaseolina* severely threatens corps production of soybean, sorghum, and several other plants. Plant-parasitic nematodes (PPNs) interact with other disease-causing pathogens resulting in the disease complexes and yield loss. Charcoal rot of maize incited by soil-borne pathogen *M. phaseolina* causes huge economic losses in most of the maize growing areas of country.

The main characteristic of *M. phaseolina* is wilting; the microsclerotia plug the vessels of xylem and chock the uptake of water generally and spread in upward direction, and symptoms appear near the flowering stage of the crops. *M. phaseolina* secretes a wide range of cell wall—degrading enzymes and toxins that aid in its pathogenesis and causes root rot, stem rot, collar rot, crown rot, and seedling blight. Although several research studies have been carried out on *M. phaseolina* infection in variety of crops, yet still it remains a persistent phytopathogen causing huge economic loss in the agriculture field.

Culture-based morphological techniques are unable to identify *M. phaseolina*; hence, nucleic acid—based molecular approaches, such as PCR-based technique, are very specific and accurate. ITS, fluorescence in situ hybridization (FISH), whole genome sequencing, genetic diversity analysis with PCR, RAPD, RFLP, metagenomics, and other molecular techniques can detect several isolates in a single attempt. Ultimately, the molecular tools help in implementing the appropriate disease management strategies against *M. phaseolina*.

M.phaseolina infects many plants or crops all over the world while survives for about 15 years in the soil. There is no generalization of predisposing factors associated with this species. The ecological niche modeling of this species has been carried out using bioclimatic and nonbioclimatic variables within India. Based on the results of these tools, four types of niches were found and mapped in India, for example, the fundamental niches, the realized niche, the potential niche, and the tolerance niche. For the first time, gains and losses of possible habitat for this species were documented across time.

M. phaseolina can also survive on many other nonhost plants due to its plurivorous nature. It survives in soil as mycelia, conidia, pycnidia, and microsclerotia. Upon decomposition of host tissues, a crucial structure called microsclerotia is released into the soil matrix. The disease's severity depends on survival of sclerotia in soil that can be controlled by various biotic, abiotic, and crop management practices. Biological control, soil solarization, bio-solarization, organic amendments, crop rotations, fumigation, uses of antibiotics, fungicides, herbicides, and plant extracts have been attempted for control survival and longevity and understanding the population dynamics of the pathogen. Resistance in some genotypes was also found to be associated with drought tolerance.

Due to its heterokaryotic mycelium, there are varied asexual sub-phases along with phenotypic, geographical, and genetic variations, making it a challenging pathogen for designing effective and long-lasting disease management. *M. phaseolina* has also been reported to be managed by various cultural practices like tillage, mixed cropping, and changing the sowing dates. In addition to essential oils, plant metabolites and plant defense elicitors are the promising antifungal substances that are inexpensive to produce and acquire easily. Various species of *Trichoderma* have also been reported for the sustainable management of *M. phaseolina*. Hence, the integrated pest management is an effective way of controlling the phytopathogens without affecting the natural conditions around the crops.

Biosurfactants are versatile, environmentally friendly, surface-active biomolecules that may be used in a variety of sectors. Biosurfactants are synthesized by microorganisms like bacteria, yeasts, and fungi that act as environment-friendly surfactants. Biosurfactants have agricultural application including their use to get rid of plant pathogens and improve the accessibility of nutrients for beneficial plant-associated microbes.

However, the control of the disease is challenging as no effective management practice is available. To overcome this problem, nanoparticles can be used as an alternative method for the management of disease as they are effective and ecofriendly method. A large number of nanoparticles are synthesized using physical, chemical, and biological method. Various metallic nanoparticles, such as Ag, Cu, Zn, ZnO, GO, TiO_2, chitosan, Cu-chitosan, and nanoform of fungicides trifloxystrobin 25% + tebuconazole 50%, have been used to control the disease under lab and greenhouse conditions.

Yield losses caused by charcoal rot, current management practices, and guidelines for the farmers and breeders to develop and produce profitable crops have been discussed in details. This will equip the essential information to farmers with disease management strategies against charcoal rot disease.

We would like to express thanks to our parents, teachers, and all the contributors/authors for their blessings and kind cooperation. The contributors did a great job in presenting a timely overview on topics of *Macrophomina* in their special fields of expertise. We also extend our sincere gratitude to Dr. Nancy Maragioglio, Ms. Kathrine Esten, and the entire team of ELSEVIER for their kind support and cooperation in timely bringing out the book.

Pankaj Kumar
Ramesh Chandra Dubey
Dehradun, Haridwar
Uttarakhand, India

Section I

The pathogen and disease

Chapter 1

Taxonomy of *Macrophomina*—traditional to molecular approaches

Barsha Poudel[1] and Niloofar Vaghefi[2]
[1]*Centre for Crop Health, Institute of Life Sciences and the Environment, University of Southern Queensland, Darling Heights, QLD, Australia;*
[2]*Faculty of Science, University of Melbourne, Parkville, VIC, Australia*

1.1 Introduction

The genus *Macrophomina* accommodates globally distributed seed-borne and soil-borne plant pathogenic species, causing charcoal rot and dry root rot on more than 500 plant species. *Macrophomina* species are well-known necrotrophic plant pathogens, carrying a suite of phytotoxins that make them some of the most destructive pathogens of agricultural and horticultural crops worldwide (Islam et al., 2012). Symptoms vary from seed rot, seedling blight, and damping off to stem, crown and root rot, and plant death (Maholay & Sohi, 1983; Adorada et al., 2018). Due to the production of microsclerotia with the ability to survive in soil for long periods of time (Baggio et al., 2019; Short et al., 1980; Zveibil et al., 2012), *Macrophomina* species are notoriously difficult and expensive to manage in agricultural systems.

Despite their global importance and the plethora of genetic and genomic studies on *Macrophomina* species (Burkhardt et al., 2019; Poudel et al., 2022; Purushotham et al., 2020; Radadiya et al., 2021; TančićŽivanov et al., 2019; Viejobueno et al., 2022; Wingfield et al., 2022), many aspects of their biology and pathogenesis remain little understood. Until 2014, *Macrophomina* was considered as a monotypic genus, with *M. phaseolina* as the only known *Macrophomina* species. However, recent molecular studies have shed light on cryptic species within this genus occurring on the same host, with almost identical morphological features (Sarr et al., 2014; Poudel et al., 2021c).

Population genetic studies have so far mostly been focused on *M. phaseolina*, reporting unexpected high genotypic diversity and genetic heterogeneity. While some studies reported specialization of *M. phaseolina* genotypes on certain hosts or association with geographical locations, others failed to identify clear relationships between *M. phaseolina* genotypes and phenotypes or host of origin (Khan et al., 2017; TančićŽivanov et al., 2019; Poudel et al., 2021a, 2021b; Viejobueno et al., 2022). It is possible that some population genetic studies conducted before the molecular taxonomic resolution of the genus in 2014 included multiple cryptic *Macrophomina* species, where sub-clustering of isolates may have represented the unknown cryptic species. This chapter outlines the historical and modern taxonomy of the genus *Macrophomina*, current knowledge on their geographical distribution and host range, and implications for future research.

1.2 History of *Macrophomina phaseolina* (Tassi) Goid

Macrophomina belongs to the family Botryosphaeriaceae within the class Dothideomycetes, which is the largest and most diverse class of Ascomycetes. Until recently, *Macrophomina* was considered as a monotypic genus, with the type species *Macrophomina phaseolina* (Tassi) Goid. The genus *Macrophomina* was originally found by Petrak upon describing *M. philippinensis* with the type specimen obtained from stems of *Sesamum orientale* from the Philippines (Petrak, 1923). Subsequently, the taxonomy of *Macrophomina* was revised several times over the years.

Ashby (Ashby, 1927) examined the type specimen of *M. philippinensis* and several other species, namely, *Dothiorella cajani* Syd. and Butl., *Macrophoma cajani* Syd. and Butl., *Macrophoma corchori* Saw., *Macrophoma sesami* Saw., *Macrophoma phaseoli* Maubl., *Rhizoctonia bataticola* (Taub.) Butl., *Rhizoctonia lamellifera* Small., and *Sclerotium bataticola* Taub., and synonymized these under the name *M. phaseoli* (Maubl.) Ashby, designating *Macrophoma phaseoli*

Macrophomina Phaseolina. https://doi.org/10.1016/B978-0-443-15443-0.00022-X

Maubl. Maublanc (1905) as the basionym (Ashby, 1927). The name *M. phaseoli* was frequently used in literature until 1947, when Goidànich revised the taxonomy of the genus and indicated that *Macrophoma phaseolina* Tassi (Tassi, 1901) was an earlier basionym, thus, proposing the combination *M. phaseolina* (Tassi) Goid (Goidànich, 1947).

Later, the name *Tiarosporella phaseolina* (Tassi) van der Aa was introduced by Von Arx (1981), who reduced *Macrophomina* to a synonym of *Tiarosporella* Höhn (Crous et al., 2006). Crous et al. (2006) induced sporulation in several *M. phaseolina* strains on pine needle agar, confirming that *M. phaseolina* can produce conidiospores with apical mucoid appendages similar to the genus *Tiarosporella*. However, *M. phaseolina* was distinguished by percurrently proliferating conidiogenous cells and conidiospores that turn to dark brown in color at maturity. Therefore, the genus *Macrophomina* and the name *M. phaseolina* (Tassi) Goid. were retained, and an amended description was proposed by Crous et al. (2006), which was further supported by phylogenetic analysis of LSU sequences separating *Tiarosporella* from *Macrophomina* (Crous et al., 2006).

1.3 Molecular taxonomy of *Macrophomina* species

Due to the high level of genetic diversity and pathogenic variation observed in *M. phaseolina* populations globally, the presence of potential cryptic species within this genus had been hypothesized (Vandemark et al., 2000; Almeida et al., 2008). However, initial molecular phylogenetic studies of *Macrophomina* based on nuclear ribosomal DNA (nrDNA) sequences, although helpful in differentiating the genus from the other morphologically similar genera like *Tiarosporella* (Crous et al., 2006; Baird et al., 2010; Saleh et al., 2010), did not provide resolution at the species level. Subsequent molecular taxonomic studies based on phylogenetic analyses of concatenated sequences data, genealogical concordance phylogenetic species recognition (GCPSR), and coalescent-based approaches using additional housekeeping loci indicated the presence of more than one species within the genus *Macrophomina*.

In 2014, Sarr et al. (2014) conducted the first systematic phylogenetic investigation on a global set of 189 *Macrophomina* isolates from 23 host species in 15 countries, using sequences of the actin (*act*), calmodulin (*cmd*), translation elongation factor one-alpha (*tef1-α*), beta tubulin (*tub2*), and internal transcribed spacers and the 5.8S region of the nrDNA (ITS). This resulted in the designation of an epitype for *M. phaseolina s. str.* and description of a new species, *M. pseudophaseolina,* which formed a sister clade to *M. phaseolina,* separated with maximum bootstrap support.

Consequently, several studies conducted on *Macrophomina* species from different regions and hosts using the same set of genetic loci described three additional species, namely, *M. euphorbiicola* (Machado et al., 2019), *M. vaccinii* (Zhao et al., 2019), and *M. tecta* (Poudel et al., 2021c). As a result, currently five *Macrophomina* species are recognized (Fig. 1.1).

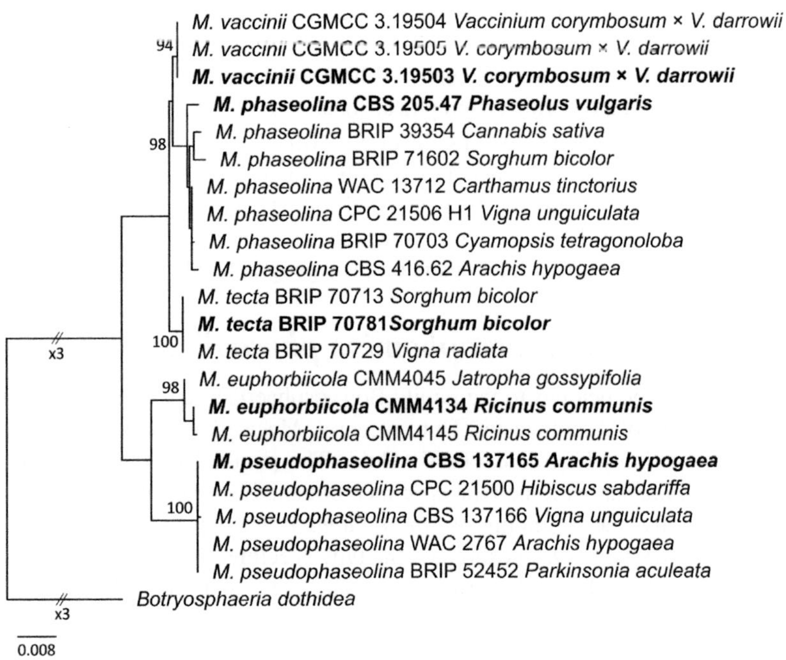

FIGURE 1.1 Phylogeny of *Macrophomina* species recognized to date. Maximum likelihood phylogeny of *Macrophomina* species based on the concatenated ITS, *cmd*, *tef1-α*, and *tub2* alignments constructed in RAxML v. 8.2.11 (Stamatakis, 2008) using the GTRGAMMA model of nucleotide substitution applied to individual partitions with 1000 pseudoreplicates. Ex-type strains are indicated in bold. Bootstrap values are given at the nodes. The tree was rooted to *Botryosphaeria dothidea* (Poudel et al., 2021c).

1.4 Host range and distribution of *Macrophomina* species

Of the five *Macrophomina* species recognized to date, *M. phaseolina* is the most widely distributed, displaying a cosmopolitan distribution on more than 500 plant species. While the cosmopolitan distribution and broad host range of this species has been confirmed in more recent molecular studies, some of the older reports of this species based on morphological studies may need to be treated with caution as this species is morphologically indistinguishable from more recently described species such as *M. tecta* (Poudel et al., 2021c) and can only be distinguished from *M. pseudophaseolina* based on morphology of the conidia (Sarr et al., 2014), which are not always readily formed in culture to allow for morphological characterization. As noted, recent molecular studies still point to the broad distribution of *M. phaseolina* on a board range of plant species. Of the 189 isolates studied by Sarr et al. (2014) in Senegal, 171 were recognized to be *M. phaseolina*. Likewise, of the 80 isolates from 28 plant species in Australia, the majority (88%) belonged to *M. phaseolina* (Poudel et al., 2021c).

Macrophomina pseudophaseolina has thus far been reported in association with 13 plant species and in six countries. These hosts include *Abelmoschus esculentus* (okra) in Senegal (Sarr et al., 2014), *Arachis hypogea* (peanut) in Australia, Brazil, and Senegal (Sarr et al., 2014; Machado et al., 2019; Poudel et al., 2021c), *Boerhavia diffusa* in Brazil (Negreiros et al., 2019), *Coleus forskohlii* in India (Mastan et al., 2019), *Gossypium hirsutum* (cotton) in Brazil (Machado et al., 2019), *Hibiscus sabdariffa* (roselle) in Senegal (Sarr et al., 2014), *Ipomoea batatas* (sweet potato) in Brazil (de Mello et al., 2021), *Jatropha curcas* in Brazil (Machado et al., 2019), *Lens culinaris* (lentil) in Algeria (Kouadri et al., 2021), *Manihot esculenta* (cassava) in Brazil (Brito et al., 2019), *Parkinsonia aculeate* a native plant species in Australia (Poudel et al., 2021c), *Phaseolus vulgaris* (bean) in Argentina (Viejobueno et al., 2022), *Ricinus communis* in Brazil (Machado et al., 2019), *Trianthema portulacastrum* in Brazil (Negreiros et al., 2019), and *Vigna unguiculata* (cowpea) in Senegal (Sarr et al., 2014).

In contrast, the remaining three species currently recognised within the genus have been recorded from a limited number of and host plants and geographical locations. *M. euphorbiicola* has thus far only been reported in Brazil on *I. batatas*, *Jatropha gossypiifolia,* and *R. communis* (Ayala-Armenta et al., 2020; de Mello et al., 2021; Machado et al., 2019). *M. vaccinii* has only been reported in China on *Pogostemon cablin* (Fang et al., 2022) and *Vaccinium* spp. (blueberry) (Ayala-Armenta et al., 2020; Zhao et al., 2019). *M. tecta* has only been recorded on *Zea mays* (corn) in Argentina (Viejobueno et al., 2022) and on *Sorghum bicolor* (sorghum) and *Vigna radiata* (mungbean) in Australia (Poudel et al., 2021c). The restricted host range could be due to the very recent recognition of these species, and further studies are warranted to investigate their potential host specialization.

1.5 Intraspecific genetic diversity and pathogenic variability

Understanding the genetic diversity and structure of pathogen populations is essential for better understanding their evolution in agricultural ecosystems to allow for sustainable disease management. Population genetic and genomic investigations provide contribution to better understanding the pathogen's biology, disease epidemiology, and co-evolution of host and pathogen, which may be used to improve breeding for disease resistance and deployment of disease mitigation strategies. Genotyping via a variety of molecular markers, such as amplified fragment length polymorphisms (AFLP), random amplified polymorphic DNA (RAPD), restriction fragment length polymorphisms (RFLP), simple sequence repeats (SSR) and single nucleotide polymorphisms (SNP), and subsequent cluster analyses have been applied to understand genetic diversity of *M. phaseolina* populations and detected substantial genetic heterogeneity in Argentina (Reznikov et al., 2019), Australia (Poudel et al., 2021a, 2021b), Brazil (Almeida et al., 2008; Sybuia et al., 2022), Europe (TančićŽivanov et al., 2019), Iran (Mahdizadeh et al., 2012), Mexico (Mayék-Pérez et al., 2001; Reyes-Franco et al., 2006) and the United States (Vandemark et al., 2000; Baird et al., 2010; Saleh et al., 2010). These studies reported separation of isolates into multiple subpopulations, with no clear association with either geographical location or host of origin.

In contrast, other studies have detected association of *M. phaseolina* genotypes and host (Su et al., 2001; Viejobueno et al., 2022). Pathogenicity and RAPD-based analysis of a collection of 45 isolates from cotton, corn, sorghum, and soybean from different geographical locations suggested an association between isolate clustering, pathogenicity, and cropping history, irrespective of geographical origin of isolates (Su et al., 2001). Similarly, in India, *M. phaseolina* isolates obtained from cotton and soybean clustered corresponding to their host of origin (Jana et al., 2005). Also, Jana et al. (2003) discovered RAPD, SSR, and URP-PCR (universal rice primer PCR) markers associated with *M. phaseolina* genotypes from certain hosts or geographical origins.

Lack of association of genotypes with pathogenicity/virulence profiles on different hosts is expected and has been reported in other fungal plant pathogens too (Barrès et al., 2008; Vaghefi et al., 2017, 2018). This is due to the fact that phylogenetic and population genomic studies usually use neutral genetic markers for population characterization, which do not necessarily represent pathogenicity and virulence phenotypes, which are under selection pressure form host populations (Vaghefi et al., 2017). On the other hand, if loci used for population characterization are linked to pathogenicity and virulence genes, the association of genotypes, and different pathogenicity/virulence profiles may be expected.

Detection of high genetic diversity in *M. phaseolina* populations, in the absence of a known sexual form, is surprising and has been attributed to parasexuality, long-distance dispersal of the isolates (genotype flow) and/or host adaptation. Parasexuality is a nonmeiotic process that produces genetically diverse progeny and is an effective mechanism to enhance genotypic diversity in asexual fungi (Pontecorvo et al., 1953). A parasexual cycle starts with the formation of heterokaryons following fusion of two genetically compatible isolates. The two haploid nuclei may fuse and form a heterozygous diploid nucleus. Upon division of the diploid nuclei, mitotic nondisjunction or mitotic recombination may take place, giving rise to recombinant haploid nuclei containing novel allelic combinations (Pontecorvo et al., 1953). Recent discovery of heterokaryon formation in *M. phaseolina* (Pereira et al., 2018) and parasexual recombination in *M. pseudophaseolina* (Pereira et al., 2018; Sybuia et al., 2022) confirms genetic exchange between vegetatively compatible isolates as a result of which mitotic recombination can be a major source of genotypic diversity in the absence of sexual recombination in *Macrophomina* populations.

Another factor underlying the high genetic diversity in *M. phaseolina* populations could be the pathogen's ability to infect a diverse range of host crops. Almeida et al. (2008) investigated how crop rotation may affect the genetic diversity of *M. phaseolina* using 89 isolates obtained from a 4-year no-tillage crop rotation paddock and an uncropped paddock. RAPD genotyping results showed that genetic diversity was higher in *M. phaseolina* population from the crop rotation paddock compared to the uncropped paddock. This led the authors to conclude that the occurrence of different hosts with different levels of resistance may have exposed the pathogen population to diversifying selection pressure leading to an increase in genetic variability.

It should be noted that some population genetic studies conducted before 2014 may have included multiple cryptic *Macrophomina* species, which may have led to inflation of genetic diversity assuming all isolates belonged to *M. phaseolina*. Population studies conducted by Poudel et al. (2021a, 2021b) using restriction site-associated DNA (RAD) sequencing have shown that sympatric populations of *M. phaseolina* and *M. tecta* occur on all surveyed sorghum crops in Australia; therefore, raising the possibility that mixed populations of *Macrophomina* spp. can also exist on other crops and in previous population genetic studies. Therefore, results from some population genomics studies that precede resolution of cryptic *Macrophomina* spp. via molecular taxonomy should be treated with caution.

1.6 Conclusions

Molecular taxonomy of the genus *Macrophomina* was resolved only very recently; therefore, many aspects of biology and epidemiology of recently described *Macophomina* species that is, *M. euphorbiicola*, *M. pseudophaseolina*, *M. tecta*, and *M. vaccinii* remain little understood. On the other hand, the extensive morphological, pathogenicity, and epidemiological studies conducted on *M. phaseolina* may need to be treated with caution since different *Macrophomina* species occur sympatrically in agricultural ecosystems (Poudel et al., 2021a, 2021b); thus, the results from studies based on morphological identification of species may have been confounded by presence of two or more cryptic species.

The very first step in successful disease management is correct identification of the causal agent. Therefore, development of reliable molecular markers to differentiate *Macrophomina* species should be a priority. To the best of our knowledge, DNA-based assays capable of distinguishing all five species of *Macrophomina* have not yet been developed. Such molecular markers may be applied for extensive surveys of *Macrophomina* species on different crops and soils, enabling generation of predictive models to understand the impact of each of these species and their interactions in agricultural ecosystems.

References

Adorada, D., Gonzales, P., Mckay, A., Vaghefi, N., & Sparks, A. (2018). A broad look at charcoal rot in the Northern Region broadacre crops through soil sampling and in-crop surveys. In *10th Australasian soilborne diseases symposium: Paddock to plates*. Adelaide: ASDS.

Almeida, Á., Sosa-Gomez, D., Binneck, E., Marin, S., Zucchi, M., Abdelnoor, R., & Rorigues, S. E. (2008). Effect of crop rotation on specialization and genetic diversity of *Macrophomina phaseolina*. *Tropical Plant Pathology, 33*, 257–264.

Ashby, S. (1927). *Macrophomina phaseoli* (Maubl.) comb. nov. the pycnidial stage of *Rhizoctonia bataticola* (Taub.) Butl. *Transactions of the British Mycological Society, 12*, 141–147.

Ayala-Armenta, Q. A., Tovar-Pedraza, J. M., Apodaca-Sánchez, M. A., Correia, K. C., Sauceda-Acosta, C. P., Camacho-Tapia, M., & Beltrán-Peña, H. (2020). Phylogeny and pathogenicity of soilborne fungi associated with wilt disease complex of tomatillo (*Physalisixo carpa*) in Northern Sinaloa, Mexico. *European Journal of Plant Pathology, 157*, 733–749.

Baggio, J. S., Cordova, L. G., & Peres, N. A. (2019). Sources of inoculum and survival of *Macrophomina phaseolina* in Florida strawberry fields. *Plant Disease, 103*, 2417–2424.

Baird, R. E., Wadl, P. A., Allen, T., McNeill, D., Wang, X., Moulton, J. K., Rinehart, T. A., Abbas, H. K., Shier, T., & Trigiano, R. N. (2010). Variability of United States isolates of *Macrophomina phaseolina* based on simple sequence repeats and cross genus transferability to related genera within Botryosphaeriaceae. *Mycopathologia, 170*, 169–180.

Barrès, B., Halkett, F., Dutech, C., Andrieux, A., Pinon, J., & Frey, P. (2008). Genetic structure of the poplar rust fungus *Melampsora laricipopulina*: Evidence for isolation by distance in Europe and recent founder effects overseas. *Infection, Genetics and Evolution, 8*, 577–587.

Brito, A. C. Q., de Mello, J. F., Michereff, S. J., de Souza-Motta, C. M., & Machado, A. R. (2019). First report of *Macrophomina pseudophaseolina* causing stem dry rot in cassava in Brazil. *Journal of Plant Pathology, 101*, 1245-1245.

Burkhardt, A. K., Childs, K. L., Wang, J., Ramon, M. L., & Martin, F. N. (2019). Assembly, annotation, and comparison of *Macrophomina phaseolina* isolates from strawberry and other hosts. *BMC Genomics, 20*, 802.

Crous, P. W., Slippers, B., Wingfield, M. J., Rheeder, J., Marasas, W. F. O., Philips, A. J. L., Alves, A., Burgess, T., Barber, P., & Groenewald, Z. J. (2006). Phylogenetic lineages in the Botryosphaeriaceae. *Studies in Mycology, 55*, 235–253.

Fang, Y., Li, J., Li, X., Chen, M., Lin, J., Huang, J., Jinhao Li, J., Tang, Y., Xiuli Song, X., & Zhang, W. (2022). *Macrophomina* vaccinii causes a basal stem and root rot of patchouli (*Pogostemon cablin*) in China. *Plant Disease, 106*, 1067.

Goidànich, G. (1947). Revisione del genere *Macrophomina* Petrak. Specie tipica: *Macrophomina phaseolina* (tassi) G. Goid. N. comb. Nec *M. phaseoli* (maubl.) Ashby. *Annali Della Sperimentazione Agaria, 1*, 449–461.

Islam, M. S., Haque, M. S., Islam, M. M., Emdad, E. M., Halim, A., Hossen, Q. M. M., Hossain, M. Z., Ahmed, B., Rahim, S., Rahman, Md S., Alam, M. M., Hou, S., Wan, X, Saito, J. A., & Alam, M. (2012). Tools to kill: Genome of one of the most destructive plant pathogenic fungi *Macrophomina phaseolina. BMC Genomics, 13*, 493.

Jana, T., Sharma, T. R., Prasad, R. D., & Arora, D. K. (2003). Molecular characterization of *Macrophomina phaseolina* and *Fusarium* species by a single primer RAPD technique. *Microbiological Research, 158*, 249–257.

Jana, T., Sharma, T. R., & Singh, N. K. (2005). SSR-based detection of genetic variability in the charcoal root rot pathogen *Macrophomina phaseolina. Mycological Research, 109*, 81–86.

Khan, A. N., Shair, F., Malik, K., Hayat, Z., Khan, M. A., Hafeez, F. Y., & Hassan, M. N. (2017). Molecular identification and genetic characterization of *Macrophomina phaseolina* strains causing pathogenicity on sunflower and chickpea. *Frontiers in Microbiology, 8*, 1309.

Kouadri, M. E. A., Zaim, S., & Bekkar, A. A. (2021). First report of *Macrophomina pseudophaseolina* infecting *Lens culinaris. Australasian Plant Disease Notes, 16*, 26.

Machado, A. R., Pinho, D. B., Soares, D. J., Gomes, A. A. M., & Pereira, O. L. (2019). Bayesian analyses of five gene regions reveal a new phylogenetic species of *Macrophomina* associated with charcoal rot on oilseed crops in Brazil. *European Journal of Plant Pathology, 153*, 89–100.

Mahdizadeh, V., Safaie, N., Goltapeh, E. M., & Mayek-Perez, N. (2012). Intraspecies diversity of *Macrophomina phaseolina* in Iran. *Archives of Phytopathology and Plant Protection, 45*, 963–976.

Maholay, M., & Sohi, H. (1983). *Macrophomina* seed rot of bottlegourd, squash and muskmelon. *Indian Journal of Mycology & Plant Pathology, 13*, 192–197.

Mastan, A., Bharadwaj, R. K. B., Kushwaha, R. K., & Babu, V. C. S. (2019). Functional fungal endophytes in *Coleus forskohlii* regulate labdane diterpene biosynthesis for elevated forskolin accumulation in roots. *Microbial Ecology, 78*, 914–926.

Maublanc, A. (1905). *Bulletin Trimestriel de la Societe Mycologique de France, 21*(1), 90.

Mayék-Pérez, N., López-Castañeda, C., González-Chavira, M., Garcia-Espinosa, R., Acosta-Gallegos, J., De La Vega, O. M., & Simpson, J. (2001). Variability of Mexican isolates of *Macrophomina phaseolina* based on pathogenesis and AFLP genotype. *Physiological and Molecular Plant Pathology, 59*, 257–264.

de Mello, J. F., Brito, A. C., Vieira, J. C., Câmara, M. P., Michereff, S. J., de Souza-Motta, C. M., & Machado, A. R. (2021). Identification and pathogenicity of Botryosphaeriaceae species associated with root and stem rot of sweet potato in Brazil. *Plant Pathology, 70*, 1601–1615.

Negreiros, A. M. P., Sales Júnior, R., León, M., Melo, N. J., Michereff, S. J., Ambrósio, M. M., De Sousa Medeiros, H. L., & Armengol, J. (2019). Identification and pathogenicity of *Macrophomina* species collected from weeds in melon fields in Northeastern Brazil. *Journal of Phytopathology, 167*, 326–337.

Pereira, T. S., Machado Esquissato, G. N., da Silva Franco, C. C., de Freitas Mathias, P. C., Soares, D. J., Bock, C. H., De Carvalho Nunes, W. M., Zanutto, C. A., & Castro-Prado, M. A. A. (2018). Heterokaryosis and diploid formation among Brazilian isolates of *Macrophomina phaseolina. Plant Pathology, 67*, 1857–1864.

Petrak, F. (1923). MykologischeNotizen VI 201-300. *Annals of Mycology, 21*, 313.

Pontecorvo, G., Roper, J. A., & Forbes, E. (1953). Genetic recombination without sexual reproduction in *Aspergillus Niger. Journal of General Microbiology, 8*, 198–210.

Poudel, B., Purushotham, N., Adorada, D., Sparks, A., Schwessinger, B., & Vaghefi, N. (2021a). Genetic diversity of *Macrophomina* species associated with sorghum in Australia. In 23rd *Australasian plant pathology society conference*. Tasmania: APPS.

Poudel, B., Purushotham, N., Adorada, D., Sparks, A., & Vaghefi, N. (2021b). Genetic diversity of *Macrophomina* species associated with sorghum in Australia. In *Proceedings of the joint CanFunNet and the great lakes mycology conference. Presented at the CanFunNet, Canada.*

Poudel, B., Purushotham, N., Jones, A., Nasim, J., Adorada, D. L., Sparks, A. H., Schwessinger, B., & Vaghefi, N. (2022). The first annotated genome assembly of *Macrophomina tecta* associated with charcoal rot of sorghum. *Genome Biologyand Evolution, 14*, evac081.

Poudel, B., Shivas, R. G., Adorada, D. L., Barbetti, M. J., Bithell, S. L., Kelly, L. A., Moore, N., Sparks, A. H., Tan, Y. P., Thomas, G., & Leur, J. V. (2021c). Hidden diversity of *Macrophomina* associated with broadacre and horticultural crops in Australia. *European Journal of Plant Pathology, 161*, 1–23.

Purushotham, N., Jones, A., Poudel, B., Nasim, J., Adorada, D., Sparks, A., Schwessinger, B., & Vaghefi, N. (2020). Draft genome resource for *Macrophomina phaseolina* associated with charcoal rot in sorghum. *MolecularPlant-Microbe Interactions, 33*, 724–726.

Radadiya, N., Mangukia, N., Antala, V., Desai, H., Chaudhari, H., Dholaria, T. L., Dholaria, D., Tomar, R. S., Golakiya, B. A., & Mahatma, M. K. (2021). Transcriptome analysis of sesame-*Macrophomina phaseolina* interactions revealing the distinct genetic components for early defense responses. *Physiology and Molecular Biology of Plants, 27*, 1675–1693.

Reyes-Franco, M. C., Hernandez-Delgado, S., Beas-Fernandez, R., Medina-Fernandez, M., Simpson, J., & Mayek-Perez, N. (2006). Pathogenic and genetic variability within *Macrophomina phaseolina* from Mexico and other countries. *Journal of Phytopathology, 154*, 447–453.

Reznikov, S., Chiesa, M. A., Pardo, E. M., De Lisi, V., Bogado, N., González, V., Ledesma, F., Morandi, E. N., Ploper, L. D., & Castagnaro, A. P. (2019). Soybean-*Macrophomina phaseolina*-specific interactions and identification of a novel source of resistance. *Phytopathology, 109*, 63–73.

Saleh, A. A., Ahmed, H. U., Todd, T. C., Travers, S. E., Zeller, K. A., Leslie, J. F., & Garrett, K. A. (2010). Relatedness of *Macrophomina phaseolina* isolates from tallgrass prairie, maize, soybean and sorghum. *Molecular Ecology, 19*, 79–91.

Sarr, P., Ndiaye, M., Groenewald, J., & Crous, P. (2014). Genetic diversity in *Macrophomina phaseolina*, the causal agent of charcoal rot. *Phytopatholgia Mediterranea, 53*, 250–268.

Short, G., Wyllie, D., & Bristow, P. (1980). Survival of *Macrophomina phaseolina* in soil and in residue of soybean. *Phytopathology, 70*, 13–17.

Stamatakis, A. (2008). RAxML version 8: A tool for phylogenetic analysis and post-analysis of large phylogenies. *Bioinformatic Application Notes, 30*, 1312–1313.

Su, G., Suh, S.-O., Schneider, R. W., & Russin, J. S. (2001). Host specialization in the charcoal rot fungus, *Macrophomina phaseolina. Phytopathology, 91*, 120–126.

Sybuia, P. A., de Castro-Prado, G., de CarvalhoNunes, W. M., Zanutto, C. A., Kaneshima, E. N., Soares, D. J., Da Silva Franco, C. C., De Freitas Mathias, P. C., & De Castro-Prado, M. A. A. (2022). Parasexual recombination in *Macrophomina pseudophaseolina* and vegetative compatibility reactions in. *M. euphorbiicola. European Journal of Plant Pathology, 163*, 937–950.

TančićŽivanov, S., Dedić, B., Dimitrijević, A., Dušanić, N., Jocić, S., Miklič, V., Kovačević, B., & Miladinovic, D. (2019). Analysis of genetic diversity among *Macrophomina phaseolina* (Tassi) Goid. isolates from Euro-Asian countries. *Journal of Plant Diseases and Protection, 126*, 565–573.

Tassi, A. (1901). Novae micromycetum species. Ser III, Bullettino del Laboratorio ed Orto botanicodella R. *Università di Siena, 4*, 7–12.

Vaghefi, N., Kikkert, J. R., Hay, F. S., Carver, G. D., Koenick, L. B., Bolton, M. D., Hanson, L. E., Secor, G. A., & Pethybridge, S. J. (2018). Cryptic diversity, pathogenicity, and evolutionary species boundaries in *Cercospora* populations associated with Cercospora leaf spot of *Beta vulgaris. Fungal Biology, 122*(4), 264–282.

Vaghefi, N., Nelson, S. C., Kikkert, J. R., & Pethybridge, S. J. (2017). Genetic structure of *Cercospora beticola* populations on Beta vulgaris in New York and Hawaii. *Scientific Reports, 7*, 1726.

Vandemark, G., Martínez, O., Pecina, V., & Alvarado, M. J. (2000). Assessment of genetic relationships among isolates of *Macrophomina phaseolina* using a simplified AFLP technique and two different methods of analysis. *Mycologia, 92*, 656–664.

Viejobueno, J., de los Santos, B., Camacho-Sanchez, M., Aguado, A., Camacho, M., & Salazar, S. M. (2022). Phenotypic variability and genetic diversity of the pathogenic fungus *Macrophomina phaseolina* from several hosts and host specialization in strawberry. *Current Microbiology, 79*, 189.

Von Arx, J. (1981). The genera of fungi sporulating in pure culture. *J. Cramer, Vaduz.*, 283–331.

Wingfield, B. D., De Vos, L., Wilson, A. M., Duong, T. A., Vaghefi, N., Botes, A., Kharwar, R. N., Chand, R., Poudel, B., Aliyu, H., Barbetti, M. J., Chen, S., De Maayer, P, Liu, F., Navathe, S., Sinha, S., Steenkamp, E. T., Suzuki, H., Tshisekedi, K. A., & Wingfield, M. J. (2022). IMA genome - F16: Draft genome assemblies of *fusarium marasasianum, huntiella abstrusa*, two *immersiporthe knoxdaviesiana* isolates, *Macrophomina pseudophaseolina, Macrophomina phaseolina, naganishia randhawae*, and *pseudocercospora cruenta. IMA Fungus, 13*(1), 3.

Zhao, L., Cai, J., He, W., & Zhang, Y. (2019). *Macrophomina vaccinii* sp. nov. causing blueberry stem blight in China. *MycoKeys, 55*, 1–14.

Zveibil, A., Mor, N., Gnayem, N., & Freeman, S. (2012). Survival, host–pathogen interaction, and management of *Macrophomina phaseolina* on strawberry in Israel. *Plant Disease, 96*, 265–272.

Chapter 2

Morphological and physiological characterization of *Macrophomina* isolates from Senegal

Mame P. Sarr[1], Mbaye Ndiaye[2] and Ndiaga Cisse[1]

[1]*Institut Sénégalais de Recherches Agricoles/Centre National de Recherches Agronomiques BP 53, Bambey, Senegal;* [2]*Centre Regional Aghrymet/ Expert in Plant Pathology, Saly, Senegal*

2.1 Introduction

Macrophomina phaseolina (Tassi) Goid. (=*Tiasrosporelle phaseolina* (Tassi) Van der Aa) is a soil-borne fungus of the class Ascomycetes (Ndiaye, 2007). This pathogen can cause yield losses on sorghum, groundnut, and cowpea (Adam, 1986; Bouhot, 1967, 1968; Diourte et al., 1995; Gray et al., 1990; Hall, 1991). It is a polyphagous fungus that can live as a saprophyte, and its propagules are preserved for several years in the soil.

M. phaseolina was first named as *Rhizoctonia bataticola* by Halsted in 1890 after observing the black nodules (sclerotia) on the roots of *Ipomoea batatas* plants. Tassi (1901) described the pycnidial stage and had named it as *M. phaseolina* and Maublanc (1905) named it *M. phaseolus*.Shaw in 1912 described it as a fungus with black sclerotia that attacked seedlings of peanut (*Arachishy pogaea* L.), and was called *Sclerotium bataticola* by Dhingra and Sinclair (1978). Mihail (1992) indicated that there was an unconfirmed report of a teleomorphic stage named *Orbilia obscura* (Ghosh et al., 1964) but there has been no confirmation of this teleomorphic stage since then.

Crous et al. (2006), based on morphological characteristics, included the pathogen in the Botryosphaeriaceae family showing a similarity between *Macrophomina* and *Tiarosporella* species. Both have conidia with a hyaline apical appendage. However, *Tiarosporella* has no microsclerotia; the conidia do not turn brown with age, and its conidiogenous cells have no visible percurrent proliferations (Crous et al., 2006).

In Senegal, *M. phaseolina* was first reported in 1964 on cowpea (*Vignaungui culata* (L.) Walp.), groundnut (*Arachis hypogaea*), bambara groundnut (*Voandzae subterranea*(L.) Verdc.), and sunflower (*Helianthus annuus*) in Diourbel and Kaolack regions (Bouhot, 1967).

From 1972 to today, the area infested by this disease has increased considerably in Louga region because of cowpea monoculture (Cissé & Hall, 2002). This practice of planting cowpeas continuously in the same field was a response to drought, since it is one of the few species able to produce under erratic and low rain rainfall (amount) conditions of the region. In addition, in 2014, a new species was found in Senegal and was named *M. pseudophaseolina* by Sarr et al. (2014).

2.2 Bio-ecology

M. phaseolina (Tassi) Goid. is a saprophytic fungus that causes damping-off, stem, and root rot on more than 500 species of cultivated and wild plants (Dhingra & Sinclair, 1977). The disease affects water balance and vigor of plants. The fungus is able to maintain itself in the soil for several years in the absence of susceptible hosts particularly with its sclerotia. Warm temperatures, around 28–35°C, and water-stress favor attacks by this pathogen *M. phaseolina.*

Microsclerotia are black bodies composed of 50–200 accumulated hyphal cells with a cement-like melanin that gives them this coloration (Gangopadhyay & Wyllie, 1974; Short & Wyllie, 1978). These microsclerotia germinate on the root

Macrophomina Phaseolina. https://doi.org/10.1016/B978-0-443-15443-0.00017-6

surface and produce numerous germ tubes. Furthermore, there is a relation between the number of cells and germ tubes and microsclerotia size (Dhingra & Sinclair, 1977). Pratt (2006) described these germ tubes and hyphae of the fungus *M. phaseolina* in a corn-based culture medium (corn meal agar diluted CMA). In this culture medium, on average, four germ tubes are produced per microsclerotia, with a length of a few micrometers to a 100 after 18–24 h of exposure in this medium (Pratt, 2006).

M. phaseolina pycnidia are obtained by alternating 12 h of light and 12 h of darkness and by growing the fungus in young plant tissue (Jimenez, 2011). They are globose to spindle-shaped, black or gray. During their initial stage, pynidia are immersed in host tissue, but eruptive at maturity with a central hyaline opening through which ellipsoid to ovoid conidia are released (Dhingra & Sinclair, 1977). The small truncate ostiole may be invisible, sometimes with a clear hyaline opening (Hartman et al., 1999; Mihail, 1992). According to Gaetàn et al. (2006), pycnidia are not produced in potato dextrose agar (PDA) medium, except under specific incubation conditions and only sometimes on host tissue for (Mihail & Taylor, 1995). In 2010, a peanut butter extract-saturated filter paper over soynut butter extact agar (PESEA) culture medium was described by Ma et al. (2010), and in this medium, there was the production of both conidia and pycnidia.

M. phaseolina is easily grown on all cell-free media from spores, sclerotia, or mycelium. According to Ndiaye (2007), out of 75 isolates of *M. phaseolina* studied at different temperatures few or no microsclerotia were formed at 37°C in the culture media after 48 h of incubation. Optimal in vitro growth temperatures for *M. phaseolina* were between 35 and 37°C for 92% of isolates, and the average growth rate was 25%.

According to Babu et al. (2011), the genus *Macrophomina* produces two asexual structures, microsclerotia and pycnidia. The sclerotia can be detected in soil and both pycnidia and sclerotia in tissues by real-time polymerase chain reaction (PCR) assay or by isolation on agar medium. Sclerotia can survive in soil up to 15 years or in plants debris for a long time (Baird et al., 2003; Cook et al., 1973; Papavizas, 1977; Short et al., 1980). When tissues begin to decompose, microsclerotia are released into the soil (Hartman et al., 1999). The primary function of microsclerotia is to ensure the survival of the fungus under adverse environmental conditions (Short et al., 1980). Seed mycelium and sclerotia in the soil on various plant debris are the main sources of primary infection (Emechebe & McDonald, 1979). Pycnidia are immersed in host tissue but erupt at maturity with a central hyaline opening through which ellipsoid to ovoid conidia are released (Dhingra & Sinclair, 1977). They are of little importance in the epidemiology of the disease in the Sahel.

2.3 Damage and economic importance of *Macrophomina*

Charcoal rot caused by the fungus *Macrophomina* is one of the most important production constraints (Adam, 1986, 1990; Paré, 1990). It is widespread in tropical, subtropical, and temperate regions (ICRISAT, 1980; Tarr, 1962).

M. phaseolina can damage cowpeas and several crops such as okra (*Hibiscus esculentus*), sesame (*Sesamum indicum*), and sorrel (*Hibiscus sabdariffa*) in dry and hot soils (Ndiaye, 2007).

In India, damage to economically important crops such as cowpea, sorghum, and groundnut has been reported. According to Shanmugan and Govindaswamy (1973), losses caused by *M. phaseolina* on groundnut crops can exceed 50%.

In the United States, charcoal rot is common in some regions of the North and is ranked fourth in economic impact of soybeans after root-knot nematode, *Phytophtora* root rot and seedling diseases (Smith & Carvil, 1997). Average yield losses in the US due to charcoal rot are estimated to be approximately 734. 670 tons from 1996 to 2009 (Wrather & Koenning, 2010).

According to Ndiaye (2007), in the West African Sahelian zone (including Burkina Faso, Niger, and Senegal), charcoal rot causes average yield losses of 10%, which is equivalent to 30,000 tons of cowpeas with an estimated value in US dollars of 146 million for Niger and Senegal alone.

In Senegal, this pest, which was considered a secondary pest of no economic importance to agricultural production, caused a considerable damage to cowpeas in 1987 (Gaikwad & Sokhy, 1987). Despite new cowpea varieties that perform well in crop rotation and under drought conditions, yields in farmers' fields remain very low (0.2–0.5 tons/ha).

Aziz and Ndiaye (2022) evaluated the effect of high seeding density in controlling charcoal rot of cowpea in the Sahel. 10 days after sowing, plant stand was significantly higher in plots sown with one seed per planting hole (83%) than in plots sown with three and four seeds per planting hole (63%). At harvest period, diseases incidence (44%) and dead plants (33%) were also higher in plots with seeding densities four and three seeds per planting hole. Also, at this density, the soil inoculum density was the highest (21,680 sclerotia/m^2). The pod and grain average yields were significantly less in plots with high seeding densities than in the one with low densities.

In the Sahelian region, the typical symptom of charcoal rot on cowpea is sudden wilting and drying of the plant with most of the leaves remaining green (Fig. 2.1). The stem and branches are then covered with black bodies (Fig. 2.2) and give the appearance of charcoal or dead plant ash. Wilt results from necrosis (roots and stems) and from mechanical press on

FIGURE 2.1 Disease symptoms of *Macrophomina* (charcoal rot) in infested fields. (A) Advanced level of infestation on cowpea (*Vigna unguiculata*) in an experimental field with a drip irrigation system (Agrhymet, Niger). (B) Charcoal rot symptoms in a sorrel (*Hibiscus sabdariffa*) field during the rainy season (Agrhymet, Niger). (C) Early infection in a young cowpea (*Vigna unguiculata*) plant, in an experimental field during the rainy season (Bambey, Senegal).

FIGURE 2.2 Brown [dark gray in print version] and black spots on the hypocotyl section of the stem of a young cowpea plant grown in a pot infested with *Macrophomina* sp. (A); cowpea plant dead from *Macrophomina* attack with a charcoal-like stem (B).

xylem vessels blocked by microsclerotia but also from production of toxin and enzymatic action (Chan & Sackston, 1973; Jones & Wang, 1997; Kuti et al., 1997). It can be observed in all stages (seedling to maturity). An infected plant can have up to 44% infected seeds (Raut, 1985).

At the seed infection stage, *M. phaseolina* invades the host very rapidly and becomes established within 24–48 h of infection. If the invaded plant survives from seedling mortality, the fungus moves into the aerial parts at a very slow rate. Typical symptoms of the disease become apparent at the seed formation stage following high temperature and low humidity. However, the range of relative humidity and temperature required may vary depending on the nature of the isolate, the climatic zone and the host cultivar (Edmunds, 1964).

2.4 Disease cycle

The fungus *M. phaseolina* causes charcoal rot on several economically important plant species. Sclerotia produced in roots and stems allows it to survive under adverse environmental conditions (Cook et al., 1973; Meyer et al., 1974; Short et al., 1980). Microsclerotia produced in host tissues are multicellular, black and spherical in structure and are composed of globular cells which sometimes germinate independently. These microsclerotia released into the soil allow the fungus to survive under adverse conditions (Hartman et al., 1999) and are the primary sources of inoculum (Abawi & Pastor-Corrales, 1990; Bouhot, 1968; Dhingra & Sinclair, 1977). Root exudates from plants induce germination of microsclerotia, which then produce numerous germ tubes that infect the host roots. Penetration is usually through appressoria formed by mature hyphae that enter the host plant through natural openings or cell walls on the epidermis (Bressano et al., 2010). This pathogen can also infect seeds. During the early stages of pathogenesis, the mycelium found is on intercellular spaces of roots. As a result, adjacent cells collapse and heavily infected plants die early due to fungal toxins production, for example, phaseolinone (Bhattacharya et al., 1994) that blocks host vessels. According to Wyllie and Scott (1988), at the flowering stage, fungal hyphae colonize intracellular tissues through the xylem and form microsclerotia that block tissues (Short & Wyllie, 1978; Mayek-Pérez et al., 2002) and destroy host cells. At this stage, penetration into the cell wall is the result of both mechanical and chemical softening pressure due to the production of pectolytic enzymes (Ammon et al., 1974). Many studies have demonstrated the importance of this phase (flowering) in the expression of disease symptoms (Blanco-Lopez & Jimenez-Diaz, 1983; Davet et al., 1986; Young & Alcorn, 1984). At the reproductive and near-mature stages, attacked plants lost vigor, and have their chlorotic and necrotic leaves remained attached to the stem. Finally, with high level infestation of roots and stems by *M. phaseolina*, discoloration appear on the epidermal and subepidermal cells of roots and stems (Ammon et al., 1974). After harvest, microsclerotia-laden roots begin to decompose and release microsclerotia (Olaya et al., 1996), which are generally distributed in clusters on soil surface and are located mainly at 0–20 cm depth of (Alabouvette, 1976; Campbell & van der Gaag, 1993; Mihail, 1989). These microsclerotia, when associated with host residues, and with moisture, temperature, and microbial population in the soil (Collins et al., 1991), can survive for 2–15 years (Baird et al., 2003; Cook et al., 1973; Papavizas, 1977; Short et al., 1980) (Fig. 2.3).

2.5 Characteristic of the isolates

Sarr et al. (2014) when studying genetic diversity of 189 isolates of *Macrophomina* coming from different hosts and continents have classified senegalese isolates from three regions (Saint-louis, Matam, Louga) and from four hosts (*Abelmoschus esculentus*, *Arachis hypogaea*, *Hibiscus sabdariffa*, and *Vigna unguiculata*) in a second species named *M. pseudophaseolina*. In this study, 05 loci were used: the Internal Transcribed Spacer regions of the nuclear rDNA operon with the primers ITS five and ITS 4 (White et al., 1990), part in (ACT) gene region using primers ACT-512F and ACT-783R (Carbone & Kohn, 1999, part of CAL gene region using primers CAL-228F and CAL-737R (Carbone & Kohn, 1999), part of the TEF-1α gene region using primers EF1-728F and EF 986R, ACT- 512F and ACT-2RD for part of the actin (ACT) gene region and T1-T22 for the partial beta-tubulin gene (O'Donnell & Cigelnik, 1997).

On this batch, 104 isolates of *Macrophomina* sp. were obtained from five plant species in Senegal: cowpea (*Vigna unguiculata* (L.) Walp.), groundnut (*Arachis hypogaea*), Sorrel (*Hibiscus sabdariffa*), sorghum (*Sorghum bicolor*) and okra (*Abelmoschus esculentus*). Physiological and morphological characterization allowed a distinction between the two species.

2.5.1 Effect of temperature on the growth rate of *Macrophomina* sp. isolates

To check if all our isolates were identical from a physiological point of view, we evaluated isolates growth rate at 6°C, 9°C, 12°C, 30°C, 36°C, and 40°C. For this purpose, to obtain young cultures, isolates were sub-cultured on a semi-selective PDA medium containing the antibiotic chloramphenicol at 0.05% and pentachloronitrobenzene (225 mg dissolved in 2 mL of pure alcohol). After incubation at 30°C for 3 days, a 4 mm explant was removed and transferred back to Petri dishes containing agar medium at a rate of three Petri dishes per isolate (thus constituting the replicates) and per temperature. The plates were then incubated at different temperatures. Measurements of mycelial diameter were made every 24 h until day 7 to determine the growth rate of the isolates (Table 2.1).

2.5.1.1 *Macrophomina phaseolina*

M. phaseolina colonies have well-defined margins with aerial and abundant flaky mycelium. Beige on PDA, they turn from winey buff to pale olive-gray with masses of dense, black sclerotia. After 2 days of incubation, the colonies do not grow at 6 and 9°C, grow at 12°C with an average diameter of 9–10 mm, have an optimal growth between 30 and 36°C with a diameter of 80 mm, and still continue to grow at 40°C (25–40 mm diameter).

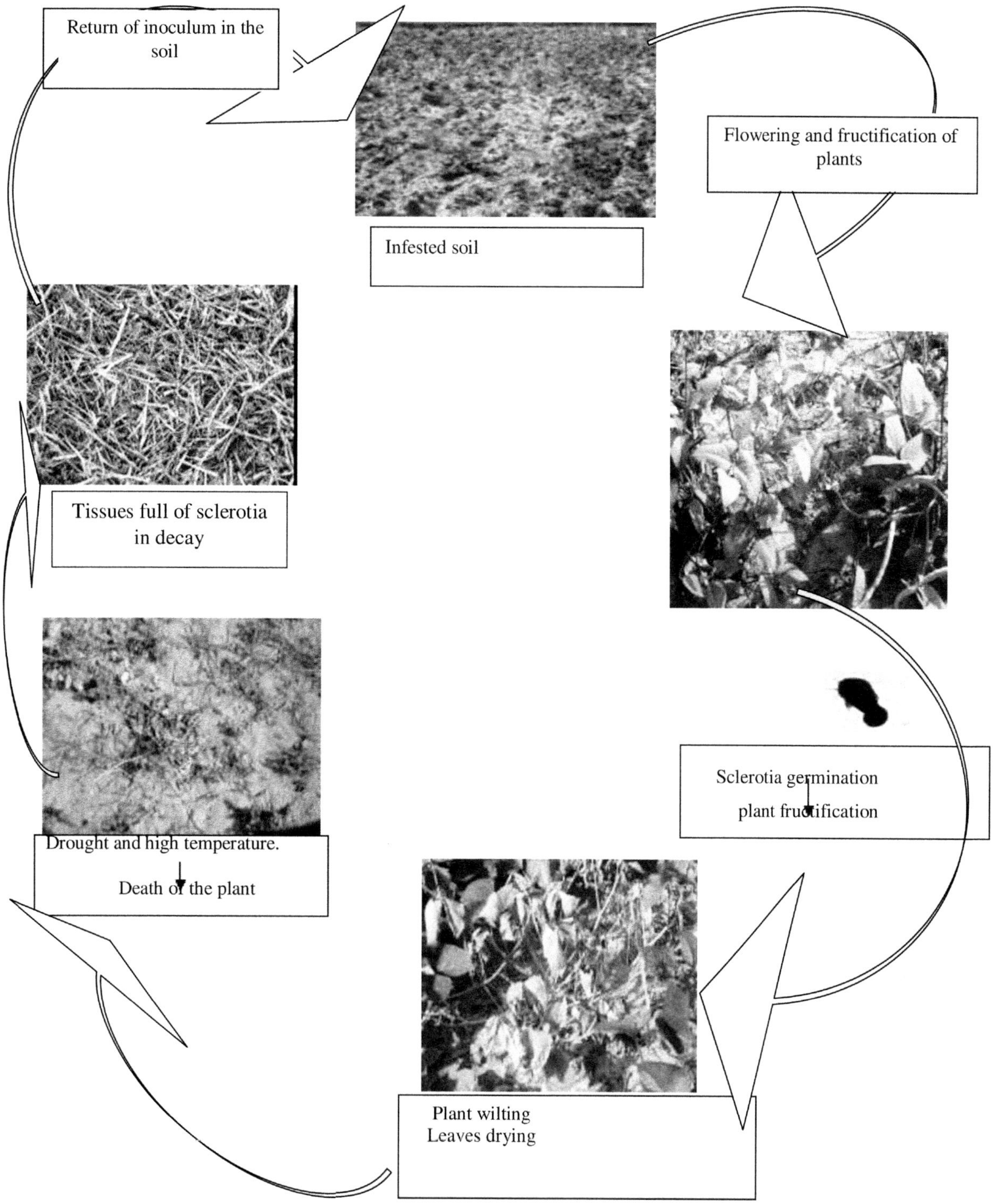

FIGURE 2.3 Disease cycle of *Macrophomina* sp. on cowpea in an infested field.

TABLE 2.1 Average colony diameter per species after 2 days of incubation at 6 and 40°C.

Temperatures (°C)	Diameter of colonies of *M. phaseolina* (mm)	Diameter of colonies of *M. pseudophaseolina* (mm)
6–9	0	0
12	9–10	4–5
30–36	80	80
40	25–40	10–15

2.5.1.2 *Macrophomina pseudophaseolina*

Colonies have well-defined margins with an abundant aerial mycelial. Initially wine-colored buff, they turn pale olive-gray with dense masses of sclerotia. Colonies do not grow at 6 and 9°C after 2 days of incubation, growing weakly at 12°C with a diameter of 4–5 mm. Optimal growth is obtained between 30 and 36°C with a diameter of 80 mm and minimal at 40°C (10–15 mm diameter).

2.5.1.3 *Comparison of the two species at 36 and 40°C*

Isolates of *Macrophomina* sp. obtained from the main cowpea production areas in Senegal were compared in terms of growth rate at two temperatures: 36 and 40°C. At 36°C, the growth rate of all isolates was rapid and similar, and the size of the fungal colonies was maximal (diameter of 9 cm petri dishes) after 3 days of incubation. In contrast, at 40°C, colony development was very low until 7 days after incubation. However, at 40°C, *M. phaseolina* isolates grew faster than *M. pseudophaseolina* isolates with an average diameter of 3.6 cm for *M. pseudophaseolina* isolates and an average diameter of 4.92 cm for *M. phaseolina* isolates in 1 week (Figs. 2.4 and 2.5).

A comparison of the mean colony diameter of all isolates regardless of species showed a faster growth of isolates at 36°C than at 40°C. At 40°C, isolates grew slowly with an average colony diameter of 4.75 cm after 7 days of incubation. At 36°C, the average is 8.9 cm after 1 week of incubation, and this diameter is reached from the third day of incubation (Fig. 2.6).

2.5.2 Phenotype of the colonies

After 07 days of incubation at different temperatures, the appearance, density, and color of the different isolates were also evaluated. For the 104 isolates from Senegal, 101 gave dense and slightly dense (sparse) mycelia, and three isolates had phenotypes of colonies with limited growth patterns.

M. phaseolina usually has a black mycelium (70% of isolates) in contrast to *M. pseudophaseolina*, which often appears as gray (88.89% of isolates). However, *M. phaseolina* may appear gray (30% of isolates) just as *M. pseudophaseolina* may appear black (11.11% of isolates) (Fig. 2.7). According to Reichert and Hellinger (1947), most of *M. phaseolina* colonies are most often gray or white with short mycelia sloping toward the direction of growth, aerial mycelia are produced scarcely.

2.5.3 Sclerotia shape

Microscopic observation was made to determine the shape of the sclerotia of both species using Optika light microscope and Optika Vision Lite software. Generally, in both species, irregular shapes were observed for most of sclerotia of the isolates. However, circular shapes were observed for some isolates of *M. phaseolina* as well as for those of *M. pseudophaseolina* (Fig. 2.8).

FIGURE 2.4 Evolution of the mean diameter of *M. phaseolina* and *M. pseudophaseolina* isolates at 36°C.

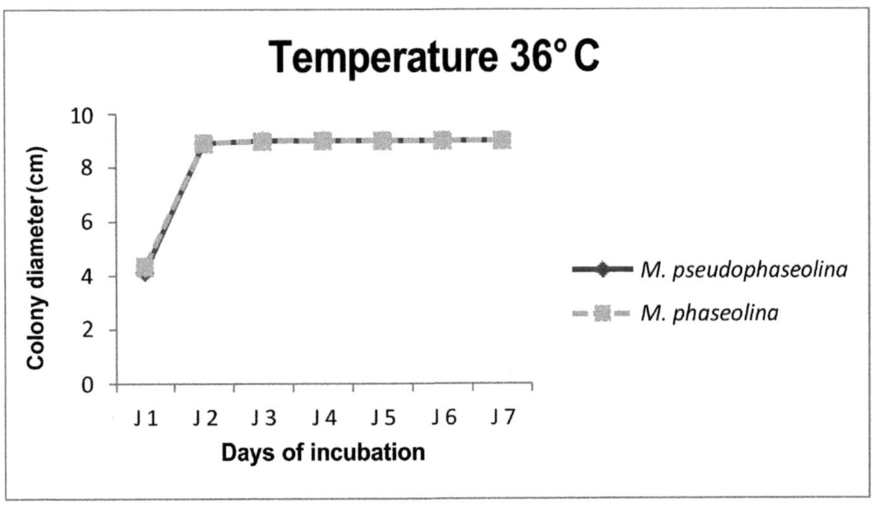

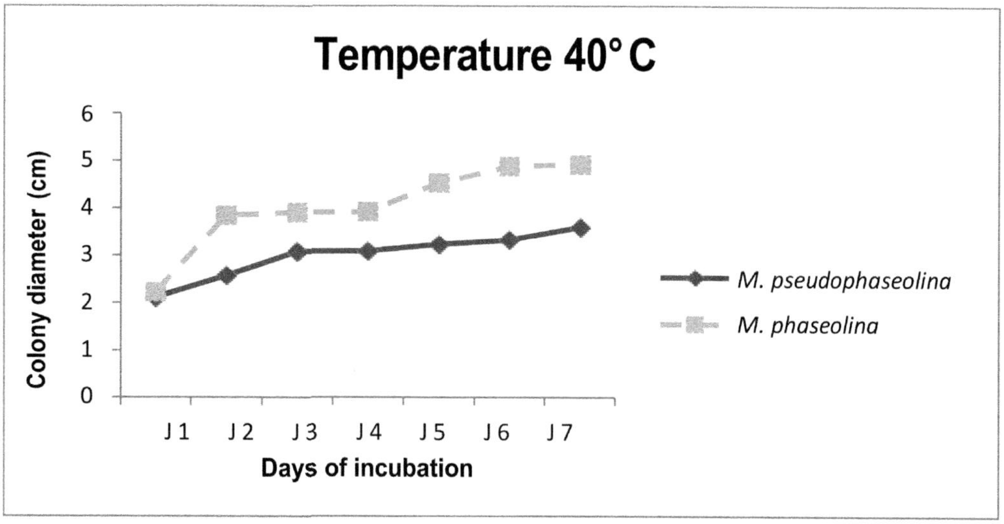

FIGURE 2.5 Evolution of the mean diameter of *M. phaseolina* and *M. pseudophaseolina* isolates at temperature 40°C.

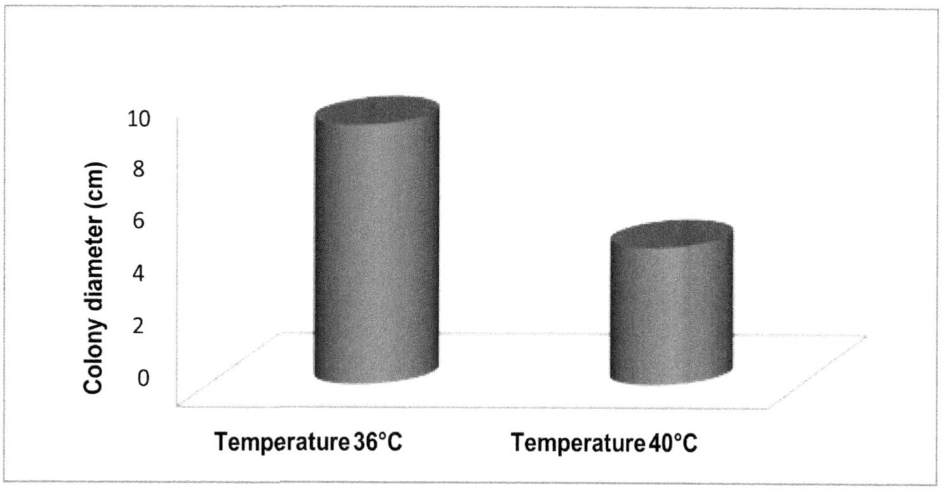

FIGURE 2.6 Color phenotypes of a sparsely growing gray *Macrophomina phaseolina* isolate (A) and a densely growing black *Macrophomina pseudophaseolina* isolate (B).

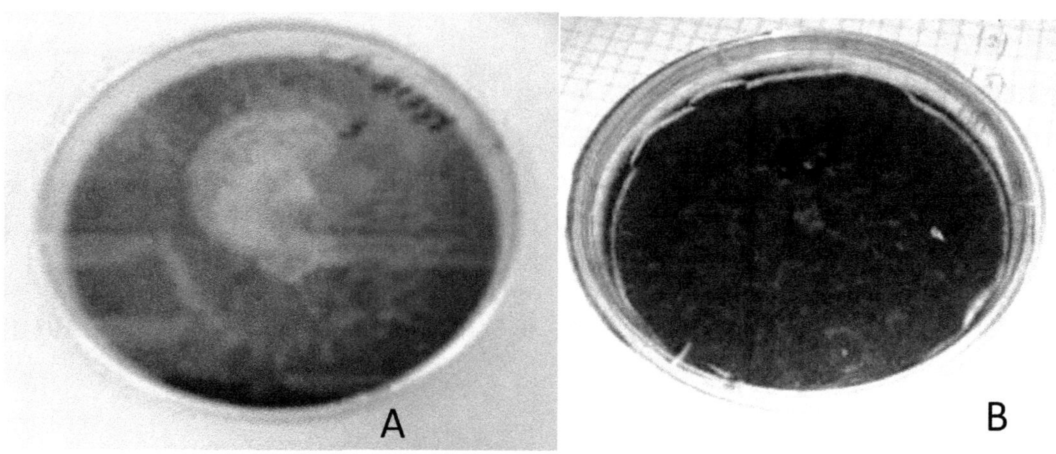

FIGURE 2.7 Color phenotype of a sparsely growing gray *Macrophomina phaseolina* isolate (A) and a densely growing black *Macrophomina pseudophaseolina* isolate (B).

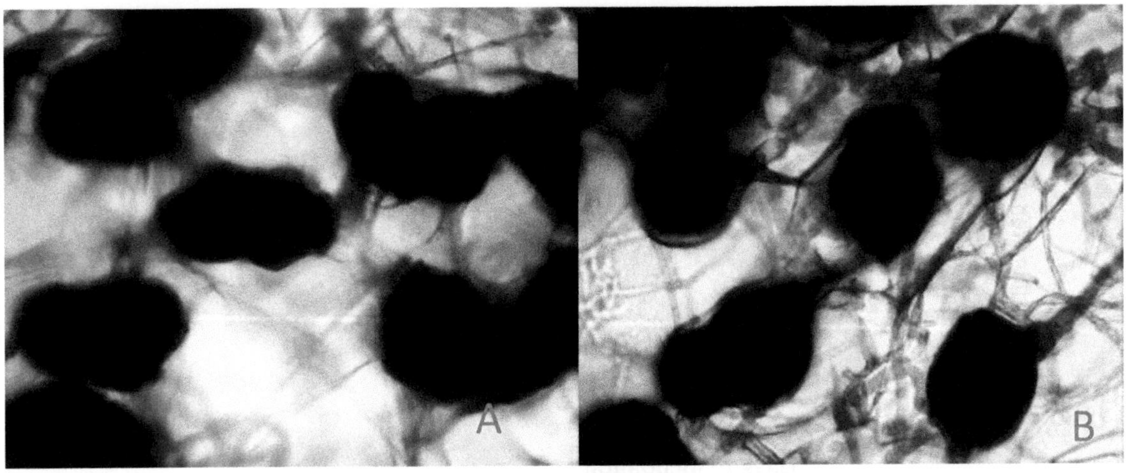

FIGURE 2.8 Sclerotia of *Macrophomina* sp. of irregular shape (A) and of circular shape (B).

2.5.4 Weight of sclerotia of both species

To determine the weight of sclerotia, isolates of *M. phaseolina* and *M. pseudophaseolina* were grown in 100 mL of soluble potato starch medium for 07 days at a rate of 03 replicates/isolate during 15 days at 32. Thus, the different isolates of the two species were classified into 03 categories: light isolates with weight less than 50 mg, medium-heavy isolates $51 \leq$ Poids ≤ 100 mg, and heavy isolates with weight greater than 100 mg. Of the 18 isolates tested, the average weight of sclerotia of the two species was 113.02 mg. Sclerotia of both species were generally heavy, only 1 M. *pseudophaseolina* isolate was light (Fig. 2.9 and Table 2.2).

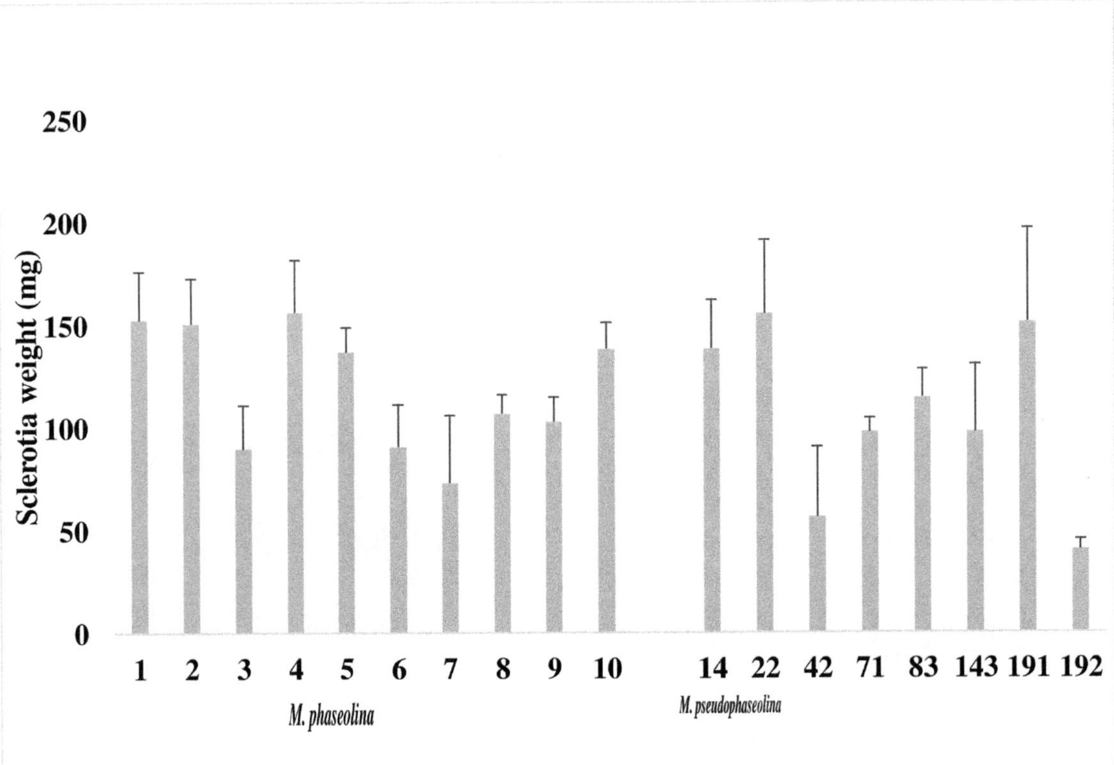

FIGURE 2.9 Sclerotia variability of 18 tested isolates according to their weight.

TABLE 2.2 Classification of the 18 isolates tested according to their weight.

Weight categories	Number of isolates
Isolats légers avec un poids inférieur à 50 mg	01
Isolats moyennement lourds 51 ≤ Poids ≤ 100 mg	06
Isolats lourds dont le poids est supérieur à 100 mg	11

2.6 Conclusion

In the Sahel, charcoal rot disease caused by *Macrophomina* sp., is wide spread and infect many economic important crops including cowpea, peanut, sorrel, okra, sorghum, etc. There were reported two species in the region; however, *M. pseudophaseolina* was only observed in Senegal. The two species have many isolates with different morphological, physiological, and pathogenic characteristics. Some factors such as high soil temperature, erratic rain fall, susceptible host, and high seeding density increase the disease incidence and the soil inoculum and reduce the pod and yield losses. The fungus can survive for more than 10 months under dry soil conditions. The study of 104 isolates collected from five main host crops: groundnut, cowpea, okra, and sorghum and in seven regions in Senegal (Saint-Louis, Thiès, Diourbel, Kaolack, Kaffrine, Tambacounda, and Louga) showed us that morphologically, *M. phaseolina* is similar to *M. pseudophaseolina*.

Physiological characterization showed a faster growth of *M. phaseolina* isolates at 40°C, indicating that *M. phaseolina* isolates are more tolerant to extreme temperature variations than *M. pseudophaseolina* isolates. Although the morphology of *M. pseudophaseolina* appears to be largely similar to that of *M. phaseolina*, its ecological attributes need further study.

References

Abawi, G. S., & Pastor-Corrales, M. A. (1990). Seed transmission and effect of fungicide seed treatments against *Macrophomina phaseolina*in dry edible beans. *Turrialba, 40*, 304–339.

Adam, T. (1986). *Contribution à la connaissance des maladies du niébé (*Vignaunguiculata *(L) Walp.) au Niger avec mention spéciale au* Macrophominaphaseolina*(Tassi) Goïd* (p. 117). Université de Renne I. Thèse de doctorat.

Adam, T. (1990). Les maladies affectant le niébé (*Vignaunguiculata*(L.) Walpers) au Niger. *Ambio, 19*, 358–359.

Alabouvette, C. (1976). Recherches sur l'écologie des champignons parasites dans le sol. VIII. - Etude écologique de *Macrophominaphaseolina*grâce à une technique d'analyse sélective. *Annales de Phytopathologie, 8*, 147–157.

Ammon, V., Wyllie, T. D., & Brown, M. F. (1974). Ultra-structural investigation of pathological alterations induced by *Macrophominaphaseolina*(Tassi) Goid in seedlings of soybean, Glycine max (L) Merrill. *Physiological Plant Pathology, 4*, 1–10.

Azize, H. T. A., & Ndiaye, M. (2022). Sustainability of endogenous farming practices in controlling plants diseases: Case of seeding density on *Macrophominaphaseolina*. *Pakistan Journal of Phytopathology, 34*, 57–63.

Babu, B. K., Mesapogu, S., Sharma, A., Somasani, S. R., & et Arora, D. K. (2011). Quantitative real-time PCR assay for rapid detection of plant and human pathogenic *Macrophominaphaseolina*from field and environmental samples. *Mycologia, 103*, 466–473.

Baird, R. E., Watson, C. E., & Scruggs, M. (2003). Relative longevity of *Macrophominaphaseolina*and associated mycobiota on residual soybean roots in soil. *Plant Disease, 87*, 563–566.

Bhattacharya, D., Dhar, T. K., Siddiqui, K. A. I., & Ali, E. (1994). Inhibition of seed germination by *Macrophominaphaseolina*is related to phaseolinone production. *Journal of Applied Bacteriolology, 77*, 129–133.

Blanco-Lopez, M. A., & et Jimenez-Diaz, R. M. (1983). Effect of irrigation on susceptibility of sunflower to *Macrophomina phaseoli*. *Plant Disease, 67*, 1214–1217.

Bouhot, D. (1967). Étude du *Macrophominaphaseoli*sur arachide. *Tropical Agriculture, 22*, 1165–1171.

Bouhot, D. (1968). Le *Macrophominaphaseoli*sur les plantes cultivées au Sénégal. *Tropical Agriculture, 23*, 1172–1181.

Bressano, M., Giachero, M. L., Luna, C. M., & Ducasse, D. A. (2010). An in vitro method for examining infection of soybean roots by *Macrophominaphaseolina*. *Physiological and Molecular Plant Pathology, 74*, 201–204.

Campbell, C. L., & van der Gaag Dirk, J. (1993). Temporal and spatial dynamics of microsclerotia of *Macrophominaphaseolina*in three fields in North Carolina over four to five years. *Phytopathology, 83*, 1434–1440.

Carbone, I., & Kohn, L. M. (1999). A method for designing primer sets for speciation studies in filamentous ascomycetes. *Mycologia, 91*, 553–556.

Chan, Y. H., & Sackston, W. E. (1973). Non specificity of the necrosis inducing toxin of Sclerotiumbataticola. *Canadian Journal of Botany, 51*, 690–692.

Cissé, N., & Hall, A. E. (2002). *Traditional Cowpea in Senegal, a case study* (p. 27). www.fao.org/ag/AGP/AGPC/doc/publicat/cowpeaCisse/cowpeacissee.html.

Collins, D. J., Wyllie, T. D., & Anderson, S. H. (1991). Biological activity of *Macrophominaphaseolina*in soil. *Soil Biology and Biochemistry, 23*, 495–496.

Cook, G. E., Boosalis, M. G., Dunkle, L. D., & Odvody, G. N. (1973). Survival of *Macrophominaphaseoli*na corn and sorghum stalk residue. *Plant Disease Reporter, 57*, 873–875.

Crous, P. W., Slippers, B., Wingfield, M. J., Rheeder, J., Marasas, W. F. O., Philips, A. J. L., Alves, A., Burgess, T., Barber, P., & Groenewald, J. Z. (2006). Phylogenetic lineages in the Botryosphaeriaceae. *Studies in Mycology, 55*, 235–253.

Davet, P., Herbach, M., & Piquemal, G. (1986). Effet de quelques facteurs intrinsèques ou extrinsèques d'affaiblissement des tournesols sur leur sensibilité au dessèchement précoce. *Agronomie, 6*, 803–810.

Dhingra, O. D., & Sinclair, J. B. (1977). *An annotated bibliography of* Macrophomina phaseoli *(1905–1975)*. Minas Gerais, Viçosa, Brazil: UniversidadeFederal de Viçosa.

Dhingra, O. D., & Sinclair, J. B. (1978). *Biology and pathology of* Macrophominaphaseolina (p. 166). Viscosa, Brazil: Universidade Federal de Viscosa.

Diourte, M., Starr, J. L., Jeger, M. J., Stack, J. P., & Rosenow, D. T. (1995). Charcoal rot (*Macrophominaphaseolina*) resistance and the effects of stress on disease development in Sorghum. *Plant Patholology, 44*, 196–202.

Edmunds, L. K. (1964). Combined relation of plant maturity, temperature, and soil moisture to charcoal stalk rot development in grain sorghum. *Phytopathology, 54*, 514–517.

Emechebe, A. M., & McDonald, D. (1979). Seed-borne pathogenic fungi and bacteria of cowpea in Northern Nigeria. *Pest Articles and News Summaries, 25*, 401–404.

Gaetàn, S. A., Fernande, L., Madia, M., et al. (2006). Occurrence of charcoal rot caused by *M. phaseolina*on canola in Argentina. *Plant Disease, 90*, 524.

Gaikwad, D. G., & Sokhy, S. S. (1987). Detection of seed rot, root rot and seedling infection in naturally infected cowpea seeds in Senegal and their control. *Plant Disease Research, 2*, 127–128.

Gangopadhyay, S., & Wyllie, T. D. (1974). Melanin-like compound in the sclerotia of *Macrophominaphaseolina*. *Indian Phytopathology, 27*, 661–663.

Ghosh, T., Mukherji, N., & Basak, M. (1964). On the occurrence of a new species of *Orbilaria Fr. Jute Bulletin, 27*, 134–141.

Gray, F. A., Kolp, B. J., & Mohamed, M. A. (1990). A disease survey of crops grown in the Bay Region of Somalia, East Africa. F. A. O. *Plant Protection Bulletin, 38*, 39–47.

Hall, R. (Ed.). (1991). *Compendium of bean diseases* (p. 73). St. Paul, MN: American Phytopathological Society.

Halsted, B. D. (1890). Some fungus diseases of the sweet potato. *New Jersey Agricultural Experiment Station Bulletin, 76*, 7–14.

Hartman, G. L., Sinclair, J. B., & Rupe, J. C. (Eds.). (1999). *Compendium of soybean diseases* (4th ed.). St. Paul, Minnesota, USA: American Phytopathological Society.

ICRISAT (Ed.). (1980). *Proceedings of the Consultants group discussion on the resistance to soil borne diseases of legumes* (p. 167). India.

Jimenez, D. R. C. (2011). *Influence of soils, nutrition, and water relations upon charcoal rot disease processes in Kansas* (p. 127). MSc. Thesis, Kansas State University, USA.

Jones, R. W., & Wang, H. Y. (1997). Immunolocalization of a beta-1,4-endoglucanase from *Macrophominaphaseolina*expressed in plant. *Canadian Journal of Microbiology, 43*, 491–495.

Kuti, J. O., Schading, R. L., Latigo, G. V., & Braford, J. M. (1997). Differential responses of guayule (Partheniumargentatum Gray) genotypes to culture filtrate and toxin from *Macrophominaphaseolina*(Tassi) Goidanich. *Journal of Phytopathology, 145*, 305–311.

Ma, J., Hill, C. B., & Hartman, G. L. (2010). Production of*Macrophominaphaseolina*conidia by multiple soybean isolates in culture. *Plant Disease, 94*, 1088–1092.

Maublanc, A. (1905). Nouvelles espèces de champignons inferieurs. *Bulletin de la Société mycologique de France, 21*, 85–95.

Mayek-Pérez, N., García-Espinosa, R., López Castañeda, C., Acosta-Gallegos, J. A., & Simpson, J. (2002). Water relations, histopathology, and growth of common bean (*Phaseolus vulgaris* L.) during pathogenesis of *Macrophominaphaseolina*under drought stress. *Physiological Plant Pathology, 60*, 185–195.

Meyer, W. A., Sinclair, J. B., & Khare, M. N. (1974). Factors affecting charcoal rot of soybean seedlings. *Phytopathology, 64*, 845–849.

Mihail, D. J. (1989). *Macrophominaphaseolina*: Spatio-temporal dynamics of inoculum and of disease in a highly susceptible crop. *Phytopathology, 79*, 848–855.

Mihail, J. D. (1992). Macrophomina. In L. L. Singleton, J. D. Mihail, & C. M. Rush (Eds.), *Methods for research on soil borne phytopathogenic fungi* (pp. 134–136). St. Paul, MN: American Phyothopathological Society.

Mihail, J. D., & Taylor, S. J. (1995). Interpreting variability among isolates of *Macrophominaphaseolina*in pathogenicity, pycnidium production, and chlorate utilization. *Canadian Journal of Botany, 73*, 1596–1603.

Ndiaye, M. (2007). *Ecology and management of charcoal rot (*Macrophominaphaseolina*) on cowpea in the Sahel*. Ph.D. thesis Wageningen University and Research Centre, Wageningen, The Netherlands, ISBN 978-90-8504-664-6.

O'Donnell, K., & Cigelnik, E. (1997). Two divergent intragenomic rDNA ITS2 types within a monophyletic lineage of the fungus *Fusarium* are no orthologous. *Molecular Phylogenetics and Evolution, 7*, 103–116.

Olaya, G., Abawi, G. S., & Barnard, J. (1996). Influence of water potential on survival of sclerotia in soil and on colonization of bean stem segments by *Macrophominaphaseolina*. *Plant Disease, 80*, 1351–1354.

Papavizas, G. C. (1977). Some factors affecting survival of sclerotia of *Macrophominaphaseolina*in soil. *Soil Biology and Biochemistry, 9*, 337–341.

Paré, D. (1990). Technique de quantification de l'inoculum et distribution géographique de *Macrophominaphaseolina*(Tassi) Goid. au Burkina Faso. Deuxième Séminaire sur la Lutte Intégrée contre les Ennemis des Cultures Vivrières dans le Sahel, 4–9 Janvier 1990. Bamako - Mali. CILSS, INSA, UCTR/PV. p. 12.

Pratt, R. G. (2006). A direct observation technique for evaluating sclerotium germination by *Macrophominaphaseolina*and effects of biocontrol materials on survival of sclerotia in soil. *Mycopathologia, 162*, 121–131.

Raut, J. G. (1985). Effect of charcoal rot caused by *Macrophominaphaseolina*on sunflower plants. *Indian Phytopathology, 38*, 345–346.

Reichert, I., & Hellinger, E. (1947). On the occurrence, morphology and parasitism of *Sclerotiumbataticola*. *Palestine Journal of Botany, 6*, 107–147.

Sarr, M. P., Ndiaye, M., Groenewald, J. Z., & Crous, P. W. (2014). Genetic diversity in *Macrophominaphaseolina*, the causal agent of charcoal rot. *Phytopathologia Mediterranea, 53*, 163–173.

Shanmugan, N., & Govindaswamy, C. V. (1973). Varietal susceptibility of groundnut (Arachis hypogeal L.) to *Macrophomina*root rot. *Madras Agricultural Journal, 60*, 591.

Shaw, F. J. F. (1912). The morphology and parasitism of *Rhizoctonia*. Memoirs of the Department of Agriculture in India. *BotanicalServey, 4*(115), 153.

Short, G. E., & Wyllie, T. D. (1978). Inoculum potential of *Macrophominaphaseolina*. *Phytopathology, 68*, 742–746.

Short, G. E., Wyllie, T. D., & Bristow, P. R. (1980). Survival of *Macrophominaphaseolina*in soil and residue of soybean. *Phytopathology, 70*, 13–17.

Smith, G. S., & Carvil, O. N. (1997). Field screening of commercial and experimental soybean cultivars for their reaction to *Macrophominaphaseolina*. *Plant Disease, 81*, 363–368.

Tarr, S. A. J. (1962). *Diseases of sorghum, Sudan grass and broom corn*. Kew, Surrey, UK: Commonwealth Mycology inst.

Tassi, A. (1901). Novae micromycetum species. *Ser III Bull Lab Orto Bot Univ Siena, 4*, 7–12.

White, T. J., Bruns, T., Lee, J., & Taylor, S. B. (1990). Amplification and direct sequencing of fungal ribosomal RNA genes for phylogenetics. In M. A. Innis, D. H. Gelfand, J. J. Sninsky, & T. J. White (Eds.), *PCR protocols: A guide to methods and applications* (pp. 315–322). San Diego, California, USA: Academic Press.

Wrather, J. A., & Koenning, S. R. (2010). *Soybean disease loss estimates for the United States, 1996–2009*. Delta Research Center: Agriculture Experiment Station. University of Missouri, College of Agriculture, Food and Natural Resources, USA.

Wyllie, T. D., & Scott, D. H. (1988). Charcoal rot of soybeans - current status. In *Soybean diseases of the north central region* (pp. 106–113). Indianapolis, Indiana: APS Press.

Young, D.,J., & Alcorn, S. M. (1984). Latent infection of *Euphorbia lathyris*and weeds by *Macrophominaphaseolina*and propagule populations in Arizona field soil. *Plant Disease, 68*, 587–589.

Chapter 3

Macrophomina phaseolina causing various diseases in different crops

Iqra Haider Khan and Arshad Javaid
Department of Plant Pathology, Faculty of Agricultural Sciences, University of the Punjab, Lahore, Pakistan

3.1 Introduction

Macrophomina phaseolina (Tassi) Goid., is a polyphagous pathogen belonging to anamorphic Ascomycetes (Ndiaye, 2007). *M. phaseolina* was the original fungal pycnidial state (Tassi, 1901) followed by *Macrophoma phaseoli* (Maublanc, 1905) and *Macrophomina phaseoli* (Ashby, 1927). Later on, it was named as *M. phaseolina* (Goidanich, 1947). It is a necrotrophic fungal pathogen that inhabits both soil and seeds. It causes various diseases such as dry weather wilt, charcoal rot, dry root rot, damping off, seedling blight, ashy stem blight, and leaf blight in over 500 plant species (Khan et al., 2021; Lodha & Mawar, 2020). It is widely distributed all around the world and present in variable soil types indicating its omnipresence. It is a pathogen of medicinal plants, weeds, and crops, such as sorghum, cotton, common bean, soybean, maize, mungbean, cluster bean, chickpea, cowpea, sesame, and peanut (Basandrai et al., 2021; Khan & Javaid, 2022a). It is also responsible for diseases in forest trees including *Cassia*, *Pseudotsuga*, *Pinus*, and *Abies* and fruit trees including, *Leucaena* spp., *Ziziphus mauritiana*, *Coffea* spp., *Cocos nucifera,* and *Citrus* spp. (Songa & Hillocks, 1996). It is also an opportunistic human pathogen (Javed et al., 2021; Uroos et al., 2022).

It persists in the form of microsclerotia in host crop debris and soil serving as a primary source of inoculum. Infected crop residues protect microsclerotia and are released after their breakdown in the soil where they remain viable for up to 3−15 years under favorable environmental conditions (Baird et al., 2003; Wagan et al., 2019). A mature sclerotium is pigmented; outermost cells are uniformly reticulated and tightly packed with no structural difference. It is a major source of inoculum. It can withstand high temperatures of dry soil, at low moisture and nutrient levels (Lodha, 1995). Nutrients stored in microsclerotia fuel fungal growth in the soil even if the level of the soil nutrients depletes for fungal competitors to grow. The soil with low nutrients and high temperature favors the *M. phaseolina* to compete with other soil-borne pathogens. Microsclerotia germinate from a few cells in close proximity or on the surface of roots. The germ tubes are formed through the induction of root exudates, which give rise to appressoria present on epidermal cell walls. These penetrate through mechanical pressure, enzymatic digestion, wounds, or natural openings (Hartman et al., 1991). Ramezani et al. (2007) observed the penetration of fungal mycelium in root epidermal cells that is primarily restricted in cortex in intracellular spaces during the initial infectious stages resulting in the collapse of adjacent cells and plantlets death. Necrotic lesions were found on infected plant branches, peduncles, and stems. Thereafter, it further spreads from pod peduncles to infect grains. At the seedling stage, multiple germinating hyphae possibly infect host plant roots and upper parts through disruption of the vascular system, water transport, and nutrients. Fungal toxins including botryodiplodia and phaseolinone result in premature plant death, with reduced seed size and poor seed setting in pods. After plant death, microsclerotia are released into the soil and the cycle continues. Plant species and crop age show variable disease expressions. In general, typical disease symptoms include leaves yellowing, sloughing of taproot, and lower stem cortical tissues with an ultimate greyish appearance, which result in premature plant death. The symptoms become more severe under dry hot conditions and when the temperature is in the range of 28−35°C (Smith & Carvil, 1997). Continuous cultivation of susceptible crops significantly increases the *M. phaseolina* populations in the soil. Moreover, the disease incidence in the field is directly related to the inoculum density of sclerotia (Sheikh & Ghaffar, 1979).

Macrophomina Phaseolina. https://doi.org/10.1016/B978-0-443-15443-0.00004-8

3.2 Morphological characterization

M. phaseolina is characterized by light or dark brown, thin-walled hyaline hyphae with septa. At the point of origin, parent hyphae produce lateral branches with constriction in the right direction. Microsclerotia are a hardened compact mass of fungal mycelium, which is light brown when young and gradually become darker with aging. These are light spherical, oval to oblong in shape. Pycnidia are often black, rough, irregular, or globose in shape and are larger than microsclerotia (Lakhran et al., 2018). Moreover, morphological heterogeneity is generally attributed to the variation in host species and fungal responses toward environmental factors (Pandey et al., 2020).

3.3 Pathogenicity of *M. phaseolina*

Pathogenicity-associated genes are encoded by *M. phaseolina* genome that plays a key role in disease expression. Around 12% of genes are encoded by the genome having identities to those involved in successful host—pathogen interactions. These genes have the potential to neutralize host defense through the production of salicylate-1-monooxygenase and enable *M. phaseolina* to adhere host tissues by cellulose-binding as well as invasion in the host epidermis. The pathogen secretes certain toxins and cell wall—degrading enzymes that gradually disrupt the host defense mechanisms with an eventual plant death (Islam et al., 2012). It also produces a wide array of cell wall—degrading hydrolytic enzymes, which directly penetrate into the host tissues. These include β-glucosidases, exocellobiohydrolases, and endoglucanases for cellulose hydrolysis, whereas lignin peroxidases, laccases, chloroperoxidases, galactose oxidases, heme peroxidases, and haloperoxidases are used to degrade lignin. Moreover, *M. phaseolina* produces cellulitic and laccases enzymes in excessive amounts in comparison to *Ustilago maydis, Cryptococcus neoformans, Phanerochaete chrysosporium, Postia placenta, Neurospora crassa, Aspergillus nidulans,* and *Saccharomyces cerevisiae* (Bandara et al., 2018). Ramos et al. (2016) have confirmed the activity and production of cell wall-degrading enzymes in plants. *M. phaseolina* also produces phytotoxic metabolites including patulin, botryodiplodin, and phaseolinone. These metabolites play a key role in necrotic lesion formations and wilt diseases (Abbas et al., 2020; Salvatore et al., 2020). In addition, the virulence mechanism is efficiently increased by phytotoxic metabolites that infect different tissues and hosts. These are adapted to different environmental conditions, which contribute to their infectivity and ubiquitous distribution (Salvatore et al., 2020). Plant defense response transformation counters through the expression of detoxifying genes including cytochrome, superoxide dismutase, and cof and P450 (Ghosh et al., 2018).

Interestingly, *M. phaseolina* genome analysis revealed the presence of those amino acids conserved with synthase-like sequences of nitric oxide (NO). Higher concentrations of NO were found in the mycelium of infected jute tissues. Therefore, Sarkar et al. (2014) suggested that NO is physiologically important in successful host—pathogen interactions.

3.4 Diseases caused by *M. phaseolina*

Diseases caused by *M. phaseolina* are summarized in Table 3.1.

3.4.1 Charcoal rot

3.4.1.1 Hosts

Charcoal rot also known as dry root rot is a disease of soybean, sunflower, alfalfa, maize, cotton, jute, sorghum, cluster bean, green gram, strawberry, and chickpea (Manikandan et al., 2022; Pastrana et al., 2022). It develops during the mid- to late summer when the plant is under stress in tropical and semi-tropical regions. It is a common disease in southern areas of the United States of America but now spreads into northern states including Illinois, Ohio, and Iowa. The pathogen incidence has also been increasing in Asian and South Asian countries. It has the ability to infect plants at both pre- and postemergence stages in different growing seasons (Manjunatha et al., 2013; Singh et al., 2008; Thilagavathi et al., 2007).

3.4.1.2 Symptoms

The disease symptoms are frequently observed in stressed field areas, hillsides, and compacted soils, whereas scattered yellow patches of infected crops can be observed over large areas. In charcoal rot, premature death of plants occurs, and for several days, infected leaves remain attached to the petioles. Charcoal rot severity is also influenced by plant maturity, and more pronounced symptoms may be observed in years with low soil moisture and high temperatures. However, the mechanism lays behind disease severity and plant maturity are not well understood. It is believed that early maturing plants

TABLE 3.1 Diseases caused by *M. phaseolina* in different host plants.

Disease	Host plant	Country	Symptoms	References
Charcoal rot	Soybean, sunflower, cotton, sorghum	Louisiana, USA	Appearance of greyish lesion on stem, black spot-on cotyledons, leaf yellowing, root discoloration, reduced vigor and plant wilt	Su et al. (2001)
	Cluster bean	India		Purkayastha et al. (2006)
	Green gram, strawberry, chickpea, Chir-pine	India, Spain, Israel		Thilagavathi et al. (2007), Avilés et al. (2008), Singh et al. (2008), Manjunatha et al. (2013)
	Mungbean, mash bean	Pakistan		Banaras et al. (2020), Javaid et al. (2018), Khan and Javaid (2020, 2022a, 2022b)
Damping-off	Cowpea	Nigeria	Seeds become soft, rotted and fail to germinate After emergence seedlings decay, wilt and die prematurely.	Adekunle et al. (2006)
	Sunflower	Egypt		Abd El-Hai et al. (2009)
	Cotton	Egypt		Aly et al. (2007)
	Common bean	Romania		Bedine et al. (2020)
Stem blight	Common bean	Africa North and South America	Characteristic symptoms include; chlorosis, premature defoliation, wilting, necrosis on stem and twigs spreading downwards resulting in early maturity or plant death.	Viteri and Linares (2017)
	Blueberries	China Serbia North and South America		Popovic et al. (2018)
	Cactus	Egypt		Abdel-Kader et al. (2010)
	Peas	South Australia		Ali and Dennis (1992)
	Cowpea	Nigeria		Amusa and Akinfenwa (2007)
	Tepary bean Common bean	Puerto Rico		Viteri and Linares (2022)
Leaf blight	Safed musli	India	Presence of water-soaked, irregular reddish-brown lesions that are surrounded by yellow holes. Later on, these spots coalesced into extended patches, become necrotic, and darker producing a complete burning of leaf areas.	Meena and Kadam (2021)
	Green gram	India		Tandel et al. (2010)
	Urdbean	India		Singh and Kumar (2002)
	White spider lilies	Malaysia		Huda-Shakirah et al. (2019)
	Kasai	Malaysia		Khoo et al. (2022)
	C. roseus	Bangladesh		Mridha and Rahman (2015)
	Blueberry	Serbia		Popovic et al. (2018)
	J. multiflorum	India		Mahadevakumar and Janardhana (2016)
Pod rot	Mungbean	Australia	Production of small galls on pods. Formation of black lesions start appearing on pods followed by rotting of internal tissue.	Fuhlbohm et al. (2013)

are less susceptible to the pathogen as they flower during cooler temperatures and show a rare contribution to symptom expression and yield loss (Mengistu et al., 2007).

Pathogen causes infections at the early vegetative stages of plant seedlings when temperature is above 35°C, and soil is dry for a continuous period of 2–3 weeks. After emergence, the initial disease symptoms include yellowing of leaves, presence of brown to black spots on cotyledons, root discoloration, and general wilting appearance with reduced vigor. While, infected hypocotyls show the presence of reddish-brown, circular to oblong lesions that become black and extend toward the stem upon maturity. The pathogen persists under hot and dry conditions, and later on, at advanced stages, foliar symptoms show premature senescence and plant death. The pathogen also infects adult plants, and the symptoms appear on pods, stems, roots, and seeds after 4 weeks of maturity. The above-ground superficial lesions are grayish to light brown in color and appear infrequently on the stem. Microsclerotia are greyish to black in appearance, formed in pith and on stem sub-epidermal tissues. The discoloration is formed first at nodes in the form of randomly distributed specks. Foliar symptoms progress from upward to the downward direction, and infected leaves turn yellow before wilting and remain smaller than normal size (Gupta & Chauhan, 2005). As the fungus spreads, lower stem and root vascular elements give reddish-brown discoloration and precede premature yellowing. Microsclerotia completely cover the mature dry pods with black bodies and penetrate into the grains and appear in the form of black diffusible spots on seeds. Sometimes microsclerotia are produced in the cracks and fissures of seed coats. After plant death, minute, pinhead-sized microsclerotia also become visible on roots and lower stem epidermal tissues. The name charcoal rot is derived due to the presence of silver or gray microsclerotia discoloration, which is usually observed on infected lower plant stem and roots.

Mungbean production in South Asia is constrained by many biotic stresses, which limits its yield. Charcoal rot of mungbean is an emerging problem of Pakistan, and it is a prevalent disease in Myanmar (Khan & Javaid, 2020). Mungbean plants are infected by the pathogen at all growth phases and have the ability to become more severe in field both at the adult and seedling stages. *M. phaseolina* spreads within infected plants vascular system and also through soil from one to the adjacent plants. At advanced stages, infected mungbean root rot results in wilting that ultimately results in plant death. Scanning electron microscopy revealed that *M. phaseolina* severely damages the cell structures in infected plant root and stem that can be prevented by application of biocontrol agents such as *Aspergillus* species (Khan & Javaid, 2022b). *M. phaseolina* particularly infects susceptible plants over the winter from soil reservoirs through the production of toxins such as phaseolinone and botryodiplodin. The disease spread has increased in Asia in recent years because of an increase in temperature, water stress, and irregular and reduced rainfall (Pandey et al., 2020). In addition to mungbean, charcoal rot is also very common in mash bean having the similar symptoms as in mungbean (Banaras et al., 2020, 2021).

Chickpea is mainly grown in Africa, India, and Middle East and is now rapidly gaining popularity in the west. It is considered the world's third most important pulse crop. *M. phaseolina* is an endemic fungus that is becoming more severe in Indian chickpea-growingareas and causes up to 25% yield losses. The disease has also been reported in European countries such as in Italy (Dell'Olmo et al., 2022). Charcoal rot symptoms in chickpeas vary according to the time of the year and are dependent on plant parts. Seedlings commonly become infected during the growing period and later on disease develops at any time in the season (Ali & Kumar, 2008). At the initial stages of infection, these show brown discoloration near the soil line extending up the stem that gradually turns black. Infected seedling foliage usually appears off-color, then turns brown, and begins to dry. In addition to this, infected leaves and stems become brittle, dry, and straw-colored in appearance. The characteristic root rot symptoms include yellowing of leaves that may drop off within two or 3 days and within a week plant may wilt. Young plants become infected under wet and cool conditions and later on in the presence of dry hot weather symptoms express and progress (Becker & Schwinn, 1993).

At advanced stages, affected tissues exhibit several scattered sclerotial bodies. *M. phaseolina* survives on or inside the seeds and persists in the form of black sclerotia in soil. These bodies are produced on infected host tissues and during tillage practices subsequently disperse in the soil in large numbers. Srivastava et al. (2001) reported *M. phaseolina* as a major constraint all around the world, especially for cotton and soybean growing areas. In these crops, the infection mechanism moves in an upward direction starting from the roots to the stem collar regions and causes charcoal color rot, foliar chlorosis, or collar region discoloration. Charcoal rot of soybean is ranked among the top five most important diseases that cause low yield and seed quality with huge economic losses (Bellaoui et al., 2021; Mengistu et al., 2011). Infected soybean plants aid *M. phaseolina* to grow rapidly and clog vascular bundles by producing sclerotia in large numbers. It results in the appearance of disease symptoms ranging from wilting, leaf yellowing, and plant death. The disease symptoms in plants often occur under drought or heat stress also referring as summer or dry-weather wilt. Moreover, the combination of stress and *M. phaseolina* causes high production losses than drought alone can do on soybean. The pathogen attacks the soybean throughout the season and often causes progressive host debilitation. After flowering, epidermal and sub-epidermal tissues develop a silvery or light gray discoloration in the lower part of the stem and taproots. The disease diagnosis symptoms can be found on the stem epidermis, and when it is peeled away, it shows numerous black small microsclerotia. These often block water flow due to their frequent production in pith and xylem vessels (Mengistu et al., 2011).

Strawberry is mainly cultivated in Spain as one of the most important crops. Its production is severely threatened by charcoal rot pathogen and becoming problematic worldwide. Strawberry charcoal rot has also been reported in Iran, Italy, Australia, Argentina, the United States, Israel, and Turkey (Angelini & Faedi, 2010; Baino et al., 2011; Golzar et al., 2007; Viejobueno et al., 2021). Koike (2008) found that in strawberries, the pathogen shows symptoms of foliage wilting, drying of older leaves that eventually collapse, and the plant dies. In addition, cut-open diseased crowns show orange to dark brown discoloration of vascular and internal cortical tissues. At mature stages, the diseased roots completely become rotted and brown. The development of charcoal rot in strawberries is increased by the consolidation of water deficit light textured soil, host reproduction-associated stress, or heat stress (Mihail, 1989). Moreover, the pathogen becomes more dominant when temperature often exceed 30°C, particularly in warmer months of the year, such as in early summer or late spring. In the field, pathogen survives in soil in the form of microsclerotia or on crop debris and infects strawberry plants, which may wilt and die rapidly. These microsclerotia allow the fungus to persist in soil for extended periods serving as the most resilient resistant structures. Charcoal rot affected strawberry fields have generally limited, small patches with symptomatic plants. However, the locations of initial infections show the appearance of small patches, which become quite large where the pathogen has developed for more than one seasons. Strawberry is an annual crop plant, and if it grows without rotation in high-temperature areas, creates optimum conditions for *M. phaseolina* development and multiplication. This situation has resulted in extensive yield losses in strawberry-producing areas of the Mediterranean basin and Europe (Santos et al., 2006).

3.4.1.3 Disease cycle

M. phaseolina mainly produces either pycnidia or sclerotia and exists as a hardened compact mass of mycelial structures called "sclerotia" in the soil. The germination of fungal hyphae from sclerotia infects plant cell walls through penetration into the host roots. The life cycle of a pathogen begins with the germination of microsclerotia. These structures survive in the soil and grow between 20 and 40°C and infect root tissues. These infect plants at both seedling and mature stages. Interestingly, dry and hot weather conditions allow rapid production of *M. phaseolina* in the stalk which multiplies in the host pith. During the initial stages of seed germination, fungal hyphae grow toward and colonize the roots and seedlings of the host plants. Initially, the pathogen remains latent, and no visible symptoms appear on aerial plant organs. Later on, disease symptoms become obvious at the end of the growing season when the pathogen approaches into its necrotic phase.

Once the infection has occurred, it colonizes and degrades plant tissues within 2–3 weeks through secretion of toxins and enzymes (Islam et al., 2012). Extracellular enzymes, such as hemicelluloses, celluloses, pectinases, lipases, and amylases, disrupt host plant cell walls and membranes. The enzymatic activities have an influence on infection process, but their correlation with fungal growth is still unknown (Schinke & Germani, 2012). *M. phaseolina*-isolated toxins like phaseolina and botryodiplodin are considerably thought to be involved in the infection process. Under favorable environmental conditions, the pathogen infects plant vascular system and interferes with normal water and nutrients transportation that cause wilting and premature plant death (Khan, 2007). When infected plants are harvested, these release microsclerotia into the soil and in the following year infection cycle starts again. In addition to this, infected seeds are also a virulent source of disease transmission.

The survival of microsclerotia in the soils is not studied under field conditions. Previous studies reported that *M. phaseolina* survives below the soil surface or on crop residues for almost 2 years. However, this finding was insufficient in describing *M. phaseolina* ability to cause infections after this duration (Short et al., 1980). According to Reis et al. (2014), microsclerotia could survive on infected crop residues for almost up to 35 months with a decrease in pathogenicity after 6 months.

3.4.2 Damping-off

3.4.2.1 Hosts

It has been reported in sunflower (Abd El-Hai et al., 2009), cotton (Aly et al., 2007), common bean (Bedine et al., 2020), and alfa alfa (Azzam et al., 2022) plants. Alfalfa (*Medicago sativa* L.), also known as lucerne, is a forage crop that is used to feed animals. Its production is largely influenced by several fungal pathogens. Among them, *M. phaseolina* is the most important one which causes damping-off disease at early stages of seed development and causes substantial yield losses (Azzam et al., 2022). Moreover, the largest cowpea producing countries are Niger and Nigeria where *M. phaseolina* also affects its growth. Damping-off is caused by the pre- and postemergence of seedlings and seeds decay (Adekunle et al., 2006).

3.4.2.2 Symptoms

Damping-off caused by *M. phaseolina* is a common problem of raising nursery seedlings. The pathogen prevents seed germination or the emergence of new seedlings and as a consequence, disease symptoms vary widely on the basis of time and stress types. The crops are severely affected at the early stages of pathogen establishment, and the seeds may rot before germination. However, seedling symptoms are dependent upon time of their appearance and become visible after the third to fifth-week of post sowing (Horst, 2013).

3.4.2.2.1 Pre-emergence symptoms

These symptoms usually appear prior to seed emergence or when germination takes place. *M. phaseolina* causes seed decay, while shoot tissues are still below ground (Crous, 2002). When the pathogen attacks, seeds become soft, rotten, and fail to germinate. Whereas, in the case of germinated seeds characteristic, water-soaked lesions are formed on affected stems below or at the soil line; as the disease progresses, these become reddish brown to black in color. Additionally, pre-emergence damping-off indicates wilt and death of seedlings (Cram, 2003).

3.4.2.2.2 Postemergence symptoms

These symptoms occur after seedling emergence and the affected plants become wilted and show premature death (Horst, 2013). In response to *M. phaseolina* attack, the plant shows symptoms that result in collapse and seedling death. Seedlings at ground level show lesions that become tough and thin often leading to reduced vigor. These symptoms may show spotting on leaves that result in complete rot. Additionally, roots and stem of the seedlings that are slightly below or above the ground level exhibit sunken, water-soaked lesions that sometimes cause plants to fall over (Filer & Peterson, 1975). If affected plants survive, they exhibit stunted and uneven growth areas.

M. phaseolina penetrates the seedlings through their roots and causes symptoms of sudden wilting and rotting of young succulent seedlings. Thereafter, it results in complete fall over, collapse, and sudden plant death. Several factors including poor soil sanitation and low temperatures result in continuous wetness, which favors the pathogen and considerably leads to disease development (Abd El-Hai et al., 2009). In general, the emergence of plant seedlings near the soil line often causes pre-emergence damping off. Whereas, those plant seedlings typically emerging above 2.5–5 cm of the soil line experience post-emergence damping-off.

3.4.2.3 Disease cycle

M. phaseolina enters and causes infections in susceptible host plants under favorable environmental conditions. The pathogen produces microsclerotia that germinate and produce hyphae. These hyphal structures enter into the seeds or seedlings through natural openings or wounds and cause decay as they grow. Microsclerotia are resistant structures that can survive for several months in their dormant state under adverse environmental conditions. They rapidly germinate in the presence of abundant moisture content and susceptible host plants. Thereafter, the infection process starts with hyphal penetration inside the host plants (Azzam et al., 2022).

3.4.3 Stem blight

3.4.3.1 Hosts

It is one of the most destructive diseases that affect the worldwide production of common beans. In susceptible cultivars, the pathogen attacks the entire cropping season on different plant parts and causes up to 60% yield losses (Viteri & Linares, 2017). The occurrence of stem blight has also been reported in peas (Ali & Dennis, 1992), cowpea (Amusa & Akinfenwa, 2007), cactus (Abdel-Kader et al., 2010), blueberries (Popovic et al., 2018), and tepary beans (Viteri & Linares, 2022).

3.4.3.2 Symptoms

The pathogen remains active on diseased plant tissues or in the soil in the form of sclerotia. Its production is higher in host tissues and disperses subsequently in the soil through decomposed plant residues and tillage practices. Moreover, sclerotial population in the soil is directly related to the disease severity. Infected seeds also contribute to the primary source of inoculum. Mycelial bodies are generally not considered as the major inoculum source in the soil. Cropping history greatly influences the inoculum densities and sclerotia populations increase annually where susceptible crops are grown continuously. Plants infected with stem blight are characterized by chlorosis, premature defoliation, wilting, vascular tissues having systemic infections, downward spread of stem, and twig necrosis followed by early maturity or death (Ali & Dennis, 1992). In blueberry plants, unusual stem blight symptoms were observed in Serbia. The infected plants initially showed yellowish-brown leaf edges followed by leaf defoliation on the infected stems. Later on, infected leaves turned reddish brown, stem tissues become discolored at the basal part and foliage was dried completely (Popovic et al., 2018).

3.4.3.3 Disease cycle

The soil colonization with *M. phaseolina* in the form of plant debris serves as primary inoculum source for subsequent cropping systems. As a necrotrophic fungus, it produces both pycnidia and sclerotia on susceptible hosts. Additionally, hyphal bodies are formed excessively in the presence of optimum temperature and dry soil conditions. On the attainment of these conditions, the fungus produces a huge number of sclerotia and pycnidia in the field on decomposed plant debris. Thereafter, a new disease cycle starts with the subsequent release of these structures. These inoculum bodies play a key role in pathogen epidemiology and a further unit of spread (Ali & Dennis, 1992).

3.4.4 Leaf blight

3.4.4.1 Hosts

A large number of valuable crops from tropical and sub-tropical countries are key hosts to this ubiquitous pathogen. It causes leaf blight in urdbean (Singh & Kumar, 2002), green gram (Tandel et al., 2010), *Catharanthus roseus* (Mridha & Rahman, 2015), *Jasminium multiflorum* (Mahadevakumar & Janardhana, 2016), blueberries (Popovic et al., 2018), white spider lilies (Huda-Shakirah et al., 2019), safedmusli (Meena & Kadam, 2021), and *Pometiapinnata* (Khoo et al., 2022).

3.4.4.2 Symptoms

The expression of leaf blight is mainly depending on the availability of inoculum, canopy and age of the crop and plant species. Initially infected plant leaves show water-soaked, irregular reddish-brown lesions surrounded by yellow holes, which gradually appear as small light brown spots. As the pathogen spreads, extended patches are formed through the fusion of spots, lesions developed, which later become necrotic, and turned darker producing a complete burning of large leaf areas that results in typical blight symptoms. Blighted leaves are usually covered with black-colored pycnidia or sclerotia. The mature leave surface area has more infection percentage when compared with younger ones. This might occur due to less nutrient availability in young leaves. Warm temperatures in the range of 28–35°C and humid growing conditions make these disease symptoms more severe (Huda-Shakirah et al., 2019).

3.4.4.3 Disease cycle

The development of *M. phaseolina* microsclerotia takes place only on diseased petioles and plant leaves. This infected plant debris remains on the soil surface, and the disease development starts again in the next cropping season. Upon suitable environmental conditions, conidia are released from infected plant debris or lesions of infected plants and dispersed through wind or rain. Thereafter, new field crops become infected by the transmission of conidial bodies. Plants become suitable hosts for *M. phaseolina*when humid and moist conditions are relatively high. Once *M. phaseolina* conidia infect healthy plant leaves, it germinates on plant tissues by forming germ tubes. The penetration of germ tube is supported through natural openings or leaves. Thereafter, the production of conidial bodies takes place on younger leaves base, and penetration is carried out through a flattened appressorium. Then, the parenchymatous leaf tissues are invaded by the fungal mycelium that penetrates into host tissues, germinates rapidly, and subsequently, begins to turn brown and collapse. The pathogen lives in the soil in the form of mycelium on infected or dead decaying plant tissues. It possibly survives as conidia or microsclerotia during dry periods and remains viable for less than 3 years (Huda-Shakirah et al., 2019).

3.4.5 Pod rot

3.4.5.1 Hosts

M. phaseolina is a ubiquitous pathogen that also causes postflowering stress. Fuhlbohm et al. (2013) found that *Macrophomina* affects the pods of mungbean plants that significantly reduced the crop yield.

3.4.5.2 Symptoms

The fungal infected pods show typical symptoms of rot including termination of flower growth and abscission of young pods. Small galls are produced on pods that proliferate in the host tissues and then gradually turned yellowish brown in color. However, mature pods are without galls that appear in the form of white specks. Afterward, infected pods develop pycnidia and microsclerotia. As the infection progresses, irregular brown to black lesions start appearing on pods. The lesions continue to expand and are completely colonized by grayish fungal mycelium. Moreover, pods start rotting on internal tissue and are smaller in size than healthy ones (Fuhlbohm et al., 2013).

3.4.5.3 Disease cycle

Microsclerotial bodies of *M. phaseolina* are either embedded in colonized plant tissues or dispersed in the soil that might dislodge during tillage operations followed by host decay (Dhingra & Sinclair, 1978). Short et al. (1980) found that initially, microsclerotia are present in the soil in higher numbers later on decrease as the infected host debris in the soil decomposes. In the late growing season, *M. phaseolina* conidia come out from pycnidia and implicate in the infection process, but their exact mechanism in pathogen epidemiology is not fully elucidated (Luttrell & Garren, 1952). The pathogen can efficiently be transmitted through seeds in mungbean plants worldwide (Sharada & Shetty, 1987), which results in poor seed germination, seedling rot, and lesion development on hypocotyls and primary leaves (Gaikwad & Sokhi, 1987). Eventually, when *M. phaseolina*-contaminated seeds disperse, it leads to the pathogen introduction in noninfected areas (Sharada & Shetty, 1987). Singh and Mathur (2004) described specific possible seed infection modes such as direct and indirect transmission. Direct infection starts from the mother plant either through vascular or nonvascular tissues, whereas indirect mediation is carried from the seed wall, ovary, or flowers. It also spreads through soil particles attached to seeds and infected debris. Several modes of *Macrophomina* infection have been reported from different seeds and host plants. Previously, the concept of *Macrophomina* systemic infection has been supported by many researchers that explained its transmission from mother plants to seed progeny of black gram, soybean, and sunflower (Sharada & Shetty, 1987). By contrast, Songa and Hillocks (1998) suggested that *Macrophomina* colonization on mungbean results in infection prior to germination from soil-borne inoculum or from preharvest pod infection. Similarly, Shama (1991) has also described *M. phaseolina* establishment within green bean pods. Excessive rainfall is usually associated with the highest infection levels, especially at the time of flowering and pod fill. However, seed colonization and pathogen transmission modes in different host plants are unclear.

3.5 Conclusion

M. phaseolina is a soil- and seed-borne fungal pathogen that generally causes charcoal rot or dry root rot in hundreds of plant species under drought stress. However, it also has the ability to cause damping off, stem blight, leaf blight, and pod rot in some susceptible host plant species.

References

Abbas, H. K., Bellaloui, N., Butler, A. M., Nelson, J. L., Abou-Karam, M., & Shier, W. T. (2020). Phytotoxic responses of soybean (*Glycine max* L.) to Botryodiplodin, a toxin produced by the charcoal rot disease fungus, *Macrophomina phaseolina. Toxins, 12*, 25.

Abd El-Hai, K. M., El-Metwally, M. A., El-Baz, S. M., & Zeid, A. M. (2009). The use of antioxidants and microelements for controlling damping-off caused by *Rhizoctonia solani* and charcoal rot caused by *Macrophomina phasoliana* on sunflower. *Plant Pathology Journal, 8*, 79–89.

Abdel-Kader, M. M., El-Mougy, N. S., Aly, M. D. E. H., & Lashin, S. M. (2010). First report of ashy stem blight caused by *Macrophomina phaseolina* on *Aeonium canariense* in Egypt. *Journal of Plant Pathology & Microbiology, 1*, 1–6.

Adekunle, A. T., Ikotun, T., Florini, D. A., & Cardwell, K. F. (2006). Field evaluation of selected formulations of *Trichoderma* species as seed treatment to control damping-off of cowpea caused by *Macrophomina phaseolina. African Journal of Biotechnology, 5*, 419–424.

Ali, S. M., & Dennis, J. (1992). Host range and physiologic specialisation of *Macrophomina phaseolina* isolated from field peas in South Australia. *Australian Journal of Experimental Agriculture, 32*, 1121–1125.

Ali, M., & Kumar, S. (2008). Productivity levels of pulses in major producing states in India. In *Survey of Indian agriculture* (pp. 44–45). Chennai: The Hindu Year Book.

Aly, A. A., Abdel-Sattar, M. A., Omar, M. R., & Abd-Elsalam, K. A. (2007). Differential interaction between isolates of *Macrophomina phaseolina* and Egyptian cotton cultivars. *Pest Technology, 1*, 127–132.

Amusa, N. O. R., & Akinfenwa, B. (2007). Reactions of cowpea to infection by *Macrophomina phaseolina* isolates from leguminous plants in Nigeria. *African Journal of Agricultural Research, 2*, 073–075.

Angelini, R., & Faedi, W. (2010). In A. R. T. La Fragola (Ed.), *Malattie e fisiopatie* (pp. 228–246). Bologna: Italy.

Ashby, S. F. (1927). *Macrophomina phaseoli* (maubl.) comb. nov., the pycnidial stage of *Rhizoctonia bataticola* (Taub.)Butl. *Transactions of the British Mycological Society, 12*, 141–147.

Avilés, M., Castillo, S., Bascon, J., Zea-Bonilla, T., Martín-Sánchez, P. M., & Pérez-Jiménez, R. M. (2008). First report of *Macrophomina phaseolina* causing crown and root rot of strawberry in Spain. *Plant Pathology, 57*, 382.

Azzam, C. R., Abd El-Naby, Z. M., Abd El-Rahman, S. S., Omar, S. A., Ali, E. F., Majrashi, A., & Rady, M. M. (2022). Association of saponin concentration, molecular markers, and biochemical factors with enhancing resistance to alfalfa seedling damping-off. *Saudi Journal of Biological Sciences, 29*, 2148–2162.

Baino, O. M., Salazar, S. M., Ramallo, A. C., & Kirschbaum, D. S. (2011). First report of *Macrophomina phaseolina* causing strawberry crown and root rot in north-western Argentina. *Plant Disease, 95*, 1477.

Baird, R. E., Watson, C. E., & Scruggs, M. (2003). Relative longevity of *Macrophomina phaseolina* and associated mycobiota on residual soybean roots in soil. *Plant Disease, 87*, 563–566.

Banaras, S., Javaid, A., & Khan, I. H. (2021). Bioassays guided fractionation of *Ageratum conyzoides* extract for the identification of natural antifungal compounds against *Macrophomina phaseolina. International Journal of Agriculture and Biology, 25*, 761–767.

Banaras, S., Javaid, A., & Shoaib, A. (2020). Non-chemical control of charcoal rot of urdbean by *Sonchus oleraceous* application. *Planta Daninha, 38*, e020216088.

Bandara, Y. M. A. Y., Weerasooriya, D. K., Liu, S., & Little, C. R. (2018). The necrotrophic fungus *Macrophomina phaseolina* promotes charcoal rotsusceptibility in grain sorghum through induced host cell-wall-degradingenzymes. *Phytopathology, 108*, 948–956.

Basandrai, A. K., Pandey, A. K., Somta, P., & Basandrai, D. (2021). *Macrophomina phaseolina*–host interface: Insights into an emerging dry root rot pathogen of mungbean and urdbean, and its mitigation strategies. *Plant Pathology, 70*, 1263–1275.

Becker, J. O., & Schwinn, S. J. (1993). Control of soil-borne pathogens with living bacteria and fungi: Status and outlook. *Pesticide Science, 37*, 355–363.

Bedine, B. M. A., Sameza, M. L., Iacomi, B., Tchameni, S. N., & Boyom, F. F. (2020). Screening, identification and evaluation of *Trichoderma* spp. for biocontrol potential of common bean damping-off pathogens. *Biocontrol Science and Technology, 30*, 228–242.

Bellaoui, N., Mengistu, A., Smith, J. R., Abbas, H. K., Accinelli, C., & Shier, W. T. (2021). Effects of charcoal rot on soybean seed composition in soybean genotypes that differ in charcoal rot resistance under irrigated and non-irrigated conditions. *Plants, 10*, 1801.

Cram, M. M. (2003). Damping-off. *Tree Planters' Notes, 50*, 1–5.

Crous, P. W. (2002). *Taxonomy and pathology of* Cylindrocladium *(Calonectria) and allied genera* (pp. 15–17). St. Paul, MN: American Phytopathological Society.

Dell'Olmo, E., Tripodi, P., Zaccardelli, M., & Sigillo, L. (2022). Occurrence of *Macrophomina phaseolina* on chickpea in Italy: Pathogen identification and characterization. *Pathogens, 11*, 842.

Dhingra, O. D., & Sinclair, J. B. (1978). *Biology and pathology of* Macrophomina phaseolina. Brazil: Universidade Federal de Vicosal.

Filer, T. H. J., & Peterson, G. W. (1975). Damping-off. In G. W. Peterson, & R. S. Smith (Eds.), *Agriculture handbook No. 470Forest nursery diseases in the United States* (pp. 6–8). Washington DC: USDA Forest Service.

Fuhlbohm, M. J., Ryley, M. J., & Aitken, E. A. B. (2013). Infection of mungbean seed by *Macrophomina phaseolina* is more likely to result from localized pod infection than from systemic plant infection. *Plant Pathology, 62*, 1271–1284.

Gaikwad, D. G., & Sokhi, S. S. (1987). Detection of seed rot, root rot and seedling infection in naturally infected cowpea seed in Senegal and their control. *Plant Disease Research, 2*, 127–128.

Ghosh, T., Biswas, M. K., Guin, C., & Roy, P. (2018). A review on characterization, therapeutic approaches and pathogenesis of *Macrophominaphaseolina. Plant Cell Biotechnology and Molecular Biology, 19*, 72–84.

Goidanich, G. (1947). A revision of the genus *Macrophomina* Petrak.type species: M. P. (Tassi) G. Goid. n. comb. M.P. (Maubl.) Ashby. *Annalidell'Istituto Sperimentale di Frutticoltura, 3*, 449–461.

Golzar, H., Phillips, D., & Mack, S. (2007). Occurrence of strawberry root and crown rot in Western Australia. *Australasian Plant Disease Notes, 2*, 145–147.

Gupta, G. K., & Chauhan, G. S. (2005). *Symptoms, identification and management of soybean diseases*. National Research Centre for Soybean, Indian Council of Agricultural Research.

Hartman, G. L., Wang, T. C., & Tschanz, A. T. (1991). Soybean rust development and the quantitative relationship between rust severity and soybean yield. *Plant Disease, 75*, 596–600.

Horst, R. K. (2013). Damping-off. *Westcott's plant disease handbook* (p. 177). Dordrecht: Springer Netherlands.

Huda-Shakirah, A. R., Kee, Y. J., Hafifi, A. B. M., Mohamad Azni, N. N., Zakaria, L., & Mohd, M. H. (2019). Identification and characterization of *Macrophomina phaseolina* causing leaf blight on white spider Lilies (*Crinum asiaticum* and *Hymenocallis littoralis*) in Malaysia. *Mycobiology, 47*, 408–414.

Islam, M. S., Haque, M. S., Islam, M. M., Emdad, E. M., Halim, A., Hossen, Q. M. M., & Alam, M. (2012). Tools to kill: Genome of one of the most destructive plant pathogenic fungi *Macrophomina phaseolina. BMC Genomics, 13*, 1–16.

Javaid, A., Khan, I. H., & Shoaib, A. (2018). Management of charcoal rot of mungbean by two *Trichoderma* species and dry biomass of *Coronopus didymus. Planta Daninha, 36*, e018182795.

Javed, S., Mahmood, Z., Khan, K. M., Sarker, S. D., Javaid, A., Khan, I. H., & Shoaib, A. (2021). Lupeol acetate as a potent antifungal compound against opportunistic human and phytopathogenic mold *Macrophomina phaseolina. Scientific Reports, 11*, 8417.

Khan, S. N. (2007). *Macrophomina phaseolina* as causal agent for charcoal rot of sunflower. *Mycopathology, 5*, 11–118.

Khan, I. H., & Javaid, A. (2020). Comparative antifungal potential of stem extracts of four quinoa varieties against *Macrophomina phaseolina. International Journal of Agriculture and Biology, 24*, 441–446.

Khan, I. H., & Javaid, A. (2022a). Histopathological changes in root and stem of mungbean exposed to *Macrophomina phaseolina* and dry biomass of *Chenopodium quinoa. Microscopy Research and Technique, 85*, 2596–2606.

Khan, I. H., & Javaid, A. (2022b). Biocontrol *Aspergillus* species together with plant biomass alter histochemical characteristics in diseased mungbean plants. *Microscopy Research and Technique, 8*, 2953–2964.

Khan, I. H., Javaid, A., & Ahmed, D. (2021). *Trichoderma viride* controls *Macrophomina phaseolina* through its DNA disintegration and production of antifungal compounds. *International Journal of Agriculture and Biology, 25*, 888–894.

Khoo, Y. W., Khaw, Y. S., Tan, H. T., Li, S. F., & Chong, K. P. (2022). First report of *Macrophomina phaseolina* causing leaf blight on *Pometia pinnata* in Malaysia. *Plant Disease*. https://doi.org/10.1094/PDIS-11-21-2478-PDN

Koike, S. T. (2008). Crown rot of strawberry caused by *Macrophomina phaseolina* in California. *Plant Disease, 92*, 1253.

Lakhran, L., Ahir, R. R., Choudhary, M., & Choudhary, S. (2018). Isolation, purification, identification and pathogenicity of *Macrophomina phaseolina* (Tassi) goid caused dry root rot of chickpea. *Journal of Pharmacognosy and Phytochemistry, 7*, 3314–3317.

Lodha, S. (1995). Soil solarization, summer irrigation and amendments for the control of *Fusarium oxysporum* f. sp. *cumini* and *Macrophomina phaseolina* in arid soils. *Crop Protection, 14*, 215–219.

Lodha, S., & Mawar, R. (2020). Population dynamics of *Macrophomina phaseolina* in relation to disease management: A review. *Journal of Phytopathology, 168*, 1–17.

Luttrell, E. S., & b Garren, K. H. (1952). Blights of snap bean in Georgia. *Phytopathology, 42*, 607–613.

Mahadevakumar, S., & Janardhana, G. R. (2016). First report of leaf blight caused by *Macrophomina phaseolina* on *Jasminium multiflorum* in India. *Journal of Plant Pathology, 98*, 177.

Manikandan, A., Jaivel, N., Johnson, I., Raghu, R., Gopal, N. O., & Anandham, R. (2022). Suppression of Macrophomina root rot, Fusarium wilt and growth promotion of some pulses by antagonistic rhizobacteria. *Physiological and Molecular Plant Pathology, 121*, 101876.

Manjunatha, S. V., Naik, M. K., Khan, M. F. R., & Goswami, R. S. (2013). Evaluation of bio-control agents for management of dry root rots of chickpea caused by *Macrophomina phaseolina. Crop Protection, 45*, 147–150. https://doi.org/10.1016/j.cropro.2012.09.003

Maublanc, A. (1905). Travaux de la Station de Pathologie Végétale – espècesnouvelles de champignons inférieurs. *Bulletin Trimestriel de la Societe Mycologique de France, 21*, 87–95.

Meena, R. P., & Kadam, V. A. (2021). Characterization of *Macrophomina phaseolina* associated with leaf blight disease on *Chlorophytum borivilianum* Santapau & RR Fern. and its fungicidal susceptibility. *Journal of Applied Research on Medicinal and Aromatic Plants, 20*, 100288.

Mengistu, A., Ray, J. D., Smith, J. R., & Paris, R. L. (2007). Charcoal rot disease assessment of soybean genotypes using a colony-forming unit index. *Crop Science, 47*, 2453–2461.

Mengistu, A., Smith, J. R., & Ray, J. D. (2011). Seasonal progress of charcoal rot and its impact on soybean productivity. *Plant Disease, 95*, 1159–1166.

Mihail, J. D. (1989). *Macrophomina phaseolina*: Spatio-temporal dynamics of inoculum and of disease in a highly susceptible crop. *Phytopathology, 79*, 848–855.

Mridha, M. A. U., & Rahman, M. M. (2015). Leaf blight of *Catharanthus roseus* (L). G. Don caused by *Macrophomina phaseolina* (Tassi) Goid. and its *in vitro* control through bio-pesticides. *Pakistan Journal of Botany, 47*, 741–745.

Ndiaye, M. (2007). *Ecology and management of charcoal rot (*Macrophomina phaseolina*) on cowpea in the Sahel*. PhD Thesis. the Netherlands, Wageningen: Wageningen University.

Pandey, A. K., Burlakoti, R. R., Rathore, A., & Nair, R. M. (2020). Morphological and molecular characterization of *Macrophomina phaseolina* isolated from three legume crops and evaluation of mungbean genotypes for resistance to dry root rot. *Crop Protection, 127*, 104962.

Pastrana, A. M., Borrero, C., Perez, A. G., & Aviles, M. (2022). Soilborne pathogens affect strawberry fruit flavor and quality. *Plant Science, 326*, 111533.

Popovic, T., Blagojević, J., Aleksić, G., Jelušić, A., Krnjajić, S., & Milovanović, P. (2018). A blight disease on highbush blueberry associated with *Macrophomina phaseolina* in Serbia. *Canadian Journal of Plant Pathology, 40*, 121–127.

Purkayastha, S., Kaur, B., Dilbaghi, N., & Chaudhury, A. (2006). Characterization of *Macrophomina phaseolina*, the charcoal rot pathogen of cluster bean, using conventional techniques and PCR-based molecular markers. *Plant Pathology, 55*, 106–116.

Ramezani, M., Shier, W. T., Abbes, H. K., Tones, J. L., Baird, R. E., & Sciumbato, G. L. (2007). Soybean charcoal rot disease fungus *Macrophomina phaseolina* in Mississippi produces the phytotoxin botryodiplodin but no detectable phaseolinone. *Journal of Natural Products, 70*, 128–129.

Ramos, A. M., Gally, M., Szapiro, G., Itzcovich, T., Carabajal, M., & Levin, L. (2016). *In vitro* growth and cell wall degrading enzyme production by Argentinean isolates of *Macrophomina phaseolina*, the causative agent of charcoal rot in corn. *Revista Argentina de Microbiología, 48*, 267–273.

Reis, E. M., Boaretto, C., & Danelli, A. L. D. (2014). *Macrophomina phaseolina*: Density and longevity of microsclerotia in soybean root tissues and free on the soil, and competitive saprophytic ability. *Summa Phytopathologica, 40*, 128–133.

Salvatore, M. M., Félix, C., Lima, F., Ferreira, V., Naviglio, D., & Salvatore, F. (2020). Secondary metabolites produced by *Macrophomina phaseolina* isolated from eucalyptus globulus. *Agriculture, 10*, 72. https://doi.org/10.3390/agriculture10030072

Santos, B. M., Gilreath, J. P., Lopez-Aranda, J. M., Miranda, L., Soria, C., Medina, J. J., & Whidden, A. J. (2006). *Performance of methyl bromide alternatives for strawberry production in Florida and Spain* (p. 119). Proceedings of the Florida State Horticultural Society.

Sarkar, T. S., Biswas, P., Ghosh, S. K., & Ghosh, S. (2014). Nitric oxide production by necrotrophic pathogen *Macrophomina phaseolina* and the host plant in charcoal rot disease of jute: Complexity of the interplay between necrotroph-host plant interactions. *PLoS One, 9*, e0107348.

Schinke, C., & Germani, J. C. (2012). Screening Brazilian *Macrophomina phaseolina* isolates for alkaline lipases and other extracellular hydrolases. *International Microbiology, 15*, 1–7.

Shama, S. M. (1991). Histopathology of bean seeds infected with *Macrophomina phaseolina*. *Alexandria Journal of Agricultural Research, 36*, 249–261.

Sharada, R., & Shetty, H. S. (1987). Location and transmission of *Macrophomina phaseolina* in black gram seeds. *Indian Phytopathology, 40*, 194–196.

Sheikh, A. H., & Ghaffar, A. (1979). Relation of sclerotial inoculum density and soil moisture to infection of field crops by *Macrophomina phaseolina*. *Pakistan Journal of Botany, 11*, 185–189.

Short, G. E., Wyllie, T. D., & Bristow, P. R. (1980). Survival of *Macrophomina phaseolina* in soil and in residue of soybean. *Phytopathology, 70*, 13–17.

Singh, J., & Kumar, K. (2002). Location, survival, transmission and control of seedborne, *Macrophomina phaseolina* causing dry root rot and leaf blight in Urdbean. *Annals of Plant Protection Sciences, 10*, 111–113.

Singh, D., & Mathur, S. B. (2004). *Histopathology of seed-borne infections*. CRC Press.

Singh, N., Pandey, P., Dubey, R. C., & Maheshwari, D. K. (2008). Biological control of root rot fungus *Macrophomina phaseolina* and growth enhancement of *Pinus roxburghii* (Sarg.) by rhizosphere competent *Bacillus subtilis* BN1. *World Journal of Microbiology and Biotechnology, 24*, 1669–1679.

Smith, G. S., & Carvil, O. N. (1997). Field screening of commercial and experimental soybean cultivars for their reaction to *Macrophomina phaseolina*. *Plant Disease, 81*, 363–368.

Songa, W., & Hillocks, R. J. (1996). Legume hosts of *Macrophomina phaseolina* in Kenya and effect of crop species on soil inoculum levels. *Journal of Phytopathology, 144*, 387–391.

Songa, W., & Hillocks, R. J. (1998). Survival of *Macrophomina phaseolina* in bean seed and crop residue. *International Journal of Pest Management, 44*, 109–114.

Srivastava, A. K., Singh, T., Jana, T. K., & Arora, D. K. (2001). Microbial colonization of *Macrophomina phaseolina* and suppression of charcoal rot of chickpea. In A. Sinha (Ed.), *Microbes and plants* (pp. 269–319). New Delhi: Vedams eBooks.

Su, G., Suh, S. O., Schneider, R. W., & Russin, J. S. (2001). Host specialization in the charcoal rot fungus, *Macrophomina phaseolina*. *Phytopathology, 91*, 120–126.

Tandel, D. H., Sabalpara, A. N., & Pandya, J. R. (2010). Efficacy of phytoextracts on *Macrophomina phaseolina* (Tassi) Goid causing leaf blight of green gram. *International Journal of Pharma Bio Sciences, 2*, 1–5.

Tassi, A. (1901). Novae micromycetum species. *Series III Bullettino del Laboratorio ed Orto Botanico, University of Siena, 4*, 7–12.

Thilagavathi, R., Saravanakumar, D., Ragupathi, N., & Samiyappan, R. (2007). A combination of biocontrol agents improves the management of dry root rot (*Macrophomina phaseolina*) in greengram. *Phytopathologia Mediterranea, 46*, 157–167.

Uroos, M., Javaid, A., Bashir, A., Tariq, J., Khan, I. H., Naz, S., Fatima, S., & Sultan, M. (2022). Green synthesis of coumarin derivatives using bronsted acidicpyridinium based ionic liquid [MBSPy][HSO4] to control anopportunistic human and a devastating plant pathogenic fungus *Macrophemina phaseolina*. *RSC Advances, 12*, 23963–23972.

Viejobueno, J., Albornoz, P. L., Camacho, M., de los Santos, B., Martínez-Zamora, M. G., & Salazar, S. M. (2021). Protection of strawberry plants against charcoal rot disease (*Macrophomina phaseolina*) induced by *Azospirillum brasilense*. *Agronomy, 11*, 195.

Viteri, D. M., & Linares, A. M. (2017). Reaction of *Phaseolus* spp. genotypes to ashy stem blight caused by *Macrophomina phaseolina*. *Euphytica, 213*, 1–8.

Viteri, D. M., & Linares, A. M. (2022). Agronomic performance of common and tepary bean genotypes and their response to ashy stem blight in Isabela and Lajas, Puerto Rico. *Legume Science, 4*, e118.

Wagan, K. H., Khaskheli, M. I., Hajano, J. D., & Lanjar, A. G. (2019). Population density and aggressiveness of *Macrophomina phaseolina* isolates from Sindh, Pakistan. *Sarhad Journal of Agriculture, 35*, 400–407.

Chapter 4

Charcoal rot: a major root and stalk disease of sorghum

Christopher R. Little[1], Ramasamy Perumal[2], Tesfaye Tesso[3], Y.M.A.Y. Bandara[4] and Afsana Noor[1]

[1]Department of Plant Pathology, Kansas State University, Manhattan, KS, United States; [2]Agricultural Research Center - Hays, Kansas State University, Hays, KS, United States; [3]Department of Agronomy, Kansas State University, Manhattan, KS, United States; [4]Department of Plant Pathology and Environmental Microbiology, Penn State University, University Park, PA, United States

4.1 Introduction

In 2020, the annual global production of sorghum, *Sorghum bicolor* (L.) Moench, reached 58.7 million metric tons (megagrams, Mg) and ranked 5th among cereal crops. During the same year, the United States produced 9.47 Mg, followed by Nigeria (6.36 Mg), Ethiopia (5.06 Mg), India (4.77 Mg), and Mexico (4.70 Mg) (Food and Agriculture Organization of the United Nations, 2022). In the United States, sorghum was planted on 5.88 million, 7.31 million, and 6.85 million acres of cropland in 2020, 2021, and 2022, respectively. Most sorghum production occurs in Kansas and Texas, which planted 48.1% and 21.9% of the total acreage for the United States (United States Department of Agriculture, National Agricultural Statistics Service, 2022). The environment in the southern and central Great Plains of the United States favors sorghum production, where the crop needs fewer inputs to be productive compared to other commodities, such as corn or soybeans (Hoffman et al., 2020; Staggenborg et al., 2008).

Sorghum has many uses, including animal feed and human food, and has recently been identified as a dedicated energy crop by the US Department of Energy (Alavi et al., 2019; DOE, 2022). Food usages occur primarily in the developing world where local landraces and varieties are used for food products, such as *roti* (an Indian unfermented bread), *injera* (an Ethiopian fermented bread), and *couscous* (a West African steamed product) (Akin et al., 2022; Galiba et al., 1988). In China, large amounts of sorghum are used for the widely consumed and traditional alcohol product *baijiu* (Xing-Lin et al., 2017). In the United States and other countries with high livestock feed demands, sorghum supplies a large portion of feed for poultry, swine, dairy, and beef cattle (McCuistion et al., 2019). In addition to grain, sorghum is a good source of cattle silage (McCary et al., 2020).

Sorghum grain is comparable to other cereals in its nutritional composition. It is rich in polyphenolic compounds that contain antioxidants known to offer numerous health benefits, including the reduction of cancer cell proliferation (Lee et al., 2021). Phytate extracted from pigmented sorghum grain has been shown to maintain blood cholesterol balance and reduce the risk of stroke and heart attack (Carr et al., 2005). Because of its slow digestion, low glycemic index, and longer gastric emptying time, sorghum foods are considered ideal for reducing blood sugar spikes and are recommended for fighting obesity and diabetes (Awika & Rooney, 2004; Cisse et al., 2018). In addition, polyphenols from pigmented sorghums have been reported to reduce oxidative stresses and thus prevent or delay the progression of degenerative diseases such as Alzheimer's (Rezaee et al., 2021). A recent systematic review of the literature on human health benefits of sorghum outlined the role of the crop in reducing chronic diseases, improving blood glucose response, reducing oxidative stress, increasing satiety, and its use for weight management (Ducksbury et al., 2021).

Sorghum is a C4, climate-resilient crop that exhibits traits suited for production under sub-optimal environments. Such features include drought and heat tolerance and an ability to grow and yield in marginal soils with moderate to poor mineral nutrition (Prasad et al., 2021). Given that the reality of climate change threatens the efficiency and productivity of the global food system, sorghum represents a crop with great potential to remain productive under such circumstances (Mundia et al., 2019).

Macrophomina Phaseolina. https://doi.org/10.1016/B978-0-443-15443-0.00018-8

A range of diseases can affect the productivity and profitability of sorghum. Foliar diseases include anthracnose (*Colletotrichum sublineolum*) and sorghum downy mildew (*Peronosclerospora sorghi*), which impact leaf photosynthate production and ultimately reduce grain filling and yields (Little, Bandara, & Perumal, 2018; Little & Perumal, 2018). Seed and panicle diseases include grain mold (numerous fungi such as *Fusarium*, *Curvularia*, and *Alternaria* spp.), ergot (*Claviceps africana*), and head smut (*Sporisorium reilianum*), which are problematic to different degrees depending on geographic location and environmental conditions (Little, Bandara, & Perumal, 2018; Little & Perumal, 2018).

Stalk rots, including charcoal rot (*Macrophomina phaseolina*) and Fusarium stalk rot (*Fusarium thapsinum* and other *Fusarium* spp.), can lead to yield reduction due to reduced photosynthate and water translocation, host senescence, and lodging (Little, Perumal, et al., 2018; Little & Perumal, 2018). Further, stalk rots are the most common sorghum diseases encountered in the central Great Plains. Some aspects of the sorghum-*M. phaseolina* pathosystem including symptomatology, pathogen biology, infection processes, disease cycle and epidemiology, and management considerations will be discussed here.

4.2 Symptoms and signs

Macrophomina phaseolina causes disease in >500 host plants. The disease first gained attention in the midwestern United States due to crop damage on corn and sorghum in the late 1930s when the corn belt states were exposed to repetitive drought for multiple years (Livingston, 1945). Severe sorghum crop damage was reported from the south-central counties of Nebraska in 1940, with statewide yield reduction and lodging occurring in corn and sorghum the following year from charcoal rot progression in the field (Livingston, 1945). Livingston (1945) also documented the severity of lodging, especially in grain sorghum, which was so damaging that the heads were not amenable to combine harvesting.

M. phaseolina infection causes seedling blight and damping-off, root rot, soft stalk, wilting, premature leaf and stalk senescence, pith degradation, and stalk breakage at the infection zone, leading to lodging (Bandara, 2017; Kapanigowda et al., 2013; Mughogho and Pande, 1984) (Fig. 4.1). Charcoal rot also reduces yields through reduced seed weight and fewer seeds per panicle (Bandara et al., 2017a, 2017c; Perumal et al., 2020) (Fig. 4.1). *M. phaseolina* infection has also been reported to decrease grain macronutrients and mineral content in susceptible sorghum varieties (Bandara et al., 2017a). Under conducive environmental conditions, infected plants often die prematurely. In addition to grain sorghum, charcoal rot affects sweet sorghum. Lodging is the primary cause of decreased harvestable yield when infected with *M. phaseolina*. In addition, infection can reduce the juice volume, sugar content, and total crop biomass (Bandara et al., 2017b).

FIGURE 4.1 Symptoms and signs of charcoal rot in sorghum. (A) Extensive field lodging in Kansas (USA) due to charcoal rot. (B) Comparison of panicles from mock-inoculated (left) and *Macrophomina phaseolina*-inoculated (right) stalks in the greenhouse. (C) Degradation of pith in susceptible sorghum stalks leaving vascular bundles colonized by numerous *M. phaseolina* microsclerotia. (D) Close-up of microsclerotia (white arrow) and sorghum stalk vascular bundles (vb).

Characteristic black-pigmented lesions in the infected areas are visible to the naked eye in longitudinally split stalks. In highly susceptible sorghum varieties, these lesions can be seen in the upper internodes and the peduncle (Tesso et al., 2012). Single, long, continuous lesions can spread across the internodes and nodes. However, a lesion can also be disjointed with marked interruptions at the nodes. Both lesion types spread vertically as the plant matures. As the infected plant dies or reaches the senescence phase, microsclerotia can be observed in the infected area and on the desiccated vascular bundles (Fig. 4.1). Apart from the black color of the lesion, densely packed microsclerotia contribute to the charcoal appearance associated with the disease (Kaur, Dhillon, Brar, Vallad, et al., 2012). As the stalks of the infected plant prematurely deteriorate or reach the senescence phase, the black pigment diminishes and turns gray (Fig. 4.1).

Seedling blight symptoms appear when young roots are infected by *M. phaseolina* (Mughogho & Pande, 1984). The tip of lower seedling leaves dry out and wilt, gradually progressing to the entire lower leaves. Eventually, seedlings turn yellow and rot in the hypocotyl region near or below the soil line. Seedling blight symptoms most often resemble water stress (Livingston, 1945). Seedling blight occurs before or during the development of secondary roots, which can also be affected. However, secondary root rot is not as common as primary root rot. The hypocotyl of the blighted seedling turns black and shrivels (Livingston, 1945). If the seedling escapes early infection and secondary roots generate; it survives and develops into a mature plant (Livingston, 1945; McLaren, 2004; Tarr, 1962). Root rot symptoms of charcoal rot vary from scattered lesions on the root surface to a dark black rot of the entire root, starting at the root base (Livingston, 1945). Early root lesions are typically brown, water-soaked and gradually darken, girdle the root surface, and progress from the cortex to the stele (Livingston, 1945). In the advanced stages of root rot, when the root stele region is affected, the root becomes nonfunctional, shrivels, and dies (Livingston, 1945).

Adult symptomatic plants appear healthy from the outside until internal colonization within the vascular bundles blocks water and nutrient translocation. The phenomenon of hindering vascular bundles with mycelia and microsclerotia can lead to lodging due to significant degradation of the pith and the stalk cortex (Fig. 4.1). Lignified fibrovascular bundles are converted into individual strands, and the stalk becomes hollow. Microsclerotia generated in the later stage of stalk maturity gives a charcoal-like appearance inside the stalk, which indicates how the disease was named (Fig. 4.1) (Mughogho & Pande, 1984). Stalk rot and microsclerotia inside the stalk vary between field and greenhouse conditions. The shredding and disintegration of stalk tissue and microsclerotia have been observed in the lower 6–8 inches of the stalk from field samples. In contrast, in the greenhouse condition, it was observed even from near the head of the plant (Livingston, 1945). The appearance of charcoal rot symptoms is distinguishable from other sorghum symptoms (seedling blight, damping off, root rot, stalk rot, and collar rot) produced by this fungus (Kaur, Dhillon, Brar, Vallad, et al., 2012). Livingston (1945) described "stalk rot symptoms in some varieties as dry and brittle and in some others as soft and spongy."

M. phaseolina causes systemic infection inside the plant, from root to xylem tissue, thus causing rot from the root to stalk. Stalk rot and lodging are usually initiated after flowering, especially at the grain-filling stage when plants are physiologically stressed (Edmunds, 1964). Lodging can also be influenced by the host's physiological status, such as stalk sugar and water level, source and sink balance, and inherent varietal nonsenescence (Das, Fakrudin, & Arora, 2008). Lodging can reach 100%, and grain yield loss can reach nearly 70% if conditions are favorable for disease development (Mughogho & Pande, 1984). Dry fodder quality is also severely deteriorated by charcoal rot disease. Fodder is as essential as grain in regions where sorghum is used in this way (Das, Fakrudin, & Arora, 2008). Seed dry matter, quality, and yield are compromised when lodging occurs during the post-flowering stages (Chadalavada et al., 2021).

4.3 Pathogen biology

Charcoal rot of sorghum is caused by the soilborne, necrotrophic, plurivorous fungus *M. phaseolina* and is of economic concern for yield losses when the host is compromised due to abiotic or biotic stresses and lodging. *M. phaseolina* (Tassi) Goid. is an ascomycete fungus belonging to the class Dothideomycetes, order Botryosphaeriales, and family Botryosphaeriaceae (Crous et al., 2006). This fungus is globally distributed and adapted to various environmental conditions due to the pathogen's physiological, morphological, genetic, and pathogenic variability (Diourte et al., 1995; Reyes-Franco et al., 2006; Reznikov et al., 2018; Su et al., 2001) (Table 4.1). The fungus is morphologically variable in terms of mycelial characteristics, color, size, microsclerotia characteristics, and the presence or absence of pycnidia (Das et al., 2015). Mycelial characteristics vary from hyaline to dark brown or black, aerial to superficial or immersed, septate, and profuse or dendroid branching (Das et al., 2015). The production of pycnidia from maize, sorghum, and soybean isolates is rare, while pycnidia formation is common in isolates from jute and garden beans (Das et al., 2015; Mihail and Taylor, 1995).

The primary inoculum for charcoal rot disease is the microsclerotia produced by *M. phaseolina* (Fig. 4.2). The fungus can survive within crop debris and in the soil as a saprophyte as microsclerotia for up to 15 years (Baird et al., 2003; Kaur,

TABLE 4.1 **A partial global distribution of charcoal rot disease in *Sorghum bicolor*.**

Country	Reference(s)
Argentina	Frezzi and Teyssandier (1980); Teyssandier (1992); Giorda (2002)
Australia	Ryley et al. (2002); Fuhlbohm et al. (2012); Purushotham et al. (2020); Poudel et al. (2021)
Benin	Marley et al. (2002)
Botswana	de Milliano (1992)
Brazil	Casela et al. (1992); Casela et al. (2002)
Burkina Faso	Zummo (1984); Marley et al. (2002)
Burundi	Halluka and Esele (1992)
Cameroon	Zummo (1984); Marley et al. (2002)
China	Indira et al. (2002)
Columbia	Teyssandier (1992)
Ethiopia	Halluka and Esele (1992)
Ghana	Zummo (1984)
Ivory coast	Zummo (1984)
India	Anahosur (1992); Indira et al. (2002)
Iran	Indira et al. (2002)
Kenya	Halluka and Esele (1992)
Mali	Zummo (1984)
Mexico	Narro et al. (1992); Stack (2002); Sarr et al. (2014)
Nigeria	Pande et al. (2008)
Pakistan	Indira et al. (2002)
Panama	Wall and Meckenstock (1992)
Philippines	Tangonan and Dalmacio (1992)
Rwanda	Halluka and Esele (1992)
Senegal	Sarr et al. (2014)
Somalia	Halluka and Esele (1992)
South Africa	de Milliano (1992)
Sudan	Halluka and Esele (1992)
Tanzania	Riley (1960); de Milliano (1992)
Thailand	Boon-Long (1992)
Uganda	Halluka and Esele (1992)
United States	Luttrell (1950); Sprague (1950); Hsi (1956); Johnston et al. (1957); Leyendecker and Nakayama (1956); Mankin (1969); Frederiksen and Duncan (1992); Su et al. (2001); Stack (2002); Bandara (2017)
Venezuela	Teyssandier (1992); Malaguti (2002); Sarr et al. (2014)
Zimbabwe	de Milliano (1992); Whiteside (1966)

Dhillon, Brar, Vallad, et al., 2012; Short et al., 1980). In soybean, *M. phaseolina* can survive as a seedborne pathogen (mycelia or microsclerotia) for up to three years (Gupta et al., 2012). However, the survival of microsclerotia could be affected by low C: N ratio, bulk density, and high soil moisture (Gupta et al., 2012). Microsclerotia are thought to germinate by sensing sorghum seedling root exudates, which infect the primary root (Little, Perumal, et al., 2018; Smith, 1969).

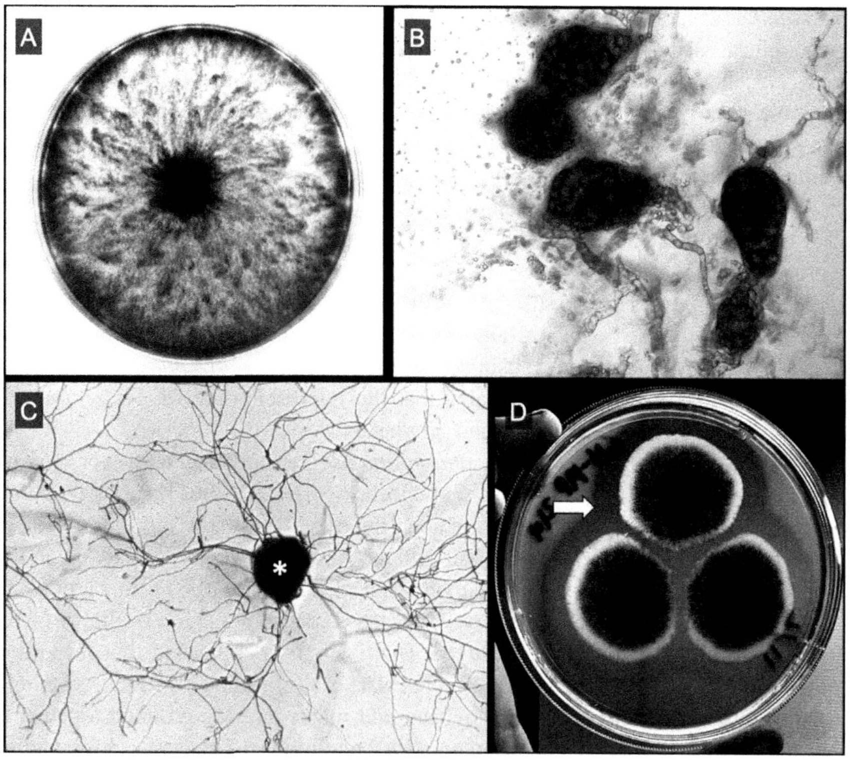

FIGURE 4.2 (A) Colony morphology of *M. phaseolina* on 1/4-strength potato dextrose agar. (B) Microscopic view of *M. phaseolina* microsclerotia (200x). (C) Microsclerotium (*) germination on water agar and hyphal development on a water agar slide after 12 h. Hyphae have been stained with methylene blue. (D) Production of proteases on casein agar by an isolate of *M. phaseolina*. Note zones of clearing in medium adjacent to *M. phaseolina* colonies (white arrow).

Microsclerotia usually measure 50–150 μm in diameter but may vary depending on the host's nutritional status or soil conditions (Short & Wyllie, 1978). Microsclerotia are composed of 50 to >200 individual cells, each containing a septal pore (Gupta et al., 2012; Short & Wyllie, 1978). Microsclerotia repeatedly germinate during a single growing season. However, fungal growth can continue beyond the cropping season as it uses stored nutrients within the microsclerotium to propagate. Thus, *M. phaseolina* competes with other soilborne fungi in nutrient-limiting soil conditions or when weather conditions are unfavorable. In the case of dry rot of chickpea caused by *M. phaseolina*, the microsclerotium germinates a few cells at a time by sensing root exudates, several germ tubes are produced near the epidermal cells of the root, and appressoria are generated (Rai et al., 2022). The appressorium penetrates through the cell walls of the root through mechanical force, enzymatic degradation, natural openings, or any wounding site. Appressoria can be produced within three days of infection by *M. phaseolina* (Ammon et al., 1975).

Invasion of the host cell wall is conferred by lectin, which is a cellulose-binding glycoprotein elicitor. Lectin interacts with the host cell wall to aid in infection and to escape the host's plasma membrane-mediated first layer of defense (Islam et al., 2012). *M. phaseolina* also uses a sophisticated mechanism to protect itself from mechanical injury during host penetration by secreting phytosphingosine and phytoceramide. Although their role in pathogenicity is unknown, these lipids play a role in pathogen-mediated programmed cell death (PCD), ROS burst, fungal germination and proliferation, and alkaline tolerance (Del Poeta et al., 2014; Magnin-Robert et al., 2015).

After penetrating through the cell wall, *M. phaseolina* grows intercellularly and remains in an endophytic or latent state for some period. This invasion stage is asymptomatic and considered a biotrophic state of the fungus (Chowdhury et al., 2017). Penetrating hyphae eventually grow intercellularly in the cortex region and progress toward the vascular bundle, where hyphal colonization blocks water and nutrient transfer. In this late stage of infection, the fungus transitions to the necrotrophic state by inducing host senescence (Chowdhury et al., 2017). Islam et al. (2012) termed this kind of transition as a "biotrophy to necrotrophy switch."

M. phaseolina produces numerous microsclerotia, toxins, and degradative enzymes that induce the stalk to lodge (Kaur, Dhillon, Brar, & Chauhan, 2012). As a result, the stalk dries and breaks at the crown or collar region (Kaur, Dhillon, Brar, & Chauhan, 2012). During the necrotroph stage of the pathogen, *M. phaseolina* induces host senescence to colonize dying or dead tissue. *M. phaseolina* uses a diverse arsenal of cell wall degrading enzymes (CWDE) to manipulate host immunity. Cell wall-degrading endoglucanases were found to be an important pathogenicity factor for *M. phaseolina* necrotrophy (Kaur, Dhillon, Brar, & Chauhan, 2012). Wang and Jones (1995) showed that *M. phaseolina* secretes similar types of endoglucanases as the plant and may provide clues as to how this fungus opportunistically parasitizes the host. *M. phaseolina* possesses two endoglucanase genes, *egl1* and *egl2*, where *egl2* is like those of other saprophytic fungi in function, but *egl1* is similar to the plant's auxin-responsive endoglucanase that functions in cell wall expansion (Hatfield & Nevins, 1987). Thus, *M. phaseolina* may use this endoglucanase to alter plant cell wall elasticity rather than the direct hydrolysis of the cell wall (Carpita & Gibeaut, 1993; Kaur, Dhillon, Brar, & Chauhan, 2012). Other hydrolytic enzymes, such as pectinases, proteases, and phosphatidases, have roles in disease development (Fig. 4.2) (Amadioha, 2000). *M. phaseolina* also secreted phosphatidases in a susceptible *Brassica juncea* plant, suggesting the role of phosphatidases in pathogenicity and host susceptibility (Srivastava & Dhawan, 1982). Kaur, Dhillon, Brar, and Chauhan (2012) compared synthesis of carbohydrate degradative enzymes using one microsclerotial *M. phaseolina* isolate and one mycelial *M. phaseolina* isolate on an artificial media and revealed that the microsclerotial *M. phaseolina* isolate produced more hydrolytic enzymes such as cellulase, hemicellulase, and amylase.

Host specialization has not been observed in *M. phaseolina* isolates per se. However, isolates recovered from a particular crop species are usually more aggressive on that species (Su et al., 2001). Cross-inoculation with isolates obtained from different crops suggested the same phenomenon (Su et al., 2001). A morphological differentiation marker using chlorate-amended media showed chlorate-resistant isolates produced dense colony phenotypes, while chlorate-sensitive isolates had a feathery or restricted phenotype (Pearson et al., 1986, 1987). In particular, sorghum and maize isolates exhibited the "dense" colony chlorate phenotype, whereas soybean (73.2%), alfalfa (100.0%), and sunflower (96.3%) isolates exhibited the feathery chlorate phenotype. However, it is suggested that these morphologically differentiated and identical isolates might not be genetically identical (Su et al., 2001). In fact, an amplified fragment length polymorphism study showed that the majority of the soybean and maize isolates grouped together and the sorghum and tallgrass prairie plants grouped together separately based on AFLP haplotypes. The dense chlorate phenotype as seen in the above study (Pearson et al., 1987) and a unique ITS-rDNA grouping were also associated with most of the sorghum isolates (Saleh et al., 2010).

M. phaseolina also produces several types of toxins. However, none of them have been studied so far concerning pathogenicity in a host except phaseolinone, and no studies have occurred in sorghum. A study of inoculating a mutagenic, nontoxigenic, and avirulent *M. phaseolina* and *Aspergillus fumigatus* in mungbean (*Phaseolus mungo*) seedlings with the presence of phaseolinone confirms the role of this phytotoxin in inducing pathogenicity (Sett et al., 2000). However, no reports on the role of phaseolinone in sorghum root infection have been found (Kaur, Dhillon, Brar, Vallad, et al., 2012).

Sexual reproduction is unknown in this fungus, and genetic variability between isolates mainly occurs through hyphal fusion, mutation, and mitotic recombination (Kaur et al., 2012). Classically, *M. phaseolinia* has been considered a monotypic, but highly variable genus, containing a single species (Echávez-Badel and Perdomo, 1991). However, two new *Macrophomina* species, *M. euphorbiicola,* and *M. tecta* sp. nov., have recently been described (Machado et al., 2019; Poudel et al., 2021). Poudel et al. (2021) indicates that *M. tecta* sp. nov. and *M. phaseolina* are associated with sorghum in Australia. However, it is possible that another species, *M. pseudophaseolina,* and *M. euphorbiicola* are associated with sorghum charcoal rot. Within the genus *Macrophomina*, morphological traits are not sufficient for species differentiation. Further, the molecular characterization of *M. phaseolina* has historically been based on *M. phaseolina*-specific primers (MpKFI and MpKRI) and ITS primers (ITS-1 and ITS-4) (Babu et al., 2007). These primers do not discriminate between *Macrophomina* species. It was only using multigene DNA data (target sequences included the internal transcribed spacer regions 1 and 2, including the 5.8S rRNA gene, translation elongation factor 1-α, actin, calmodulin, and β-tubulin) that the two new species were segregated from isolates that were previously identified as *M. phaseolina* (Santos et al., 2020). Therefore, multilocus phylogenetic analysis of tentative *M. phaseolina* isolates from sorghum could potentially unravel the occurrence of *M. pseudophaseolina* and *M. euphorbiicola*, as well as other cryptic species that could be present under the name *M. phaseolina*.

4.4 Infection process, disease cycle, and epidemiology

Charcoal rot disease is a major yield-limiting disease in all dry sorghum growing regions of Africa, America, Asia, and Australia (Little, Perumal, et al., 2018; Pandey & Basandrai, 2021). *M. phaseolina* can cause substantial yield losses under high temperatures and low soil moisture (Marquez et al., 2021; Perumal et al., 2020).

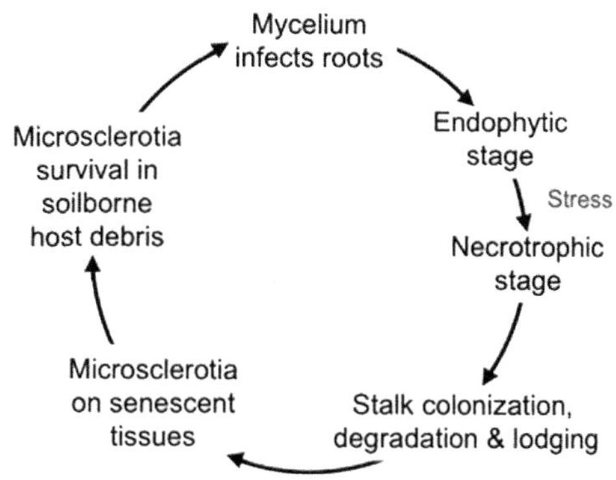

FIGURE 4.3 General disease cycle of *M. phaseolina* in sorghum.

M. phaseolina survives in the form of microsclerotia and mycelia in crop residue and these serve as the initial inoculum to incite disease (Fig. 4.3) (Little, Perumal, et al., 2018). Pycnidia production is a feature that depends on the *M. phaseolina* isolate and the host species (Kaur, Dhillon, Brar, Vallad, et al., 2012). Therefore, the epidemiological role of conidia in the charcoal rot disease cycle is determined by the ability to produce pycnidia. In sorghum, evidence of pycnidia formation is lacking. Consequently, conidia are thought to play no role in sorghum charcoal rot. On the other hand, microsclerotia (highly compacted and melanized masses of fungal mycelia) can remain viable in soil and crop residues for up to 15 years (Marquez et al., 2021).

Microsclerotia germination occurs when the soil temperatures are 30−35°C and in the presence of host root exudates (Smith, 1969). Mycelia produced from germinating microsclerotia can adhere to the root epidermis using transglutaminase-like and cellulose-binding elicitor lectin (CBEL) proteins (Marquez et al., 2021). Although properly adhered mycelia form germ tubes, appressoria, and penetration pegs, they are less prominent structures as compared to those produced by biotrophic fungi. Once the penetration peg breaches the root epidermis, rapid colonization of the internal root tissues occurs. During penetration, *M. phaseolina* deploys molecules, such as salicylate-1-monooxygenase, to neutralize the initial host defense. Typically, adherence to the root epidermis, penetration, and subsequent spread of mycelia within the roots and stalk occurs when plants undergo temperature and moisture stresses (Odvody & Dunkle, 1979). In addition, post-anthesis stress conditions, such as drought, heat, and nutrient imbalances, promote *M. phaseolina* infection. Further, lesion nematodes can enhance charcoal rot incidence by providing entry points for the pathogen (Norton, 1958).

Once in the host, *M. phaseolina* secretes cell wall-degrading enzymes (CWDEs), such as cellulolytic enzymes and laccases, as well as a suite of phytotoxins, including botryodiplodin, patulin, and phaseolinone (Di Matteo et al., 2006; Esquerre-Tugaye et al., 2000; Islam et al., 2012). However, no evidence of phytotoxins as pathogenicity factors has been found in the sorghum-*M. phaseolina* pathosystem. In other pathosystems, these secretions result in host cell death and disease establishment. Moreover, gene expression and biochemical studies of sorghum have revealed that *M. phaseolina* infection increases the biosynthesis of host CWDEs and reactive oxygen and nitrogen species in certain charcoal rot susceptible sorghum varieties (Bandara et al., 2018b, 2019c).

Mycelia promptly spreads to the above-ground basal stalk portions upon root invasion and begins colonizing the lower internodes. Depending on the degree of sorghum susceptibility, growing mycelia can internally spread and invade the upper internodes and the peduncle (Tesso et al., 2012). Microsclerotia are formed within the infected tissues (on vascular stands and pith) as susceptible, infected plants prematurely deteriorate or as moderately susceptible infected plants reach the latter stages of their life cycle. As the infected plants continue to degrade, microsclerotia in root and stem debris amass in the soil and contribute to infections in subsequent seasons (Fig. 4.3). Within the same cropping season, microsclerotia do not contribute to new secondary disease cycles. Further, the vertical transmission of *M. phaseolina* from the infected plant to seeds has not been reported in sorghum.

Pith, cortical, and vascular tissue disintegration is characteristic of charcoal rot disease in both the stalk and root systems. The disintegration of the pith tissues retards water and mineral translocation and storage capacity of the pith and the lateral movement of water and nutrients between the pith and vascular tissues. Degraded pith weakens stalk strength and can contribute to lodging (Zummo, 1984). In addition, damaged cortical and vascular tissues reduce water and mineral uptake and their vertical movement, which results in disrupted photosynthetic activity. Scanning electron microscopy reveals that *M. phaseolina* damages the xylem and phloem tissues. Therefore, besides hindering water and mineral uptake and translocation, the infection also impedes the movement of photosynthates (Hundekar & Anahosur, 1994).

4.5 Disease control

4.5.1 Host resistance

Although host resistance is the primary strategy for controlling charcoal rot, evidence for R-gene or major gene-based resistance for charcoal rot is currently lacking. Charcoal rot resistance in sorghum has historically been attributed to the staygreen trait or delayed post-flowering senescence. Pathogens with aggressive necrotrophic stages, such as *M. phaseolina* can fulfill their energy requirements, colonize, and spread in rapidly senescing tissues. Consequently, staygreen sorghums delay charcoal rot development and severity. For example, leaf greenness and photosystem II photochemical efficiency of two staygreen sorghum parental lines, SC599 and SC35, were not found to be significantly affected after *M. phaseolina* inoculation as compared to mock-inoculated control treatment (Bandara et al., 2019d). In addition, charcoal rot resistance and staygreen did not co-segregate in populations derived from the resistant sorghum variety SC599. This finding suggests that charcoal rot resistance and the staygreen trait in SC599 are independent (Tenkouano et al., 1993). Furthermore, some nonstaygreen lines exhibit charcoal rot resistance. Consequently, nonstaygreen parental lines should remain in breeding programs that target the development of charcoal rot-resistant sorghum hybrids.

Four major QTL blocks are known to be associated with the staygreen trait in sorghum (*stg1*, *stg2*, *stg3*, and *stg4*). Using 14 genotypic groups derived from the Tx642 × Tx7000 recombinant inbred line (RIL) population carrying a combination of *stg* QTLs, it was shown that sorghum varieties that harbor either *stg1* or *stg3,* or both, exhibit high levels of resistance to *M. phaseolina* (Adeyanju et al., 2016). Neither *stg2* nor *stg4* retards charcoal rot disease. Further, the original staygreen QTL donor parent Tx642 is shown to have significantly lower disease severity than the RILs carrying any combination of the four QTL blocks (Adeyanju et al., 2016). This indicates the possible existence of additional QTLs that contribute to charcoal rot resistance.

A genome-wide association study (GWAS) conducted using a sorghum diversity panel comprising 300 genotypes investigated the association between charcoal rot stalk lesion length and single nucleotide polymorphic markers (SNPs) (Adeyanju et al., 2015). This study aimed to unravel the genetic basis of charcoal rot resistance at the reproductive stage. Plants were inoculated with *M. phaseolina* two weeks after anthesis, and lesion measurements were recorded four weeks after inoculation. The analysis revealed ten experiment location-specific SNPs. A combined analysis of data collected across different experimental locations revealed an SNP on sorghum chromosome 9 (position = 57222599) strongly associated with total and major lesion length. *Sb09g028280.1*, which encodes an ROP GTPase protein, was identified as the candidate gene associated with this SNP (Adeyanju et al., 2015). ROP GTPase proteins are signaling molecules that regulate plant hormone responses and disease resistance.

Other studies have used variables, such as pre- and post-emergence damping-off, and plant survival rates as charcoal rot resistance-related phenotypes to perform QTL analysis. By measuring these traits using a sorghum population consisting of landraces and genotyping the same population, SSR markers significantly associated with QTLs conferring resistance to charcoal rot have been identified. Further, several QTL-linked SSR markers were shown to be associated with more than one resistance-related parameter. In silico analysis of four major QTLs (SB2437, SB3258, SbAGB03, and Xtxp287) revealed 75 putative gene homologs that could be considered candidate genes conferring resistance against charcoal rot (Mahmoud et al., 2018). These genes confer resistance to plant pathogens and biotic stressors.

Sorghum breeding lines and hybrids have been developed with resistance or tolerance to charcoal rot. In Texas, Frederiksen and Rosenow (1979) indicated SC599-6 (R 9188, R 9247; IS 17459 derivatives), IS12568C (SC56-14), SC56-6 (IS125 derivative), SC170-6-17 (IS12661 derivative), RT584 (SC56 × SC33), SC35-6 (IS12555 derivative), B4R (B406 × Rio), 1790E (SC56 × SC33), 1790L (SC56 × SC33), 1778 (SC56 × SC33), and NSA440. Further, hybrids (A35 × SC56-14, A4R × NSA440, and A399 × 1790L) were evaluated involving these lines as parents and revealed several combinations that showed low levels of lodging (<5%) in Lubbock and Halfway, Texas in at least one of the test years (Rosenow, 1980). Nonsenescence phenotypes were associated with the charcoal rot resistant materials in these breeding studies. Diourte et al. (1995) found that B35-6 and SC265-14E were resistant to charcoal rot as assessed under conditions of low soil moisture.

In India, a major effort took place at the International Crops Research Institute for the Semi-Arid Tropics (ICRISAT) to identify sources of charcoal rot resistance in Indian sorghums. Rao et al. (1980) found 27 entries (out of 517 tested) that exhibited no soft stalks or diseased nodes crossed after inoculation in the field. Of these, seven (21-78, 25-98, 1-52, 1-30, SC120-14, 4-45, and 6-38) showed <2 cm of lesion length within the stalk. One B-line, 10248-B, exhibited <2 cm of lesion length, but 9.4% of plants had soft stalk symptoms. Further, E36-1 is resistant to charcoal rot, whereas A1, Afzalapur local, SPV249, SPV428, and SPV488 were tolerant (Anahosur, 1992; Anahosur & Patil, 1983). Anahosur (1992) reported that IS3443, IS3547, IS6365, IS8383, and SB1085 were resistant to charcoal rot and IS2328, IS18758,

IS14332, MR750, SPV386, SPV462, SPV736, and SPV775 were moderately resistant. Desal (1998) found that SPV655 and 1055 were resistant to charcoal rot. Indira et al. (2002) indicated that BJ111, CSH6, E36-1, PQ16-2, PQ54, PQ60-2, PQ76, Q101, Q102, Q103, and Q104 were resistant to a complex of stalk rot pathogens including *M. phaseolina*. In a study of defense responses to *M. phaseolina*, Sharma et al. (2013) indicated that PJ-1430 is resistant to charcoal rot.

One parental line developed in Kansas, KS135A/B, was not significantly different than the resistant check, SC599, after artificial inoculation with *M. phaseolina*, for lesion length, diseased nodes crossed, or tolerance rating (100-seed weight divided by lesion length) (Perumal et al., 2015). Perumal et al. (2019) released two Kansas R lines, KS144R and KS146R, that did not appear to differ significantly from mock-inoculated control for relative lesion length (lesion length divided by plant height) or disease nodes crossed after artificial inoculation with *M. phaseolina*. Among the sorghum nested association mapping (NAM) parents, Macia, Ajabsido, and SC35 produced significantly higher relative lesion length than the resistant check, SC599. The other NAM parents, including SC265, SC283, SC971, SC1103, SC1345, and P898012 were not significantly different that the resistant check (Perumal et al., 2021).

4.5.2 Defense responses and molecular mechanisms of resistance

A genome-wide transcriptome study conducted using RNA-sequencing revealed *M. phaseolina*'s ability to significantly upregulate cell wall-degrading enzymes (CWDEs), such as cellulase-, endoglucanase-, pectin methylesterase-, polygalacturonase-, and glycosyl hydrolase encoding genes in a charcoal rot-susceptible sorghum variety (Tx7000) but not in a resistant variety (SC599) (Bandara, Weerasooriya, et al., 2018). Further, functional biochemical assays were conducted using crude enzyme mixtures from sorghum stalk tissues of the same two charcoal rot susceptible sorghum varieties: Tx7000 and BTx3042. These varieties exhibited significantly increased pectin methylesterase, polygalacturonase, and cellulose-degrading enzyme activity after *M. phaseolina* inoculation as compared to mock-inoculated control treatments. On the other hand, *M. phaseolina* did not significantly increase the activity of the same enzymes extracted from two resistant sorghum varieties, SC599 and SC35. These findings revealed *M. phaseolina*'s ability to manipulate host CWDE, which may induce stalk cell wall degradation in charcoal rot-susceptible sorghum varieties (Bandara, Weerasooriya, et al., 2018). Although the host cell wall is a significant barrier to pathogen entrance, plant pathogens, especially pathogens with significant necrotrophic stages such as *M. phaseolina*, can breach this barrier using several mechanisms, such as producing CWDE (Di Matteo et al., 2006; Esquerré-Tugayé et al., 2000). However, as a defensive measure, plants protect their cell walls from fungal CWDEs by producing fungal CWDE-inhibiting proteins, such as polygalacturonase inhibitors, pectin methylesterases inhibitors, pectin lyase inhibitors, xylanase inhibitors, and xyloglucan endoglucanase inhibitors (Juge, 2006). Therefore, *M. phaseolina*-infection-driven upregulation of the host CWDEs can be viewed as a mechanism of enhanced charcoal rot disease susceptibility in sorghum. It also appears to be an efficient virulence strategy of *M. phaseolina*.

Several studies have found sets of sorghum genes that are upregulated in response to *M. phaseolina* infection. Sharma et al. (2013) found that chitinase (*SbChit*) and stilbene synthase (*SbSTS1*) were induced within 24 h after *M. phaseolina* inoculation in a charcoal rot resistant line, PJ-1430, and at a much higher level than in the susceptible line, SU-1080. In a sorghum *bmr12* (COMT, caffeic-O-methyltransferase) mutant, Khasin et al. (2021) found that genes associated with several GO-terms were enriched in response to *M. phaseolina* infection including RNA transport (eukaryotic translation initiation factors), spliceosome (DEAD-box RNA helicases), and protein turnover (E3 ubiquitin ligases and proteasome subunits) as soon as three days after infection. Additionally, these authors showed that two "yieldins" were expressed in response to *M. phaseolina* infection. Yieldins modulate the mechanical characteristics of plant cell walls, especially cell wall loosening, and have been discovered previously in sorghum (Rai et al., 2016).

In another study, RNA sequencing revealed many differentially expressed genes related to host oxidative stress (Bandara et al., 2019c). Among these, genes encoding NADPH oxidase (catalyzes the synthesis of $O_2^{\bullet-}$ [superoxide radical] in the apoplast) and NADH dehydrogenase (catalyzes the synthesis of $O_2^{\bullet-}$ in the mitochondria) were found to be significantly upregulated in *M. phaseolina*-inoculated Tx7000. Similarly, five genes involved in the nitrate reduction I pathway were significantly upregulated in pathogen-inoculated Tx7000 and encoded for isozymes nitric oxide (NO)-forming nitrite reductase. These findings provided transcriptional-level evidence of enhanced reactive oxygen species (ROS) and NO biosynthesis in the charcoal rot susceptible sorghum variety after *M. phaseolina* inoculation. Transcriptional data were corroborated with follow-up functional experiments. For example, compared to the control treatment, greater ROS, and reactive nitrogen species (RNS) were detected in stalk extracts of two charcoal rot susceptible sorghum varieties, Tx7000 and BTx3042, but not in the two resistant varieties, SC599 and SC35. This was further confirmed by increased malondialdehyde content (an indicator of ROS- and RNS-mediated lipid peroxidation) in the stalk extracts of two susceptible varieties. Furthermore, NO over-accumulation in susceptible variety stalk tissues was confirmed using a

NO-specific fluorescent probe (DAF-FM DA) and confocal microscopy (Bandara et al., 2019c). These findings suggested that *M. phaseolina*'s ability to promote a host-derived oxidative stress response in susceptible sorghum genotypes could contribute to induced cell death-associated disease susceptibility to *M. phaseolina*

In the same study, after *M. phaseolina* inoculation, several differentially expressed genes related to the host antioxidant system were reported (Bandara et al., 2019a). Compared to the mock-inoculated control treatment, *M. phaseolina* significantly upregulated the peroxidase, glutathione S-transferase (GST), glutathione peroxidase (GPX), glutathione reductase (GR), glutathione synthetase, and glutamate-cysteine ligase genes in Tx7000, but not in SC599. Biochemical experiments conducted using stalk tissue extracts showed a significantly lower GSH (glutathione, reduced): GSSG (glutathione disulfide, oxidized) ratio in Tx7000 and BTx3042 indicating the strong oxidative stress faced by charcoal rot susceptible sorghum varieties after *M. phaseolina* infection. However, the GSH: GSSG ratio was not significantly different between *M. phaseolina* and control treatments in the resistant varieties SC599 and SC35. Furthermore, *M. phaseolina* infection significantly increased the enzymatic activity of GST, GPX, and GR in Tx7000 and BTx3042 but not in SC599 and SC35 (Bandara et al., 2019a). Glutathione and its related enzymes play an integral role in cellular detoxification and redox buffering (Noctor & Foyer, 1998). In fact, charcoal rot disease severity can be reduced with the exogenous application of glutathione (Fig. 4.4). Therefore, charcoal rot susceptible sorghum varieties mitigate the *M. phaseolina* infection-driven oxidative burst by increased biosynthesis and activity of different antioxidative enzymes. In charcoal rot-resistant sorghum varieties, *M. phaseolina* inoculation does not result in enhanced oxidative stress.

A lipidome study conducted using automated direct infusion electrospray ionization-triple quadrupole mass spectrometry (ESI-MS/MS) quantitatively analyzed the lipid changes that occur in charcoal rot resistant (SC599) and susceptible (Tx7000) sorghum varieties in response to *M. phaseolina* inoculation (Bandara et al., 2019b). This study showed that *M. phaseolina* inoculation significantly decreased phosphatidylserine, phytosterol, and ox-lipid content while significantly increasing the monogalactosyldiacylglycerol (MGDG): digalactosyldiacylglycerol (DGDG) and stigmasterol: sitosterol ratios in Tx7000 stalk tissues. Phosphatidylserine is a key constituent in mitochondrial and cell membranes (Horvath & Daum, 2013). It also plays an essential role in cell death signaling, vesicular trafficking, lipid-protein interactions, and membrane lipid metabolism (Vance, 2008). Phytosterols such as stigmasterol, sitosterol, and campesterol represent the most abundant sterols in plants (Benveniste, 2004). They are integral elements of membrane lipid bilayers and regulators of membrane fluidity and permeability, and influence membrane properties, functions, and structure (Roche et al., 2008; Schaller, 2003). Phytosterols also play a vital role in plant innate immunity against phytopathogens (Wang

FIGURE 4.4 (A) A charcoal rot resistant line (SC599; *top panel*, mock-inoculated; *upper middle panel, M. phaseolina*-inoculated) and susceptible line (Tx7000; *lower middle panel*, mock-inoculated; *lower panel, M. phaseolina*-inoculated). (B) Greenhouse-grown and *M. phaseolina*-inoculated sorghum plants after daily (left two plants) and every 3 days (right two plants) watering regimes. Resistant (SC599, leftmost and center-right plants) and susceptible (Tx7000, center-left and rightmost) plants grew through an inoculum layer of *M. phaseolina*-colonized Japanese millet in the upper portion of the pot to mimic natural infection conditions. (C) Effect of exogenous addition of glutathione after artificial stalk inoculation with *M. phaseolina*. Mock-inoculated (top stalk), *M. phaseolina*-inoculated (middle stalk), and *M. phaseolina*-inoculated + glutathione-treated (bottom stalk) treatments are shown. (D) Zones of clearing on *M. phaseolina* cultures formed by streptomycete isolates obtained from south Texas sorghum field soils.

et al., 2012). Ox-lipids function as signaling molecules that initiate stress responses in plants (Andersson et al., 2006). Plants can produce ox-lipids in response to various stresses including pathogen attacks (Thoma et al., 2003).

MGDG does not form bilayers in mixtures with water, while DGDG is a bilayer-forming lipid (Webb & Green, 1991). The ratio of non-bilayer-forming to bilayer-forming lipids is critical for protein folding and membrane insertion and intracellular protein trafficking (Bogdanov & Dowhan, 1999; Dörmann & Benning, 2002; Gounaris & Barber, 1983). Therefore, the MGDG: DGDG ratio in chloroplasts must be tightly regulated to maintain a stable physical phase and for the proper functioning of the thylakoid membranes. Decreased MGDG: DGDG ratios enhance membrane stability under various stresses (Chen et al., 2006; Ge et al., 2022; Hazei & Williams, 1990; Moellering et al., 2010; Welti et al., 2002). Therefore, the changes observed in sorghum lipids suggest the lethal impacts of *M. phaseolina* inoculation on plastid-, mitochondrial-, and cell-membrane integrity and the lipid-based signaling capacity of the charcoal rot susceptible sorghum variety, Tx7000 (Bandara et al., 2019b). Further, the conversion of sitosterol to stigmasterol is triggered through the perception of pathogen-associated molecular patterns, such as flagellin and lipopolysaccharides, and the generation of ROS and RNS (Griebel & Zeier, 2010). Therefore, the increased stigmasterol: sitosterol ratio corroborates the strong oxidative stress experienced by Tx7000 after *M. phaseolina* inoculation (Bandara et al., 2019b).

4.5.3 Cultural controls and nutrition

M. phaseolina exhibits a broad host range (Burkhardt et al., 2019; Marquez et al., 2021). Further, the microsclerotia they produce can remain viable in the soil for extended periods (Pandey & Basandrai, 2021). Because of this, agronomic practices such as fallowing, crop rotation, and intercropping are less helpful in controlling charcoal rot (Lodha & Mawar, 2019). Disease incidence and severity can be reduced by conserving soil moisture and minimizing water deficit conditions, particularly during the post-flowering stages (Little, Perumal, et al., 2018). Consequently, any conservation tillage practice such as no-till, ridge-till, and mulch-till can contribute to reduced charcoal rot disease. High soil moisture content reduces microsclerotia survival in soil and reduces the initial field inoculum density (Lodha & Mawar, 2019; Marquez et al., 2021). Avoiding excessive N fertilizer application and maintaining a balance between N and K fertilizers are reported to decrease charcoal rot incidence and severity (Little, Perumal, et al., 2018).

Plant stress and competition for resources promote charcoal rot development in most hosts that have been studied. Mughogho and Pande (1984) indicated that the greatest amount of lodging occurred in plant populations of 266,700 plants ha^{-1} while under moisture stress and high N fertilization (Fig. 4.5). Pande et al. (1990) showed that high plant populations led to lodging when moisture stress regimes were imposed. Lodging was greatest when irrigation was stopped at flag leaf visible in the whorl stage, the boot stage, and at 50% flowering. However, plant density did not impact lodging when irrigation was stopped at physiological maturity.

On a related note, varying responses to row spacing, which influences plant populations, have been observed in sorghum. For example, Brown et al. (1964) observed that narrow spacing (50 cm [~20 in]) resulted in greater levels of charcoal rot-associated lodging than wider row spacing (100 cm [~40 in]). On the other hand, Bond et al. (1964) observed

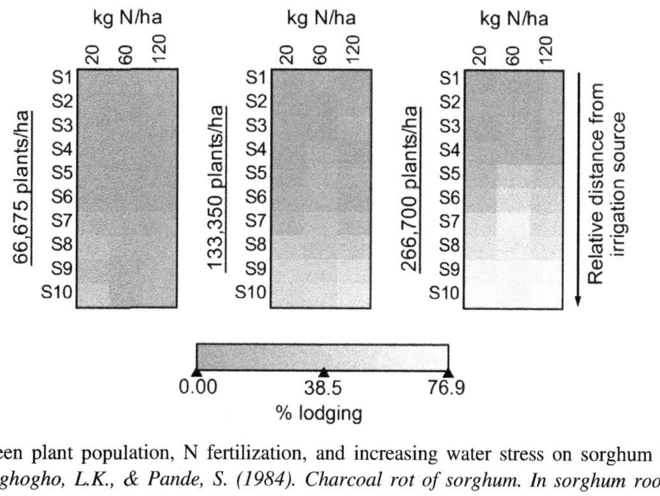

FIGURE 4.5 The interaction between plant population, N fertilization, and increasing water stress on sorghum lodging caused by *M. phaseolina*. *Modified from data presented in Mughogho, L.K., & Pande, S. (1984). Charcoal rot of sorghum. In sorghum root and stalk rots, a critical review: Proceedings of the consultative group discussion on research needs and strategies for control of sorghum root and stalk rot diseases (pp. 11–24), 27 Nov to 2 Dec 1983, Bellagio, Italy.*

the opposite response in Texas dryland production, where wider row spacing exhibited greater levels of lodging. This is possible if plant height and stalk diameter increase, without a concomitant increase in vascular bundle numbers or pith density.

The response of sorghum charcoal rot to NPK has exhibited varied results. Perhaps the most consistent is nitrogen fertilizer, which promotes charcoal rot and lodging in sorghum. Mote and Ramshe (1980) showed that nitrogen application increased charcoal rot incidence in several Rabi sorghum cultivars. Dodd (1980) indicated that high N rates should be avoided in sorghum production where lodging is a concern. Further, Pande et al. (1989) demonstrated that lodging increased in high N environments. Although root infection by *M. phaseolina* did not appear to be influenced by N, Cloud and Rupe (1994) found that disease severity (measured as lesion length) was significantly affected by the interaction between N treatment and growth stage. In general, increased N led to increased lesion length. In corn, Singh and Kaiser (1993) showed a similar response to sorghum. In this case, charcoal rot disease severity increased as N rates were increased. Maximum infection and severity occurred at an N rate of 160 kg ha^{-1}.

Limited information is available on the influence of phosphorus and potassium on charcoal rot in sorghum. In corn, Singh and Kaiser (1993) indicate that P addition reduced *M. phaseolina* growth in vitro. However, it is unclear how that response translates to disease progression in the host under a similar P regime. On the other hand, Gopalakrishnan et al. (2012) suggest that P-solubilizing rhizobacteria could indirectly improve plant health and lodging resilience by making greater levels of plant-useable P available for sorghum plants.

Murphy (1975) added potassium to K-deficient soils, which reduced stalk rot and lodging. Further, Dodd (1980) indicates that low K soils should be avoided in sorghum production where environmental stresses are such that lodging could be a problem. As with phosphorus, K-amended media reduced the in vitro growth of *M. phaseolina* (Singh & Kaiser, 1993).

Numerous weedy plant species that are associated with sorghum production and agricultural lands are also hosts for *M. phaseolina* (Table 4.2 for Poaceae). Monocot weed hosts include *Bouteloua curtipendula* (sideoats grama), *Panicum virgatum* (switchgrass), *Setaria italica* (foxtail), *Sorghum halepense* (Johnsongrass), and *Sorghum* × *drummondii* (shatter cane) (Boewe, 1963; Fuhlbohm et al., 2012; Ghimire et al., 2011; Sprague, 1950, p. 538). Dicot weed hosts include *Abutilon theophrasti* (velvetleaf), *Amaranthus* spp. (pigweeds), *Ambrosia artemisiifolia* (common ragweed), *Asclepias* spp. (milkweeds), *Bidens bipinnata* (Spanish needles), *Boerhavia diffusa* (red spiderling), *Chenopodium album* (lamb's quarters), *Cirsium arvense* (Canada thistle), *Datura stramonium* (jimsonweed), *Helianthus annuus* (annual sunflower), *Malva parviflora* (little mallow), *Malvastrum americanum* (American false mallow), *Melilotus albus* (white sweet clover),

TABLE 4.2 Hosts of *Macrophomina phaseolina* in the Poaceae.

Tribe	Genus	Citation(s)
Andropogoneae	*Saccharum officinarum*	Sarr et al. (2014)
	Sorghum bicolor	*M. phaseolina* (Hsi, 1956; Johnston et al., 1957; Leyendecker & Nakayama, 1956; Luttrell, 1950; Poudel et al., 2021; Saleh et al., 2010; Sarr et al., 2014; Vandemark et al., 2000); *M. tecta* (Poudel et al., 2021)
	S. halepense	Fuhlbohm et al. (2012)
	Zea mays	Johnston et al. (1957); Young and Kucharek (1977); Windham and King (1983); Singh and Kaiser (1993); Baird et al. (2010); Saleh et al. (2010); Sarr et al. (2014); Lopez-Nicora et al. (2016)
Cynodonteae	*Bouteloua curtipendula*	Sprague (1950)
	Muhlenbergia mexicana	Sprague (1950)
	M. schreberi	Lenne (1990)
Oryzeae	*Oryza sativa*	Adikaram and Yakandawala (2020)
Paniceae	*Cenchrus americanus*	Boewe (1963); Lenne (1990)
	Panicum sp.	Lenne (1990)
	Panicum virgatum	Saleh et al. (2010)
Poeae	*Avena sativa*	Boewe (1963)

Oenothera biennis (common evening-primrose), *Physalis minima* (pygmy ground cherry), *Ricinus communis* (weedy castorbean), *Sida spinosa* (prickly sida), *Solanum carolinense* (Carolina horsenettle), *Sonchus oleraceus* (annual sow thistle), *Tribulus terrestris* (puncture vine), *Trifolium repens* (white clover), and *Xanthium* spp. (cockleburs) (Boewe, 1963; Fuhlbohm et al., 2012; Hepperly et al., 1980; Kaur, Dhillon, Brar, Vallad, et al., 2012; Machado et al., 2019; McGlohon, 1959; Nikandrow et al., 1990; Poudel et al., 2021). Therefore, weed management and well-kempt production fields may serve as a management strategy to reduce overall soilborne and residue-borne *M. phaseolina* microsclerotia populations.

4.5.4 Fungicides

The mycelial growth of *M. phaseolina* has been reduced by fungicides, including under in vitro conditions (Marquez et al., 2021). Fungicides could protect sorghum seedlings from *M. phaseolina*-associated pre- and post-emergence damping-off under conducive environmental conditions. However, sorghum seed treatment fungicides do not appear to protect against charcoal rot under field conditions given that disease occurs during the sorghum reproductive stages and microsclerotia endure in the soil. It is currently unknown if the foliar application of systemic fungicides in fields with a history of charcoal rot disease reduces the disease incidence and severity.

Among the most effective fungicides with direct activity against *M. phaseolina* are the methyl benzimidazole carbamates (FRAC Code 1; Table 4.3), which affect tubulin polymerization in the fungus. Ilyas et al. (1976) showed that benomyl, thiophanate-methyl, and thiabendazole decreased microsclerotial viability in fungicide-treated soils. Papavisas (1977) showed that benomyl reduced *M. phaseolina* soil populations. Jaiman and Jain (2010) used the fungicide Bavistin (50% carbendazim) to reduce pre-emergent seed rot, post-emergent damping-off, and adult plant charcoal rot disease severity in cotton. Likewise, these authors tested Topsin (45% thiophanate-methyl) in vitro, which could restrict *M. phaseolina* growth. Tonin et al. (2013) also showed that carbendazim had activity against *M. phaseolina*. However, the methyl benzimidazole carbamates are "high-risk" fungicides because β-tubulin gene mutations have led to fungicide resistance in many pathogens. Also, there is evidence to suggest that some of these fungicides, such as benomyl and thiabendazole, have carcinogenic and teratogenic activities (Kara et al., 2020; Sharma et al., 2022).

Among the lower-risk fungicides with activity against *M. phaseolina* is captan (FRAC Code M04; Table 4.3). Thus far, there has been no evidence of fungicide resistance to captan emerging in fungal pathogen populations. Papavisas (1977) indicated that captan reduced *M. phaseolina* populations. Although not as effective as the methyl benzimidazole carbamates tested, Ilyas et al. (1976) found that captan also decreased microsclerotial viability.

Cohen et al. (2012) found that azoxystrobin (FRAC Code 11; Table 4.3) exhibited activity against charcoal rot and vine disease in melons. Azoxystrobin is a quinone outside inhibitor fungicide that acts on the ubiquinol oxidase cytochrome bc_1 at the Q_o site. Azoxystrobin (and the related strobilurin fungicides) are considered high risk for developing fungicide resistance in pathogen populations and this has been observed in various fungal species. Such resistance occurs due to mutation in the *cyt b* gene that encodes the cytochrome protein. Mahmoud et al. (2006) found that another respiration inhibitor, flutolanil (FRAC Code 7, phenyl-benzamide; Table 4.3), a succinate dehydrogenase inhibitor, could inhibit the growth of 45% of *M. phaseolina* isolates to varying degrees. As in the case of azoxystrobin, flutolanil use has a high risk of mutations in the *sdh* gene, which leads to fungicide resistance.

TABLE 4.3 Fungicide active ingredients with activity against *Macrophomina phaseolina*.

Active ingredient	FRAC code, chemical group	Citations
Azoxystrobin	11, quinone outside inhibitors	Cohen et al. (2012)
Benomyl[a]	1, methyl benzimidazole carbamates	Ilyas et al. (1976); Papavisas (1977)
Captan	M04, pthalimide	Papavisas (1977)
Carbendazim	1, methyl benzimidazole carbamates	Jaiman and Jain (2010); Tonin et al. (2013)
Flutolanil	7, succinate-dehydrogenase inhibitors	Mahmoud et al. (2006)
Thiabendazole[a]	1, methyl benzimidazole carbamates	Ilyas et al. (1976)
Thiophanate-methyl	1, methyl benzimidazole carbamates	Ilyas et al. (1976); Jaiman and Jain (2010)

[a]*Possible carcinogen and teratogen.*

4.5.5 Biocontrol

Numerous biocontrol agents have been reported to suppress *M. phaseolina*. These include bacteria, streptomycetes, and fungi (Table 4.4).

Gopalakrishnan et al. (2011) found several bacterial species, including *Acinetobacter tandoii* and *Bacillus altitudinis*, that could decrease visual charcoal rot disease ratings and decrease sorghum root infection by *M. phaseolina*. Other *Bacillus* spp., including *B. amyloliquefaciens*, *B. siamensis*, and *B. subtilis*, have been shown to inhibit *M. phaseolina*. Torres et al. (2016) found that *B. amyloliquefaciens* and *B. siamensis* produced surfactin lipopeptides could damage *M. phaseolina* microsclerotia. In contrast, these authors hypothesized that *B. subtilis* uses a resource competition strategy to inhibit *M. phaseolina*. *Pseudomonas fluorescens* produces siderophores and inhibits *M. phaseolina* (Simonetti et al., 2015). Das, Indira, et al. (2008) tested 126 *P. fluorescens* strains against *M. phaseolina* in dual culture assays found 30.5–76.5% inhibition for 10 of 126 strains. Three isolates significantly reduced microsclerotial germination (SRB129,

TABLE 4.4 Partial summary of studies examining microorganisms with biocontrol potential for charcoal rot in sorghum.

Organism group and species	Reference(s)
Bacteria	
Acinetobacter tandoii	Gopalakrishnan et al. (2011)
Bacillus altitudinis	Gopalakrishnan et al. (2011)
B. amyloliquefaciens	Torres et al. (2016)
B. siamensis	Hussain and Khan (2020); Torres et al. (2016)
B. subtilis	Torres et al. (2016)
Brevibacterium antiquum	Gopalakrishnan et al. (2011)
Enterobacter ludwigii	Gopalakrishnan et al. (2011)
Pseudomonas chlororaphis	Das, Indira, et al. (2008)
P. fluorescens	Jahagirdar et al. (2001); Das, Indira, et al. (2008)
P. monteilii	Gopalakrishnan et al. (2011)
P. plecoglossicida	Gopalakrishnan et al. (2011)
Unidentified rhizobacteria	Das et al. (2007)
Actinomycetes	
Amycolatopsis sp.	Gopalakrishnan et al. (2019)
Streptomyces spp.	Alekhya et al. (2016)
S. albus	Gopalakrishnan et al. (2020); Gopalakrishnan et al. (2021)
S. aureofaciens	Shrivastava et al. (2017)
S. cavourensis	Gopalakrishnan et al. (2021)
S. griseoplanus	Vijayabharathi et al. (2018)
Fungi	
Penicillium citrinum	Sreevidya and Gopalakrishnan (2016)
Trichoderma atroviride	Larralde-Corona et al. (2007)
T. citrinoviride	Larralde-Corona et al. (2007)
T. koningiopsis	Larralde-Corona et al. (2007)
T. longibrachiatum	Larralde-Corona et al. (2007)
T. reesei	Larralde-Corona et al. (2007)

-127, -288). Further, SRB127, -129 when applied as a seed dressing, reduced charcoal rot and lodging compared to the control.

Blanco et al. (2007) found that sorghum fields in south Texas had a high richness and diversity of inhibitory phenotypes among the streptomycetes found as compared to cucurbit production fields. Therefore, regionally native streptomycete communities associated with sorghum may be a source of biocontrol agents with effective inhibitory activity against soilborne pathogens (Fig. 4.4). Gopalakrishnan et al. (2020) characterized a strain of *Streptomyces albus* (CAI-21) that produced an antifungal organophosphate metabolite. This strain reduced *M. phaseolina* seedling root disease in a blotter test and reduced charcoal rot stalk disease severity in a greenhouse assay. Further, Gopalakrishnan et al. (2021) tested five other *Streptomyces* strains of the species *S. griseus* and *S. cavourensis*. All five strains decreased visual root colonization in a blotter paper assay. In greenhouse experiments, *S. albus* (CAI-17 and KAI-27), *S. cavourensis* (SAI-14), and *S. griseus* (KAI-26) decreased the number of diseased internodes and decreased lesion length compared to an inoculated control plants. In one year of the field experiment, *S. albus* (CAI-17) and *S. griseus* (MMA-32) reduced the same severity measures. Silver nanoparticles synthesized with extracellular extracts of *Streptomyces griseoplanus* SAI-25 also showed antifungal activity against *M. phaseolina* (Vijayabharathi et al., 2018).

Fungi such as *Penicillium* and *Trichoderma* spp. have been found to inhibit *M. phaseolina* in vitro or control charcoal rot in various plant assays. Sreevidya and Gopalakrishnan (2016) found that *P. citrinum* strain VFI-57 could reduce charcoal rot in a seedling blotter paper assay by 80% and in an adult plant disease severity assay by 75%. Larralde-Corona et al. (2007) tested several strains of *Trichoderma* spp. and found that TCBG-2 (*Trichoderma* sp.) and TCBG-8 (*T. koningiopsis*) showed the greatest reduction of fungal growth and production of N-acetylhexosamidase and β-1, 3-glucanase among the strains tested. Further, the authors demonstrated the ability of these two strains to hyperparastize *M. phaseolina*.

4.6 Conclusions

Charcoal rot is a significant disease in all dry sorghum-growing regions of Africa, America, Asia, and Australia. Sorghum charcoal rot can reduce yield, increase host senescence, and cause stalk lodging. *Macrophomina phaseolina* damages the pith, cortical, and vascular tissues, which weakens the stalk. Further, this disease can decrease macronutrients and mineral content in grain sorghum and reduce sugar quality and content in sweet sorghum. Although host resistance is the primary strategy for controlling charcoal rot, evidence for major gene-based resistance is currently lacking. Because *M. phaseolina* has a wide host range, agronomic practices such as fallowing, crop rotation and intercropping, are less helpful in controlling the disease.

Given that plant stress and competition for resources promote charcoal rot in most hosts that have been studied, available moisture and balanced nutrition can reduce charcoal rot incidence and severity. Since sorghum is a water-use-efficient C4 crop, it will remain productive under future climate change scenarios and increased stress-associated disease issues. Using fungicides as seed treatments for sorghum is unlikely to be economically feasible. In addition to improved host resistance and climate resilience, using biological control options is a more environmentally sustainable approach to charcoal rot disease control in sorghum.

Acknowledgments

This book chapter is Publication No. 23-123-B from the Kansas Agricultural Experiment Station, Manhattan, Kansas, USA.

References

Adeyanju, A., Little, C., Yu, J., & Tesso, T. (2015). Genome-wide association study on resistance to stalk rot diseases in grain sorghum. *G3: Genes Genomes Genetics, 5*, 1165–1175.

Adeyanju, A., Yu, J., Little, C., Rooney, W., Klein, P., Burke, J., & Tesso, T. (2016). Sorghum RILs segregating for stay-green QTL and leaf dhurrin content show differential reaction to stalk rot diseases. *Crop Science, 56*, 2895–2903.

Adikaram, N. K. B., & Yakandawala, D. M. D. (2020). A checklist of plant pathogenic fungi and Oomycota in Sri Lanka. *Ceylon Journal of Science, 49*, 93–123.

Akin, P. A., Demirkesen, I., Bean, S. R., Aramouni, F., & Boyaci, I. H. (2022). Sorghum flour application in bread: Technological challenges and opportunities. *Foods, 11*, 2466.

Alavi, S., Mazumdar, S. D., & Taylor, J. R. N. (2019). Modern convenient sorghum and millet food, beverage and animal feed products, and their technologies. In J. R. N. Taylor, & K. G. Duodu (Eds.), *Sorghum and millets: Chemistry, technology and nutritional attributes* (2nd ed., pp. 293–329). Amsterdam, Netherlands: Elsevier.

Alekhya, G., Sharma, R., & Gopalokrishnan, S. (2016). *Streptomyces* spp., a potential biocontrol agent of charcoal rot of sorghum caused by *Macrophomina phaseolina*. *Indian Journal of Plant Protection, 44*, 222–228.

Amadioha, A. C. (2000). The production and activity of extracellular amylase by *Rhizoctonia bataticola*. *Archives of Phytopathology and Plant Protection, 33*, 1–9.

Ammon, V., Wyllie, T. D., & Brown, M. F. (1975). Investigation of the infection process of *Macrophomina phaseolina* on the surface of soybean roots using scanning electron microscopy. *Mycopathologia, 55*, 77–81.

Anahosur, K. H. (1992). Sorghum diseases in India: Knowledge and research needs. In W. A. J. de Milliano, R. A. Frederiksen, & G. D. Bengston (Eds.), *Sorghum and millets diseases: A second world review* (pp. 45–56). Patancheru, India: ICRISAT.

Anahosur, K. H., & Patil, S. H. (1983). Evaluation of sorghum genotypes for charcoal rot resistance. *Indian Phytopathology, 36*, 407–408.

Andersson, M. X., Hamberg, M., Kourtchenko, O., Brunnstrom, A., McPhail, K. L., Gerwick, W. H., Göbel, C., Feussner, I., & Ellerström, M. (2006). Oxylipin profiling of the hypersensitive response in *Arabidopsis thaliana*. Formation of a novel oxo-phytodienoic acid-containing galactolipid, arabidopside E. *Journal of Biological Chemistry, 281*, 31528–31537.

Awika, J. M., & Rooney, L. W. (2004). Sorghum phytochemicals and their potential impact on human health. *Phytochemistry, 65*, 1199–1221.

Babu, B. K., Saxena, A. K., Srivastava, A. K., & Arora, D. K. (2007). Identification and detection of *Macrophomina phaseolina* by using species-specific oligonucleotide primers and probe. *Mycologia, 99*, 797–803.

Baird, R. E., Wadl, P. A., Allen, T., McNeill, D., Wang, X., Moulton, J. K., Rinehart, T. A., Abbas, H. K., Shier, T., & Trigiano, R. N. (2010). Variability of United States isolates of *Macrophomina phaseolina* based on simple sequence repeats and cross genus transferability to related genera within Botryosphaeriaceae. *Mycopathologia, 170*, 169–180.

Baird, R. E., Watson, C. E., & Scruggs, M. (2003). Relative longevity of *Macrophomina phaseolina* and associated mycobiota on residual soybean roots in soil. *Plant Disease, 87*, 563–566.

Bandara, Y. M. A. Y. (2017). *Dissecting the molecular responses of* Sorghum bicolor *to* Macrophomina phaseolina *infection*. Ph.D. dissertation. Kansas State University, 271 pp.

Bandara, Y. M. A. Y., Tesso, T. T., Bean, S. R., Dowell, F. E., & Little, C. R. (2017a). Impacts of fungal stalk rot pathogens on physicochemical properties of sorghum grain. *Plant Disease, 101*, 2059–2065.

Bandara, Y. M. A. Y., Tesso, T. T., Zhang, K., Wang, D., & Little, C. R. (2018). Charcoal rot and Fusarium stalk rot diseases influence sweet sorghum sugar attributes. *Industrial Crops and Products, 112*, 188–195.

Bandara, Y. M. A. Y., Weerasooriya, D. K., Liu, S., & Little, C. R. (2018). The necrotrophic fungus *Macrophomina phaseolina* promotes charcoal rot susceptibility in grain sorghum through induced host cell-wall-degrading enzymes. *Phytopathology, 108*, 948–956.

Bandara, A. Y., Weerasooriya, D. K., Liu, S., & Little, C. R. (2019a). Dynamics of host glutathione and glutathione related enzymes in *Macrophomina phaseolina-Sorghum bicolor* interaction. *bioRxiv*, 853986. https://doi.org/10.1101/853986

Bandara, A. Y., Weerasooriya, D. K., Liu, S., & Little, C. R. (2019b). Host lipid alterations after *Macrophomina phaseolina* infection contribute to charcoal rot disease susceptibility in grain sorghum: Evidence from transcriptomic and lipidomic data. *bioRxiv*, 854067.

Bandara, A. Y., Weerasooriya, D. K., Liu, S., & Little, C. R. (2019c). The necrotrophic fungus *Macrophomina phaseolina* induces oxidative stress-associated genes and related biochemical responses in charcoal rot susceptible sorghum genotypes. *bioRxiv*, 853846.

Bandara, Y. M. A. Y., Weerasooriya, D. K., Tesso, T. T., & Little, C. R. (2017b). Stalk rot diseases impact sweet sorghum biofuel traits. *BioEnergy Research, 10*, 26–35.

Bandara, A. Y., Weerasooriya, D. K., Tesso, T. T., & Little, C. R. (2019d). Stalk rot resistant sorghum genotypes are resilient to pathogen-mediated photosystem II quantum yield retardation. *Crop Protection, 124*, 104852.

Bandara, Y. M. A. Y., Weerasooriya, D. K., Tesso, T. T., Prasad, P. V. V., & Little, C. R. (2017c). Stalk rot fungi affect grain sorghum yield components in an inoculation stage-specific manner. *Crop Protection, 94*, 97–105.

Benveniste, P. (2004). Biosynthesis and accumulation of sterols. *Annual Review of Plant Biology, 55*, 429–457.

Blanco, E. M., Little, C. R., & Davlos Baines, A. L. (2007). Variation in antibiotic inhibitory abilities among streptomycetes from south Texas agricultural soils. *Soil Biology and Biochemistry, 39*, 268–275.

Boewe, G. H. (1963). Host plants of charcoal rot disease in Illinois. *Plant Disease Reporter, 47*, 753–755.

Bogdanov, M., & Dowhan, W. (1999). Lipid-assisted protein folding. *Journal of Biological Chemistry, 274*, 36827–36830.

Bond, J. J., Army, T. J., & Lehman, O. R. (1964). Row spacing, plant populations and moisture supply as factors in dryland grain sorghum production. *Agronomy Journal, 56*, 3–6.

Boon-Long, T. (1992). Sorghum diseases in Thailand. In W. A. J. de Milliano, R. A. Frederiksen, & G. D. Bengston (Eds.), *Sorghum and millets diseases: A second world review* (pp. 41–43). Patancheru, India: ICRISAT.

Brown, A. R., Cobb, C., & Wood, E. H. (1964). Effects of irrigation and row spacing on grain sorghum in the Piedmont. *Agronomy Journal, 56*, 506–509.

Burkhardt, A. K., Childs, K. L., Wang, J., Ramon, M. L., & Martin, F. L. (2019). Assembly, notation, and comparison of *Macrophomina phaseolina* isolates from strawberry and other hosts. *BMC Genomics, 20*, 802.

Carpita, N. C., & Gibeaut, D. M. (1993). Structural models of primary cell walls in flowering plants: Consistency of molecular structure with the physical properties of the walls during growth. *The Plant Journal, 3*, 1–30.

Carr, T. P., Weller, C. L., Schlegel, V. L., Cuppett, S. L., Guderian, D. M., Jr., & Johnson, K. R. (2005). Grain sorghum lipid extract reduces cholesterol absorption and plasma non-HDL cholesterol concentration in hamsters. *Journal of Nutrition, 135*, 2236–2240.

Casela, C. R., Ferreira, A. S., Santos, F. G., & Guimarães, F. B. (2002). Sorghum diseases in Brazil. In J. F. Leslie (Ed.), *Sorghum and millets diseases* (pp. 379–382). Ames, Iowa, United States: Iowa State Press.

Casela, C. R., Ferreira, A. S., & Schaffert, R. E. (1992). Sorghum diseases in Brazil. In W. A. J. de Milliano, R. A. Frederiksen, & G. D. Bengston (Eds.), *Sorghum and millets diseases: A second world review* (pp. 57–62). Patancheru, India: ICRISAT.

Chadalavada, K., Kumari, B. D. R., & Kumar, T. S. (2021). Sorghum mitigates climate variability and change on crop yield and quality. *Planta, 253*, 113.

Chen, J., Burke, J. J., Xin, Z., Xu, C., & Velten, J. (2006). Characterization of the *Arabidopsis* thermo sensitive mutant atts02 reveals an important role for galactolipids in thermotolerance. *Plant, Cell and Environment, 29*, 1437–1448.

Chowdhury, S., Basu, A., & Kundu, S. (2017). Biotrophy-necrotrophy switch in pathogen evoke differential response in resistant and susceptible sesame involving multiple signaling pathways at different phases. *Scientific Reports, 7*, 1–17.

Cisse, F., Erickson, D. P., Hayes, A. M. R., Opekun, A. R., Nichols, B. L., & Hamaker, B. R. (2018). Traditional Malian solid foods made from sorghum and millet have markedly slower gastric emptying than rice, potato, or pasta. *Nutrients, 10*, 124.

Cloud, G. L., & Rupe, J. C. (1994). Influence of nitrogen, plant growth stage, and environment on charcoal rot of grain sorghum caused by *Macrophomina phaseolina* (Tassi) Goid. *Plant and Soil, 158*, 203–210.

Cohen, R., Omari, N., Porat, A., & Edelstein, M. (2012). Management of *Macrophomina* wilt in melons using grafting or fungicide soil application: Pathological, horticultural and economical aspects. *Crop Protection, 35*, 58–63.

Crous, P. W., Slippers, B. W., Wingfield, M. J., Rheeder, J., Marasas, W. F. O., Philips, A. J. L., Alves, A., Burgess, T., Barber, P., & Groenewald, J. Z. (2006). Phylogenetic lineages in the Botryosphaeriaceae. *Studies in Mycology, 55*, 235–253.

Das, I. K., Fakrudin, B., & Arora, D. K. (2008). RAPD cluster analysis and chlorate sensitivity of some Indian isolates of *Macrophomina phaseolina* from sorghum and their relationships with pathogenicity. *Microbiological Research, 163*, 215–224.

Das, I. K., Indira, S., Annapurina, A., & Seetharama, N. (2008). Biocontrol of charcoal rot in sorghum by fluorescent pseudomonads associated with the rhizosphere. *Crop Protection, 27*, 1407–1414.

Das, I. K., Indira, S., & Annapurna, A. (2007). Early growth promotion and charcoal rot suppression in sorghum by plant growth promoting rhizobacteria. *Journal of Biological Control, 21*, 137–144.

Das, I. K., Kishore, B. B., Srivastava, A. K., Kumar, S., & Arora, D. K. (2015). Ecology, biology and management of *Macrophomina phaseolina*: An overview. *Agriculturally Important Microorganisms, 1*, 193–206.

Del Poeta, M., Nimrichter, L., Rodrigues, M. L., & Luberto, C. (2014). Synthesis and biological properties of fungal glucosylceramide. *PLoS Pathogens, 10*, e1003832.

deMilliano, W. A. J. (1992). Sorghum diseases in southern Africa. In W. A. J. de Milliano, R. A. Frederiksen, & G. D. Bengston (Eds.), *Sorghum and millets diseases: A second world review* (pp. 9–19). Patancheru, India: ICRISAT.

Department of Energy (DOE).(2022). Biomass resources: Bioenergy Technologies Office. https://www.energy.gov/eere/bioenergy/biomass-resources. Accessed: 20 Oct 2022.

Desal, S. A. (1998). Evaluation of advanced sorghum varieties against charcoal rot disease in Karnataka. *Karnataka Journal of Agricultural Science, 11*, 104–107.

Di Matteo, A., Bonivento, D., Tsernoglou, D., Federici, L., & Cervone, F. (2006). Polygalacturonase-inhibiting protein (PGIP) in plant defence: A structural view. *Phytochemistry, 67*, 528–533.

Diourte, M., Starr, J. L., Jeger, M. J., Stack, J. P., & Rosenow, D. T. (1995). Charcoal rot (*Macrophomina phaseolina*) resistance and the effects of water stress on disease development in sorghum. *Plant Pathology, 44*, 196–202.

Dodd, J. L. (1980). The photosynthetic stress translocation balance concept of sorghum stalk rot. In *Sorghum diseases: A world review. Proceedings of an international workshop at ICRISAT* (pp. 300–305). Hyderabad, India: ICRISAT Press.

Dörmann, P., & Benning, C. (2002). Galactolipids rule in seed plants. *Trends in Plant Science, 7*, 112–118.

Ducksbury, C., Neale, E. P., & Stefoska-Needham, A. (2021). The effect of sorghum consumption on markers of chronic disease: A systematic review. *Critical Reviews in Food Science and Nutrition, 21*, 1–19.

Echávez-Badel, R., & Perdomo, F. (1991). Characterization and comparative pathogenicity of two *Macrophomina phaseolina* isolates from Puerto Rico. *Journal of Agriculture of the University of Puerto Rico, 75*, 419–421.

Edmunds, L. K. (1964). Combined relation of plant maturity, temperature and soil moisture to charcoal stalk rot development in grain sorghum. *Phytopathology, 54*, 514–517.

Esquerré-Tugayé, M.-T., Boudart, G., & Dumas, B. (2000). Cell wall degrading enzymes, inhibitory proteins, and oligosaccharides participate in the molecular dialogue between plants and pathogens. *Plant Physiology and Biochemistry, 38*, 157–163.

Food and Agriculture Organization of the United Nations. (2022). FAOSTAT Statistical Database (Rome). https://www.fao.org/statistics/en/. Accessed 30 November 2022.

Frederiksen, R. A., & Duncan, R. R. (1992). Sorghum diseases in North America. In W. A. J. de Milliano, R. A. Frederiksen, & G. D. Bengston (Eds.), *Sorghum and millets diseases: A second world review* (pp. 85–88). Patancheru, India: ICRISAT.

Frederiksen, R. A., & Rosenow, D. T. (1979). Breeding for disease resistance in sorghum. In M. K. Harris (Ed.), *Biology and breeding for resistance to arthropods and pathogens in agricultural plants* (pp. 137–167). College Station: Texas, USA: Texas Agricultural Experiment Station, Texas A&M University.

Frezzi, M., & Teyssandier, E. F. (1980). Summary and historical review of sorghum diseases in Argentina. In *Sorghum diseases, A world review: Proceedings of the international workshop on sorghum diseases* (pp. 11–15). Patancheru, India: ICRISAT.

Fuhlbohm, M. F., Ryley, M. J., & Aitken, E. A. B. (2012). New weed hosts of *Macrophomina phaseolina* in Australia. *Australasian Plant Disease Notes, 7*, 193–195.

Galiba, M., Waniska, R. D., Rooney, L. W., & Miller, F. R. (1988). Couscous quality of sorghum with different kernel characteristics. *Journal of Cereal Science, 7*, 183–193.

Ge, S., Liu, D., Chu, M., Liu, X., Wei, Y., Che, X., Zhu, L., He, L., & Xu, J. (2022). Dynamic and adaptive membrane remodeling in leaves of sorghum under salt stress. *The Crop Journal, 10*, 1557–1569.

Ghimire, S. R., Charlton, N. D., Bell, J. D., Krishnamurthy, Y. L., & Craven, K. D. (2011). Biodiversity of fungal endophyte communities inhabiting switchgrass (*Panicum virgatum* L.) growing in the native tallgrass prairie of northern Oklahoma. *Fungal Diversity, 47*, 19–27.

Giorda, L. M. (2002). Sorghum diseases in Argentina. In J. F. Leslie (Ed.), *Sorghum and millets diseases* (pp. 389–391). Ames, Iowa, United States: Iowa State Press.

Gopalakrishnan, S., Humayun, P., Kiran, B. K., Kannan, I. G. K., Vidya, M. S., Deepthi, K., & Rupela, O. (2011). Evaluation of bacteria isolated from rice rhizosphere for biological control of charcoal rot of sorghum caused by *Macrophomina phaseolina* (Tassi) Goid. *World Journal of Microbiology and Biotechnology, 27*, 1313–1321.

Gopalakrishnan, S., Sharma, R., Srinivas, V., Naresh, N., Mishra, S. P., Ankati, S., Pratyusha, S., Govindaraj, M., Gonzalez, S., Nervik, S., & Simic, N. (2020). Identification and characterization of a *Streptomyces albus* strain and its secondary metabolite organophosphate against charcoal rot of sorghum. *Plants, 9*, 1727.

Gopalakrishnan, S., Srinivas, V., Naresh, N., Alekhya, G., & Sharma, R. (2019). Exploiting plant growth-promoting *Amycolatopsis* sp. for bio-control of charcoal rot of sorghum (*Sorghum bicolor* L.) caused by *Macrophomina phaseolina* (Tassi) Goid. *Archives of Phytopathology and Plant Protection, 52*, 543–559.

Gopalakrishnan, S., Srinivas, V., Naresh, N., Pratyusha, S., Ankati, S., Madhuprakash, J., Govindaraj, M., & Sharma, R. (2021). Deciphering the antagonistic effect of *Streptomyces* spp. and host-plant resistance induction against charcoal rot of sorghum. *Planta, 253*, 1–12.

Gopalakrishnan, S., Upadhyay, H. D., Vadlamudi, S., Humayun, P., Vidya, M. S., Alekhya, G., Singh, A., Vijayabharathi, R., Bhimineni, R. K., Seema, M., Rathore, A., & Rupela, O. (2012). Plant growth-promoting traits of biocontrol potential bacteria isolated from rice rhizosphere. *SpringerPlus, 1*, 71.

Gounaris, K., & Barber, J. (1983). Monogalactosyldiacylglycerol: The most abundant polar lipid in nature. *Trends in Biochemical Sciences, 8*, 378–381.

Griebel, T., & Zeier, J. (2010). A role for β-sitosterol to stigmasterol conversion in plant-pathogen interactions. *The Plant Journal, 63*, 254–268.

Gupta, G. K., Sharma, S. K., & Ramteke, R. (2012). Biology, epidemiology and management of the pathogenic fungus *Macrophomina phaseolina* (Tassi) Goid with special reference to charcoal rot of soybean (*Glycine max* (L.) Merrill). *Journal of Phytopathology, 160*, 167–180.

Halluka, M., & Esele, J. P. E. (1992). Sorghum diseases in eastern Africa. In W. A. J. de Milliano, R. A. Frederiksen, & G. D. Bengston (Eds.), *Sorghum and millets diseases: A second world review* (pp. 21–24). Patancheru, India: ICRISAT.

Hatfield, R. D., & Nevins, D. J. (1987). Hydrolytic activity and substrate specificity of an endoglucanase from *Zea mays* seedling cell walls. *Plant Physiology, 83*, 203–207.

Hazei, J. R., & Williams, E. E. (1990). The role of alterations in membrane lipid composition in enabling physiological adaptation of organisms to their physical environment. *Progress in Lipid Research, 29*, 167–227.

Hepperly, P. R., Kirkpatrick, B. L., & Sinclair, J. B. (1980). *Abutilon theophrasti*: Wild host for three fungal parasites of soybean. *Phytopathology, 70*, 307–310.

Hoffman, A. L., Kemanian, A. R., & Forest, C. E. (2020). The response of maize, sorghum, and soybean yield to growing-phase climate revealed with machine learning. *Environmental Research Letters, 15*, 094013.

Horvath, S. E., & Daum, G. (2013). Lipids of mitochondria. *Progress in Lipid Research, 52*, 590–614.

Hsi, C. H. (1956). Stalk rots of sorghum in eastern New Mexico. *Plant Disease Reporter, 40*, 369–371.

Hundekar, A. R., & Anahosur, K. H. (1994). Pathogenicity of fungi associated with sorghum stalk rot. *Karnataka Journal of Agriculture Science, 7*, 291–295.

Hussain, T., & Khan, A. A. (2020). Determining the antifungal activity and characterization of *Bacillus siamensis* AMU03 against *Macrophomina phaseolina* (Tassi) goid. *Indian Phytopathololgy, 73*, 507–516.

Ilyas, M. B., Ellis, M. A., & Sinclair, J. B. (1976). Effect of soil fungicides on *Macrophomina phaseolina* sclerotium viability in soil and in soybean stem pieces. *Phytopathology, 66*, 355–359.

Indira, S., Xu, X., Iamsupasit, N., Shetty, H. S., Vasanthi, N. S., Singh, S. D., & Bandyopadhyay, R. (2002). Diseases of sorghum and millet in Asia. In J. F. Leslie (Ed.), *Sorghum and millets diseases* (pp. 393–402). Ames, Iowa, United States: Iowa State Press.

Islam, M. S., Haque, M. S., Islam, M. M., Emdad, E. M., Halim, A., Hossen, Q. M. M., et al. (2012). Tools to kill: Genome of one of the most destructive plant pathogenic fungi *Macrophomina phaseolina*. *BMC Genomics, 13*, 1–16.

Jahagirdar, S., Patil, M. S., & Indira, S. (2001). Biological control of charcoal rot of sorghum caused by *Macrophomina phaseolina*. *Agricultural Science Digest, 21*, 153–156.

Jaiman, R. K., & Jain, S. C. (2010). Effect of fungicides on root rot of cluster bean caused by *Macrophomina phaseolina*. *Environment and Ecology, 28*, 1138–1140.

Johnston, C. O., Hansing, E. D., Sill, W. H., Jr., Haskett, W. C., & Rogerson, C. T. (1957). Kansas Phytopathological notes: 1955. *Transactions of the Kansas Academy of Science, 60*, 148–154.

Juge, N. (2006). Plant protein inhibitors of cell wall degrading enzymes. *Trends in Plant Science, 11*, 359–367.

Kapanigowda, M. H., Perumal, R., Djanaguiraman, M., Aiken, R. M., Tesso, T., Prasad, P. V. V., & Little, C. R. (2013). Genotypic variation in sorghum [*Sorghum bicolor* (L.) Moench] exotic germplasm collections for drought and disease tolerance. *SpringerPlus, 2*, 650.

Kara, M., Oztas, E., Ramazanogullari, R., Kouretas, D., Nepka, C., Tsatsakis, A. M., & Veskoukis, A. S. (2020). Benomyl, a benzimidazole fungicide, induces oxidative stress and apoptosis in neural cells. *Toxicology Reports, 7*, 501–509.

Kaur, S., Dhillon, G. S., Brar, S. K., & Chauhan, V. B. (2012). Carbohydrate degrading enzyme production by plant pathogenic mycelia and microsclerotia isolates of *Macrophomina phaseolina* through koji fermentation. *Industrial Crops and Products, 36*, 140–148.

Kaur, S., Dhillon, G. S., Brar, S. K., Vallad, G. E., Chand, R., & Chauhan, V. B. (2012). Emerging phytopathogen *Macrophomina phaseolina*: Biology, economic importance and current diagnostic trends. *Critical Reviews in Microbiology, 38*, 136–151.

Khasin, M., Bernhardson, L. F., O'Neill, P. M., Palmer, N. A., Scully, E. D., Sattler, S. E., & Funnell-Harris, D. L. (2021). Pathogen and drought stress affect cell wall and phytohormone signaling to shape host responses in a sorghum COMT *bmr12* mutant. *BMC Plant Biology, 21*, 391.

Larralde-Corona, C. P., Santiago-Mena, M. R., Sifuentes-Rincón, A. M., Rodríguez-Luna, I. C., Rodríguez-Pérez, M. A., Shirai, K., & Narváez-Zapata, J. A. (2007). Biocontrol potential and polyphasic characterization of novel native *Trichoderma* strains against *Macrophomina phaseolina* isolated from sorghum and common bean. *Applied Microbiology and Biotechnology, 80*, 167–177.

Lee, S.-H., Lee, H.-S., Lee, J., Amarakoon, D., Lou, Z., Noronha, L. E., Herald, T. J., Perumal, R., & Smolensky, D. (2021). Polyphenol containing sorghum brans exhibit an anti-cancer effect in Apc Min/+ mice treated with dextran sodium sulfate. *International Journal of Molecular Sciences, 22*, 8286.

Lenne, J. M. (1990). World list of fungal diseases of tropical pasture species. *Phytopathological Papers, 31*, 1–162.

Leyendecker, P. J., & Nakayama, R. M. (1956). 1955 plant disease summary for New Mexico. *Plant Disease Reporter, 40*, 159–161.

Little, C. R., Bandara, Y. M. A. Y., & Perumal, R. (2018). Diseases and their management in sorghum cultivation: Seedling, seed, panicle, and foliar diseases. In W. L. Rooney (Ed.), *Achieving sustainable cultivation in sorghum: Genetics, breeding, and production techniques*. Burleigh Dodds Science Publishing, Ltd.

Little, C. R., & Perumal, R. (2018). The biology and control of sorghum diseases. In I. Ciampitti, & P. V. V. Prasad (Eds.), Sorghum: State of the art and future perspectives agronomy monographs, *No. 58*. Madison, WI, USA: ASA and CSSA.

Little, C. R., Perumal, R., Bandara, Y. M. A. Y., & Todd, T. C. (2018). Diseases and their management in sorghum cultivation: Stalk, root, and other diseases. In W. L. Rooney (Ed.), *Achieving sustainable cultivation in sorghum: Genetics, breeding, and production techniques*. Burleigh Dodds Science Publishing, Ltd.

Livingston, J. E. (1945). Charcoal rot of corn and sorghum in Nebraska. *Plant Disease Reporter, 34*, 45–52.

Lodha, S., & Mawar, R. (2019). Population dynamics of *Macrophomina phaseolina* in relation to disease management: A review. *Journal of Phytopathology, 168*, 1–17.

Lopez-Nicora, H. D., Simon, A. C. M., Dossman, B. C., Paul, P. A., Dorrance, A. E., Lindsey, L. E., & Niblack, T. L. (2016). Distribution and abundance of *Heterodera glycines* and *Macrophomina phaseolina* in Ohio. *Plant Health Progress, 17*, 35–41.

Luttrell, E. S. (1950). Grain sorghum diseases in Georgia—1949. *Plant Disease Reporter, 34*, 45–52.

Machado, A. R., Pinho, D. B., Soares, D. J., Gomes, A. A. M., & Pereira, O. L. (2019). Bayesian analyses of five gene regions reveal a new phylogenetic species of *Macrophomina* associated with charcoal rot on oilseed crops in Brazil. *European Journal of Plant Pathology, 153*, 89–100.

Magnin-Robert, M., Le Bourse, D., Markham, J., Dorey, S., Clément, C., Baillieul, F., & Dhondt-Cordelier, S. (2015). Modifications of sphingolipid content affect tolerance to hemibiotrophic and necrotrophic pathogens by modulating plant defense responses in *Arabidopsis*. *Plant Physiology, 169*, 2255–2274.

Mahmoud, A. F., Abou-Elwafa, S. F., & Shehzad, T. (2018). Identification of charcoal rot resistance QTLs in sorghum using association and in silico analyses. *Journal of Applied Genetics, 59*, 243–251.

Mahmoud, Y. A.-G., Aly, A. A., Omar, M. R., & Ismail, A.-W. (2006). Variation in sensitivity among some isolates of *Macrophomina phaseolina* isolates from cotton roots to flutolanil fungicide. *Mycobiology, 34*, 99–103.

Malaguti, G. (2002). Current status of sorghum diseases in Venezuela. In J. F. Leslie (Ed.), *Sorghum and millets diseases* (pp. 415–417). Ames, Iowa, United States: Iowa State Press.

Mankin, C. J. (1969). *Diseases of grasses and cereals in South Dakota* (Vol 35, pp. 1–27). South Dakota State University Technical Bulletin.

Marley, P. S., Diourté, M., Neya, A., Nutsugah, S. K., Sérémé, P., Katilé, S. O., Hess, D. E., Mbaye, D. F., & Ngoko, Z. (2002). Sorghum and pearl millet diseases in west and central Africa. In J. F. Leslie (Ed.), *Sorghum and millets diseases* (pp. 419–425). Ames, Iowa, United States: Iowa State Press.

Marquez, N., Giachero, M. L., Declerck, S., & Ducasse, D. A. (2021). *Macrophomina phaseolina*: General characteristics of pathogenicity and methods of control. *Frontiers of Plant Science, 12*, 634397.

McCary, C. L., Vyas, D., Faciola, A. P., & Ferraretto, L. F. (2020). *Graduate student literature review*: Current perspectives on whole-plant sorghum silage production and utilization by lactating dairy cows. *Journal of Dairy Science, 103*, 5783–5790.

McCuistion, K. C., Selle, P. H., Liu, S. Y., & Goodband, R. D. (2019). Sorghum as a feed grain for animal production. In J. R. N. Taylor, & K. G. Duodu (Eds.), *Sorghum and millets: Chemistry, technology and nutritional attributes* (2nd ed., pp. 355–391). Amsterdam, Netherlands: Elsevier.

McGlohon, N. E. (1959). Survey of fungi associated with white clover stolons. *Plant Disease Reporter, 43*, 22–24.

McLaren, N. W. (2004). Effect of soil nutrient status on severity of seedling diseases and root rot of sorghum (*Sorghum bicolor*). *South Africa Journal of Plant and Soil, 21*, 263–265.

Mihail, J. D., & Taylor, S. J. (1995). Interpreting variability among isolates of *Macrophomina phaseolina* in pathogenicity, pycnidium production, and chlorate utilization. *Canadian Journal of Botany, 73*, 1596–1603.

Moellering, E. R., Muthan, B., & Benning, C. (2010). Freezing tolerance in plants requires lipid remodeling at the outer chloroplast membrane. *Science, 330*, 226–228.

Mote, U. N., & Ramshe, D. G. (1980). Nitrogen application increases the incidence of charcoal rot in Rabi sorghum cultivars. *Sorghum Newsletter, 23*, 129.

Mughogho, L. K., & Pande, S. (1984). Charcoal rot of sorghum. In Sorghum root and stalk rots, a critical review: Proceedings of the consultative group discussion on research needs and strategies for control of sorghum root and stalk rot diseases*(pp 11–24), 27 Nov to 2 Dec 1983, Bellagio, Italy*.

Mundia, C. W., Secchi, S., Akamani, K., & Wang, G. (2019). A regional comparison of factors affecting global sorghum production: The case of North America, Asia and Africa's Sahel. *Sustainability, 11*, 2135.

Murphy, L. S. (1975). Fertilizer efficiency for corn and grain sorghum. In *Proceedings of the 30th annual corn and sorghum research conference* (Vol 30, pp. 49–72).

Narro, J., Betancourt, V. A., & Aguirre, J. I. (1992). Sorghum diseases in Mexico. In W. A. J. de Milliano, R. A. Frederiksen, & G. D. Bengston (Eds.), *Sorghum and millets diseases: A second world review* (pp. 75–84). Patancheru, India: ICRISAT.

Nikandrow, A., Weidemann, G. J., & Auld, B. A. (1990). Incidence and pathogenicity of *Colletotrichum orbiculare* and a *Phomopsis* sp. on *Xanthium* sp. *Plant Disease, 74*, 796–799.

Noctor, G., & Foyer, C. H. (1998). Ascorbate and glutathione: Keeping active oxygen under control. *Annual Review of Plant Physiology and Plant Molecular Biology, 49*, 249–279.

Norton, D. C. (1958). The association of *Pratylenchus hexincisus* with charcoal rot of sorghum. *Phytopathology, 48*, 355–358.

Odvody, G. N., & Dunkle, L. D. (1979). Charcoal stalk rot of sorghum: Effect of environment on host-parasite relations. *Phytopathology, 69*, 250–254.

Pande, S., Harikrishnan, R., Alegbejo, M. D., Mughogho, L. K., Karunakar, R. I., & Ajayi, O. (2008). Prevalence of sorghum diseases in Nigeria. *International Journal of Pest Management, 39*, 297–303.

Pande, S., Mughogho, L. K., & Karunakar, R. I. (1990). Effect of moisture stress, plant population density and pathogen inoculation on charcoal rot of sorghum. *Annals of Applied Biology, 116*, 221–232.

Pande, S., Mughogho, L. K., Seetharama, N., & Karunakar, R. I. (1989). Effects of nitrogen, plant density, moisture stress, and artificial inoculation with *Macrophomina phaseolina* on charcoal rot incidence in grain sorghum. *Journal of Phytopathology, 126*, 343–352.

Pandey, A. K., & Basandrai, A. K. (2021). Will *Macrophomina phaseolina* spread in legumes due to climate change? A critical review of current knowledge. *Journal of Plant Diseases and Protection, 128*, 9–18.

Papavisas, G. C. (1977). Some factors affecting survival of *Macrophomina phaseolina* in soil. *Soil Biology and Biochemistry, 9*, 337–341.

Pearson, C. A. S., Leslie, J. F., & Schwenk, F. W. (1986). Variable chlorate resistance in *Macrophomina phaseolina* from corn, soybean, and soil. *Phytopathology, 76*, 646–649.

Pearson, C. A. S., Leslie, J. F., & Schwenk, F. W. (1987). Host preference correlated with chlorate resistance in *Macrophomina phaseolina. Plant Disease, 71*, 828–831.

Perumal, R., Tesso, T. T., Kofoid, K. D., Aiken, R. M., Prasad, P. V. V., Bean, S. R., Wilson, J. D., Herald, T. J., & Little, C. R. (2019). Registration of six grain sorghum pollinator (R) lines. *Journal of Plant Registrations, 13*, 113–117.

Perumal, R., Tesso, T. T., Kofoid, K. D., Prasad, P. V. V., Aiken, R. M., Bean, S. R., Wilson, J. D., Herald, T. J., & Little, C. R. (2015). Registration of nine grain sorghum seed parent (A/B) lines. *Journal of Plant Registrations, 9*, 244–248.

Perumal, R., Tesso, T. T., Morris, G. P., Jagadish, S. V. K., Little, C. R., Bean, S. R., Yu, J., Prasad, P. V. V., & Tuinstra, M. R. (2021). Registration of the sorghum nested association mapping (NAM) population in RTx430 background. *Journal of Plant Registrations, 15*, 395–402.

Perumal, R., Tomar, S. S., Bandara, A., Maduraimuthu, D., Tesso, T. T., Prasad, P. V., Upadhyaya, H. D., & Little, C. R. (2020). Variation in stalk rot resistance and physiological traits of sorghum genotypes in the field under high temperature. *Journal of General Plant Pathology, 86*, 350–359.

Poudel, B., Shivas, R. G., Adorada, D. L., Barbetti, M. J., Bithell, S. L., Kelly, L. A., Moore, N., Sparks, A. H., Tan, Y. P., Thomas, G., Van Leur, J., & Vaghefi, N. (2021). Hidden diversity of *Macrophomina* associated with broadacre and horticultural crops in Australia. *European Journal of Plant Pathology, 161*, 1–23.

Prasad, V. B. R., Govindaraj, M., Djanaguiraman, M., Djalovic, I., Shailani, A., Rawat, N., Singla-Pareek, S. L., Pareek, A., & Prasad, P. V. V. (2021). Drought and high temperature stress in sorghum: Physiological, genetic, and molecular insights and breeding approaches. *International Journal of Molecular Sciences, 22*, 9826.

Purushotham, N., Jones, A., Poudel, B., Nasim, J., Adorada, D., Sparks, A., Schwessinger, B., & Vaghefi, N. (2020). Draft genome resource for *Macrophomina phaseolina* associated with charcoal rot in sorghum. *Molecular Plant-Microbe Interactions, 33*, 724–726.

Rai, A., Irulappan, V., & Senthil-Kumar, M. (2022). Dry root rot of chickpea: A disease favored by drought. *Plant Disease, 106*, 346–356.

Rai, K. M., Thu, S. W., Balasubramanian, V. K., Cobos, C. J., Disasa, T., & Mendu, V. (2016). Identification, characterization, and expression analysis of cell wall related genes in *Sorghum bicolor* (L.) Moench, a food, fodder, and biofuel crop. *Frontiers of Plant Science, 7*, 1287.

Rao, K. N., Reddy, V. S., Williams, R. J., & House, L. R. (1980). The ICRISAT charcoal rot resistance program. In *Sorghum diseases: A world review* (pp 306–314). In *Proceedings of an international workshop at ICRISAT.* Hyderabad, India: ICRISAT Press.

Reyes-Franco, M. C., Hernández-Delgado, S., Beas-Fernández, R., Medina-Fernandez, M., Simpson, J., & Mayek-Pérez, N. (2006). Pathogenic and genetic variability within *Macrophomina phaseolina* from Mexico and other countries. *Journal of Phytopathology, 154*, 447–453.

Rezaee, N., Fernando, W. M. A. D. B., Hone, E., Sohrabi, H. R., Johnson, S. K., Gunzburg, S., & Martins, R. N. (2021). Potential of sorghum polyphenols to prevent and treat Alzheimer's disease: A review article. *Frontiers in Aging Neuroscience, 13*, 729949.

Reznikov, S., Vellicce, G. R., Mengistu, A., Arias, R. S., Gonzalez, V., Lisi, V. D., Gabriela, G. M., Lourdes, R. C. M., Pardo, E. M., Castagnaro, A. P., & Ploper, L. D. (2018). Disease incidence of charcoal rot (*Macrophomina phaseolina*) on soybean in north-western Argentina and genetic characteristics of the pathogen. *Canadian Journal of Plant Pathology, 40*, 423–433.

Riley, E. A. (1960). A revised list of plant diseases in Tanganyika Territory. *Mycological Papers, 75*, 1–42.

Roche, Y., Gerbeau-Pissot, P., Buhot, B., Thomas, D., Bonneau, L., Gresti, J., Mongrand, S., Perrier-Cornet, J.-M., & Simon-Plas, F. (2008). Depletion of phytosterols from the plant plasma membrane provides evidence for disruption of lipid rafts. *The FASEB Journal, 22*, 3980–3991.

Rosenow, D. T. (1980). Stalk rot resistance breeding in Texas. In *sorghum diseases: A world review* (pp. 306–314). In *Proceedings of an international workshop at ICRISAT*. Hyderabad, India: ICRISAT Press.

Ryley, M. J., Persley, D. M., Jordan, D. R., & Henzell, R. G. (2002). Status of sorghum and pearl millet diseases in Australia. In J. F. Leslie (Ed.), *Sorghum and millets diseases* (pp. 441–448). Ames, Iowa, United States: Iowa State Press.

Saleh, A. A., Ahmed, H. U., Todd, T. C., Travers, S. E., Zeller, K. A., Leslie, J. F., & Garrett, K. A. (2010). Relatedness of *Macrophomina phaseolina* isolates from tallgrass prairie, maize, soybean, and sorghum. *Molecular Ecology, 19*, 79–91.

Santos, K. M., Lima, G. S., Barros, A. P. O., Machado, A. R., Souza-Motta, C. M., Correia, K. C., & Michereff, S. J. (2020). Novel specific primers for rapid identification of *Macrophomina* species. *European Journal of Plant Pathology, 156*, 1213–1218.

Sarr, M. P., Ndiaye, M., Groenewald, J. Z., & Crous, P. W. (2014). Genetic diversity in *Macrophomina phaseolina*, the causal agent of charcoal rot. *Phytopathologia Mediterranea, 53*, 250–268.

Schaller, H. (2003). The role of sterols in plant growth and development. *Progress in Lipid Research, 42*, 163–175.

Sett, S., Mishra, S. K., & Siddiqui, K. I. (2000). Avirulent mutants of *Macrophomina phaseolina* and *Aspergillus fumigatus* initiate infection in *Phaseolus mungo* in the presence of phaseolinone; levamisole gives protection. *Journal of Biosciences, 25*, 73–80.

Sharma, I., Kumari, N., & Sharma, V. (2013). Defense gene expression in *Sorghum bicolor* against *Macrophomina phaseolina* in leaves and roots of susceptible and resistant cultivars. *Journal of Plant Interactions, 9*, 315–323.

Sharma, M., Maheshwari, N., Khan, F. H., & Mahmood, R. (2022). Carbendazim toxicity in different cell lines and mammalian tissues. *Journal of Biochemical and Molecular Toxicology*, e23194.

Short, G. E., & Wyllie, T. D. (1978). Inoculum potential of *Macrophomina phaseolina. Phytopathology, 66*, 188–192.

Short, G. E., Wyllie, T. D., & Bristow, P. R. (1980). Survival of *Macrophomina phaseolina* in soil and in residue of soybean. *Phytopathology, 70*, 13–17.

Shrivastava, P., Kumar, R., & Yandigeri, M. S. (2017). *In vitro* biocontrol activity of halotolerant *Streptomyces aureofaciens* K20: A potent antagonist against *Macrophomina phaseolina* (Tassi) goid. *Saudi Journal of Biological Sciences, 24*, 192–199.

Simonetti, E., Viso, N. P., Montecchia, M., Zilli, C., Balestrasse, K., & Carmona, M. (2015). Evaluation of native bacteria and manganese phosphite for alternative control of charcoal root rot of soybean. *Microbiological Research, 180*, 40–48.

Singh, R. D. N., & Kaiser, S. A. K. M. (1993). Effect of agronomic practices on the growth and spread of charcoal rot pathogen (*Macrophomina phaseolina*) infecting maize. *Crop Research, 6*, 499–508.

Smith, W. H. (1969). Germination of *Macrophomina phaseoli* sclerotia as affected by *Pinus lambertiana* root exudates. *Canadian Journal of Microbiology, 15*, 1387–1391.

Sprague, R. (1950). *Diseases of cereals and grasses in North America*. New York: Ronald Press Company.

Sreevidya, M., & Gopalakrishnan, S. (2016). *Penicillium citrinum* VFI-51 as biocontrol agent to control charcoal rot of sorghum (*Sorghum bicolor* (L.) Moench). *African Journal of Microbiology Research, 10*, 669–674.

Srivastava, S. K., & Dhawan, S. (1982). Phosphatidase activity in *Brassica juncea* plants infected with isolates of *Macrophomina phaseolina* and its role in pathogenesis. *Bulletin of the Torrey Botanical Club, 109*, 508–512.

Stack, J. P. (2002). Recurring and emerging diseases in North America. In J. F. Leslie (Ed.), *Sorghum and millets diseases* (pp. 449–456). Ames, Iowa, United States: Iowa State Press.

Staggenborg, S. A., Dhuyvetter, K. C., & Gordon, W. B. (2008). Grain sorghum and corn comparisons: Yield, economic, and environmental responses. *Agronomy Journal, 100*, 1600–1604.

Su, G., Suh, S. O., Schneider, R. W., & Russin, J. S. (2001). Host specialization in the charcoal rot fungus, *Macrophomina phaseolina. Phytopathology, 91*, 120–126.

Tangonan, N. G., & Dalmacio, S. C. (1992). Diseases of sorghum in the Phillipines. In W. A. J. de Milliano, R. A. Frederiksen, & G. D. Bengston (Eds.), *Sorghum and millets diseases: A second world review* (pp. 35–40). Patancheru, India: ICRISAT.

Tarr, S. A. J. (1962). *Diseases of sorghum, Sudan grass, and broom corn*. Kew, Surrey, England: Commonwealth Mycological Institute.

Tenkouano, A., Miller, F. R., Frederiksen, R. A., & Rosenow, D. T. (1993). Genetics of nonsenescence and charcoal rot resistance in sorghum. *Theoretical and Applied Genetics, 85*, 644–648.

Tesso, T., Perumal, R., Little, C. R., Adeyanju, A., Radwan, G. L., Prom, L. K., & Magill, C. W. (2012). Sorghum pathology and biotechnology—a fungal disease perspective: Part II. Anthracnose, stalk rot, and downy mildew. *The European Journal of Plant Science and Biotechnology, 6*, 31–44.

Teyssandier, E. (1992). Sorghum diseases in south America. In W. A. J. de Milliano, R. A. Frederiksen, & G. D. Bengston (Eds.), *Sorghum and millets diseases: A second world review* (pp. 63–66). Patancheru, India: ICRISAT.

Thoma, I., Loeffler, C., Sinha, A. K., Gupta, M., Krischke, M., Steffan, B., Roitsch, T., & Mueller, M. J. (2003). Cyclopentenone isoprostanes induced by reactive oxygen species trigger defense gene activation and phytoalexin accumulation in plants. *The Plant Journal, 34*, 363–375.

Tonin, R. F. B., Avozani, A., Danelli, A. L. D., Reis, E. M., Zoldan, S. M., & Garcés-Fiallos, F. R. (2013). *In vitro* mycelial sensitivity of *Macrophomina phaseolina* to fungicides. *Pesquisa Agropecuária Tropical, 43*, 460–466.

Torres, M. J., Brandan, C. P., Petroselli, G., Erra-Balsells, R., & Audisio, M. C. (2016). Antagonistic effects of *Bacillus subtilis* subsp. *subtilis* and *B. amyloliquefaciens* against *Macrophomina phaseolina*: SEM study of fungal changes and UV-MALDI-TOF MS analysis of their bioactive compounds. *Microbiological Research, 182*, 31–39.

United States Department of Agriculture, National Agricultural Statistics Service. (2022). https://www.nass.usda.gov. Accessed 30 November 2022.

Vance, J. E. (2008). Phosphatidylserine and phosphatidylethanolamine in mammalian cells: Two metabolically related aminophospholipids. *Journal of Lipid Research, 49*, 1377–1387.

Vandemark, G. J., Martinez, O., Pecina, V., & Alvarado, M. (2000). Assessment of genetic relationships among isolates of *Macrophomina phaseolina* using a simplified AFLP technique and two different methods of analysis. *Mycologia, 92*, 656–664.

Vijayabharathi, R., Sathya, A., & Gopalakrishnan, S. (2018). Extracellular biosynthesis of silver nanoparticles using *Streptomyces griseoplanus* SAI-25 and its antifungal activity against *Macrophomina phaseolina*, the charcoal rot pathogen of sorghum. *Biocatalysis and Agricultural Biotechnology, 14*, 166–171.

Wall, G. C., & Meckenstock, D. H. (1992). Sorghum diseases in Central America and the Caribbean basin. In W. A. J. de Milliano, R. A. Frederiksen, & G. D. Bengston (Eds.), *Sorghum and millets diseases: A second world review* (pp. 67–73). Patancheru, India: ICRISAT.

Wang, H., & Jones, R. W. (1995). A unique endoglucanase-encoding gene cloned from the phytopathogenic fungus *Macrophomina phaseolina*. *Applied and Environmental Microbiology, 61*, 2004–2006.

Wang, K., Senthil-Kumar, M., Ryu, C. M., Kang, L., & Mysore, K. S. (2012). Phytosterols play a key role in plant innate immunity against bacterial pathogens by regulating nutrient efflux into the apoplast. *Plant Physiology, 158*, 1789–1802.

Webb, M., & Green, B. R. (1991). Biochemical and biophysical properties of thylakoid acyl lipids. *Biochimica et Biophysica Acta, 1060*, 133–158.

Welti, R., Li, W., Li, M., Sang, Y., Biesiada, H., Zhou, H. E., Rajashekar, C. B., Williams, T. D., & Wang, X. (2002). Profiling membrane lipids in plant stress responses. Role of phospholipase D alpha in freezing-induced lipid changes in *Arabidopsis*. *Journal of Biological Chemistry, 277*, 31994–32002.

Whiteside, J. O. (1966). A revised list of plant diseases in Rhodesia. *Kirkia, 5*, 87–196.

Windham, M. T., & King, S. B. (1983). Mycoflora of roots of maize plants at seedling and silking stages in Mississippi. *Plant Disease, 67*, 1366–1368.

Xing-Lin, H., De-Liang, W., Wu-Jiu, Z., & Shi-Ru, J. (2017). The production of Chinese baijiu from sorghum and other cereals. *Journal of the Institute of Brewing, 123*, 600–604.

Young, T. R., & Kucharek, T. A. (1977). Succession of fungal communities in roots and stalks of hybrid field corn grown in Florida. *Plant Disease Reporter, 61*, 76–80.

Zummo, N. (1984). *Sorghum diseases in western Africa*. Washington DC 20250, USA: US-AID. 32 pp. USDA.

Chapter 5

Recent advancement in management of soybean charcoal rot caused by *Macrophomina phaseolina*

Laxman Singh Rajput[1,4], Sanjeev Kumar[1], V. Nataraj[2], M. Shivakumar[2], Kriti Pathak[1], Sapna Jaiswal[1,3], Saloni Mandloi[1,3], Nisha Agrawal[2], M.B. Ratnaparkhe[2], Giriraj Kumawat[2], Hemant Singh Maheshwari[2] and Vibha Pandey[3]

[1]*Division of Crop Protection, ICAR-Indian Institute of Soybean Research (IISR), Indore, Madhya Pradesh, India;* [2]*Division of Crop Improvement, ICAR-Indian Institute of Soybean Research (IISR), Indore, Madhya Pradesh, India;* [3]*Division of Plant Pathology, Jawaharlal Nehru Krishi Vishwa Vidyalaya, Jabalpur, Madhya Pradesh, India;* [4]*Division of Plant Improvement and Pest Management, ICAR-Central Arid Zone Research Institute, Jodhpur, Rajasthan, India*

5.1 Introduction

Soybean is an essential oilseed crop that produces about two-thirds of global protein and a quarter of global edible oil (Pagano & Miransari, 2016). *Macrophomina phaseolina* is a devastating necrotroph pathogen causing charcoal rot disease in soybean. Soybean is vulnerable to various fungal pathogens among which *M. phaseolina* is fourth most important pathogens, having economical impact on soybean production. *M. phaseolina* has diverse host range including several economically important crops such as soybean, sorghum, and maize (Singh, Chatterjee, Rajput, Kumar, et al., 2021). Drought or drought-like conditions along with high soil and air temperatures (30−35°C) under nutrient-deficient soils foster the high incidence of charcoal rot disease (Nataraj et al., 2019). In India, an estimate of about 77% of yield loss due to this disease has been reported (Sharma et al., 2014).

In India, soybean megavarieties, such as JS 335, JS 93−05, and JS 95−60, have become sensitive to charcoal rot. The absence of durable resistance in the genetic material may be related to the prevalence of variability in the *M. phaseolina* population and the multidimensional mode of pathogenicity (Nataraj et al., 2019). Looking at the importance of this disease in soybean, current chapter has been outlined under the following five sections.

5.2 General information on soybean charcoal rot disease

It has been dealt with the following three subsections.

5.2.1 Symptomatology

Pathogenic infection on soybean begins through the biotrophic phase, with no visual signs. With the advent of maturity and under a favorable environment (drought/drought-like conditions), the pathogen aggravates through the establishment of necrotic relationship with host (Twizeyimana et al., 2012). This phase is characterized by typical symptoms, *i.e.*, yellowing, wilting, and dropping of leaves attributed to the fungal mechanical obstruction of host-vascular tissues and necrosis induced by phytotoxins and enzymatic activity (Luna et al., 2017). During the reproductive phase, a light gray or silver discoloration occurs on the epidermal and subepidermal tissues of the taproot and the lower section of the stem. The growth of microsclerotia becomes so abundant that the stem takes on a greyish-black appearance. The split-open taproot shows black stripes in the woody section of the crown (Singh, Chatterjee, Rajput, Kumar, et al., 2021). Symptoms and sign

Macrophomina Phaseolina. https://doi.org/10.1016/B978-0-443-15443-0.00023-1

of soybean charcoal rot some time get confused; symptoms arise due to drought stress, and early senescence and infection of soybean cyst nematode (*Heterodera glycines*). Prematurely death of plant with attached of leaves to petiole is main characteristic symptom of soybean charcoal rot (Mengistu et al., 2007). Appearance of grayish and silvery discoloration on lower part of plant that resulted development of microsclerotia called as charcoal rot (Gupta et al., 2012). Charcoal rot signs are more common in stressed sections of the field, such as field margins, compacted soils, and slopes. Disease is aggravated with high temperatures and water deficit conditions of soil. The severity of charcoal rot is also affected by plant maturity (Nataraj et al., 2019). The mechanism behind and the impact of plant age on disease severity are not understood. However, it is believed that blossoming of the early maturing plants happen at lower temperature that make them less susceptible of high-temperature stress, which is the main region of symptom expression and yield loss (Smith & Carvil, 1997). Until recently, thin black dark lines on the cortex of stem caused by *Diaporthe longicolla completely confuses the researchers and leads to misidentification of soybean charcoal rot* (Olson et al., 2015).

5.2.2 Disease cycle and epidemiology

Black and hard structures produced by *M. phaseolina in soil and plant* are known as microsclerotia (Crous et al., 2006). At a temperature range of 20–40°C, these microsclerotia germinate and establish host pathogen relationship in root tissue. The pathogen can infect the soybean plant staring from seedling to mature plant (Collins et al., 1991). *M. phaseolina* releases cell wall degrading hydrolytic enzymes, such as hemicellulose, cellulase, amylase, pectinase. and lipase (Kaur, 2012), and toxins that breakdown root tissue, and once infected, root and stem tissues are colonized within in 2–3 weeks (Islam et al., 2012). The microsclerotia can persist over 2 years on residue at or below the soil surface (Short et al., 1980). The sexual status of *M. phaseolina* is yet to be determined. Pycnidia production has been seen in a variety of host plants (Dhingra & Sinclair, 1978) as well as in artificial media such as potato dextrose agar (Gaetan et al., 2006) and water-agar-leaf media (Ma et al., 2010). The relevance of pycnidia in disease expansion is unclear. Hyaline, ellipsoid to ovoid conidia were produced by *M. phaseolina* with diameters ranging from 14–30 × 5–10 μm (Crous et al., 2006). Infection of soybean seedlings by conidia has been proved in laboratory (Ma et al., 2010), but their involvement in the disease cycle in soybean remains uncertain.

5.2.3 Economic impact

M. phaseolina can infect nearly 500 other plant species, including field crops, horticulture crops, and trees, and cause nearly 50% of yield loss (Luna et al., 2017). During 1949, soybean charcoal rot was identified in the United States of America (Young, 1949). Later, it was observed throughout North America and Asia (Luna et al., 2017). As soybean charcoal rot is influenced by environmental stress, yield loss also varies yearly. In 1994, a total of 1.234 Mt yield loss was estimated to be caused by soybean charcoal rot disease in the top 10 soybean-harvesting countries (Wrather et al., 1997). During 1998, 2.01 Mt of soybean yield loss due to charcoal rot disease was observed in the top 10 soybean-producing countries (Wrather et al., 2001). During 2001–02, the climatic condition of Argentina was favorable for charcoal rot disease development, and it caused nearly 100% yield loss in some farmer fields. Argentina's maximum soybean growing area suffered from varying charcoal rot disease severity (Ploper et al., 2001). During 2003 and 2012, soybean charcoal rot was second in causing yield loss. It causes 1.98 Mt and 2.0 Mt yield loss in the respective years (Wrather et al., 2010). From 2003 to 2005, soybean charcoal rot increased and corresponding to that yield loss were also increased. It was 1.98 Mt in 2003, 2.75 Mt in 2004 and 4.95 Mt in 2005 in the United States (Wrather & Koenning, 2006). In 2006, soybean charcoal rots caused a 4.2% yield loss worldwide (Wrather et al., 2010). Under rain-fed conditions, the yield loss estimate caused by charcoal rot is quite tricky due to the confounding effect of environmental stress. Under irrigated conditions, 6%–33% of yield loss was reported in soybean in the United States (Mengistu et al., 2011). Soybean charcoal rot secured the fifth position in the top five yield loss-causing disease for soybean in the United States from 2010 to 2013, and the incidence of soybean charcoal rot in the United States is increasing in the past 30 years (Luna et al., 2017). Due to soybean charcoal rot, a maximum soybean yield loss of 73.36 Mt was reported in 2012 in the United States and Canada. In 2010 to 2014, soybean charcoal rot caused 43.8 Mt yield loss per year in the United States and Canada (Allen et al., 2017). Mengistu et al. (2018) reported that for every 1000 colony-forming units of soybean charcoal rot at the R7 stage of the soybean plant, leaded 11.5 kg yield loss per hector. Soybean charcoal rot disease causes 6.13% of yield loss (Jordaan et al., 2019). Bandara et al. (2020) reported that soybean charcoal rot secured the fourth position for loss in monitory terms ($7484 million). In 1997, Guna district of India faced an epidemic, causing an 80% yield loss in soybean (Gupta & Chauhan, 2005).

5.3 Screening for charcoal rot resistance

5.3.1 Field screening

Field-level screening is performed by creating hot spot conditions or naturally in field plots with a history of normal disease incidence. Field-level screening is often based on percent mortality. For that 0–9 scale was used predominantly, where 0 (highly resistance) stands for no mortality, 1 (resistance) for 1% mortality, 3 (moderate resistance) for 1.1%–10% mortality, 5 (moderate susceptible) for 10.1%–25% mortality, 7 (susceptible) for 25%–50% mortality, and 9 (highly susceptible) for more than 50% mortality (Amrate et al., 2019). At the field level, soybean genotypes are also evaluated through foliar symptoms on leaves at the R7 stage (Fehr et al., 1971). For that 0–11 scale (Horsefall-Barratt scale) was used, where 0 is no symptoms, 1 is 0%–3% symptoms, 2 is 3%–6% symptoms, 3 is 6%–12% symptoms, 4 is 12%–25% symptoms, 5 is 25%–50% symptoms, 6 is 50%–75% symptoms, 7 is 75%–87% symptoms, 8 is 87%–94% symptoms, 9 is 94%–97% symptoms, 10 is 97%–100% symptoms, and 11 is 100% symptoms. According to this scale, genotypes were categorized as 0 for resistance, one to four for moderate resistance, five to eight for moderately susceptible, and more than eight for susceptible (James, 1974). Another method is the "percent height of stem discoloration" (PHSD), which was developed based on stem discoloration at the R7 stage of genotypes through the production of microsclerotia. PHSD is calculated as: 100 × theinternal vascular necrosis length (cm)/plant height (cm) (Mengistu et al., 2007). Soybean genotypes are also evaluated through the root stem severity (RSS) index at the R7 stage. In that, the diseased plants are uprooted with the entire root system and the root and stem portion of the plant at the collar region is split longitudinally. Production of microsclerotia in the pith region was estimated in 1–5 scale. Score 1 is given when stem of soybean plant is not showing any microsclerotia, 2 was given when vascular tissue get discoloration and very few microsclerotia observe in the pith, vascular tissue, or beneath the stem epidermis, if partially discoloration happen in vascular tissue with microsclerotia partially distributed in stem than 3 score was provided, along with this if numerous microsclerotia was distributed in under the outer stem and root epidermis than 4 score was provided and if dark color was observed in both inner and outer surface of root and stem tissues along with numerous microsclerotia than 5 score was provided. Soybean genotypes are categorized based on the RSS scale, where 1 is for resistance, more than 1 and less than 2 is moderately resistant, more than 2 and less than 3 is moderately susceptible, and 3–5 is susceptible (Paris et al., 2006). Based on a 0–9 scale of disease incidence or 0 to 11 scale of foliar symptoms, or 1–5 scales of RSS, the Percent Disease Index (PDI) can be measured. PDI was calculated as the summation of individual scale×100 divided by maximum disease scale x full scale (Rajput & Harlapur, 2015). Based on PDI, soybean genotypes were categorized as highly resistance (0), resistance (0.01–11.11), moderated resistance (11.12–33.33), moderated susceptible (33.34–55.55), susceptible (55.56–77.77) and highly susceptible (77.78–100). PDI based on foliar symptoms calculated at different stages of soybean genotypes—R1, R3, R5, and R7 stage is used to calculate the area under the disease progress curve (AUDPC) with the subsequent formula (Dangi et al., 2019).

$$\text{AUDPC} = \sum_{i=1}^{n-1} \left(\frac{P_{i+1} + P_i}{2} \right) \times (t_{i+1} - t_i)$$

Where P_i is PDI at initial time, P_{i+1} is PDI at ith observation, t_i is time for initiation time, t_{i+1} is time at ith observation and $n =$ the number of observations of foliar symptoms.

All the screening methods of genotypes are based on visual observation, so there may be the change of high error. Smith and Carvil (1997) identified a new method of disease measurement called the colony forming unit (CFU) assay. CFU assay was performed by gathering of the lower stem and roots of soybean plants at the R7 stage and dried naturally. The dried sample was surface sterilized with 0.5% NaOCl for 3 min. The sample was incubated on sterilized selective PDA media containing rifampicin (100 mg L^{-1}) and tergitol (0.1 mL) antibiotics to avoid bacterial contamination. The plate incubated at 30°C for 3 days to develop cottony mycelium. CFU was counted and converted into per gram of sample after 3 days of incubation. This CFU is converted into colony forming unit index (CFUI) by dividing the CFU of tested genotypes by the highest CFU value obtained by other tested genotypes or susceptible check (Mengistu et al., 2007). Based on CFUI value, the genotypes were denoted as resistant to having CFUI 0 to <10, moderately resistant to having CFUI 10 to ≤ 30, moderately susceptible to having CFUI >30 to 60, and susceptible to having CFUI >60 (Mengistu et al., 2007). Estimation of CFU is a time-consuming, expensive, and irreproducible process. The data of CFU vary across time, locations, and environment as disease severity varies. Another limitation is that evaluation of significant genotypes is challenging through the CFU method. This inconsistent result leads to ambiguity in categorizing resistance and susceptible genotypes.

5.3.2 Artificial screening

Different artificial screening techniques were developed to avoid inconsistent results for resistance screening. For consistent and reliable results for identification of soybean charcoal resistance genotype, the cut-stem inoculation technique was developed against *M. phaseolina* (Twizeyimana et al., 2012). At the V2 stage of the soybean plant, the apex of the soybean stem was cut nearly 2.5 cm using a sterilized sharp razor blade. The cut should be smooth enough so that pathogen can penetrate equally. A 9-day young culture was used to inoculate *M. phaseolina*. The sterile 200 μL micropipette tip was filled with the young culture of *M. phaseolina* and placed on the blunt end of the soybean plant. The upper end of the micropipette tip was removed slightly for better aeration of the pathogen. The inoculation must be placed on the tip of the blunt end of the soybean plant so that the micropipette tip can be pushed further. The lesion lengths were measured at regular intervals, especially 3 days after inoculation to 15 days after inoculation. The area under disease progress was calculated through the trapezoidal method through lesion length according to the formula $AUDPC = \sum_{t=1}^{n-1}\left(\frac{L_i+L_{i+1}}{2}\right)(t_{i+1} - t_i)$, where Li lesion length (cm) at the ith observation, t_i is time (days) in ith observation, and n is total number of observation (Dangi et al., 2019). According to lesion length at 12th days after inoculation, soybean genotypes were categorized as resistance (0–1 cm), moderate resistance (1–3 cm), moderate susceptible (3–10 cm), and susceptible (>10 cm) (Amrate et al., 2019). Da Silva et al. (2019) developed a seed plate assay to identification of the soybean genotypes for resistance against charcoal rot. A young grown culture of *M. phaseolina* was inoculated on 2% water agar media in a sterilized Petri plate. The water agar media was covered with the 2g of sterilized vermiculite. Seeds of soybean genotypes were thoroughly washed and surface sterilized with 70% ethanol for 3 min and dried on sterilized paper towel. Equal seeds were evenly placed on the media, and entire Petri plate was covered with aluminum foil to avoid light penetration and incubated at normal room temperature (22°C). After 7 days of incubation, the number of germination seeds (radical growth should be > 1 cm) was counted. Paper towel method (Konde et al., 2017) was vastly used for screening of various phytopathogens in cereals, pulsed, and oilseed crop. The seeds of soybean genotypes were surfaced sterilized by using 1% sodium hypochlorite for 1 min and rinsed in sterile water for 2 min to wash off NaOCl. These seeds were sown in pots filled with sterilized soil and nutrient sources. The seedlings were carefully removed from the pot after 8 days of sowing. The roots of seedlings were rinsed properly with autoclaved water, and the roots were also surface sterilized with 1% NaOCl for 2 min and washed in sterile water for 3 min. Seedlings of soybean genotypes were inoculated with a culture solution of *M. phaseolina* for 2 min. Equal inoculated seedlings of soybean genotypes were kept in blotter paper, and the blotter paper was folded. The entire assembly was incubated at 35°C with 12 h of photoperiod under the growth chamber. The water was regularly sprinkled to ensure adequate moisture. The one to nine severity scale was used as suggested by Pande et al. (2012) based on the damage caused by the pathogen. In this scale, if no root infection observed than score 1 (resistant) was provided, if root having very few lesion than score 3 (moderately resistant) was provided, if root showing clear lesion but new and small root is infection free than score 5 (moderately susceptible) was provided, if root lesion was increased and new root is infection free than score 7 (susceptible) was provided and if root was completely infected and discolored than that was treated as score 9 (highly susceptible).

Another interesting method is the tooth prick method, where the soybean plants at the R6 stage are inoculated with the pathogen using a toothpick (Ghorbanipour et al., 2019). The disease progress was measured in terms of change in stem color and rotting of the stem as the method was given by Mengistu et al. (2007).

5.4 Management of soybean charcoal rot

Most of the charcoal rot disease management methods aim to diminish microsclerotia production. Management of soybean charcoal rot disease is a slightly formidable task as the phytopathogen *M. phaseolina* shows a high degree of pathogenic variation, diminished to the changing environments, persistence in soil, and wider host range (Gulya et al., 2002). Management of soybean charcoal rot with single approach is highly impractical and uneconomical. A combination of cultural, biological, and chemical methods can minimize the charcoal rot disease severity in soybean. Depending upon survival ability and adaptation to the environment and stage of the crop, the management of soybean charcoal rot varies (Khamari et al., 2019).

5.4.1 Biological control

Management strategies of *M. phaseolina* include using biocontrol agents to prevent host infection or inhibit the growth of the pathogen is an eco-friendly approaches (Ndiaye et al., 2007). Chauhan et al. (2022) identified two soybean endophytic strains of *Bacillus subtilis* for effective glasshouse management of charcoal rot. *B. subtilis* suppress the growth of *M. phaseolina* by the production of lipophytic secondary metabolite and hydrolytic enzymes such as cellulase, chitinase, and β,one to three endoglucanase that disintegrate the cell wall of phytopathogen. It also secretes siderophores, which allow the pathogen to compete for iron, which is essential for the growth and development of the pathogen. Seed inoculation of *B. subtilis* showed plant growth promotion activity; it also triggered the defense system of the plant against *M. phaseolina* (Chauhan et al., 2022). Similarly, Lodha and Mawar (2020) summarized in their review that seed treatment with *Bacillus firmus* could reduce the contact of *M. phseolina* with oil seeds crops (Lodha & Mawar, 2020). Another important group of bacteria called rhizobia has the main function of establishing symbiotic relationships to provide nitrogen to soybean plants. Apart from plant growth promotion activity, they are also reported to inhibit soybean charcoal rot pathogen by the production of antibiotics, HCN, and siderophores. Seed treatment with these rhizobia leads to enhances the accumulation of phenolic compounds, which leads to the production of phytoalexin and also enhances essential defense enzymes, such as polyphenol oxidase (PPO), L-phenylalanine ammonia lyase (PAL), peroxidase (POX), and chalcone synthase (CHS) (Das et al., 2017). Similarly, *Rhizobium japonicum* is also reported to reduce *M. phaseolina* pathogen infection in soybean in both field and pot conditions (Al-Ani et al., 2012). Another well-studied biocontrol agent genus is *Pseudomonas* for managing plant diseases. Two bacterial strains, *B. subtilis* 54 and *P. fluorescens* nine showed a significant reduction of charcoal rot disease in both fields and lab conditions with excellent PGPR activities (Simonetti et al., 2015). Mufti and Bano (2019) identified two strain of *Pseudomonas,* i.e., *P. stutzeri* and *P. putida*. Both strains had the capacity to enhance plant growth and elicited induce systemic resistance (ISR) mechanism in soybean plants against *M. phaseolina*. Recently, Yasmin et al. (2020) identified two biocontrol agents *P. putida* and *B. clausii*, effective against soybean charcoal rot. They manage the disease through the production of HCN and siderophore. Application of this *B. clausii* enhanced the proline content and plant defense enzymes viz., SOD, PO, CAT, and ascorbate peroxidase in both root and shoots of charcoal rot infected plant. It also induced salicylic acid and jasmonic acid in charcoal rot infected plants, which activated systemic resistance in the soybean plant. Apart from bacteria, a mycorrhiza *Rhizophagus intraradices* also reduced the severity of charcoal rot in soybean by the colonization of roots and rapidly providing nutrients to soybean for enhanced growth of plant height and root length (Spagnoletti et al., 2020). Analysis of transcriptomes obtained from *R. intraradices* inoculated soybean roots revealed that after infection of *M. phaseolina,* GmPR1 genes induce a defense system against the pathogen in the root. It also induces genes related to defense enzymes such as super oxide dismutase (SOD), catalyases (CAT), and guaiacol peroxidase (GPOX) were increased in leaves and roots, except GPOX that decreased in roots after being challenged by *M. phaseolina* (Spagnoletti et al., 2016). Another mycorrhizal fungus, *R. irregularis* colonized soybean roots and induced defense against *M. phaseolina* through induction of plant defense-related compounds such as hormones, PR protein, resistance protein, and another secondary metabolite compound (Marquez et al., 2018). The combination of fungi and bacteria also reduces the disease severity drastically. Seed treatment with *Trichoderma viride* @ 700 mL/kg of seeds along with *B. subtilis* @ 1 mL/kg of seed reduced charcoal rot of soybean (Reznikov et al., 2016). Similarly, Thakkar and Saraf (2015) developed consortia of three bacterial strains—*Pseudomonas aeruginosa*, *B. cereus,* and *B. amyloliquefaciens,* and one fungus *T. citrinoviride* for effective management of soybean charcoal rot through the antimicrobial compound ammonia, which directly kills the pathogen and also induces hydrolytic enzymes. The microbial consortia were also shown to enhance the growth of the soybean plant. Seed treatment with *Pseudomonas aeruginosa* along with *T. citrinoviride* increased the germination percentage of soybean along with vigor index and chlorophyll content in charcoal rot infected soybean plant. Khaledi and Taheri (2016) identified that the soil treatment of *T. harzianum* enhanced root and shoot mass in charcoal rot infected soybean plants. It also induces defense enzymes in the soybean root system against *M. phaseolina*. Parmar and Patel (2020) identified *T. asperellum*, inhibiting the growth of *M. phaseolina*. Trichoderma can kill plant pathogens through the production of antibiotics. The use of *T. harzianum* as seed treatment and soil application significantly reduced soybean charcoal rot incidence by 23.50% and 60.83%. Seed treatment and soil application also enhanced soybean grain yield (Choudhary et al., 2018). The biocontrol agents used to manage the disease are summarized here under (Table 5.1).

5.4.2 Chemical control

Earlier researchers focused on the management of soybean charcoal rot through fungicides that can diminish the microsclerotia production in soil and plant (Luna et al., 2017). The fungicides used to manage the disease are summarized in Table 5.2. Seed dressing chemicals, such as carbendazim, thiram, thiram + carboxin, thiram + carbendazim, and captan were used extensively to manage soybean charcoal rot. These chemicals reduce *M. phaseolina* severity in the roots of soybean seedlings by reducing the pathogen population and its ability to produce microsclerotia in soybean (Singh et al, 1990; Kumar & Singh, 2000; Gupta & Chauhan, 2005). However, the management of soybean charcoal rot through

TABLE 5.1 List of biocontrol agents used for management of soybean charcoal rot.

Name of organism	Mechanism of disease control	Mode of study	References
Bacillus subtilis	Production of secondary metabolite and hydrolytic enzymes	In vitro and glasshouse	Chauhan et al. (2022)
Rhizobia	Induction of defense compounds and enzymes.	In vitro and glasshouse	Das et al. (2017)
Rhizobium japonicum	Induce PGPR activities	In vitro, glass-house and field	Al-Ani et al. (2012)
B. subtilis 54 and *P. fluorescens* 9	Induce PGPR activities	In vitro, glass-house and field	Simonetti et al. (2015)
Pseudomonas stutzeri and *P. putida*	Enhance PGPR and induce ISR	In vitro and glasshouse	Mufti and Bano (2019)
P. putida and *B. clausii*	Induction of ISR	In vitro and glasshouse	Yasmin et al. (2020)
Rhizophagus intraradices	Induce defense gene GmPR1 and defense enzymes	In vitro and glasshouse	Spagnoletti et al. (2016)
Rhizophagus irregularis	Induction of ISR	In vitro and glasshouse	Marquez et al. (2018)
Consortia of *P. aeruginosa, B. cereus, B. amyloliquefaciens, and Trichoderma citrinoviride*	Mycoparasitism, induces NH_3, hydrolytic enzymes and PGPR activity	In vitro and glasshouse	Thakkar and Saraf (2015)
T. viride and along with *B. subtilis*	-	Field	Reznikov et al. (2016)
P. aeruginosa along with *T. citrinoviride*	Enhance PGPR activities	Glasshouse	Thakkar and Saraf (2015)
T. harzianum	Enhance PGPR activities	In vitro and glasshouse	Khaledi and Taheri (2016)
T. harzianum	-	In vitro, glass-house and field	Choudhary et al. (2018)
T. asperellum	Production of antibiotics and mycoparasitism	In vitro	Parmar and Patel (2020)

TABLE 5.2 List of fungicide used for management of soybean charcoal rot.

Name of fungicide	Mode of application	References
Pyraclostrobin + thiophanate methyl	Seed treatment	Reznikov et al. (2016)
Pyraclostrobin, thiophanate methyl, and Carbendazim + mancozeb	Foliar application	Sanjay et al. (2020)
Fluopyram	Seed treatment	Padgett (2018)
Thiram + carboxin	Seed treatment	Rahman et al. (2021)
The nanoparticle of trifloxystrobin 25%+ tebuconazole 50% (75 WG)	Seed treatment	Kumar et al. (2016)

fungicide is inconsistent and challenging due to the necrotrophic nature of pathogens (Reznikov et al., 2016). Reznikov et al. (2016) reported that the seed treatment with pyraclostrobin + thiophanate methyl @ 1 mL/kg seed reduced the incidence of soybean charcoal rot. Besides disease management, seed treatment with pyraclostrobin + thiophanate methyl also increases emergence percentage, seed yield, and 1000 seed weight in charcoal rot infected soybean plant.

Further, foliar application of Pyraclostrobin, Thiophanate Methyl, and Carbendazim + Mancozeb reduced the infection of soybean charcoal rot and enhanced grain yield (Sanjay et al., 2020). Rahman et al. (2021) identified that a combination of thiram + carboxin and *T. harzianum* was the best seed dressing fungicide in managing soybean charcoal rot. Padgett (2018) identified the impact of Fluopyram on soybean charcoal rot pathogen's ability to colonize soybean plants and the impact of Fluopyram (a succinate dehydrogenase inhibitor) on soybean grain yield in charcoal rot infected plants. The effect of seed treatment on grain yield and plant height was found nonsignificant. The nanoparticle (particle size<100 nm) of the fungicide trifloxystrobin 25%+ tebuconazole 50% (75 WG) found effective in reduction of hypal growth and production of microsclerotial of *M. phaseolina* (Kumar et al., 2016). The fungicide trifloxystrobin and pyraclostrobin were, belonging to the strobilurin group, inhibited the electron transfer in an electron chain reaction from cytochrome *b* to cytochrome *c*1 that, leads to the block of energy production in fungus and ultimately death of the pathogen (Feng et al., 2020).

Similarly, fungicides such as thiophanate methyl and carbendazim belonging to the benzimidazoles group can stop the synthesis of β-tubulin protein that is essential for the production of spindle fiber during cell division, resulting in a reduction of growth and reproduction of fungi. Fungicides mancozeb and thiram belonging to dithiocarbamate fungicides have abilities to inhibit various enzymes, which are metal dependent and enzymes having active sulfhydryl group that are essential for growth and metabolism of the pathogen (Yang et al., 2011). Besides this, carboxin and fluopyram inhibit the succinate dehydrogenase enzyme, which is essential for oxidative phosphorylation in the tricarboxylic acid cycle in fungi (Di et al., 2021).

5.4.3 Botanicals or plant extracts

Plants have unique mechanisms to inhibit the growth and reproduction of fungi through their metabolites (Ushiki et al., 1996). The botanicals used to manage the disease are summarized in Table 5.3. Botanicals consist of plant extract and their volatile oil, which can reduce pathogen growth and act as a substitute for fungicides. Secondary metabolites of botanicals consist of terpenes, phenolics, nitrogen, and sulfur-enriched compounds with the antifungal ability (Zaynab et al., 2018). Extract of leaves or plant part of *Allium sativum, Parthenium hysterophorus, Polyalthia longifolia, Citrus limon, Azadirachta indica, Ricinus communis*, and *Allium cepa* were screened for inhibition of soybean charcoal rot pathogen *M. phaseolina. A. sativum* was found most effective in both field and lab condition (Sanjay et al., 2020). The bulb of *A. sativum* has various metabolites viz., alliin, allicin, ajoenes, vinyldithiins, and quercetin, which can reduce the growth of this fungus (El-Saber Batiha et al., 2020). Apart from fungicidal ability, botanicals also induce a defense system against phytopathogen. Foliar application of rosemary crude extract on soybean charcoal rot inoculated plant induces plant defense enzymes such as PPO, PO, and PAL (Lorenzetti et al., 2018). Seed treatment with neem oil combined with foliar application of salicylic acid @ 5 mM reduced charcoal rot severity in glasshouse conditions and also enhanced the root and shoot weights of charcoal rot inoculated soybean plant (Mostafanezhad et al., 2012, p. 250). The inhibitory action of neem oil through azadirachtin and monoterpenes alkaloids reduces the growth of fungus (Rajput et al., 2016). The previous researcher also found that dipping soybean seeds infected with extract of ginger, garlic, and neem for 5 min reduced the charcoal rot disease severity (Hossain et al., 1999). *A. sativum* and *A.* were found to be effective in the management of soybean charcoal rot pathogen in lab condition (Hingole & Kendre, 2021).

5.4.4 Elicitors

The natural and synthetic compounds that induce systemic acquired resistance (SAR) and induce systemic resistance (ISR) in plants against phytopathogen after the application are called elicitors. The elicitors used to manage the disease are summarized in Table 5.4. Pawlowski et al. (2016) identified that foliar application of four elicitors, benzothiadiazole (BTH), chitosan (CHT), phenylalanine (PHE), and salicylic acid (SA) reduced soybean charcoal rot severity and also

TABLE 5.3 List of botanical used for management of soybean charcoal rot.

Name of botanical	Mechanism of disease management	Mode of study	References
A. sativum	Secondary metabolites	Glasshouse and field	Sanjay et al. (2020)
Rosemary	Induce plant defense enzyme	Glasshouse	Lorenzetti et al. (2018)
Neem oil with salicylic acid	Secondary metabolites	Glasshouse	Mostafanezhad et al. (2012)
A. sativum and *A. cepa*	Secondary metabolites	Glasshouse	Hingole and Kendre (2021)

TABLE 5.4 List of botanical used for management of soybean charcoal rot.

Name of elicitor	Mechanism of disease management	Method of application	References
BTH, CHT, PHE, and SA	Induce plant defense system	Foliar application in glasshouse	Pawlowski et al. (2016)
Chitosan, SA, BABA and JA	Induction of ISR	Foliar application and seed treatment in glasshouse	Agale et al. (2020)
Sodium silicate	Create physical barrier to the pathogen and may induce systemic resistance	Soil and foliar application in glasshouse	El-Abdean et al. (2020)
Silica	Improve physiology of plant	foliar application in glasshouse	Hussain et al. (2021)

enhanced plant defense system. They also observed that the management of soybean charcoal rot depends upon soybean genotypes, pathogens population, and microenvironmental conditions. Foliar application and seed treatment of chitosan, SA, β-amino butyric acid (BABA), and jasmonic acid (JA) reduced soybean charcoal rot severity by induction of ISR against *M. phaseolina* (Agale et al., 2020). Soil and foliar application of sodium silicate reduced the charcoal rot severity by nearly 50% as compared to control and also enhanced the thickness of cortex and pith of root, stem, and leaves. The mechanism of silica in disease management is not well understood, but researchers believe that silica may act as a physical barrier to the pathogen or may induce ISR or SAR mechanism or may act as fungicidal to pathogen (El-Abdean et al., 2020). Foliar application of silica also improved soybean plant health through an increase in the rate of net photosynthesis, stomatal conductance, rate of transpiration, and a decrease in the concentration of intercellular carbon dioxide in a soybean plant (Hussain et al., 2021). The presence of phenol, isoflavones, and lignin in the seed is also reported to provide resistance to charcoal rot.

Similarly, high concentrations of Ca, K, P, Zn, Mn, Cu, and B plant nutrient defense against soybean charcoal rot (Bellaloui et al., 2014). Interestingly, the foliar application of calcium and magnesium was not effective in soybean charcoal rot management (Wilkerson, 2017). Induce resistance (IR) system is a sustainable way to manage phytopathogen by understanding the natural phenomenon, and IR can be an essential method for disease management, mainly where the pathogen is highly variable and management of the disease is tedious. Adapting this IR system to manage the disease helps minimize the fungicide load on plants. The use of elicitors for disease management may impact negatively on plant growth. Therefore, one should use it consciously.

5.4.5 Cultural management

The ability to produce abundant inoculum in the soil and plant and a high genetic variation make this pathogen more dangerous and cause a high disease and yield loss in soybean crops. The cultural methods were used by various researchers to manage the disease are summarized in Table 5.5. Tillage changes physical properties of soil, which may change the inoculum density of the pathogen. Baird et al. (2003) observed that the reduced charcoal rot disease severity due to an increase in antagonistic microbes that decompose the microsclerotia. Mengistu et al. (2009) observed CFU reduction in no-tillage treatment as compared to the conventional tillage treatment indicating that no-tillage is unfavorable for the growth and development of *M. phaseolina*. The reduction of CFU was due to a decline in soil temperature because of the mulching effect of crop residues on the soil surface.

Similarly, Perez-Brandán et al. (2012) conducted a long term field experiment to investigate the impact of tillage on the incidence of *M. phaseolina* on soybean. A higher microbial population was observed in direct seedling compared to no-tillage and conventional tillage, but fungal biomass in tillage did not show any significant difference. This indicates that the direct seedling system has abundant and active microbial communities compared to no-tillage and conventional tillage. Due to the wide host range, management of soybean charcoal rot through crop rotation is unreliable. On the contrary, Perez-Brandán et al. (2014) identified high disease severity of soybean charcoal rot in the monoculture system of soybean rather than rotated with maize for 15 or 4 years. Crop rotation enhances potential biocontrol agents, such as *Gliocladium* spp., *Trichoderma* spp., and fluorescent strain of *Pseudomonas*. Previously researchers found that the number of viable microsclerotia from soybean charcoal rot is directly proportional to the number of crop rotation years within the soybean maize cropping system (Short et al., 1980). In contrast, after 2 years of crop rotation (with maize, wheat, and pasture), the disease severities of soybean charcoal rot were increased (Maier et al., 2018).

TABLE 5.5 List of cultural method used for management of soybean charcoal rot.

Name of cultural practices	Mechanism of disease control	References
Tillage	Increase antagonist microbes and decompose the microsclerotia	Baird et al. (2003)
No tillage or soybean followed by a fallow system of cultivation for 3 years	Decline in soil temperature make unfavorable environment	Mengistu et al. (2009)
Direct seedling	Enhanced active microbial population	Perez-Brandán et al. (2012)
Crop rotation with maize for 15 or 4 years	Enhances potential biocontrol agents such	Perez-Brandán et al. (2014)
Long term crop rotation with maize, wheat, and pasture	Increase biocontrol agents'population	Yuan et al. (2021)
Soil incorporation of mustard with biofumigant	Producing antimicrobial compound glucosinolate	Sassenrath et al. (2017)
Soil solarization for 1400 h	Reduction of the sclerotia population	Marquez et al. (2021)
Soil solarization for 30 days	Reduced the lived sclerotial population	Dubey et al. (2009)
Irrigation during the reproductive stage	Delaying charcoal rot infection	Mengistu et al. (2011)
Irrigation during seedling stage	Reduce colonization at harvesting stage	Jordaan et al. (2019)
Application of phosphorus fertilizer	-	Spagnoletti et al. (2018)
Application of sulfur along with Planomicrobium	Induction of defense enzyme and PGPR activity	Ijaz et al. (2021)
Variety with high protein and oleic acid	-	Bellaloui et al. (2021)
Variety with high concentration of Ca, K, P, Zn, Mn, Cu and B in seed	-	Bellaloui et al. (2014)
Application of sodium silicate and potassium silicate	Create physical barrier and act as a fungicide	El-Abdean et al. (2020)
Application of herbicide dinoseb, EPTC, fluorodifen, fluometuron, and alachlor	reduced the viable population and act as fungicide	Filho and Dhingra (1980)
Application of glyphosate herbicide at mid-vegetative stages	Fungicidal activity	Mengistu et al. (2013)
Application of quizalophop-p-ethyle, propaquizafop, oxyfluorfen, and oxadiargyl	Reduced the microsclerotia formation and growth	Ingole and Zade (2020)
Application of trifluralin	Induce ISR and lignin content in seedlings	Canaday et al. (1986)
Application of B. amyloliquefaciens and B. firmus	*Kill the SCN*	Roth, Jacobs, et al. (2020)

Yuan et al. (2021) identified that the continuous cropping system drastically increased the beneficial biocontrol agents' population viz., *Pseudomonas*, *Nitrospira,* and *Streptomyces* as compared to the rotation and short-term continuous cropping system. This indicated that the continuous cropping system of soybean changes microbial community to provide good health to the soybean plant. Mengistu et al. (2009) found that soybean followed by a fallow system of cultivation for 3 years reduced the population of *M. phaseolina* in soybean field. Sassenrath et al. (2017) reported that soil incorporation of mustard with biofumigant properties in soybean fields reduced the charcoal rot population in soil and soybean plants by producing antimicrobial compound glucosinolate. Soil solarization is also the most effective technique to minimize soil population density of soil-borne phytopathogen. Soil solarization has been shown to be effective in the reduction or optimize the population of *M. phaseolina* sclerotia in soil (Marquez et al., 2021). Lodha and Mawar (2020) summarized that in the United States condition, the rising temperature of 63°C reduces the population of *M. phaseolina* microsclerotia considerably, especially in the top 15 cm depth. The US researcher found that 50°C for 48 h required the killing the pathogen; but in Indian, subcontinental conditions temperature rising to 58–69°C for 1400 h through soil solarization is sufficient to reduce the primary inocula of *M. phaseolina* in topsoil (Lodha & Mawar, 2020). The inoculum level in the lower depth of soil cannot be ruled out through soil solarization; some amount of sclerotia population survives there and may cause disease (Lodha & Mawar, 2020). Due to heat-tolerant nature of *M. phaseolina*, the soil solarization method has been used to control disease at different levels. Dubey et al. (2009) identified that 30 days of soil solarization along with neem cake and neem leaf powder reduced the live sclerotial population of soybean charcoal rot pathogen. Mengistu et al. (2011) well studied the management of soybean charcoal rot through irrigation in 2011. They identified that the severity of soybean charcoal rot is directly associated with water availability at the reproductive stage of soybean. Providing irrigation under drought conditions during the reproductive stage controls charcoal rot, but 6%–30% of yield loss is still observed. They also observed that a higher yield was observed at irrigated soybean than nonirrigated soybean and also severity of disease was less in irrigated soybean. Jordaan et al. (2019) identified that during the seedling stage under irrigated conditions, disease severity was similar to that of water deficient conditions. Interestingly at the harvest stage, the colonization of *M. phaseolina,* below 1 cm of soil, was reduced sharply under the irrigated condition as compared to the unirrigated condition. The study concluded that soil moisture condition had no significant impact on delaying charcoal rot infection at the seedling stage, but the colonization of pathogen on the stem was favored by less soil moisture condition at the maturity stage. The water level at field capacity, saturation, and permanent wilting point did not affect the population of microsclerotia of *M. phaseolina* in soybean (Jimenez, 2011). Contrary to the previous reports, Kendig et al. (2000) identified that infection of charcoal rot seedlings in water stress conditions was more as compared to irrigated conditions. They also observed that at the vegetative stage, the viable sclerotia population reached a stable level under irrigated conditions than under low-moisture stress conditions, indicating irrigation inversely proportion to microsclerotia formation at the vegetative and reproductive stage of soybean. This study also demonstrated that soybean charcoal rot could be managed through irrigation. The researcher also found no correlation between drought-tolerant soybean genotypes and disease severity of charcoal rot (Mengistu et al., 2018; Wrather et al., 2008). Mengistu et al. (2016) studied the impact of phosphorus and potassium on soybean charcoal rot severity and found that the application of phosphorus and potassium fertilizer did not affect the charcoal rot severity. On the contrary, the application of phosphorus fertilization reduced the soybean charcoal rot severity, while the application of nitrogen increased the infection of *M. phaseolina* in soybean (Spagnoletti et al., 2018). It may be because high nitrogen leads to reduced phenol compound production, which has fungistatic property, and reduced production of lignin that leads to increased susceptibility to pathogens (Spagnoletti et al., 2020). Soybean charcoal rot is also influenced by soil texture. Jimenez (2011) identified an increased inhabitation of *M. phaseolina* in silt loam soils. The minimum population of sclerotia was observed in sandy soil due to nutrient deprivation and soil microbes, indicating an optimum nutrient requirement to cause infection of charcoal rot in soybean. Recently, researchers found that seed quality also impacts charcoal rot disease severity. As the protein and oleic acid content increases or linolenic acid content decreases, the severity of soybean decreases (Bellaloui et al., 2021). Therefore, the selection of a variety with high protein and oleic acid and low linolenic acid may help in disease management through genetic resistance. Similarly, the same group of researchers found that moderate resistance genotypes possess high seed phenol, total isoflavones, and lignin. Seeds of moderate resistance genotypes also had a high concentration of Ca, K, P, Zn, Mn, Cu, and B compared to susceptible genotypes (Bellaloui et al., 2014). EL-Abdean et al. (2020) applied sodium silicate and potassium silicate to manage soybean charcoal rot in the glasshouse and field conditions. They identified sodium silicate @ 10 mM concentration reduced 37.8% of charcoal rot severity when applied as a soil drenching with irrigation water. Applying sodium silicate induces vigor in plants and increases thickness in cortex and pith of root, stem, and leave of soybean plant. Applying Si can reduce disease severity through a physical barrier, induction of resistance mechanism, or by acting as a fungicidal on phytopathogens (El-Abdean et al., 2020). Application of sulfur along with *Planomicrobium* sp. strain significantly reduced charcoal rot disease in sunflower through induction of defense enzyme and plant growth (Ijaz et al., 2021). A similar management strategy can be followed in soybean to manage charcoal rot. Charcoal rot pathogen management through herbicide application was started long ago. At the same time, weed provides competition to soybean plants for water and nutrient that increase the susceptibility against charcoal rot in soybean (Mengistu et al., 2015). Some herbicides restrict *M. phaseolina* from colonizing soybean plant or reduce the population of *M. phaseolina*. Initially, Filho and Dhingra (1980) identified that the herbicides—dinoseb, EPTC, fluorodifen, fluometuron, and alachlor reduced the viable count of *M. phaseolina* in sandy clay loam and sandy loam soil. The application of dinoseb herbicide reduced the population of *M. phaseolina* sharply in soybean. This could happen either through the fungicidal effect of herbicide or restricting the fungus to establish host–pathogen relationship. Mengistu et al. (2013) observed that a single application of glyphosate herbicide at mid-vegetative stages (V3 and V6) in soybean reduced the severity of soybean charcoal rot. Ingole and Zade (2020) observed that quizalophop-p-ethyle, propaquizafop, oxyfluorfen, and oxadiargyl reduced the mean mycelia growth of soybean charcoal rot pathogen and also reduced the microsclerotia formation. In contrast, soil application of chloramben and 2, 4-D herbicide increases the sclerotia production and colonization of *M. phaseolina* to soybean, whereas application of alachlor reduced the colonization of pathogen (Bowman et al., 1986). Similarly, another group of researchers identified that trifluralin reduced the colonization of *M. phaseolina* to soybean plants through induction of resistance in seedling and increased the lignin content in seedlings (Canaday et al., 1986). Foliar application of lactofen at the flowering stage and azoxystrobin at the pod initiation or filling stage is ineffective in managing soybean charcoal rot (Cross et al., 2012).

Soybean charcoal rot is also aggravated by the presence of soybean cyst nematode (SCN), so management of SCN is also an essential tool for managing soybean charcoal rot. The nature of damage to the host is different for both SCN (biotroph) and *M. phaseolina* (necrotroph); therefore, management of disease requires different approaches. Luna et al. (2017) summarized that some researchers found SCN increased colonization of *M. phaseolina*, whereas the other group found no interaction in both organisms. Still state department USDA recommended that farmers grow resistance variety that carries resistance against both pathogens. Mengistu et al. (2021) identified 12 moderate resistance genotypes having resistance against one or two races of SCN. Fluopyram has both fungicidal and nematicidal abilities. It inhibited the SCN population in roots through reduction of reproduction in soybean plant (Roth, Jacobs, et al., 2020), whereas same time Fluopyram was also inhibited the growth and development of soybean charcoal rot pathogen (Padgett, 2018). Fluopyram was also reported to reduce SCN motility and the ability to cause infection in soybean (Beeman et al., 2019). Roth, Webster, et al. (2020) concluded that *B. amyloliquefaciens* and *B. firmus* also could manage both the pathogen SCN and *M. phaseolina* in soybean crops. *Recently*, Strom et al. (2020) found that crop rotation with soybean and maize decrease the SCN population due to the increased population of the other fungal community, which secretes toxins that are effective against SCN. Crop rotation of soybean followed by maize was also reported as effective in the management of charcoal rot (Maier et al., 2018).

5.4.6 Genetic resistance

Soybean charcoal rot pathogen is necrotrophic and has various mechanisms to establish the host−parasite relationship; therefore, growing of resistant/tolerant cultivars, apart from different management strategies, is most important in the management of soybean charcoal rot. Recently identified sources of resistance/tolerance for soybean charcoal rot are presented in Table 5.1. The mechanism of disease resistance in soybean charcoal rot was not fully understood whether soybean charcoal rot resistance is monogenic, polygenic, or oligogenic with high or low heritability (da Silva et al., 2019). The heritability was reported as 0.06 and 0.45, indicating the environmental influence on the expression of resistance genes (Coser et al., 2017; da Silva et al., 2019). The disease resistance was reported as polygenic in nature through continuous distribution of genotypes in RIL population from high susceptibility to highly resistance (Talukdar et al., 2009). Two minor QTLs for charcoal rot resistance were identified with 9.9% variation (Jones, 2017). The QTL mapping of soybean charcoal rot resistance in mapping the population PI 567,562A (R)/PI 567,437 (S) revealed one QTL on chromosome 15 and two other QTL on chromosome 16 (da Silva et al., 2019).

One single-nucleotide polymorphism (SNPs) Gm15_01,842,053and Gm15_03,051,337 were identified through fine mapped of QTL on chromosome 15 with a 5.25 LOD value and 29.4% of R2 value on 1209 kb interval. After fine mapping of the first QTL on chromosome 16, another SNPGm16_28,961,127 and Gm16_30,493,887 were identified with 4.32 LOD value and 25.4% of R2 value on 1533 kb interval, whereas another SNP Gm16_35,973,543 and Gm16_37,078,478 were identified with 3.6 LOD value and 8.84% of R2 value on 1105 kb interval after mapping of second QTL (da Silva et al., 2019). Coser et al. (2017) attempted for identification of charcoal rot resistance gene in soybean and studied the architecture of resistance gene in 456 germplasm lines. Field evaluation of these 459 germplasm lines collected from the core collection of soybean germplasm of USDA revealed five SNP and putative candidate genes governing soybean charcoal rot resistance. Glasshouse screening of the identical genotypes provided information on eight candidate genes loci from different gene families controlling soybean charcoal rot resistance.

Interestingly, no common gene or marker was identified in both glass house and field level screening methods. This indicated the complexity of resistance and the influence of the environment on resistance (Coser et al., 2017). Bulked-segregant analysis of soybean charcoal rot resistance genotype PI 567,562A and charcoal rot susceptible genotype PI 567,437 was performed through the QTL-sequencing approach. Three genomic regions were identified on chromosomes 5, 8, and 14, which have highly correlated values in Δ SNP-index and differ in resistance and susceptible bulk in QTL governing soybean charcoal rot resistance (da Silva et al., 2020). Genome-wide association analysis revealed that six SNPs and three SNP markers, Gm08_44,422,211_T_C, Gm08_18,909,193_A_G, and Gm19_34,320,762_A_C, were associated with soybean charcoal rot resistance (Vinholes et al., 2019). The major QTL *Mac-2* was identified in soybean governing charcoal rot resistance. Comparative genome analysis of the QTL *Mac-2* revealed that a pectin esterase inhibitor gene, and two copies of pectin esterase were located on chromosome number 8 (Muchero et al., 2011). Association analysis for soybean charcoal rot resistance in 130 different soybean lines through the general linear model and the mixed linear model revealed that the Sat_252 marker was associated with the severity of charcoal rot disease, whereas Satt190 and Sat_169 markers were associated with the population of microsclerotia on soybean stem. Another marker, Satt_359 was associated with the population of microsclerotia on soybean stem and incidence and severity of charcoal rot, maker Sat_416, and Satt_460 were also associated with the population of microsclerotia on soybean stem and severity of charcoal rot (Ghorbanipour et al., 2019). These studies indicate soybean charcoal rot resistance governs by multiple loci.

Upadhyay et al. (2020) developed 13 lines of soybean charcoal rot resistance/tolerance lines from charcoal rot susceptible varieties JS 9560 and JS 335 through tissue culture. Hemmati et al. (2018) identified that genotypes with adventitious roots encircled to the crown region of soybean provide a barrier to *M. phaseolina* for colonization and restrict the pathogen from completing its life cycle. Resistance factors restrict the pathogen growth within the host tissue due to a decrease in the concentration of amino acids such as proline and asparagines in a soybean plant (Lodha & Mawar, 2020). The resistance/tolerance soybean genotype against charcoal rot disease is summarized in Table 5.6.

5.5 Diagnosis of soybean charcoal rot and utilization in resistance identification

Early and accurate phytopathogen detection is essential for managing plant diseases. Charcoal rot disease is often diagnosed through typical symptoms, such as yellowing of leaves, dropping of petioles attached leaves, blacking or gray appearance of plant tissue present inside the soil or near the soil, production of abundance microsclerotia on these tissues, and ultimately death of plant (Gupta et al., 2012). Babu et al. (2007) developed a species-specific oligonucleotide PCR-based primer, and probes can rapidly detect *M. phaseolina* through PCR reaction and hybridization. Similarly, Santos et al. (2020) developed a species-specific oligonucleotide PCR-based primer to detect *M. phaseolina* and *M. pseudophaseolina* that help in evaluation of diversity and distribution of *M. phaseolina* in particular geographical areas. Further, Babu et al. (2011) developed a real-time qPCR assay to detect and quantify *M. phaseolina* in soil and plant tissue. Later on, Fletcher et al. (2014) observed a monotonic relationship between light reflectance through disease tissue and the microsclerotia population in diseased tissue. The resistance levels of six genotypes were differentiated through hyperspectral reflectance measured with 400−2490 nm of light wavelengths. Singh, Chatterjee, Rajput, Kumar, et al. (2021) developed a novel method for identification of soybean charcoal rot resistance through a laser biospeckle-based sensor. Later on, the same group identified soybean anthracnose disease in seeds through machine learning and a vector support system (Singh, Chatterjee, Rajput, Rana, et al., 2021). Al-Ahmadi et al. (2018) identified the effect of phaseolin toxin on soybean charcoal rot inoculate plant through hyperspectral spectroscopy. Recently, the same group identified different response of soybean

TABLE 5.6 List of charcoal rot resistance/tolerance soybean genotypes.

Resistance/Tolerant genotypes	References
PI 548178, PI 548414 and PI 548302	Pawlowski et al. (2015)
DT99-16864	Gillen et al. (2016)
PI379559D	Coser et al. (2017)
PI567241	Coser et al. (2017)
Y 227−1	Smith et al. (2018)
Munasqa RR	Reznikov et al. (2019)
AMS MB 5−18, AMS MB 5−19 AMSs - 77 and AMS—1001	Chavan et al. (2019)
CPLRC5663DK4866, JTN-4307,, Osage, R01-581F, JTN-5208, and CPLRC5007	da Silva et al. (2019)
JS 20−98, JS 20−34, MACS 450 and PS 1225	Amrate et al. (2019)
GDM151029 and BMX Elite IPRO	Ishikawa et al. (2019)
DS-880, DT97−4290 and R07-7232	Bellaloui et al. (2021)

charcoal rot toxin in resistance and susceptible cultivars through hyperspectral spectroscopy (900–2400 nm) (Al-Ahmadi et al., 2021). Similarly, Nagasubramanian et al. (2018) also developed a hyperspectral imaging process for detecting soybean charcoal rot and identifying the resistant cultivars. These contemporary techniques can be used to identify resistance sources in soybean.

5.6 Host–pathogen interaction

Compared to other phytopathogens, the molecular and cellular mechanism of soybean charcoal rot infection was less concluded, despite its potential to cause huge yield loss and broad host range. Ammon et al. (1975) studied the early stage of soybean charcoal rot infection through scanning electron microscopy. They identified *M. phaseolina* entry to soybean through the generation of mechanical pressure and secretion of enzymes that lead to breaking the plant cell wall. The entry of *M. phaseolina* inside to the soybean root required only 24 h, later on rapid colonization occurred for 2 days (Bressano et al., 2010). *M. phaseolina* infected soybean plant revealed that 3 days after inoculation, only *M. phaseolina* entry to soybean plant. Prepenetration steps, such as germination of microsclerotia and development of hyphae occur the same time in both resistance and susceptible cultivars, indicating no association between prepenetration steps and resistance in postpenetration steps, such as root colonization and degeneration of root tissues of *M. phaseolina* in soybean differed in resistant and susceptible plants (Hemmati et al., 2018).

M. phaseolina possesses transglutaminase-like proteins and cellulose binding elicitor lectin genes that allow adhering to plant tissue and challenge plant defense system through inhibition of salicylate-1-monooxygenase that allows penetrating to the epidermis of host tissue. Pathogen secrets various cell wall degradation enzymes to degrade cell wall polysaccharides, i.e., endoglucanases, exocellobiohydrolases, and β-glucosidases. To degrade lignin pathogen can secrete lignin peroxidases, galactose oxidases, laccases, haloperoxidases, chloroperoxidases, and heme peroxidase (Islam et al., 2012). *M. phaseolina* has a tremendous amount of cytochrome P450s genes, superoxide dismutase, and Cofto counter the plant defense system. The plant defense was also encountered by the major facilitator superfamilies genes, which included glycosidases, type membrane transporters, transposases, and production of secondary metabolites (Islam et al., 2012). Secretome analysis of *M. phaseolina* revealed that cell wall degradation enzymes (xylanase, endoglucanase, amylase, and lignocellulolytic) contributed 40% of total secreted material, protease contributed 12%, and signaling protein contributed 14.7% (Sinha et al., 2022). Charcoal rot infected jute revealed that *M. phaseolina* produced nitric oxide that led to kill the host tissue (Sarkar et al., 2014). Similarly, toxins produced by the pathogen lead to yellowing symptoms of soybean charcoal rot infected plants due to the production of botryodiplodin toxin by *M. phaseolina* (Ramezani et al., 2007). Abbas et al. (2019) also detected botryodiplodin but not phaseolinone in soybean charcoal rot pathogen. This botryodiplodin toxin was affecting the root system of the soybean plant. It inhibited the lateral- and tap-root growth and disintegrated the root cap and tip (Abbas et al., 2020). The killing of the meristematic cell near the root cap and tip through botryodiplodin toxin provides a path for *M. phaseolina* to enter the vascular system of soybean plant, which leads to colonization of *M. phaseolina*. Through the leaf disc assay, it was identified that in a susceptible soybean cultivar, botryodiplodin toxin caused browning and necrosis of the soybean leaf (Bellaloui et al., 2012). Radwan et al. (2014) concluded that phaseolinone toxin was not reported in soybean, but it was reported in other crop's charcoal rot pathogen. This was due to the *M. phaseolina* isolates of soybean evolving differently in both continents, Asia and North America, which may result in different gene pools to regulate phytotoxin production, or either *M. phaseolina* can produce two types of toxin Radwan et al. (2014). Khambhati et al. (2020) identified different mycotoxins from soybean charcoal rot pathogen, i.e., (−)-botryodiplodin, orsellinic acid, kojic acid, mellein, cyclo (L-proline-L-tyrosine), and moniliformin. They also identified emodin, cordycepin, citrinin, endocrocin, infectopyron, gliocladic acid, monocerin, methylorsellinic acid, and N-benzoyl-L-phenylalanine. All of these toxins have specific role in the pathogenicity of *M. phaseolina*. Mullein inhibited the other fungi from invasion, whereas similar property was reported by cyclo (L-proline-L-tyrosine). Kojic acid inhibited tyrosinase enzyme, which requires the synthesizing plant defense secondary metabolites. Moniliformin inhibited pyruvate enzymes and α-ketoglutarate oxidation, leading to inhibiting mitochondrial respiration. Orsellinic acid has little or no phytotoxicity (Khambhati et al., 2020). Radwan et al. (2014) summarized that *M. phaseolina* could produce encode polyketide synthases, which synthesize cyclic and toxins. *M. phaseolina* genome encoded genes for nonribosomal peptide synthetases that produce iron chelating protein siderophores. The siderophores act as a virulent factor for pathogenic fungi.

The resistance mechanism is better understood through functional omics strategies, such as genomics, proteomics, and transcriptomics. This can be done through analysis of the interaction between soybean pathogen and several cultivars. Reznikov et al. (2019) studied the interaction between soybean cultivars and *M. phaseolina*. Four soybean genotypes, Munasqa RR, CRIA 4, DM 6.2i RR, and DT 97–4290 having moderate resistance to susceptible reaction toward *M. phaseolina*, was inoculated with seven isolates of *M. phaseolina*, i.e., Mp15, Mp17, Mp18, Mp32, Mp37, Mp42, and Mp4 in the field as well as in glasshouse condition. They identified significant genotype × isolate interactions in both field and glass house conditions, where in-field CFU ($P = .0025$) was measured and in glasshouse AUDPC ($P = .0277$) was measured. The minimum value of 33.3 CFU/g of dry weight was observed with Munasqa RR × MP 15, whereas the maximum value of 2366.7 CFU/g of dry weight was observed with DM6.2iRR × MP15 in field conditions. Under glasshouse condition, Munasqa RR × MP 17 was provided a minimum AUDPC value (4.21), whereas DM6.2iRR × MP37 (233.3) was provided the highest AUDPC value (233.3) (Reznikov et al., 2019).

The molecular basis of host-pathogen interaction, especially for soybean charcoal rot, was not much understood. Deshmukh and Tiwari (2021a) recently studied transcriptomics from charcoal rot infected plant in susceptible and resistant genotypes. They identified the induction of several defense-related signal genes in both genotypes. The observed defense pathway was up-regulated in the R6 stage, and Ca^{2+} ion-mediated defense pathway was mainly responsible for defense. Several WRKY transcription factors for resistance genes were unregulated that regulate phenylpropanoid metabolism, jasmonic acid, abscissic acid, salicylic acid, and secondary metabolite production (Deshmukh & Tiwari, 2021b). Genomics analysis revealed that this happened due to induction of jasmonic acid and ethylene pathway. Transcriptomic analysis revealed that charcoal rot induces infection through increased indole-three-acetic acid (IAA) (Gaige et al., 2010; Mah et al., 2012). Mutant of ethylene and jasmonic acid of *M. phaseolina* revealed that ethylene and jasmonic acid enhanced susceptibility of *M. phaseolina*. That indicates this pathway is vital to counter plant defense by *M. phaseolina* (Schroeder et al., 2019). Proteomics and transcriptomics analysis must be performed with charcoal rot infected soybean plant. Overexpression of two antifungal genes, stilbene synthase and chitinase, lead to enhanced resistance to sorghum charcoal rot pathogen in the first hour of infection. That indicates both genes have a role in disease resistance (Sharma et al., 2014). In another study, the researcher identified that overexpression of chitinase and β-1, 3-glucanase genes also increase resistance to groundnut charcoal disease (Iwuala et al., 2020). Transcriptomics and microRNA analysis of the RIL population for jute charcoal rot resistance identified increased expression of several defense genes such as phenylpropanoid metabolism, abscisic acid, jasmonic acid, cell wall biosynthesis, salicylic acid, and proteolysis (Biswas et al., 2014). They also identified an increase in specific 22-nt miRNA, which trigger systemic acquired resistance (Biswas et al., 2014). Further, they identified that these novel microRNAs provided multilayer resistance against *M. phaseolina* through resistance gene having nucleotide binding site and leucine-rich repeat motifs and regulating the production of reactive oxygen species (Dey et al., 2016). Trancriptomics analysis of charcoal rot resistance plant revealed that after infection with *M. phaseolina*, soybean plant induces plant defense genes related to signaling pathway, transcriptional factor, resistance gene and resistance protein, PR proteins, and genes related to secondary metabolism (Marquez et al., 2018). Gain and loss functional analysis revealed that glyceollin-mutant soybean plant might increase susceptibility to *M. phaseolina* infection. The phytoalexin glyceollin was accumulated in root tissue and provided an innate defense to soybean against *M. phaseolina* (Lygin et al., 2013).

5.7 Conclusions

The interaction of *M. phaseolina* with microbes below the soil is very complex. A harsh environment makes it challenging to manage soybean charcoal rot with any single strategy. The necrotrophic nature and high degree of genetic variation in *M. phaseolina* entail extreme difficulty in understanding the host—pathogen interaction and identifying the stable resistance sources. Under these circumstances, a strategic management approach should be designed to encompass every component of disease management. Biocontrol management mode is found effective under suitable agronomic practice and with moderately resistant varieties. Therefore, integration of agronomical practice, resistance/tolerance variety, and biological and fungicidal application should be integrated and made sustainable.

References

Abbas, H. K., Bellaloui, N., Accinelli, C., Smith, J. R., & Shier, W. T. (2019). Toxin production in soybean (*Glycine max* L.) plants with charcoal rot disease and by *Macrophomina phaseolina,* the fungus that causes the disease. *Toxins, 11*(11), 645. https://doi.org/10.3390/toxins11110645

Abbas, H. K., Bellaloui, N., Butler, A. M., Nelson, J. L., Abou-Karam, M., & Shier, W. T. (2020). Phytotoxic responses of soybean (*Glycine max* L.) to Botryodiplodin, a Toxin Produced by the charcoal rot disease fungus, *Macrophomina phaseolina*. *Toxins, 12*(1), 25. https://doi.org/10.3390/toxins12010025

Agale, R. C., Suryawanshi, A. P., & Patil, A. G. (2020). Role of resistance inducing chemicals against soybean dry root rot (*R. bataticola*) disease. *International Journal of Chemical Studies, 8*(6), 1308—1310. https://doi.org/10.22271/chemi.2020.v8.i6s.10940

Al-Ahmadi, A., Subedi, A., Choudhary, R., Fakhoury, A., & Watson, D. G. (2021). Hyperspectral spectroscopy to detect different responses of two soybean (*Glycine max*) cultivars to charcoal rot (*Macrophomina phaseolina*) toxin. *Engenharia Agrícola, 41*, 78—86. https://doi.org/10.1590/1809-4430-Eng.Agric.v41n1p78-86/2021

Al-Ahmadi, A. H., Subedi, A., Wang, G., Choudhary, R., Fakhoury, A., & Watson, D. G. (2018). Detection of charcoal rot (*Macrophomina phaseolina*) toxin effects in soybean(*Glycine max*) seedlings using hyperspectral spectroscopy. *Computers and Electronics in Agriculture, 150*, 188–195. https://doi.org/10.1016/j.compag.2018.04.013

Al-Ani, R. A., Adhab, M. A., Mahdi, M. H., & Abood, H. M. (2012). *Rhizobium japonicum* as a biocontrol agent of soybean root rot disease caused by *Fusarium solani* and *Macrophomina phaseolina*. *Plant Protection Science, 48*(4), 149–155. https://doi.org/10.17221/16/2012-PPS

Allen, T. W., Bradley, C. A., Sisson, A. J., Byamukama, E., Chilvers, M. I., Coker, C. M., & Wrather, J. A. (2017). Soybean yield loss estimates due to diseases in the United States and Ontario, Canada, from 2010 to 2014. *Plant Health Progress, 18*(1), 19–27. https://doi.org/10.1094/PHP-RS-16-0066

Ammon, V., Wyllie, T. D., & Brown, M. F. (1975). Investigation of the infection process of *Macrophomina phaseolina* on the surface of soybean roots using scanning electron microscopy. *Mycopathologia, 55*(2), 77–81.

Amrate, P. K., Shrivastava, M. K., & Bhale, M. S. (2019). Resistance in soybean varieties against charcoal rot disease caused by *Macrophomina phaseolina*. *Plant Disease Research, 34*(2), 124–128. https://doi.org/10.5958/2249-8788.2019.00021.0

Babu, B. K., Mesapogu, S., Sharma, A., Somasani, S. R., & Arora, D. K. (2011). Quantitative real-time PCR assay for rapid detection of plant and human pathogenic *Macrophomina phaseolina* from field and environmental samples. *Mycologia, 103*(3), 466–473. https://doi.org/10.3852/10-181

Babu, B. K., Saxena, A. K., Srivastava, A. K., & Arora, D. K. (2007). Identification and detection of *Macrophomina phaseolina* by using species specific oligonucleotide primers and probe. *Mycologia, 99*(6), 797–803. https://doi.org/10.3852/mycologia.99.6.797

Baird, R. E., Watson, C. E., & Scruggs, M. (2003). Relative longevity of *Macrophomina phaseolina* and associated mycobiota on residual soybean roots in soil. *Plant Disease, 87*(5), 563–566.

Bandara, A. Y., Weerasooriya, D. K., Bradley, C. A., Allen, T. W., & Esker, P. D. (2020). Dissecting the economic impact of soybean diseases in the United States over two decades. *PLoS One, 15*(4), e0231141. https://doi.org/10.1371/journal.pone.0231141

Beeman, A. Q., Njus, Z. L., Pandey, S., & Tylka, G. L. (2019). The effects of ILeVO and VOTiVO on root penetration and behavior of the soybean cyst nematode *Heterodera glycines*. *Plant Disease, 103*, 392–397.

Bellaloui, N., Mengistu, A., Smith, J. R., Abbas, H. K., Accinelli, C., & Shier, W. T. (2021). Effects of charcoal rot on soybean seed composition in soybean genotypes that differ in charcoal rot resistance under irrigated and non-irrigated conditions. *Plants, 10*(9), 1801.

Bellaloui, N., Mengistu, A., Zobiole, L. H. S., Abbas, H. K., & Kassem, M. A. (2014). Soybean seed phenolics, sugars, and minerals are altered by charcoal rot infection in mg iii soybean cultivars. *Food and Nutrition Sciences, 5*(19), 1843–1859.

Bellaloui, N., Mengistu, A., Zobiole, L. H. S., & Shier, W. T. (2012). Resistance to toxin-mediated fungal infection: Role of lignins, isoflavones, other seed phenolics, sugars, and boron in the mechanism of resistance to charcoal rot disease in soybean. *Toxin Reviews, 31*(1–2), 16–26.

Biswas, C., Dey, P., Karmakar, P. G., & Satpathy, S. (2014). Next-generation sequencing and micro RNAs analysis reveal SA/MeJA1/ABA pathway genes mediated systemic acquired resistance (SAR) and its master regulation via production of phased, trans-acting siRNAs against stem rot pathogen *Macrophomina phaseolina* in a RIL population of jute (*Corchorus capsularis*). *Physiological and Molecular Plant Pathology, 87*, 76–85.

Bowman, J. E., Hartman, G. L., Mcclary, R. D., Sinclair, J. B., Hummel, J. W., & Wax, L. M. (1986). Effects of weed control and row spacing in conventional tillage reduced tillage and non-tillage on soybean seed quality. *Plant Disease, 70*, 673–676.

Bressano, M., Giachero, M. L., Luna, C. M., & Ducasse, D. A. (2010). An in vitro method for examining infection of soybean roots by *Macrophomina phaseolina*. *Physiological and Molecular Plant Pathology, 74*(3–4), 201–204.

Canaday, C. H., Helsel, D. G., & Wyllie, T. D. (1986). Effects of herbicide induced stress on root colonization of soybeans by *Macrophomina phaseolina*. *Plant Disease, 70*, 863–866.

Chauhan, P., Bhattacharya, A., Giri, V. P., Singh, S. P., Gupta, S. C., Verma, P., & Mishra, A. (2022). *Bacillus subtilis* suppresses the charcoal rot disease by inducing defence responses and physiological attributes in soybean. *Archives of Microbiology, 204*(5), 1–18.

Chavan, S. V., Jadhav, P. V., Mane, S. S., & Nichal, S. S. (2019). Morphological characterizations of soybean genotypes in response to charcoal rot disease under sick plot conditions. *Journal of Pharmacognosy and Phytochemistry, 8*(5), 989–992.

Choudhary, K., Meena, A. K., Dhakar, H., & Kumar, M. (2018). Management of charcoal rot of sesame [*Macrophomina phaseolina* (Tassi.) Goid.]by different bio-agent and fungicide. *International Journal of Chemical Studies, 6*(6), 1556–1559.

Collins, D. J., Wyllie, T. D., & Anderson, S. H. (1991). Biological activity of *Macrophomina phaseolina* in soil. *Soil Biology and Biochemistry, 23*, 495–496.

Coser, S. M., Chowda Reddy, R. V., Zhang, J., Mueller, D. S., Mengistu, A., Wise, K. A., & Singh, A. K. (2017). Genetic architecture of charcoal rot (*Macrophomina phaseolina*) resistance in soybean revealed using a diverse panel. *Frontiers of Plant Science, 8*, 1626.

Cross, C., Wrather, A., Fothergill, K., Shannon, G., Li, S., Shumway, C., & Rupe, J. (2012). Effect of lactofen, azoxystrobin, and genotypes on charcoal rot, *Phomopsis* seed decay, and pod and stem blight in soybean. *Plant Disease, 96*(8), 1154–1158.

Crous, P. W., Slippers, B., Wingfield, M. J., Rheeder, J., Marasas, W. F., Philips, A. J., Alves, A., Burgess, T., Barber, P., & Groenewald, J. Z. (2006). Phylogenetic lineages in the botryosphaeriaceae. *Studies in Mycology, 55*(1), 235–253.

Dangi, R., Sinha, P., Islam, S., Gupta, A., & Kumar, A. (2019). Screening of onion accessions for Stemphylium blight resistance under artificially inoculated field experiments. *Australasian Plant Pathology, 48*(4), 375–384.

Das, K., Prasanna, R., & Saxena, A. K. (2017). Rhizobia: A potential biocontrol agent for soilborne fungal pathogens. *Folia Microbiologica, 62*(5), 425–435.

Deshmukh, R., & Tiwari, S. (2021a). Analysis of soybean transcriptome uncovers differential genomic regulations in response to charcoal rot in resistant and susceptible genotypes. *Physiological and Molecular Plant Pathology, 115*, 101672.

Deshmukh, R., & Tiwari, S. (2021b). Molecular interaction of charcoal rot pathogenesis in soybean: A complex interaction. *Plant Cell Reports, 40*(10), 1799–1812.

Dey, P., Biswas, C., & Karmakar, P. G. (2016). Identification and characterization of differentially expressed novel miRNAs (21–24 nt) in a *Macrophomina phaseolina* resistant RIL line of jute (*Corchorus capsularis* L.). *Physiological and Molecular Plant Pathology, 94*, 62–66.

Dhingra, O. D., & Sinclair, J. B. (1978). *Biology and pathology of* Macrophomina phaseolina. Minas Gerais, Brazil: Universidade Federal de Vicosa.

Di, S., Wang, Z., Cang, T., Xie, Y., Zhao, H., Qi, P., Xu, H., & Wang, X. (2021). Enantioselective toxicity and mechanism of chiral fungicide penflufen based on experiments and computational chemistry. *Ecotoxicology and Environmental Safety, 222*, 112534.

Dubey, R. C., Kumar, H., & Pandey, R. R. (2009). Combined effect of soil solarization and neem amendment on survival of *Macrophomina phaseolina* sclerotia and growth of soybean. *Nature and Science, 7*(11), 52–57.

El-Abdean, W. Z., Abo-Elyousr, K. A., Hassan, M. H., & El-sharkawy, R. M. (2020). Effect of silicon compounds against *Macrophomina phaseolina* the causal agent of soybean charcoal rot disease. *Archives of Phytopathology and Plant Protection, 53*(19–20), 983–998.

El-Saber Batiha, G., Magdy Beshbishy, A., G. Wasef, L., Elewa, Y. H., A. Al-Sagan, A., Abd El-Hack, M. E., Taha, A. E., Abd-Elhakim, Y. M., & Devkota, H. P. (2020). Chemical constituents and pharmacological activities of garlic (*Allium sativum* L.): A review. *Nutrients, 12*(3), 872.

Fehr, W. R., Caviness, C. E., Burmood, D. T., & Pennington, J. S. (1971). Stage of development descriptions for soybeans, *Glycine Max* (L.) Merrill. *Crop Science, 11*(6), 929–931.

Feng, Y., Huang, Y., Zhan, H., Bhatt, P., & Chen, S. (2020). An overview of strobilurin fungicide degradation: Current status and future perspective. *Frontiers in Microbiology, 11*, 389.

Filho, E. S., & Dhingra, O. D. (1980). Population changes of *Macrophomina phaseolina* in amended soils. *Transactions of the British Mycological Society, 74*(3), 471–481.

Fletcher, R. S., Smith, J. R., Mengistu, A., & Ray, J. D. (2014). Relationships between microsclerotia content and hyperspectral reflectance data in soybean tissue infected by Macrophomina phaseolina. *American Journal of Plant Sciences, 5*(25), 3737.

Gaetan, S. A., Fernandez, L., & Madia, M. (2006). Occurrence of charcoal rot caused by *Macrophomina phaseolina* on canola in Argentina. *Plant Disease, 90*(4), 524-524.

Gaige, A. R., Ayella, A., & Shuai, B. (2010). Methyl jasmonate and ethylene induce partial resistance in Medicago truncatula against the charcoal rot pathogen *Macrophomina phaseolina*. *Physiological and Molecular Plant Pathology, 74*(5–6), 412–418.

Ghorbanipour, A., Rabiei, B., Rahmanpour, S., & Khodaparast, S. A. (2019). Association analysis of charcoal rot disease resistance in soybean. *Plant Pathology Journal, 35*(3), 189–199.

Gillen, A. M., Mengistu, A., Smith, J. R., & Paris, R. L. (2016). Registration of DT99-16864 soybean germplasm line with moderate resistance to charcoal rot [*Macrophomina phaseolina* (tassi) goid.]. *Journal of Plant Registrations, 10*(3), 309–315.

Gulya, T. J., Krupinsky, J., Draper, M., & Charlet, L. D. (2002). First report of charcoal rot (Macrophomina phaseolina) on sunflower in North and South Dakota. *Plant Disease, 86*(8), 923-923.

Gupta, G. K., & Chauhan, G. S. (2005). *Symptoms, identification and management of soybean diseases.Technical Bulletin 10*. Indore, India: National Research Centre for Soybean.

Gupta, G. K., Sharma, S. K., & Ramteke, R. (2012). Biology, epidemiology and management of the pathogenic fungus *Macrophomina phaseolina* (Tassi) Goid. with special reference to charcoal rot of soybean (*Glycine max* (L.) Merrill). *Journal of Phytopathology, 160*(4), 167–180.

Hemmati, P., Zafari, D., Mahmoodi, S. B., Hashemi, M., Gholamhoseini, M., Dolatabadian, A., & Ataei, R. (2018). Histopathology of charcoal rot disease (*Macrophomina phaseolina*) in resistant and susceptible cultivars of soybean. *Rhizosphere, 7*, 27–34.

Hingole, D. G., & Kendre, A. H. (2021). *In-vitro* evaluation of fungicides, antagonist and plant extracts on management of charcoal rot of soybean caused by *Macrophomina phaseolina* (TASSI.) Goid. *Journal of Plant Disease Sciences, 16*, 11–16.

Hossain, I., Suratuzzaman, M., & Khalil, M. I. (1999). Seed health of soybean and control of seed borne fungi with botanicals. *Bangladesh Journal of Training and Development, 12*, 99–105.

Hussain, S., Maryam, M., Manzoor, S., Shuxian, L., Ahmed, I., Skalicky, M., Brestic, M., Rastogi, A., Ulhassan, Z., Shafiq, I., Allakhverdiev, S. I., Khurshid, H., & Liu, W. (2021). Foliar application of silicon improves growth of soybean by enhancing carbon metabolism under shading conditions. *Plant Physiology and Biochemistry, 159*, 43–52.

Ijaz, M., Sattar, A., Sher, A., Ul-Allah, S., Mansha, M. Z., Khan, K. A., Shahzad, M. A., Al-Sadi, A. M., Arif, M., Aljuaid, B. S., El-Shehawi, A. M., & Farooq, S. (2021). Sulfur application combined with *Planomicrobium* sp. strain MSSA-10 and farmyard manure biochar helps in the management of charcoal rot disease in sunflower (*Helianthus annuus* L.). *Sustainability, 13*(15), 8535.

Ingole, B. M., & Zade, S. B. (2020). Bioassay of different agrochemicals and fertilizers against Root rot (*Macrophominaphaseolina*) of soybean in *in vitro*. *International Journal of Current Microbiology and Applied Sciences, 9*(04), 582–589.

Ishikawa, M. S., Ribeiro, N. R., Almeida, A. A., & Balbi-Peña, M. I. (2019). Identification of soybean genotypes resistant to charcoal rot by seed inoculation with *Macrophomina phaseolina*. *Journal of Agricultural Science, 11*(9), 213–219.

Islam, M. S., Haque, M. S., Islam, M. M., Emdad, E. M., Halim, A., Hossen, Q. M. M., Hossain, M. Z., Ahmed, B., Rahim, S., & Rahman, M. S. (2012). Tools to kill: Genome of one of the most destructive plant pathogenic fungi *Macrophomina phaseolina*. *BMC Genomics*, 1–16.

Iwuala, E., Odjegba, V., Unung, O., & Alam, A. (2020). Expression of stress responsive β-1, 3-glucanase and chitinase genes in *Arachis hypogaea* seedlings against *Macrophomina phaseolina*. *Gene Reports, 20*, 100693.

James, W. C. (1974). Assessment of plant disease losses. *Annual Review of Phytopathology, 12*, 27–48.

Jimenez. (2011). Influence of soils, nutrition, and water. M.Sc. Thesis. *Relations upon charcoal rot disease processes in Kansas*. Kansas: Kansas State University.

Jones, S. (2017). *Identification of spectral disease signatures and resistant QTL for charcoal rot infection in soybean.* M.Sc. Thesis. Iowa: Iowa State University.

Jordaan, E., van der Waals, J. E., & McLaren, N. W. (2019). Effect of irrigation on charcoal rot severity, yield loss and colonization of soybean and sunflower. *Crop Protection, 122*, 63–69.

Kaur, S. (2012). Carbohydrate degrading enzyme production by plant pathogenic mycelia and microscleroia isolates of *Macrophomina phaseolina* through koji fermentation. *Industrial Crops and Products, 36*, 140–148.

Kendig, S. R., Rupe, J. C., & Scott, H. D. (2000). Effect of irrigation and soil water stress on densities of *Macrophomina phaseolina* in soil and roots of two soybean cultivars. *Plant Disease, 84*, 895–900.

Khaledi, N., & Taheri, P. (2016). Biocontrol mechanisms of *Trichoderma harzianum* against soybean charcoal rot caused by *Macrophomina phaseolina. Journal of Plant Protection Research, 56*(1), 21–31.

Khamari, B., Satapathy, S. N., & Patra, C. (2019). Effect of inoculum load and duration of exposure to *Macrophomina phaseolina* on disease incidence of sesame. *International Journal of Chemical Studies, 7*(1), 1728–1730.

Khambhati, V. H., Abbas, H. K., Sulyok, M., Peterson, M. T., & Shier, W. T. (2020). First report of the production of mycotoxins and other secondary metabolites by *Macrophomina phaseolina* (Tassi) Goid.isolates from soybeans (*Glycine max* L.) symptomatic with charcoal rot disease. *Journal of Fungi, 6*, 332.

Konde, S., Kothikar, R., Koche, M., & Surpam, A. (2017). Studies on varietal resistance in soybean against root rot and collar rot diseases. *Innovative Farming, 2*(2), 108–110.

Kumar, D. G., Natarajan, N., & Nakkeeran, S. (2016). Antifungal activity of nano fungicide Trifloxystrobin 25% + Tebuconazole 50% against *macrophominaphaseolina. African Journal of Microbiology Research, 10*, 100–105.

Kumar, K., & Singh, J. (2000). Location, survival, transmission and management of seed-borne *Macrophomina phaseolina*, causing charcoal rot in soybean. *Annals of Plant Protection Sciences, 8*, 44–46.

Lodha, S., & Mawar, R. (2020). Population dynamics of Macrophomina phaseolina in relation to disease management: A review. *Journal of Phytopathology, 168*(1), 1–17.

Lorenzetti, E., Stangarlin, J. R., Kuhn, O. J., & Portz, R. L. (2018). Induction of resistance to *Macrophomina phaseolina* in soyben treated with rosemary extract. *Summa Phytopathologica, 44*, 45–50.

Luna, M. P. R., Mueller, D., Mengistu, A., Singh, A. K., Hartman, G. L., & Wise, K. A. (2017). Advancing our understanding of charcoal rot in soybeans. *Journal of Integrated Pest Management, 8*(1), 1–8.

Lygin, A. V., Zernova, O. V., Hill, C. B., Kholina, N. A., Widholm, J. M., & Hartman, G. L. (2013). Glyceollin is an important component of soybean plant defense against *Phytophthora sojae* and *Macrophomina phaseolina. Phytopathology, 103*, 984–994.

Ma, J., Hill, C. B., & Hartman, G. L. (2010). Production of *Macrophomina phaseolina* conidia by multiple soybean isolates in culture. *Plant Disease*, 1088–1092.

Mah, K. M., Uppalapati, S. R., Tang, Y., Allen, S., & Shuai, B. (2012). Gene expression profiling of Macrophomina phaseolina infected Medicago truncatula roots reveals a role for auxin in plant tolerance against the charcoal rot pathogen. *Physiological and Molecular Plant Pathology, 79*, 21–30.

Maier, M., Souza, C. A., & Casa, R. T. (2018). Cropping systems on root rot and soybean seed yield. *Ciência Rural, 48*(7), 1–8.

Marquez, N., Giachero, M. L., Declerck, S., & Ducasse, D. A. (2021). *Macrophomina phaseolina*: General characteristics of pathogenicity and methods of control. *Frontiers of Plant Science, 12*, 634397.

Marquez, N., Giachero, M. L., Gallou, A., Debat, H. J., Cranenbrouck, S., & Di Rienzo, J. A. (2018). Transcriptional changes in mycorrhizal and non-mycorrhizal soybean plants upon infection with the fungal pathogen *Macrophomina phaseolina. Molecular Plant-Microbe Interactions, 31*, 842–855.

Mengistu, A., Bellaloui, N., & Arelli, P. (2021). Resistance to charcoal rot identified within soybean cyst nematode resistant accessions. *Plant Health Progress, 22*(4). https://doi.org/10.1094/PHP-01-21-0004-RS

Mengistu, A., Bond, J., Nelson, R., Rupe, J., Shannon, G., Arelli, P., & Wrather, A. (2013). Identification of soybean accessions resistant to *Macrophomina phaseolina* by field screening and laboratory validation. *Plant Health Progress, 14*(1), 25.

Mengistu, A., Ray, J. D., Smith, J. R., Arelli, P. R., Bellaloui, N., & Chen, P. (2018). Effect of charcoal rot on selected putative drought tolerant soybean genotypes and yield. *Crop Protection, 105*, 90–101.

Mengistu, A., Ray, J. D., Smith, J. R., & Paris, R. L. (2007). Charcoal rot disease assessment of soybean genotypes using a colony-forming unit index. *Crop Science, 47*, 2453–2461.

Mengistu, A., Reddy, K. N., Zablotowicz, R. M., & Wrather, A. J. (2009). Propagule densities of *Macrophomina phaseolina* in soybean tissue and soil as affected by tillage, cover crop, and herbicide. *Plant Health Progress, 10*(1), 28.

Mengistu, A., Rupe, J. C., & Wrather, J. A. (2015). Charcoal rot. In GL Hartman, JC Rupe, EJ Sikora, LL Domier, KL Steffey, & JA Davis (Eds.), *Compendium of Soybean Diseases and Pests* (5th, pp. 29–31). St. Paul, MN: American Phytopathological Society. In this issue.

Mengistu, A., Smith, R. J., Ray, J. D., & Bellaloui, N. (2011). Seasonal progress of charcoal rot and its impact on soybean productivity. *Plant Disease, 95*, 1159–1166.

Mengistu, A., Yin, X., Bellaloui, N., McClure, A. M., Tyler, D. D., & Reddy, K. N. (2016). Potassium and phosphorus have no effect on severity of charcoal rot of soybean. *Canadian Journal of Plant Pathology, 38*, 174–182.

Mostafanezhad, H., Mostafanezhad, N., Sahebani, N., Khaledi, & Hosseini, M. (2012). *Control of the charcoal rot of soybean (*Macrophominaphaseolina*) by salicylic acid and neem oil.20th Iranian Plant ProtectionCongress.*

Muchero, W., Ehlers, J. D., & Close, T. J. (2011). Genic SNP markers and legume synteny reveal candidate genes underlying QTL for *Macrophomina phaseolina* resistance and maturity in cowpea [*Vigna unguiculata* (L) Walp.]. *BMC Genomics, 12*, 8.

Mufti, R., & Bano, A. (2019). PGPR-induced defense responses in the soybean plant against charcoal rot disease. *European Journal of Plant Pathology, 155*(3), 983–1000.

Nagasubramanian, K., Jones, S., Sarkar, S., Singh, A. K., Singh, A., & Ganapathy subramanian, B. (2018). Hyperspectral band selection using genetic algorithm and support vector machines for early identification of charcoal rot disease in soybean stems. *Plant Methods, 14*, 86–87.

Nataraj, V., Kumar, S., Kumawat, G., Shivakumar, M., Rajput, L. S., Ratnaparkhe, M. B., Ramteke, R., Gupta, S., Satpute, G. K., Rajesh, V., Kamble, V., & Chandra, S. (2019). Charcoal rot resistance in soybean: Current understanding and future perspectives. In S. H. Wani (Ed.), *Disease resistance in crop plants* (pp. 241–259). Cham: Springer.

Ndiaye, M., Termorshuizen, A. J., & VanBruggen, A. H. C. (2007). Combined effects of solarization and organic amendment on charcoal rot caused by *Macrophomina phaseolina. The Sahelian Phytoparasitica, 35*, 392–400.

Olson, T. R., Gebreil, A., Micijevic, A., Bradley, C. A., Wise, K. A., Mueller, D. S., Chilvers, M. I., & Mathew, F. M. (2015). Association of Diaporthe longicolla with black zone lines on mature soybean plants. *Plant Health Progress, 16*(3), 118–122.

Padgett, A. C. (2018). *The effects of Fluopyram on Macrophomina phaseolina and charcoal rot of soybean.* Southern Illinois University at Carbondale.

Pagano, M. C., & Miransari, M. (2016). The importance of soybean production worldwide. In M. Miransari (Ed.), *Abiotic and biotic stresses in soybean production* (pp. 1–26). Academic Press, Elsevier.

Pande, A. K., Burlakoti, R. R., Rathore, A., & Nair, R. M. (2012). Morphological and molecular characterization of *Macrophomina phaseolina* isolated from three legume crops and evaluation of mungbean genotypes for resistance to dry root rot. *Crop Protection, 127*, 104962.

Paris, R. L., Mengistu, A., Tyler, J. M., & Smith, J. R. (2006). Registration of soybean germplasm line DT97-4290 with moderate resistance to charcoal rot. *Crop Science, 46*(5), 2324.

Parmar, R. G., & Patel, P. S. (2020). Efficacy of bioagents against *Macrophomina phaseolina* causing root rot of soybean *in vitro*. *Journal of Pharmacognosy and Phytochemistry, 9*(6S), 196–198.

Pawlowski, M. L., Bowen, C. R., Hill, C. B., & Hartman, G. L. (2016). Responses of soybean genotypes to pathogen infection after the application of elicitors. *Crop Protection, 87*, 78–84.

Pawlowski, M. L., Hill, C. B., & Hartman, G. L. (2015). Resistance to charcoal rot identified in ancestral soybean germplasm. *Crop Science, 55*, 1230–1235.

Perez -Brandán, C., Huidobro, J., Grümberg, B., Scandiani, M. M., Luque, A. G., Meriles, J. M., & Vargas-Gil, S. (2014). Soybean fungal soil-borne diseases: A parameter for measuring the effect of agricultural intensification on soil health. *Canadian Journal of Microbiology, 60*(2), 73–84.

Perez-Brandán, C., Arzeno, J. L., Huidobro, J., Grümberg, B., Conforto, C., Hilton, S., & Vargas-Gil, S. (2012). Long-term effect of tillage systems on soil microbiological, chemical and physical parameters and the incidence of charcoal rot by *Macrophomina phaseolina* (Tassi) Goid. in soybean. *Crop Protection, 40*, 73–82.

Ploper, L. D., Gonzalez, V., de-Ramallo, N. V., Galvez, R., Devani, M., & de-Ramallo, N. V. (2001). Presence of black stem rot on soybeans in central and northeast Argentina. *Advance Agroindustrial, 22*, 30–34.

Radwan, O., Rouhana, L. V., Hartman, G. L., & Korban, S. S. (2014). Genetic mechanisms of host–pathogen interactions for charcoal rot in soybean. *Plant Molecular Biology Reporter, 32*(3), 617–629.

Rahman, M. T., Rubayet, M., Khan, A. A., & Bhuiyan, M. (2021). Integrated management of charcoal rot disease of soybean caused by *Macrophomina phaseolina. Egyptian Journal of Agricultural Research, 99*(1), 10–19.

Rajput, L. S., & Harlapur, S. I. (2015). Evaluation of fungicides and biocontrol agents for suppression of banded leaf and sheath blight of maize (*Zea mays*). *Indian Phytopathology, 68*(2), 149–155.

Rajput, L. S., Harlapur, S. I., Venkatesh, I., Aggarwal, S. K., & Choudhary, M. (2016). *In-Vitro* study of botanicals and biocontrol agents against *Rhizoctonia solani* f. sp. sasakii causing banded leaf and sheath blight of maize. *International Journal of Agriculture Sciences, 8*(53), 2777–2779.

Ramezani, M., Shier, W. T., Abbas, H. K., Tonos, J. L., Baird, R. E., & Sciumbato, G. L. (2007). Soybean charcoal rot disease fungus *Macrophomina phaseolina* in Mississippi produces the phytotoxin (−)-botryodiplodin but no detectable phaseolinone. *Journal of Natural Products, 70*(1), 128–129.

Reznikov, S., Chiesa, M. A., Pardo, E. M., De Lisi, V., Bogado, N., & González, V. (2019). Soybean-*Macrophomina phaseolina*-Specific interactions and identification of a novel source of resistance. *Phytopathology, 109*, 63–73.

Reznikov, S., Vellicce, G. R., González, V., de Lisi, V., Castagnaro, A. P., & Ploper, L. D. (2016). Evaluation of chemical and biological seed treatments to control charcoal rot of soybean. *Journal of General Plant Pathology, 82*(5), 273–280.

Roth, M. G., Jacobs, J. L., Napieralski, S., Byrne, A. M., Stoffer-Hopkins, A., Warner, F., & Chilvers, M. I. (2020). Fluopyram suppresses population densities of *Heterodera glycines* in field and greenhouse studies in Michigan. *Plant Disease, 104*, 1305–1311.

Roth, M. G., Webster, R. W., Mueller, D. S., Chilvers, M. I., Faske, T. R., Mathew, F. M., Bradley, C. A., Damicone, J. P., Kabbage, M., & Smith, D. L. (2020). Integrated management of important soybean pathogens of the United States in changing climate. *Journal of Integrated Pest Management, 11*(1), 17.

Sanjay, Kumar, S., & Chaudhary, B. K. (2020). Potential of few fungicides and plant extracts for managing charcoal rot of soybean caused by *Macrophomina phaseolina* (Tassi) Gold. in Madhya Pradesh, India. *Journal of Applied and Natural Science, 12*(3), 388–393.

Santos, K. M., Lima, G. S., Barros, A. P., Machado, A. R., Souza-Motta, C. M., Correia, K. C., & Michereff, S. J. (2020). Novel specific primers for rapid identification of Macrophomina species. *European Journal of Plant Pathology, 156*(4), 1213–1218.

Sarkar, T. S., Biswas, P., Ghosh, S. K., & Ghosh, S. (2014). Nitric oxide production by necrotrophic pathogen *Macrophomina phaseolina* and the host plant in charcoal rot disease of jute: Complexity of the interplay between necrotroph–host plant interactions. *PLoS One, 9*(9), e107348.

Sassenrath, G. F., Little, C. R., Hsiao, C. J., Shoup, D. E., & Lin, X. (2017). Cover crop system to control charcoal rot in soybeans. *Kansas Agricultural Experiment Station Research Reports, 3*(2), 14.

Schroeder, M. M., Lai, Y., Shirai, M., Alsalek, N., Tsuchiya, T., Roberts, P., & Eulgem, T. (2019). A novel Arabidopsis pathosystem reveals cooperation of multiple hormonal response-pathways in host resistance against the global crop destroyer *Macrophomina phaseolina*. *Scientific Reports, 9*(1), 1–14.

Sharma, I., Kumari, N., & Sharma, V. (2014). Defense gene expression in Sorghum bicolor against *Macrophomina phaseolina* in leaves and roots of susceptible and resistant cultivars. *Journal of Plant Interactions, 9*(1), 315–323.

Short, G. E., Wyllie, T. D., & Bristow, P. R. (1980). Survival of *Macrophomina phaseolina* in soil and in residue of soybean. *The American phytopathological Society, 7*(13), 17.

da Silva, M. P., Pereira, A., Rupe, J. C., Bluhm, B. H., Gbur, E. E., Wood, L., Mozzoni, L. A., & Chen, P. (2019). Effectiveness of a seed plate assay for evaluating charcoal rot resistance in soybean and the relationship to field performance. *Plant Disease, 103*(8), 1947–1953.

da Silva, M. P., Zaccaron, A. Z., Bluhm, B. H., Rupe, J. C., Wood, L., Mozzoni, L. A., & Pereira, A. (2020). Bulked segregant analysis using next-generation sequencing for identification of genetic loci for charcoal rot resistance in soybean. *Physiological and Molecular Plant Pathology, 109*, 101440.

Simonetti, E., Viso, N. P., Montecchia, M., Zilli, C., Balestrasse, K., & Carmona, M. (2015). Evaluation of native bacteria and manganese phosphite for alternative control of charcoal root rot of soybean. *Microbiology Research, 180*, 40–48. https://doi.org/10.1016/j.micres.2015.07.004

Singh, S. N., Srivastava, S. K., Bhargava, P. K., & Khare, M. N. (1990b). Chemical control of seedling mortality in cv. JS 72-44 and bacterial pustule in cv. Punjab-1 of soybean (*Glycine max* (L.) Merrill). *Legume Research, 13*(1), 17–20.

Singh, P., Chatterjee, A., Rajput, L. S., Kumar, S., Nataraj, V., Bhatia, V., & Prakash, S. (2021). Biospeckle-based sensor for characterization of charcoal rot[*Macrophomina phaseolina* (Tassi) Goid.] disease in soybean [*Glycine Max* (L.) Merr.]crop. *IEEE Access, 9*, 31562–31574.

Singh, P., Chatterjee, A., Rajput, L. S., Rana, S., Kumar, S., Nataraj, V., Bhatia, V., & Prakash, S. (2021). Development of an intelligent laser biospeckle system for early detection and classification of soybean seeds infected with seed-borne fungal pathogen (*Colletotrichum truncatum*). *Biosystems Engineering, 212*, 442–457.

Sinha, N., Patra, S. K., & Ghosh, S. (2022). Secretome analysis of *Macrophomina phaseolina* identifies an array of putative virulence factors responsible for charcoal rot disease in plants. *Frontiers in Microbiology, 13*, 847832. https://doi.org/10.3389/fmicb.2022.847832

Smith, G. S., & Carvil, O. N. (1997). Field screening of commercial and experimental soybean cultivars for their reaction to *Macrophomina phaseolina*. *Plant Disease, 81*(4), 363–368.

Smith, J. R., Ray, J. D., & Mengistu, A. (2018). Genotypic differences in yield loss of irrigated soybean attributable to charcoal rot. *Journal of Crop Improvement, 32*(6), 781–800. https://doi.org/10.1080/15427528.2018.1516262

Spagnoletti, F. N., Balestrasse, K., Lavado, R. S., & Giacometti, R. (2016). Arbuscular mycorrhiza detoxifying response against arsenic and pathogenic fungus in soybean. *Ecotoxicology and Environmental Safety, 133*, 47–56.

Spagnoletti, F. N., Cornero, M., Chiocchio, V., Lavado, R. S., & Roberts, I. N. (2020). Arbuscular mycorrhiza protects soybean plants against *Macrophominaphaseolina* even under nitrogen fertilization. *European Journal of Plant Pathology, 156*, 839–849. https://doi.org/10.1007/s10658-020-01934-w

Spagnoletti, F. N., Leiva, M., Chiocchio, V., & Lavado, R. S. (2018). Phosphorus fertilization reduces the severity of charcoal rot (*Macrophomina phaseolina*) and the arbuscular mycorrhizal protection in soybean. *Journal of Plant Nutrition and Soil Science, 181*, 855–860. https://doi.org/10.1002/jpln.201700569

Strom, N., Hu, W., Haarith, D., Chen, S., & Bushley, K. (2020). Interactions between soil properties, fungal communities, the soybean cyst nematode, and crop yield under continuous corn and soybean monoculture. *Applied Soil Ecology, 147*, 103388.

Talukdar, A., Verma, K., & Gowda, D. S. S. (2009). Molecular breeding for charcoal rot resistance in soybean I. Screening and mapping population development. *Indian Journal of Genetics and Plant Breeding, 69*, 367–370.

Thakkar, A., & Saraf, M. (2015). Development of microbial consortia as a biocontrol agent for effective management of fungal diseases in *Glycine max* L. *Archives of Phytopathology and Plant Protection, 48*(6), 459–474.

Twizeyimana, M., Hill, C. B., Pawlowski, M., Paul, C., & Hartman, G. L. (2012). A cut stem inoculation technique to evaluate soybean for resistance to *Macrophomina phaseolina*. *Plant Disease, 96*, 1210–1215.

Upadhyay, S., Singh, A. K., Tripathi, M. K., Tiwari, S., Tripathi, N., & Patel, R. P. (2020). *In Vitro*selection for resistance against charcoal rot disease of soybean [*Glycine max*(L.) Merrill] caused by *Macrophomina phaseolina* (Tassi) Goid. *Legume Research, LR-4440*. https://doi.org/10.18805/LR-4440

Ushiki, J., Hayakawa, Y., & Tadano, T. (1996). Medicinal plants for suppressing soil-borne plant diseases: I. Screening for medicinal plants with antimicrobial activity in roots. *Journal of Soil Science and Plant Nutrition, 42*, 423–426. https://doi.org/10.1080/00380768.1996.10415116

Vinholes, P., Rosado, R., & Roberts, P. (2019). Single nucleotide polymorphism-based haplotypes associated with charcoal rot resistance in Brazilian soybean germplasm. *Agronomy Journal, 111*, 182–192.

Wilkerson, T. H. (2017). *Managing charcoal rot of soybean through supplementing secondary nutrients*. Ph.D. Thesis. Mississippi: College of Agriculture and Life Sciences, Mississippi State University.

Wrather, J. A., Anderson, T. R., Arsyad, D. M., Gai, J., Ploper, L. D., Porta-Puglia, A., Ram, H. H., & Yorinori, J. T. (1997). Soybean disease loss estimates for the top 10 soybean producing countries in 1994. *Plant Disease, 81*, 107–110.

Wrather, J. A., Anderson, T. R., Arsyad, D. M., Tan, Y., Ploper, L. D., Porta-Puglia, A., Ram, H. H., & Yorinori, J. T. (2001). Soybean disease loss estimates for the top ten soybean-producing countries in 1998. *Canadian Journal of Plant Pathology, 23*, 115–121.

Wrather, J. A., & Koenning, S. R. (2006). Estimates of disease effects on soybean yields in the United States 2003 to 2005. *Journal of Nematology, 38*(2), 173–180.

Wrather, A., Shannon, G., Balardin, R., Carregal, L., Escobar, R., Gupta, G. K., Ma, Z., Morel, W., Ploper, D., & Tenuta, A. (2010). Effect of diseases on soybean yield in the top eight producing countries in 2006. *Plant Health Progress.* https://doi.org/10.1094/PHP-2010-0125-01-RS

Wrather, J. A., Shannon, J. G., Carter, T. E., Bond, J. P., Rupe, J. C., & Almeida, A. M. R. (2008). Reaction of drought-tolerant soybean genotypes to *Macrophomina phaseolina. Plant Health Progress.* https://doi.org/10.1094/PHP-2008-0618-01-RS

Yang, C., Hamel, C., Vujanovic, V., & Gan, Y. (2011). Fungicide: Modes of action and possible impact on nontarget microorganisms. *International Scholarly Research Notices, 8.* https://doi.org/10.5402/2011/13028s

Yasmin, H., Naz, R., Nosheen, A., Hassan, M. N., Ilyas, N., Sajjad, M., Anjum, S., Gao, X., & Geng, Z. (2020). Identification of new biocontrol agent against charcoal rot disease caused by *Macrophomina phaseolina* in soybean (*Glycine max* L.). *Sustainability, 12*(17), 6856. https://doi.org/10.3390/su12176856

Young, P. A. (1949). *Charcoal rot of plants in east Texas*. College Station, TX: Texas Agricultural Experiment Station.

Yuan, M., Yu, T., Shi, Q., Han, D., Yu, K., Wang, L., Wang, S., Xiang, H., Wen, R., Nian, H., & Lian, T. (2021). Rhizosphere soil bacterial communities of continuous cropping-tolerant and sensitive soybean genotypes respond differently to long-term continuous cropping in mollisols. *Frontiers in Microbiology, 13*(12), 729047. https://doi.org/10.3389/fmicb.2021.729047

Zaynab, M., Fatima, M., Abbas, S., Sharif, Y., Umair, M., & Zafar, M. H. (2018). Role of secondary metabolites in plant defense against pathogens. *Microbial Pathogenesis, 124*, 198–202.

Chapter 6

Management of nematode-charcoal rot disease complexes in legumes

Zaki Anwar Siddiqui and Manzoor Raiees Khan
Section of Plant Pathology and Nematology, Department of Botany, Aligarh Muslim University, Aligarh, Uttar Pradesh, India

6.1 Introduction

Plant-parasitic nematodes (PPNs) are generally microscopic pests, which feed on all plant parts including seeds, flowers, leaves, stems, and roots, although most species are feeding on plant roots. For penetrating plant cells and feeding, PPNs produce/secrete effector molecules with pharyngeal glands connected to protrusible stylet facilitate penetration and migration inside the roots (Jones et al., 2013; Mejias et al., 2019). PPNs are categorized generally as ecto-, semiendo-, and endo-parasites, but sedentary endoparasites, i.e., *Meloidogyne* spp., *Globodera* spp., and *Heterodera* spp. cause severe crop losses (Jones et al., 2013). They are called hidden enemies of the farmers responsible for an average 12.6% annual yield loss amounting $215.77 billion globally (Abd-Elgawad & Vagelas, 2015; Holbein et al., 2016).

Macrophomina phaseolina (Tassi) Goid. is a soil-borne fungus belonging to the family Botryosphaeriaceae. No subspecies/physiological races of *M. phaseolina* have been identified on morphological or genomic characterizations (Crous et al., 2006; Marquez et al., 2021). The fungus is characterized by brown (light/dark) thin-walled hyaline hyphae with septa. The hyphal branches are formed at right angles on parent hyphae with a constriction at the point of origin. Microsclerotia are oblong, oval, or spherical, which change the color from light brown to brown or black with aging. Pycnidia are larger in comparison to microsclerotia usually brown or black colored, globose or irregular, ostiolate, beaked, and rough (Lakhran et al., 2018). On synthetic media, *M. phaseolina* showed wide heterogeneity in color of mycelium, distribution of microsclerotia, and formation of pycnidia. Isolates of this fungus belong to the single species (Marquez et al., 2021).

Microsclerotia of *M. phaseolina* are the infective source (primary) and can survive in soil for years (Gupta et al., 2012; Marquez et al., 2021). At the seedling stage, it infects the roots of host plants, and once inside the roots, it disrupts the transport of nutrients and water to the upper portions of plants, finally resulting in premature death of infected plants (Fig. 6.1). The disease symptoms includes senescence and yellowing of leaves, sloughing of tissues (cortical) from taproot and lower stem, and graying of tissues due to presence of microsclerotia (Marquez et al., 2021; Sinclair & Backman, 1989; Smith & Carvil, 1997).

PPNs interact with fungi in different ways resulting in disease complexes (Bergeson, 1972). PPNs may serve as agents or vectors for transmission of pathogens, induce necrotic infection courts, provide portals of entry, modify the host plant physiology, and break the resistance of the host plants to other pathogens, etc. Moreover, in the process of penetration and feeding, PPNs frequently wound plants, which become subject to infection by fungal pathogens requiring aid in the penetration of their host. PPNs also modify plant tissues and make them a better substrate for the fungal pathogens and thus enhance their growth detrimental to the host plant. They may also induce qualitative and quantitative changes in root exudates which stimulate the growth, germination, and reproduction of fungi present in the rhizosphere. PPNs may also induce drought stress and influence some plant diseases by restricting development of host roots. They are also known to exert influence on the reproduction of their own and also the host plants. Co-infection may stimulate or inhibit the nematode reproduction depending on the nematode and fungal species (Bergeson, 1972).

Mechanisms of nematode—fungal interactions and the development of disease complexes by the interplay of mechanisms of interactions are crucial because of the ultimate effect on disease control. It may also enhance the understanding

Macrophomina Phaseolina. https://doi.org/10.1016/B978-0-443-15443-0.00001-2

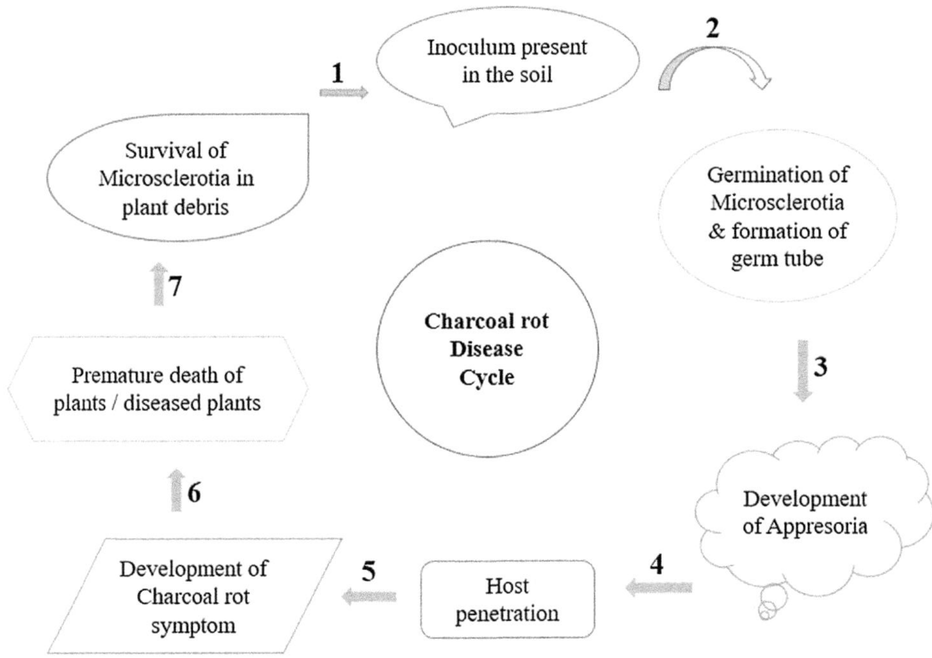

FIGURE 6.1 Disease cycle of *Macrophomina phaseolina* causing charcoal rot in legume crops.

of the factors that regulate the interactions between soil-borne pathogens in the rhizosphere (Riedel, 1988). Management of diseases involving more than one pathogen is more complex than one might anticipate. Brodie (1970) highlighted the significance of the multipathogenic scenario and emphasized the importance of their management. Various methods for the management of nematode-fungus disease complexes in grain legumes have been discussed in this chapter and also the difficulties in their management.

6.2 Disease complexes of legumes

Legumes provide nutritious food for both humans and animals. They also provide fibers and raw material for various purposes, edible oil, while the edible seeds are a source of high protein and many essential amino acids. The legume family includes bean, chickpea, cowpea, lentil, pea, peanut, soybean, etc., and these plants are susceptible to pests and diseases than cereals. This increases the risk for smallholder farmers and constraints for the successful cultivation of these crops. Therefore, the adoption of improved and resistant cultivars, and the judicious use of biological and chemical control methods are required. Important disease complexes of legume crops that have been studied by various workers are summarized in Table 6.1.

6.3 Management of disease complexes

Disease complexes involving PPNs and other plant pathogens under field condition are not uncommon (Powell, 1971). Nematode–fungal disease complexes are most important and cause serious losses to agriculture (Siddiqui & Mahmood, 1999). The prohibitive cost of chemicals and also the adverse ecological effects are the reasons for searching alternative methods of disease management. Shifting priorities of disease control from chemicals to cultural and biological and integrated management demonstrated are changing the scientific ideas with time. The urgent need is to work on management strategies for disease complexes because they cause severe losses to legumes and other crops. Management of disease complexes by using various methods has been discussed.

Use of chemicals for the management of one of the interacting organisms and thus prevent the occurring of disease complex is the most promising solution. Knowledge of mechanisms and interactions is vital for the control of disease complexes, even where low densities of both pathogens, that is, fungi or nematodes are present in a disease complex (Bowers et al., 1996; Saeed et al., 1998). Therefore, problem of the interaction may not be resolved by reducing one pathogen. Therefore, populations of all the pathogens should be reduced, but the main emphasis must be on the reduction

TABLE 6.1 Nematode–root rot disease complexes in legumes.

Nematode	Fungus	Host	References
Meloidogyne javanica	*Rhizoctonia bataticola*	Cowpea	Kanwar et al. (1987, 1988)
M. incognita	*Macrophomina phaseolina*	French bean	Al-Hazmi (1985)
M. javanica	*R. bataticola, F. solani*	Groundnut	Sakhuja and Sethi (1986)
M. javanica	*R. bataticola*	Chickpea	Goel and Gupta (1986)
H. glycines	*M. phaseolina*	Soybean	Todd et al. (1987)
M. inocognita race3	*M. phaseolina*	Chickpea	Siddiqui and Husain (1991a)
M. inocognita race3	*M. phaseolina, Bradyrhizobium* sp.	Chickpea	Siddiqui and Husain (1992)
M. incognita	*M. phaseoli*	Soybean	Agarwal and Goswami (1973)
Heterodera glycines	*M. phaseolina*	Soybean	Winkler et al. (1994)
H. cajani	*M. phaseolina*	Black gram	Latha and Narasimhan (2000)
M. javanica	*M. phaseolina*	Green been	Imani et al. (2014)
M. javanica	*M. phaseolina*	Green beans	Al-Hazmi et al. (2017)
M. inocgnita	*M. phaseolina*	Chickpea	Sumbul and Mahmood (2020)
H. glycines	*M. phaseolina*	Soybean	Lopez-Nicora et al., 2020

of the nematode population because nematodes generally predispose the host to fungal pathogens, which typically aggravate the disease.

6.3.1 Chemical control

Generally, the growers have knowledge about adverse effects of the disease complex, but there is no rationale behind the application of higher rates of nematicides. The field trial was conducted to observe the effect of chemicals on the disease complex of *M. phaseolina* and *H. cajani* on black gram (*Vigna mungo*). Combination of fungicide, carbendazim (seed treatment @ 2g kg^{-1} of seed), and nematicide carbofuran (soil application @ 3.3 kg ha^{-1}) was most effective in increasing the yield and in reducing the root-rot incidence and nematode population (Latha et al., 2000). The basic aim behind the use of fungicide and nematicide in combination is to reduce the population of both pathogens, but cost of both pesticides may not be feasible for farmers and sustainable agriculture.

The occurrence of co-infected diseases and the application of fumigants against nematodes and also effectiveness of fumigants against fungi have resulted in the foundation of a multiple control strategy. According to Brodie (1970), the use of selective pesticides mixtures to achieve multiple pest control on a field appears most promising. Therefore, formulated mixtures can be prepared from technical materials with exact ratios of toxicants necessary for effective control. The "super" granule with nematicide, herbicide, and fungicide can be formulated and may be applied for specific crops and against disease complexes under a particular condition (Brodie, 1970). Kinloch and Schenck (1978) reported that DBCP possesses both nematicidal and fungicidal properties. Therefore, use of DBCP with sodium azide was effective against the nematode–fungus disease complex of soybean.

6.3.2 Physical methods

Commonly used physical methods to control plant diseases include temperature, unfavorable light lengths, and types of radiation. Physical treatments involve hazards for plants and operators, but there are fewer residual problems. Physical methods generally include soil sterilization, hot water treatment (propagative organs), soil pasteurization by solar heating, field burning, desiccation, hot air treatment (storage organs), drying storage of grains, refrigeration, radiation and mechanical seed cleaning, etc.

Out of physical control methods used, solarization and hot water treatments are generally effective against plant pathogens (Yuliar et al., 2015). The soil solarization by transparent plastic mulches for 60 days before planting reduced the incidence of diseases (Vinh et al., 2005). Mechanism of soil solarization by which plant diseases are reduced was studied by Baptista et al.(2006, 2007). Soil solarization did not significantly affect soil chemical properties but reduced microbial biomass and microbial respiration in soil. Soil solarization needs careful application because increase in temperature/the release of volatile compounds affects practical feasibility of the field. Overall, physical methods for disease complexes have rarely been used with less success.

6.3.3 Breeding for resistance

Numbers of alternative strategies have been investigated for disease complexes management. Breeding programs for legumes should be designed in a way as to include potential pathogens of the region. Breeding of legumes cultivars for the multipathogenic condition is more challenging and has numerous impediments, but it relieves farmers from extra capital investments. Multipathogenic resistance generally requires the pooling of all eco-pathogenic information of pathogens involved in disease complexes, but field populations of nematodes and fungi consist of combination of different races/pathotypes. Therefore, incorporation of genes providing resistance to pathogens in cultivars with high yields and desirable agronomical attributes is a challenging task and requires strenuous efforts.

6.3.4 Biological control

Due to public concern and scrutiny, biocontrol of plant diseases is a suitable alternative for the management of plant diseases (Cook, 1993). Biocontrol curtails the growth, infection, or reproduction of pathogenic organism by another beneficial organism (Baker, 1987; Cook, 1993). It is an ecologically safe alternative, and, in few cases, this is the only option available to protect the plants from diseases (Cook, 1993).

Root-rot disease complex (*M. incognita* plus *M. phaseolina*) was studied using leaf extracts on chickpea. Increasing doses of leaf extracts were increasingly beneficial to increase plant growth of infected plants by reducing nematode multiplication. *Cymbopogon citratus* was most effective for the disease complex of chickpea, followed by *Eichhornia crassipes* and *Ipomea carnea* (Siddiqui & Husain, 1990). Similarly, culture filtrates of *Paecilomyces lilacinus, A. flavus, A. niger, Alternaria brassicicola, A. triticina,* and *Fusarium solani* were used for the management of root-rot disease complex of chickpea. Culture filtrate of *P. lilacinus* was most effective, followed by *A. niger, A. flavus, A. brassicicola, A. triticina,* and *F. solani* (Siddiqui & Husain, 1991b). In another study, *P. lilacinus* was the best biocontrol agent for the chickpea root-rot disease complex, while *B. licheniformis* and *Acrophialophora fusispora* were equally effective (Siddiqui & Mahmood, 1992a). Similarly, inoculation of *Bradyrhizobium* sp., 10 days prior to *M. incognita* and *M. phaseolina,* resulted in a reduced disease index of chickpea (Siddiqui & Husain, 1992). An arbuscular mycorrhizal (AM) fungus *Glomus fasciculatum* caused a greater increase in plant growth and pod number of plants having root-rot disease complex, followed by *G. intaradices, G. constrictum, Sclerocystis*sp., *Gigaspora margarita,* and *Acaulospora* sp. (Siddiqui & Akhtar, 2006), while another study revealed that *Pseudomonas putida* provided significant disease control of chickpea disease complex (Akhtar & Siddiqui, 2009a). Use of *Rhizobium* sp. Resulted in a higher yield of chickpea than the use of *Pseudomonas putida/P. aeruginosa,* or *G. intraradices.* The root nodules/root systems in plants inoculated with *Rhizobium* sp. were significantly higher than the plants without *Rhizobium* sp. Use of *P. putida* resulted in higher reduction in root-knot galling, while inoculation of *A. awamori* the least. However, inoculation of *Rhizobium* sp. resulted in a similar reduction in root galling to that by *P. putida* (Akhtar & Siddiqui, 2009b).

Use of *P. lilacinus* with seed treatment or ascorbic acid foliar spray was equally effective against chickpea root-rot disease complex, while the combination of foliar spray and seed treatment of ascorbic acid the least (Siddiqui & Mahmood, 1992b). Use of *P. lilacinus* plus *Bacillus subtilis* caused a greater increase in dry shoot weight when chickpea plants were either with *M. incognita* or *M. phaseolina* or with both (Siddiqui & Mahmood, 1993a). The use of *Cymbopogan citratus* as green manure with culture filtrate of *P. lilacinus* provided better biocontrol than the combination of *P. lilacinus* and *A. niger* culture filtrates in the management of chickpea disease complex (Siddiqui & Mahmood, 1993b). Increase in plant growth and reduction in nematode multiplication, galling, and disease indices of chickpea were greater when plants inoculated with both pathogens were treated with *B. subtilis* plus *A. niger* (Siddiqui & Mahmood, 1995). The greatest decrease in gall formation, nematode reproduction, and root-rot indices was obtained by inoculation of *A. awamori* with *G. intraradices* and *P. aeruginosa* Pa28 (Akhtar & Siddiqui, 2006). Inoculation of *G. fasciculatum* plus *Rhizobium* sp. resulted in higher increase in plant growth and also reduced the disease indices of chickpea greater than by either of them. Both pathogens adversely affect root colonization by *G. fasciculatum* (Akhtar & Siddiqui, 2007a). Similarly, greater reduction in galling and nematode multiplication was caused by *P. aeruginosa* followed by *A. awamori* and *G. intraradices.* Individual inoculation of these biocontrol agents caused lesser increase in plant growth and also lesser reduction in disease indices of chickpea than when the mixture of all three biocontrol agents was used (Siddiqui & Akhtar, 2007). Inoculation of plants with mixture of *G. intraradices*, *P. putida*, and *Paenibacillus polymyxa* caused the highest reduction in galling, nematode multiplication, and disease indices of chickpea (Akhtar & Siddiqui, 2007b). Combined use of *Rhizobium* sp., *G. intraradices,* and *P. straita*, or use of *Rhizobium* sp. plus *P. straita* provided substantial biocontrol of chickpea root-rot disease complex (Akhtar & Siddiqui, 2008a). Application of *P. alcaligenes* plus *B. pumilus* with *G. intraradices* resulted in the greatest reduction in root-rot indices, galling, and nematode multiplication of chickpea (Akhtar & Siddiqui, 2008b).

The effect of antagonists, i.e., *P. lilacinus* with *Trichoderma viride* and *P. fluorescens* on the root-rot disease complex caused by *M. phaseolina* and *Heterodera cajani* were studied in a field trial on blackgram (*Vigna mungo*). Seed treatment with *P. lilacinus* plus *T. viride* and *P. fluorescens* resulted in lesser root-rot incidence, nematode population, and higher pod yield (Latha et al., 2000).

6.3.5 Culture practices

The activities of humans aim to control disease or disease complexes through the cultural manipulation of plants. These practices have their greatest importance for the control of diseases in the large areas and crops of low unit value. High labor costs may prohibit the use of these cultural practices in modern agriculture. The success of these practices for disease control is dependent on knowledge of pathogens biology and the host plant response to infection. These understanding facilitate the decisions of management of pathogens at vulnerable stages of their life cycles. The farming operations are performed well in advance than pathogen attack; therefore, assessment of their effectiveness is difficult.

Knowledge of cropping history of the land site for a crop is necessary so that the crop to be planted may not be infected by pathogens of previous crop. Growers can reduce the initial inoculum levels of pathogens or reduce the spread rate of established pathogens to prevent/control disease outbreaks. Cultural practices generally destroy the inoculum or inhibit the spread of plant pathogens. This includes destruction of crop residues, removal of living plants that have pathogens, crop rotation, tillage practices, sowing and harvesting practices, intercropping, mulching and soil amendments, flooding, irrigation, rouging, fertilizer application and plant nutrition, strip farming, trap and decoy crops, etc. Cultural practices, such as multiyear cropping regimes (Chen et al., 1995), reduce population densities of *Verticillium dahliae* and *Pratylenchus penetrans* with variable success.

Crop rotations involve changing of crop grown in one area regularly. Growing of alternate groups of plants, such as roots, cereals, brassicas, and legumes generally prevent pests and diseases for building up and also add to soil fertility. Some plants add nutrients to the soils, while others use up nutrients. Observance of bare fallow periods or crop rotation with other crops may serve as effective alternative for the management of diseases of field crops and low-cost practices (Stone et al., 2004; Vilich & Sikora, 1998). *Zea mays and Glycine* max rotation is superior based on economic, energy than monoculture of soybean/corn (Helmers et al., 2001; Howard et al., 1998; West et al., 1996).

Plant-parasitic nematodes and other plant pathogens are widely managed by organic amendments. Decomposing organic materials release toxic compounds, which results in rapid declines of pathogen population levels, while increases in pathogen antagonists may be its long-term effects. Improved crop nutrition and plant growth may result in increased tolerance to plant pathogens. Materials used, processing/composting of material, application rate, test arena, crop rotation and agronomic practices, soil type, climate, and other environmental factors also affect the results of cultural practices. To explain the beneficial effects of organic amendments on plants several mechanisms includes release of nematicidal compounds from decomposing organic materials, stimulation of natural enemies of pathogens, improved plant growth, and tolerance to plant pathogens (Akhtar & Malik, 2000; Thoden et al., 2011). It is not easy to distinguish which mechanism is most important because multiple mechanisms operate simultaneously (Akhtar & Malik, 2000).

Organic amendments for the management of plant diseases appear to be one of the most promising for its low cost and the limited environmental impact compared to the pesticides. Although organic amendments have great potential in the management of diseases but sometimes give inconsistent results and may also increase disease severity and phytotoxicity. The benefits of organic amendments no doubt far outweigh their drawbacks, but the impact on disease suppression remains unpredictable.

6.3.6 Integrated management

Integrated pest management consists of sneering with the timely application of action plan and steps of action. These may include use of resistant cultivars, alternation of planting practices, modification of environment, selection of site and its preparation, use of pesticides, pruning and thinning of plants, etc. The monitoring of temperature, moisture, soil pH, nutrients, etc., k nowledge of economic thresholds and disease forecasting are also are essential for the integrated management. All these measures should be applied in a coordinated, integrated, and harmonized manner for maximum benefits. For example, adequate use of fertilizers with irrigation practices helps promote healthy plants. Whatever the measures used, they must be compatible with the cultural practices essential for the crop being managed.

6.4 Conclusions

Under field conditions, synergistic interactions of nematodes and charcoal rot fungus result in serious crop losses. Several strategies for the management are in practice, but breeding for resistant varieties against plant disease complexes is the most acceptable form of disease management. Resistant varieties are generally vulnerable to new races, and continuous use of a particular variety may lead to the development of a new race or shift in the pest population. Chemical control is also effective and efficient means of managing disease complexes, but problems like toxicity, pollution, bioaccumulation, etc. have limited their application. Available cultural methods are generally least accepted by farmers. A number of biocontrol agents are available to manage disease complexes, but their application needs a greater knowledge of the complex interactions among plants, pathogens, biocontrol agents, and the environment. Therefore, specifically targeted biocontrol agents for disease complexes need to be searched. Control of diseases with a single biocontrol agent is generally not consistent under field conditions. Therefore, use of a mixture of biocontrol agents against disease complexes is required. Application of biocontrol agents in combination generally increases the competitiveness and efficacy. Mostly the mixture of biocontrol agents adapts better to the environmental changes, protects greater in the growing season, and protects greater against pathogens (Siddiqui & Futai, 2009). This mixture may increase the genetic diversity in the rhizosphere, utilize an array of mechanisms, and may persist longer (Pierson & Weller, 1994; Siddiqui & Futai, 2009). Therefore, biocontrol agent's mixture may be promising for management of disease complexes of nematodes and fungi.

References

Abd-Elgawad, M. M., & Vagelas, I. K. (2015). Nematophagous bacteria: Field application and commercialization. In T. H. Askary, & P. R. P. Martinelli (Eds.), *Biocontrol agent of phytonematodes* (pp. 276–309). Wallingford, UK: CAB International.

Agarwal, D. K., & Goswami, B. K. (1973). Interrelationship between a fungus, *Macrophomina phaseoli* (Maubl.) Ashby and root-knot nematode, *Meloidogyne incognita* (kofoid and white, 1991) chitwood, 1949 in soybean (*Glycine max* L.) Merill. *Proceedings of the Indian National Science Academy - Part B: Biological Sciences, 39*, 701–704.

Akhtar, M., & Malik, A. (2000). Roles of organic soil amendments and soil organisms in the biological control of plant-parasitic nematodes: A review. *Bioresource Technology, 74*, 35–47.

Akhtar, M. S., & Siddiqui, Z. A. (2006). Effects of phosphate solubilizing microorganisms on the growth and root-rot disease complex of chickpea. *Mikologiai Fitopatologia, 40*, 246–254.

Akhtar, M. S., & Siddiqui, Z. A. (2007a). Effects of *Glomus fasciculatum* and *Rhizobium* sp. on the growth and root-rot disease complex of chickpea. *Archives of Phytopathology and Plant Protection, 40*, 37–43.

Akhtar, M. S., & Siddiqui, Z. A. (2007b). Biocontrol of a chickpea root-rot disease complex with *Glomus intraradices, Pseudomonas putida* and *Paenibacillus polymyxa. Australian Plant Pathology, 36*, 175–180.

Akhtar, M. S., & Siddiqui, Z. A. (2008a). Biocontrol of a root-rot disease complex of chickpea by *Glomus intraradices, Rhizobium* sp. and *Pseudomonas straita. Crop Protection, 27*, 410–417.

Akhtar, M. S., & Siddiqui, Z. A. (2008b). *Glomus intraradices, Pseudomonas alcaligenes, Bacillus pumilus* as effective biocontrol agents for the root-rot disease complex of chickpea *Cicer arietinum* L. *Journal of General Plant Pathology, 74*, 53–60.

Akhtar, M. S., & Siddiqui, Z. A. (2009a). Use of plant growth promoting rhizobacteria for the biocontrol of root- rot disease complex of chickpea. *Australasian Plant Pathology, 38*, 44–50.

Akhtar, M. S., & Siddiqui, Z. A. (2009b). Effects of phosphate solubilizing microorganisms and *Rhizobium* sp. on the growth, nodulation,yield and root-rot disease complex of chickpea under field condition. *African Journal of Biotechnology, 8*, 3489–3496.

Al-Hazmi, A. S. (1985). Interaction of *Meloidogyne incognita* and *Macrophomina phaseolina* in a root-rot disease complex of French bean. *Journal of Phytopathology, 113*, 311–316.

Al-Hazmi, A. S., Dawabah, A. A., Al-Nadhary, S. N., Al-Yahya, F. A., & Lafi, H. A. (2017). Influence of soil texture and moisture on the interaction of *Meloidogyne javanica* and *Macrophomina phaseolina* on green beans. *Journal of Experimental Biology and Agricultural Sciences, 5*, 148–154.

Baker, K. F. (1987). Evolving concepts of biological control of plant pathogens. *Annual Review of Phytopathology, 25*, 67–85.

Baptista, M. J., de Souza, R. B., Pereira, W., Lopes, C. A., & Carrijo, O. A. (2007). Effect of soil solarization and biofumigation on tomato bacterial wilt incidence. *Horticultura Brasileira, 24*, 161–165.

Baptista, M. J., Lopes, C. A., de Souza, R. B., & Furumoto, O. (2006). Effect of soil solarization and biofumigation during autumn on bacterial wilt incidence and potato yield. *Horticultura Brasileira, 24*, 99–102.

Bergeson, G. B. (1972). Concepts of nematode–fungus associations in plant disease complexes: A review. *Experimental Pathology, 32*, 301–314.

Bowers, J. H., Nameth, S. T., Riedel, R. M., & Rowe, R. C. (1996). Infection and colonization of potato roots by *Verticillium dahliae* as affected by *Pratylenchus penetrans* and *P. crenatus*. *Phytopathology, 86*, 614–621.

Brodie, B. B. (1970). Use of nonselective and mixtures of selective pesticides for multiple pest control. *Phytopathology, 60*, 12–15.

Chen, J., Bird, G. W., & Mather, R. L. (1995). Impact of multi-year cropping regimes on *Solanum tuberosum* tuber yields in the presence of *Pratylenchus penetrans* and *Verticillium dahliae*. *Journal of Nematology, 27*, 654–660.

Cook, R. J. (1993). Making greater use of introduced microorganisms for biological control of plant pathogens. *Annual Review of Phytopathology, 31*, 53–80.

Crous, P. W., Slippers, B., Wingfield, M. J., Rheeder, J., Marasas, W. F. O., Phillips, A. J. L., Alves, A., Burgess, T., Barber, P., & Groenewald, J. Z. (2006). Phylogenetic lineages in the botryosphaeriaceae. *Studies in Mycology, 55*, 235–253.

Goel, S. R., & Gupta, D. C. (1986). Interaction of *Meloidogyne javanica* and *Rhizoctonia bataticola* on chickpea (*Cicer arietinum* L.). *Indian Journal of Nematology, 16*, 133–134.

Gupta, G. K., Sharma, S. K., & Ramteke, R. (2012). Biology, epidemiology and management of the pathogenic fungus *Macrophomina phaseolina*(Tassi) Goid.with special reference to charcoal rot of soybean (*Glycine max* (L.) Merrill). *Journal of Phytopathology, 160*, 167–180.

Helmers, G. A., Yamoah, C. F., & Varvel, G. E. (2001). Separating the impacts of crop diversification and rotations on risk. *Agronomy Journal, 93*, 1337–1340.

Holbein, J., Grundler, F. M., & Siddique, S. (2016). Plant basal resistance to nematodes: An update. *Journal of Experimental Botany, 67*, 2049–2061.

Howard, D. D., Chambers, A. Y., & Lessman, G. M. (1998). Rotation and fertilization effects on corn and soybean yields and soybean cyst nematode populations in a no-tillage system. *Agronomy Journal, 90*, 518–522.

Imani, S., Moosavi, M. R., & Basirnia, T. (2014). Interaction of *Macrophomina phaseolina* and *Meloidogyne javanica* on green bean. *Journal of Research in Plant Pathology, 2*(3), 41–50.

Jones, J. T., Haegeman, A., Danchin, E. G. J., Gaur, H. S., Helder, J., Jones, M. G. K., Kikuchi, T., Manzanilla-López, R., Palomares-Rius, J. E., Wesemael, W. M. L., & Perry, R. N. (2013). Top 10 plant-parasitic nematodes in molecular plant pathology. *Molecular Plant Pathology, 14*(9), 946–961.

Kanwar, R. S., Gupta, D. C., & Walia, K. K. (1987). Interaction of *Meloidogyne javanica* and *Rhizoctonia solani* on cowpea. *Nematologia mediterranea, 15*, 385–386.

Kanwar, R. S., Gupta, D. C., & Walia, K. K. (1988). Effect of nematode inoculations at different intervals and soil types on interactions between *Meloidogyne javanica* and *Rhizoctonia solani* on cowpea. *Indian Journal of Nematology, 18*(1), 112–113.

Kinloch, R. A., & Schenck, N. C. (1978). Nematodes and fungi associated with soybean grown under various pesticide regimes. *Proceedings - Soil & Crop Science Society of Florida, 37*, 224–227.

Lakhran, L., Ahir, R. R., Choudhary, M., & Choudhary, S. (2018). Isolation, purification, identification and pathogenicity of *Macrophomina phaseolina* (Tassi) Goid. caused dry root rot of chickpea. *Journal of Pharmacognosy and Phytochemistry, 7*(3), 3314–3317.

Latha, T. K. S., & Narasimhan, V. (2000). Interaction between *Macrophomina phaseolina* and *Heterodera cajani* in root disease complex of black gram. *Madras Agricultural Journal, 87*, 98–101.

Latha, T. K. S., Rajeswari, E., & Narasimhan, V. (2000). Management of root-rot disease complex through antagonists and chemicals. *Indian Phytopathology, 53*, 216–218.

Lopez-Nicora, H. D., Carr, J. K., Paul, P. A., Dorrance, A. E., Ralston, T. I., Williams, C. A., & Niblack, T. L. (2020). Evaluation of the Combined Effect of *Heterodera glycines* and *Macrophomina phaseolina* on soybean yield in naturally infested fields with spatial regression analysis and in greenhouse studies. *Phytopathology, 110*, 406–417.

Marquez, N., Giachero, M. L., Declerck, S., & Ducasse, D. A. (2021). *Macrophomina phaseolina*: General characteristics of pathogenicity and methods of control. *Frontiers of Plant Science, 12*, 634397.

Mejias, J., Truong, N. M., Abad, P., Favery, B., & Quentin, M. (2019). Plant proteins and processes targeted by parasitic nematode effectors. *Frontiers of Plant Science, 10*, 970. https://doi.org/10.3389/fpls.2019.00970

Pierson, E. A., & Weller, D. M. (1994). Use of mixtures of fluorescent pseudomonads to suppress take-all and improve the growth of wheat. *Phytopathology, 84*, 940–947.

Powell, N. T. (1971). Interactions between nematodes and fungi in disease complexes. *Annual Review of Phytopathology, 9*, 253–274.

Riedel, R. M. (1988). Interactions of plant-parasitic nematodes with soil-borne plant pathogens. *Agriculture, Ecosystems & Environment, 24*, 281–292.

Saeed, I. A. M., MacGuidwin, A. E., & Rouse, D. I. (1998). Effect of initial nematode population density on the interaction of *Pratylenchus penetrans* and *Verticillium dahlia* on 'Russet Burbank' potato. *Journal of Nematology, 30*, 100–107.

Sakhuja, P. K., & Sethi, C. L. (1986). Multiplication of *Meloidogyne javanica* as affected by *Fusarium solani* and *Rhizoctonia bataticola* on groundnut. *Indian Journal of Nematology, 16*(1), 1–3.

Siddiqui, Z. A., & Akhtar, M. S. (2006). Biological control of root-rot disease complexof chickpea by AM fungi. *Archives of Phytopathology andPlant Protection, 39*, 389–395.

Siddiqui, Z. A., & Akhtar, M. S. (2007). Biocontrol of a chickpea root-rot disease complex with phosphate solubilizing microorganisms. *Journal of Plant Pathology, 89*, 67–77.

Siddiqui, Z. A., & Futai, K. (2009). Biocontrol of *Meloidogyne incognita* using antagonistic fungi, plant growth-promoting rhizobacteria and cattle manure on tomato. *Pest Management Science, 65*, 943–948.

Siddiqui, Z. A., & Husain, S. I. (1990). Herbal control of root-knot and root-rot disease of chickpea. I. Effect of plant extracts. *New Agriculturist, 1*, 1–6.

Siddiqui, Z. A., & Husain, S. I. (1991a). Interaction of *Meloidogyne incognita* race 3 and *Macrophomina phaseolinain* a root-rot disease complex of chickpea. *Nematologia mediterranea, 19*, 237–239.

Siddiqui, Z. A., & Husain, S. I. (1991b). Control of *Meloidogyne incognita* and *Macrophomina phaseolina* in chickpea by fungal filtrates. *Pakistan Journal of Nematology, 9*, 131–137.

Siddiqui, Z. A., & Husain, S. I. (1992). Interaction between *Meloidogyne incognita* race 3, *Macrophomina phaseolina* and *Bradyrhizobium* sp. in the root rot disease complex of chickpea, *Cicer arietinum. Fundamental and Applied Nematology, 15*, 491–494.

Siddiqui, Z. A., & Mahmood, I. (1992a). Biological control of root-rot disease complex of chickpea caused by *Meloidogyne incognita* race 3 and *Macrophomina phaseolina. Nematologia mediterranea, 20*, 199–202.

Siddiqui, Z. A., & Mahmood, I. (1992b). Control of *Meloidogyne incognita* race 3 and *Macrophomina phaseolina* by ascorbic acid alone and in combination with *Paecilomyces lilacinus* on chickpea. *Indian Journal of Plant Pathology, 10*, 45–52.

Siddiqui, Z. A., & Mahmood, I. (1993a). Biological control of *Meloidogyne incognita* race-3 and *Macrophomina phaseolina* by *Paecilomyces lilacinus* and *Bacillus subtilis* alone and in combination of chickpea. *Fundamental and Applied Nematology, 16*, 215–218.

Siddiqui, Z. A., & Mahmood, I. (1993b). Integrated control of root-rot disease complex of chickpea by fungal filtrates and green manuring. *Nematologia mediterranea, 21*, 161–164.

Siddiqui, Z. A., & Mahmood, I. (1995). Management of *Meloidogyne incognita* race 3 and *Macrophomina phaseolina* by fungus culture filtrates and *Bacillus subtilis* on chickpea. *Fundamental and Applied Nematology, 18*, 71–76.

Siddiqui, Z. A., & Mahmood, I. (1999). The effect of inoculations of *Heterodera cajani* and *Meloidogyne incognita* with *Fusarium udum* and *Bradyrhizobium japonicum* on the wilt disease complex of pigeon pea. *Indian Phytopathology, 52*, 66–70.

Sinclair, J. B., & Backman, P. A. (1989). *Compendium of soybean diseases* (3rdEdn.). St.Paul, MN: APS Press.

Smith, G. S., & Carvil, O. N. (1997). Field screening of commercial and experimental soybean cultivars for their reaction to *Macrophomina phaseolina. Plant Disease, 81*, 363–368.

Stone, A. G., Scheuerell, S. J., & Darby, H. M. (2004). Suppression of soil-borne diseases in field agricultural systems: Organic matter management, cover cropping, and other cultural practices. In F. Magdoff, & R. Weil (Eds.), *Soil organic matter in sustainable agriculture* (pp. 131–177). Boca Raton, FL: CRC Press.

Sumbul, A., & Mahmood, I. (2020). Interactive effect of *Meloidogyne incognita* and *Macrophomina phaseolina* on the development of root–rot disease complex in relation to growth and physiological attributes of chickpea. *Hellenic Plant Protection Journal, 13*, 13–23.

Thoden, T. C., Korthals, G. W., & Termorshuizen, A. J. (2011). Organic amendments and their influences on plant-parasitic and free-living nematodes: A promising method for nematode management? *Nematology, 13*, 133–153.

Todd, T. C., Pearson, C. A. S., & Schwenk, F. W. (1987). Effect of *Heterodera glycines* on charcoal rot severity in soybean cultivars resistant and susceptible to soybean cyst nematode. *Journal of Nematology, 1*, 35–40.

Vilich, V., & Sikora, R. A. (1998). Diversity in soilborne microbial communities: A tool for biological system management of root health. In G. J. Boland, & L. D. Kuykendall (Eds.), *Plant–Microbe interactions and biological control* (pp. 1–14). New York, NY: Marcel Dekker Inc.

Vinh, M. T., Tung, T. T., & Quang, H. X. (2005). Primary bacterial wilt study on tomato in vegetable areas of Ho Chi Minch city, Vietnam. In C. Allen, P. Prior, & A. Hayward (Eds.), *Bacterial wilt disease and the Ralstonia solanacearum species complex* (pp. 177–184). St Paul, MN: American-Phytopathological Society Press.

West, T. D., Griffith, D. R., Steinhardt, G. C., Kladivko, E. J., & Parsons, S. D. (1996). Effect of tillage and rotation on agronomic performance of corn and soybean: Twenty-year study on dark silty clay loam soil. *Journal of Production Agriculture, 9*, 241–248.

Winkler, H. E., Hetrick, B. A. D., & Todd, T. C. (1994). Interactions of *Heterodera glycines, Macrophomina phaseolina*, and mycorrhizal fungi on soybean in Kansas. *Supplement Journal of Nematology, 26*(4S), 675–682.

Yuliar, Nion, Y. A., & Toyota, K. (2015). Recent Trends in control methods for bacterial wilt diseases caused by *Ralstonia solanacearum. Microbes and Environments, 30*(1), 1–11.

Chapter 7

Epidemiology and management of charcoal rot of maize caused by *Macrophomina phaseolina* (Tassi.) Goid

S.I. Harlapur[1], M.K. Khokhar[2] and K.S. Hooda[3]

[1]*ICAR-All India Coordinated Maize Improvement Project, University of Agricultural Sciences, Dharwad, India;* [2]*ICAR-National Research Centre for Integrated Pest Management, Pusa Campus, New Delhi, India;* [3]*ICAR-National Bureau of Plant Genetic Resources, Pusa Campus, New Delhi, India*

7.1 Introduction

The soil-borne pathogen *Macrophomina phaseolina* (Tassi) Goid. causing charcoal rot of maize *(Zea mays* L.) is an important pathogen that severely hampered maize production (Shekhar & Kumar, 2010). Sustainable maize production is thwarted by this pathogen, which causes quantitative and qualitative losses in yield at grain filling stage. There have been reports of yield losses of up to 70% due to this disease from various maize growing regions of the world. High plant density, heavy nitrogenous fertilizers, and insect damage provide congenial conditions for charcoal rot development in hot and dry seasons. In India, this disease outbreak appeared since 1960 till recent time in different parts of the country. Payak and Renfro (1969) reported it for the first time from Kashmir region and later on from Hyderabad and Pantnagar during *Kharif season.* Since then, it is regularly present in states of Himachal, Uttarakhand, Punjab, Telangana, Delhi, Haryana, Karnataka, Rajasthan, Andhra Pradesh, Madhya Pradesh, Uttar Pradesh, Bihar, Jharkhand, Chattisgarh, Tamil Nadu, and West Bengal (Khokhar et al., 2014).

7.2 Symptoms

Initial symptoms of charcoal rot in maize plants can be observed after flowering with the unusual drying of plant upper leaves, breaking and lodging of stalk, and finally premature withering and drying of the entire plant (Fig. 7.1). When plants reach to maturity, the lower stalk nodes (usually lower five nodes) of plants show a typical charcoal, gray-black discoloration. The distinguishing character of this disease is the presence of black sclerotia that are pin-head like structures on the pith of the infected stalks (Khokhar et al., 2014). These sclerotia play important roles in the disease cycle of this menacing pathogen. Internal disintegration and discoloration of the infected stalk tissues are distinctive symptoms of this disease. As a result, translocation of water and nutrients is blocked, leading to drastic reduction of the yield and consequently lodging of the plants with wind movement. Ultimately, plants are dead. Roots of infected plants are also invaded and show black sclerotia in the disorganized tissue which later on serve as the source of primary inoculum for the next season.

7.3 Economic importance

The charcoal rot fungus has the ability to cause huge economic yield loss in the range of 25%–32.2% along with deterioration in fodder and foliage quality (Payak & Sharma, 1978; Krishna et al., 2013). Satyanarayana and Begum (1996) reported significantly more yield in the protected plots as compared to unprotected plots in maize crop due to the incidence of charcoal rot. This pathogen has caused 90% yield loss under favorable conditions, and it attacks on roots and stem, and cob. Harris (1962) noted that rotting caused early leaf drying, plant lodging, and tassel dropping by spreading from diseased internodes to the healthy nodes. Charcoal rot pathogen has the ability to cause 39.5% yield reduction in susceptible genotypes of maize (Payak & Sharma, 1978), though the highest yield losses of 63.5% (AICRP, 2014) have been observed in recent years.

Macrophomina Phaseolina. https://doi.org/10.1016/B978-0-443-15443-0.00003-6

FIGURE 7.1 (A) Premature wilting and (B) severely infected maize plants with charcoal rot.

7.4 The pathogen

M. phaseolina (Tassi) Goid., an anamorphic ascomycete fungus of the family Botryosphaeriaceae, causes charcoal rot on a variety of plants in different locations of the world where maize is grown (Reichert & Hellinger, 1947). Over the time, its taxonomy has stagnated due to the absence of its perfect stage. However, a systematic phylogenetic molecular study of 113 isolates of the family *Botryosphaeriaceae* using ribosomal DNA sequences helped in recognition of two separate genera *Macrophomina* and *Tiarosporella* (Crous et al., 2006). According to the ICBN, the binomial system of nomenclature, *M. phaseolina* was the real name for the pycnidial stage of *Rhizoctonia bataticola*. Its hypha is hyaline in color with septa. On parent hyphae, branches of hypha often develop at a right angle with constriction at the point of origin (Fig. 7.2). Numerous black sclerotia on infected plant parts are produced and are globular to irregular in shape. The strains of this fungus can be differentiated based on the sclerotial size and the presence or absence of pycnidia (Malcom, 1980). Microsclerotia are round, oval, or oblong, first appearing light brown at young stage and becoming darker (brown to black) with age (Lakhran et al., 2018; Marquez et al., 2021).

7.5 Host range

The pathogen *M. phaseolina* is known to infect nearly 500 plants species in different parts of the world (Ghaffar, 1988). This fungus has wide host range including *Glycine* max (soybean), *Gossypium* (cotton), *Helianthus annuus* (sunflower), *Phaseolus vulgaris* (common bean), *Cajanus cajan* (pigeon Pea), *Cyamopsis tetragonoloba* (guar), *Sesamum indicum* (til), *Sorghum bicolor* (sorghum), *Zea mays* (maize), *Vigna sinensis* (cow pea), *Nicotiana tabacum* (tobacco), and a range of vegetable crops.

7.6 Epidemiology

Charcoal rot disease is more visible in water stressed plants and further favored by high soil temperature (28−42°C), low soil moisture, and soil pH (5.4−6.0) (Shekhar & Kumar, 2010). The sclerotia and mycelia serve as the primary source of inoculum for this disease (Meyer et al., 1974). The explosive pathogenicity of *M. phaseolina* in favorable conditions and the ability of its sclerotia to withstand adverse conditions allow it to be a successful pathogen of maize. In maize, the charcoal rot disease is aggravated rapidly under favorable conditions of high temperature and low relative humidity (McLean & Cook, 1965; Shekhar & Kumar, 2010). The conditions of severe water stress and high soil temperature at post-flowering stage are highly conducive for developing the charcoal rot symptoms (Smith & McLaren, 1997). Grain filling stage is more prone to infection of this disease.

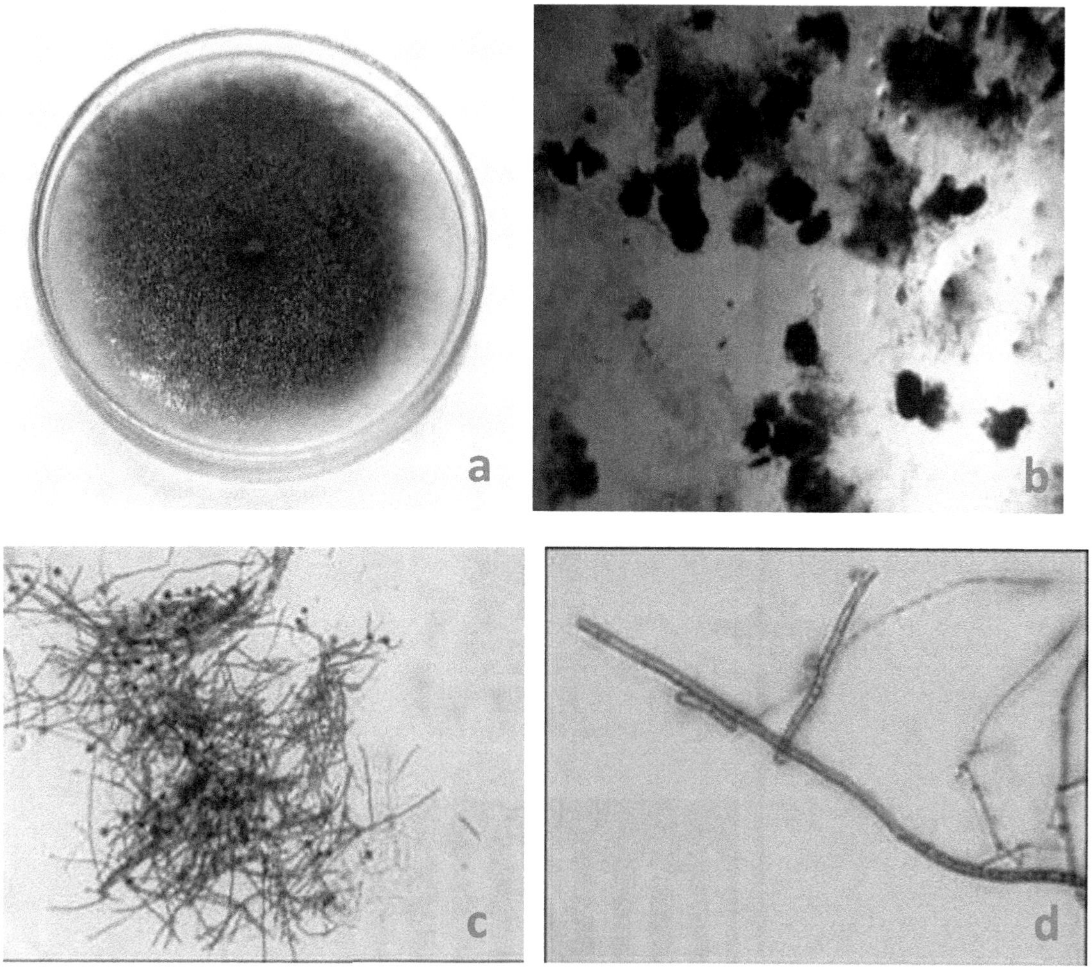

FIGURE 7.2 (A) pure culture of pathogen, (B) sclerotia, (C) mycelium, and (D) mycelium branches of *M.phaseolina*.

7.7 Disease cycle

The charcoal rot pathogen is seed and soil-borne, and crop residues serve as major inoculum source (Malcom, 1980; Payak & Sharma, 1985; Sandra et al., 2008). Sheikh and Ghaffar (1978) reported that *M. phaseolina* persisted in the soil in the form of large number of black sclerotia in infected host tissue which are later spread in soil during field intercultural operations. Savita et al. (1996) reported its survival upto 12 months in maize seeds and soil. The fungus primarily penetrates through sclerotia released into the soil after decaying of host tissues. Sclerotial population was immensely enhanced if susceptible maize cultivars were grown every year (Meyer et al., 1974). It forms micro-sclerotia in senescing shoot tissues and survive in the soil for a longer period. Multiple germinating hyphae infect the roots of the seedlings (Fig. 7.3). As a result of blocking of xylem and phloem vessels due to infection of *M. phaseolina,* the minerals, water, and nutrients transport to the top foliage parts of the plants is disrupted. Normally the age of plant at this stage is 60–65 days depending on maturity group, but infection may occur before and after 5 days.

7.8 Phytopathometry

The disease rating of charcoal rot usually kept as one to nine because of its internal symptoms, which can be judged only by splitting the stalk across the internodes (Fig. 7.4). Up to rating of more than five is considered as susceptible (Table 7.1). If disease is restricted to one internode, it indicates that the biochemical defense of plant is effective.

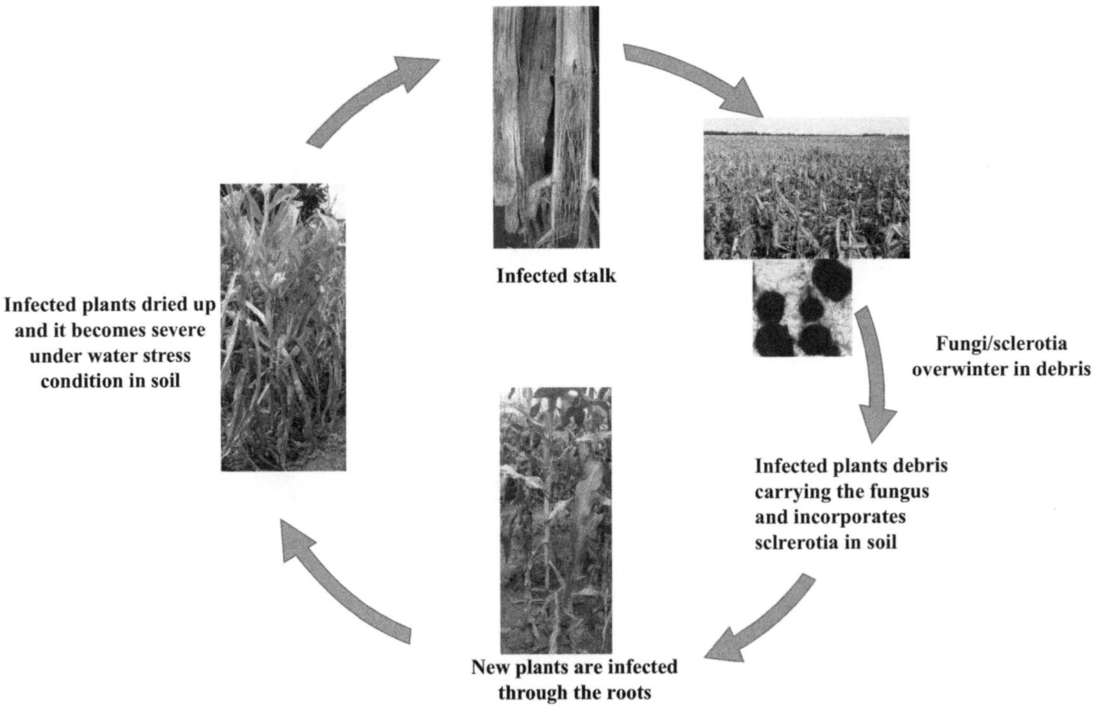

FIGURE 7.3 Disease cycle of charcoal rot of maize.

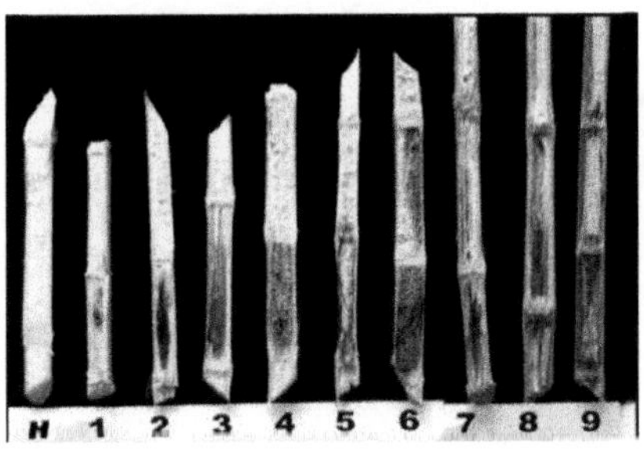

FIGURE 7.4 Disease rating scale of charcoal rot on one to nine scale.

TABLE 7.1 Measurement of disease on standard rating scale giving severity and disease reaction against each score.

S. No.	Severity (%)/percent incidence	Score on standard rating scale[a]	Disease reaction (R/MR/S/HS)
1	0–20 of first internode	1	HR
2	21–50 of first internode	2	HR
3	50–75 of first internode	3	R
4	First internode completely showing fungal growth	4	R
5	Second internode half covered with fungal growth	5	MR
6	Second internode fully covered with fungal growth	6	MS
7	Third internode fully covered with fungal growth	7	S
8	Fourth internode fully covered with fungal growth	8	HS
9	All the subsequent internodes covered with fungal growth	9	HS

[a]R = resistant, MR = medium resistant, S=susceptible, and HS= highly susceptible.

7.9 Disease management

Fungicidal control of charcoal rot is not much effective because of its soil-borne nature. Moreover, it appears at or after flowering stage when any of management measures has little or no effect. Soil treatment/fumigation with biostimulants or chemical fungicides though effective but not viable economically (Cohen et al., 2012; Tonin et al., 2013; Khokhar et al., 2014; Chamorro et al., 2015; Parmar et al., 2017; Lokesh et al., 2020). The use of crop rotation with non-host crops helps to disrupt pathogen cycle. In the absence of resistant host, the population of *M. phaseolina* in soil will increase when susceptible hosts are grown in successive years. Rotation with a poor host such as cotton or nonhost plants will need two or more years to reduce inoculum levels in soil.

The common methods of disease management are cultural practices, use of biocontrol agents, breeding for resistant varieties, and use of chemical control. Disease-resistant specific breeding for this disease has not shown particularly encouraging outcomes (Krishna et al., 2013). Due to continuously increasing cost of fungicides and their hazardous impact on environment and human body, the world has moved toward biological control. Scientists are becoming more and more interested in biological control as a disease management alternative that is also ecologically sustainable, economical, and safe for humans and mammals. *Trichoderma, Pseudomonas,* and *Bacillus* spp. are effective biocontrol agents for several soil-borne fungal plant pathogens including *M. phaseolina* (Bastakoti et al., 2017). Manczinger et al. (2002) reported that the members of the genus *Trichoderma (T. viride, T. polysporum* and *T. harzianum*) have antagonistic properties against charcoal rot pathogen. Bhantangar and Bansal (2003) observed that *T. polysporum* was highly antagonistic against *M. phaseolina* causing dry root rot in cowpea. The production of *M. phaseolina's* microsclerotia is inhibited by *T. harzianum*, according to Gajera et al. (2012). Hyder et al. (2017) have also noted improvements in plant health and a decrease in disease incidence with seed treatment of *Trichoderma viride*. Sridharan et al. (2020) reported that seed coating with *T. viride* is beneficial for raising the shoot dry weight, grains, and nodules and enhancing plant growth. The seed treatment with *T. viride* (8 g/kg), thiram + carbendazim (seed treatment 2.0 + 1.0 g/kg), and vitavax 75 WP (2.5 g/kg) against root rot of sunflower significantly reduced the disease incidence significantly (Patil et al., 2003).

The antagonistic effects of several biocontrol agents (*Trichoderma, Bacillus, Pantoea, Pseudomonas, Stenotrophomonas*, and *Serratia*) against *M. phaseolina* have been identified and studied by several researchers (Sajeena et al., 2004; Kumar et al., 2007; Shekhar & Kumar, 2010; Vasebi et al., 2013; Torres et al., 2016; Mmbaga et al., 2018; Sanjeevkumar et al., 2020; Hussain & Khan, 2020).

Kulkarni and Anahosur (2011) observed that the application of biocontrol agents individually either as soil application or seed treatment also showed good germination with reduced mortality due to dry root rot of chickpea as compared to treatments. Among bioagents, *T. viride* (Tv-R) isolate showed the maximum inhibition of mycelial growth by recording 78.50 % followed by *T. viride* (Tv −32) with inhibition of 74.92%. *M. phaseolina* sclerotial population was reduced after a period of 15 days when poultry manure was applied @ 1%, 3%, and 5% (Sheikh & Ghaffar, 2004). Rajkumar and Raja (2007) reported that adding poultry litter is effective against *M. phaseolina* at 30, 15, and 10 days before planting.

Disease management of *M. phaseolina* through combining cultural practices, host plant resistance, with biological control and chemicals has been reported, but no clear results could be drawn and need further investigations for management of this menace in maize (Reznikov et al., 2016; Swamy et al., 2018; Adhikary et al., 2019; Iqbal & Mukhtar, 2020; Lokesh et al., 2020). Therefore, integrated disease management viz., plant extracts, organic amendments, and antagonists are required to minimize environmental hazards and to increase crop yield by suppressing or destroying the pathogen and protecting plants against infection.

7.10 Conclusion

Charcoal rot is one of the most serious and destructive maize disease causing significant yield loss in maize production. The disease causes yellowing and drooping of leaves, discoloration, and lodging of stalks and tilting of cobs. Most of the maize growing areas of India particularly in rain-fed areas are prone to charcoal rot. Severity of the disease aggravates in case of water stress at or after flowering stage of the crop. To develop an integrated disease management strategy for charcoal rot, it is vitally necessary to focus efforts on isolating and identifying efficient bio-control agents. Information on use of fungicides in disease management of charcoal rot is meagre. Being soil-borne pathogen, management of *M. phaseolina* is a challenge. Therefore, it is the need of hours to develop cost effective management consisting of agronomical practices, balance nutrition, host plant resistance, biocontrol agents, and fungicides. Thus, we can reduce the yield losses due to charcoal rot and can grow the maize successfully.

References

Adhikary, N. K., Chowdhury, M. R., Begum, T., & Mallick, R. (2019). Integrated management of stem and root rot of sesame (*Sesamumindicum* L.) caused by *Macrophominaphaseolina* (Tassi) Goid. *International Journal of Current Microbiology and Applied Sciences, 8*, 804–808. https://doi.org/10.20546/ijcmas.2019.804.089

AICRP. (2014). *Annual report of AICRP maize pathology udaipur center.*

Bastakoti, S., Belbase, S., Manandhar, S., & Arjyal, C. (2017). *Trichoderma*species as Biocontrol agent against soil borne fungal pathogens. *Nepal Journal of Biotechnology, 5*, 39–45. https://doi.org/10.3126/njb.v5i1.18492

Bhantangar, K., & Bansal, R. K. (2003). *Trichoderma polysporum* a new antagonist against *Macrophomina phaseolina* causing dry root in cowpea. *Journal of Mycology and Plant Pathology, 33*(2), 331.

Chamorro, M., Miranda, L., Domínguez, P., Medina, J. J., Soria, C., & Romero, F. (2015). Evaluation of biosolarization for the control of charcoal rot disease (*Macrophomina phaseolina*) in strawberry. *Crop Protection, 67*, 279–286. https://doi.org/10.1016/j.cropro.2014.10.021

Cohen, R., Omari, N., Porat, A., & Edelstein, M. (2012). Management of *macrophomina* wilt in melons using grafting or fungicide soil application: Pathological, horticultural and economical aspects. *Crop Protection, 35*, 58–63. https://doi.org/10.1016/j.cropro.2011.12.015

Crous, P. W., Slippers, B., Wingfield, M. J., Rheeder, J., Marasas, W. F. O., Philips, A. J. L., Alves, A., Burgess, T., Barber, P., & Groenewald, J. Z. (2006). Phylogenetic lineages in the Botryosphaeriaceae. *Studies in Mycology, 55*, 235–253.

Gajera, H., Bambharolia, R., Patel, S., Khatrani, T., & Goalkiya, B. (2012). Antagonism of *Trichoderma* spp. against *Macrophomina phaseolina*: Evaluation of coiling and cell wall degrading enzymatic activities. *Journal of Plant Pathology & Microbiology, 3*, 149. https://doi.org/10.4172/2157-7471.1000149

Ghaffar, A. (1988). *Soil borne disease research centre research report 1* (Karachi, Pakistan).

Harris, E. (1962). Diseases of Guinea corn. *Sernaru Technical Notes, 2*, 1–13.

Hussain, T., & Khan, A. A. (2020). Determining the antifungal activity and characterization of *Bacillus siamensis AMU03* against *macrophominaphaseolina* (tassi) goid. *Indian Phytopathology, 73*, 507–516. https://doi.org/10.1007/s42360-020-00239-6

Hyder, S., Inam-ul-haq, M., Bibi, S., & Humayun, A. (2017). Novel potential of *Trichoderma* Spp. as biocontrol agent. *Journal of Entomology and Zoology Studies, 5*, 214–222.

Iqbal, U., & Mukhtar, T. (2020). Inhibitory effects of some fungicides against *Macrophomina phaseolina* causing charcoal rot. *Pakistan Journal of Zoology, 52*, 709–715. https://doi.org/10.17582/journal.pjz/20181228101230

Khokhar, M. K., Hooda, K. S., Sharma, S. S., & Singh, V. (2014). Post flowering stalk rot complex of maize - present status and future prospects. *Maydica, 59*, 226–242.

Krishna, K. M., Chikkappa, G. K., & Manjulatha, G. (2013). Components of genetic variation for *Macrophomina phaseolina* resistance in maize. *Journal of Research AN-GRAU, 41*(3), 12–15.

Kulkarni, S., & Anahosur, K. H. (2011). Integratedmanagement of dry stalk rot disease of maize. *Journal of Plant Disease Sciences, 6*(2), 99–106.

Kumar, V., Kumar, A., & Kharwar, R. N. (2007). Antagonistic potential of fluorescent pseudomonads and control of charcoal rot of chickpea caused by *Macrophomina phaseolina. Journal of Environmental Biology, 28*, 15–20.

Lakhran, L., Ahir, R. R., Choudhary, M., & Choudhary, S. (2018). Isolation, purification, identification and pathogenicity of *Macrophominaphaseolina* (Tassi) goid caused dry root rot of chickpea. *Journal of Pharmacognosy and Phytochemistry, 7*, 3314–3317.

Lokesh, R., Rakhollya, K. B., & Thesiya, M. R. (2020). Evaluation of different fungicides against *Macrophominaphaseolina* (Tassi) goid. causing dry root rot of chickpea (*Cicerarietinum* L.) *in vitro. International Journal of Current Microbiology and Applied Sciences, 9*, 1–11. https://doi.org/10.20546/ijcmas.2020.907

Malcom, C. S. (1980). *Compendium of corn diseases.* Minnesota: APS Press.

Manczinger, L., Antal, Z., & Kredics, L. (2002). Ecophysiology and breeding of mycoparasitic*Trichoderma* studies (a review). *Acta Microbiologica et Immunologica Hungarica, 49*(1), 1–14.

Marquez, N., Giachero, M. L., Declerck, S., & Ducasse, D. A. (2021). *Macrophomina phaseolina*: General characteristics of pathogenicity and methods of control. *Frontiers of Plant Science, 12*, 634397. https://doi.org/10.3389/fpls.2021.634397

McLean, R. C., & Cook, W. R. I. (1965). *Plant science formula.* London: McMillan and Co. Ltd.

Meyer, W. A., Sinclair, J. B., & Khare, M. N. (1974). Biology of *Macrophomina phaseolina* in soil studied with selective media. *Phytopathology, 64*, 845–849.

Mmbaga, M. T., Mackasmiel, L. M., & Mrema, F. A. (2018). Evaluation of biological agents for control of *macrophomina* root and powdery mildew in flowering dogwood (*Cornus Florida* L.). *Horticulture Science, 53*, 1461–1466. https://doi.org/10.21273/HORTSCI13071-18

Parmar, H. V., Kapadiya, H. J., & Bhaliya, C. M. (2017). Efficacy of different fungicides against *Macrophomina phaseolina* (Tassi) Goid causing castor root rot. *International Journal of Chemical Studies, 5*, 1807–1809.

Patil, R. K., Goyal, S. N., Patel, B. A., Patel, R. G., Patel, D. J., Singh, R. V., & Gour, H. S. (2003). Integrated management of stalk rot disease and phytonematodes in Rabi maize. In *Proceedings of national symposium on biodiversity and management of nematodes in cropping systems for sustainable agriculture, Jaipur* (pp. 250–254).

Payak, M. M., & Renfro, B. L. (1969). *Final technical report (1963–1968) Research on diseases of maize with special reference to* Erwinia Cartovoravarzeae *CMI. Grant No FG in 146.* New Delhi: Indian Council of Agricultural Research.

Payak, M. M., & Sharma, R. C. (1978). *Final technical reptrort . April 1969–March1975* (pp. 60–68). New Delhi: ICAR.

Payak, M. M., & Sharma, R. C. (1985). Maize diseases and approaches to their management in India. *Tropical Pest Management, 31*, 302–310.

Rajkumar, V., & Raja, J. (2007). Effect of poultry litter soil amendmenton*Macrophominaphaseolina* in sunflower. *Journal of Mycology and Plant Pathology, 37*(1), 171–177.

Reichert, I., & Hellinger, E. (1947). On the occurrence, morphology and parasitism of *Sclerotium bataticola*. *Palestine Journal of Botany, 6*, 107–147.

Reznikov, S., Vellicce, G. R., González, V., de Lisi, V., Castagnaro, A. P., & Ploper, L. D. (2016). Evaluation of chemical and biological seed treatments to control charcoal rot of soybean. *Journal of General Plant Pathology, 82*, 273–280. https://doi.org/10.1007/s10327-016-0669-4

Sajeena, A., Rajan, F., Seetharaman, K., & Mohan Babu, R. (2004). Evaluation of biocontrol agents against dry root of blackgram (*Vigna mungo*). *Journal of Mycology and Plant Pathology, 34*(2), 341–344.

Sandra, G. A., Carolina, V. F., Katy, D. P., José, H. C., Matilde, U. M., & Eugenio, S. V. (2008). Identification and biological characterization of isolates with activity inhibitive against *Macrophomina phaseolina* (Tassi) goid. *Chilean Journal of Agricultural Research, 69*(4), 526–533.

Sanjeevkumar, K., Kumar, K. S., Balabaskar, P., Sivakumar, T., Kannan, R., & Saravanan, K. R. (2020). Efficacy of seed plus soil application of Bacillus cereus on the root rot (*Macrophominaphaseolina* (tassi.) Goid.) incidence and plant growth of groundnut. *Plant Archives, 20*, 1217–1221.

Satyanarayana, E., & Begum, H. (1996). Relative efficacy of fungicides (seed dressers) and irrigation schedule for the control of late wilt in maize. *Current Research University of Agricultural Sciences Bangalore, 25*(4), 59–60.

Savita, P., Jain, B. L., & Pareek, S. (1996). Survival of *Macrophomina phaseolina* on seed and diseased stem pieces of maize. *Journal of Phytology Research, 9*, 75–76.

Sheikh, A. H., & Gaffar, A. (1978). Relation of sclerotial inoculum density and soil moisture to infection of field crops by *Macrophomina phaseolina*. *Pakistan Journal of Botany, 11*, 185–189.

Sheikh, A. H., & Gaffar, A. (2004). Effect of poultry manure and sawdust on survival of sclerotia of *Macrophomina phaseolina* in soil. *Pakistan Journal of Botany, 36*(2), 425–428.

Shekhar, M., & Kumar, S. (2010). Potential biocontrol agents for the management of *Macrophomina phaseolina,* incitant of charcoal rot in maize. *Archives of Phytopathology and Plant Protection, 43*, 379–383.

Smith, E., & McLaren, M. (1997). Effect of water stress on colonization of maize roots by root infecting fungi. *African Journal of Plant Protection, 3*, 47–51.

Sridharan, A., Thankappan, S., Karthikeyan, G., & Uthandi, S. (2020). Comprehensive profiling of the VOCs of *Trichoderma longibrachiatum* EF5 while interacting with *sclerotium rolfsii* and *macrophomina phaseolina*. *Microbiology Research, 236*, 126436. https://doi.org/10.1016/j.micres.2020.126436

Swamy, C., Naik, M. K., Amaresh, Y. S., & Jayalakshmi, S. K. (2018). Evaluation of fungicides and bio-agents under *in vitro* condition against *Macrophomina phaseolina* causing stem canker of pigeonpea. *International Journal of Current Microbiology and Applied Sciences, 7*, 811–819. https://doi.org/10.20546/ijcmas.2018.701.099

Tonin, R. F. B., Avozani, A., Durante Danelli, A. L., Reis, E. M., Zoldan, S. M., & Garcés-Fiallos, F. R. (2013). *In vitro* mycelial sensitivity of *Macrophomina phaseolina* to fungicides ISensibilidademicelial in vitro de *macrophomina phaseolina* a fungicidas. *Pesquisa Agropecuária Tropical, 43*, 460–466. https://doi.org/10.1590/S1983-40632013000400014

Torres, M. J., Brandan, C. P., Petroselli, G., Erra-Balsells, R., & Audisio, M. C. (2016). Antagonistic effects of *Bacillus subtilis* subsp. subtilis and B. Amyloliquefaciens against macrophomina phaseolina: SEM study of fungal changes and UV-MALDI-TOF MS analysis of their bioactive compounds. *Microbiology Research, 182*, 31–39. https://doi.org/10.1016/j.micres.2015.09.005

Vasebi, Y., Safaie, N., & Alizadeh, A. (2013). Biological control of soybean charcoal root rot disease using bacterial and fungal antagonists *in vitro* and greenhouse condition. *Journal of Crop Protection, 2*, 139–150.

Chapter 8

Spatial distribution of *Macrophomina phaseolina* (Tassi) Goid. in the Carpathian Basin and its damage on sunflowers

Kitti Csüllög, Gyula Tóth, Cintia Vartek, Alexandra Piti, Antal Nagy, András Csótó, Péter Riczu, Györgyi Biró and Gábor Tarcali

University of Debrecen, Faculty of Agricultural and Food Sciences, Environmental Management Plant Protection Institute, Debrecen, Hungary

8.1 Introduction

One of the most significant limiting factors in crop production is plant diseases accounting for 10%–30% yield loss worldwide (Strange & Scott, 2005). A very aggressive and widely distributed fungal pathogen is *Macrophomina phaseolina* (syn: *Rhizoctonia bataticola*), the cause of charcoal rot disease. *M. phaseolina* is a ubiquitous soil-borne fungal pathogen, infecting more than 500 plants (Dhingra & Sinclair, 1977; Marquez et al., 2021). *M. phaseolina* causes various diseases in a multitude of crops, namely damping off, stem canker, dry root rot, and seedling blight (Romero Luna et al., 2017).

Two stages are known—the pycnidial stage (*M. phaseolina*) and the sclerotial stage (*Rhizoctonia bataticola*), in which the fungus survives in the form of microsclerotia. There is no known teleomorph stage of the fungus (Premamalini et al., 2012). According to Haig (1928), the pycnidia always form microsclerotia, whereas the microsclerotia usually do not form pycnidia. This finding is also supported by a later study (Singh & Kaiser, 1995). The main morphological characteristics of the pathogen are the pycnidia, which are dark brown, thick walled, slightly flattened, without setae, and formed on the leaves and stem of the plant. The size of the pycnidia is 100–200 µm. Conidiophores are hyaline, cone-shaped, and $5-13 \times 4-6$ µm in size. Pycnoconidia are smoke-colored, unicellular, and ovoid and are $14-30 \times 5-10$ µm in size (Horváth, 1995). Microsclerotia are black and hard with a smooth surface and are found in the roots, stem, and leaves. The shape is spherical or ovoid, the hyphae branch at right angles (Bánhegyi et al., 1985). Haig has distinguished three types of the pathogen based on the size of microsclerotia. Consequently, the microsclerotia size of group "A" is approximately 1 mm, but the average size of microsclerotia in group 'B' is 200 µm, while it is around 120 µm in group "C." Békési (1970) was the first in Hungary to describe a Haig group "C" type of the pathogen on sunflower with microsclerotia of 80–160 µm in diameter. Dhingra and Sinclair (1973) detected microsclerotia of 44–291 µm, whereas Ilyas and Sinclair (1974) measured microsclerotia between 80 and 360 µm in size. In a narrower size interval than the previous ones, microsclerotia sizes were measured from Italian isolates (72.8–127.8 µm) by Manici et al. (1992) and from Indian isolates (44.65–126.2 µm) by Purkayastha et al. (2004).

The sclerotial form enables the fungus to survive in the soil and in crop debris for 15 years (Ghosh et al., 2018). Microsclerotia are the primary source of inoculum and can germinate in the vicinity of roots within 2 days but are capable of infection throughout the entire vegetation period (Chowdhury et al., 2014). Infection usually occurs in the early phenological stages (Almeida et al., 2003; Romero Luna et al., 2017). Seedlings, usually under additional stress, are infected by the fungus and it remains in the plant tissues until conditions become optimal for the development of the disease (Gupta et al., 2012). The symptoms usually appear following flowering, at the end of the season (Ijaz et al., 2012; Kaur et al., 2012; Romero Luna et al., 2017). The disease is characterized by symptoms including stunted growth,

Macrophomina Phaseolina. https://doi.org/10.1016/B978-0-443-15443-0.00009-7

chlorosis, and wilting. It is also responsible for early aging and reduced yields (Kaur et al., 2012; Romero Luna et al., 2017). The fungus obstructs the nutrient and water transport of the plant by infecting the vascular system, resulting in wilting, decreased vigor, and ultimately, premature plant death. Due to the formation of microsclerotia, the lower parts of the stem develop gray or silver discoloration (Gupta & Chauhan, 2005). It secretes a diverse arsenal of cell wall–degrading enzymes (Islam et al., 2012) as well as toxins (Khambhati et al., 2020) enabling it to penetrate plant cells and destroy host tissues. Khambhati et al. (2020) studied the toxins and secondary metabolites secreted by the fungus and found that in addition to phaseolinone, mellein, and botryodiplodin, the fungus also secretes kojic acid, moniliformin, orsellinic acid, and cyclo[L-proline-L-tyrosine]. They also discovered other metabolites of the fungus, including citrinin, cordycepin, emodin, endocrocin, gliocladic acid, infectopyron, methylorsellinic acid, monocerin, and N-benzoyl-L-phenylalanine.

8.1.1 Spread of *M. phaseolina* in europe

This fungus is located in Belgium (Hunt, 1952), Bulgaria (Alexandrov, 1999), Czech Republic (Kudlíková et al., 2002), the United Kingdom (Scholefield & Griffin, 1979), Germany (Müller & Grill, 1991), France (Alabouvetto & Bremeersch, 1976), the Netherlands (Turkensteen & Lablans, 1988), Italy (Zazzerini, 1980), Hungary (Békési et al., 1970), Portugal (De Barros, 1985), Romania (Bontea, 1985), Spain (Jimenéz-Diaz et al., 1983), Slovakia (Bokor, 2007), Croatia (Marič et al., 1988, pp. 1–227) Serbia (Aćimović, 1998), Denmark (Dell'Olmo et al., 2022), Malta (Haleem et al., 2016), Slovenia (Žerjav et al., 2017), Turkey (Onan et al., 1992), and Ukraine (Tančić-Živanov et al., 2019) (Fig. 8.1).

8.1.2 Yield losses caused by *Macrophomina phaseolina*

This widespread disease causes considerable economic damage in numerous economically significant plants. Soy is one of the most researched and investigated of them. *M. phaseolina* causes charcoal rot disease in soy. It was ranked as one of the

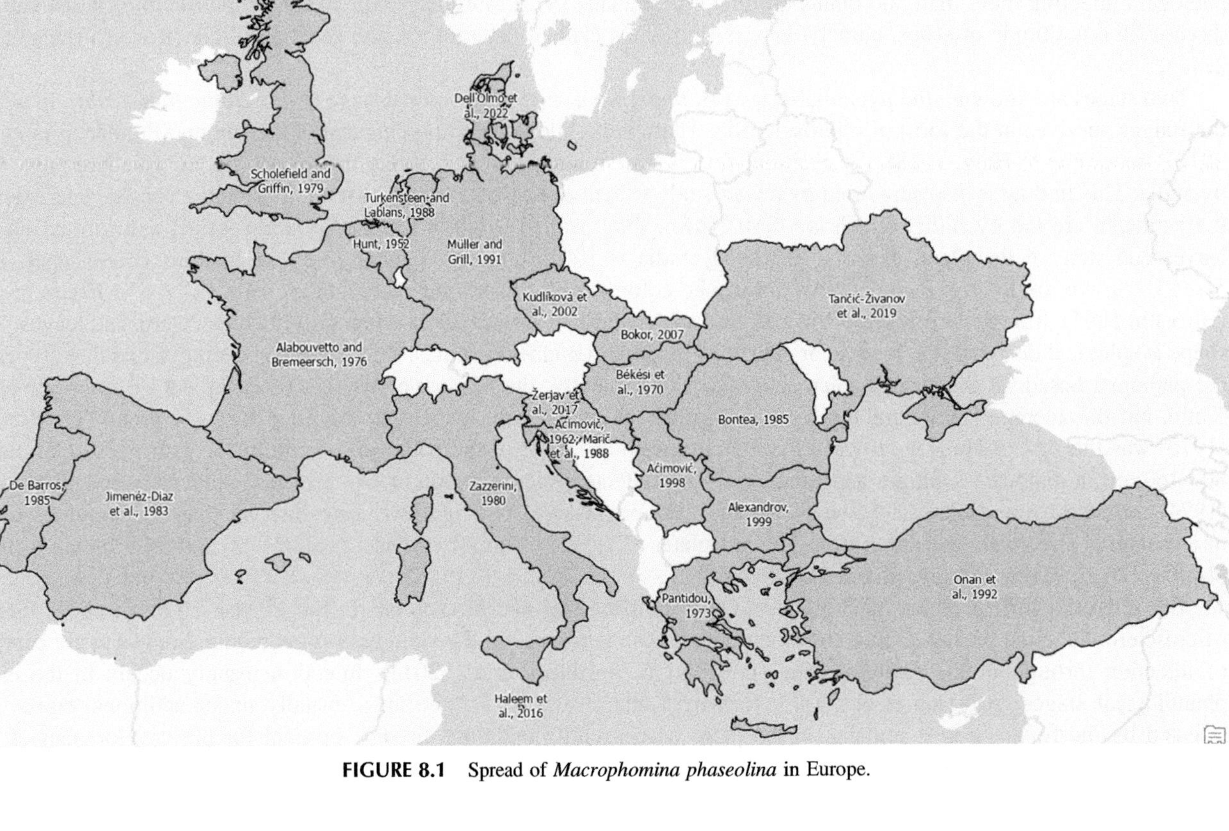

FIGURE 8.1 Spread of *Macrophomina phaseolina* in Europe.

top five diseases of soy in 1998. It causes massive yearly losses in the 10 largest producers of soy (USA, China, India, Brazil, Argentina, Paraguay, Indonesia, Canada, Italy, and Bolivia) (Wrather et al., 2001).

In the United States, charcoal rot was the second most prevalent disease of soy in 2003 and 2012, where the estimated yield losses amounted to 1.9 million and 2.0 million tons. In the United States, charcoal rot was in the top five diseases causing the most yield loss between 2010 and 2013 (Romero Luna et al., 2017). Though the distribution of charcoal rot in the Northern parts of the United States has expanded in the past 30 years (Birrenkott et al., 1984; Bradley & Río, 2003; Cummings & Bergstrom, 2013; Yang & Navi, 2005), yet most research related to this disease was conducted in the Southern regions due to its higher prevalence and annual significance in this region (Gangopadhyay et al., 1970; Shokes et al., 1977; Short et al., 1978; Mengistu et al., 2007).

The occurrence of charcoal rot disease is on the rise in soy and sunflower in South Africa. Soybean and sunflower oil used in protein cakes is the primary source of feed in South Africa. Charcoal rot caused by *M. phaseolina* is not generally regarded as a major disease of soy and sunflower, but it is becoming a greater concern in the industry due to increasing drought conditions. Most South African oilseed producing regions reach the average temperature required for the development of charcoal rot in the vegetation period. Due to changes in the climate, the average temperature is expected to rise by 2°C, whereas average annual rainfall will decrease (Van der Waals et al., 2013). This could increase the frequency of optimal conditions for charcoal rot. Currently, information is limited in connection with yield loss caused by the disease in South Africa.

The bioecological factors of arid and semiarid regions of India contribute to the development of the disease. An 80% incidence was recorded for cowpea (Lodha et al., 1986). In the case of sunflower, a 30% loss of seed weight was documented (Raut, 1981). Infection rate of ~40% was detected in chickpea and urdbean in India (Indira & Gayatri, 2003; Lakhran et al., 2018). The fungus caused 100% infection rate and a yield loss of 23–64% in sorghum (Harris, 1962). Cultivars and hybrids that revolutionized Indian sorghum production have proven to be highly susceptible to the disease and, in severe cases, shown a 100% infection rate (Rao & Balasubramanian, 1983). Mughogho and Pande (1983) measured a 35% reduction in the weight of sorghum seeds. In addition, annual yield loss in soybean can reach 80% (Gupta & Chauhan, 2005).

The sole host plant of silkworm (*Bombyx mori* L.) in India is mulberry (*Morus* spp., Moraceae), a fast-growing perennial tree with deep roots. The tree can adapt to a wide range of agricultural climates. Approximately, 60% of the cost of silk cocoon production is reported to be spent on mulberry leaf production (Anonymous, 1997). Thus, mulberry production is a major aspect of farming. The disease is becoming more alarming due to its incidence of epidemic proportions and its tendency to cause the death of the entire plant (Chowdary, 2006).

In Israel since 2015, the number of plant species from which this pathogen has been isolated has increased from 5 to 23 by 2019. This encompasses a wide range of horticultural crops, including annual and perennial flowers, herbs, lawns, vegetables, and fruit tree species (Cohen et al., 2022). In this country, *M. phaseolina* has become the most common pathogenic fungus in the last 20 years (Cohen et al., 2002; Pickel et al., 2020).

Harris (1962) reported a significant yield loss in sorghum caused by charcoal rot in Kano, Nigeria. The pathogen infected common bean, pigeon pea, soy, and cowpea in Kenya; however, the percentage of disease occurrence and economic damages due to yield loss were not estimated. Being the most severe disease of common bean, it causes up to 70% yield loss in semiarid areas of Kenya (Songa & Hillocks, 1996).

8.1.3 European countries

In Spain, Jiménez-Diaz et al. (1983) examined the incidence and damage of *M. phaseolina* in sunflower in the years 1976, 1977, and 1978. The occurrence varied greatly between sampling years and regions, being the most frequent in areas where low precipitation and high temperatures coincided with the flowering period. The highest incidence was observed in 1976 when the disease was present in 76 parcels out of 163 and on 25 parcels at least 25% of plants were infected. *M. phaseolina* is a common pathogen of sunflower throughout Italy (Zazzerini et al., 1985). Tosi et al. (1995) recorded an infection index of 87% infection of sunflowers in an Italian field. Between 1990 and 1996, the fungus caused 16–42.8% of damage to sunflowers in Bulgaria (Alexandrov, 1999). In Slovakia, it was reported that this fungus caused high infection rates in sunflower in 2005 and 2006 (Bokor, 2007).

The pathogen was first described in Hungary in 1970 (Békési, 1970). The first isolates were obtained from sunflower plants, and it still causes the most economic damage to this crop today. In 2002, Békési (2007) reported a 90% infection rate of sunflowers causing a yield loss of about 302,007) 35%. Further in 2007, *M. phaseolina* was already the predominant pathogen of sunflower domestically.

8.1.4 Oil content

Infection by the pathogen results in yield loss due to reduced head diameter and seed weight (Tikhonov et al., 1976). Oil content declines to 30–80 g/kg and fatty acid composition changes as well, although this change varies greatly between sunflower cultivars (Conte et al., 1989). Furthermore, oil extracted from the infected plants has a higher amount of free fatty acids and is darker in color (Lakshmidevi et al., 1992).

8.1.5 Environmental conditions for disease development

Several ecological factors influence the persistence and survival of the microsclerotia of the pathogen in different soil types. The most prominent of these factors are temperature, moisture, and soil quality.

8.1.6 Temperature

M. phaseolina is widely distributed in arid and semiarid areas. The temperature optimum of the fungus is between 30 and 35°C (Manici et al., 1995). Akhtar et al. (2011) examined the growth of the fungus in laboratory conditions at 10, 15, 20, 25, 30, 35, and 40°C. They found that after 24 h, there was no observable growth of the fungus at 25, 30, 35, and 40°C. At 10°C, the fungus did not develop mycelia even after 96 h. The formation of microsclerotia was detectable after 72 h in cultures incubated at 20, 25, 30, and 35°C. Formation of microsclerotia was not detectable after 96 h in cultures incubated at 15°C and 40°C. Similarly, Manici et al. (1995) found that the optimal temperature between 30°C and 35°C for the growth of the fungus and its microsclerotia formation.

8.1.7 Soil moisture

Ghaffar and Erwin (1969) determined that low soil moisture is a greater predisposing factor for disease development, and the damage is more severe in higher temperatures. Infection by *M. phaseolina* occurs at a considerably greater scale when the soil is dry at the seedling stage (Lodha & Mawar, 2019). Based on field observations by Meyer et al. (1973), the microsclerotia density in the soil was directly correlated with the severity of the disease, and the fungal infection is primarily influenced by soil moisture. The disease developed with low soil moisture; furthermore, increasing soil moisture the susceptibility of the host plants to infection was decreased. In light of the results of the in vitro research conducted by Dhingra and Sinclair (1974), it can be established that at higher soil moisture levels, bacteria may parasitize the microsclerotia. Hartman et al. (2015) found that in wet soil mycelia cannot survive for more than 7 days, and microsclerotia lose viability in 7–8 weeks. In practice, an efficient control method could be to keep the moisture level at 60% MHC when the soil temperature is consistently around 30°C for 2–3 weeks (Lodha & Mawar, 2019). Dhingra and Sinclair (1974) examined the survivability of the pathogen under different moisture levels in sandy loam and silty clay loam soils and found that with the increase in moisture level and incubation time, the population of the pathogen declined sharply in both soil types. Research by Kending et al. (2000) in Arkansas determined that in the case of soy grown in irrigated and nonirrigated conditions, microsclerotia appeared in the roots 6 weeks after sowing. The number of microsclerotia was proportionally lower under irrigated conditions, and this number increased sharply as soon as irrigation was stopped. The study has shown that irrigation limits but does not prevent root infection. According to Lodha and Mawar (2019), it can be deduced that soil moisture stress is the primary abiotic factor influencing population dynamics and disease development. Arid and semi-arid climates are characterized by low and fluctuating precipitation, which leads to unpredictable short and long-term moisture load. Further research is necessary to study the effects of using drought-resistant and tolerant genotypes in these regions as well as to find ways to maintain soil moisture at appropriate levels.

8.1.8 Host plant and previous cropping

This invasive fungus has a wide host range. It has more than 500 host plant, including numerous field crop, vegetable, and fruit species worldwide, such as soybean (Preston, 1945, pp. 1–168), maize (Maneval, 1937), sunflower (Mackie, 1932), lucerne (Hoffmaster et al., 1943), tobacco (Reichert & Hellinger, 1947), bean, tomato (Small, 1928), potato (Ashby, 1927), sweet potato (Taubenhaus, 1913), apple, pear (Reichert–Hellinger, 1947), and pineapple (Thompson & Johnston, 1953). It was also found on weeds, for example common ragweed (Boewe, 1963) and cocklebur species (Nikandrow et al., 1990), as well as ornamental plants.

8.1.9 Crop rotation

Crop rotation is hardly possible considering the wide range of hosts and the long-term viability of microsclerotia. Francl et al. (1988) investigated the effects of crop rotation on the population density of microsclerotia. The highest microsclerotia accumulation was shown in the case of soy monoculture. If soy was present in the crop rotation every 2–3 years, high numbers of microsclerotia were observed. Lower microsclerotia density was found when soy was grown every 3 years than when it was grown every 2 years. Their results also showed that microsclerotia density in the soil was reduced significantly if soy was followed by maize, cotton, and sorghum in the crop rotation as compared to soy monoculture. According to the findings of Singh et al. (1990), growing susceptible host plants, such as sorghum and pigeon peas, resulted in a higher microsclerotia density in most cropping systems. Similarly, an increase in microsclerotia density was observed when susceptible corn and soybean cultivars were grown continuously on the same field for 3 years (Dhingra & Sinclair, 1978). The production of susceptible plants, for instance, soybean, pine, and jute, likewise increases the inoculum density of *M. phaseolina* in the soil (Hodges, 1962; Lodha & Singh, 1985; Tiwari & Shroff, 1982).

8.1.10 Tillage

A decrease in disease incidence was observed in uncultivated fields where straw remaining on the soil surface contributed to decreased soil temperature and increased moisture (Almeida et al., 2001). Almeida et al. (2003) investigated a 20 years-old crop rotation and no-tillage system to determine its effects on the density of *M. phaseolina* propagules. Their results showed that soil samples from conventional cropping systems contained larger amounts of propagules and had a higher prevalence of infected roots than soil from the no-tillage system. One possible explanation was the significant amount of crop debris on the soil surface, which could reduce soil temperature and increase moisture, two factors that suppress charcoal rot (Olaya & Abawi, 1996) and could also lead to the improvement of the conditions facilitating mycoparasitism (Chet et al., 1981).

8.1.11 Irrigation

According to Nischwitz et al. (2004), *M. phaseolina* did not cause economic damage in melon production up until the early 2000s, but disease incidence increased drastically due to growers switching from furrow irrigation to drip irrigation. Edaphic properties, such as soil moisture, salinity, pH, and temperature differ between drip and furrow irrigated soils. The same characteristics also affect the survival of *M. phaseolina* and the incidence of charcoal rot in other crops as well (Dhingra & Sinclair, 1978; Olaya & Abawi, 1996; Reuveni et al., 1982). In the case of furrow irrigation, soil moisture is uniformly distributed in the soil even after 2 days, whereas for subterranean drip irrigation, soil moisture decreases constantly with the increase of distance from the nozzle (Subbarao et al., 1997).

Nischwitz et al. (2004) compared two fields where one was irrigated with furrow irrigation and the other with drip irrigation. The fields were close to one another and had similar crop rotations and soil types. The results showed that drip irrigation could contribute to a more frequent occurrence of the disease in melons in Arizona. The low soil moisture of the field is highly conducive to *M. phaseolina* infection. The two methods differ from each other in several ways, and the culmination of these effects facilitates the accumulation of microsclerotia in the soil and subsequently, damage to the crops.

It was established by Lodha et al. (1997) that naturally sun-warmed moist soil and crucifer stem debris mixed with polyethylene reduces the number of microsclerotia in the soil. Previous research by Sharma et al. (1995) confirmed that with sufficient temperature and incubation time (28 ± 2°C for 30 days) the amount of fungus decreased in the soil by 100% when mustard pellets were mixed into the soil. In the study conducted by Lodha et al. (1997), the top 7.5 cm layer of moist soil was exposed to 50°C for an hour and a half, causing the pathogen to reach lethal temperatures. This high soil temperature could promote the release of toxic, sulfur-containing volatile compounds from crucifer residues. According to a publication by Gamliel and Stapleton (1993), under humid conditions and high soil temperature, toxic volatile compounds, such as dimethyl sulfide, mercaptan, and isothiocyanate, are released from the remains of crucifers. The latter does not occur at low soil temperatures (Lewis & Papavizas, 1970). The toxic chemical materials are mainly responsible for the inhibition or reduction of soil-borne pathogens (Angus et al., 1994).

Lodha et al. (1997) observed only low mortality while investigating the viability of microsclerotia in dry soil at high temperatures. Temperatures were lower in the deeper layers of the soil (20–30 cm), and thus, the mortality of microsclerotia was proportionally low as well. Hot and moist environments negatively influence the population of *M. phaseolina* in the soil, and therefore, the viability of microsclerotia may be reduced by irrigation during the summer (Lodha, 1995).

8.1.12 Chemical control

Lokesh et al. (2020) evaluated the efficiency of different systemic (azoxystrobin, carbendazim, hexaconazole propiconazole, tebuconazole, thiophanate methyl, difenoconazole) and contact (copper oxychloride, mancozeb, chlorothalonil, propineb, thiram, zineb, sulfur) fungicides at different concentrations under *in vitro* conditions against *M. phaseolina*. It was found that the growth inhibition efficiency of *M. phaseolina* mycelia increased with the increase in concentrations of systemic fungicides. As compared to control, carbendazim produced significantly the highest percent growth inhibition at 50 ppm (82%), followed by thiophanate methyl with an efficacy of 79%. At 500 ppm, carbendazim produced the highest mycelial growth inhibition of 92%, and the second most effective was thiophanate methyl with 85% percent inhibition. These results are supported by similar findings published by Chattopadhyay & Kalpana (2002), Khan and Gangopadhyay (2008), Kumari and Shekhawat (2012) and Ebenezar and Wesely (2000) according to them carbendazim completely inhibits the growth of *M. phaseolina* at a concentration of 0.1%.

The contact fungicides significantly inhibited the growth of the fungus. The highest percent growth inhibition at 1000 ppm was observed for mancozeb (83%). At the same concentration, chlorothalonil (79%), zineb (71%), thiram (65%), and propineb (42%) were moderately effective on fungal growth inhibition. Sulfur (12%) had the least efficiency in growth inhibition. Also at 2500 ppm concentration, mancozeb (90%) was the most effective fungicide in inhibiting fungal growth followed by chlorothalonil with 88% growth inhibition. The other fungicides were moderately effective (zineb 78%, thiram 76%, propineb 49%).

Ravichandran & Hedge (2017) made similar observations as they found that chlorothalonil and mancozeb at 0.2% level effectively inhibited fungal growth (Lokesh et al., 2020). Savitha (2017) reported that at 0.3% concentration, mancozeb, and thiram produced 100% growth inhibition. Based on the results of Sangappa and Mallesh (2016), thiophanate methyl showed 100% mycelial growth inhibition at concentrations of 0.1%, 0.2%, and 0.3%. A number of pesticides have been shown to be clearly effective against the pathogen, but the continued ban on these active substances means a reduction in the number of effective active substances available. The ban on the use of the soil fumigant methyl bromide coincided with a surge in losses in the case of various crops due to charcoal rot. Soil fumigation with methylbromide allowed strawberry producers worldwide to exclude soil-borne pathogens for years (Chamorro et al., 2016; Holmes et al., 2020). Since the gradual ban of the fumigant, *M. phaseolina* has become a massive issue in several strawberry-producing countries including Israel (Pickel et al., 2020).

8.1.13 Biocontrol agents

Biological control is one of the most promising strategic measures to decrease disease incidence and increasing agricultural yield. Beneficial microbes with antagonistic properties against plant pathogens in the soil are used in the biological control of diseases. Plant extracts and microorganisms or the materials they secrete can be used safely and effectively against pathogens (Sreevidya & Gopalakrishnan, 2012). Among rhizobacteria, the species of the *Pseudomonas* and *Bacillus* have proven to be efficient biocontrol agents (BCA) against soil-borne pathogens. *Pseudomonas* and *Bacillus* species produce a wide range of antibiotic compounds that have a negative effect on *M. phaseolina* (Dave et al., 2021). Members of the genus *Pseudomonas* involve a large number of biocontrol strains producing significant metabolites to varying volumes, such as pyrrolnitrin, which is effective against *Fusarium*, *R. bataticola,* and other fungal pathogens (Ligon et al., 2000). Anis et al. (2010) examined the antagonistic effect of *Bacillus subtilis* in sunflower and found that the bacterium effectively inhibits radial growth of *M. phaseolina*. Kumar et al. (2015) determined that *B. subtilis* isolated from the rhizosphere of mungbean inhibited the growth of *M. phaseolina* by 24% in vitro. Similar results were obtained in greenhouse conditions, decreasing disease incidence to 46,67%. Several *Trichoderma* species can be used as potential biocontrol agents and growth enhancers in various crops (Bai et al., 2008; Savazzini et al., 2009; Verma et al., 2007). There are several antagonist species in the sizable *Trichoderma* genus. This effect is achieved either by direct (competition for nutrients and space) or direct (mycoparasitism) means (Benitez et al., 2004). These biocontrol products are effective against soil- and/or seed-borne fungal pathogens of various plants (Kubicek et al., 2001). Kehri and Chandra (1991) were the first to report the antagonistic activity of *T. viride* against *M. phaseolina* and its application against the root rot of *Vigna mungo* in greenhouse conditions. Raguchander et al. (1997) determined that seeds coated with *T. viride* effectively decrease the damage of *M. phaseolina*. Bandopadhyay et al. (2008) also suggested seed coating of jute with *T. viride* as effective. An investigation by Sankar and Sharma (2001) explicitly shows the efficacy of *T. viride in vitro* as well as in vivo mitigated the disease caused by *M. phaseolina*. Studies by Sreedevi et al. (2011) revealed that isolates of *T. harzianum* and *T. viride* are successful against *M. phaseolina* in peanuts. Results of the study conducted by Pastrana et al. (2016) show that the application of commercial formulation of *T. asperellum* based on T18 strain decreases the damage caused by

M. phaseolina in strawberry. Elad et al. (1986) were the first to conclude that *T. harzianum* is an excellent biocontrol of *M. phaseolina*. The use of *T. harzianum* isolates in the interaction studies with *M. phaseolina* showed high phytopathogenic inhibition ability through several mechanisms, such as hyperparasitism and production of volatile compounds (Khalili et al., 2016). BCAs can be effective control solutions to reduce *M. phaseolina* damage in many plant crops.

8.1.14 Climate change

Changes that occurred in the past decade have created a process that permanently affected all organic life on Earth. Drought is the chief issue with soil moisture as it directly influences plant growth and subsequently agricultural yield (Ciais et al., 2005). The rapid decrease of soil moisture levels or irregular low soil moisture conditions combined with short warm surface temperatures can trigger flash droughts (Otkin et al., 2018). In parallel, long-term drought and thus lack of soil moisture led to the reduction of groundwater or surface reservoir levels (Naumann et al., 2018). The 2003 European heatwave, combined with the preceding drought period, caused significant yield loss in certain parts of South Europe, adding up to almost 15 billion euros in expenses (García-Herrera et al., 2010). In 2018, North and Central Europe faced extraordinarily hot weather conditions, leading to record temperatures and drought. According to the media, the drought of 2018 caused the most severe problems in the past 40 years regarding vegetable production in the EU. The increase in temperature due to global warming decreases yields in most cases and plants under such stress conditions are more prone to infection by pests and diseases (Ghini et al., 2012). It is expected that this process will continue, and these conditions will become more frequent making the conditions more favorable for *M. phaseolina*.

8.1.15 Hungarian climate

Climatic characteristics of the Carpathian Basin are mainly determined by its situation on the globe, which influences the solar incidence angle. The country is situated between 45 degrees4548 degrees35′ North latitudes and 16 degrees5−22 degrees58′ East longitudes. Its distance from the seas is approximately the same, which is why typical elements of wet oceanic climate with balanced temperature and wet continental climate with more extreme temperatures are present in the climate. According to the Trewartha climate, small regions of Hungary classification is one of the most frequently used systems; the Hungarian climate belongs to the warm summer (wet) continental type with major contrasts in temperature. Its main trait is the distinguishable four seasons, although the manifestation of the seasons is occasionally extreme. The summer is at least 3 months long with an average temperature of over 18°C, and diurnal maximum temperatures of over 35°C are not a rare occurrence. Winter temperatures can be erratic as severe and mild periods. However, there are no more than 3 days with an average temperature below 0°C (Mezősi, 2011). Hungary consists of several large and small regions (Fig. 8.2). Table 8.1 shows the climatic conditions of some small regions. In Hungary, we selected the survey areas within the small regions.

8.2 Materials and methods

8.2.1 Collection of isolates

The distribution of the pathogen was surveyed in the years 2019 and 2022 between August and November. During this process, a sunflower field was chosen randomly, and 50 randomly selected plants were examined. The lower third of the stem was cut with an ax. Once the pathogen was found, no more plants were examined. The lowest third of the stems were stored in dark at 25°C. From each sample, tissues containing microsclerotia were put on potato dextrose agar medium and incubated in dark conditions at 30°C. Hyphal ends of *M. phaseolina* cultures were taken and inoculated to a PDA medium and incubated in dark conditions at 30°C. Pure cultures were made of the isolates, which were permanently stored at −80°C in 50% glycerol.

8.2.2 Size of microsclerotia

Average diameter of microsclerotia taken from pure cultures was measured using a light microscope. This was carried out for the isolates collected on each surveyed site (n = 50).

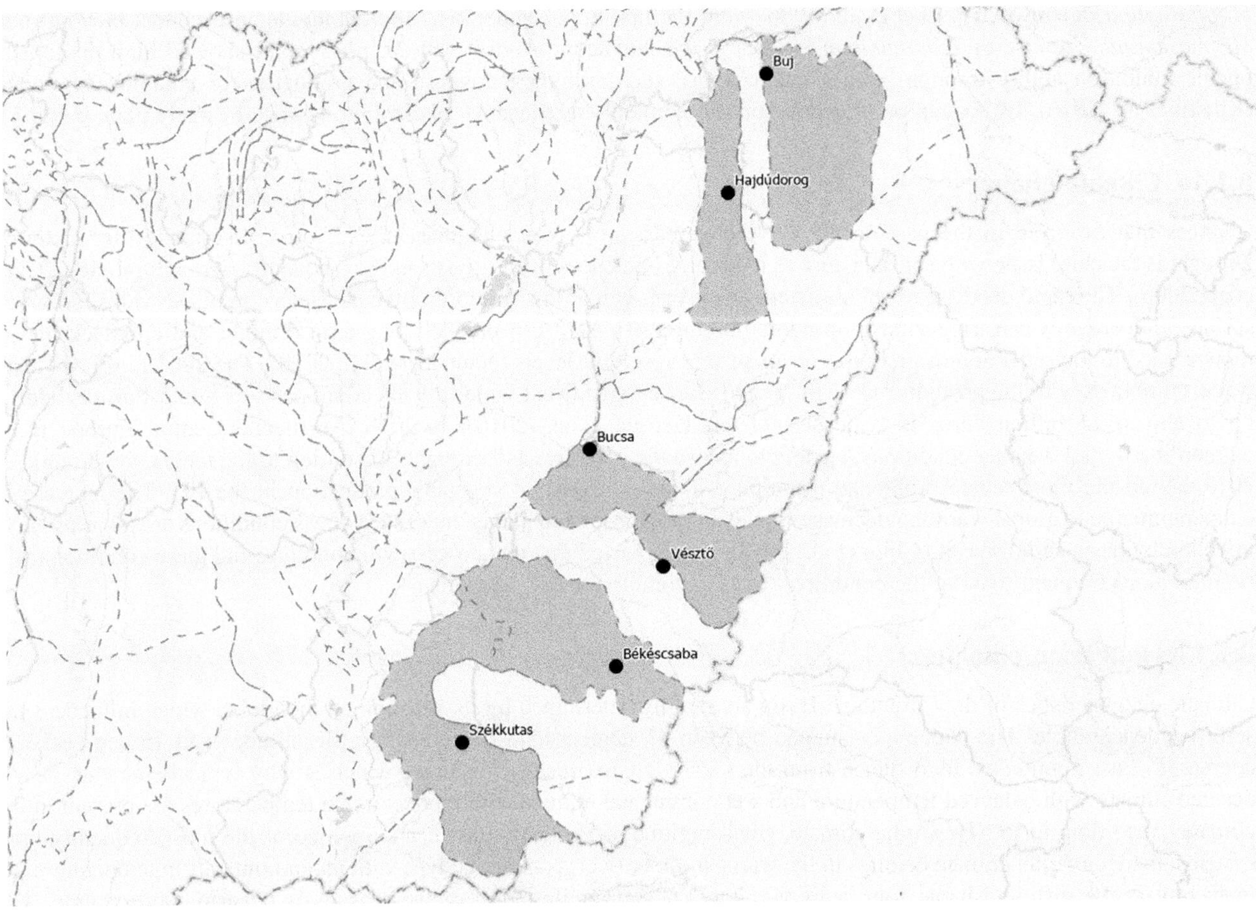

FIGURE 8.2 Small regions of Hungary.

TABLE 8.1 Climatic data of examined areas following Dövényi (2010).

Region	Csongrádi-sík	Békési-sík	Kis-sárrét	Dévaványai-sík	Hajdúhát	Közép-nyírség
Site	**Székkutas**	**Békéscsaba**	**Vésztő**	**Bucsa**	**Hajdúdorog**	**Buj**
Sunshine duration per year (h)	2000–2020	2000–2020	2000–2020	1980–2000	1850–980	1850–1900
Average mean temperature per year (°C)	10.3–10.6	10.2–10.4	10.2–10.3	10.3–10.4	9.7–10.0	9.4–9.7
Mean temperature of vegetation period (°C)	17.4–17.6	17.3–17.5	17.2–17.4	17.2–17.4	16.8–17.1	16.6–16.9
Average of annual maximum temperatures (50 yrs) (°C)	>34.0	34.0	34.0–34.5	>34.0	>34.0	34.0–34.5
Average of total annual precipitation (50 yrs) (mm)	500–570	500–550	540–570	510–540	520–550	530–580
Average precipitation in the vegetation period (mm)	290–350	320–330	320–330	310–320	310–350	350

8.2.3 Symptomatology

During the survey, the symptoms caused by *M. phaseolina* were also observed. Outward appearance of the symptoms are often different, thus how these symptoms present were also examined.

8.2.4 The survey

Infection of the sunflower hybrid by *M. phaseolina* was studied in six areas in the eastern region of the country (Hungary) (Fig. 8.2). During the surveys, 100 randomly selected plants were examined from the study area. The lower one-third of the stem (up to 50–60 cm in height) was cut in half and the presence of the disease was recorded. The heads were removed, and the average diameters were measured based on two diameters perpendicular to each other. The heads were threshed one by one and stored separately. To achieve the desired moisture, they were dried in an oven to a moisture content of 4%. The oil content of cypsela from infected and healthy plants was also examined. We randomly selected 5–5 sachets containing infected and healthy cypsela per area and measured their oil content. Isolates were collected from all areas and cultured. 100 microsclerotia from pure cultures were weighed.

8.2.5 Description of sunflower hybrid

It has high yield potential and is characterized by excellent stability in different habitats and weather while also having high oil content. Its 1000 seed weight is exceptionally high as well as its hectoliter weight. Its oil content can be 6–8% above average. Tolerance against *Sclerotinia sclerotiorum* is slightly lower than average, and the same is true for *Diaporthe helianthi.* It shows a less than average tolerance against the pathogen *M. phaseolina,* especially in heat stress. It is resistant against all pathotypes (100, 700, 730, 710, 330) of *Peronospora* (*Plasmopara halstedii*) described in Hungary, resistant against the “E” pathotype of broomrape. Weed control property is based on Clearfield technology (imidazolinone tolerance). It is characterized by a shorter than average growing season. In Hungary, the recommended sowing time is between 10th April and 10th May. Recommended plant number is 55.000–55.800. Highly sensitive to drought and heat stress to which it reacts with early maturation. Plant height is low, the hybrid is early flowering, and the heads face down in maturation. There is no ripening on living stem.

8.2.6 Weather data

Data of the closest meteorological station to the sites were used. The average temperature, number of heat days, and the distribution of precipitation of the vegetation period (April–September) were also analyzed.

8.2.7 Data analysis

To characterize the distribution of the studied pathogen published and our unpublished distribution, data were collected, and an actual distribution map was made for Hungary and some surrounding regions. The infection was studied at 12 sampling sites and was characterized with morphological parameter of plants (head diameter) and parameters of yield (seed weight, 1000-seed weight, and oil content). The effect of cultivation was characterized by the frequency of crop rotation (two or 3 years rotation of sunflower), while the weather conditions were characterized with total precipitation, and mean temperature of the vegetation periods and the number of heat days in the studied years. The economic loss was also estimated using country wide yield data of the Hungarian central Statistical Office (KSH).

The values of infested and healthy populations were compared with t–test if our data fulfill the assumptions of the parametric tests. The variance of homogeneity was checked with Levene-test, while the normal distribution was studied with Q-Q plots. In the other cases, nonparametric Mann–Whitney U-test was used at $P < .5$ significance level. The correlation between the weather parameters and the infection rate was analyzed with linear regression. Statistica seven software was used for calculation.

8.3 Results

8.3.1 The survey of *M. phaseolina* in Hungary and in the Carpathian Basin

A total of 73 sites were surveyed in the Carpathian Basin in 2019 and 2020. 63 of the 73 sites were found to be infected with the fungus. Lack of symptoms or presence of the fungus *M. phaseolina* was mainly observed in colder areas at higher altitudes in the duration of the survey. Unsuccessful surveys do not mean that the pathogen is not present at the site; simply that the survival structures of the pathogen were not found on the host plants during the period of the survey (Table 8.2).

Sites in Hungary were surveyed in early September, since sunflower is harvested in September in the southern regions of the country. Water content of sunflower plants examined in the northern regions was well over the harvestable amount. In these regions, harvest usually starts mid to late September. In Romania, sites were also surveyed mid-September. Sunflower fields selected for sampling did not show symptoms of drought damage. Due to the close proximity of the Carpathians, this region is considered relatively rainy. Most pathogens observed were *Sclerotinia sclerotiorum* and *Alternaria* species, both of which prefer rainy and humid weather. In Slovenia, maize fields were surveyed as sunflower was not sown in the area. Compared to sunflower, maize is less susceptible to the pathogen. It causes significant damage in maize during drought years. Symptoms of *M. phaseolina* or microsclerotia were not found on the sites surveyed in Slovenia.

The prevalence of *M. phaseolina* in Hungary according to the data published so far and the isolates we found are given in Fig. 8.3 and Table 8.3.

8.3.2 The size of microsclerotia

Our results are supported by the findings of Békési (1970) that the pathogen produces microsclerotia are of 80–160 μm in diameter (Table 8.4). The cultures were made from the isolates collected on each site. The fungus produced microsclerotia on the PDA medium.

Based on the average size of microsclerotia formed in our cultures (Fig. 8.4), they belong to group "C." Bokor (2007) first described the pathogen in Slovakia in 2007. According to Bokor, the size of microsclerotia is 88 μm ± 20 μm. Gulya (2002) took samples from sunflower plants in North and South Dakota, measured the microsclerotia taken from pure cultures, and found that the diameter was between 80 and 90 μm. Bradley and Río (2003) found that microsclerotia formed in and on outside of soybean plants grown in North Dakota were 76 ± 28 μm in diameter.

8.3.3 Symptoms on sunflower

In Hungary, infection develops in the first 4–6 weeks after sowing, especially during dry Aprils when plant emergence is slow and the plants are under stress. After the infection has occurred, symptoms cannot be observed until pollination. However, the pathogen can be detected in samples taken mid-June (Fig. 8.5) and early August (Fig. 8.6) that were dyed with lactophenol plus cotton blue. In the case of dry and hot July and August, the symptoms appear drastically and the disease progresses rapidly. To the layman, the symptoms may be mistaken for drought damage, as the plants wilt and turn brown, then show signs of early maturation (Fig. 8.7).

Symptoms start with a loss of turgor in the leaves. The leaves then turn yellow and brown and constantly dry out. After a few days, the leaves shrink drastically and dry out. The typical dusty black discoloration of the stem is not always present. Similar discoloration often appears on the bottom third of the stem in the case of infection by *Alternaria* species or completely healthy plants as well. By the end of the vegetation period, more discolored stems can be observed. Symptoms

TABLE 8.2 Unsuccessful surveys in the Carpathian Basin.

Site	Year	Country
Sátoraljaújhely	2020	Hungary (HU)
Gönc	2021	Hungary (HU)
Edelény	2020	Hungary (HU)
Turda	2021	Romania (RO)
Cluj napoca	2021	Romania (RO)
Târgu mureş	2021	Romania (RO)
Lendava	2020	Slovenia (SL)
Lipovci	2020	Slovenia (SL)
Lucova	2020	Slovenia (SL)

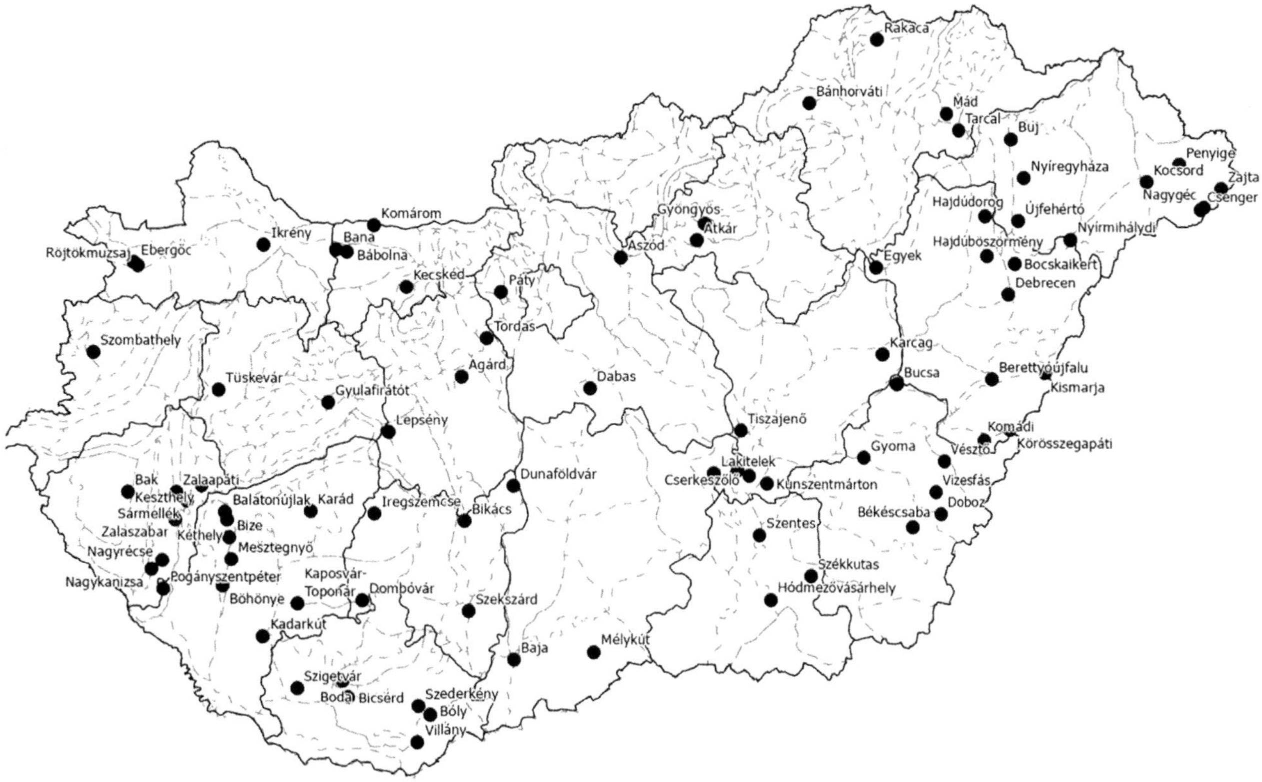

FIGURE 8.3 *M. phaseolina* sampling sites in Hungary.

appear 3–4 weeks later in the northern parts of the country. One telltale sign on sunflower may be the detached epidermis at the bottom of the stem. Microsclerotia of the pathogen often appear on the outside of the stem (Fig. 8.8) and the taproot (Fig. 8.9). The pathogen appears in patches in a field. The pathogen appears especially in heterogenous areas where elevations can be found. As rainwater flows down from these elevated patches, the plants suffer from water deficiency, thus creating favorable conditions for infection and damage by the pathogen. However, the pathogen is also present in areas with unfavorable conditions where the water accumulates, and the plants are not under drought stress. Disease incidence may exceed 60% even in fields that appear healthy. The most reliable symptom is the mass presence of microsclerotia inside the stem from the root collar to 40–50 cm heights. The fungus may completely consume the transport system and the pith (Fig. 8.10 and 8.11) and is capable of forming several thousand microsclerotia. This causes the symptoms that can be mistaken for drought damage.

Sunflower diseases, such as *Alternaria* sp. (Fig. 8.12) and *Diaporthe helianthi,* can cause similar symptoms as *M. phaseolina* (Fig. 8.13). Sunflower leaf infection by *Alternaria* is characterized by dark necrotic lesions with brownish–gray center and a chlorotic edge on the leaves, the stem, and leaf stalks (Kim & Mathur, 2006). These lesions are irregular in shape and size with dark edges and gray center. On younger plants, this may be surrounded by a yellow halo. Lesions on the leaves may fuse together and the leaves may wither. Symptoms on the stem initially present as dark lesions that become long and thin as they grow larger. These lesions often fuse, forming large blackened areas on the plant, which often leads to the breaking of the stem (Bhargav & Meena, 2014). The fungi *M. phaseolina* and *Diaporthe helianthi* may also cause similar symptoms on the stem (Perron et al., 1990). In order to differentiate between these diseases, the stem must be cut in half, and a magnifying glass should be used. Pure cultures are necessary in order to ensure accurate identification. The morphological characteristics of the cultures make the identification simple.

According to Mukhtar (2010), Su et al. (2001) and Mahmoud and Budak (2011), we detected similar symptoms. Mukhtar (2010) recorded that the early symptoms of the disease are not visible. However, the most common symptom is the rapid wilting of the plants, usually after pollination (Mukhtar, 2010). Symptoms initially observed on plants nearing physiological maturity consisted of the premature death of the plant and decreased head diameter (Su et al., 2001). Infected plants ripen early and spots appear on the stems ranging from brown to dark brown in color, which is why the disease is called "charcoal rot." The stem is split and black microsclerotia appear in the pith (Mukhtar, 2010). The pith is either

TABLE 8.3 *M. phaseolina* **sampling sites in Hungary and some surrounded areas. 1–62: sites sampled by Csüllög, K., 63–100: published data of Csöndes (2009), 101–108: published data of Békési et al. (1970). 1–12: study sites for evaluation of the infection in 2019–20.**

No	Sites	Year	No	Sites	Year	No	Sites	Year
1	Békéscsaba 1	2019	37	Keszthely	2021	73	Hódmezővásárhely	2005
2	Békéscsaba 2	2020	38	Dabas	2021	74	Iregszemcse	2005
3	Buj 1	2019	39	Vizesfás	2020	75	Kadarkút	2005
4	Buj 2	2020	40	Békéscsaba	2019	76	Kaposvár- toponár	2006
5	Bucsa 1	2019	41	Újfehertó	2019	77	Karcag	2005
6	Bucsa 2	2020	42	Bak	2020	78	Keszthely	2005
7	Hajdúdorog 1	2019	43	Aszód	2020	79	Kéthely	2006
8	Hajdúdorog 2	2020	44	Gyöngyös	2020	80	Kunszentmárton	2005
9	Székkutas 1	2019	45	Balatonújlak	2021	81	Lakitelek	2005
10	Székkutas 2	2020	46	Tiszajenő	2020	82	Lepsény	2005
11	Vésztő 1	2019	47	Ebergőc	2020	83	Marcali	2006
12	Vésztő 2	2020	48	Tüskevár	2020	84	Mesztegnyő	2005
13	Bana	2019	49	Tarcal	2020	85	Nagykanizsa	2005
14	Bábolna	2019	50	Mád	2020	86	Nyíregyháza	2005
15	Röjtökmuzsaly	2019	51	Bocskaikert	2021	87	Nagyrécse	2006
16	Berettyóújfalu	2019	52	Atkár	2021	88	Pogányszentpéter	2005
17	Hajdúböszörmény	2020	53	Páty	2021	89	Röjtökmuzsaj	2005
18	Csenger	2020	54	Bánhorváti	2020	90	Sármellék	2005
19	Szombathely	2019	55	Komárom	2021	91	Szederkény	2005
20	Kocsord	2020	56	Veľký Meder (SK)	2021	92	Székkutas	2005
21	Kismarja	2019	57	Čiližská Radvaň (SK)	2021	93	Szentes	2005
22	Kakasszék	2020	58	Čantavir (SRB)	2021	94	Szigetvár	2005
23	Nyírmihálydi	2020	59	Bačka Topola (SRB)	2021	95	Tiszakürt	2005
24	Nagygéc	2019	60	Horgoš (SRB)	2021	96	Zalaapáti	2006
25	Zajta	2019	61	Orohovo (SRB)	2021	97	Zalaszabar	2006
26	Rakaca	2019	62	Tileagd (RO)	2021	98	Tordas	2005
27	Körösszegapáti	2019	63	Balatonújlak	2005	99	Szekszárd	1984
28	Komádi	2019	64	Bize	2005	100	Bicsérd	1984
29	Doboz	2021	65	Boda	2005	101	Röjtökmuzsaj	1970
30	Penyige	2019	66	Bóly	2005	102	Ikrény	1970
31	Obid (SK)	2020	67	Böhönye	2006	103	Tordas	1970
32	Egyek	2019	68	Cserkeszőlő	2005	104	Agárd	1970
33	Bikács	2020	69	Debrecen	2005	105	Iregszemcse	1970
34	Mélykút	2020	70	Dombóvár	2006	106	Villány	1970
35	Karád	2020	71	Dunaföldvár	2005	107	Baja	1970
36	Kecskéd	2021	72	Gyulafirátót	2005	108	Gyoma	1970

TABLE 8.4 The average diameter of microsclerotia.

Site	Average size of microsclerotia (μm) (n = 50)	Group. According to Haig (1929)
Székkutas 1	100	C
Székkutas 2	98	C
Békéscsaba 1	121	C
Békéscsaba 2	120	C
Vésztő 1	99	C
Vésztő 2	105	C
Bucsa 1	110	C
Bucsa 2	105	C
Hajdúdorog 1	115	C
Hajdúdorog 2	122	C
Buj 1	98	C
Buj 2	105	C

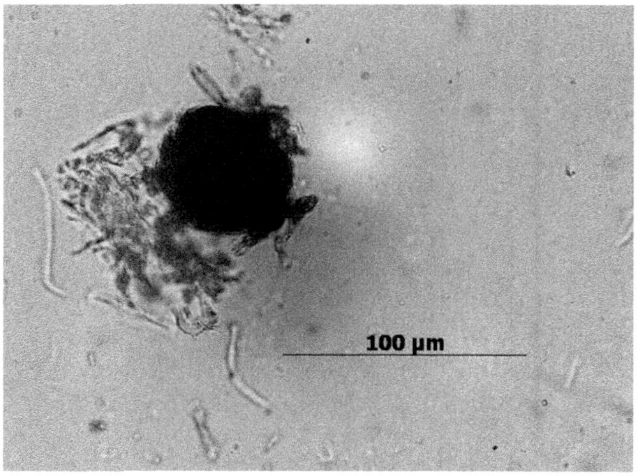

FIGURE 8.4 Size of microsclerotium.

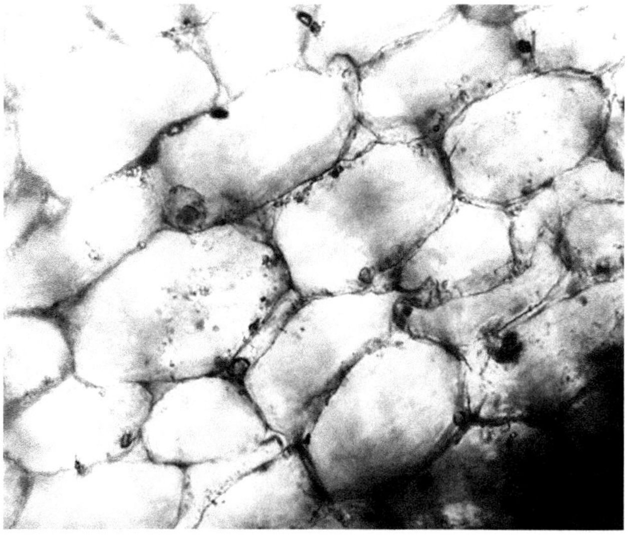

FIGURE 8.5 Mycelia in the sunflower pith.

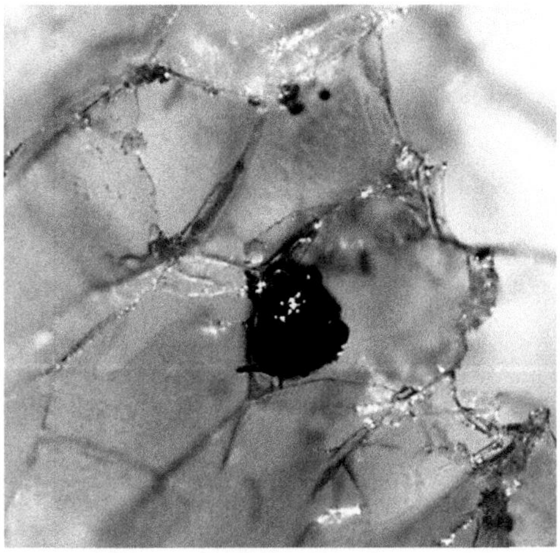

FIGURE 8.6 Microsclerotium in the sunflower pith.

FIGURE 8.7 Symptoms of *M. phaseolina* on the field.

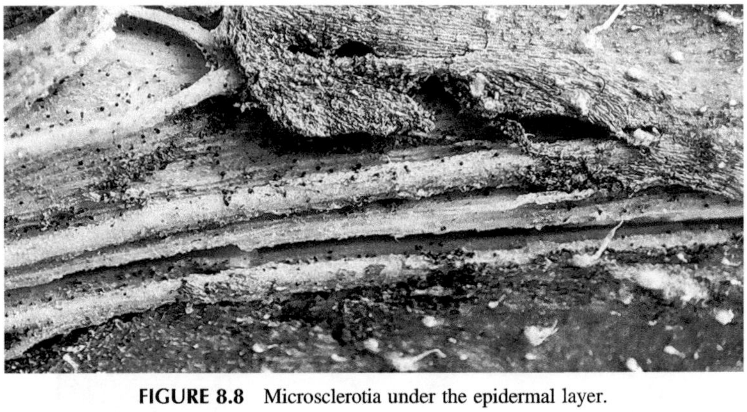

FIGURE 8.8 Microsclerotia under the epidermal layer.

FIGURE 8.9 Microsclerotia on the outside (root collar).

FIGURE 8.10 Microsclerotia in the fibers.

FIGURE 8.11 Microsclerotia in the pith.

FIGURE 8.12 *Alternaria* sp. infection in sunflower stem.

completely missing in the lower stem or it is compressed into horizontal layers. Spherical, black microsclerotia can be observed inside the lower stem, directly under the epidermis, and the outer surface of the taproot (Su et al., 2001). Mukhtar (2010) noted similar symptoms and found black microsclerotia inside the stem. The plants become weaker and dry out. The stem shows light gray or ashy black discoloration. The heads bear fewer seeds and the stems may completely dry out and become discolored. The seed falls apart; vascular fibers appear segmented. According to Mahmoud and Budak (2011), symptoms can be first observed in sunflower plants approaching physiological maturity. The stem turned a silvery grey color near ground level. Compared to noninfected plants, the head diameter was reduced and the plant withered earlier. Black, spherical microsclerotia were observed on the lower part of the stem, under the epidermis, and on the outside of the taproot.

8.3.4 Yield loss caused in Hungary

The *M. phaseolina* appeared in both studied years in each sampling sites (Table 8.5). The southernmost surveyed site was Székkutas. In this site, disease incidence of over 70% was measured 73% in 2019 and 71% in 2020. Békéscsaba is situated 46 km north of Székkutas. In this area, a disease incidence of 77% was recorded in 2019% and 68% in 2020. Vésztő is located 30 km north of Békéscsaba, where similarly to the latter; in 2019, a disease incidence of 78% was recorded in 2019 and 68% in 2020. Bucsa is located 83 km north of Székkutas. The disease incidence in Bucsa was 70% in 2019 and 65% in 2020. Hajdúdorog is the second northernmost surveyed site (163 km from Székkutas), although it was the most infected site in both 2019 (83%) and 2020 (74%). The northernmost area was around Buj (200 km from Székkutas). This was the second most infected site in 2019, whereas in 2020, only nine infected plants were found on the surveyed site.

Previous crops greatly influence the amount of inoculum. In Hungary, the main hosts include sunflower, maize, and oilseed rape. It has no host plants among cereals, with the exception of maize. Incidence of the pathogen and crop rotation are closely connected, and the effect of crop rotation was significant on the infection value in the studied sites (Mann–Whitney U-test: $P = .0061$). In areas where 2 years have passed between sowing, sunflower showed a disease incidence of 77.34% (1.94 ± SE), whereas this value was 67.00% (1.10 ± SE) in areas where 3 years pass between sowing sunflowers.

FIGURE 8.13 *M. phaseolina* infection in the stem.

TABLE 8.5 Disease incidence percentages (DIs%) measured in the sampling sites studied in 2019–20 and crop rotation data of the sites. underlined: known host plants in Hungary, bold underlined: primary host plant in Hungary.

	Di%	Crop rotation			
Site	**2019**	**2018**	**2017**	**2016**	**2015**
Székkutas 1	73	Winter wheat	Sunflower	Winter wheat	Maize
Békéscsaba 1	77	Winter wheat	Sunflower	Maize	Winter wheat
Vésztő 1	78	Winter wheat	Sunflower	Maize	Winter wheat
Bucsa 1	70	Winter wheat	Maize	Sunflower	Maize
Hajdúdorog 1	83	Winter wheat	Sunflower	Maize	Sunflower
Buj 1	82	Triticale	Sunflower	Maize	Winter wheat
	2020	**2019**	**2018**	**2017**	**2016**
Székkutas 2	71	Winter barley	Sunflower	Winter wheat	Maize
Békéscsaba 2	68	Winter wheat	Maize	Sunflower	Winter wheat
Vésztő 2	68	Winter wheat	Maize	Sunflower	Winter wheat
Bucsa 2	65	Winter wheat	Maize	Sunflower	Winter wheat
Hajdúdorog 2	64	Maize	Winter wheat	Sunflower	Winter wheat
Buj 2	9	Winter wheat	Maize	Oilseed rape	Winter wheat

This is also supported by the findings on Buj in 2020, where alfalfa was grown for 4 years, and sunflower was not sown for an additional 4 years after clearing of land.

The main source of inoculum of the pathogen is microsclerotia in the soil that can survive for many years. Crop rotation is difficult to maintain due to the large number of host plants and the longevity of microsclerotia and the high acreage of sunflower in Hungary. Francl et al. (1988) observed the highest accumulation of microsclerotia in soy monoculture. Microsclerotia accumulation was shown to increase when susceptible host plants were present in crop rotation. Lower microsclerotia density was recorded in areas where soy was present every 3 years as compared to every 2 years. Therefore, it can be established that the more frequently a susceptible host plant appears in the crop rotation, the higher the risk of infection will be. Our research did not include sunflower monoculture.

8.3.5 Effect of weather conditions on infection

In 2019, the average precipitation in the six areas was only 366 mm (Table 8.6). In 2019, the driest areas were Bucsa, Hajdúdorog, and Buj. There were records of highest number of hot days out of all the studied sites in 2019 (Table 8.7). Buj and Hajdúdorog had the highest disease incidence % in 2019 (82–83%), while the arid Bucsa only 70% disease incidence was detected. In Bucsa, 297 mm of precipitation occurred during the vegetation period. Vegetation precipitation in 2019 was the requested in Vésztő and Székkutas (449 and 432 mm), which resulted in high and medium (73%) infection. Infection index was 77% in Békéscsaba. This high index is due to the low precipitation of June, July, and August and the number of hot days. Vésztő and Székkutas had a lower number of hot days than the above mentioned areas with only 35 and 32 days, respectively. The precipitation in these two areas was the highest in the vegetation period of 2019 with 449 and 432 mm, respectively. In Vésztő, the infection index was 78%, while in Székkutas, it was 73%.

In 2020, the average amount of precipitation in the six areas was similar to 2019 (Table 8.6). The driest area was Békéscsaba (271 mm). The number of hot days was 31 (Table 8.7). The infection rate in this area was high at 68%. Vésztő and Székkutas were areas with moderate precipitation (320 and 336 mm), where the infection rate was 68% and 71%. In terms of hot days, there were 49 days in Vésztő and 43 days in Székkutas. The areas with the most precipitation in 2020 were Hajdúdorog and Buj (456 mm), where the disease infection was 64% and 9%, respectively. The number of hot days in the two areas in 2020 was the lowest of 25 days. The low infection rate could be due to several factors.

The weather conditions of both the sampling sites and the studied years showed large differences and overview of neither precipitation nor temperature data could lead obvious results. The effect of the aridity of the weather characterized by the total precipitation of the vegetation period was negative but not significant effect in the studied period (2019–20),

TABLE 8.6 Precipitation in studied areas.

			Precipitation (mm)						
	Site	Di%	Veg. per.	April	May	June	July	August	Sept.
2019	Székkutas 1	73	432.4	37.1	121.9	138.5	78.0	17.4	39.5
	Békéscsaba 1	77	353.4	59.4	117.6	31.6	75.2	13.1	56.5
	Vésztő 1	78	449.1	76.68	146.2	86.0	79.8	15.8	44.6
	Bucsa 1	70	297.9	43.3	87.5	19.9	94.1	10.7	42.4
	Hajdúdorog 1	83	333.6	21.0	64.2	123.4	83.8	18.6	22.6
	Buj 1	82	333.6	21.0	64.2	123.4	83.8	18.6	22.6
2020	Székkutas 2	71	336.3	10.4	36.9	138.0	60.5	85.5	5.0
	Békéscsaba 2	68	271.5	12.2	17.3	108.5	74.3	18.8	40.4
	Vésztő 2	68	320.8	0.0	6.1	118.7	94.8	62.2	39.0
	Bucsa 2	65	361.9	12.0	20.1	142.7	113.5	40.0	33.6
	Hajdúdorog 2	64	456.2	3.5	37.3	218.1	82.5	60.8	54.0
	Buj 2	9	456.2	3.5	37.3	218.1	82.5	60.8	54.0

TABLE 8.7 Mean temperature in studied areas.

					Mean temperature (°C)					
	Site	Di%	Hd	Veg. per.	April	May	June	July	August	Sept.
2019	Székkutas 1	73	32	18.3	12.9	14.7	22.5	22.1	23.8	18.0
	Békéscsaba 1	77	42	18.1	12.8	14.6	22.7	21.8	23.7	17.8
	Vésztő 1	78	35	18.8	13.6	15.4	23.7	22.6	23.6	17.8
	Bucsa 1	70	45	17.8	12.7	14.4	23.1	21.5	23.2	17.0
	Hajdúdorog 1	83	46	17.6	12.6	14.2	23.2	20.9	22.9	16.7
	Buj 1	82	46	17.6	12.6	14.2	23.2	20.9	22.9	16.7
2020	Székkutas 2	71	43	17.8	11.9	14.9	20.1	21.7	23.3	18.9
	Békéscsaba 2	68	31	17.4	11.7	15.1	20.4	21.4	22.5	18.4
	Vésztő 2	68	49	19.4	18.5	14.9	20.6	22.0	23.6	19.2
	Bucsa 2	65	20	17.06	11.04	14.62	20.21	21.07	22.41	17.88
	Hajdúdorog 2	64	25	17.13	11.61	14.42	20.16	21.25	22.82	17.73
	Buj 2	9	25	17.13	11.61	14.42	20.16	21.25	22.82	17.73

while analysis of the monthly values of the precipitation did not show obvious trend. The effect of monthly precipitation in spring was positive, but this effect was not significant. In the summer months, this effect was significantly negative in case of June and August, while in July, there was not any effect. In autumn, the effect of the precipitation was negative but not significant. Based on results, lower precipitation led to higher infestation (Fig. 8.14).

The mean temperature of the vegetation period had similar but also not significant effect; however, increasing temperature caused relatively higher infestation (Fig. 8.15).

The number of heat days had similar effect as temperature had. Higher number of heat days resulted in higher infestation, which was correlated marginally significant at $P < .1$ level (Fig. 8.16).

Under favorable conditions, 90% incidence of charcoal rot has been reported in major sunflower-producing areas of Sindh and Punjab in India (Rana, 1999, pp. 1−13). An incidence of 25% was recorded in 1998 in western North Dakota (Gulya et al., 2005). In the Eskişehir province of Turkey, the disease was detected in 70% of the surveyed 2 month-old oil sunflower crops in June 2009. The disease was observed in 70% of the surveyed areas, with a prevalence ranging from 10%−50% (Mahmoud & Budak, 2011). In Hungary, Békési (2007) reported 90% disease incidence in the drought year 2007.

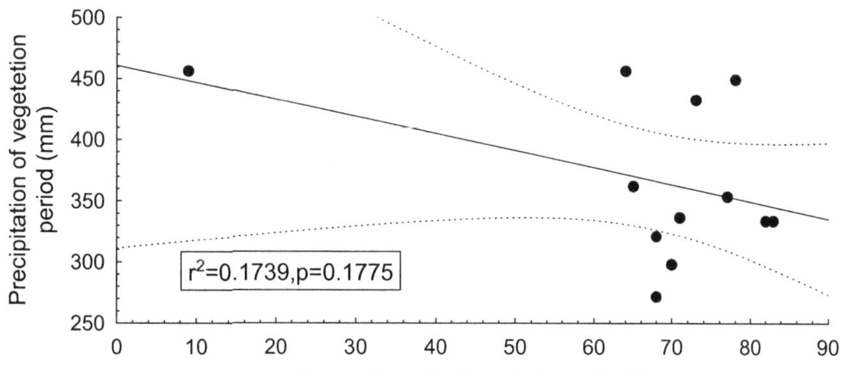

FIGURE 8.14 The effect of the precipitation.

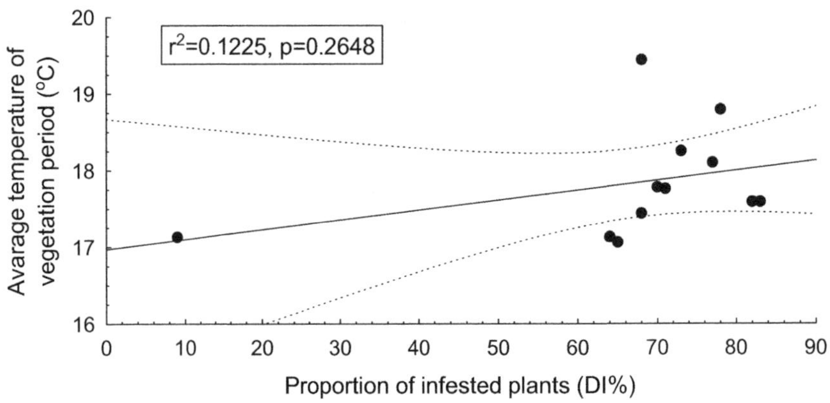

FIGURE 8.15 The effect of the mean temperature.

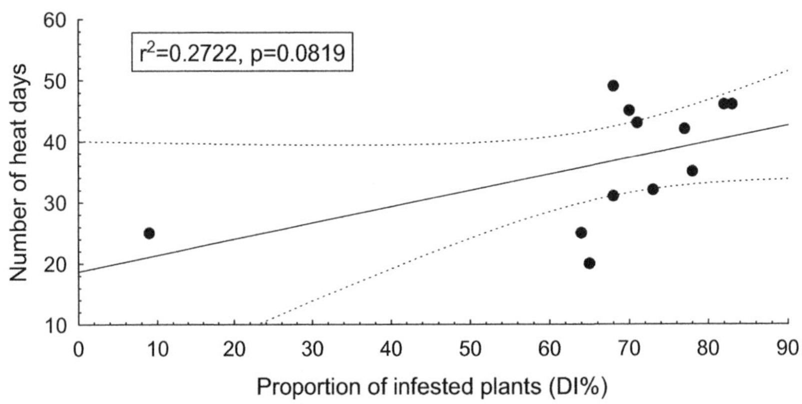

FIGURE 8.16 The effect of number of heat days.

8.3.6 Effect yield loss

The *M. phaseolina* infection caused significant yield loss showed by decrease of head diameter, seed weight, and 1000 seed weight of the studied plants (Table 8.8). The lowest effect was detected in case of the head diameters that decreased in both studied years with 27.76% and 15.86%, respectively. The weight loss of seeds was higher since it was 42.36% in 2019 and 25.56% in 2020. The decrease of 1000 seed weight showed the same trend in the two studied years. Although the yield loss of the 2 years differed, yet it was significant in each period.

The oil content showed opposite trend since seeds grown by infested plants have significantly higher oil content than healthy ones (Table 8.8). The increase of oil content was remarkably lower in both the years than the decrease of seed weight; thus, its increase could not compensate the yield loss. In favorable conditions, yield loss can reach 60–90% in the case of certain crops, such as sunflower (Khan, 2007). Recently, infection of epidemic proportions and 46–74% yield loss were observed on cowpea in several bean producing areas of Nigeria due to charcoal rot caused by *M. phaseolina* (Amusa et al., 2007).

8.3.7 Economic damage

In 2019, sunflower yield broke a record with an average of 3.0 tonnes/hectare. The harvested area was 50.000 ha less (8.1%) than in the previous year, although the amount of harvested product was only 7.0% less than that of the previous year. The total harvest of 2019 in tonnes: 1,706,851. Purchase price was around 100 ft/kg (35 cent) in the past 3 years (KSH).

Sunflower production has been steadily decreasing since 2018 in Hungary. In 2020, only 1.7 million tonnes were harvested, which is approximately 2.4% less than the harvest of 2019. This is primarily due to the fact that yield per hectares was 2.8 tonnes, which is a 9.2% decrease compared to the previous year. The 605 1000 hectares harvested were a

TABLE 8.8 Effect of infection on head diameter, seed weight, and 1000 seed weight of sunflower in the two studied years (2019 and 2020). Small letters refer to significant differences between healthy and infested plants, based on Mann–Whitney U-test and t-test in case of oil content ($P < .05$).

	2019	2020	2019	2020
Head diameter (cm±SE)			**Yield loss (%)**	
Healthy	25.07 (±0.20)a	24.27 (±0.14)a		
Infected	18.11 (±0.10)b	20.42 (±0.10)b	−27.76	−15.86
Seed weight (g±SE)				
Healthy	133.85 (±1.50)a	128.41 (±1.08)a		
Infected	77.15 (±1.00)b	95.59 (±0.90)b	−42.36	−25.56
1000 seed weight (g±SE)				
Healthy	84.44 (±0.66)a	82.38 (±0.40)a		
Infected	58.64 (±0.66)b	68.30 (±0.55)b	−30.55	−17.09
Oil content (%)				
Healthy	42.67 (±0.24)a	41.89 (±0.24)a		
Infected	44.98 (±0.22)b	44.57 (±0.19)b	5.41	6.40

7.3% increase contrasted with the previous year. In 2020, the total harvest was 1,697,961 tonnes. Partly, due to the lower harvest, the purchase price was 21% higher than the previous year at 122 forints (40 cents) per kilogram (KSH). If the susceptible Neoma hybrid was sown throughout Hungary, *M. phaseolina* would cause a yield loss of 35% (919,073 t), which at the average purchase price and national average yield of the 2019 crop year, would amount to a loss of USD 325 912 500. If the susceptible Neoma hybrid were sown throughout Hungary, *M. phaseolina* would cause a yield loss of 14.72% (293,142 t), which at the average purchase price and national average yield of the 2019 crop year, would amount to a loss of USD 118 815 000.

8.4 Conclusion

M. phaseolina is a widespread pathogen. The 12 isolates studied belonged to group "C" of the three groups described by Haig (1929). This is supported by the findings of Békési (1970) in Hungary and Bokor (2007) in Slovakia regarding microsclerotia size. Numerous studies have been made on the damage this pathogen. Significant yield losses are attributed to the fungus across the Europe. In Hungary, it has become a major pathogen of sunflower, causing considerable economic damage since its first recorded appearance in 1970. It is found in almost all regions of Hungary. In addition to the 62 isolates, we have found the pathogen has been recorded in a number of other areas in Hungary (Békési, 1970; Csöndes, 2009). It is considered to be widespread in nearly the entire country. In order to determine its distribution more accurately, it is necessary to examine 3–4 sunflower fields per region.

The spatial distribution of the pathogen is influenced by a combination of factors. In addition to the analysis of weather data, the importance of crop rotation should be highlighted. In both the years, high infection indices were recorded in all surveyed areas. The only outlier was Buj in 2020, where sunflower had not been in the crop rotation for 8 years prior to the survey. Therefore, although environmental conditions were favorable for a high or medium infection index, the previous cropping had a greater impact on the incidence of the pathogen. In areas where farmers used at least 3 years of crop rotation, the incidence of the fungus was significantly less. In fields where a 2-year crop rotation was used and congenial environmental conditions (high number of hot days, low precipitation) were present, a high incidence of the pathogen was detected. If the microsclerotia of the pathogen are present in high levels in the field and the environmental conditions are favorable, high infection index and yield losses can be expected. Sunflower should be present in the crop rotation every 6 years. The adherence to crop rotation is essential not only because of the threat *M. phaseolina* poses but also because of other diseases of sunflower that cause great economic damage, such as *Plasmopara halstedii, Sclerotinia sclerotiorum, Alternaria* sp., etc.

The pathogen had a major impact on the diameter of the sunflower heads, as well as on the total seed and 1000 seed weights. Infection with *M. phaseolina* caused significant yield losses. The lowest impact was observed on head diameter, which decreased in both studied years by 27.76% and 15.86%, respectively. Seed weight loss was higher as it was 42.36% in 2019 and 25.56% in 2020. The decrease in 1000 seed weight showed a similar trend. Although yield losses in the 2 years were different, they were significant in both years.

Oil content showed an opposite trend, with seeds from infected plants having a significantly higher oil content than healthy ones (Table 8.8). In both years, the increase in oil content was significantly lower than the decrease in seed weight, and therefore, the increase in oil content could not compensate for the yield loss. The economic damage caused by this deterioration could be enormous. A 35% yield loss in 2019 could result in a loss of 919,073 t, or USD 325 912 500. In 2020, an estimated loss of 293,000 t due to a rise in prices would amount to a loss of USD 815 000.

The fungus is expected to increasingly spread north due to global warming and its extraordinary adaptability. As of yet, no infection has been reported in Poland, Norway and Finland, but it is only a matter of time. The continuing rise in average temperatures and the lack of rainfall in the past few years all contribute to the wide distribution and damage caused by the pathogen. A few pieces of soil, plant residue or seeds contaminated with microsclerotia are sufficient for a new infection. Once the pathogen is introduced into the field, it is almost impossible to make it free of microsclerotia again.

Acknowledgment

We would like to thank the landowners for providing their lands for the survey:Laura Zsuzsa Oláh, László Oláh, Péter Borgula, Tamás Faluházi, Tamás Baukó, Sándor Dráguly, Tamás Dráguly, László Harkály, and Zsolt Baranyi.

References

Aćimović, M. (1998). Charcoal root and stem rot. In *Sunflower diseases*. Novi Sad, Serbia: Scientific Institute of Field and Vegetable Crops.

Akhtar, K. P., Sarwar, G., & Arshad, H. M. I. (2011). Temperature response, pathogenicity, seed infection and mutant evaluation against *Macrophomina phaseolina* causing charcoal rot disease of sesame. *Archives of Phytopathology and Plant Protection, 44*(4), 320–330. https://doi.org/10.1080/03235400903052945

Alabouvetto, C., & Bremeersch, P. (1976). *Macrophomina phaseolina* parasite due to tournesol.et el. actual de nos. connaissances et orientation des. researches. *Centre Tech. Interprob, Oleagineusx, Melrop. Inform. Tech., 52*, 13.

Alexandrov, V. (1999). Incidence of charcoal rot of sunflower caused by *Sclerotium bataticola* Taub. in Bulgaria. *Bulgarian Journal of Agricultural Science, 5*(6), 867–870.

Almeida, Á. M., Amorim, L., Filho, A. B., Torres, E., Farias, J. R., Benato, L., Pinto, M. C., & Valentim, N. (2003). Progress of soybean charcoal rot under tillage and no-tillage systems in Brazil. *Fitopatologia Brasileira, 28*, 131–135. https://doi.org/10.1590/S0100-41582003000200002

Almeida, A. M. R., Ferreira Saraiva, O., Bouças Farias, J. R., Almeida Gaudêncio, C., & Torres, E. (2001). Survival of pathogens on soybean debris under no-tillage and conventional tillage systems. *PesquisaAgropecuariaBrasileira, 36*(10), 1231–1238.

Amusa, N. A., Okechukwu, R. U., & Akinfenwa, B. (2007). Reactions of cowpea to infection by *Macrophomina phaseolina* isolates from leguminous plants in Nigeria. *African Journal of Agricultural Research, 2*, 73–75.

Angus, J. F., Gardner, P. A., Kirkegaard, J. A., & Desmarchelier, J. M. (1994). Biofumigation: Isothiocyanates released from Brassica roots inhibit growth of the take-all fungus. *Plant and Soil, 162*, 107–112.

Anis, M., Abbasi, W. M., & Zaki, M. J. (2010). Bioefficacy of microbial antagonists against *Macrophomina phaseolina* on sunflower. *Pakistan Journal of Botany, 42*(4), 2935–2940.

Anonymous. (1997). *New illustrated sericulture reader*. Bangalore, India: Central Silk Board.

Ashby, S. F. (1927). *Macrophomina phaseoli* (Maubl.) comb.now. The pycnidial stage of *Rhizoctonia bataticola* (Taub.)Butl. *Transactions of the British Mycological Society, 12*, 141–147.

Bai, Z., Jin, B., Li, Y., Chen, J., & Li, Z. (2008). Utilization of winery wastes for *Trichoderma viride* biocontrol agent production by solid state fermentation. *Journal of Environmental Sciences, 20*, 353–358.

Bandopadhyay, A., Bandopadhyay, A. K., & Samajpati, N. (2008). *In vitro* antifungal activity of some biocontrol fungi against jute pathogen *Macrophomina phaseolina*. *Indian Phytopathology, 61*(2), 204–211.

Bánhegyi, J., Tóth, S., Ubrizsy, G., & Vörös, J. (1985). *Magyarországmikroszkópikusgombáinakhatározókönyve*. Budapest: *Akadémiai*Kiadó.

Békési, P. (1970). A *Macrophomina phaseoli* (Maubl.)Ashby magyarországimegjelenéseéskártételenapraforgón. *Növényvédelem, 7*, 304–307.

Békési, P. (2007). A napraforgónövénykórtaniállapota 2007-ben. *GyakorlatiAgrofórum, 18*(11), 17–19.

Békési, P., Vörös, J., & Calvert, O. H. (1970). *Macrophomina phaseolina* in Hungary damaging sunflower. *Plant Disease Reporter, 54*(4), 286–287.

Benitez, T., Rincon, A. M., Limon, M. C., & Condon, A. C. (2004). Biocontrol mechanisms of *Trichoderma* strains. *International Microbiology, 7*(4), 249–260.

Bhargav, D. K., & Meena, H. P. (2014). Alternaria blight: A chronic disease in sunflower. *Popular Kheti, 2*(1), 146–153.

Birrenkott, G. L., Mengistu, A., & Grau, C. R. (1984). First report of charcoal rot caused by *Macrophomina phaseolina* on soybeans in Wisconsin. *Plant Disease, 68*, 628.

Boewe, G. H. (1963). Host plants of charcoal rot disease in Illinois. *Plant Disease Reporter, 47*, 753–755.

Bokor, P. (2007). *Macrophomina phaseolina* causing a charcoal rot of sunflower through Slovakia. *Biologia, 62*(2), 136–138.

Bontea, V. (1985). *Ciuperci parazite şi saprofite din republica populara romāna* (Volume I). Bucharest, Romania: Editura,AcademieiRepublicii SocialisteRomânia.

Bradley, C. A., & Río, L. E. D. (2003). First report of charcoal rot on soybean caused by *Macrophomina phaseolina* in North Dakota. *Plant Disease, 87*, 601–601.

Chamorro, M., Seijo, T. E., Noling, J. C., De los Santos, B., & Peres, N. A. (2016). Efficacy of fumigant treatments and inoculum placement on control of *Macrophomina phaseolina* in strawberry beds. *Crop Protection, 90*, 163–169.

Chattopadhyay, C., & Kalpana Sastry, R. (2002). Combining viable disease control tools for management of sesame stem-root rot caused by *Macrophomina phaseolina* (Tassi) Goid. *IndianJournal Protection, 30*(2), 132–138.

Chet, I., Harman, G. E., & Baker, R. (1981). *Trichoderma hamatum*: Its hyphal interactions with *Rhizoctonia solani* and *Pythium* spp. Microbial Ecol*ogy, 7*, 29–38.

Chowdary, N. B. (2006). *Studies on root rot disease of mulberry (*Morus *spp.) and its management with special reference to the antagonistic microbes*. Ph.D. Thesis. Mysore, India: submitted to the University of Mysore.

Chowdhury, S., Basu, A., Raychaudhuri, T., & Kundu, S. (2014). *In-vitro* characterization of the behaviour of *Macrophomina phaseolina (Tassi)*Goid at the rhizosphere and during early infection of roots of resistant and susceptible cultivars of sesame. *European Journal of Plant Pathology, 138*, 361–375. https://doi.org/10.1007/s10658-013-0335-z

Ciais, P., Reichstein, M., Viovy, N., Granier, A., Ogée, J., Allard, V., Aubinet, M., Buchmann, N., Bernhofer, C., Carrara, A., Chevallier, F., De Noblet, N., Friend, A. D., Friedlingstein, P., Grünwald, T., Heinesch, B., Keronen, P., Knohl, A., Krinner, G., Loustau, D., & Valentini, R. (2005). Europe-wide reduction in primary productivity caused by the heat and drought in 2003. *Nature, 437*(7058), 529–533. https://doi.org/10.1038/nature03972

Cohen, R., Elkabetz, M., Paris, H. S., Gur, A., Dai, N., Rabinovitz, O., & Freeman, S. (2022). Occurrence of *Macrophomina phaseolina* in Israel: Challenges for disease Management and crop germplasm enhancement. *Plant Disease, 106*(1), 15–25. https://doi.org/10.1094/PDIS-07-21-1390-FE

Cohen, R., Shreiber, S., & Graph, S. (2002). Vine decline and yield reduction in watermelon: The causal agents and evaluating chemical control efficacy. *Phytoparasitica, 30*, 16–17.

Conte, L., Zazzerini, A., & Tosi, L. (1989). Changes in composition of sunflower oil extracted from achenes of *Sclerotium bataticola* infected plants. *Journal of Agricultural and Food Chemistry, 37*, 36–38.

Csöndes, I. (2009). *Investigation of the factors affecting damage of* Macrophomina phaseolina *in different hostparasite relationships*. PhD thesis. Keszthely, Hungary: Pannon University, 120 pp.

Cummings, J. A., & Bergstrom, G. C. (2013). First report of charcoal rot caused by *Macrophomina phaseolina* in soybean in New York. *Plant Disease, 97*, 1506–1506.

Dave, K., Gothalwal, R., Singh, M., & Joshi, N. (2021). Facets of rhizospheric microflora in biocontrol of phytopathogen *Macrophomina phaseolina* in oil crop soybean. *Archives of Microbiology, 203*(2), 405–412. https://doi.org/10.1007/s00203-020-02046-z

De Barros, L. M. (1985). Disease Complex (*Fusarium oxysporum* and *Macrophomina phaseolina*) responsible for sunflower wilt in Portugal. In *Proc. 11th International Sunflower Conference, Mar del Plata, Argentina* (pp. 445–448). Paris, France: International Sunflower Association.

Dell'Olmo, E., Tripodi, P., Zaccardelli, M., & Sigillo, L. (2022). Occurrence of *Macrophomina phaseolina* on chickpea in Italy: Pathogen identification and characterization. *Pathogens, 11*(8), 842. https://doi.org/10.3390/pathogens11080842

Dhingra, O. D., & Sinclair, J. B. (1973). Variation among isolates of *Macrophomina phaseoli (Rhizoctonia bataticola*) from different regions. *Phytopathologische Zeitschrift, 76*, 200–204.

Dhingra, O. D., & Sinclair, J. B. (1974). Effect of soil moisture and carbon: Nitrogen ratio on survival of *Macrophomina phaseolina* in soybean stems in soil. *Plant Disease Reporter, 58*, 1034–1037.

Dhingra, O. D., & Sinclair, J. B. (1977). *An annotated bibliography of Macrophomina phaseolina*. Vicosa, MG, Brasil: Universidade Federal de Vicosa Press.

Dhingra, O. D., & Sinclair, J. B. (1978). *Biology and pathology of* Macrophomina phaseolina. Minas Gerais, Brazil: Universidade Federal de Vicosa.

Dövényi, Z. (2010). Magyarországkistájainakkatasztere. In *MTA FöldrajztudományiKutatóintézet* (Budapest).

Ebenezar, E. G., & Wesely, E. G. (2000). Effect of fungicides on *Macrophomina phaseolina* causing root rot in greengram. *Pestology, 24*(3), 12–14.

Elad, Y., Zvieli, Y., & Chet, I. (1986). Biological control of *Macrophomina phaseolina* (tassi) Goid.by*Trichoderma harzianum*. *Crop Protection, 5*, 288–292.

Francl, L. J., Wyllie, T. D., & Rosembrok, S. M. (1988). Influence of crop rotation on population density of *Macrophomina phaseolina* in soil infested with *Heteroderaglycines*. *Plant Disease, 72*, 760–764.

Gamliel, A., & Stapleton, J. J. (1993). Characterization of antifungal volatile compounds evolved from solarized soil amended with cabbage residue. *Phytopathology, 83*, 899–905.

Gangopadhyay, S., Wyllie, T. D., & Luedders, V. D. (1970). Charcoal rot disease of soybean transmitted by seeds. *Plant Disease Reporter, 54*, 1088–1091.

García-Herrera, R., Díaz, J., Trigo, R., Luterbacher, J., & Fischer, E. (2010). A review of the European summer heat wave of 2003. *Critical Reviews in Environmental Science and Technology, 40*(4), 267–306. https://doi.org/10.1080/10643380802238137

Ghaffar, A., & Erwin, D. C. (1969). Effect of soil water stress on root rot of cotton caused by *Macrophomina phaseoli*. *Phytopathology, 59*, 795–797.

Ghini, R., Hamada, E., Angelotti, F., Costa, L. B., & Bettiol, W. (2012). Research approaches, adaptation strategies, and knowledge gaps concerning the impacts of climate change on plant diseases. *Plant Pathology, 37*, 5–24.

Ghosh, T., Biswas, M. K., Guin, C., & Roy, P. (2018). A review on characterization, therapeutic approaches and pathogenesis of *Macrophomina phaseolina*. *Plant Cell Biotechnology and Molecular Biology, 19*(3–4), 72–84.

Gupta, G. K., & Chauhan, G. S. (2005). *Symptoms, identification and management of soybean diseases*. TechnicalBulletin 10. Indore, India: National Research Centre for Soybean.

Gupta, G. K., Sharma, S. K., & Ramteke, R. (2012). Biology, epidemiology and management of the pathogenic fungus *Macrophomina phaseolina* (Tassi) Goid.withspecial reference to charcoal rot of soybean (*Glycine max* (L.) Merrill). *Journal of Phytopathology, 160*, 167–180. https://doi.org/10.1111/j.1439-0434.2012.01884.x

Haig, J. C. (1928). *Macrophomina phaseoli* (Maubl.)Ashby.The picnidial stage of *Rhizoctonia bataticola* (taub.) butler. *Tropical Agriculturist, 70*, 77–79.

Haig, J. C. (1929). *Macrophomina phaseoli*(Maubl.) ashby and *Rhizoctonia bataticola*(taub.) butl. *TropicalAgriculture, 23*, 3–4.

Haleem, R., Saedo, K., & Shareef, B. (2016). Fungi colonized the roots of seedlings in forest nurseries. *Mycopath, 14*(1), 9–13.

Harris, E. (1962). *Disease of gluiheacorh*. Samaru, Technical Notes, 2,13.

Hartman, G. L., Rupe, J. C., Sikora, E. J., Domier, L. L., Davis, J. A., & Steffey, K. L. (2015). *Compendium of soybean diseases and pests*. The American Phytopathological Society, APS Press.

Hodges, C. S. (1962). Black root rot of pine seedlings. *Phytopathology, 52*, 210–219.

Hoffmaster, D. F., Mclaughlin, J. H., Ray, W. W., & Chester, K. S. (1943). The problem of dry rot caused by *Macrophomina phaseoli*. *Phytopathology, 33*, 1113.

Holmes, G. J., Mansouripour, S. Y., & Hewavitharana, S. S. (2020). Strawberries at the crossroads: Management of soil-borne diseases in California without methyl bromide. *Phytopathology, 110*, 956–968.

Horváth, J. (1995). *A szántóföldinövényekbetegségei*. Budapest: MezőgazdaKiadó.

Hunt, J. (1952). List of intercepted plant pests, 1951 *Service and Regulatory announcements*. U.S. Department of Agriculture.Agricultural Research Service, Plant Quarantine Division.

Ijaz, S., Sadaqat, H. A., & Khan, M. N. (2012). A review of the impact of charcoal rot *(Macrophomina phaseolina)* on sunflower. *Journal of Agriculture Science, 151*, 222–227.

Ilyas, M. B., & Sinclair, J. B. (1974). Effects of plant age upon development of necrosis and occurrence of intraxylem sclerotia in soybean infected with *Macrophomina phaseolina*. *Phytopathology, 64*, 156–157.

Indira, N., & Gayatri, S. (2003). Management of blackgram root rot caused by *Macrophomina phaseolina* by antagonistic microorganisms. *Madras Agriculture Journal, 90*, 490–494.

Islam, M. S., Haque, M. S., Islam, M. M., Emdad, E. M., Halim, A., Hossen, Q. M., Hossain, M. Z., Ahmed, B., Rahim, S., Rahman, M. S., Alam, M. M., Hou, S., Wan, X., Saito, J. A, & Alam, M. (2012). Tools to kill: Genome of one of the most destructive plant pathogenic fungi*Macrophomina phaseolina*. *BMC Genomics, 13*, 493. https://doi.org/10.1186/1471-2164-13-493

Jimenéz-Diaz, R. M., Blanco-López, M. A., & Sackston, W. E. (1983). Incidence and distribution of charcoal rot of sunflower caused by *Macrophomina phaseolina* in Spain. *Plant Disease, 63*, 1033–1036.

Kaur, S., Dhillon, G. S., Brar, S. K., Vallad, G. E., Chad, R., & Chauhan, V. B. (2012). Emerging phytopathogen *Macrophomina phaseolina*: Biology, economic importance and current diagnostic trends. *Critical Reviews in Microbiology, 38*, 136–151.

Kehri, K. H., & Chandra, S. (1991). Antagonism of *Trichoderma viride* to *Macrophomina phaseolina* and its application in the control of dry root-rot of mung. *Indian Phytopathology, 44*(1), 60–63.

Kending, S. R., Rupe, J. C., & Scott, H. D. (2000). Effect of irrigation and soil water stress on densities of *Macrophomona phaseolina* in soil and roots of two soybean cultivars. *Plant Disease, 84*, 895–900.

Khalili, E., Javed, M. A., Huyop, F., Rayatpanah, S., Jamshidi, S., & Wahab, R. A. (2016). Evaluation of *Trichoderma* isolates as potential biological control agent against soybean charcoal rot disease caused by *Macrophomina phaseolina*. *Biotechnology & Biotechnological Equipment, 30*(3), 479–488.

Khambhati, V. H., Abbas, H. K., Sulyok, M., Tomaso-Peterson, M., & Shier, W. T. (2020). First report of the production of mycotoxins and other secondary metabolites by *Macrophomina phaseolina* (tassi) Goid.Isolates from soybeans (*Glycine max* L.)Symptomatic with charcoal rot disease. *Journal of Fungi, 6*(4), 332. https://doi.org/10.3390/jof6040332

Khan, S. N. (2007). *Macrophomina phaseolina* as causal agent for charcoal rot of sunflower. *Mycopath, 5*, 111–118.

Khan, M. A., & Gangopadhyay, S. (2008). Efficacy of *Pseudomonas fluorescens* in controlling root rot of chickpea caused by *Macrophomina phaseolina*. *Indian Journal of Mycology & Plant Pathology, 38*(3), 580–587.

Kim, W. G., & Mathur, S. B. (2006). Detection of *Alternaria* spp. in carrot seeds and effect of the fungi on seed germination and seedling growth. *Journal of Plant Pathology, 22*(1), 11–15. https://doi.org/10.5423/PPJ.2006.22.1.011

KSH (https://www.ksh.hu/stadat_files/mez/hu/mez0078.html).

Kubicek, C. P., Mach, R. L., Peterbauer, C. K., & LoritoKubicek, M. (2001). *Trichoderma*: From genes to biocontrol. *Journal of Plant Pathology, 83*(2), 11–23.

Kudlíková, I., Šárová, J., Žalud, Z., & Veverka, K. (2002). Neue sonnenblumenkrankheiten in der Tschechischenrepublik. In *53. Pflanzenschutztagung DPG, bonn*, 17.–19.9.356–357.

Kumar, P., Gaur, V. K., Meena, A. K., Paliwal, S. K., & Meena, A. K. (2015). Evaluation of antagonists isolated from rhizosphere of Mungbean against *Macrophomina phaseolina*. *International Journal of Tourism Anthropology, 33*(4), 2531–2536.

Kumari, R., & Shekhawat, K. S. (2012). Integrated management against root-rot of mungbean (*Vigna radiata* L. Wilczek) incited by *Macrophomina phaseolina. Indian Journal Plant Pathology Microbiology, 3*(5), 215–747.

Lakhran, L., Ahir, R. R., Choudhary, M., & Choudhary, S. (2018). Isolation, purification, identification and pathogenicity of *Macrophomina phaseolina* (Tassi) Goid. caused dry root rot of chickpea. *Journal of Pharmacognosy and Phytochemistry, 7*, 3314–3317.

Lakshmidevi, N., Prakash, H. S., & Shetty, H. S. (1992). Effect of seed-borne*Macrophomina phaseolina* on physicochemical properties of sunflower oil. *International Journal of Tropical Diseases, 10*, 79–84.

Lewis, J. A., & Papavizas, G. C. (1970). Evolution of volatile sulfur-containing compounds from decomposition of crucifers in soil. *Soil Biology and Biochemistry, 2*, 239–246.

Ligon, J. M., Hill, D. S., Hammer, P. E., Torkewitz, N. R., Hofmann, D., Kempf, H. J., & Van Pee, K. H. (2000). Natural products with antifungal activity from *Pseudomonas* biocontrol bacteria. *Pest Management Science, 56*, 688–695.

Lodha, S. (1995). Soil solarization, summer irrigation and amendments for the control of *Fusarium oxysporum* f. sp. *cumini* and *Macrophomina phaseolina* in arid soils. *Crop Protection, 14*, 215–219.

Lodha, S., Gupta, G. K., & Singh, R. S. (1986). Crop disease situation and some new records in Indian arid zone. *Annals of Arid Zone, 25*, 311–320.

Lodha, S., & Mawar, R. (2019). Population dynamics of *Macrophomina phaseolina* in relation to disease management: A review. *Journal of Phytopathology, 168*(1), 1–17. https://doi.org/10.1111/jph.12854

Lodha, S., Sharma, S. K., & Aggarwal, R. K. (1997). Solarization and natural heating of irrigated soil amended with cruciferous residues for improved control of. *Macrophomina phaseolina.Plant Pathology, 46*, 186–190. https://doi.org/10.1046/j.1365-3059.1997.d01-223.x

Lodha, S., & Singh, M. (1985). Quantitative determination of *Macrophomina phaseolina* (Tani) Goid.in gram-legume intercropping systems. *Annals of Arid Zone, 23*, 259–261.

Lokesh, R., Rakholiya, K. B., & Thesiya, M. R. (2020). Evaluation of different fungicides against *Macrophomina phaseolina* (Tassi) Goid.causingdry root rot of chickpea (*Cicer arietinum* L.) invitro. *International Journal of Current Microbiology and Applied Sciences, 9*, 901–911.

Mackie, W. W. (1932). A hitherto unreported disease of maize and beans. *Phytopathology, 22*, 637–644.

Mahmoud, A., & Budak, H. (2011). First report of charcoal rot caused by *Macrophomina phaseolina* in sunflower in Turkey. *Plant Disease, 95*, 223.

Maneval, W. E. (1937). *A list of the Missouri fungi* (Vol 12). University of Missouri Studied Science Ser.

Manici, L., Caputo, F., & Cerato, C. (1995). Temperature responses of isolates of *Macrophomina phaseolina* from different climatic regions of sunflower production in Italy. *Plant Disease, 79*, 834–838. https://doi.org/10.1094/PD-79-0834

Manici, L. M., Cerato, C., & Caputo, F. (1992). Pathogenic and biologic variability of *Macrophomina phaseolina* (Tassi) Goid.isolates in different areas of sunflower cultivation in Italy. In *Proceedings of the 13th international sunflower Conference.Vol. 1. Pisa, Italy* (pp. 779–784).

Marič, A., Camprag, D., & Masirevič, S. (1988). *Bolesti i stetocinesuncokreta i njihhovosuzbijanje*. Beograd: Nolit.

Marquez, N., Giachero, L., Declerck, S., & Ducasse, D. (2021). *Macrophomina phaseolina:*General characteristics of pathogenicity and methods of control. *Frontiers of Plant Science, 12*, 634397. https://doi.org/10.3389/fpls.2021.634397

Mengistu, A., Ray, J. D., Smith, J. R., & Paris, R. L. (2007). Charcoal rot disease assessment of soybean genotypes using a colony-forming unit index. *Crop Science, 47*, 2453–2461.

Meyer, W. A., Sinclair, J. B., & Khare, M. N. (1973). Factors affecting charcoal rot of soybean seedlings. *Phytopathology, 64*, 845–849.

Mezősi, G. (2011). *Magyarország természetföldrajza. Akadémia Kiadó.*

Mughogho, L. K., & Pande, S. K. (1983). Charcoal rot of sorghum. In *A critlcal review: Proceedings of the consultative group discussion on research needs and strategies for control of sorghum root and stalk rot diseases, 27 Nov - 2 Dec 1983, Bellagio, Italy.*

Mukhtar, I. (2010). *Sunflower Disease and insect Pests in Pakistan:* A review. *African Crop Science Journal, 17*(2) (ISSN: 1021-9730).

Müller, D., & Grill, E. (1991). *Macrophomina phaseolina* on sonnenblumen in mitteldeutschland. *Nachrichtenblatt des Deutschen Pflanzenschutzdienstes, 43*, 61–62.

Naumann, G., Alfieri, L., Wyser, K., Mentaschi, L., Betts, R. A., Carrão, H., Spinoni, J., Vogt, J., & Feyen, L. (2018). Global changes in drought conditions under different levels of warming. *Geophysical Research Letters, 45*, 3285–3296.

Nikandrow, A., Weidemann, G. J., & Auld, B. A. (1990). Incidence and pathogenicity of *Colletotrichum orbiculare* and a *Phomopsis* sp. on *Xanthium* sp. *Plant Disease, 74*, 796–799.

Nischwitz, C., Olsenm, M., & Rasmussen, S. (2004). Effect of irrigation type on inoculum density of *Macrophomina phaseolina* in melon fields in Arizona. *Journal of Phytopathology, 152*(3), 133–137. https://doi.org/10.1111/j.1439-0434.2004.00813.x

Olaya, G., & Abawi, G. S. (1996). Effect of water potentialon mycelial growth and on production and germination of sclerotia of *Macrophomina phaseolina. Plant Disease, 80*, 1347–1350.

Onan, E., Çimen, M., & Karcilioğlu, A. (1992). Fungal diseases of sunflower in Aegean region of Turkey. *Journal ofTurkish Phytopathology, 21*(2–3), 101–107.

Otkin, J. A., Svoboda, M., Hunt, E. D., Ford, T. W., Anderson, M. C., Hain, C., & Basara, J. B. (2018). Flash droughts: A review and assessment of the challenges imposed by rapid-onset droughts in the United States. *Bulletin of the American Meteorological Society, 99*, 911–919.

Pastrana, A. M., Basallote-Ureba, M. J., Aguado, A., Akdi, K., & Capote, N. (2016). Biological control of strawberry soil-borne pathogens *Macrophomina phaseolina* and *Fusarium solani*, using *Trichoderma asperellum* and *Bacillus* spp. *Phytopathologia Mediterranea, 55*(1), 109–120. http://www.jstor.org/stable/43871859.

Perron, G., Delos, M., Hariot, J., & Germillon, C. (1990). Symptoms on sunflower stem: Beware of confusions. *Phytoma, 419*, 31–33.

Pickel, B., Dai, N., Maymon, M., Elazar, M., Tanami, Z., Frenkel, O., Toamy, M. A., Mor, N., & Freeman, S. (2020). Development of a reliable screening technique for determining tolerance to *Macrophomina phaseolina* in strawberry. *European Journal of Plant Pathology, 157*, 707–718.

Premamalini, T., Ambujavalli, B. T., Vijayakumar, R., Rajyoganandh, S. V., Kalpana, S., & Kindo, A. J. (2012). Fungal keratitis caused by *Macrophomina phaseolina* - a case report. *Medical Mycology Case Reports, 1*(1), 123–126. https://doi.org/10.1016/j.mmcr.2012.10.007

Preston, D. A. (1945). *Host index of Oklahoma plant diseases*. Oklahoma Agricultural and Mechanical College Agricultural Experiment Station. Technical Bulletin.

Purkayastha, S., Kaur, B., Dilbaghi, N., & Chaudhury, A. (2004). Cultural and pathogenic variation in the charcoal rot pathogen from clusterbean. *Annals of Agri Bio Research, 9*(2), 217–221.

Raguchander, T., Rajappan, K., & Samiappan, R. (1997). Evaluating methods of application of biocontrol agent in the control of mungbean root rot. *Indian Phytopathology, 50*(2), 229–234.

Rana, M. A. (1999). *Oilseed crops in Pakistan, status, constrains and strategy*. Islamabad: Ministry of Food Agriculture and Livestock, Government of Pakistan.

Rao, T. G. N., & Balasubramanian, K. A. (1983). Use of rank correlation coefficient in deciding charcoal rot expression parameters in Sorghum.(En). *Indian Phytopathology, 36*(2), 363–364.

Raut, J. G. (1981). Anthesis and seed development stage of sunflower in relation to seed-borne infection of *Macrophomina phaseolina* and comparison of charcoal rot affected plants with healthy plants (abstract). In IARI (Ed.), *Third international symposium on plant pathology, New Delhi, 14–18 december, 1981*. New Delhi, India: Indian Agricultural Research Institute.

Ravichandran, S., & Hegde, Y. R. (2017). Management of dry root rot of chickpea caused by *Rhizoctonia bataticola* through fungicides. *International Journal of Current Microbiology and Applied Sciences, 6*(7), 59416.

Reichert, I., & Hellinger, E. (1947). On the occurrence, morphology, and parasitism of Sclerotium bataticola. *Journal of Palestine, 6*, 107–147.

Reuveni, R., Kirkun, J., Nachmias, A., & Shlevin, E. (1982). The role of *Macrophomina phaseolina* in a collapse of melon plants in Israel. *Phytoparasitica, 10*, 51–56.

Romero Luna, M. P., Mueller, D., Mengistu, A., Singh, A. K., Hartman, G. L., & Wise, K. A. (2017). Advancing our understanding of charcoal rot in soybeans. *Journal of Integrated Pest Management, 8*, 1–8. https://doi.org/10.1093/jipm/pmw020

Sangappa, G., & Mallesh, S. B. (2016). Effect of bioagents and chemicals for the management of aerial blight and dry root rot of black gram incited by *Rhizoctonia bataticola*. *International Journal of Plant Protection, 9*(2), 424429.

Sankar, P., & Sharma, R. C. (2001). Management of charcoal rot of maize with *Trichoderma viride*. *Indian Phytopathology, 54*(3), 390–391.

Savazzini, F., Longa, C. M. O., & Pertot, I. (2009). Impact of the biocontrol agent *Trichoderma atroviride* SC1 on soil microbial communities of a vineyard in northern Italy. *Soil Biology and Biochemistry, 41*, 1457–1465.

Savitha, A. S. (2017). *In vitro* efficacy of fungicides and bioagents against dry root rot of pigeonpea caused by *Rhizoctonia bataticola* (Taub.) Butler. *International Journal of Pure and Applied Sciences, 5*(6), 1341–1347.

Scholefield, S. M., & Griffin, M. J. (1979). Charcoal rot (*Macrophomina phaseolina*) on mung bean. *Plant Pathology, 28*(3), 155–156.

Sharma, S. K., Aggarwal, R. K., & Lodha, S. (1995). Population changes of *Macrophomina phaseolina* and *Fusarium oxysporum* f. sp. *cumini* in oil cake and crop residue-amended sandy soils. *Applied Soil Ecology, 2*, 281–284.

Shokes, F. M., Lyda, S. D., & Jordan, W. R. (1977). Effect of water potential on the growth and survival of *Macrophomina phaseolina*. *Phytopathology, 67*, 239–241.

Short, G. E., Wyllie, T. D., & Ammon, V. D. (1978). Quantitative enumeration of *Macrophomina phaseolina* in soybean tissues. *Phytopathology, 68*, 736–741.

Singh, R. D. N., & Kaiser, S. A. K. M. (1995). Occurrence of pycnidial stage of charcoal rot pathogen *Macrophomina phaseolina* on maize stalk. *Advances in Plant Species, 8*(1), 189–193.

Singh, S. K., Nene, Y. L., & Reddy, M. V. (1990). Influence of cropping systems on *Macrophomina phaseolina* in soil. *Plant Disease, 74*, 812–814.

Small, W. (1928). On *Rhizoctonia bataticola* (Taub.)Butler as a cause of root disease in the tropics. *Transactions of the British Mycological Society, 13*, 40–68.

Songa, W., & Hillocks, R. J. (1996). Legume hosts of *Macrophomina phaseolina* in Kenya and effect of crop species on soil inoculum level. *Journal of Phytopathology, 144*, 387–391.

Sreedevi, B., Charithadevi, M., & Saigopal, D. V. R. (2011). Induction of defense enzymes in *Trichoderma harzianum* treated groundnut plants against *Macrophomina phaseolina*. *Journal of Biological Control, 25*(1), 67–73.

Sreevidya, M., & Gopalakrishnan, S. (2012). Bacteria and actinomycetes as biocontrol agents for the control of fungal pathogens of chickpea and sorghum. In *International conference proceeding* (p. 263). Hyderabad, India: ICPHM.

Strange, R., & Scott, P. (2005). Plant disease: A threat to global food security. *Annual Review of Phytopathology, 43*, 83–116. https://doi.org/10.1146/annurev.phyto.43.113004.133839

Subbarao, K. V., Hubbard, J. C., & Schulbach, K. F. (1997). Comparison of lettuce diseases and yield under subsurface drip and furrow irrigation. *Phytopathology, 87*, 877–883.

Su, G., Suh, S. O., Schneider, R. W., & Russin, J. S. (2001). Host specialization in the charcoal rot fungus. *Macrophomina phaseolina. Phytopathology, 91*, 120126.

Tančić-Živanov, S., Dedić, B., Dimitrijević, A., Dušanić, N., Jocić, S., Miklič, V., Kovačević, B., & Miladinović, D. (2019). Analysis of genetic diversity among *Macrophomina phaseolina* (Tassi) Goid.isolates from Euro-Asian countries. *Journal of Plant Diseases and Protection, 126*, 565–573. https://doi.org/10.1007/s41348-019-00260-6

Taubenhaus, J. J. (1913). The black rots of the sweet potato. *Phytopathology, 3*, 159–166.

Thompson, A., & Johnston, A. (1953). *A host list of plant diseases in Malaya*. Kew, Surrey UK: Commonwealth Mycological Institute. Mycological Papers, 1953 No. 52 38 pp.

Tikhonov, O. J., Nedelko, Y. K., & Perestova, T. A. (1976). Infection period and spreading pattern of *Sclerotium bataticola* in sunflower tissue. In *Proc. 7th int. Sunflower conf., kransnodar, USSR.27 lune-3 july* (p. 134). Paris, France: Int. Sunflower Assoc.

Tiwari, A., & Shroff, V. N. (1982). Differential reaction of cotton genotypes to root rot. *Indian Phytopathology, 35*, 514–515.

Tosi, L., Zazzerini, A., & Monotti, M. (1995). Comparative phytopathological surveys on varieties of sunflower conducted in 1994. *InformatoreAgrario, 51*, 71–73.

Turkensteen, L. J., & Lablans, W. N. (1988). Meteorological aspects of disease management in potato crops. *Acta Horticulturae, 214*, 157–163.

Van der Waals, J. E., Krüger, K., Franke, A. C., Haverkort, A. J., & Steyn, J. M. (2013). Climate change and potato production in contrasting South African agro-ecosystems. 3. Effects on relative development rates of selected pathogens and pests. *Potato Research, 56*, 67–84.

Verma, M., Brar, S. K., Tyagi, R. D., Sahai, V., Prévost, D., Valéro, J. R., & Surampalli, R. Y. (2007). Bench-scale fermentation of *Trichoderma viride* on wastewater sludge: Rheology, lytic enzymes and biocontrol activity. *Enzyme and Microbial Technology, 41*(6), 764–771.

Wrather, J. A., Anderson, T. R., Arsyad, D. M., Tan, Y., Ploper, L. D., Porta-Puglia, A., Ram, H. H., & Yorinori, J. T. (2001). Soybean disease loss estimates for the top ten soybean-producing countries in 1998. *Canadian Journal of Plant Pathology, 23*(2), 115–121. https://doi.org/10.1080/07060660109506918

Yang, X. B., & Navi, S. S. (2005). First report of charcoal rot epidemics caused by *Macrophomina phaseolina* in soybean in Iowa. *Plant Disease, 89*, 526–526.

Zazzerini, A. (1980). Sunflower diseases in Italy. *Helia, 3*, 45–46.

Zazzerini, A., Monotti, M., Buonaurio, R., & Pirani, V. (1985). Effects of some environmental and agronomical factors on charcoal rot of sunflower. *Helia, 8*, 45–49.

Žerjav, M., Rodič, K., Zemljič, M. U., & Schroer, H. J. (2017). Soybean diseases caused by fungi in *Slovenia*. In *13th Slovenian conference on plant protection with international participation RimskeToplice, March 7-8, Slovenia*.

Section II

Resistance and pathogenicity

Chapter 9

Candidate genes of *Macrophomina phaseolina* involved in the pathogenicity

Shulbhi Verma[1], Narendra Kumar[2], Ravita Tadiya[3], Garima Kumari[4] and Sunita Rawat[3]
[1]Department of Plant Breeding and Genetics, C.P College of Agriculture, S.D Agricultural University, S.K Nagar, Gujarat, India; [2]Department of Agriculture, Doon (PG) College of Agriculture Science and Technology, Dehradun, Uttarakhand, India; [3]Plant Physiology Discipline, Genetics and Tree Improvement Division, Forest Research Institute, Dehradun, Uttarakhand, India; [4]Forest Ecology and Climate Change Division, ForestResearch Institute, Dehradun, Uttarakhand, India

9.1 Introduction

Soil- and seed-borne *Macrophomina phaseolina* (Tassi) Goid. is a polyphagous fungus that infects plants on almost all the stages of their growth. Their growth is encouraged through soil, seed, and plant residues. The microsclerotia of fungus tolerate the adverse condition and survive in soil for long time, and they flourish by root exudate and infect the host plants. Plant defoliation and wilting occur through formation of dark mycelia and black microsclerotia. They appear on roots and stems, causing red to brown lesions. The black color microsclerotia production in the infected plant tissues causes charcoal rot. Colonization of charcoal rot pathogen results in the seedling death due to the impediment of xylem vessels (Sarr et al., 2014). The pathogen transmit via infected roots to stems, obstructing root system (Kaur et al., 2012b), and reduction in the seed germination causes seedling rots. Throughout the world, *M. phaseolina* infests many species and causes charcoal rot disease (Srivastava et al., 2001). The economically important crops, such as oilseed plants, bean (*Phaseolus vulgaris* L.), cowpea (*Vigna unguiculata*), corn (*Zea mays* L.), cotton (*Gossypium herbaceum* L.), sesame (*Sesamum indicum* L.), sorghum (*Sorghum bicolor* [L.]Moench), soybean (*Glycine* max L.), castor (*Ricinus communis*), jatropha (*Jatropha gossypifolia*), groundnut (*Arachis hypogaea*), and sunflower (*Helianthus annuus* L.), suffer the great losses. *M. phaseolina* have the genetic and morphological variations due to its adaption in variable environments (Iqbal & Mukhtar, 2014), physiological (Sanchez et al., 2018), pathogenic (Iqbal & Mukhtar, 2014), and genetic (Khan et al., 2017) diversity has been reported. *M. phaseolina* shows higher virulence and pathogenicity on crops due to existence of pathogenic architect of enzymes, protein, and genes for disease development during pathogenic invasion. In aspect of *M. phaseolina* behavior, recognition techniques related to the identification and detection become more important for disease management in the crops. Advance technologies that are utilized for characterization of *M. phaseolina* deepen the knowledge of the fungus pathogenicity (Burkhardt et al., 2018; Islam et al., 2012; Yousef, 2021).

9.2 Morphology characterization and life cycle of *M. Phaseolina*

9.2.1 Morphological characteristics

M. phaseolina has hyaline hyphae designed in thin walls with septa, and its texture varies from light brown to dark according to age. Microsclerotia may be spherical, oval, or oblong compact masses of hardened fungal mycelium. The size of pycnidia is larger than microsclerotia; their color varies from deep brown to black; surface is bulbous, rough, irregular, beaked, and etiolated, and are rarely seen in nature (Lakhran et al., 2018). The variation in color of the fungal mycelia, microsclerotia distribution, and pycnidia production occur due to the pathogen. Besides, the internal transcribed spacers (ITS) amplification revealed the similarity in the isolates and indicates that all belongs to same species (Almomani et al., 2013). The discrimination on the basis of morphological polymorphism of the fungus has been imputing to the fungal

Macrophomina Phaseolina. https://doi.org/10.1016/B978-0-443-15443-0.00016-4

responses to surrounding environmental conditions or changes occurring in their host species (Pandey et al., 2020; Tok, 2019).

Similarly, Tok (2019) found a strong link between virulence and their phenotype. Individual sclerotium is made up of 50–200 or more individual cells that are connected by each other through a gelatinous material and have capacity to germinate and revive itself multiple times in the crop growing season. The size of the microsclerotia varies according to the nutrients availability in the substrate, the place of propagules occurrence (Marquez et al., 2021).

9.2.2 Genetic diversity

From different countries, *M. phaseolina* populations are used for comparative analysis on the basis of their genetic approaches; few markers like universal random amplified polymorphic DNA, restriction site–based restriction fragment length polymorphism, PCR-based specific amplified fragment length polymorphism, and rDNA sequencing have been successfully utilized (Ghosh et al., 2018; Vandemark et al., 2020). Molecular marker–based cluster analysis has investigated the genetic variation in the isolates of *M. phaseolina*. As *M. phaseolina* does not reproduce sexually, still the data showed significant genetic variation across isolates of this disease. Most likely, parasexualism occurs for genetic variation because in the mechanism, cell gets united from distinct hyphae, resulting in heterokaryons for variations (Marquez et al., 2021).

9.2.3 Life cycle of *M. phaseolina*

The disease cycle of the *M. phaseolina* is monocyclic (Figs. 9.1 and 9.2). The sclerotia which are the primary pathogenic source of *M. phaseolina* survive not less than 15 years in the soil (Gupta et al., 2012). For survival, microsclerotia require soil moisture, but when they do not get appropriate condition, their population significantly reduces within 7–8 weeks and be in dormant condition. These microsclerotia produce hyphae again during the favorable conditions. Generally, temperatures between 28 and 35°C are required for the growth of microsclerotia to germinate throughout the growing season. Microsclerotia germinate on the roots surface, and germ tubes on the sclerotial ends generate appressoria, which use turgor pressure or natural holes to enter the epidermal cell walls of the hosts. Additionally, the fungus can be carried in the seed coats of infected seeds which do not allow germinating or generating seedlings and dying shortly after emergence. The fungus mainly invades the root xylem resulting in the yellowness and senescence in leaves and cortical tissues sloughing from the stems lower part and root, and the microsclerotia cause dull appearance, somewhat gray in the tissues, and lead the host plant's death (Ghosh et al., 2018) (Fig. 9.3).

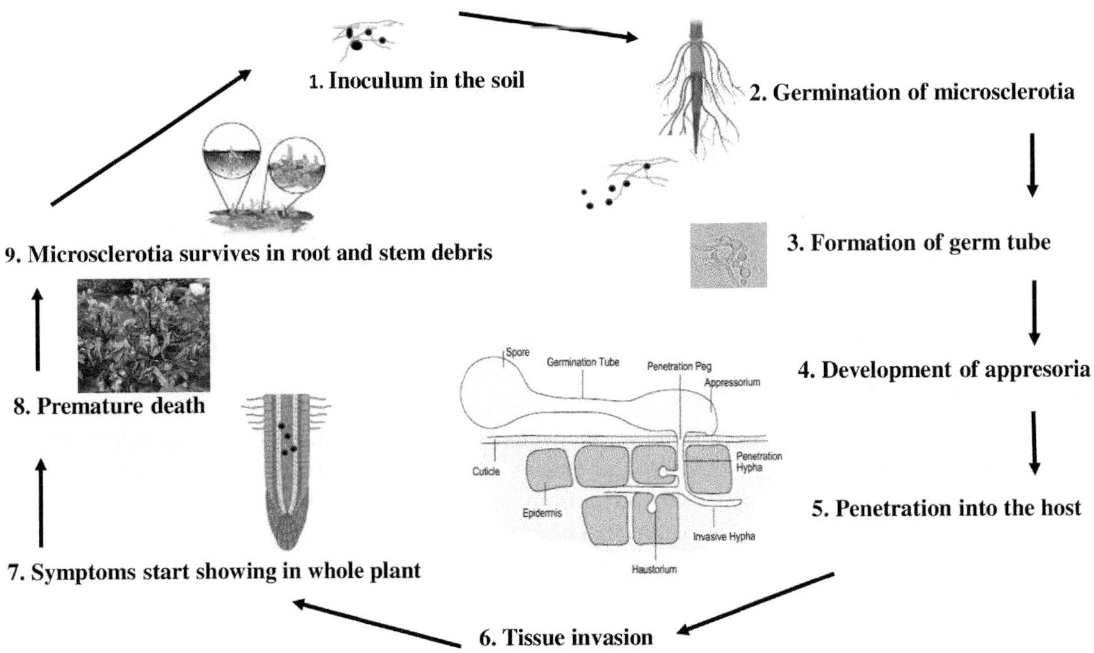

FIGURE 9.1 Diagrammatic representation of life cycle of *M. phaseolina*.

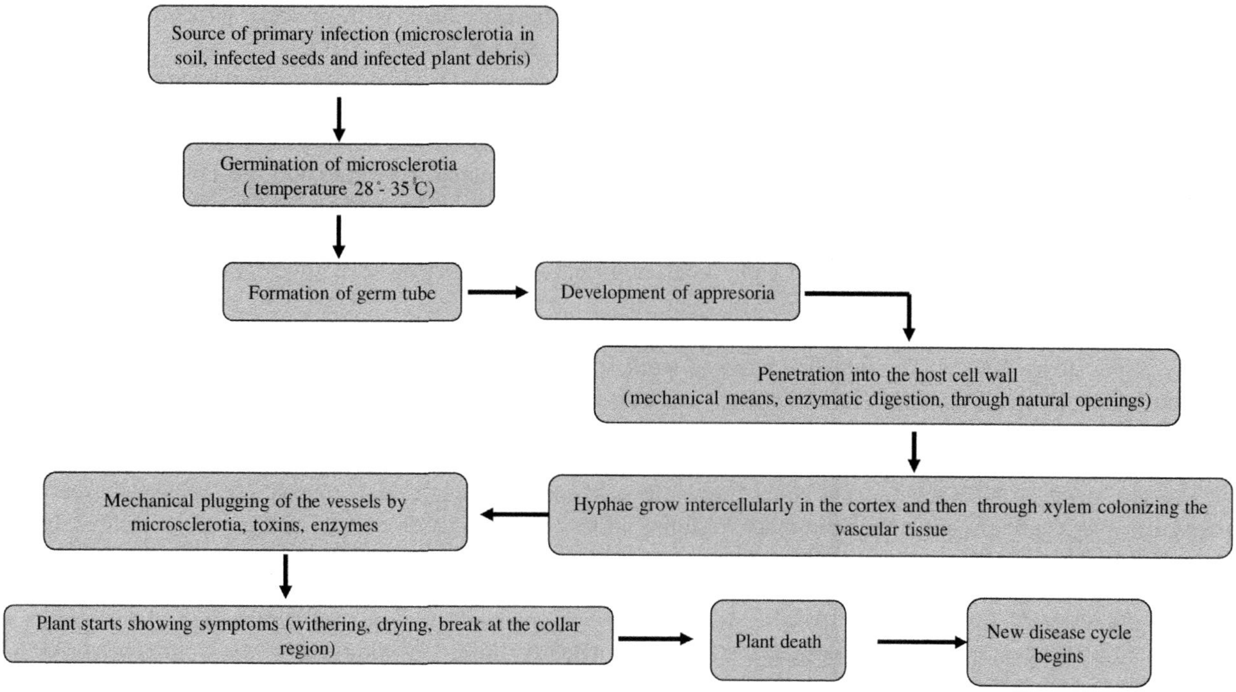

FIGURE 9.2 Flowchart representation of the disease cycle caused by *M. phaseolina*.

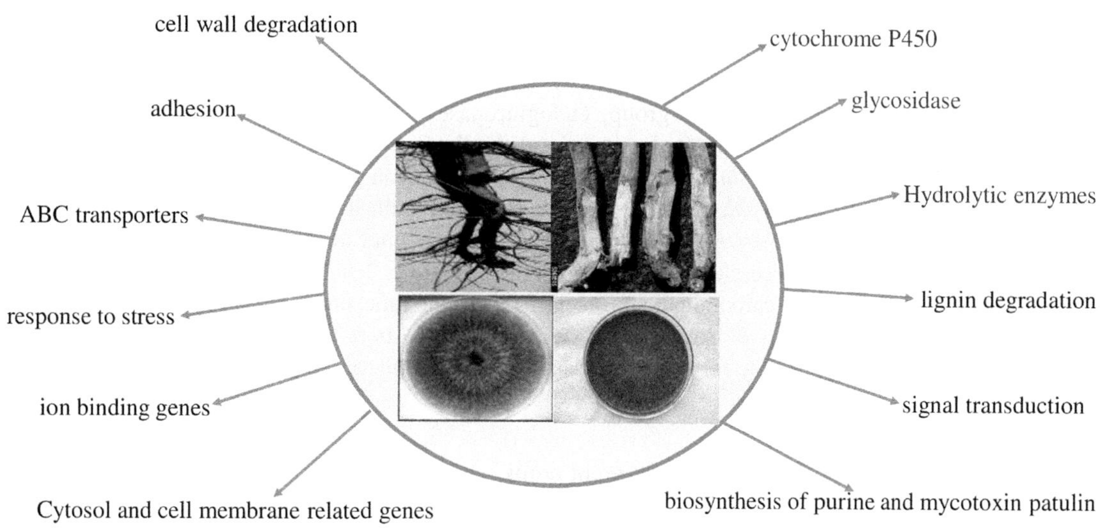

FIGURE 9.3 Major genes related to pathogenicity in *M. phaseolina*.

9.3 Crops and *M. phaseolina* interaction for disease development

M. phaseolina resides in the soil and seeds more than years in the form of microsclerotia, which pose worldwide severe hazard to crops and crop productivity (Ghosh et al., 2018). The environment-tolerant sclerotia especially in heat can live in soil or root debris for 2–15 years, and as host tissues disintegrate, they are discharged in clusters above the soil surface, where both living and dead tissues colonize. On the basis of behavior, its life cycle has been differentiated in four stages: germination, penetration, parasitic, and saprophytic (Lodha & Mawar, 2020; Pandey & Basandrai, 2021).

During the parasitic phase, *M. phaseolina* invades the living root tissues. In the advances of disease progresses, the fungus enters a saprophytic phase where host cells breakdown, and the fungus decomposes the host tissue for nutrition. This condition in host results in necrosis, and the black discoloration appears in host roots (Gupta et al., 2012). The host tissue necrosis ultimately causes root tissue fragility and insufficient nutrient and water uptake, resulting in plant growth stop and declined yield. Penetrating hyphae can completely block root channels in severely infected plants, limiting nutrition and water intake and eventually leading to plant death (Lodha & Mawar, 2020).

M. phaseolina causes diseases between 30 and 35°C temperatures with 60% below soil moisture and significant production losses in sorghum and soybean crops. Similarly, in the case of groundnut, losses occur at early infestation of the pathogen. Infection of two or more phytopathogens in the same host cause more severity than single infection (Bashir, 2017; Cohen et al., 2016). Tremendous range of plant species occurs; among more than 500 and families approximately 100 are infected by *M. phaseolina* including major green gram, mung bean, groundnut, cowpea, urdbean pulse crops, maize and sorghum food crops, and sesame, sunflower, soybean, and castor oil crops, tomato, potato, onion, and garlic vegetable crops, tomato, potato, onion, and garlic fiber crops and cotton (Degani et al., 2022; Lamini et al., 2020), and strawberry (Gomez et al., 2020).

M. phaseolina spreads through roots and generally symptoms appear later in the season, around the flowering stage, and are aggravate by dryness. Through their mode of work, it comes to light that pathogen obstructs water supply as the growing season develops, causing dehydration and yield decrease. *M. phaseolina* produces phytotoxins, which enhance its pathogenicity. Its control becomes difficult because of high survival rate in the soil and host tissues (Pandey & Basandrai, 2021).

The pathogen has ability to infect and colonizing in the seed coat, endosperm, and embryo. The pathogen's seed-borne propagules attack urdbean and mungbean seeds, resulting in loss in germination of seed and in their viability (Sarita & Singh, 2014). The infection causes mungbean seeds deterioration in the storage condition; degeneration of 2%–36% has been documented in stored mungbean from several South Asian countries where its consumption is high, including India, Pakistan, Bangladesh, and Thailand. Likewise, *M. phaseolina* significantly lowers seed germination of mungbean and results in 12.3% protein loss. The pathogen also affects the cotyledons during seedling emergence of mungbean (Ashwini & Giri, 2014). Disease caused by *M. phaseolina* in different crops mentioned in Table 9.1

9.4 *M. Phaseolina* enzymes and their role in pathogenicity

M. phaseolina causes severe damage due to pathogenicity, which is governed by the abundance amount of enzymes directing the degradation of plant cell wall. The most common enzymes are polygalacturonase and polymethylgalacturonase that belong to pectinases group, endoglucanase belongs to cellulases, endoxylanase belongs to hemicellulases group, and others like xylanases, laccases, and the ligninolytic enzyme laccase are also involved in the degradation.These enzymes of *M. phaseolina* can be used for bioconversion process (Ghosh et al., 2018; Ramos et al., 2016; Tonukari, 2003). As soon as *M. phaseolina* enters the plant cell wall, it releases an extensive range of plant cell wall–degrading enzymes that depolymerize the plant cell wall and penetrate to utilize the nutrients from plant tissues, which are the basic requirement of successful infection (Javaid & Saddique, 2012). Pectinase is an enzyme that collapses pectin from the substrate for the use of carbon in the host cell. At the same time, endo- and exo-polygalacturonase enzymes are synthesized during the early stages of pathogenesis and intrusion of host tissues (Murad & Azzaz, 2011). Lipase enzymes of *M. phaseolina* are responsible for the hydrolysis of lipids, mono- and diglycerides (Kakde & Chavan, 2011).

TABLE 9.1 Disease caused by *M. phaseolina* in different crops.

Crop	Disease	References
Soybean	Charcoal Rot	Ghosh et al. (2018)
Strawberry	Charcoal Rot	Peres et al. (2018); Gomez et al. (2020); Nelson et al. (2021)
Sesame	Charcoal Rot	Ibrahim and Abdel-Azeem (2015); Marquez et al. (2021)
Sorghum	Stalk Rot	Madhusudhana (2019)
Cowpea	Charcoal Rot	Sales Junior et al. (2020)
Cowpea	Root Rot	Lamini et al. (2020)
Castor	Root Rot	Verma et al. (2018)
Castor	Charcoal Rot	Claudino and Soares (2014)
Groundnut	Charcoal Rot	Marquez et al. (2021)
Sunflower	Charcoal Rot	Siddique et al. (2021)

In order to enter the host tissue, *M. phaseolina* has an abundance of hydrolytic enzymes that can degrade the polysaccharides and lignocelluloses found in the cell wall. Hydrolytic enzymes play an essential role in plant pathogenicity by ease the tissue colonization in the host (Bandara et al., 2018; Marquez et al., 2021). In addition, it produces toxins, such as phaseolinone and botryodiplodin, which assist in spreading the disease (Bressano et al., 2010; Ramezani, 2008). In some cases, it has been suggested that the synthesis of hydrolytic enzymes is essential for expanding various infections (Kaur et al., 2012b) (Fig. 9.3).

9.4.1 Enzymes produced by *M. phaseolina*

A combination of hydrolytic enzymes, as carbohydrate-active enzymes (CAZymes), is secreted by pathogenic fungi, responsible for destroying the cell wall, while entering into the host tissue for disease development. According to Islam et al. (2012), studies on *M. phaseolina* transcriptome analysis revealed that it encodes 362 putative CAZymes, including glycoside hydrolases (GH), glycosyltransferases (GT), carbohydrate esterases (CE), carbohydrate binding modules (CBM), and polysaccharide lyases (PL) distributed on the whole genome. These carbohydrate esterases are very important in host invasion and pathogenesis. In *M.* phaseolina, GHs are more abundant than GTs and CEs. CEs consist of sterol esterase, chitin, and cutinases required to enter into the host initially (Islam et al., 2012). It has been demonstrated that the cellulolytic activity of *M. phaseolina* is substantially higher than other fungal species, which indicates its high pathogenic potential (Kaur et al., 2012a).

The lignin depolymerization mechanism in *M. phaseolina* comprises a number of enzymes, including laccases, galactose oxidases, lignin peroxidases, haloperoxidases, chloroperoxidases, and heme peroxidases; all of which are essential components of the system (Ghosh et al., 2018; Himo et al., 2000). Seven halo peroxidases, six chloroperoxidases, and six extracellular class II heme peroxidases altogether contribute to the ligninolytic action of the enzymes. Furthermore, a considerable number of enzyme related to glucose-methanol-choline oxidoreductase have been identified, which include alcohol oxidases and cellobiose dehydrogenases for disease development (Martinez et al., 2004). These enzymes are directly involved in the breakdown of lignocelluloses. In addition, *M. phaseolina* synthesizes various phytotoxic metabolites, including phaseolin, patulin, and botryodiplodin, which participate in the initial phase of the infection and cause seed wilt and necrotic lesions on plant parts especially upper part on leaves and lower parts on roots (Abbas et al., 2020; Ramezani et al., 2007; Salvatore et al., 2020). The high fungal tolerance has variation due to range in ecological circumstances, which contributes its dissemination and infection in plant (Islam et al., 2012). This results in more virulence effect in *M. phaseolina* leading to infect wide range of hosts and tissues with high efficacy and precision. List of enzymes produced by *M. phaseolina* with their act of mechanism. mentioned in Table 9.2

9.5 Molecular techniques to identify the pathogen specific genes

Researchers are much interested in gathering the information of its genes from very long time due to devastating nature of *M. phaseolina* fungus. Earlier markers, probes and other molecular diagnostic techniques are utilized for diagnosis and early detection of *M. phaseolina*. Further, these techniques cannot be utilized to correlate the relationships and genetically variations among the populations of *M. phaseolina*.Furthermore, the emergence of new technologies, such as qPCR detect and quantify pathogen and their gene expression; genomics revealed the genes related to invasion into the host and pathogenicity (Islam et al., 2012), proteomic data narrates the protein expression (Zaman et al., 2020), transcriptome reveals the insight of genome (Verma et al., 2018) and metabolic analysis of metabolites induce the disease establishment (Salvatore et al., 2020).

9.5.1 Molecular approach for identification of pathogenicity related genes

Plant disease management necessitates precise detection and diagnosis of pathogen at very early stage. In addition, peculiar primers and probes, in vitro technology, such as the polymerase chain reaction and in situ hybridization technology, are applied to detect *M. phaseolina* in a short duration (Ghosh et al., 2018; Santos et al., 2020).

9.5.1.1 Markers and probes

Genetically, *M. phaseolina* has variation in their nature so it varies in its different isolates, which were identified by molecule markers and probes18S rRNA, ITS-1, ITS-2, and rDNA. The findings based on RAPD markers and 18SrRNA narrated the genetic variation of isolates collected from different sources. Isolates of *M. phaseolina* were identified by 18SrRNA gene sequencing and the target fragment sequenced and followed by BLAST for annotation (Brandfass & Karlovsky, 2008). Another research works on six OPA series of RAPD marker utilized for variation detection among the

TABLE 9.2 List of enzymes produced by *M. phaseolina* with their act of mechanism.

Name of enzymes	Mechanism	References
Carbohydrate-active enzymes (cazymes)		
Glycoside hydrolases (GH)	Hydrolyze glycosidic bonds	Graebin et al. (2016)
Carbohydrate esterases (CE)	Catalyze de-O or de-N-acylation by eliminating ester modification	Nakamura et al. (2017)
Glycosyltransferases (GT)	Catalyze the formation of glycosidic bonds utilizing sugar donors with a lipid phosphate or nucleoside phosphate leaving group.	Lairson et al. (2008)
Polysaccharide lyases (PL)	Create an unsaturated hexenuronic acid residue and a new reducing end from uronic acid-containing polysaccharide chains via a β-elimination mechanism	Michaud et al. (2003)
carbohydrate-binding modules (CBM)	Targeting added catalytic modules to substrate and concentrating appended catalytic modules on the surface of substrate	Abbott and Boraston (2012)
Lignin depolymerization		
Laccases	Catalyze the polymerization or depolymerization of the lignin model compounds by the abstraction of one electron from the phenolic hydroxyl group	Youn et al. (1995)
Galactose oxidases	Catalyzes the oxidation of many primary alcohols to their respective aldehydes, as well as the reduction of molecular oxygen to hydrogen peroxide	Himo et al. (2000)
Lignin peroxidases	Oxidises lignin using hydrogen peroxide as a catalyst	Falade et al. (2017)
Haloperoxidases	In the presence of halide ions and peroxides, catalyzes the halogenation of organic compounds	Franssen (1994)
Heme peroxidases	Catalyze the oxidation of a range of substrates using hydrogen peroxide as a catalyst	Raven (2013)

M. phaseolina (Prabhu et al., 2012). In the sequence of markers study, 18 RAPD and 15 ISSR primers were used for amplification of *M. phaseolina* DNA (Yousef, 2021). Other than restricted fragment length polymorphism [RAPD] marker, amplified restriction fragment length polymorphism) molecular markers were also used on *M. phaseolina* DNA used to evaluate genetic diversity among the *M. phaseolina* (Vandemark et al., 2020).

9.5.1.2 Genomic approach

Genomic approaches are used to identify the structural changes generally associated with a genotype. In the genome analysis, transcriptome analysis is used to identify a locus of *M. phaseolina* infected on strawberry (Burkhardt et al., 2018). Other researchers have revealed the pathogenic nature of genes existing in the genome through transcriptome analysis of *M. phaseolina* (Islam et al., 2012). He edited the complete genome of this fungus first ever and provided information related to genes in the recent decade. Use of wide range of plants enzymatic and toxin tools in the study of investigation how plants become infected at the cytological and molecular levels, the *M. phaseolina* genomes included isolates from jute, strawberry, alfalfa, and sorghum, using different molecular techniques (Purushotham et al., 2020), and the genomics helps us to understand the pathogenic specific genes and their regulatory network that can be further studied by using PCR-based techniques. RNA seq analysis of microsclerotia of *M. phaseolina* reveals that the ROS related genes, such as glutathione S-transferase, glutaredoxin, catalase, superoxide dismutase I, NADH ubiquinone oxidoreductase, and lignin peroxidase present at different stage of microsclerotia development (Falade et al., 2017). The NADPH oxidase showed highest expression at hyphal stage. On other side, antioxidant-related genes have negative impact on the sclerotia growth. The cytochrome *c* related transcript exhibits the highest expression at initial phase of sclerotia development (Liu et al., 2022).

9.5.1.3 Quantitative PCR

Real-time PCR technique used on *M. phaseolina* nucleic acids to detect and quantify its abundance in the host cell. Both TaqMan probes and SYBR dye are used for the detection (Babu et al., 2011). TaqMan assay has been done on *M.*

phaseolina infected on strawberry, and the same locus was used to detect also from recombinase polymerase amplification (RPA) assay with the modified primers and probes sequence (Burkhardt et al., 2018). Same qPCR technique was used on *M. phaseolina* infected in cotton crop for detection (Degani et al., 2022).

9.5.1.4 Proteomic approach

Proteomics is a very powerful technique in identifying the protein and gene function analysis because protein formation participates very importantly in disease development. Sadat et al. (2021) revealed the F box of *M. phaseolina*. Proteins related to F box are the important components in controlling the protein degradation and its consumption in the disease development process and their molecular mechanism. In *M. phaseolina,* 31 F box proteins were identified throughout the genome during the research. *M. phaseolina* whole proteome was studied by using the quantitative proteomics technique iTRAQ LC-MS/MS, which reveals the activity related to carboxylic metabolic process, biological related process, transport machinery and protein metabolic process (Zaman et al., 2020). The secondary metabolite assists in the disease establishment in *M. phaseolina* because during growth host tissues presence at the time (Salvatore et al., 2020).

9.5.1.5 CRISPER/CAS

CRISPER technology can be adapted to engineer fungal genome. Utilization of CRISPER/CAS technology assists in exploring the basic fungal biology, which can be utilized for screening purposes as well as characterization of fungal virulence related genes which enhance the opportunity for antifungal therapy (Krappmann, 2017; Dort et al., 2020).

9.5.1.6 Secretome analysis

In secretome analysis, secretory protein was isolated and analyzed on SDS-PAGE followed by LC-ESI-MS/MS for protein identification. These *M. phaseolina* secretome proteins are analyzed through bioinformatics tools. These secretome protein analysis based on whole genome sequencing and proteome analysis of secretome of pathogen confirms the pathogenicity of the virulence genes of *M. phaseolina* (Sinha et al., 2022).

9.6 *M. phaseolina* genes and their role in pathogenicity

Among RAPD-PCR amplification profiles, primer series of OPA-13 amplified all the isolates of *M. phaseolina* genomic classified into nine major groups with 64% similarity. In *in vitro* condition, the pathogenicity of isolates was evaluated at seedling stage in various crops, such as soybean, sunflower, and maize plants. These isolates have different effect on different crop as they are more virulent on soybean than sunflower and no pathogenic effect on corn. These results indicated a considerable variability within the Iranian isolates of *M. phaseolina* with the aspect of pathogenicity and genetics (Siavosh et al., 2012).

The *M. phaseolina* possesses the abundant protein related to, glycosidase, cytochrome P450, and secondary metabolite and slighter proteases. Due to these proteins and metabolites, *M. phaseolina* had capability to vanquish the defense response of the host plant. *M. phaseolina* consists of major components of the lignin degradation including laccases, chloroperoxidases, haloperoxidases, lignin peroxidases, galactose oxidases, and heme peroxidases (Martinez et al., 2004). Several pathogenicity-associated genes occur in *M. phaseolina* genome, and these genes play a role in the fungus pathogenesis. Cellulose-binding elicitor lectin and transglutaminase-like proteins utilized for attachments and adherence to host tissue; this diverse gene assemblage neutralize the initial host defense of *M. phaseolina* with the help of salicylate-1-monooxygenase and penetrates and invades plant external most layer called the epidermis (Islam et al., 2012).

M. phaseolina is prominent for its abundance of hydrolytic enzymes capable of degrading polysaccharides and lignocelluloses of cell wall, which helps the host cell invasion. The gene category in fungus includes different exocellobiohydrolases, endoglucanases, glucosidases hydrolysis of cellulose, and laccases, galactose oxidases, lignin, chloro, halo, and heme peroxidases utilize for degradation of lignin. Furthermore, the highest number of laccases and cellulolytic enzymes occurs in *M. phaseolina* in comparison to other fungal genomes (Islam et al., 2012).

It is believed that *M. phaseolina* secretes three toxic metabolites, phaseolinone, botryodiplodin, and patulin that contribute in the early stages of infection resulting in seedling stage wilting and necrotic lesions on the plants leaves and root. In order to counteract the plant defense response, cytochrome P450, C of, and superoxide dismutase genes are expressed after the interaction of host and *M. phaseolina* (Ghosh et al., 2018). This way increases virulence to infect wide range of host's tissues. The fungal ability to thrive in a wide range of environments is also a factor in its widespread distribution and ability to infect plants (Salvatore et al., 2020).

M. phaseolina consists of sufficient number of pathogen–host interaction genes almost 537 PHI genes identified (Winnenburg et al., 2008). They are related to signal transduction, cell wall degradation, adhesion, biosynthesis of purine, and mycotoxin patulin; ABC transporters assist pathogen from host. Apart from the pathogenesis, genes detoxifying enzymes are present to assist *M. phaseolina* pathogenesis that include cytochrome P450, superoxide dismutase, Cu/Zn binding, and beta ketoacyl synthases (Morschhauser, 2010). Signals are very important for the pathogenicity in *M. phaseolina* genome, which consists of 44 putative PTH11 like GPCRs protein kinases for signaling the fungal development and infection process (Li et al., 2007). *M. phaseolina* genome encodes 839 transporter genes, which comprise of 106 families of transporter genes. Pathogenic fungi survival and establishment depend on host; therefore, 782 transporter genes are related to pathogen–host interaction followed by major facilitator superfamily (MFS) 270 transporter genes, ABC superfamily have 59 transporter genes, amino acid polyamine organocation (APC) possess 40 transporter genes and 54 transporter genes related to the amino acid. Transporter genes related to the membrane may be required for the host–pathogen interaction; amino acid transporter requires accessing the protein degradation products from the host. ABC superfamily transporters act as transporters as well as receptors to utilize the energy generated by ATP hydrolysis (Islam et al., 2012). *M. phaseolina* whole proteome was studied by using the quantitative proteomics technique iTRAQ LC-MS/MS. In that, 2204 different protein were identified with 137 of them being found to be differentially regulated. Among those, relative higher number of enzymes belongs to hydrolases, cellular metabolic process, oxidoreductase, and transferase (Zaman et al., 2020). In KEGG, pathways carbohydrate metabolism (26.11%) and amino acid metabolism pathways (19.91%) results showing similarity as in the transcriptomics result (Zaman et al., 2020). In genome sequencing analysis of *M. phaseolina,* susceptible castor genotype 48-1 several genes were identified related to the pathogenicity of *M. phaseolina*. To reveal the function of genes, we performed the GO analysis under the sub-category of GO analysis in biological process; the maximum genes expressed are related to response to stress, which is required for pathogenic fungi for the survival in the hostile environments encountered in the host followed by biosynthetic process, which deepenes our understanding of the different process and reveals bioactive compound and mechanism adopted by *M. phaseolina*. In cellular component, the highest numbers of genes are related to cytosol followed by plasma membrane. Host-mediating infection senses after the entry of the pathogen through pathogen carry associated molecular patterns (PAMPs) into the cytosol or damage induction, which causes the cytosolic liberation of danger associated molecular patterns (DAMPs) in the host cell. Through different mechanisms, microbes induce inflammasome and in the functional consequence, inflammasome activation in infectious disease. The wall is the external cellular component that plays important role in the interaction with host cell. The cell membrane of pathogenic microorganism carries important antigenic determination and adhesion to the host tissues, essential for the colonization, and causes disease. In biological process, ion binding genes are in abundance; it shows that ion binding plays an important role in fungal pathogenesis (Verma et al., 2018).

9.7 Conclusion

Knowledge of pathogenic genes of *M. phaseolina* assists in the modification in the fungus as well as in plants against the harmful pathogenic effect of *M. phaseolina* genes. Modification in fungus is little difficult, whereas genetically modification in plants against the fungi is very practical and easy. As soon as fungus attacks, crops' immune system gets activated, but it may not be strong enough to cope with the infection. To enhance the protection capacity of plant, modern technology such as genetic engineering and CRISPER-CAS can be implemented for more yield, production and reduce crop loss.

References

Abbas, H. K., Bellaloui, N., Butler, A. M., Nelson, J. L., Abou-Karam, M., & Shier, W. T. (2020). Phytotoxic responses of soybean (*Glycine max* L.) to botryodiplodin, a toxin produced by the charcoal rot disease fungus, *Macrophominaphaseolina*. *Toxins, 12*(1), 25. https://doi.org/10.3390/toxins12010025

Abbott, D. W., & Boraston, A. B. (2012). Quantitative approaches to the analysis of carbohydrate-binding module function. *Methods in Enzymology, 510*, 211–231.

Almomani, F., Alhawatema, M., & Hameed, K. (2013). Detection, identification and morphological characteristic of *Macrophominaphaseolina*: The charcoal rot disease pathogens isolated from infected plants in northern Jordan. *Archives of Phytopathology and Plant Protection, 46*(9), 1005–1014.

Ashwini, C., & Giri, G. (2014). Control of seed borne fungi in green gram and black gram through bio-agents. *International Journal of Applied Biology and Pharmaceutical Technology, 5*, 68–170.

Babu, B. K., Mesapogu, S., Sharma, A., Somasani, S. R., & Arora, D. K. (2011). Quantitative real-time PCR assay for rapid detection of plant and human pathogenic *Macrophominaphaseolina*from field and environmental samples. *Mycologia, 103*(3), 466–473. https://doi.org/10.3852/10-18

Bandara, Y. M. A. Y., Weerasooriya, D. K., Liu, S., & Little, C. R. (2018). The necrotrophic fungus *Macrophominaphaseolina* promotes charcoal rot susceptibility in grain sorghum through induced host cell-wall-degrading enzymes. *Phytopathology, 108*, 948–956. https://doi.org/10.1094/PHYTO-12-17-0404-R

Bashir, M. R. (2017). Impact of global climate change on charcoal rot of sesame caused by *Macrophominaphaseolina*. *Journal of Horticulture, 4*(1).

Brandfass, C., & Karlovsky, P. (2008). Upscaled CTAB-based DNA extraction and real-time PCR assays for *Fusariumculmorum* and *F. graminearum* DNA in plant material with reduced sampling error. *International Journal of Molecular Sciences, 9*, 2306–2321. https://doi.org/10.3390/ijms9112306

Bressano, M., Giachero, M. L., Luna, C. M., & Ducasse, D. A. (2010). An in vitro method for examining infection of soybean roots by *Macrophominaphaseolina*. *Physiological and Molecular Plant Pathology, 74*(3–4), 201–204. https://doi.org/10.1016/j.pmpp.2009.12.003

Burkhardt, A., Ramon, M. L., Smith, B., Koike, S. T., & Martin, F. (2018). Development of molecular methods to detect *Macrophominaphaseolina* from strawberry plants and soil. *Phytopathology, 108*, 1386–1394.

Claudino, M. R., & Soares, D. J. (2014). Pathogenicity and aggressiveness of *Macrophominaphaseolina* isolates to castor (*ricinuscommunis*). *Tropical Plant Pathology, 39*, 453–456.

Cohen, R., Elkabetz, M., & Edelstein, M. (2016). Variation in the responses of melon and watermelon to *Macrophominaphaseolina*. *Crop Protection, 85*, 46–51.

Degani, O., Becher, P., & Gordani, A. (2022). Pathogenic interactions between *Macrophominaphaseolina* and *Magnaporthiopsismaydis* in mutually infected cotton sprouts. *Agriculture, 12*(2), 255.

Dort, E. N., Tanguay, P., & Hamelin, R. C. (2020). CRISPR/Cas9 gene editing: An unexplored frontier for forest pathology. *Frontiorin Plant Science, 11*, 1126. https://doi.org/10.3389/fpls.2020.01126

Falade, A. O., Nwodo, U. U., Iweriebor, B. C., Green, E., Mabinya, L. V., & Okoh, A. I. (2017). Lignin peroxidase functionalities and prospective applications. *Microbiology Open, 6*(1), e00394.

Franssen, M. C. R. (1994). Haloperoxidases: Useful catalysts for halogenation and oxidation reactions. *Catalysis Today, 22*(3), 441–457.

Ghosh, T., Biswas, M. K., Guin, C., & Roy, P. (2018). A review on characterization, therapeutic approaches and pathogenesis of *Macrophominaphaseolina*. *Plant Cell Biotechnology and Molecular Biology, 19*, 72–84.

Gomez, A. O., De Faveri, J., Neal, J. M., Aitken, E. A., & Herrington, M. E. (2020). Response of strawberry cultivars inoculated with *Macrophominaphaseolina* in Australia. *International Journal of Fruit Science, 20*(Suppl. 2), 164–177.

Graebin, N. G., Schöffer, J. D. N., Andrades, D. D., Hertz, P. F., Ayub, M. A., & Rodrigues, R. C. (2016). Immobilization of glycoside hydrolase families GH1, GH13, and GH70: State of the art and perspectives. *Molecules, 21*(8), 1074.

Gupta, G. K., Sharma, S. K., & Ramteke, R. (2012). Biology, epidemiology and management of the pathogenic fungus *Macrophominaphaseolina* (Tassi) Goid. with special reference to charcoal rot of soybean (*Glycine max* (L.) Merrill). *Journal of Phytopathology, 160*(4), 67–180.

Himo, F., Eriksson, L. A., Maseras, F., & Siegbahn, P. E. (2000). Catalytic mechanism of galactose oxidase: A theoretical study. *Journal of the American Chemical Society, 122*(33), 8031–8036.

Ibrahim, M. E., & Abdel-Azeem, A. M. (2015). Management of sesame (*Sesamumindicum* L.) charcoal rot caused by *Macrophominaphaseolina* (Tassi) Goid. through the application of different control measure. *Journal of Pure and Applied Microbiology, 9*(1), 141–149.

Iqbal, U., & Mukhtar, T. (2014). Morphological and pathogenic variability among *Macrophominaphaseolina* isolates associated with Mungbean (*Vignaradiata* L.) Wilczek from Pakistan. *The Scientific World Journal, 950175*, 1–9.

Islam, M., Haque, M., Islam, M., Emdaa, E., Halim, A., Hossen, Q. M., Hossain, M. Z., Ahmed, B., Rahim, S., Rahman, M. S., Alam, M. M., Hou, S., Wan, X., Saito, J. A., & Alam, M. (2012). Tools to kill: Genome of one of the most destructive plant pathogenic fungi Macrophominaphaseolina. *BMC Genomics, 13*, 493.

Javaid, A., & Saddique, A. (2012). Control of charcoal rot fungus *Macrophominaphaseolina* by extracts of daturametel. *Natural Product Research, 26*, 1715–1720. https://doi.org/10.1080/14786419.2011.605363

Kakde, R. B., & Chavan, A. M. (2011). Extracellular lipase enzyme production by seed-borne fungi under the influence of physical factors. *International Journal of Biology, 3*, 94.

Kaur, S., Dhillon, G. S., Brar, S. K., & Chauhan, V. B. (2012a). Carbohydrate degrading enzyme production by plant pathogenic mycelia and microsclerotia isolates of *Macrophomina phaseolina* through koji fermentation. *Industrial Crops and Products, 36*, 140–148. https://doi.org/10.1016/j.indcrop.2011.08.020

Kaur, S., Dhillon, G. S., Brar, S. K., Vallad, G. E., Chand, R., & Chauhan, V. B. (2012b). Emerging phytopathogen *Macrophomina phaseolina*: Biology, economic importance and current diagnostic trends. *Critical Review in Microbiology, 38*, 136–151. https://doi.org/10.3109/1040841X.2011.640977

Khan, A. N., Shair, F., Malik, K., Hayat, Z., Khan, M. A., Hafeez, F. Y., & Hassan, M. N. (2017). Molecular identification and genetic characterization of *Macrophominaphaseolina* strains causing pathogenicity on sunflower and chickpea. *Frontiers in Microbiology, 8*, 309.

Krappmann, S. (2017). CRISPR-Cas9, the new kid on the block of fungal molecular biology. *Medical Mycology, 55*(1), 16–23. https://doi.org/10.1093/mmy/myw097

Lairson, L. L., Henrissat, B., Davies, G. J., & Withers, S. G. (2008). Glycosyltransferases: Structures, functions, and mechanisms. *Annual Review of Biochemistry, 77*, 521–555.

Lakhran, L., Ahir, R. R., Choudhary, M., & Choudhary, S. (2018). Isolation, purification, identification and pathogenicity of *Macrophominaphaseolina* (Tassi) goid caused dry root rot of chickpea. *Journal of Pharmacognosy and Phytochemistry, 7*(3), 3314–3317.

Lamini, S., Cornelius, E. W., Kusi, F., Danquah, A., Attamah, P., Mukhtaru, Z., Awuku, J. F., & Mensah, G. (2020). Prevalence, incidence and severity of a new root rot disease of cowpea caused by *Macrophominaphaseolina* (Tassi) goid in Northern Ghana. *West African Journal of Applied Ecology, 28*(2), 140–154.

Liu, H. H., Huang, C. C., Lin, Y. H., Tseng, M. N., & Chang, H. X. (2022). Superoxide initiates the hyphal differentiation to microsclerotia formation of Macrophominaphaseolina. *Microbiology Spectrum, 10*(1). https://doi.org/10.1128/spectrum.02084-21

Li, L., Wright, S. J., Krystofova, S., Park, G., & Borkovich, K. A. (2007). Heterotrimeric G protein signaling in filamentous fungi. *Annual Review of Microbiology, 61*, 423–452.

Lodha, S., & Mawar, R. (2020). Population dynamics of *Macrophominaphaseolina* in relation to disease management: A review. *Journal of Phytopathology, 168*(1), 1–17.

Madhusudhana, R. (2019). Marker-assisted breeding in sorghum. In C. Aruna, K. B. R. S. Visarada, B. V. Bhat, & V. A. Tonapi (Eds.), *Breeding sorghum for diverse end uses* (pp. 93–114). Woodhead Publishing series in Food Science, Technology and Nutrition.

Marquez, N., Giachero, M. L., Declerck, S., & Ducasse, D. A. (2021). *Macrophominaphaseolina:* General characteristics of pathogenicity and methods of control. *Frontiers of Plant Science, 12*, 634397. https://doi.org/10.3389/fpls.2021.634397

Martinez, D., Larrondo, L. F., Putnam, N., Gelpke, M. D., Huang, K., Chapman, J., Helfenbein, K. G., Ramaiya, P., Detter, J. C., Larimer, F., Coutinho, P. M., Henrissat, B., Berka, R., Cullen, D., & Rokhsar, D. (2004). Genome sequence of the lignocellulose degrading fungus *Phanerochaetechrysosporium* strain RP78. *Nature Biotechnology, 22*, 695–700.

Michaud, P., Da Costa, A., Courtois, B., & Courtois, J. (2003). Polysaccharide lyases: Recent developments as biotechnological tools. *Critical Reviews in Biotechnology, 23*(4), 233–266.

Morschhauser, J. (2010). Regulation of multidrug resistance in pathogenic fungi. *Fungal Genetics and Biology, 47*, 94–106.

Murad, H., & Azzaz, H. (2011). Microbial pectinases and ruminant nutrition. *Research Journal of Microbiology, 6*, 246–269. https://doi.org/10.3923/jm.2011.246.269

Nakamura, A. M., Nascimento, A. S., & Polikarpov, I. (2017). Structural diversity of carbohydrate esterases. *Biotechnology Research and Innovation, 1*(1), 35–51.

Nelson, J. R., Verma, S., Bassil, N. V., Finn, C. E., Hancock, J. F., Cole, G. S., Knapp, S. J., & Whitaker, V. M. (2021). Discovery of three loci increasing resistance to charcoal rot caused by *Macrophominaphaseolina* in octoploid strawberry. *G3(Bethesda), 11*(3), 37.

Pandey, A. K., & Bansadrai, A. K. (2021). Will *Macrophominaphaseolina* spread in legumes due to climate change? A critical review of current knowledge. *Journal of Plant Diseases and Protection, 128*, 9–18.

Pandey, A. K., Burlakoti, R. R., Rathore, A., & Nair, R. M. (2020). Morphological and molecular characterization of *Macrophominaphaseolina* isolated from three legume crops and evaluation of Mungbean genotypes for resistance to dry root rot. *Crop Protection, 127*, 104962.

Peres, N. A., Baggio, J. S., & Mertely, J. C. (2018). Charcoal rot of strawberries caused by Macrophominaphaseolina. In *Plant pathology* (p. 33598). Wimauma, Florida: Department, UF/IFAS Extension, Gulf Coast Research and Education Center.

Prabhu, H. V., Adiver, S., Bhat, R., Narayana, Y., Jahagirdar, S., & Parameshwarappa, K. (2012). Genetic variability in *Macrophominaphaseolina* (Tassi.) goid., causal agent of charcoal rot of sorghum. *Karnataka Journal of Agricultural Science, 25*, 72–76.

Purushotham, N., Jones, A., Poudel, B., Nasim, J., Adorada, D., Sparks, A., Schwessinger, B., & Vaghefi, N. (2020). Draftgenome resource for *Macrophominaphaseolina* associated with charcoal rot in sorghum. *MolecularPlant-Microbe Interaction, 33*, 724–726. https://doi.org/10.1094/MPMI-12-19-0356-A

Ramezani, H. (2008). Biological control of root-rot of eggplant caused by *Macrophominaphaseolina. American-Eurasian Journal of Agricultural and Environmental Sciences, Dubai, 4*, 218–220.

Ramezani, M., Shier, W. T., Abbas, H. K., Tonos, J. L., Baird, R. E., & Sciumbato, G. L. (2007). Soybean charcoal rot disease fungus *Macrophominaphaseolina* in Mississippi produces the phytotoxin (−)-botryodiplodin but no detectable phaseolinone. *Journal of Natural Products, 70*(1), 128–129.

Ramos, A. M., Gally, M., Szapiro, G., Itzcovich, T., Carabajal, M., & Levin, L. (2016). *In vitro* growth and cell wall degrading enzyme production by Argentinean isolates of *Macrophominaphaseolina*, the causative agent of charcoal rot in corn. *Revista Argentina de Microbiología, 48*(4), 267–273.

Raven, E. L. (2013). Heme peroxidases. In G. C. K. Roberts (Ed.), *Encyclopedia of biophysics* (pp. 1–73). Berlin, Heidelberg: Springer. https://doi.org/10.1007/978-3-642-16712-6_52

Sadat, M. A., Ullah, M. W., Bashar, K. K., Hossen, Q. M. M., & Islam, S. (2021). Genome-wide identification of F-box proteins in *Macrophominaphaseolina* and comparison with other fungus. *Journal of Genetic Engineering and Biotechnology, 19*(46), 1–14.

Sales Junior, R., Silva Neto, A. N. D., Negreiros, A. M. P., Gomes, T. R. R., Ambrósio, M. M. D. Q., & Armengol, J. (2020). Pathogenicity of Macrophomina species collected from weeds in Cowpea. *RevistaCaatinga, 33*, 395–401.

Salvatore, M. M., Felix, C., Lima, F., Ferreira, V., Naviglio, D., Salvatore, F., Duarte, A. S., Alves, A., Andolfi, A, & Esteves, A. C. (2020). Secondary metabolites produces by *Macrophominaphaseolina* isolate from *eucalyptus globulus. Agriculture, 10*, 72. https://doi.org/10.3390/agriculture10030072

Sanchez, S., Grez, J., Contreras, E., Gil, P. M., & Gambardella, M. (2018). Physiological response and susceptibility of strawberry cultivars to the charcoal rot caused by *Macrophominaphaseolina* under drought stress condition. *Journal of Berry Research.* https://doi.org/10.3233/JBR-180329

Santos, K. M., Lima, G. S., Barros, A. P. O., Machado, A. R., Souza-Motta, C. M., Correia, K. C., & Michereff, S. J. (2020). Novel specific primers for rapid identification of *Macrophomina* species. *European Journal of Plant Pathology, 156*, 1213–1218.

Sarita, B. A. K., & Singh, R. (2014). Study of seed borne mycoflora of mungbean treated with potassium nitrate during storage. *Journal of Advanced and Applied Science Research, 5*, 11–13.

Sarr, M., Ndiaye, M., Groenewald, J., & Crous, P. (2014). Genetic diversity in *Macrophominaphaseolina*, the causal agent of charcoal rot. *Phytopathology, 53*, 250–268. https://doi.org/10.14601/Phytopathol_Mediterr 13736

Siavosh, R., Siranoush, G. N., Seyed, V. A., Mohammad, R., & Abbas, G. M. (2012). Pathogenic and genetic diversity among Iranian isolated of *Macrophominaphaseolina. Chilean Journal of Agricultural Research, 72*(1), 40–44.

Siddique, S., Shoaib, A., Khan, S. N., & Mohy-Ud-Din, A. (2021). Screening and histopathological characterization of sunflower germplasm for resistance to *Macrophominaphaseolina. Mycologia, 113*(1), 92–107.

Sinha, N., Patra, S. K., & Ghosh, S. (2022). Secretome analysis of *Macrophominaphaseolina*identifies an array of putative virulence factors responsible for charcoal rot disease in plants. *Frontiers in Microbiology, 13*, 847832. https://doi.org/10.3389/fmicb.2022.847832

Srivastava, A. K., Singh, T., Jana, T. K., & Arora, D. K. (2001). Microbial colonization of *Macrophominaphaseolina* and suppression of charcoal rot of chickpea. In A. Sinha (Ed.), *Microbes and plants* (pp. 269–319). New Delhi, India: Vedam eBooks.

Tok, F. M. (2019). Relationship between morphologic, phenotypic and pathogenic characteristics in *Macrophominaphaseliona* Isolates from cucumber plants. *International Journal of Innovative Approaches in Agricultural Research, 3*, 651–660.

Tonukari, N. J. (2003). Enzymes and fungal virulence. *Journal of Applied Sciences & Environmental Management, 7*(1), 8.

Vandemark, G., Martinez, O., Pecina, V., & Alvarado, M. (2020). Assessment of genetic relationships among isolated of *Macrophominaphaseolina*using a simplified AFLP techniques and two different methods of analysis. *Genetics and Molecular Biology, 92*, 656–664.

Verma, S., Tomar, R. S., Rathode, V., Thakker, J., Shubham, Bhagwaat, N., Raval, S., Antala, T., Jogia, Z., & Golakiya, B. A. (2018). Genome sequencing analysis of *Macrophominaphaseolina* resistant and susceptible Castor genotype. *Bioscience Biotechnology Research Asia, 15*(1), 195–215.

Winnenburg, R., Urban, M., Beacham, A., Baldwin, T. K., Holland, S., Lindeberg, M., Hansen, H., Rawlings, C., Hammond-Kasack, K. E., & Kohler, J. (2008). PHI-Base update: Additions to the pathogen host interaction database. *Nucleic Acids Research, 36*, 572–576.

Youn, H. D., Hah, Y. C., & Kang, S. O. (1995). Role of laccase in lignin degradation by white-rot fungi. *FEMS Microbiology Letters, 132*(3), 183–188.

Yousef, H. (2021). Pathogenic variation and molecular characterization of *Macrophominaphaseolina*, the cause of sesame charcoal rot. *Egyptian Journal of Phytopathology, 49*(1), 151–165.

Zaman, N. R., Kumar, B., Nasrin, Z., Islam, M. R., Maiti, T. K., & Khan, H. (2020). Proteome analyses reveal *Macrophominaphaseolina* 's survival tools when challenged by *Burkholderiacontaminans* NZ. *ACS Omega, 5*, 1352–1362. https://doi.org/10.1021/acsomega.9b01870

Chapter 10

Mechanism of disease development in host plants by *Macrophomina phaseolina*

Harshvardhan Chaudhary[1], Pankaj Kumar[2] and Ramesh Chandra Dubey[1]
[1]*Department of Botany and Microbiology, Gurukula Kangri (Deemed to be University), Haridwar, Uttarakhand, India;* [2]*Department of Microbiology, Dolphin (PG) Institute of Biomedical and Natural Sciences, Dehradun, Uttarakhand, India*

10.1 Introduction

Macrophomina phaseolina (Tassi) Goid. is one of the most ruinous necrotrophic fungi which is known to infect more than 700 plant species including major food, pulse and oil crops, such as maize, sorghum, common bean, soybean, sunflower, sesame, etc. It is a member of the family *Botryosphaeriaceae* and also known by several synonyms like *Sclerotium bataticola* (Taub.), *M. phaseoli* (Maubl.) Ashby, *M. phillipinensis* Petrak, *M. carchori* Sawada, *M. sesame* Sawada, etc. Basically, *M. phaseolina* is the pycnidial stage (perfect stage) and *Rhizoctonia bataticola* (Taub.) Butler is the sclerotial stage (imperfect stage). But the name *Macrophomina* has become popular and used by most of the researchers. Earlier there was no report on physiological races or subspecies based on morphological or genetic characterizations (Crous et al., 2006; Dhingra & Sinclair, 1978). Recently, two new species of *Macrophomina* have been reported, such as *M. pseudophaseolina* isolated from okra (*Abelmoschus esculentus*), groundnut (*Arachis hypogea*), roselle (*Hibiscus sabdariffa*), and cowpea (*Vigna unguiculata*) (Sarr et al., 2014), and a new phylogenetic species, *M. euphorbia* is known to cause charcoal rot in castor bean (*Ricinus communis*) and bellyache bush (*Jatropha gossypifolia*) (Machado et al., 2019).

M. phaseolina is characterized by light to dark brown septate hyphae having thick wall. Branches from the primary hyphae are framed at right angles on parent hyphae, narrowing at the starting point. Microsclerotia (the small-sized sclerotia) are the closely packed mass of numerous fungal mycelia; these are round, oval, or elliptical, light brown in the initial stage becoming more obscure (brown to dark) with the attainment of maturity (Fig. 10.1). Sclerotia are multicellular structures where each cell comprises of 1−3 nuclei, mitochondria, lipid droplets, etc. and are usually dark brown or black in color with barrel-shaped cells (Wyllie & Brown, 1970). An individual sclerotium usually comprises of 50−200 or even more cells that are united by a septal pore present in each cell. All the cells are tightly cemented together to form a microsclerotium that have variable size (50−150 μm) depending on the availability of nutrients in the substrate on which the propagules are produced (Short & Wyllie, 1978). A single sclerotium of *M. phaseolina* germinates and forms numerous hyphae, which in turn also form secondary sclerotia, thus increases its number in soil (Fig. 10.1). *Macrophomina* causes a wide range of diseases in plants such as charcoal rot, root rot, collar rot, stem rot, etc., with a favored temperature range of 28−35°C and low moisture content of soil (Sandhu et al., 1999). Pycnidia, which are seldom seen under normal circumstances, are bigger than microsclerotia, dim brown to dark, rough, globose or sporadic, bent, and ostiolate (Lakhran et al., 2018).

In addition, humans are also susceptible to a variety of fungal infections, and they have increased considerably over the last few decades, more particularly invasive mycosis. People who are at higher risk of developing these fungal infections are the ones that usually have a compromised immune system. There are some reports of *Macrophomina* infection in humans also. The first case of *M. phaseolina* infection in humans was reported in a Sri Lankan-born Canadian man who went through a renal transplant in India.

Pathogenesis of *M. phaseolina* in plants involves the action of several cell wall-degrading enzymes (CWDEs) and toxins. *M. phaseolina* comes in action by making contact with the host plants and initiates infection by adhering to plant cell surface followed by invasion of host epidermis through appressoria formation being established in different plant tissues.

Macrophomina Phaseolina. https://doi.org/10.1016/B978-0-443-15443-0.00010-3

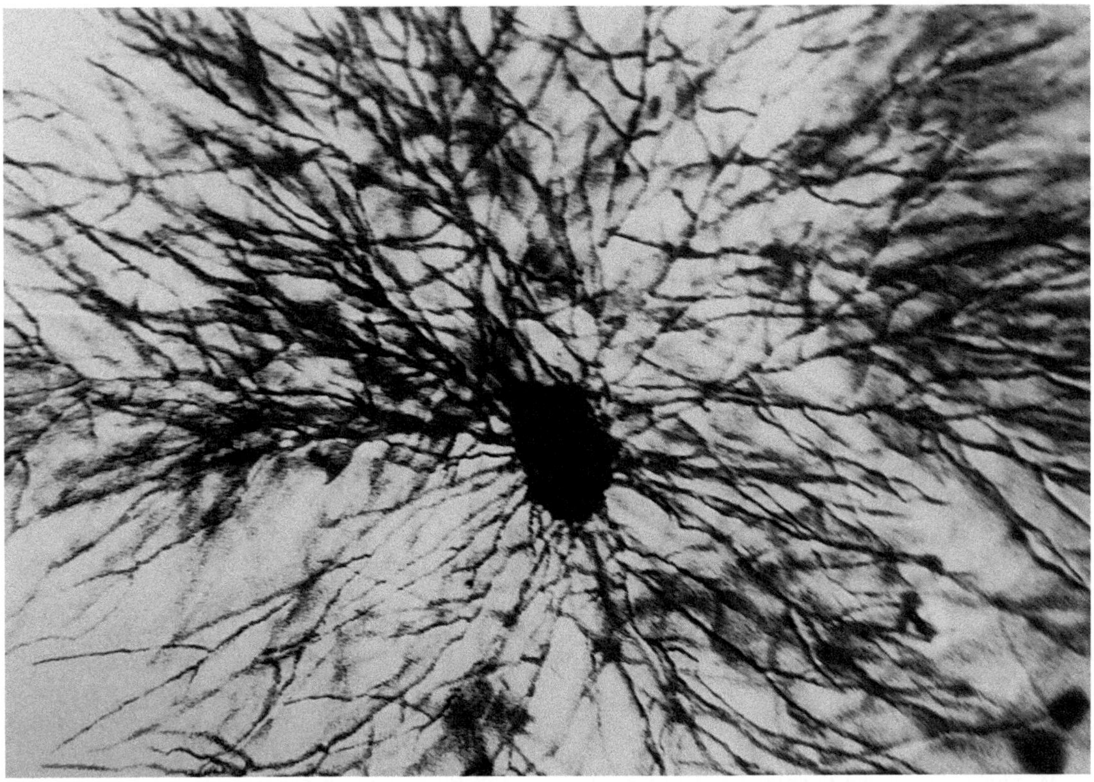

FIGURE 10.1 Germination of a single sclerotium of *M. phaseolina* (40 X).

CWDEs are among the most important parameters of *Macrophomina* infection as these enzymes play an important role in overcoming the first true barrier of plant defense system, i.e., the cell wall. These enzymes show specificity toward varied cell wall components of plant cell wall. Toxins such as (−)-botryodiplodin and phaseolinone are another important virulence factors of *M. phaseolina* associated with disease development in hosts. However, this field requires further research to understand clearly the role and impact of these toxins in infection. All the steps leading to disease development are also governed at molecular level including the production of CWDEs and toxins as revealed in the genomic study carried out by Islam et al. (2012).

10.2 Survival of *M. phaseolina*

The main modes of survival of *M. phaseolina* in soil or in plant debris are mycelium and microsclerotia. Meyer et al. (1973) reported that mycelia do not survive for than 7 days in soil, but the survival time varied from 8 to 12 days at different pH levels. Under dry and saturated nonsterile soil conditions, the hyphae of *M. phaseolina* did not survive for more than a day (Ghaffar & Erwin, 1969). Microsclerotia are referred to as the primary source of inoculum and main surviving propagules present in soil (Abawi & Pastor-Corrales, 1990; Dhingra & Sinclair, 1977). They are formed by mycelial aggregation either on dead substrates saprophytically or in the infected host tissue during pathogenesis. Further, duration of survival of microsclerotia of *M. phaseolina* in soil and in host tissues varies from 3 months to up to 16 years (Ghaffar & Erwin, 1969; Young et al., 1982).

The survivability and duration of survival of *M. phaseolina* for a longer time depend on various biotic and abiotic factors in soil. In potato tubers and corn residues, the duration of survival of *M. phaseolina* has been recorded as 18 months followed by 16 months in sorghum residues (Cook et al., 1973; Watanabe, 1973). Papavizas (1977) evaluated the survival of free sclerotia in soil, different soil factors including temperature, moisture, pH, fungicides, organic and inorganic amendments. Dubey and Dwivedi (1988a) studied the effects of heavy metals, viz., Cd, Co, Ni, and Zn on growth and survival of *M. phaseolina* and found that Cd, Co, and Ni (at 500 μg/mL) inhibited the growth, while 100 μg/mL concentration of Zn enhanced the yield of mycelial dry weight. Furthermore, Dubey (1992) studied the survival of *M. phaseolina* in fungicide amended and unamended soils along with the effect of soil solarization.

10.3 Genetic diversity of *M. phaseolina*

Although *Macrophomina* has a very vast range of hosts, it is a monotypic genus with very high degree of genetic variability among different isolates even though the pathogen lacks sexual reproduction. Almeida et al. (2003) hypothesized the possibility of parasexualism being the reason for this variability by formation of heterokaryons. Kaur et al. (2012) also emphasized on mutation, hyphal fusion, and mitotic recombination being responsible for such genetic variability. Amplified fragment length polymorphism (AFLP) was implemented by Mayék-Pérez et al. (2001) to understand the relation between pathogenesis of different isolates and their genotype, origin, and geographical location. Almeida et al. (2003) analyzed the genetic diversity of 55 isolates of *M. phaseolina* from soybean roots of different regions of Brazil through random amplified polymorphic DNA (RAPD) studies. More divergence was seen in isolates from regions with a single crop as compared to the isolates of areas where crop rotation was carried out. Furthermore, variability among Brazilian isolates of *M. phaseolina* was observed that showed the presence of more than one haplotype in a single root.

Jana et al. (2003) and Babu et al. (2010) carried out RAPD-based molecular analysis of 99 *Macrophomina* isolates infecting different plant hosts of various regions. They found out that primers OPA-13 (5′-CAGCACCCAC-3′) and OPB-08 (5′-GTGAGCTAGG-3′) induced the maximum polymorphism that could be used for identification and differentiation of different isolates of *M. phaseolina*. Furthermore, they also reported monomorphic bands in all isolates, indicating the origin of all isolates of *M. phaseolina* from a common ancestor. Recently, simple-sequence repeat (SSR)-based markers were used for characterization of strawberry and soybean infecting isolates of *M. phaseolina* (Reznikov et al., 2019; Sánchez et al., 2017).

Some studies associate the genetic diversity of *M. phaseolina* with the type of host plant and/or geographical regions (Babu et al., 2010; Jana et al., 2005; Mahdizadeh et al., 2012), whereas some others were not able to clearly differentiate between the isolates on the basis of host and geographical regions (Mahdizadeh et al., 2011; Reznikov et al., 2018, 2019). However, various studies reported the independency of *M. phaseolina* isolates on sampling sites and hosts (Khan et al., 2017; TančićŽivanov et al., 2019). Some researchers have given the idea of categorizing the isolates of *Macrophomina* on the basis of their ability to reduce chlorate, an analogue of nitrate, to chlorite through nitrate reductase pathway. When grown on a minimal medium amended with 120 mM potassium chlorate, the isolates of *M. phaseolina* from soybean roots and field soils only produced sensitive phenotypes, while the corn isolates produced only the resistant ones; hence, it could be used for distinguishing different *Macrophomina* isolates (Pearson et al., 1986). However, Das et al. (2008) contrasted these findings in their studies of characterizing sorghum stalks isolates of *M. phaseolina*. Variation in the pathogenicity of different isolates from different crop fields, such as soybean, cotton, sorghum, and maize, has also been reported mainly due to the cropping history along with some degree of host specialization by *M. phaseolina* (Su et al., 2001). Studies carried out on different isolates of soybean, sorghum, sunflower, cowpea, corn, and wheat along with soil samples of native areas, reported the reduction in host specialization levels in *M. phaseolina* isolates due to crop rotation (Almeida et al., 2003).

Several studies have been carried out by implementing RAPD and RFLP of ITS region to analyze the genetic variability among the *Macrophomina* isolates. Purkayastha et al. (2006) distinguished the isolates on the basis of chlorate utilization and host origin based on their studies on the use of RAPD patterns of *M. phaseolina*. SSR loci have also be used for studying the gene flow as they can be transferred from one genus to another closely related genus and the same is also being used in other genera, such as *Botryosphaeria* and *Diplodia*, which are closely related to *Macrophomina* (Crous et al., 2006). Jana et al. (2005) suggested the use of frequently interspersed repeats known as microsatellite core sequence repeats, such as (CAC) 5, (ACTG) 4, and (GACAC), as potent markers that can be implemented for studies related to epidemiology, genetics as well as population of *M. phaseolina*.

10.4 Microbial interactions in rhizosphere

Rhizosphere region is unique because of continuous flow of secretion of root exudates that contain a variety of chemicals, such as carbohydrates, organic acids, amino acids, hormones, vitamins, and other compounds (enzymes, flavones, nucleotides, HCN, saponines, glycosides, auxins, scopoletin). These chemicals are utilized by the microorganisms residing in rhizosphere. However, both beneficial and harmful microorganisms utilize these food resources through different types of interactions (Fig. 10.2). Moreover, increased microbial activities in the rhizosphere region result in different types of microbial interactions viz., antibiosis, competition, parasitism, etc. that lead toward biological control of plant pathogens. Root exudation varies with different stages of plant growth and can specifically modulate the microbiome of the rhizosphere (Chaparro et al., 2013). The soil microbiome consists of bacteria, fungi, ascomycetes, nematodes, etc. Before initiation of infection, *M. phaseolina* should first survive under the aforesaid competitive environment.

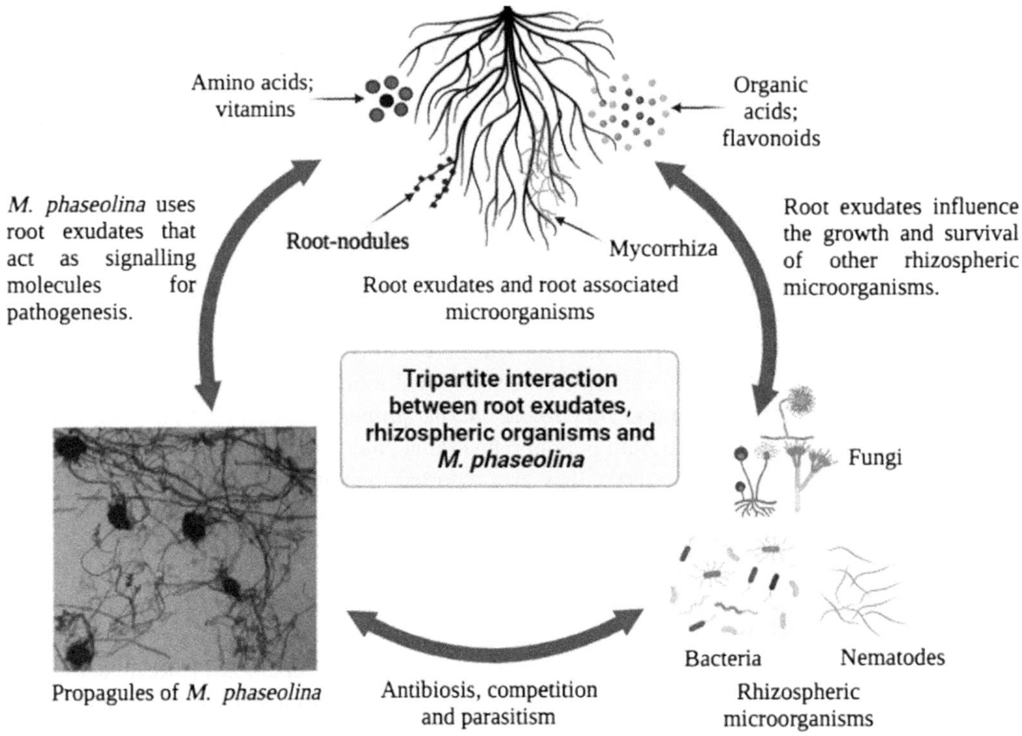

FIGURE 10.2 Interactions of *M. phaseolina* with rhizosphere microorganisms and host root.

10.4.1 Interaction with bacteria

Plant growth promoting rhizobacteria (PGPR) have shown considerable amount of antibiosis potential against *M. phaseolina*. Increase in bacterial numbers in rhizosphere reduces disease incidence in hosts due to reduction in numbers and inhibition in invasion by pathogens on host plants. Singh et al. (2008) have found an increase in population of antagonistic *Bacillus subtilis* BN1 by 1.5×10^4 cfu/g root after 1 month was further increased to 4.5×10^4 cfu/g root within 3 months even in the presence of *M. phaseolina*, which possibly interfered the infection of host by the pathogen. In a study carried out by Dubey et al. (2011), *Azotobacter chroococcum* AZO2 isolated from rhizosphere of *Sesamum indicum* showed a strong antagonistic activity against *M. phaseolina* causing 81% inhibition of its growth. Kumar et al. (2012, 2020) also reported the antagonistic activities of *Bacillus* against *M. phaseolina*. Furthermore, Dubey et al. (2012), while studying the antagonistic potential of *Bradyrhizobium* sp. VR1 and VR2 against *M. phaseolina*, reported fragmentation, shrinkage, and lysis of hyphae along with cytoplasm vacuolation and inability of *M. phaseolina* to form sclerotia with the loss of mycelial pigmentation. The efficiency and virulence of *M. phaseolina* also get declined possibly due to aggressive behavior of antagonistic microorganisms. Further, *Pseudomonas putida* was reported to reduce charcoal rot incidence caused by *M. phaseolina* in *Vigna radiata* up to 66.66% (Sharma & Dubey, 2017). In a study carried out by Kumar et al. (2021), the biosurfactant cyclic siloxane, produced by *B. cereus* BS14 from *V. mungo*, inhibited the growth of *M. phaseolina* causing deformities at cellular level in the mycelia of the pathogen. *B. amyloliquefaciens*, through secretion of various diffusible compounds, such as bacillomycin D, macrolactin, iturin A, surfactin, and fengycin, inhibited the growth and germination of sclerotia of *M. phaseolina* along with growth promotion in cowpea (Rangel-Montoya et al., 2022).

10.4.2 Interaction with fungi

Several fungi as well as fungal metabolites are known to inhibit the growth and germination of several phytopathogens. However, *Trichoderma* species are known to elicit plant defense response through an array of mechanism comprising of nutrient competition, antibiotic production and mycoparasitism. It has also been reported that *Trichoderma* spp. induce systemic resistance in host plant against phytopathogens (Hyder et al., 2017; Martinez-Medina et al., 2016). Gajera et al. (2012) examined the antagonistic potential of 7 *Trichoderma* spp. namely, *T. harzianum* NBAII Th 1, *T. viride* NBAII Tv 23,

T. virens NBAII Tvs 12, *T. koningi* MTCC 796, *T. pseudokoningi* MTCC 2048, *T. hamatun* NBAII Tha-1 and a local *T. harzianum* against *M. phaseolina*. *T. koningi* MTCC 796, and *T. harzianum* NBAII Th 1 showed the maximum inhibition of *M. phaseolina* by 74.3% and 61.4%, respectively on seventh days after inoculation (DAI). They also observed an increase in the total phenolic content along with induction of chitinase and β-1, 3-glucanase, thus aiding in inhibition of *Macrophomina*. Introduction of *T. virens* in soil through seed bacterization of peanut resulted in the increased seed germination (80%−86%) and seedling emergence (60%−68%) after 7 days of sowing in *M. phaseolina*−infested soil. *T. virens* produced volatile and nonvolatile organic compounds that also inhibited the growth of *M. phaseolina* (Maheshwari et al., 2001). Sridharan et al. (2020) reported that volatile organic compounds (VOCs) produced by *T. longibrachiatum*, inhibited the growth of *M. phaseolina*. These VOCs included terpenoids along with some newly expressed compounds like, limonene, azulene, 3-methyl-1-butanol, styrene, salicylaldehyde, undecane, and 3-methylphenol that might be involved in microbe-microbe signaling along with a potential of being antimicrobial agents. Nafady et al. (2019) delineated the role of a growth-promoting yeast, *Brettanomyces naardensis*, as a biocontrol agent against *M. phaseolina*. The pathogen−pathogen interaction between *Magnaporthiopsis maydis* and *M. phaseolina* (both infected cotton sprouts) was investigated by Degani et al. (2022). They reported that consortium of these two fungi decreased the severity of cotton charcoal rot disease (CRD) with an increase in plant growth and yield. These results delineate toward the antagonistic relation between the two fungi.

10.4.3 Interaction with mycorrhizal fungi

Mycorrhiza, especially arbuscular mycorrhizal fungi (AMF), have also been reported to provide the host with protection against varied phytopathogens through a range of mechanisms ranging from competition for space and nutrition with the soil-borne pathogens to elicitation of innate plant defense responses by reprogramming the plant gene expression and metabolisms (Dey & Ghosh, 2022). Jalali et al. (1990) reported antagonistic activity of a vesicular-arbuscular mycorrhizal (VAM) fungus, *Glomus mosseae,* towards *M. phaseolina* and its restricted root colonization of mungbean. In a study to know the effect of *Rhizophagus intraradices* (basionym *Glomus intraradices*) on protection of soybean against infection of *M. phaseolina* reported a reduction in AMF colonization when amended with nitrogen (N) fertilizer but arbuscules percentage was not affected. When *Macrophomina* infected soybean plants were co-inoculated with AMF, there was an increase in the biomass as well as greenness index of the infected plants. The severity of *Macrophomina* infection was directly proportional to N fertilization while inversely to mycorrhizal symbiosis (Spagnoletti et al., 2020).

10.4.4 Interaction with nematodes

A wide range of parasitic organisms reside in soil including nematodes that interact with other soil residing microbes including bacteria, fungi, etc. Synergistic interaction of these phytophagous nematodes and soil-borne fungi have resulted in an increase in damage to plants and crops, but in some studies, antagonistic interactions has also been reported. Due to nematode infestation, there is a decrease in plants resistance to pathogens, and as a result, it facilitates fungal infection. However, this tripartite interaction depends on all the three participating partners, i.e., nematodes, soil-borne fungi, and the host plant (Al-Ani et al., 2022). While feeding or entering through plant roots, several phytophagous nematodes tend to create a variety of wounds/openings, through which the entry of phytopathogenic fungi is facilitated. Endo-parasitic nematodes, e.g., *Pratylenchus* sp. are far more disruptive as compared to ecto-parasitic nematodes, such as *Trichodorus* sp. and *Tylenchorhynchus* sp. Furthermore, some nematodes may induce physiological changes in host plant that trigger changes in the population dynamics of fungal pathogens causing them to increase their population which in turn increases the pathogenicity (Back et al., 2002). Winkler et al. (1994) studied interactions of *Heterodera glycines*, *M. phaseolina,* and mycorrhizal fungi on soybean and reported that *H. glycines* infested soil caused an eightfold increase in *M. phaseolina* colonization with 28% decrease in colonization by mycorrhizal fungi. Al-Hazmi (1984) examined the effect of *Meloidogyne incognita* and *M. phaseolina* in root-rot of French bean and reported that *M. incognita* enhanced the severity of *M. phaseolina* infection after simultaneous inoculation. Gupta and Metha (1989) reported a decrease in nematode galls in mungbean and delineated that production of metabolites by *M. phaseolina* has negative effect on *Meloidogyne javanica*. Sharma and Khan (2010) reported that *M. phaseolina* affected the life cycle of *M. javanica* by delaying it 8 days. While studying the combined effect of *H. glycines* and *M. phaseolina* on Soybean, Lopez-Nicora et al. (2020) reported the negative impact of *M. phaseolina* on *H. glycines* population with no effect on soybean tissue-colonization by *M. phaseolina*. However, it is still unclear that how these two organisms interact with each other.

10.5 Pathogenesis by *M. phaseolina*

Pathogenesis is defined as the chain of events that leads to the establishment of a disease. Disease cycle of *M. phaseolina* is initiated at the moment when the soil residing microsclerotia come in the contact of the host tissues, usually root hairs followed by invasion of host tissues to establishment of disease leading to the death of the host plant (Fig. 10.3). Pathogenesis by *M. phaseolina* is administered at genetic as well as biochemical levels. The genome of *M. phaseolina* consists of a large number of pathogen–host interaction (PHI) genes. Islam et al. (2012) carried out genome analysis of *M. phaseolina* and reported that 12% of the genes are PHI genes. A total of 537 putative PHI genes have been identified in the genome of *M. phaseolina* that are responsible for pathogenesis.

10.5.1 Role of inoculum in pathogenesis

Pathogenesis by any fungal pathogen starts with germination on root surface of its propagules that reside in the soil. The germination of these propagules, microsclerotia in case of *M. phaseolina*, starts after coming in contact with the host tissues. But before infecting the host, there must be a considerable number (density) of fungal propagules present in soil that can compete with varied biotic and abiotic factors to initiate infection process.

10.5.1.1 Sclerotia density and germination

The primary infective source of *M. phaseolina* is microsclerotia, which germinate repeatedly during the growing season of crop. The most favored temperature for germination of microsclerotia of *M. phaseolina* ranges between 28 and 35°C

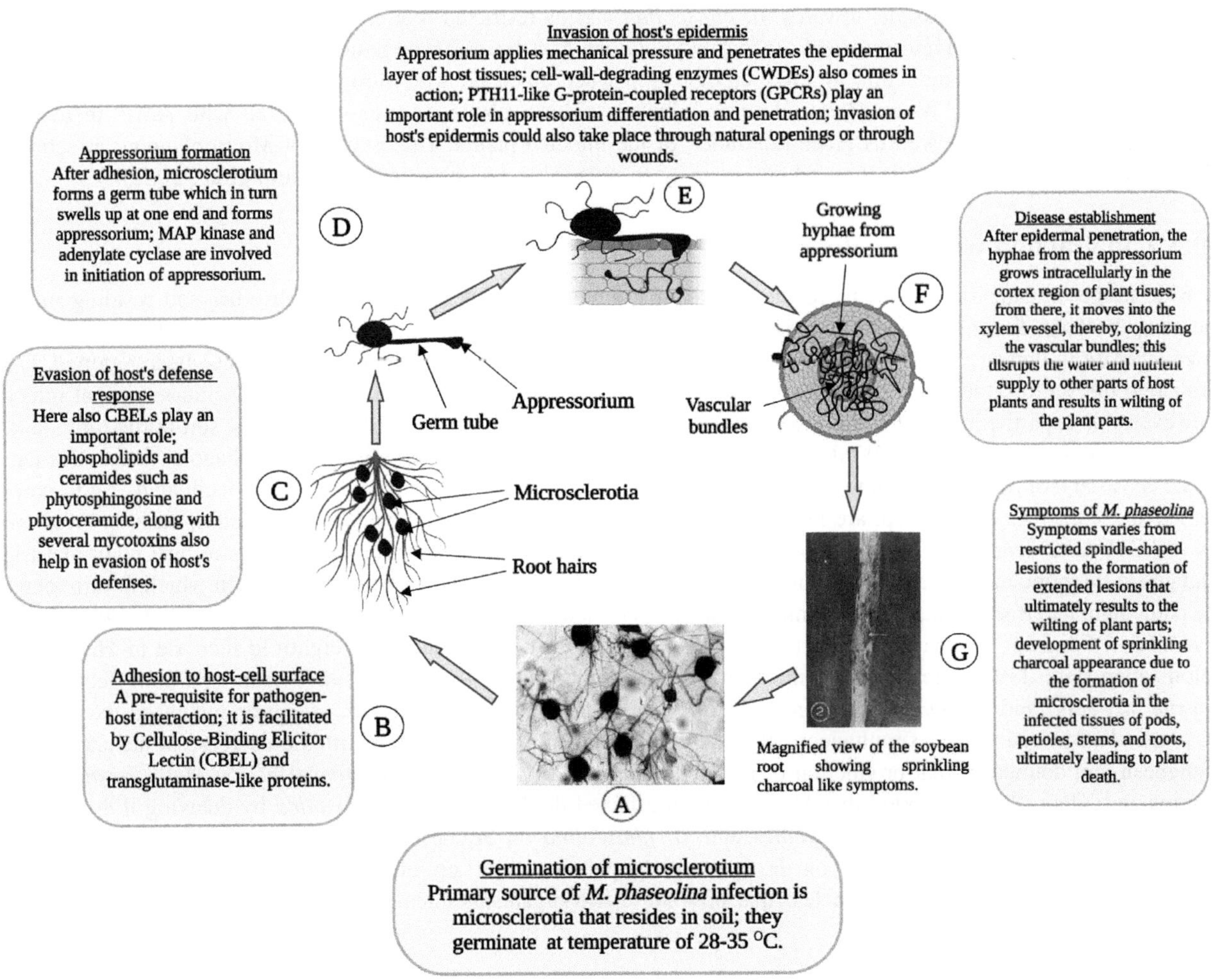

FIGURE 10.3 Disease cycle of *M. phaseolina* causing charcoal rot in soybean.

(Mihail, 1989). The highest sclerotial density is usually found in top layer of soil to a depth of about 30 cm (Bruton & Reuveni, 1985). Lodha et al. (1990) while studying sclerotia survival of *M. phaseolina* reported the maximum sclerotial population at a depth of 0–5 cm followed by 20–30 cm. Taliei et al. (2013) surveyed 172 soybean fields for 2 years for population density of *M. phaseolina*. In the two sampling years, sclerotia recovery from the fields was 73.33% and 93.57%, respectively, with an average population density ranging from 0.65 to 14.31 g^{-1} and 4.7–16.9 g^{-1} soil, respectively.

10.5.1.2 Factors affecting sclerotia germinations

Germination of sclerotia is influenced by a number of factors, such as inoculum potential, temperature, and soil moisture. The inoculum potential of the infective propagules greatly influences the severity of disease caused by *M. phaseolina*. In soybean fields, sclerotia germination is favored by dry soil conditions along with high soil C:N ratio of amendments, low bulk density (Dhingra & Sinclair, 1975; Gangopadhyay et al., 1982), and more oxygen concentrations (Todd et al., 1987).

10.5.1.3 Temperature

Temperature plays an important role in pathogenesis of any phytopathogen. If the temperature is not suitable, the propagules of the pathogen will not germinate. *M. phaseolina* has a germination range of 28–35°C, but studies have shown variation in these claims. An increase or decrease in temperature range can hinder the germination process of sclerotia. Dhingra and Chagas (1981) reported the maximum colonization of wheat stems by *M. phaseolina* at 15 and 20°C, while bean stem colonization was maximum at 15 and 25°C. Soil solarization is another important criterion that has been studied to evaluate the effect of higher temperature on sclerotia germination and survival. Dubey (1992) reported 78% reduction in inoculum density after solarization period of 45 days with the maximum temperature of 55.5°C in tarped soil. Dubey (1992) further studied the effect of increase in solarization in fungicide–mixed and nonmixed natural soils and found that 15 days of solarization reduced inoculum density to below detectable levels in soil–mixed with 250 μg/g of Bavistin. Yildiz et al. (2010) carried out experiments to understand the effects soil solarization using different plastics films to control *M. phaseolina* infesting strawberry and reported up to 66% reduction in inoculum density at 5 cm depth but no change in inoculum density at depths of 10 cm. All these studies delineate toward the finding that higher temperature is unfavorable for sclerotia germination and considerably decreases the inoculum density of *M. phaseolina* in soil.

10.5.1.4 Soil moisture

Soil moisture is among the main predisposing factor that influences sclerotia germination of *Macrophomina*. Soil moisture has an inverse relationship with *M. phaseolina*-induced diseases in plants (Ghaffar, 1968). Dhingra and Sinclair (1975) examined the effect of different moisture gradients on survival of pathogen and reported a decrease in population dynamics of *Macrophomina* with an increase in soil moisture and duration of incubation. Sheikh and Ghaffar (1979) reported the variation in susceptibility to *Macrophomina* infection of test plants. At 25% water holding capacity, 50% infection in black gram was recorded at a density of 5 sclerotia/g dry soil, 20 sclerotia/g dry soil for okra and guar, and 40 sclerotia/g dry soil for cotton. More than 50% infection could not be achieved at an inoculum density of 40 sclerotia/g dry soil plus high soil moisture (100% water holding capacity).

10.5.1.5 Organic amendments

Addition of organic matter, such as crop residues, in soil can also help in lowering of population density of phytopathogens by changing the general microbial balance of soil. Lukade and Rane (1994) reported a decrease in root rot incidence in sunflower when soil was amended with farm yard manure (FYM). A considerable reduction of 75%–96% inoculum density was observed after 15 days of solarization of soil amended with cruciferous residues and soil amended with mustard pod residues followed by solarization, caused propagules elimination even upto the depth of 30 cm (Israel et al., 2005; Lodha et al., 2003). Sudha and Prabhu (2008) also reported a decrease in the incidence of charcoal rot by 13.3% and 15.0%, respectively as compared to control (63.3%) in sunflower with soil amendment by FYM (12.5 t/ha) and neem cake (250 kg/ha), with an increase in the total bacterial population in the rhizosphere. This increase in rhizosphere microbiome aided in control of *Macrophomina* as it housed such microbes that showed antagonism toward *M. phaseolina*.

10.5.1.6 Inorganic amendments

Several workers have investigated the effect of inorganic soil amendments on germination and survival of sclerotia. Filho and Dhingra (1980) found a significant reduction in counts of *M. phaseolina* from 640 to 2–10 propagules in soil supplemented with 3–7 mg N/g nitrogen (N) in the form of sodium nitrate or ammonium sulfate.

10.5.1.7 Chemicals fungicides

Presence of chemicals that dissipate in the soil can also hinder the germination of sclerotia. Dwivedi and Dubey (1987) reported that application of bavistin (250 μg/g) completely eradicated the sclerotial as well as mycelial propagules in soil after 30 days, whereas dithane M-45 and brassicol completely eradicated at 1000 μg/g concentration in soil. Parmar et al. (2017) examined the efficacy of various systemic as well as nonsystemic fungicides against *M. phaseolina* causing root rot in castor. Among the nonsystemic ones, propineb and mancozeb showed the maximum mycelial growth inhibition (99.97%) followed by captan (85.63%), but no sclerotia germination was observed in these treatments. In case of systemic fungicides, carbendazim was the most effective followed by difenoconazole and thiophanate methyl, showing mycelial growth inhibition by 95.23%, 75.47%, and 69.49%, respectively. Carbendazim (50 ppm) inhibited the growth of mycelia and sclerotia germination of *M. phaseolina* that causes dry root rot in chickpea. With the increase in concentration of systemic fungicide, the inhibitory effect of fungicides increased (Lokesh et al., 2020).

10.5.2 Identification of host and adhesion

Adhesion of the fungal propagules to the host cell surface is a prerequisite for pathogen–host interaction leading to disease development. The genome of *M. phaseolina* encodes eight homologs of cellulose-binding elicitor lectin (CBEL), a cell wall localized protein that aids in adhesion to host wall components (Gaulin et al., 2002) along with a class protein II hydrophobic protein that enables the fungus to maintain contact and communication with its environment (Whiteford & Spanu, 2001). In addition, *M. phaseolina* also possess three transglutaminase–like proteins comprising of 13-amino acid motif (Pep-13) that stimulate plant defense response. The genome of *M. phaseolina* also contains certain heterotrimeric G-proteins (MPH_02326, MPH_07066, MPH_00770, MPH_05860, and MPH_09815), which are known to contribute to the signaling pathways for pathogenesis (Islam et al., 2012).

10.5.3 Induction of defense mechanism in host

Local root colonization is sufficient to trigger induced systemic resistance (ISR). A reaction in the roots is triggered by the PGPR creating a signal that spreads throughout the plant. This signal activates the defense mechanisms re-enforcing the plant cell wall to produce antimicrobial phytoalexins and to synthesize the pathogen related proteins (PRP). The main defense mechanisms used by ISR are the lignification of host tissues and the oxidation of phenols in response to harmful phytopathogens. A consortium of *P. putida* CRN-09 and *B. subtilis* CRN-16 activated ISR in *V. radiata* against *M. phaseolina* through enhancement in high level expression of peroxidase (PO), polyphenol oxidase (PPO), phenylalanine ammonia lyase (PAL), β–1,3 glucanase and chitinase, playing a greater role in the development of plant immunity (Sharma et al., 2018).

10.5.4 Countering of host defense responses

During the course of pathogenesis, *M. phaseolina* must also evade the defensive responses of the host. These defense mechanisms get activated at the moment when the fungi get adhered to the host cell surfaces. Cellulose-binding elicitor lectin, along with aiding in adhesion to host cells, also aids in the evasion of host defenses (Bhowal et al., 2010). In addition, the secretion of phytosphingosine and phytoceramide by *M. phaseolina* also helps in overcoming the mechanical damage caused to its cell membrane by the host cell as a measure for entry barrier defense mechanisms. Phytosphingosine and phytoceramide aid in the pathogenicity by inciting pathogen-induced apoptosis along with ROS accumulation in host. These are also known to assist in fungal cell division and germination along with providing the fungus the ability to tolerate alkaline conditions (del Poeta et al., 2014; Magnin-Robert et al., 2015). Besides, the fungus also secretes certain secondary metabolites especially mycotoxins, such as patulin, phaseolinone, etc. that suppress the host defenses (Islam et al., 2012; Thakur & Jana, 2017).

10.5.5 Invasion

It begins with the physical contact of microsclerotium with the plant tissues resulting in a germ tube formation that extends further and forms an appressorium. Ammon et al. (1975) reported appressoria formation at the tip of the primary hyphae on the root surface within 3 days of inoculation. Appressoria are the specialized structures produced by fungal pathogens at the tip of a hypha to infect host plants. MAP kinase *PMK1* (MPH_065596) and adenylate cyclase *MAC1* (MPH_00975) are involved in appressoria initiation and invasive growth of *M. phaseolina* (Islam et al., 2012; Xu & Hamer, 1996).

The appressoria thus formed apply turgor and mechanical pressures and penetrate the host's epidermal cell walls through the secretion of CWDEs.The G-protein-coupled receptors (GPCRs) are the most abundant class of cell surface receptors present in fungus, which are responsible for fungal development and pathogenesis. *M. phaseolina* genome encodes 44 putative PTH11-like GPCRs in comparison to other pathogenic fungi that average to 34. PTH11-like GPCRs regulate the differentiation of appressorium, typical to plant pathogenic fungi that aid in penetration of epidermis in response to the plant surface (Islam et al., 2012). In some cases, indirect penetration of sclerotia through wounds or natural openings has also been reported (Bowers & Russin, 1999; Mayek-PÉrez et al., 2002).

10.5.6 Establishment of disease

After invading the epidermal layer, the hyphae from the appressoria grow intracellularly in the cortex region and then into the xylem vessels thus colonizing the vascular tissues (Abawi & Corrales, 1990). After colonization of vascular tissues, the fungus disrupts the water and nutrient supply to the other parts of the plant and results in wilting of the plant, which is characterized by silver and gray color development in tissues of pods, stem, petioles, and roots in soybean and other crop plants (Wyllie, 1988). *M. phaseolina* is known to produce certain phytotoxic metabolites, such as phaseolinone, botryodiplodin, and patulin that also play an important role in initial stages of infection and cause wilting of seedlings and the formation of necrotic lesions on leaves and roots (Abbas et al., 2020; Bhattacharya et al., 1987; Ramezani et al., 2007; Salvatore et al., 2020). Role of toxins in pathogenesis is described in detail in the preceding section.

10.5.7 Symptoms caused by *M. phaseolina*

The emergence of spindle-shaped lesions and the production of pycnidial bodies are among the first signs of *M. phaseolina* in woody plants. The symptoms of *Macrophomina* infection vary from restricted spindle-shaped lesions to the formation of extended lesions that ultimately results in wilting of plant parts. Appearance of deep and irregular necrotic lesions have been observed in soybean, chickpea, and sorghum that tend to extend toward the hypocotyls and root surfaces (Ammon & Wyllie, 1972; Pedgaonkar & Mayee, 1990; Singh & Mehrotra, 1982). When these lesions coalesce together, they form larger patches on branches and on entire plant leading to early senescence ultimately leading to plant death (Dhingra & Sinclair, 1978). It has been reported that *Macrophomina* colonizes epidermal and cortical cells followed by vascular cambium and phloem cells in chickpea, garlic, maize, and guava (Dwivedi et al., 1994; Suárez et al., 1998). Typical symptoms observed in different crops infected by *M. phaseolina* are given in Table 10.1. The symptoms of *Macrophomina* infection usually start during the flowering season but collapsing of seedlings due to the development of thin brown to black streaks has also been observed in *Phaseolina vulgaris* and *Vigna mungo* (Chandra et al., 1995; Machado, 1987). In the infected tissues of pods, petioles, stems and roots of soybean, there was appearance of gray or silver color due to the germination and multiplication of microsclerotia (Wyllie, 1988). After flowering, charcoal rot symptoms usually appear at the end of the season. *M. phaseolina* isolates from one host usually will infect and more-or-less cause disease in another host (Su et al., 2001). Initial infection of host by *M. phaseolina* cannot be prevented by an increase in soil moisture, but colonization of the stem can significantly be reduced at maturity (Jordaan et al., 2019).

10.6 Role of cell wall-degrading enzymes (CWDEs) of *M. phaseolina* in pathogenesis

All the plants bear a rigid cell wall, which acts as the first barrier to the entry of pathogen. In order to overcome the cell wall barrier, *M. phaseolina* encodes a large repertoire of hydrolytic enzymes required for the degradation of almost all major components of plant cell wall and cuticle including cellulose, hemicellulose, pectin, lignin, etc. (Table 10.2). Several CWDEs produced by *M. phaseolina* are the main virulence factor.

10.6.1 Pectin-degrading enzymes

Pectin is a complex heteropolysaccharide composed of linear chains of α-D-galacturonic acid having high percentage of methyl esterification. Plant pathogens usually degrade pectin by the collaborative action of several pectinolytic enzymes involving two enzymatic processes, lyases splitting α-1,4- glycosidic bonds between galacturonic acid residues and by hydrolytic cleavage catalyzed by polygalactouronase (Sakai et al., 1993). The action of these pectinolytic enzymes is a prerequisite for activity of other CWDEs; therefore, pectinases are the first CWDEs secreted by most fungal pathogens

TABLE 10.1 Common symptoms of diseases caused by *M. phaseolina*.

Disease	Symptoms	Host	References
Charcoal rot	Infection in the growth stages of the crop leading to colonization through the roots followed by blocking of xylem vessels and disintegration of tissues.	Sorghum	Parmeshwarappa et al. (1976), Gray et al. (1991), Seetharama et al. (1987), Bandara et al. (2018)
	Reduction in seed emergence due to development of black spots on the infected seed coat; most favorable temperature for disease development was 30–38°C.	*Glycine* max (Soybean)	Gangopadhyay et al. (1974)
	Prevalence of disease 32%–82%; wilting and browning of leaves along with the base of the crown showing water-soaked lesions along with amber to dark brown exudates.	Cantaloupe melon	Jacob et al. (2013)
	Appearance of symptoms at unifoliate leaf stage; charcoal-colored spots of the size of a pinhead starts to appear; these spots or lesions expand in size and form necrotic lesions ultimately leading to the death of the plant.	*Vigna unguiculata* (L.) Walp. (Cowpea)	Adam (1986)
Wilting or dry root rot	High prevalence during postflowering stage in hot dry climate; dryness occurs in tap root which becomes devoid of fine roots; development of numerous pinhead-shaped, dark sclerotia underneath the bark.	*Cicer arietinhum*	Leach and Garber (1970)
	Development of small gray-black sclerotia in the root pith resulting in tap root becoming brittle; plant height is stunted; disease is more prevalent in summer.	*Cajanus cajan*	Noble and Richardson (1968)
Damping off/Seedling blight	Infection at seedling stage; development of pinhead-shaped spots that later on gets merged together forming larger, necrotic lesions; collapsing of seedlings due to the development of thin streaks of brown to black color in the collar region.	*Phaseolus vulgaris* and *Vigna mungo*	Chandra et al. (1995)
	Infection at seedling stage with leaves turning yellowish brown to black in color; upper part of the stem gets infected by necrotic lesions and breaking of stem from the collar region is also seen.	Eucalyptus	Soni and Dadwal (1985)
Root rot	During the early stages of development, the root system becomes rusty brown followed by dropping and wilting.	*Phaseolus vulgaris* and *Vigna mungo*	Jhooty and Brains (1972)
Crown rot	Disease occurred in the form of discrete patches; upon dissection of plant crowns, internal vascular and cortical tissued appears dark brown to orange brown in color; other symptoms include wilting of foliage with stunted plant growth eventually leading to collapse and death of the plant.	*Fragaria ananassa* (Strawberry)	Koike (2008)
	Cortical and vascular tissues of taproot and crown exhibited discoloration of red-brown to dark-brown; decaying of root and crown; wilting of shoots in severly afflicted plants.	*Medicago sativa* (Alfalfa)	Wang et al. (2020)

TABLE 10.2 Cell wall-degrading enzymes produced by *M. phaseolina*.

Category of enzyme	Enzymes	Mode of cleavage/action	References
Pectin-degrading	Pectin methyl-esterase (PME); polymethylgalacturonase (PMG); polygalacturonase (PG)	Splitting of α-1,4- glycosidic bonds between galacturonic acid residues and by hydrolytic cleavage.	Madkour and Aly (1981), Dubey & Dwivedi (1988), Sakai et al. (1993), Ramos et al. (2016)
Cellulose-hydrolyzing	Endoglucanases	Cleavage of β-1,4- glycosidic bonds and release of reducing and nonreducing chain ends.	Wang and Jones (1995), Dashtban et al. (2009)
	Exocellobiohydrolases	Works on the reducing and nonreducing ends and releases a disaccharide cellobiose.	Islam et al. (2012), Couturier and Berrin (2013)
	B-glucosidase	Cleaves the cellobiose further into glucose units.	Saha et al. (1979), Kracher and Ludwig (2016)
Lignin-degrading	Chloroperoxidases; lignin peroxidases; laccase; glucose-methanol-choline (GMC) oxidases	Catalyzes the oxidation of phytoalexins, tannins and phenolic substrates.	Mayer and Staples (2002), Islam et al. (2012), Ali et al. (2014)
Protein-degrading	Proteinases and peptidases	Breakdown proteins into polypeptides followed by hydrolyzation to small peptides and amino acids which also serve as nitrogen source for pathogens.	Dubey & Dwivedi (1988), Sinha et al. (2022)
Hemicellulose-degradation	Xylanase (endo- β −1,4-xylanase)	Depolymerization of xylan to xylose.	Reddy and Reddy (1985), Kaur et al. (2012), Kubicek et al. (2014)

followed by hemicellulases and cellulases, and as a result, pectin degradation is must for a pathogen in order to initiate its disease cycle (Herrera et al., 2009).

Ali et al. (1969) revealed the defoliation and wilting of the jute plants by the action of pectinolytic enzymes produced by *M. phaseolina*. Madkour and Aly (1981) delineated high pectolytic enzyme activity, specifically pectin methyl-esterase (PME), polymethylgalacturonase (PMG), and polygalacturonase (PG), in the culture filtrates of *M.phaseolina* infecting strawberry plants. Dubey and Dwivedi (1988b) studied in vitro production of PG, PME, and PME along with cellulase, amylase, and protease by *M. phaseolina* in seed extract of soybean. They reported that an increase in production of PG and PMG with the increase in culture age, recording the maximum production after 12 days of incubation. Endo-PG production was higher in comparison to endo-PMG production. PME production was determined only in 8-day-old culture filtrates, which showed the similar trait of increase in production with the course of time. Complete saccharification of pectin is possible by *M. phaseolina* through complimentary action of pectin lyases (PL1, PL3, and PL4), pectin hydrolases, and pectin esterases (Islam et al., 2012). Ramos et al. (2016) reported in vitro production of polymethylgalacturonase (PMG) as well as polygalacturonase (PG) by six isolates of *M. phaseolina*. Bandara et al. (2018) provided evidence for enhanced polygalacturonan degradation in genotypes of sorghum that were susceptible to charcoal rot through upregulated PME and PG activity.

10.6.2 Cellulose-hydrolyzing enzymes

Cellulose is a primary constituent of plant cell walls as well as the most abundant macromolecule on earth. Cellulose fibers aggregate together and form bundles known as microfibrils. Fungi produce a wide range of glycoside hydrolase (GH) enzymes which are often referred to as cellulases. These enzymes typically hydrolyze/cleave the β-1,4-glycosidic bonds connecting the glucose units in the cellulose fiber. Wang and Jones (1995) reported the presence of a unique pathogen-specific endoglucanase (*egl1*) gene, in addition to the endoglucanase gene (*egl2*) commonly found in *M. phaseolina*. *Egl2* is known for showing activity against carboxymethylcellulose (CMC) along with cleaving cellodextrin. *M. phaseolina* possesses a higher number of GHs as compared to other plant pathogenic fungi. Furthermore, Islam et al. (2012) reported about the encoding of approximately 219 GHs in the genome of *M. phaseolina*. Out of these GHs, 25 codes for putative endoglucanases (GH5, GH12, GH45 and GH61), 7 for exocellobiohydrolases (GH6, GH7, GH81), and 28 for β-glucosidases (GH1, GH3, GH17) are required for cellulose hydrolysis.

Endoglucanases show cellulolytic activity by cleaving β-1,4-glycosidic bonds internally in the amorphous regions of cellulose and releases the reducing and nonreducing sugar chain ends (Dashtban et al., 2009). Exocellobiohydrolases work on the reducing and nonreducing ends of the cleaved amorphous regions of cellulose and release a disaccharide cellobiose (Couturier & Berrin, 2013). The cellobiose is further cleaved into glucose units by the action of another GH β-glucosidase (Kracher & Ludwig, 2016). Madkour and Aly (1981) reported high cellulolytic enzyme activities in the culture filtrates of *M. phaseolina*. Endoglucanases are known as the most prominent cellulolytic enzymes involved in pathogenesis by *M. phaseolina* (Cooper & Wood, 1973; Heiler et al., 1993). The jute infecting *M. phaseolina* showed considerable cellulolytic activity through the production of endo-β-1,4-glucanase and two forms of β-glucosidase (Saha et al., 1979). An increase in the susceptibility of sorghum to charcoal rot due to increased cellulose-degrading enzyme activity was also reported (Bandara et al., 2018).

10.6.3 Lignin-degrading enzymes

Lignin is a complex aromatic polymer component of the secondary cell walls in plants that acts as a barrier for fungal infections. Genomic analysis of *M. phaseolina* revealed the presence of a high number of ligninolytic enzymes comprising of 22 laccases, 7 haloperoxidases, 7 galactose oxidases, 6 chloroperoxidases, 6 class II hemeperoxidases, and 3 lignin peroxidases. Additionally, 40 glucose-methanol-choline (GMC) oxidases including alcohol oxidases and cellobiose dehydrogenases have also been reported (Islam et al., 2012). By destroying the harmful chemicals like phytoalexins, tannins, and other phenolic compounds, laccase shields phytopathogenic fungi from the plant defense mechanism initiated in response to a microbial infection (Mayer & Staples, 2002). Ali et al. (2014) carried out in silico analysis of chloroperoxidases and lignin peroxidases of *M. phaseolina* and delineated their physio-chemical, structural, and phylogenetic characteristics. Ramos et al. (2016) reported high laccase activity in *M. phaseolina* isolates from Argentina and recorded the production of CWDEs by submerged fermentation technique for the first time. Thus, *M. phaseolina* weakens the cell wall through the secretion of lignin-degrading enzymes.

10.6.4 Protein-degrading enzymes

Proteolytic enzymes play an important role in the hydrolysis of proteins present in the cell wall and protoplasts of plant cells. Several studies have reported the role of these proteolytic enzymes secreted by fungi in degradation of protein molecules during pathogenesis (Chouhan & Nagar, 1979). Proteolytic enzymes of the pathogens degrade proteins, which ultimately yield amino acids and aminides as the degraded products, in turn serve as nitrogen source for the pathogens. Dubey and Dwivedi (1988b) studied production of proteolytic enzymes in vitro and reported the production of protease enzyme in soybean seed extract (SSE) broth. They also reported an increase in the protease activity with increase in culture age, but protease production was relatively lower as compared to other enzymes. Sinha et al. (2022) carried out secretome analysis of *M. phaseolina* and reported a number of proteases in the solid-state culture of *M. phaseolina*. These proteases are sought to play an important role as putative virulence factors in plant–pathogen interactions. Proteolytic cleavage is usually dependent on proteases that are classified into two types: (i) proteinases which breakdown proteins into polypeptides and (ii) peptidases that hydrolyze polypeptides into small peptides and amino acids.

10.6.5 Hemicellulose-degrading enzymes

Another group of plant cell wall polysaccharides is the hemicellulose that comprises of xyloglucans, xylans, and galactomannans. Along with other carbohydrates and pectin, it surrounds the cellulosic fibers of plant cells. It has random, amorphous structure with little strength as compared to cellulose, which is crystalline and strong. The hemicellulolytic activity of several fungi is due to the presences of xyloglucanases and xylanases. The major constituent of hemicellulose is xylan, and its hydrolysis is carried out by the action of two enzymes, endo-β-1,4-xylanase and β-xylosidase. Reddy and Reddy (1985) reported the production of glucanase and xylanase by *M. phaseolina* causing seed-rot in *Sesamum indicum*. Depolymerization of hemicellulose fraction, containing mostly xylan, xylose, and other sugars, is carried out by xylanases. Xylanases are the members of the family GH10 and GH11, whereas GH10 family xylanases are more abundant in fungi (Kubicek et al., 2014). Both the mycelial and sclerotial isolates of *M. phaseolina* showed high xylanase activity with the latter showing higher activity than the former (Kaur et al., 2012). All six isolates of *M. phaseolina* of Argentinean soil were able to grow on medium supplemented with xylan as a carbon source and produced xylanolytic enzymes (Ramos et al., 2016).

Furthermore, several other enzymes have also been reported in the secretome of *M. phaseolina* that may also play an important role in its pathogenesis. Sinha et al. (2022) reported the secretion of neuraminidase, an exoglycosidase that can cleave glycoconjugates and release terminal sialic acid residues, from *M. phaseolina* secretome collected from wheat bran matrix. The secretion of neuraminidase corroborated with the findings of Bhowal et al. (2005) who also showed neuraminidase production by *M. phaseolina* during sialic acid pathway for host recognition and infection.Furthermore, the study also reveals the production of other enzymes, such as malate dehydrogenase, feruloyl esterases, triose phosphate isomerase, thioredoxin, etc.

10.7 Role of toxins in pathogenesis by *M. phaseolina*

M. phaseolina produces a wide range of secondary metabolites that are phytotoxic in nature. These metabolites play an important role in establishment of disease in their hosts by aiding the fungus in initial stages of infection including adhesion and evasion of defense responses of host leading to wilting and necrosis in different plant parts. Siddiqui et al. (1979) first time identified phytotoxin in the culture filtrates of *M. phaseolina*, whose structure was later described as an eremophilane sesquiterpenoid, specifically an epoxidized analogue of phomenone. Phaseolinone, botryodiplodin, and patulin are among the major toxins produced by *M. phaseolina*; however, some studies reported the production of other phytotoxins, such as mullein, moniliformin, tyrosol, phomenone, azelaic acid, asperlin, etc. (Dhar et al., 1982; Mahato et al., 1987; Salvatore et al., 2020).

A wide variety of fungi produce phaseolinone, which is a heat-resistant and nonhost specific exotoxin. Very low concentration of phaseolinone reduces seed germination of *Phaseolus mungo* and also known to cause seedling wilting and necrotic lesions in leaves of variety of hosts (Bhattacharya et al., 1987; Bilgrami et al., 1979). Phaseolinone has an eremophilane structure and is an ll,12-epoxide of phomenone (Fig. 10.4). Bhattacharya et al. (1992) found 50% inhibition in seed germination in *P. mungo* at 0.60 μg/g toxin concentration of wet tissue. Furthermore, phaseolinone has also been reported to suppress the induced resistance in host plants that make them susceptible to infection by fungal pathogens. When avirulent and nontoxigenic strains of *M. phaseolina* and *Aspergillus fumigatus* were inoculated in *Phaseolus mungo* seedlings in the presence of phaseolinone, there was initiation of infection. It was found that phaseolinone was necessary

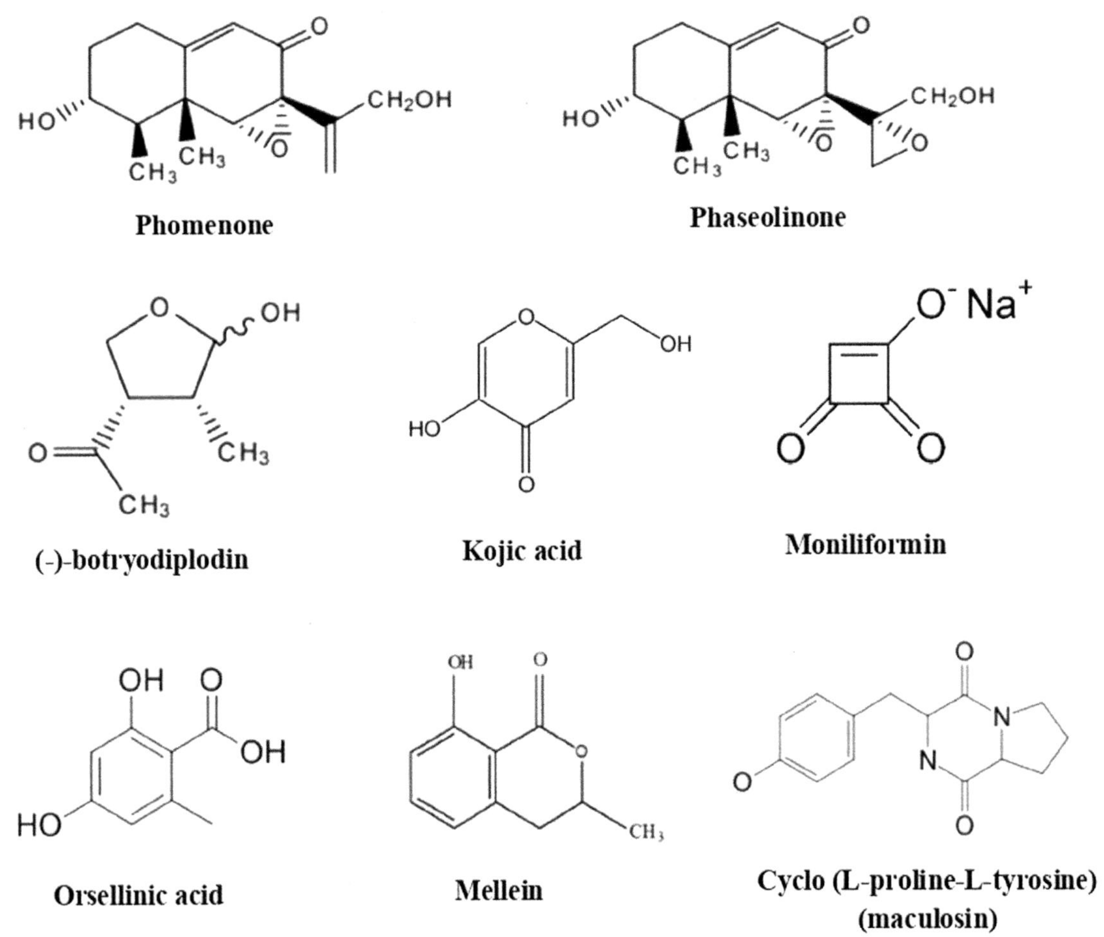

FIGURE 10.4 Chemical structure of different toxins produced by *M. phaseolina*.

for disease development in *P. mungo* (Sett et al., 2000). However, the role of phaseolinone in the development of disease is still not very clear.

The (−)-botryodiplodin is another mycotoxin secreted by *M. phaseolina*. For the first time, it was isolated from the culture filtrates of *Botryodiplodia theobromae* Pat. (syn. *Lasiodiplodia theobromae* [Pat.] Griffon & Maubl.), which is a cellulolytic fungus infecting a wide range of crops in tropical and sub-tropical regions around the world (San Gupta et al., 1966). Production of a significant amount of (−)-botryodiplodin was reported in soybean cultivators of Mississippi infected by *M. phaseolina*, while there was no production of phaseolinone (Ramezani et al., 2007). Mycotoxins including (−)-botryodiplodin help the fungus in the initial stages of infection as they cause necrosis in plant tissues and facilitate in the entry of fungi. Shier et al. (2012) investigated the effects of (−)-botryodiplodin on soybean seedlings and reported that (−)-botryodiplodin selectively killed the growing meristematic tissues at root tips. Abbas et al. (2020) reported growth inhibition of lateral and tap root with abnormal cellular structure of root tips in soybean seedlings due to phytotoxic activity of (−)-botryodiplodin.

Khambhati et al. (2020) carried out LC-MS/MS analysis of cell-free culture filtrates of 89 isolates of *M. phaseolina* causing charcoal rot in soybean for identification and quantification of secondary metabolites. They reported a total of 15 metabolites including six metabolites having frequency greater than 19%, such as kojic acid, (−)-botryodiplodin, moniliformin, orsellinic acid, mullein, and cyclo (L-proline-L-tyrosine) (i.e., maculosin) with frequencies of84.3%, 82%, 61.8%, 49.4%, 28.31%, and 19.1% of isolates, respectively (Fig. 10.4), while the others were in small frequency of <19%, namely cordycepin, emodin, endocrocin, citrinin, gliocladic acid, infectopyron, methylorsellinic acid, monocerin, and N-benzoyl-phenylalanine. In addition, other nine metabolites were also reported, but they had a relatively low frequency (<5%). All these metabolites are believed to be involved in pathogenesis of *M. phaseolina,* but further work is required to strengthen this statement.

Draft genome sequencing of *M. phaseolina* revealed a very high number (75) of putative secondary metabolite backbone genes. These genes regulate the production of several secondary metabolites including mycotoxin that are required for pathogenesis of several phytopathogenic fungi. Fifteen nonribosomal peptide synthesases (NRPS) and 13 NRPS like gene families regulating toxin production were reported in the genome of *M. phaseolina*. The number of these secondary metabolite backbone genes present in *M. phaseolina* was much higher in comparison to the other fungal pathogens (Islam et al., 2012). However, there has been an inconsistency regarding the type of toxins produced by different isolates of *M. phaseolina* in Asia and in North America. This variation could have been due to the divergent evolution on the two continents that dates back to 255 million years and resulted in the accretion of different set of genes that regulate mycotoxin synthesis (Shier et al., 2007). Since there are contradictory reports regarding toxin production by *M. phaseolina*, comprehensive research must be carried out in future to elucidate toxins of *M. phaseolina* along with its analogs.

10.8 *Macrophomina*—host plants interaction

Several genomic studies, such as proteomics and transcriptomics, have been carried out over the past years to understand the type of interactions between *M. phaseolina* and its host plants. These interaction studies were carried out to better understanding the underlying mechanisms of defense in host plants against *M. phaseolina*. Biswas et al. (2014) carried out micro-RNA analysis and transcriptomic studies to understand the levels of resistance in a recombinant inbred line (RIL) population of jute against *Macrophomina* infection. From the 177 RIL lines, a significant number of defense-triggering genes were identified that were related to cell wall biosynthesis, reactive oxygen species (ROS), salicylic acid (SA), ethylene, jasmonic acid (JA), abscisic acid (ABA), hormone signaling, hypersensitive response (HR), and programmed cell death (PCD) pathways. Furthermore, in microRNA analysis, it was revealed that the highly abundant 22-nt miRNA families possessing an innate ability for triggering phased small RNA cascades in SA/JA/ABA-mediated natural SAR resistance, thereby imparting resistance to *M. phaseolina* infection in jute. Dey et al. (2016) suggested that miR-845b and miR-166 superfamilies are attributed to provide a multilayered defense against *M. phaseolina* in jute through nucleotide binding site (NBS) and leucine-rich repeat (LRR) along with ROS mediated defense.

Studies with respect to sorghum cultivators reported that during the early hours of infection by *M. phaseolina*, overexpression of antifungal genes regulating chitinase (*SbChit*) and stilbene synthase production (*SbSTS1*) occurs with maximum expression levels as observed at 72 and 24 h post inoculation, respectively. However, gene expression was even higher in resistant varieties of sorghum (Sharma et al., 2014). Iwuala et al. (2020) carried out gene expression analysis of two antifungal genes viz., chitinase (*AhChit*) and β-1,3-glucanase (*AhGlu*), in two genotypes of groundnut (*Arachis hypogea* L.) viz., SAMNUT 21 and SAMNUT 23. The expression of these two genes was found to be higher in SAMNUT 21 in comparison to SAMNUT 23.

Soybean is among the most prominent agronomic crop grown all over the world. Charcoal rot of soybean is the most common and widely studied disease in the world. As a result, a good amount of work has been carried out for understanding the interactions of *M. phaseolina* with soybean. Bellaloui et al. (2012) studied to understand how total phenolics of seeds, along with lignin, isoflavones, sugars, and boron impart resistance in soybean against toxin-mediated *Macrophomina* infection. The study was carried out under irrigated and nonirrigated conditions using susceptible (S) as well as moderately resistant (MR) genotypes of soybean. Under irrigated, nonirrigated, experimentally infested, and noninfested conditions, MR genotypes showed significantly increased levels of total phenolics, lignin, isoflavones, and total boron as compared to S genotypes. These results indicate toward the role of these substances in imparting resistance to *M. phaseolina* infection in soybean. Phenylpropanoids, such as phytoalexins, are among the secondary metabolites that are known to interact with pathogens in plants and have strong antagonistic and suppressive effects on them. The measurement of effect of *M. phaseolina* colonization was carried out using these genes of transgenic soybean lines that suppresses glyceollin (the collective name for soybean phytoalexins) biosynthesis. These transformed soybeans significantly reduced glyceollin producing capacity of roots and increased susceptibility to infection by pathogen, thus indicating the glyceollin accumulation as an important compound that may trigger innate defense response in soybean (Lygin et al., 2013). Genome-wide association studies (GWAS) provided useful information for understanding the genes responsible for imparting resistance and in the dissection of the genetic architecture of resistance. GWAS study was carried out for identifying the genes and molecular markers responsible for imparting resistance to charcoal rot in soybean germplasm using field and greenhouse experiments (Coser et al., 2017).

Marquez et al. (2018) carried out studies regarding the transcriptional changes occurring in mycorrhizal as well as nonmycorrhizal soybean plants after being infected by *M. phaseolina* in *vitro*. Numerous plant defense genes associated with hormone signaling, such as PR proteins, disease-resistance proteins, transcription factors, and genes involved in secondary metabolism, were revealed to have increased expression. The study also showed that premycorrhized plantlets have lower frequency of upregulation of defense-related genes as compared to the nonmycorrhized ones, post inoculation with *Macrophomina*. Since, *Macrophomina* infection being highly dependable on environmental conditions, the application of different temperatures for better characterization of the resistance levels allows the selection of plant cultivars most appropriately for different climatic conditions (de Sousa Linhares et al., 2020).

10.9 Conclusion

When we talk about the pathogenesis of *M. phaseolina*, several environmental as well as virulence factors affect disease cycle of the fungus. The environmental factors, such as temperature, soil moisture, etc., play an important role in epidemiology and germination of microsclerotia. On the other hand, virulence factors are responsible for the events that take place when the fungus enters inside host tissues. These virulence factors include the genes controlling the production of several CWDEs along with toxins produced by *M. phaseolina*. CWDEs, such as pectin methyl-esterase (PME), polymethylgalacturonase (PMG) and polygalacturonase (PG), endoglucanase, laccases, etc. that are responsible for the entry of *Macrophomina* inside plant tissues by overcoming the cell wall barrier of host plant. Apart from these enzymes, various studied have shown the presence of several toxins, such as phaseolinone and (−)-botryodiplodin in parts of plants infected by *M. phaseolina* and delineated toward the possible role of these toxins in occurrence of symptoms of *Macrophomina* infection. Further research is to be carried out in order to deduce the effects of toxins on plant growth and germination. In addition to these factors, the interaction of *M. phaseolina* with the microbes residing in the rhizosphere also influences its pathogenesis. But some of these micro- and macro-organisms show antagonistic activity against *Macrophomina*, while some have showed synergistic relationship as well. Apart from the interaction with rhizosphere entities, studies should also be focused to understand the underlying signaling mechanisms between *M. phaseolina* and the host plant responsible for entry of the fungus with emphasis on host specificity.

References

Abawi, G. S., & Corrales, P. (1990). Seed transmission and effects of fungicide seed treatments against *Macrophomina phaseolina* in dry edible beans. *Turrialba, 40*, 334–339.

Abawi, G. S., & Pastor-Corrales, M. A. (1990). *Root rots of beans in Latin America and Africa: Diagnosis, research methodologies, and management strategies*. Cali, CO: Centro Internacional de Agricultura Tropical (CIAT).

Abbas, H. K., Bellaloui, N., Butler, A. M., Nelson, J. L., Abou-Karam, M., & Shier, W. T. (2020). Phytotoxic responses of soybean (*Glycine max* L.) to Botryodiplodin, a toxin produced by the charcoal rot disease fungus, *Macrophomina phaseolina*. *Toxins, 12*(1). https://doi.org/10.3390/toxins12010025

Adam, T. (1986). *Contribution à la connaissance des maladies du niébé (*Vigna unguiculata *(L.)Walp.) au Niger avec mention spéciale au* Macrophomina phaseolina *(Tassi) Goïd*. Université de Renne I.

Al-Ani, L. K. T., Soares, F. E. D. F., Sharma, A., de los Santos Villalobos, S., Valdivia-Padilla, A. V., & Aguilar-Marcelino, L. (2022). Strategy of nematophagous fungi in determining the activity of plant parasitic nematodes and their prospective role in sustainable agriculture. *Frontiers in Fungal Biology, 3*, 863198. https://doi.org/10.3389/ffunb.2022.863198

Al-Hazmi, A. S. (1984). Host-index of plant parasitic nematodes in the Kingdom of Saudi Arabia. *Journal of the College of Agriculture, King Saud University, 6*, 69–85.

Ali, M. M., Sayem, A. Z. M., Alam, S., & Ishaque, M. (1969). Relationship of pectic enzyme of *Macrophomina phaseoli* with stem-rot disease and retting of jute. *Mycopathologia et Mycologia Applicata, 38*(3), 289–298. https://doi.org/10.1007/BF02052681

Ali, Y., Sikder, M. O. F., Marzan, L. W., Sharmenb, F., Khan, M. M., & Hosain, M. A. (2014). In silico analysis of chloroperoxidase and lignin peroxidase of pathogenic fungus *Macrophomina phaseolina*. *International Journal of Biological & Medical Research, 5*(3), 4246–4257.

Almeida, Á. M. R., Abdelnoor, R. V., Arias, C. A. A., Carvalho, V. P., Filho, D. S. J., et al. (2003). Genotypic diversity among Brazilian isolates of *Macrophomina phaseolina* revealed by RAPD. *SciELOBrasil, 28*(3), 279.

Ammon, V., & Wyllie, T. D. (1972). Penetration and host parasite relationships of *Macrophomina phaseolina* on *Glycine max* L. *Phytopathology, 62*(7), 743–744.

Ammon, V., Wyllie, T. D., & Brown, M. F. (1975). Investigation of the infection process of *Macrophomina phaseolina* on the surface of soybean roots using scanning electron microscopy. *Mycopathologia, 55*(2), 77–81. https://doi.org/10.1007/BF00444275

Babu, B. K., Reddy, S. S., Yadav, M. K., Sukumar, M., Mishra, V., Saxena, A. K., et al. (2010). Genetic diversity of *Macrophomina phaseolina* isolates from certain agro-climatic regions of India by using RAPD markers. *Indian Journal of Microbiology, 50*(2), 199–204. https://doi.org/10.1007/s12088-010-0033-x

Back, M. A., Haydock, P. P. J., & Jenkinson, P. (2002). Disease complexes involving plant parasitic nematodes and soilborne pathogens. *Plant Pathology, 51*(6), 683–697. https://doi.org/10.1046/J.1365-3059.2002.00785.X

Bandara, Y., Weerasooriya, D., Liu, S., & Little, C. R. (2018). The necrotrophic fungus *Macrophomina phaseolina* promotes charcoal rot susceptibility in grain Sorghum through induced host cell-wall-degrading enzymes. *American Phytopathological Society, 108*(8), 948–956. https://doi.org/10.1094/PHYTO-12-17-0404-R

Bellaloui, N., Mengistu, A., Zobiole, L. H. S., & Shier, W. T. (2012). Resistance to toxin-mediated fungal infection: Role of lignins, isoflavones, other seed phenolics, sugars, and boron in the mechanism of resistance to charcoal rot disease in soybean. *Toxin Reviews, 31*(1–2), 16–26. https://doi.org/10.3109/15569543.2012.691150

Bhattacharya, D., Dhar, T. K., & Ali, E. (1992). An enzyme immunoassay of phaseolinone and its application in estimation of the amount of toxin in *Macrophomina phaseolina*-infected seeds. *Applied and Environmental Microbiology, 58*(6), 1970–1974. https://doi.org/10.1128/AEM.58.6.1970-1974.1992

Bhattacharya, G., Dhar, T. K., Bhattacharyya, F. K., & Siddiqui, K. A. L. (1987). Mutagenic action of phaseolinone, a mycotoxin isolated from *Macrophomina phaseolina. CSIRO Publishing, 40*, 349–353.

Bhowal, J., Ghosh, S., Guha, A. K., & Chaterjee, B. P. (2010). Infection of jute seedlings by the phytopathogenic fungus *Macrophomina phaseolina* mediated by endogenous lectin. *Research Journal of Microbiology, 5*, 657–666.

Bhowal, J., Guha, A. K., & Chatterjee, B. P. (2005). Purification and molecular characterization of a sialic acid specific lectin from the phytopathogenic fungus *Macrophomina phaseolina. Carbohydrate Research, 340*(12), 1973–1982.

Bilgrami, K. S., & Jamaluddin, R. M. A. (1979). *The fungi of India: Part I.* New Delhi: Today and Tomorrow's Printers and Publishers.

Biswas, C., Dey, P., Karmakar, P., & Satpathy, S. (2014). Next-generation sequencing and micro RNAs analysis reveal SA/MeJA1/ABA pathway genes mediated systemic acquired resistance (SAR) and its master regulation. *Physiological and Molecular Plant Pathology, 87*, 76–85. https://doi.org/10.1016/j.pmpp.2014.07.003

Bowers, G. R., & Russin, J. S. (1999). Soybean disease management. In L. G. Heatherly, & H. F. Hodges (Eds.), *Soybean production in the midsouth* (pp. 231–271). Boca Raton, FL: CRC Press.

Bruton, B. D., & Reuveni, R. (1985). Vertical distribution of microsclerotia of *Macrophomina phaseolina* under various soil types and host crops. *Agriculture, Ecosystems & Environment, 12*(2), 165–169.

Chandra, S., Kehri, H. K., & Maheshwari, S. (1995). *Macrophomina* dry root rot of leguminous crops and its management through VAM biopesticides. In J. P. Verma, A. Varma, & D. Kumar (Eds.), *Detection of plant pathogens and their management* (pp. 357–364). New Delhi: Angkor Publishers (P) Ltd.

Chaparro, J. M., Badri, D. V., Bakker, M. G., Sugiyama, A., Manter, D. K., & Vivanco, J. M. (2013). Root Exudation of phytochemicals in *Arabidopsis* follows specific patterns that are developmentally programmed and correlate with soil microbial functions. *PLoS One, 8*(2). https://doi.org/10.1371/JOURNAL.PONE.0055731

Chouhan, S. K., & Nagar, M. L. (1979). Study of pectolytic and cellulolytic enzymes secreted by *Fusarium moniliforme* var. *subglutinans* and *Mucor hemalis* pathogenic to maize (*Zea mays* L.). In G. P. Agrawal, & K. S. Biligrami (Eds.), *Physiology of parasitism* (pp. 171–178). New Delhi: Today and Tomorrow, Publisherand Printer.

Cook, G. E., Boosalis, M. G., Dunkle, L. D., & Odvody, G. N. (1973). Survival of *Macrophomina phaseoli* in corn and sorghum stalk residue. *Plant Disease Reporter, 57*(10), 873–875.

Cooper, R. M., & Wood, R. K. S. (1973). Induction of synthesis of extracellular cell-wall degrading enzymes in vascular wilt fungi. *Nature, 246*, 309–311.

Coser, S. M., Reddy, R. V. C., Zhang, J., Mueller, D. S., Mengistu, A., Wise, K. A., et al. (2017). Genetic architecture of charcoal rot (*Macrophomina phaseolina*) resistance in soybean revealed using a diverse panel. *Frontiers of Plant Science, 8*, 1626. https://doi.org/10.3389/FPLS.2017.01626

Couturier, M., & Berrin, J. G. (2013). The saccharification step: The main enzymatic components. In V. Faraco (Ed.), *Lignocellulose conversion* (pp. 93–110). Berlin: Heidelberg: Springer. https://doi.org/10.1007/978-3-642-37861-4_5

Crous, P. W., Slippers, B., Wingfild, M. J., Rheeder, J., Marasas, W. F. O., Philips, A. J. L., et al. (2006). Phylogenetic lineages in the botryosphaeriaceae. *Studies in Mycology, 55*, 235–253. https://doi.org/10.3114/sim.55.1.235

Das, I. K., Fakrudin, B., & Arora, D. K. (2008). RAPD cluster analysis and chlorate sensitivity of some Indian isolates of *Macrophomina phaseolina* from sorghum and their relationships with pathogenicity. *Microbiological Research, 163*(2), 215–224.

Dashtban, M., Schraft, H., & Qin, W. (2009). Fungal bioconversion of lignocellulosic residues; opportunities & perspectives. *International Journal of Biological Sciences, 5*(6), 578–595.

Degani, O., Becher, P., & Gordani, A. (2022). Pathogenic interactions between *Macrophomina phaseolina* and *Magnaporthiopsis maydis* in mutually infected cotton sprouts. *Agriculture, 12*(2). https://doi.org/10.3390/agriculture12020255

Dey, P., Biswas, C., & Karmakar, P. G. (2016). Identification and characterization of differentially expressed novel miRNAs (21–24 nt) in a *Macrophomina phaseolina* resistant RIL line of jute (*Corchorus capsularis* L.). *Physiological and Molecular Plant Pathology, 94*, 62–66. https://doi.org/10.1016/j.pmpp.2016.04.005

Dey, K., & Ghosh, S. (2022). Arbuscular mycorrhizae in plant immunity and crop pathogen control. *Rhizosphere, 22*, 100524. https://doi.org/10.1186/1471-2164-13-493

Dhar, T. K., Siddiqui, K. A., & Ali, E. (1982). Structure of phaseolinone, a novel phytotoxin from *Macrophomina phaseolina. Tetrahedron Letters, 23*(51), 5459–5462.

Dhingra, O. D., & Chagas, D. (1981). Effect of soil temperature, moisture, and nitrogen on competitive saprophytic ability of *Macrophomina phaseolina. Transactions of the British Mycological Society, 11*(1), 15–20.

Dhingra, O. D., & Sinclair, J. B. (1975). Survival of *Macrophomina phaseolina* sclerotia in soil: Effects of soil moisture, carbon: Nitrogen ratios, carbon sources, and nitrogen concentration. *Phytopathology, 65*, 236–240.

Dhingra, O. D., & Sinclair, J. B. (1977). *An annotated bibliography of* Macrophomina phaseolina *1905–1975* (p. 244). Urbana, USA Brazil: University of Illinois. Universidade Federal de Vicosa.

Dhingra, O. D., & Sinclair, J. B. (1978). *Biology and pathology of macrophomina phaseolina*. Minas Gerais: Universidade Federal de Viçosa: Impresa Universitaria.

Dubey, R. C. (1992). Effects of soil solarization on the survival of of *Macrophomina phaseolina* in fungicide-amended and unamended soils. *Tropical Science, 32*(3), 275–279.

Dubey, R. C., & Dwivedi, R. S. (1988a). Effect of heavy metals on growth and survival of *Macrophomina phaseolina* (Tassi) Goid. *Biology and Fertility of Soils, 6*(4), 311–314. https://doi.org/10.1007/BF00261019

Dubey, R. C., & Dwivedi, R. S. (1988b). *In vitro* production of pectolytic, cellulolytic, amylolytic and proteolytic enzymes by *Macrophomina phaseolina* (Tassi) Goid. *Acta Botanica Indica, 16*(2), 175–181.

Dubey, P., Gupta, G. P., & Dubey, R. C. (2012). Culture filtrates of plant growth promoting *Bradyrhizobium* sp. (*Vigna*) strains VR1 and VR2 inhibit growth and sclerotia germination of *Macrophomina phaseolina in vitro*. *New York Science Journal, 5*, 1–9.

Dubey, R. C., Maheshwari, D. K., Kumar, V., & Pandey, R. R. (2011). Growth enhancement of *Sesamum indicum* L. by rhizosphere-competent *Azotobacter chroococcum* AZO2 and its antagonistic activity against *Macrophomina phaseolina*. *Archives of Phytopathology and Plant Protection, 45*(4), 437–454. https://doi.org/10.1080/03235408.2011.587979

Dwivedi, R. S., & Dubey, R. C. (1987). Effect of fungicides on survival of *Macrophomina phaseolina* in soil and in soybean stem in soil. *International Journal of tropical Plant Diseases, 5*(2), 147–152.

Dwivedi, A. K., Singh, T., & Singh, D. (1994). Histopathology of garlic bulbs infected with *Rhizoctonia bataticola*. *Phytomorphology, 44*, 3–4.

Filho, S. E., & Dhingra, O. D. (1980). Effect of herbicides on survival of *Macrophomina phaseolina* in soil. *Transactions of the British Mycological Society, 74*(1), 61–64.

Gajera, H. P., Bambharoliam, R. P., Patel, S. V., Khatrani, T. J., & Goalkiya, B. A. (2012). Antagonism of *Trichoderma* spp. against *Macrophomina phaseolina*: Evaluation of coiling and cell wall degrading enzymatic activities. *Journal of Plant Pathology & Microbiology, 3*(7), 149. https://doi.org/10.4172/2157-7471.1000149

Gangopadhyay, S., Agarwal, D. K., Sarbhoy, A. K., & Wadhi, S. R. (1974). Charcoal rot disease of soybean in India. *Indian Phytopathology, 26*, 730–732.

Gangopadhyay, S., Wyllie, T. D., & Teague, W. R. (1982). Effect of bulk density and moisture content of soil on the survival of *Macrophomina phaseolina*. *Plant and Soil, 68*(2), 241–247. https://doi.org/10.1007/BF02373710

Gaulin, E., Jauneau, A., Villalba, F., Rickauer, M., Esquerré-Tugayé, M. T., & Bottin, A. (2002). The CBEL glycoprotein of *Phytophthora parasiticavar-nicotianae* is involved in cell wall deposition and adhesion to cellulosic substrates. *Journal of Cell Science, 115*(23), 4565–4575. https://doi.org/10.1242/jcs.00138

Ghaffar, A. (1968). Interactions of soil fungi with *Macrophomina phaseoli* (Maubl.) Ashby, the cause of root rot of cotton. *Mycopathologia et Mycologia Applicata, 34*(2), 196–201. https://doi.org/10.1007/BF02051428

Ghaffar, A., & Erwin, D. C. (1969). Effect of soil water stress on root rot of cotton caused by *Macrophomina phaseoli*. *Phytopathology, 59*(6), 795–797.

Gray, F. A., Mihail, J. D., Lavigne, R. J., & Porter, P. M. (1991). Incidence of charcoal rot of sorghum and soil populations of *Macrophomina phaseolina* associated with sorghum and native vegetation in Somalia. *Mycopathologia, 114*(3), 145–151. https://doi.org/10.1007/BF00437203

Gupta, D. C., & Metha, N. (1989). Interaction studies between different levels of *Meloidogyne javanica* and *Rhizoctonia* spp. in mung bean (*Vigna radiata* L.) Wilczek. *Indian Journal of Nematology, 19*, 138–143.

Heiler, S., Mendgen, K., & Deising, H. (1993). Cellulolytic enzymes of the obligately biotrophic rust fungus *Uromyces viciae-fabae* are regulated differentiation-specifically. *Mycological Research, 97*(1), 77–85.

Herrera, M. D. V., Toro, M. E., de Figueroa, L. I., & Vazquez, F. (2009). Extracellular hydrolytic enzymes produced by phytopathogenic fungi. In J. M. Walker, J. F. T. Spencer, & A. L. Ragout de Spencer (Eds.), *Environmental microbiology: Vol. 16.Methods in biotechnology* (pp. 299–321). Humana Press. https://doi.org/10.1385/1-59259-765-3:299

Hyder, S., Inam-ul-Haq, M., Bibi, S., Humayun, A., Ghuffar, S., & Iqbal, S. (2017). Novel potential of *Trichoderma* spp. as biocontrol agent. *Journal of Entomology and Zoology Studies, 5*(4), 214–222.

Islam, M. S., Haque, M. S., Islam, M. M., Emdad, E. M., Halim, A., Hossen, Q. M. M., et al. (2012). Tools to kill: Genome of one of the most destructive plant pathogenic fungi *Macrophomina phaseolina*. *BMC Genomics, 13*(1), 1–16.

Israel, S., Mawar, R., & Lodha, S. (2005). Soil solarisation, amendments and bio-control agents for the control of *Macrophomina phaseolina* and *Fusarium oxysporum* f.sp. *cumini* in aridisols. *Annals of Applied Biology, 146*(4), 481–491. https://doi.org/10.1111/J.1744-7348.2005.040127.X

Iwuala, E., Odjegba, V., Unung, O., & Alam, A. (2020). Expression of stress responsive β-1, 3-glucanase and chitinase genes in *Arachis hypogaea* seedlings against *Macrophomina phaseolina*. *Gene Reports, 20*, 100693.

Jacob, C. J., Krarup, C., Díaz, G. A., & Latorre, B. A. (2013). A severe outbreak of charcoal rot in cantaloupe melon caused by *Macrophomina phaseolina* in Chile. *Plant Disease, 97*(1), 141. https://doi.org/10.1094/PDIS-06-12-0588-PDN

Jalali, B. L., Chhabra, M. L., & Singh, R. P. (1990). Interaction between vesicular-arbuscular mycorrhizal endophyte and *Macrophomina phaseolina* in mungbean. *Indian Phytopathology, 43*(4), 527–530.

Jana, T., Sharma, T. R., Prasad, R. D., & Arora, D. K. (2003). Molecular characterization of *Macrophomina phaseolina* and *Fusarium* species by a single primer RAPD technique. *Microbiological Research, 158*(3), 249–257.

Jana, T. K., Singh, N. K., Koundal, K. R., & Sharma, T. R. (2005). Genetic differentiation of charcoal rot pathogen, *Macrophomina phaseolina*, into specific groups using URP-PCR. *Canadian Journal of Microbiology, 51*(2), 159–164. https://doi.org/10.1139/W04-122

Jhooty, J. S., & Brains, S. S. (1972). Evaluation of different systemic and non-systemic fungicides for the control of damping-off of moong (*Phaseolus aureus*) caused by *Rhizoctonia solani*. *Indian Phytopathology, 25*, 509–512.

Jordaan, E., van der Waals, J. E., & McLaren, N. W. (2019). Effect of irrigation on charcoal rot severity, yield loss and colonization of soybean and sunflower. *Crop Protection, 122*, 63–69.

Kaur, S., Dhillon, G. S., Brar, S. K., Vallad, G. E., Chand, R., & Chauhan, V. B. (2012). Emerging phytopathogen *Macrophomina phaseolina*: Biology, economic importance and current diagnostic trends. *Critical Reviews in Microbiology, 38*(2), 136–151. https://doi.org/10.3109/1040841X.2011.640977

Khambhati, V. H., Abbas, H. K., Sulyok, M., Tomaso-Peterson, M., & Shier, W. T. (2020). First report of the production of mycotoxins and other secondary metabolites by *Macrophomina phaseolina* (Tassi) Goid. Isolates from soybeans (*Glycine max* L.). *Journal of Fungi, 6*(4), 332.

Khan, A. N., Shair, F., Malik, K., Hayat, Z., Khan, M. A., Hafeez, F. Y., et al. (2017). Molecular identification and genetic characterization of *Macrophomina phaseolina* strains causing pathogenicity on sunflower and chickpea. *Frontiers in Microbiology, 8*, 1309. https://doi.org/10.3389/FMICB.2017.01309/FULL

Koike, S. T. (2008). Crown rot of strawberry caused by *Macrophomina phaseolina* in California. *Plant Disease, 92*(8), 1253. https://doi.org/10.1094/PDIS-92-8-1253B

Kracher, D., & Ludwig, R. (2016). Cellobiose dehydrogenase: An essential enzyme for lignocellulose degradation in nature - a review. *Die Bodenkultur: Journal of Land Management, Food and Environment, 67*(3), 145–163. https://doi.org/10.1515/BOKU-2016-0013

Kubicek, C. P., Starr, T. L., & Glass, N. L. (2014). Plant cell wall-degrading enzymes and their secretion in plant-pathogenic fungi. *Annual Review of Phytopathology, 52*(1), 427–451.

Kumar, S., Dheeman, S., Dubey, R. C., Maheshwari, D. K., & Baliyan, N. (2021). Cyclic siloxane biosurfactant-producing *Bacillus cereus* BS14 biocontrols charcoal rot pathogen *Macrophomina phaseolina* and induces growth promotion in *Vigna mungo* L. *Archives of Microbiology, 203*(8), 5043–5054. https://doi.org/10.1007/S00203-021-02492-3

Kumar, P., Dubey, R. C., & Maheshwari, D. K. (2012). *Bacillus* strains isolated from rhizosphere showed plant growth promoting and antagonistic activity against phytopathogens. *Microbiological Research, 167*(8), 493–499. https://doi.org/10.1016/j.micres.2012.05.002

Kumar, P., Pahal, V., Gupta, A., Vadhan, R., Chandra, H., & Dubey, R. C. (2020). Effect of silver nanoparticles and *Bacillus cereus* LPR2 on the growth of Zea mays. *Scientific Reports, 10*(1), 20409.

Lakhran, L., Ahir, R. R., Choudhary, M., & Choudhary, S. (2018). Isolation, purification, identification and pathogenicity of *Macrophomina phaseolina* (Tassi) Goid.caused dry root rot of chickpea. *Journal of Pharmacognosy and Phytochemistry, 7*(3), 3314–3317.

Leach, L. D., & Garber, R. H. (1970). Control of Rhizoctonia. In J. R. Parmeter (Ed.), *Rhizoctonia solani, biology and pathology* (pp. 189–198). Berkeley: University of California Press.

Lodha, S., Mathur, B. K., & Solanki, K. R. (1990). Factors influencing population dynamics of *Macrophomina phaseolina* in arid soils. *Plant and Soil, 125*(1), 75–80. https://doi.org/10.1007/BF00010746

Lodha, S., Sharma, S. K., Mathur, B. K., & Aggarwal, R. K. (2003). Integrating sub-lethal heating with *Brassica* amendments and summer irrigation for control of *Macrophomina phaseolina*. *Plant and Soil, 256*(2), 423–430. https://doi.org/10.1023/A:1026181009751

Lokesh, R., Rakholiya, K. B., & Thesiya, M. R. (2020). Evaluation of different fungicides against *Macrophomina phaseolina* (Tassi) Goid. causing dry root rot of chickpea (*Cicer arietinum* L.) *in vitro*. *International Journal of Current Microbiology and Applied Sciences, 9*(7), 901–911. https://doi.org/10.20546/ijcmas.2020.907.105

Lopez-Nicora, H. D., Carr, J. K., Paul, P. A., Dorrance, A. E., Ralston, T. I., Williams, C. A., et al. (2020). Evaluation of the combined effect of *Heteroderaglycines* and *Macrophomina phaseolina* on Soybean yield in naturally infested fields with spatial regression. *American Phytopathological Society, 110*(2), 406–417. https://doi.org/10.1094/PHYTO-03-19-0087-R

Lukade, C. M., & Rane, M. S. (1994). Effect of organic and inorganic soil amendments on pre and post emergence of root rot and yield of safflower. *Madras Agricultural Journal, 81*(1), 3–4.

Lygin, A. V., Zernova, O. V., Hill, C. B., Kholina, N. A., Widholm, J. M., Hartman, G. L., et al. (2013). Glyceollin is an important component of soybean plant defense against *Phytophthora sojae* and *Macrophomina phaseolina*. *American Phytopathological Society, 103*(10), 984–994. https://doi.org/10.1094/PHYTO-12-12-0328-R

Machado, C. C. (1987). *Macrophomina phaseolina: Biological behavior of isolates, spatial pattern of microsclerotia in the soil, and incidence on soybeans*. University of Illinois at Urbana-Champaign.

Machado, A. R., Pinho, D. B., Soares, D. J., Gomes, A. A. M., & Pereira, O. L. (2019). Bayesian analyses of five gene regions reveal a new phylogenetic species of *Macrophomina* associated with charcoal rot on oilseed crops in Brazil. *European Journal of Plant Pathology, 153*(1), 89–100. https://doi.org/10.1007/s10658-018-1545-1

Madkour, M. A., & Aly, M. H. (1981). Cell wall-degrading enzymes produced during pathogenesis of *Macrophomina phaseolina* on strawberry plants. *Journal of Phytopathology, 100*(1), 36–43. https://doi.org/10.1111/J.1439-0434.1981.TB03288.X

Magnin-Robert, M., le Bourse, D., Markham, J., Dorey, S., Clément, C., Baillieul, F., et al. (2015). Modifications of sphingolipid content affect tolerance to hemibiotrophic and necrotrophic pathogens by modulating plant defense responses in *Arabidopsis*. *Plant Physiology, 169*(3), 2255–2274.

Mahato, S. B., Siddiqui, K. A. I., Bhattacharya, G., Ghosal, T., Miyahara, K., Sholichin, M., & Kawasaki, T. (1987). Structure and stereochemistry of phaseolinic acid: A new acid from *Macrophomina phaseolina*. *Journal of Natural Products, 50*(2), 245–247. https://doi.org/10.1021/np50050a024

Mahdizadeh, V., Safaie, N., & Goltapeh, E. M. (2011). Diversity of *Macrophomina phaseolina* based on morphological and genotypic characteristics in Iran. *Plant Pathology Journal, 27*(2), 128–137.

Mahdizadeh, V., Safaie, N., & Goltapeh, E. M. (2012). Genetic diversity of sesame isolates of *Macrophomina phaseolina* using RAPD and ISSR markers. *Trakia Journal of Sciences, 10*(2), 65–74.

Maheshwari, D. K., Dubey, R. C., & Sharma, V. K. (2001). Biocontrol efects of *Trichoderma virens* on *Macrophomina phaseolina*causing charcoalrot of peanut. *Indian Journal of Microbiology, 41*(4), 251–256.

Marquez, N., Giachero, M. L., Gallou, A., Debat, H. J., Cranenbrouck, S., di Rienzo, J. A., et al. (2018). Transcriptional changes in mycorrhizal and nonmycorrhizal soybean plants upon infection with the fungal pathogen *Macrophomina phaseolina. Molecular Plant-Microbe Interactions, 31*(8), 842–855. https://doi.org/10.1094/MPMI-11-17-0282-R

Martinez-Medina, A., Pozo, M. J., Cammue, B., & Vos, C. M. (2016). Belowground defence strategies in plants: The plant–*Trichoderma*dialogue. In M. F. V. Christine, & K. Kazan (Eds.), *Belowground defence strategies in plants* (pp. 301–327). Cham: Springer. https://link.springer.com/chapter/10.1007/978-3-319-42319-7_13.

Mayek-Pérez, N., GarcÍa-Espinosa, R., López-Castañeda, C., Acosta-Gallegos, J. A., & Simpson, J. (2002). Water relations, histopathology and growth of common bean (*Phaseolus vulgaris* L.) during pathogenesis of *Macrophomina phaseolina* under drought stress. *Physiological and Molecular Plant Pathology, 60*(4), 185–195.

Mayék-Pérez, N., López-Castañeda, C., González-Chavira, M., Garcia-Espinosa, R., Acosta-Gallegos, J., de la Vega, O. M., & Simpson, J. (2001). Variability of Mexican isolates of *Macrophomina phaseolina* based on pathogenesis and AFLP genotype. *Physiological and Molecular Plant Pathology, 59*(5), 257–264.

Mayer, A. M., & Staples, R. C. (2002). Laccase: New functions for an old enzyme. *Phytochemistry, 60*(6), 551–565.

Meyer, W. A., Siclair, J. B., & Khare, M. N. (1973). Biology of *Macrophomina phaseoli* in soil studied with selective media. *Phytopathology, 63*, 613–620.

Mihail, J. D. (1989). *Macrophomina phaseolina*: Spatio-temporal dynamics of inoculum and of disease in a highly susceptible crop. *Phytopathology, 79*(8), 848–855.

Nafady, N. A., Hashem, M., Hassan, E. A., Ahmed, H. A., & Alamri, S. A. (2019). The combined effect of arbuscular mycorrhizae and plant-growth-promoting yeast improves sunflower defense against *Macrophomina phaseolina* diseases. *Biological Control, 138*, 104049.

Noble, M., & Richardson, M. J. (1968). *An annotated list of seed-borne diseases* (p. 191). Commonwealth Mycological Institute. Phytopathological Papers No. 8.

Papavizas, G. C. (1977). Some factors affecting survival of sclerotia of *Macrophomina phaseolina* in soil. *Soil Biology and Biochemistry, 9*(5), 337–341. https://doi.org/10.1016/0038-0717(77)90006-2

Parmar, H. V., Kapadiya, H. J., & Bhaliya, C. M. (2017). Efficacy of different fungicides against *Macrophomina phaseolina* (Tassi) Goid causing castor root rot. *International Journal of Communication Systems, 5*(5), 1807–1809.

Parmeshwarappa, R., Kajjari, N. B., Patil, C. S. P., Thimmiaah, H. C., & Bettsur, S. R. (1976). Charcoal rot incidence recorded at regional research station, Dharwad (Karnataka) during kharif. *Sorghum News, 19*, 1–37.

Pearson, C. A. S., Leslie, J. F., & Schwenk, F. W. (1986). Variable chlorate resistance in *Macrophomina phaseolina* from corn, soybean, and soil. *Phytopathology, 76*(6), 646–649.

Pedgaonkar, S. M., & Mayee, C. D. (1990). Stalk water potential in relation to charcoal rot of sorghum. *Indian Phytopathology, 43*(2), 192–196.

del Poeta, M., Nimrichter, L., Rodrigues, M. L., & Luberto, C. (2014). Synthesis and biological properties of fungal glucosylceramide. *PLoS Pathogens, 10*(1), e1003832. https://doi.org/10.1371/JOURNAL.PPAT.1003832

Purkayastha, S., Kaur, B., Dilbaghi, N., & Chaudhury, A. (2006). Characterization of *Macrophomina phaseolina*, the charcoal rot pathogen of cluster bean, using conventional techniques and PCR-based molecular markers. *Plant Pathology, 55*(1), 106–116. https://doi.org/10.1111/j.1365-3059.2005.01317.x

Ramezani, M., Shier, W. T., Abbas, H. K., Tonos, J. L., Baird, R. E., & Sciumbato, G. L. (2007). Soybean charcoal rot disease fungus *Macrophomina phaseolina* in Mississippi produces the phytotoxin (−)-botryodiplodin but no detectable phaseolinone. *Journal of Natural Products, 70*(1), 128–129. https://doi.org/10.1021/np060480t

Ramos, A. M., Gally, M., Szapiro, G., Itzcovich, T., Carabajal, M., & Levin, L. (2016). *In vitro* growth and cell wall degrading enzyme production by Argentinean isolates of *Macrophomina phaseolina*, the causative agent of charcoal rot in corn. *Revista Argentina de Microbiología, 48*(4), 267–273. https://doi.org/10.1016/j.ram.2016.06.002

Rangel-Montoya, E. A., Delgado-Ramírez, C. S., Sepulveda, E., & Hernández-Martínez, R. (2022). Biocontrol of *Macrophomina phaseolina* using *Bacillus amyloliquefaciens* strains in cowpea (*Vigna unguiculata* L.). *Agronomy, 12*(3), 676. https://doi.org/10.3390/agronomy12030676

Reddy, A., & Reddy, S. (1985). Production of extracellular cellulase-hemicellulase complex by *Macrophomina phaseolina*, the seed-rot fungus of sesamum (*Sesamum indicum*). *Journal of the Indian Botanical Society, 64*(6), 377–380.

Reznikov, S., Chiesa, M. A., Pardo, E. M., de Lisi, V., Bogado, N., González, V., et al. (2019). Soybean-*Macrophomina phaseolina* -specific interactions and identification of a novel source of resistance. *Phytopathology, 109*(1), 63–73. https://doi.org/10.1094/PHYTO-08-17-0287-R

Reznikov, S., Vellicce, G. R., Mengistu, A., Arias, R. S., Gonzalez, V., Lisi, V. D., et al. (2018). Disease incidence of charcoal rot (*Macrophomina phaseolina*) on soybean in north-western Argentina and genetic characteristics of the pathogen. *Canadian Journal of Plant Pathology, 40*(3), 423–433. https://doi.org/10.1080/07060661.2018.1484390

Saha, S. C., Dube, S., Sanyal, A., Dube, D. K., & Sinha, S. N. (1979). Production of extra-cellular enzymes by jute pathogenic fungus *Macrophomina phaseolina. Current Science, 48*(11), 484–485.

Sakai, T., Sakamoto, T., Hallaert, J., & Vandamme, E. J. (1993). Pectin, pectinase, and protopectinase: Production, properties, and applications. *Advances in Applied Microbiology, 39*, 213–294. https://doi.org/10.1016/S0065-2164(08)70597-5

Salvatore, M. M., Félix, C., Lima, F., Ferreira, V., Naviglio, D., Salvatore, F., et al. (2020). Secondary metabolites produced by *Macrophomina phaseolina* isolated from *Eucalyptus globulus*. *Agriculture, 10*(3), 72. https://doi.org/10.3390/agriculture10030072

San Gupta, R., Chandran, R. R., & Divekar, P. V. (1966). Botryodiplodin–a new antibiotic from *Botryodiplodiatheobromae*. II. Production, isolation and biological properties. *Indian Journal of Experimental Biology, 4*(3), 152–153.

Sánchez, S., Chamorro, M., Henríquez, J. L., Grez, J., Díaz, I., Santos, B. D. L., et al. (2017). Genetic and biological characterization of *Macrophomina phaseolina* (Tassi) Goid.causing crown and root rot of strawberry. *Chilean Journal of Agricultural Research, 77*(4), 325–331. https://doi.org/10.4067/S0718-58392017000400325

Sandhu, A., Singh, R. D., & Sandhu, A. (1999). Factors influencing susceptibility of cowpea to *Macrophomina phaseolina*. *Journal of Mycology and Plant Pathology, 29*, 421–424.

Sarr, M. P., Diaye, M. N., Groenewald, J. Z., & Crous, P. (2014). Genetic diversity in*Macrophomina phaseolina*, the causal agent of charcoal rot. *Phytopathologia Mediterranea, 53*(2), 250–268.

Seetharama, N., Bidinger, F. R., Rao, K. N., Gill, K. S., & Mulgund, M. (1987). Effect of pattern and severity of moisture deficit stress on stalk rot incidence in Sorghum I. Use of line source irrigation technique, and the effect of time of inoculation. *Field Crops Research, 15*(3–4), 289–308. https://doi.org/10.1016/0378-4290(87)90018-9

Sett, S., Mishra, S. K., & Siddiqui, K. I. (2000). Avirulent mutants of *Macrophomina phaseolina* and *Aspergillus fumigatus* initiate infection in *Phaseolus mungo* in the presence of phaseolinone; levamisole gives protection. *Journal of Biosciences, 25*(1), 73–80. https://doi.org/10.1007/BF02985184

Sharma, C. K., & Dubey, R. C. (2017). Plant growth promoting and antagonistic properties of *Pseudomonas putida* CRN-09 against *Macrophomina phaseolina* (Tassi) Goid. *Indian Journal of Scientific Research, 13*(2), 1–5.

Sharma, S., & Khan, T. A. (2010). Effect of root-rot fungus (*Macrophomina phaseolina*) on the life cycle of root-knot nematode (*Meloidogyne javanica*) Race-2 on balsam. *Archives of Phytopathology and Plant Protection, 43*(14), 1387–1391. https://doi.org/10.1080/03235400802476518

Sharma, I., Kumari, N., & Sharma, V. (2014). Defense gene expression in *Sorghum bicolor* against *Macrophomina phaseolina* in leaves and roots of susceptible and resistant cultivars. *Journal of Plant Interactions, 9*(1), 315–323. https://doi.org/10.1080/17429145.2013.832425

Sharma, C. K., Vishnoi, V. K., Dubey, R. C., & Maheshwari, D. K. (2018). A twin rhizospheric bacterial consortium induces systemic resistance to a phytopathogen *Macrophomina phaseolina* in mung bean. *Rhizosphere, 5*, 71–75. https://doi.org/10.1016/j.rhisph.2018.01.001

Sheikh, A. H., & Ghaffar, A. (1979). Relation of sclerotial inoculum density and soil moisture to infection of field crops by *Macrophomina phaseolina*. *Pakistan Journal of Botany, 11*(2), 185–189.

Shier, W. T., Abbas, H. K., Baird, R. E., Ramezani, M., & Sciumbato, G. L. (2007). (-)-Botryodiplodin, a unique ribose-analog toxin. *Toxin Reviews, 26*(4), 343–386. https://doi.org/10.1080/15569540701741866

Shier, W. T., Nelson, J., Abbas, H. K., & Baird, R. E. (2012). Root toxicity of the mycotoxin botryodiplodin in soybean seedlings. *Toxicon, 60*, 163.

Short, G. E., & Wyllie, T. D. (1978). Inoculum potential of *Macrophomina phaseolina* [causes of root and stem rot in seed crops]. *Phytopathology, 68*, 742–746.

Siddiqui, K. A. I., Gupta, A. K., Paul, A. K., & Banerjee, A. K. (1979). Purification and properties of a heat-resistant exotoxin produced by*Macrophomina phaseolina* (Tassi) Goid in culture. *Experientia, 35*(9), 1222–1223. https://doi.org/10.1007/BF01963302

Singh, P. J., & Mehrotra, R. S. (1982). Influence of soil moisture and temperature on *Rhizoctonia bataticola* infection of gram [*Cicer arietinum* L.]. *Indian Phytopathology, 35*, 327–329.

Singh, N., Pandey, P., Dubey, R. C., & Maheshwari, D. K. (2008). Biological control of root rot fungus *Macrophomina phaseolina* and growth enhancement of *Pinus roxburghii* (Sarg.) by rhizosphere competent *Bacillus subtilis* BN1. *World Journal of Microbiology and Biotechnology, 24*(9), 1669–1679. https://doi.org/10.1007/s11274-008-9680-z

Sinha, N., Patra, S. K., Sarkar, T. S., & Ghosh, S. (2022). Secretome analysis identified extracellular superoxide dismutase and catalase of *Macrophomina phaseolina*. *Archives of Microbiology, 204*(1), 1–11. https://doi.org/10.1007/s00203-021-02631-w

Soni, K. K., & Dadwal, V. S. (1985). Charcoal root and stem rot of Eucalyptus. *Forest Pathology, 15*(7), 397–401. https://doi.org/10.1111/j.1439-0329.1985.tb00895.x

de Sousa Linhares, C. M., Ambrósio, M. M. Q., Castro, G., Torres, S. B., Esteras, C., de Sousa Nunes, G. H., et al. (2020). Effect of temperature on disease severity of charcoal rot of melons caused by *Macrophomina phaseolina*: Implications for selection of resistance sources. *European Journal of Plant Pathology, 158*(2), 431–441. https://doi.org/10.1007/S10658-020-02083-W

Spagnoletti, F. N., Cornero, M., Chiocchio, V., Lavado, R. S., & Roberts, I. N. (2020). Arbuscular mycorrhiza protects soybean plants against *Macrophomina phaseolina* even under nitrogen fertilization. *European Journal of Plant Pathology, 156*(3), 839–849. https://doi.org/10.1007/s10658-020-01934-w

Sridharan, A. P., Thankappan, S., Karthikeyan, G., & Uthandi, S. (2020). Comprehensive profiling of the VOCs of *Trichoderma longibrachiatum* EF5 while interacting with *Sclerotium rolfsii* and *Macrophomina phaseolina*. *Microbiological Research, 236*, 126436. https://doi.org/10.1016/j.micres.2020.126436

Suárez, Z., Rosales, L. C., Rondón, A., & González, M. S. (1998). Histopatología de raíces de *Psidium guajava*atacadasporelnematodo*Meloidogyne incognita*raza 1 y loshongos*Macrophomina phaseolina* y *Fusarium oxysporum*. *Fitopathol Venezolana, 11*, 44–47.

Sudha, A., & Prabhu, S. (2008). Evaluation of organic soil amendments against *Macrophomina phaseolina* (Tassi). *Journal of Biopestcides, 1*(2), 143–145.

Su, G., Suh, S. O., Schneider, R. W., & Russin, J. S. (2001). Host specialization in the charcoal rot fungus, *Macrophomina phaseolina*. *Phytopathology, 91*(2), 120–126. https://doi.org/10.1094/PHYTO.2001.91.2.120

Taliei, F., Safaie, N., & Aghajani, M. A. (2013). Spatial distribution of *Macrophomina phaseolina* and soybean charcoal rot incidence using geographic information system (a case study in northern Iran). *Journal of Agricultural Science and Technology A, 15*(7), 1523–1536.

TančićŽivanov, S., Dedić, B., Dimitrijević, A., Dušanić, N., Jocić, S., Miklič, V., Kovačević, B., & Miladinović, D. (2019). Analysis of genetic diversity among *Macrophomina phaseolina* (Tassi) Goid.isolates from Euro-Asian countries. *Journal of Plant Diseases and Protection, 126*(6), 565–573. https://doi.org/10.1007/s41348-019-00260-6

Thakur, S., & Jana, T. (2017). Computational analysis reveals industrial natural products and toxins in charcoal rot pathogen, *Macrophomina phaseolina*. *Current Science, 112*(2), 322–331.

Todd, T. C., Pearson, C. A. S., & Schwenk, F. W. (1987). Effect of *Heteroderaglycines* on charcoal rot severity in soybean cultivars resistant and susceptible to soybean cyst nematode. *Journal of Nematology, 19*(Annals 1), 35–40.

Wang, H., & Jones, R. W. (1995). A unique endoglucanase-encoding gene cloned from the phytopathogenic fungus *Macrophomina phaseolina*. *Applied and Environmental Microbiology, 61*(5), 2004–2006. https://doi.org/10.1128/aem.61.5.2004-2006.1995

Wang, Z., Tian, H., Zhang, C. X., Fang, X. L., & Nan, Z. B. (2020). Occurrence of *Macrophomina phaseolina* causing root and crown rot on alfalfa (*Medicago sativa*) in China. *Plant Disease, 104*(9), 2521. https://doi.org/10.1094/PDIS-03-20-0555-PDN

Watanabe, T. (1973). Survivability of *Macrophomina phaseoli* (Maubl.) Ashby in naturally-infested soils and longevity of the sclerotia formed *in vitro*. *Japanese Journal of Phytopathology, 39*(4), 333–337. https://doi.org/10.3186/jjphytopath.39.333

Whiteford, J. R., & Spanu, P. D. (2001). The Hydrophobin HCf-1 of *Cladosporium fulvum* is required for efficient water-mediated dispersal of conidia. *Fungal Genetics and Biology, 32*(3), 159–168. https://doi.org/10.1006/fgbi.2001.1263

Winkler, H. E., Hetrick, B. A. D., & Todd, T. C. (1994). Interactions of *Heteroderaglycines, Macrophomina phaseolina*, and mycorrhizal fungi on soybean in Kansas. *Journal of Nematology, 26*(4S), 675–682.

Wyllie, T. D. (1988). Charcoal rot of soybeans: Current status. In T. D. Wyllie, & S. DH (Eds.), *Soybean diseases of the North central region* (pp. 106–113). St. Paul, MN: APS Press.

Wyllie, T. D., & Brown, M. F. (1970). Ultrastructural formation of sclerotia of Macrophomina phaseoli. *Phytopathology, 60*(3), 524–528.

Xu, J. R., & Hamer, J. E. (1996). MAP kinase and cAMP signaling regulate infection structure formation and pathogenic growth in the rice blast fungus *Magnaporthe grisea*. *Genes & Development, 10*(21), 2696–2706. https://doi.org/10.1101/gad.10.21.2696

Yildiz, A., Benlioğlu, S., Boz, Ö., & Benlioğlu, K. (2010). Use of different plastics for soil solarization in strawberry growth and time–temperature relationships for the control of *Macrophomina phaseolina* and weeds. *Phytoparasitica, 38*(5), 463–473. https://doi.org/10.1007/s12600-010-0123-7

Young, D. J., Gilbertson, R. L., & Alcorn, S. M. (1982). A new record for longevity of *Macrophomina phaseolina* sclerotia [Soil-inhabiting Fungi]. *Mycologia, 74*(3), 504–505.

Chapter 11

Molecular tools for detection and management of *Macrophomina phaseolina*

Ramesh Nath Gupta and Ram Niwas
Department of Plant Pathology, Bihar Agricultural University, Bhagalpur, Bihar, India

11.1 Introduction

Macrophomina phaseolina (Tassi) Goid. is an economically important soil- and seed-borne pathogen. It may infect approximately 500 plant species globally and is found in most farmed soils. Pulses (green gram, chickpea, and cowpea), cereals (sorghum, maize, and wheat), oilseeds (soybean, sunflower, groundnut, and castor), vegetables (potato, brinjal, okra, squash, and bottle gourd), fruits (watermelon and strawberry), fiber crops (jute and cotton), medicinal plants (coleus), and many others are affected by this pathogen. Pathogen causes an increased damage in warm, temperate, and tropical climates. The biology and control of the pathogen have previously been examined by Sharma et al. (2015). This investigation provides a recent research discovery with a focus on the ecology, biology, and control pathogen. Genetic diversity, pathogenic variants, host specialization, molecular and biochemical identification, and contemporary management techniques of this disease have been explored. *Macraphomina phaseolina* (Tassi) Goid. belongs to the Deuteromycotina, class Coelomycetes, order Sphaeropsidales, and family Sphaeropsidaceae. *Rhizoctoniabataticola* Taub. is the imperfect state of the fungus, whereas *M. phaseolina* is the perfect state. *M. phaseoli*, *M. Philippines*, *M. crochori*, *M. cajani*, *M. sesamae*, *Rhizoctonia bataticola*, *R. lamellifera*, and *Dothiorella cajani* are the synonyms of *M. phaseolina* (Das et al., 2015). The size of sclerotia and the presence or absence of pycnidia in *M. phaseolina* vary widely. The mycelium is hyaline to brown, septate, abundantly branched, or dendroid. Sclerotia are a loose form of fungus and varying in brown to black in color, scale-like appearance, irregular in shape, with size varying greatly among isolates. The sclerotia of *M. phaseolina* do not have a well-defined layer of tissues like those of *Sclerotium rolfsii*. The pycnidial stage is usually prevalent in jute and garden beans but not so often in soybean, maize, and sorghum (Raj & Rathi, 2016).

11.2 Symptomatology

M. phaseolina causes extensive damage in terms of wider host range or various plant parts affected. Infection causes a diverse range of symptoms, depends upon host and growth stages of the crop. Damping off and seedling blight (jute, okra, groundnut), dry root rot (most pulses), charcoal rot (sorghum, maize, soybean, groundnut, cotton, potato, sweet potato), and stem rot (sorghum, maize, soya, groundnut, cotton, potato, sweet potato), and seed rot (pigeon pea, chickpea, cowpea, sunflower, muskmelon, squash, bottle gourd, pumpkin) are some common symptoms. Plant stand, vigor, grain yield, and dry fodder quality are adversely affected by the disease (Table 11.1).

Infected small spots may coalesce and eventually build up a big necrotic lesion, causing death of the plant. *M. phaseolina* causes infection in roots resulting in the development of necrotic lesion (Ghosh et al., 2018). *Macrophomina* infections are systemic in origin, and the pathogen travels up to xylem bundles. The entire host–defense system is weakened under abiotic stresses (moisture, temperature, and photosynthetic stressors), and pathogenic activity rises several fold, resulting in fast and widespread rotting of plant tissues. Many black sclerotia are generated on the infected plant tissue, giving the diseased areas a burned look; hence, the term "charcoal rot" is commonly used to characterize the disease

Macrophomina Phaseolina. https://doi.org/10.1016/B978-0-443-15443-0.00005-X

TABLE 11.1 Symptoms produced by *M. phaseolina* on different host.

Symptoms	Host	References
Damping off and seedling blight	Cowpea (*Vigna unguiculata*)	Muchero et al. (2011)
	Jute (*Corehorus* spp.)	Dey et al. (2016)
Stalk rot	Sorghum (sorghum bicolor)	Mahmoud et al. (2018)
Reduction in shoot length, root length, photosynthetic pigments	Rockcress (*Arabidopsis thaliana*)	Saima & Wu (2019)
Induction in secondary metabolism, hormone metabolism and stress	Soybean (*Glycine max*)	Marquez et al. (2018)
Charcoal rot	Strawberry (*Fragaria ananassa*)	Gomez et al. (2020)

symptoms. Severe losses in cereals, such as sorghum and maize, are caused by decaying stalk tissue resulting in a loss of both quantity and quality of dry feed followed by lodging of the diseased plant (loss in quantity and quality of grain). Sorghum has been reported to have lodging of up to 100% and grain yield losses of 23–64% (Badigannavar et al., 2018). Many economically important crops are infected with the pathogen, which can reduce seed germination and seedling vigor. *M. phaseolina* generate at least six phytotoxins, the most common of which is phaseolina. Phaseolinone (LDW 0.98 g/kg body weight) can induce infection (Das et al., 2015) observed phytotoxin component (−)botryodiplodin in the cultures of pathogenic *M. phaseolina* identified in Mississippi in recent investigations. Continuous feeding inebriated (charcoal rotting) sorghum and maize stover to domestic animals might be harmful to their health.

11.3 Causal organism

M. phaseolina is a common soil- and seed-borne plant pathogenic fungus. Here, it is characterized as anamorphic ascomycetes and includes the creation of sclerotia and pycnidia in culture medium and host tissues (Fig. 11.1A). The development of *M. phaseolina* favors at temperatures ranging from 15 to 40°C and produces a grayish, black colony on the PDA medium. However, the best development of pathogen shows at 30–35°C. Some isolates can be distinguished based on their morphological features and thermophilic behavior (Pandey et al., 2019). Sclerotia are variable in nature except few are circular to oblong (Fig. 11.1B).

The septate mycelium is 1.5–2.5 mm broad, at first hyaline and later turns into black in color. Microsclerotia are black in color and vary in size on affected plants and culture medium. Sclerotia are hyaline or light brown, and 89 μm size has been recorded under laboratory conditions, while it measures 50–120 μm under field conditions (Ghosh et al., 2018).

11.4 Disease cycle

Pycnidia of *M. phaseolina* appear in host plants near the close of the wet season, although their epidemiological impact appears to be small. The pycnidiospores are produced on the early infected leaf and stem tissues and considered as

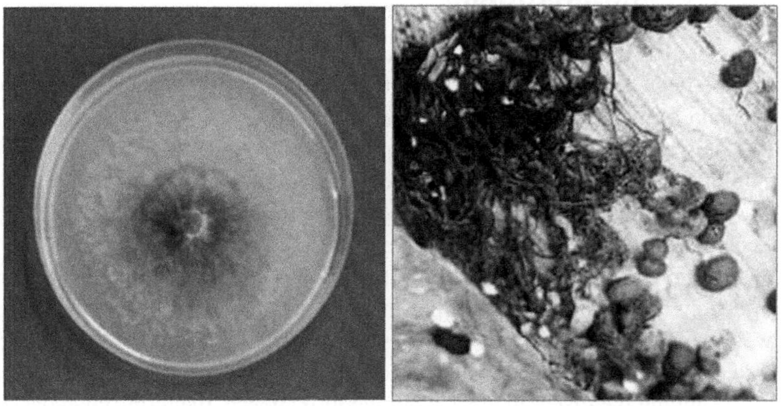

FIGURE 11.1 *M. phaseolina;* (A)-culture; (B)-microsclerotia.

secondary source of disease transmission. Primary inoculum comes from soil microsclerotia, infected seeds, or host tissues. Microsclerotia germination and root infection of hosts are induced by root exudates. Infectious hyphae penetrate inside plant through wounds or root epidermal cells. During early stages of pathogenesis, mycelium enters into epidermis of the root and is mostly confined in intercellular space of root cortex (Fig. 11.2).

As a result, neighboring cells disintegrate, and diseased plants may die extensively. The fungal hypha develops intracellular in xylem and produces microsclerotia, which clog vessels and damage host cells (Kumar et al., 2016). Necrotic lesions can be seen on peduncles, stems, and branches of the diseased plants. Further, the pathogen spreads from the peduncles to pods and ultimately invades the grains. Marquez et al. (2021) reported the development of fungal toxin, such as phaseolinone, formation of fungal tissue which blocks the host vasculature and causes severe infection, finally causing premature plant death. Microsclerotia development in soybean is influenced by blooming and pod setting and leading to death of the host plant (Forster, 2019).

11.5 Molecular identification of the pathogen

Significant progress has been achieved regarding the development in various molecular techniques and technologies in the recent years. They have been utilized to investigate the plant diseases. The genetic differences and taxonomic connections in fungus have been studied by using DNA sequences which encode for RNAs (Liao et al., 2014). Conserved sequences of the pathogen have been identified by small and large subunits of genes which are utilized in investigating diverse connections between related fungi (Edel-Hermann & Lecomte, 2019). Intragenic spacers (IGS) between gene clusters are more diverse than the subunits of internal transcribed spacers (ITS).

Sequencing the ITS1 and ITS2 sections of 18S rRNA gene, isolates of *M. phaseolina* from soybean, mungbean, and urdbean have been identified. Pandey et al. (2020) applied buffer (10% CTAB) along with salt (1.5 M) for heat lysis at 65°C for extracting genomic DNA from every isolate. The amount of genomic DNA, quantified by using Thermo Scientific 1000 spectrophotometer, is used for measuring absorbance at 260 nm wavelength. Purity of the extracted genomic

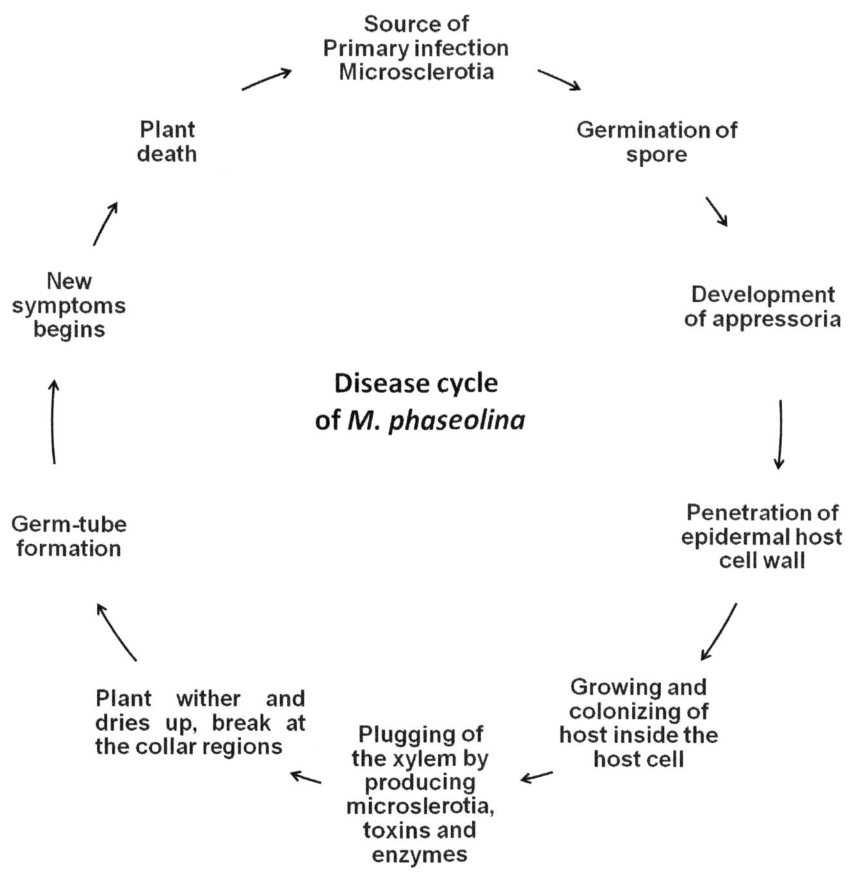

FIGURE 11.2 Disease cycle of *M. phaseolina.*

DNA is determined by electrophoresis with 0.8% agarose gel and measuring absorbance ratio at wavelength 260:280 nm. ITS1, ITS2 rDNA, and 5.8S rDNA clustering genes have amplified by primer 50-TCCGTAGGTGAACCTGCGG-30 for the ITS1 and another primer 50-GCTGCGTTCTTCATCGATGC-30 used for the ITS 2 (White et al., 1990) and an amount of 5–10 ng DNA using for the primer ITS1 and ITS2 in each combination of PCR reaction. KPC Life Sciences used SUPERZymTaq DNA Polymerase for the PCR experiment. In total volume of 50, reactions are conducted by buffer in accordance with the protocol of manufacturer. Initial denaturation temperature 95°C for the 5 min is followed by denaturation at temperature 95°C for 30 s with 30 cycles, annealing at temperature 50.2°C for the 30 s and extension at temperature 72°C for 30 s along with final extension temperature at 72°C for 10 min in the thermo cycler (BioRad). By using buffer 0.5× TAE, the PCR-amplified products have fractionated in 2% agarose gel. ITS amplified band of the interest by agarose gel is purified with the help of UniPro Gel Extraction Kit (KPC Life Sciences). Macrogen Inc., Korea, sequenced the gel-purified DNA utilized in the sequencing procedure.

The ITS region of the fungal isolates genomes has been sequenced in both sense and antisense ways. To find the nearest homolog, sequences were annotated and processed using BLAST. BLAST search in database GenBank is used to validate the sequence identity (NCBI) and preparation of phylogenetic tree.

11.6 Internal transcribed spacer

Forward primers ITS1, ITS5, and reverse primer ITS4 are applied to amplify ITS regions one and two for the gene 5.8S rRNA of *M. phaseolina* isolate (Sarr et al., 2014). Three sections of ITS region including ITS1, gene 5.8 S rRNA, and ITS2 for the *M. phaseolina,* and associated species of fungi are shown to be conserved among *M. phaseolina* isolates. From ITS-1 to ITS-2, they separated the whole sequence into five sections. The region four of aligned sequences of the *M. phaseolina* isolates, which demonstrated the variability could not considered for further study. The 5.8 S RNA genes are sequenced likewise left out of the investigation. Regions one and five are chosen for construction of species specific primers of *M. phaseolina* following rearranging, editing the matched sequences and comparing sequences with close genus of the fungus. Two primers MpKFI and MpKRI along with an oligonucleotide probe, the MpKH1 (19 mer) designed for the conserved region adjoining 5.8 S gene that revealed in region 3. From all tested *M. phaseolina* isolates, chosen primers produced only single 350bp amplified product. In recent investigations, variations in the restriction patterns of ITS region are observed that include a portion of 25 SrDNA (Purkayastha et al., 2006). Sequencing of eight isolates obtained from distinctive hosts and the diverse ecological settings since ITS-RFLP, which does not perceive changes within different isolates in the investigation. Alignment of the sequences with two GenBank sequences revealed a diversity of variable and conserved regions in the *M. phaseolina* ITS sequences. Amplified ITS sequences almost separated into five regions for better understanding of these regions. Region 4 (not shown in the picture) is removed because it contains sequences that differed among *M. phaseolina* isolates. Within the ITS region, an oligonucleotides probe (19 mer) is also constructed. With little or no background, a DIG-labeled probe may identify the target sequence in diverse concentration. The probe is also found *M. phaseolina* specific, with no results recorded by using nonspecific target ITS sequence. It can also be used to track the *M. phaseolina* ecology in more complicated environments. This is first report for designing of the specific primer and probe for identification and detection of *M. phaseolina* isolates. (Khalid & Ramadan, 2018) also reported sequencing and amplification of ITS employed for the identification of *Trichoderma* isolates that exhibited 99–100% similarity with different species of *Trichoderma viz., T. asperellum, T. longibrachiatum,* and *T. harzianum.*

11.7 PCR-finger printing techniques

Molecular fingerprinting methods including AFLP, RAPD, and SSR are used to demonstrate genetic heterogeneity in this soil-borne filamentous fungus. Pathogenic variability and genetic diversity of *M. phaseolina* have been studied extensively with the help of molecular fingerprinting methods. In the genus *Macrophomina,* only a single species has been identified, despite significant levels of pathogenicity variation. *M. phaseolina* showed morphologically, pathogenically, physiologically, and genetically variations among the isolates. These variations help the pathogen for the adaptation and survival in diverse environments. Several therapeutic approaches have been applied against *M. phaseolina,* which will help to increase the production of jute (Ghosh et al., 2018).

11.8 Genetic diversity

Genetic diversity among isolates of *M. phaseolina* is used for investigation; total seven of nine isolates have successfully amplified the SSR loci that showed polymorphism. Total 54 alleles have been identified, which ranged from 2 to 14 per locus.

The value of PIC was ranging between 0.38 (MP31) and 0.85 (MP85). Six of the seven polymorphic loci have PIC values more than 0.5, showing highly edifying. A dendrogram based on matrix of the genetic distance pairings is generated by using data received from 40 isolates and the seven specified primers. There are two primary recognized groups. Twenty two Chilean isolates obtained from the strawberry plant growing in San Pedro-Melipilla region formed the Group I. Group II has 16 isolates which include 12 isolates from strawberry field in Huelva (Spain), two isolates from pine and two isolates from the Chilean melon. A distinct group is discovered that includes two isolates, one from Chilean melon and the other from Spanish strawberry. A primary coordinate analysis revealed the existence of the two major groups, one strawberry isolate clearly distinguishable from Chile and Spain although melon and the pine isolate genetically closer to the Spanish isolates. The use of Bayesian inference to analyze the population structure resulted in identification of the two distinct populations. Initial population has 22 strawberry isolates from Chile and one from the Spain. The presence of a third population is ruled out by the presence of the Spanish strawberry isolates, which combined with the Chilean pine and melon isolates. The genetic distances revealed are constant, allowing the formation of considerable disparities among the examined populations, according to the two primary groupings determined from various SSR studies. A global set of 189 *M. phaseolina* isolates has been collected from 23 diverse hosts along with 30 soil samples from the 15 countries. The multigene DNA analysis was performed for all five loci *viz*., ITS, TEF, ACT, CAL, and TUB. Based on the findings, two precise clusters were delineated; one structure corresponding to *M. phaseolina* for this an appropriate epitype is selected. The other clade corresponds to *M. pseudophaseolina* occurring on *Vigna unguiculata*, *Abelmoschus esculentus*, and *Arachis hypogaea*. Correlation was not observed among hosts, genotypes, geographic locations, although these species might be present on the identical host at identical location (Sarr et al., 2014).

11.9 SSR markers

Simple-sequence repeat (SSR) is also called as microsatellite, a short tandem repeated motif that may vary in the number of repeat at a given locus. With the help of primers, namely, SSR 2, SSR five, and SSR 10, different degrees of polymorphism and bands are produced. For each isolate, produced bands ranged between y250 and y3500 bp. By using the SSR two primer, just a few bands changes are identified in isolates of the *M. phaseolina* collected from the roots of soybean. Isolates of the *M. phaseolina* collected from roots of soybean included DNA bands of 1300, 1200, and 700 bp that are not found in isolates obtain from cotton roots. Isolate of the *M. phaseolina* derived from roots of cotton plant, DNA bands *viz*., y2000, 1500, 1000, and y900 bp are amplified but not isolates of soybean roots. On the other side, SSR five formed multiple DNA band; one of them is common to each *M. phaseolina* isolates, regardless of hosts or geographical origin. SSR produced five DNA bands with various sizes, e.g., 2000, 1300, 1200, 800 and 600 bp, which are similar to *M. phaseolina* isolates taken from roots of soybean but not with cotton roots. Similarly, cotton isolates have DNA bands with various sizes *viz*., y1000, 800, 600, and 400 bp. Effective multiplex ratio is an estimation of polymorphic loci in the set of isolates of fungi that were analyzed in the investigation, ranging from 0.70 (SSR two and SSR 5) to 0.79 (SSR 10) used in SSR markers, and marker index varied between 0.43 (SSR 2) and 0.61 (SSR 10) in the applied SSR markers. *M. phaseolina* isolated from cotton and soybean roots are genetically unique. The observed distinctness across *M. phaseolina* isolates from a host population is substantial, apparently owing to differences in the isolates derived from the various hosts. Hamer et al. (1989) observed host species-specific conservation in family of repetitive sequences of DNA in the genome of fungal pathogens of plants. Confirmation of the isolates of *M. phaseolina* was done with specific primer and genetic diversity analyzed by RAPD and SSR markers (Tables 11.2 and 11.3).

Correlation between RAPD analysis of the *M. phaseolina* isolates and their pathogenicity was estimated. The cluster analysis placed the isolates in two diverse clusters and showed clear correlation with the pathogenicity. Analysis of the ISSR profiles revealed the distinct genetic diversity among various isolates, which showed different clusters according to geographical regions, in which related geographic origins be inclined to the nearly group (Yousef, 2021).

11.10 RFLP

RFLP is a technique used to exploit the variations in homologous DNA sequences. ITS region of various isolates is amplified from genomic DNA, yielding a 650-bp fragment. Cleaving fragments with various restriction enzymes and diversity within amplified areas are also investigated. Restriction enzymes, namely, Alu I, EcoR I, Cla I, and Sau3A I, produced fragments of the different sizes from a mono incise for all isolates. In ITS amplified product, i.e., enzymes, namely, Msp I, Hap II, and Taq I show several sites of cleavage that created pieces lesser than 100 bp. Although Sac I, Apa I, and Hind III are unable to split the ITS region. There are many examples of RFLP for spacer regions to differentiate the closely related species among the fungal genera; ITS region of *M. phaseolina* has revealed no differences between pattern

TABLE 11.2 RAPD primer with sequence used for characterization of *M. phaseolina* (Yousef, 2021).

Sl. No.	Primer	Sequences
1	RAPD-1	(5′- GGTGCGGGAA-3′)
2	RAPD-2	(5′- GTTTCGCTCC-3′)
3	RAPD-3	(5′- GTAGACCCGT-3′)
4	RAPD-4	(5′- AAGAGCCCGT-3′)
5	RAPD-5	(5′- AACGCGCAAC-3′)
6	RAPD-6	(5′- CCCGTCAGCA-3′)
7	RAPD-7	(5′- CAGGCCCTTC-3′)
8	RAPD-8	(5′-AGGTGACCGT-3′)
9	RAPD-9	(5′-AGTCAGCCAC-3′)
10	RAPD-10	(5′-GTGATCGCAG-3′)
11	RAPD-11	(5′-CAATCGCCGT-3′)
12	RAPD -12	(5′-TCGGCGATAG-3′)
13	RAPD-13	(5′-TCTGTGCTGG-3′)
14	RAPD-14	(5′-TTCCGAACCC-3′)
15	RAPD-15	(5′-CCGCATCTAC-3′)
16	RAPD-16	(5′-GAACGGACTC-3′)
17	RAPD-17	(5′-AGGGGTCTTG-3′)
18	RAPD-18	(5′-GAAACGGGTG-3′)

TABLE 11.3 ISSR primer with sequence used for characterization of *M. phaseolina*(Yousef, 2021).

Sl. No.	Primer	Sequence
1	ISSR-1	TC(GACA)4
2	ISSR-2	(AG)8T
3	ISSR-3	(CT)8T
4	ISSR-4	(CA)8G
5	ISSR-5	(TG)8A
6	ISSR-6	(CAG)5AT
7	ISSR-7	(CAC)3 GC
8	ISSR-8	(GT)6CC
9	ISSR -9	(AG)8C
10	ISSR-10	(AC)8CT
11	ISSR-11	(CT)8G
12	ISSR-12	(GGAT)3AG
13	ISSR -13	(ATG)6
14	ISSR-14	(AC)8T
15	ISSR-15	(GAA)6AA

of restriction of the DNA fragment that amplified with PCR and cover the ITS regions of 5.8S rRNA and some part of the 25S rRNA among the isolates (Mangrauthia et al., 2018). Deviation in the patterns of restriction in ITS region included a portion of the 25 SiDNA, which showed that analysis of RFLP is inadequate for detecting genetic diversity of the *M. phaseolina* (Purkayastha et al., 2006).

RAPD markers and PCR-RFLP were used for the analysis of genetic diversity in *M. phaseolina* isolates. ITS regions were amplified by using primers LR5 and ITS5 which formed 1600 bp long DNA fragment. Findings of investigation showed genetic variability among *M. phaseolina* isolates with the help of RAPD marker. The RAPD data based on randomly primers tested on 16 isolates showed the presence of high degree of a polymorphism in unrelated isolates (Farzi et al., 2019).

11.11 Genome sequencing

RNA sequencing allows for gene quantification and mapping with the reference genome. RNA sequencing (RNA seq) is applied to measure genes, findings of novel transcripts, and analyzing data within the sample (Conesa et al., 2016). To determine the state of castor plant's reaction to infection by *M. phaseolina*, transcriptome of the diseased plant is sequenced by the RNA seq technique and further compared with transcriptome of a castor genotype, which shows resistance against infection. Number of matched and nonmatched read to a reference genome in the resistant genotype of castor JI357 is 1,749, 311 and 901, 800, respectively, whereas number of matched and nonmatched reads in the susceptible castor genotype 48-1 is 44,739 and 80, 121, respectively.

11.12 Metagenomics

Metagenomics is another potential method for detecting the microbes on a wide scale from a variety of sources. It is direct genetic analysis of the genomes contained with an environmental sample. It is the study of genetic content of microorganisms in their natural habitats, rather than isolating and growing specific species (Bell et al., 2016). As a result, metagenomics allows researchers to study both culturable and nonculturable species in their natural habitat. These novel sequencing tools, which employ the metagenomics methodology, allow researchers to promptly and economically recognize the species present in a complicated samples (plant tissues, soil or irrigation water) without earlier information. Such metagenomics approach for detection of pathogen should allow for the quick recognition and quantification of different pathogens prevalence in the sample, as well as implementation of appropriate disease management techniques long before pathogen populations attain to critical injury threshold. A 454 GS FLX system is ideal for metagenomics because its lengthy reads allow for precise detection of the pathogenic strains in a sample. This platform is frequently used by researchers to count the gene tags and detect abundance of diverse microbial species in different samples of the isolates. Khan et al. (2017) also identified isolates of *M. phaseolina* by 18S rRNA and analyzed their genetic diversity by using RAPD markers. Finding based on the analysis of 18S rRNA sequence and RAPD markers clearly reveals genetic variations in the different *M. phaseolina* isolates obtained from various hosts.

11.13 Fluorescence in situ hybridization

FISH is a microbial detection technique, which combines ease of the microscope viewing with the specificity of hybridization (Frickmann et al., 2017). It is a new technique used for the identification of plant pathogenic microbes (Kliot & Ghanim, 2016) based on hybridization of DNA probes for species-specific ribosome's sections in a microbe. The rRNA has a functional sequence and is distinctive for the individual species; a FISH identifies the precise information. A probe can hybridize by linear RNA sequence and also with three-dimensional protein/RNA structures. The sensitivity of FISH is comparable with the other amplification methods and may theoretically identify the single cell. FISH conducted under highly stern hybridization condition, wherever a change in a single nucleotide of a 15−20 oligonucleotides probe, which is sufficient for distinguishing the binding, and enormous sensitivity is due to high selectivity and affinity of the DNA probes. Furthermore, a FISH preserves the structural integrity of microorganisms by containing the chemicals in a single tiny vessel, with one probe binding to each of the 1−5. The potential capacity of this technique to attain single-cell sensitivity is due to its exceedingly high signal. DNA- and protein-based pathogen identification techniques like DNA fingerprinting, isothermal amplification techniques, serology, and biochemical assays are gaining significance in the quick soil-borne pathogen detection by high degree of specificity for distinguishing the closely related microbes at various taxonomic levels (Ghosh et al., 2019).

11.14 Target sequences used for diagnostic assay development

Selection of nucleic acid sequence can be done for identification of microbes is initial step for construction of a diagnostic test. For bacteria and fungi, there are some strategies that can be applied universally, but viruses require a distinct approach. Their genomes are quite short, and sequencing databases often have information on whole viral genomes. Coat protein gene is perhaps the most popular target for the viral diagnostics; however, any section of genome might be relevant for dependence on sequence of data and compare with the virus in similar areas of genome is available (Hariharan & Prasannath, 2021).

11.15 Conclusion and future prospects

Genomics research is producing rapidly expanding datasets that may be utilized to construct molecular tests for instant detection of a large number of pathogens, as well as innovative platforms with unparalleled multiplexing, throughput and mobility capabilities. Routine detection, diagnosis, monitoring, and management of plant diseases should become increasingly prevalent as these novel molecular technologies achieve widespread adoption. The oligonucleotides which are manufactured directly on the micro-spotted or solid surface are being used to make microarray chips. Similar to 454 and SOLiD, next-generation sequencing technologies may sequence numerous microbial genomes in a single run. Plant disease diagnosis based on molecular tools is becoming a field reality, advent of inexpensive and the moveable real-time PCR devices (Tarim et al. (2021). It is a simplified method which helps in precise management strategies. Recent breakthrough technology has made routine detection of many agricultural diseases achievable in 1 day or less. Along with high throughput and lower sample cost, it makes assays more appealing for the applications of plant protection measures. Combining microarrays of DNA with the other molecular techniques will definitely speed up the attempts for describing function of unknown genome regions of the fungus. Resultant database allows the thorough examination of the fungal developmental stages as well as revealing the molecular basis of pathogenicity. Complete genome sequences of *forma specialis* for *M. phaseolina* and other phytopathogenic fungi may be available in the next few years and combination of the DNA microarrays and other genomic approaches will undoubtedly speed up attempt for defining the fungus. The convergence of genomics, biology, and bioinformatics makes easy to identify the genetic resources of *M. phaseolina* and their potential use for precise disease management. Early detection and reliable identification of *M. phaseolina* and its isolates ensure appropriate implementation of the disease management strategies. Seed treatment with *Trichoderma* spp., organic amendment of soil and cultural practices are eco-friendly ways for the management of *M. phaseolina*.

References

Badigannavar, A., Girish, G., Jayalakshmi, J., & Ganapathi, T. R. (2018). Gamma ray induced genetic improvement of sorghum landraces for grain yield and charcoal rot tolerance. *Electronic Journal of Plant Breeding, 9*(3), 894–898.

Bell, R. L., Jarvis, K. G., Ottesen, A. R., McFarland, M. A., & Brown, E. W. (2016). Recent and emerging innovations in *Salmonella* detection: A food and environmental perspective. *Microbial Biotechnology, 9*(3), 279–292.

Conesa, A., Madrigal, P., Tarazona, S., Gomez-Cabrero, D., Cervera, A., McPherson, A., & Mortazavi, A. (2016). A survey of best practices for RNA-seq data analysis. *Genome Biology, 17*(1), 1–19.

Das, I. K., Kishore, B. B., Srivastava, A. K., Kumar, S., & Arora, D. K. (2015). Ecology, biology and management of *Macrophominaphaseolina*: An overview. *Agriculturally Important Microorganisms, 1*, 193–206.

Dey, P., Biswas, C., & Karmakar, P. G. (2016). Identification and characterization of differentially expressed novel miRNAs (21-24 nt) in a *Macrophomina phaseolina*resistant RIL line of jute (*Corchoruscapsularis* L.). *Physiology and Molecular Plant Pathology, 94*, 62–66. https://doi.org/10.1016/j.pmpp.2016.04.005

Edel-Hermann, V., & Lecomte, C. (2019). Current status of *Fusariumoxysporum*formaespeciales and races. *Phytopathology, 109*(4), 512–530.

Farzi, N., Behzad, C., Hasani, Z., Alebouyeh, M., Zojaji, H., & Zali, M. R. (2019). Characterization of *Clarithromycin heteroresistance* among *Helicobacter pylori* strains isolated from the antrum and corpus of the stomach. *Folia Microbiologica, 64*(2), 143–151.

Forster, H. (2019). *Suppression of growth of the phytopathogenic fungus* Macrophomina phaseolina *through application of exogenous sirnas* (Doctoral Dissertation). Wichita State University (U.S.A.).

Frickmann, H., Zautner, A. E., Moter, A., Kikhney, J., Hagen, R. M., Stender, H., & Poppert, S. (2017). Fluorescence in situ hybridization (FISH) in the microbiological diagnostic routine laboratory: A review. *Critical Reviews in Microbiology, 43*(3), 263–293.

Ghosh, T., Biswas, M. K., Guin, C., & Roy, P. (2018). A review on characterization, therapeutic approaches and pathogenesis of *Macrophomina phaseolina. Plant Cell Biotechnology and Molecular Biology, 19*, 72–84.

Ghosh, R., Tarafdar, A., Chobe, D. R., SharathChandran, U. S., Rani, S., & Sharma, M. (2019). Diagnostic techniques of soil borne plant diseases: Recent advances and next generation evolutionary trends. *Biological Forum—An International Journal, 11*(2), 1–13.

Gomez, A. O., De Faveri, J., Neal, J. M., Aitken, E. A. B., & Herrington, M. E. (2020). Response of strawberry cultivars inoculated with *Macrophomina phaseolina* in Australia. *International Journal of Fruit Science, 20*, 1−14.

Hamer, J. E., Farral, L., Orbach, M. J., Valent, B., & Chumley, F. G. (1989). Host species-specific conservation of a family of repeated DNA sequences in the genome of a fungal plant pathogen. *Proceedings of the National Academy Sciences, USA, 86*, 9981−9985.

Hariharan, G., & Prasannath, K. (2021). Recent advances in molecular diagnostics of fungal plant pathogens: A mini review. *Frontiers in Cellular and Infection Microbiology, 10*, 600234.

Khalid, S. A. L., & Ramadan, A. B. (2018). Internal transcribed spacers (ITS) based identification of Trichoderma isolates and biocontrol activity against *Macrophomina phaseolina, Aspergillus niger* and *Meloidogyne incognita*. *African Journal of Microbiology Research, 12*(30), 715−722.

Khan, A. N., Shair, F., Malik, K., Hayat, Z., Khan, M. A., Hafeez, F. Y., & Hassan, M. N. (2017). Molecular identification and genetic characterization of *Macrophomina phaseolina* strains causing pathogenicity on sunflower and chickpea. *Frontiers in Microbiology, 8*, 1309.

Kliot, A., & Ghanim, M. (2016). Fluorescent in situ hybridization for the localization of viruses, bacteria and other microorganisms in insect and plant tissues. *Methods, 98*, 74−81.

Kumar, P., Gupta, V. K., Tiwari, A. K., & Kamle, M. (Eds.). (2016). *Current trends in plant disease diagnostics and management practices*. Cham: Springer International Publishing.

Liao, H. L., Chen, Y., Bruns, T. D., Peay, K. G., Taylor, J. W., Branco, S., & Vilgalys, R. (2014). Metatranscriptomic analysis of ectomycorrhizal roots reveals genes associated with *PilodermaPinus* symbiosis: Improved methodologies for assessing gene expression in situ. *Environmental Microbiology, 16*(12), 3730−3742.

Mahmoud, A. F., Abou-Elwafa, S. F., & Shehzad, T. (2018). Identification of charcoalrot resistance QTLs in sorghum using association and in silico analyses. *Journal of Applied Genetics, 59*, 243−251.

Mangrauthia, S. K., Sailaja, B., Pusuluri, M., Jena, B., Prasanth, V. V., Agarwal, S., & Voleti, S. R. (2018). Deep sequencing of small RNAs reveals ribosomal origin of microRNAs in Oryza sativa and their regulatory role in high temperature. *Gene Reports, 11*, 270−278.

Marquez, N., Giachero, M. L., Declerck, S., & Ducasse, D. A. (2021). *Macrophominaphaseolina:* General characteristics of pathogenicity and methods of control. *Frontiers of Plant Science, 12*, 634397. https://doi.org/10.3389/fpls.2021.634397

Marquez, N., Giachero, M. L., Gallou, A., Debat, H. J., Cranenbrouck, S., & Di Rienzo, J. A. (2018). Transcriptional changes in mycorrhizal and nonmycorrhizal soybean plants upon infection with the fungal pathogen *Macrophomina phaseolina*. *Molecular Plant Microbe Interaction, 31*, 842−855.

Muchero, W., Ehlers, J. D., Close, T. J., & Roberts, P. A. (2011). Genic SNP markers and legume synteny reveal candidate genes underlying QTL for *Macrophomina phaseolina* resistance and maturity in cowpea [*Vignaunguiculata* (L) Walp.]. *BMC Genomics, 12*, 8.

Pandey, A. K., Burlakoti, R. R., Rathore, A., & Nair, R. M. (2020). Morphological and molecular characterization of *Macrophomina phaseolina* isolated from three legume crops and evaluation of mungbean genotypes for resistance to dry root rot. *Crop Protection, 127*, 104962.

Pandey, A., Jain, R., Sharma, A., Dhakar, K., Kaira, G. S., Rahi, P., & Shouche, Y. S. (2019). 16S rRNA gene sequencing and MALDI-TOF mass spectrometry based comparative assessment and bioprospection of psychrotolerant bacteria isolated from high altitudes under mountain ecosystem. *SN Applied Sciences, 1*(3), 1−12.

Purkayastha, S., Kaur, B., Dilbaghi, N., & Chaudhury, A. (2006). Characterization of *Macrophomina phaseolina*, the charcoal rot pathogen of cluster bean, using conventional techniques and PCR-based molecular markers. *Plant Pathology, 55*(1), 106−116.

Raj, K., & Rathi, A. S. (2016). *Diseases of rapeseed-mustard and their management*. New Delhi, India: Indian Council of Agricutural Research.

Saima, S., & Wu, G. (2019). Effect of *Macrophomina phaseolina* on growth and expression of defense related genes in *Arabidopsis thaliana*. *Journal of the Natural Science Foundation of Sri Lanka, 47*, 113.

Sarr, M. P., Ndiaye, M., Groenewald, J. Z., & Crous, P. W. (2014). Genetic diversity in *Macrophomina phaseolina*, the causal agent of charcoal rot. *Phytopathologia Mediterranea, 53*, 250−268.

Sharma, I., Kumari, N., & Sharma, V. (2015). Sorghum fungal diseases. In E. Lichtfouse, & A. Goyal (Eds.), *Vol 16. Sustainable agriculture reviews*. Cham: Springer.

Tarim, E. A., Karakuzu, B., Oksuz, C., Sarigil, O., Kizilkaya, M., Al-Ruweidi, M. K. A., & Tekin, H. C. (2021). Microfluidic-based virus detection methods for respiratory diseases. *Emergent Materials, 4*(1), 143−168.

White, T. J., Bruns, T., Lee, S. J. W. T., & Taylor, J. (1990). Amplification and direct sequencing of fungal ribosomal RNA genes for phylogenetics. *PCR protocols: A Guide to Methods and Applications, 18*(1), 315−322.

Yousef, H. (2021). Pathogenic variation and molecular characterization of *Macrophomina phaseolina*, the cause of sesame charcoal rot. *Egyptian Journal of Phytopathology, 49*(1), 151−165.

Chapter 12

Genetics of plant resistance against *Macrophomina phaseolina*

Adikshita Sharma[1], Monica Sharma[2] and Kishor Sharma[3]

[1]*Department of Plant Pathology, Yashwant Singh Parmar University of Horticulture and Forestry, Solan, Himachal Pradesh, India;* [2]*Department of Plant Pathology, Yashwant Singh Parmar University of Horticulture and Forestry, Hamirpur, Himachal Pradesh, India;* [3]*Department of Plant Pathology, Yashwant Singh Parmar University of Horticulture and Forestry, Mandi, Himachal Pradesh, India*

12.1 Introduction

Macrophomina phaseolina (Tassi) Goid. (*Tiarosporella phaseolina* (Tassi) Vander Aa) is an important destructive necrotrophic soil and seed-borne fungus causing dry root rot (Dhingra & Sinclair, 1978; Ghosh et al., 2018). This pathogen attacks a number of hosts and distributed very widely in all the geographic regions. Major cultivated hosts include: *Beta vulgaris* (Beet), *Pinus* spp., *Brassica oleracea* (cabbage), *Arachis hypogea* (peanut), *Capsicum annuum* (pepper), *Corchorus* sp., *Fragaria* sp., *Cucumis* spp., *Phaseolus* spp., *Prunus* spp., *Citrus* spp., *Gossypium* sp., *Helianthus annuus* (sunflower), *Glycine* max (soybean), *Cicer arietinum* (chickpea), *Zea mays* (corn), *Sorghum bicolor* (sorghum), *Ipomoea batatas* (sweet potato), *Vigna unguiculata* (bean), *Medicago sativa* (alfalfa), and groundnut in the world (Aly et al., 2007; Mayek-Perez et al., 2001; Sobti & Sharma, 1992; Su et al., 2001). It shows the monotypic nature instead having a wide range of hosts (Purkayastha et al., 2006). It infects all the growth stages of plants. Crop yields are reduced significantly, e.g., soybean and sorghum when subjected to soil moisture below 60% and temperatures are high, i.e., 30−35°C (Kaur et al., 2012). Therefore, a number of physiological and ecological factors, such as low moisture contents, high temperatures, and heat may increase its occurrence. The disease intensity is related to the quantity of the sclerotia, which are viable and present in the soil. In the worst example, when disease initially appeared during the pre-emergence stage, groundnut cultivars experienced 100% yield reductions (Sharma & Bhowmik, 1986). According to studies, *M. phaseolina* numbers varied significantly in other parts of the world showing variations morphologically (Mayek-Perez.,1997), physiologically (Mihail & Taylor, 1995), pathogenically (Mayek-Perez et al., 2001; Su et al., 2001; Reyes-Franco et al., 2006), and genetically (Vandemark et al., 2000; Jana, Singh, et al., 2005). These variations allow the pathogen to adapt and flourish in a variety of conditions. There had been a lot of work done for its management by various scientists, but to control the devastating effect on various crops pathogen, the different strategies are not easy to lay into practice. This pathogen develops as result of interactions between the host plant, the pathogen, and the biotic and abiotic factors of the environment. Genetic resistance is also one of the good management practices to control the pathogen infection. These approaches reduce the need for pesticides particularly fungicides, having very negative effect on humans, soil, water, plants, and every part of environment. *M. phaseolina* resistance is often described as adult plant resistance or all phases resistance (seedling resistance). *M. phaseolina* have the ability to develop new races.

Few years back, high-throughput genome annotation, gene expression profiling, and more experiments linked to resistance genes and genomes have been made as a result of rapid next-generation sequencing. The color of mycelia, distribution of microsclerotia, sequence of pycnidia generation, and phenotypes of chlorate of the fungus can differ significantly between isolates on synthetic media. The morphological heterogeneity of the pathogen will be explained by the fungal interaction with the environmental conditions and differences between the hosts (Pandey et al., 2020; Tok, 2019). Consequently, our goal in this work is to define the characteristic of *M. phaseolina* in general and report the most recent information on the pathogenic and genetic variations, genetic resistance, and management strategies.

Macrophomina Phaseolina. https://doi.org/10.1016/B978-0-443-15443-0.00011-5

12.2 Pathogen background

M. phaseolina (Tassi) Goid. (syns. *M. phaseolina* (Maubl.) Ashby, *Rhizoctonia bataticola* (Taub.) Briton-Jones, *Sclerotium bataticola* Taub., and *Botryodiplodia phaseoli* (Maubl.) Thrium.) falls in Deuteromycetes and class Coelomycetes. It is anamorphic Ascomycetes and produces pycnidia and sclerotia in the host tissues and in laboratory condition on the media. At the start, the pycnidial state was known as *M. phaseolina* (Tassi, 1901) and *M. phaseoli* (Maublanc, 1905). In 1927, Ashby named it *M. phaseoli*, while Goidanich (1947) suggested for *M. phaseolina*. In 1981 Van der Aa used *Tiarosporella phaseolina* for the species first. Mihail (1992) got a first and last report as published by Ghosh et al. (1964) on teleomorph of *M. phaseolina* known as *Orbilia obscura*. *Rhizoctonia bataticola* (Taub.). Butler was sclerotial state on *Ipomoea batatas* was given by Halsted in 1890. Dhingra and Sinclair (1978) reported that this fungus from cowpea was isolated in 1912 in India by Shaw and named it as *Sclerotium bataticola*. *M. phaseolina* is produced hyaline hyphae, which have thin walls they appear to be light brown or dark brown in color having septations in between. Main hyphal branches are produced at right angle at parent hyphae having constriction at its origin. Initially, pycnidia are submerged in host tissue, after maturity they emerge out. They are membranous to sub-carbonaceous, globose, or flattened globose or flattened globose, 100–200 m in diameter, dark to greyish, turning black with age, and contain a subtly or overtly truncate ostiole. It produces rod-shaped and simple conidiophores of the pycnida that are 10–15 mm long (Lakhran et al., 2018). *M. phaseolina* microsclerotia have a jet-black appearance and a rectangular to irregular form. Among isolates, microsclerotia differ in size, shape, and substrate type. Hyaline, single-celled conidia have dimensions of 14–33 × 6–12 m and are elliptic to oval in shape. It produces jet-black microsclerotia that can be oblong or irregular in shape, as well as smooth and round. Size, morphology, and substrate of microsclerotia vary among the isolates. Individual microsclerotium is made up of 50–200 hyphal cells and cemented by a melanin substance. The microsclerotia serve as the principal inoculum source (Dhingra & Sinclair, 1978; Gahlot, 2018). On synthetic media, the fungus can exhibit significant variation between isolates in terms of microsclerotia distribution, mycelium color, pycnidial formation, and chlorate phenotypes. The production of aerial mycelium in culture with either fully or partially suppressed growth is possible. Concentric growth rings may arise on certain isolates.

12.3 Disease cycle

Drought and heat favor dry root rot. The pathogen attacks when exposed to moisture stress. There are four stages in its life cycle, i.e., germination, penetration, parasitism, and saprophysis (Dhingra & Sinclair, 1978). *M. phaseolina* enters and colonizes living root tissues during the parasitic phase of its disease cycle. The pathogen enters a saprophytic phase as the disease progresses during which host cells begin to degrade.

Later, the pathogen consumes or ingests these host cells (Gupta et al., 2012) causing necrosis and the characteristic black discoloration of the host roots. In addition to weakening the root tissues and preventing the host from effectively absorbing nutrients and water, the necrosis of the host tissues also hinders plant growth and productivity (Lodha & Mawar, 2020). Penetrating hyphae may completely clog the root capillaries in plants with severe infection, limiting nutrition, and water intake until the plant dies. In host tissue, the fungal saprophytic phase manifests as mycelia, pycnidia, and microsclerotia. Heavenly infected plants produce lot of microsclerotia. As the microsclerotia grow, black hyphae are formed. The microsclerotia are discharged in soil after the decomposition of plant debris.These multicellular structures enable the fungus to survive in harsh environments like high soil temperatures and low soil nitrogen levels. In damp soils, mycelia can only last for 7 days, and microsclerotial survival is reduced significantly, lasting no more than seven to 8 weeks. Microsclerotia survive in the soil for about 3 years. The fungus was present in the seed coat of the seed. A contaminated seed does not develop into the sprout or give birth to seedlings that perish shortly after emerging. When the temperature is between 28 and 35°C, microsclerotia germinate. The germ tube, which is produce by appressoria, enters the host's epidermal cell walls through various procedures like mechanical pressure, enzymatic digestion, or natural apertures. The hyphae cultivate intracellularly by the xylem to colonize the vascular tissue after first growing intercellularly in the cortex. *M. phaseolina* enters the vascular tissue and spreads throughout the plant's taproot and lower stem, creating microsclerotia that block the vessels (Gupta & Chauhan, 2005; Khan, 2007). Disease is brought on by toxin generation, enzymatic activity, and mechanical pressure during penetration and mechanical blockage of the xylem vessels by microsclerotia. Therefore, infected plants frequently wilt and die too soon because of this systemic infection.

12.4 Symptomatology

It shows early signs of infection in woody plants in the form of spindle-shaped lesions with blotchy pycnidial bodies having gray centers and brown borders. After a successful infection, *Macrophomina* typically causes a variety of symptoms ranging from small lesions in the shape of spindles on the stem to larger lesions that cause the plant to wilt. Premature senescence and plant death result from lesions coalescing into larger areas on branches or the entire plant (Dhingra & Sinclair, 1978). These lesions have numerous black spots, resembling pycnidia in the early stages of the disease. Later, numerous pycnidia can be seen on the majority of the plant's afflicted portions. Infected plants shrivel up and their roots seem shredded and decaying (Ilyas & Sinclair, 1974). Typical symptoms are caused by *M. phaseolina* in various crops.

12.5 Pathogenic and genetic variability

There are reports from other parts of the world unambiguously demonstrate that populations of *M. phaseolina* exhibit morphological (Mayek-Perez et al., 1997), pathological (Mayek-Perez et al., 2001; Su et al., 2001), physiological (Mihail & Taylor, 1995), and genetic (Almeida et al., 2003; Jana et al., 2003; Jana, Singh, et al., 2005; Su et al., 2001; Vandemark et al., 2000) variation. These variations aid the pathogen to adapt and survive in diverse environments. A significant amount of intraspecific morphological and pathogenic variation prevented *M. phaseolina* from being classified into separate forms, subspecies, or physiological races (Purkayastha et al., 2003). Several studies have been carried out to understand the variation based on the regional source (Reyes-Franco et al., 2006). The pathogenic diversity of *Macrophomina* isolates from soybean, sunflower, peanuts, mulberry, and common beans has been demonstrated by various scientists (Dhingra & Sinclair, 1978; Sobti & Sharma, 1992; Sowmya et al., 2022). Compared to isolates from the Asteraceae, Euphorbiaceae, or Fabaceae families, *M. phaseolina* isolates from the Poaceae family exhibit less virulent pathogenicity on the sunflower cv. *Mammoth* and *Euphorbia lathyris*. From Mexico and various countries, nine isolates (Japan, Italy, Brazil, USA, and Australia) were gathered, but they did not exhibit any symptoms on bean (Reyes-Franco et al., 2006). According to the history of cropping, distinct *M. phaseolina* isolates growing in different fields with cotton, soybean, sorghum, and maize exhibited variations in pathogenic symptoms, proving that the fungus has some degree of host specificity (Su et al., 2001).

M. phaseolina has a wide range of hosts and can leave in two asexual forms that improve its ability to survive (Cloud & Rupe, 1988; Dhingra & Sinclair, 1978; Mihail & Taylor, 1995). *M. phaseolina* isolates from Pakistan (Iqbal & Mukhtar, 2014) and India (Kumar et al., 2017) and from a variety of hosts, including mungbean, demonstrated pathogenic variability. Variability was also linked by some researchers to the *M. phaseolina* phenomenon of host specialization. Setting up this technique within a particular host requires a lot of time. According to Mihail and Taylor (1995), *M. phaseolina* is heterogenic, making it impossible to divide it into several subgroups based on pathogenicity and morphology. Host-plant resistance is specifically influenced by pathogenesis and genetic diversity. Due to the absence of a sexual phase, *M. phaseolina* genetic diversity depends on hyphal fusion, mutation, and mitotic recombination. Isolates with comparable morphology are not always genetically identical; there may be some variations. Using other molecular methods, it has been demonstrated that the varied genetic pattern leads to variance in morphology and disease (Reyes-Franco et al., 2006; Raj Kumar et al., 2007; Sarr et al., 2014). Many researches used SSR markers for showing genetic diversity, among *M. phaseolina* isolates particularly from soybean (Baird et al., 2010; Jana, Singh, et al., 2005). In maize, Su et al. (2001) discovered host specialization based on pathogenic, genetic and physiological variations. Cloud and Rupe conducted similar research on host specialization in soybean. High genetic diversity was confirmed by the high degree of polymorphism seen across *M. phaseolina* isolates taken from the same geographic region (Rajkumar et al., 2007). According to some researchers (Babu et al., 2010; Jana, Singh, et al., 2005; Mahdizadeh et al., 2011), genetic variation among different isolates has been associated to the source where from the host plant has been developed and their geographic area related to it, while according to other researchers (Mahdizadeh et al., 2011; Reznikov et al., 2018, 2019) they separate the different isolates very easily on the basis of isolates morphological characteristics, pathogenicity, their hosts, or the areas where from they belong. Various investigations have shown that the genotype distribution of *M. phaseolina* is unaffected by sampling site and host (Khan et al., 2017; Tančić Živanov et al., 2019).

Several studies employing molecular markers and cluster analysis have been done on the genetic diversity of *M. phaseolina* isolates. A variety of genetic techniques have been used for the comparative genomics among *M. phaseolina* population in different countries. These techniques are loop-mediated isothermal amplification (LAMP)-based detection, rDNA sequencing, restriction fragment length polymorphism (RFLP), amplified fragment length polymorphism (AFLP), and RFLP (Almeida et al., 2003; Babu et al., 2010; Ghosh et al., 2017; Jana, Singh, et al., 2005; Khan et al., 2017). The other regions of the pathogen's genome basically housekeeping genes and conserved genes tubulin and translation elongation factor 1 was also give good information (Gautam et al., 2014). Identification and characterization of fungal plant diseases were effectively done by multilocus sequence analysis of housekeeping genes, that is, calmodulin, histone H3, and translation elongation factor 1-alpha genes (Iqbal & Mukhtar, 2014). For the SYBR green and TaqMan test, sets of particular primers have been developed (Babu et al., 2011; Burkhardt et al., 2018). Different isolates from the different hosts sunflower, beans, peanut, sorghum, cowpea, soybean, and chickpea give different genetic diversity as shown by using various molecular techniques by various scientist (Aboshosha et al., 2007; Das et al., 2008; Ghosh et al., 2017; Mayek-Pérez et al., 2002; Jana, Singh, et al., 2005; Muchero et al., 2011). *M. phaseolina* isolates from a variety of host was subjected to RAPD-based molecular analysis and was found that the primers of one of the mungbean isolates, i.e., primers OPA-13 (5′-CAGCACCCAC-3′) and OPB-08 (5′-GTGAGCTAGG-3′) were used for comparative studies and identification of *M. phaseolina* because of showing most of the polymorphism. Monomorphic bands were shown by the different isolates of the *M. phaseolina*. Thus, various researchers came to the conclusion that they had come out from the single same ancestor. Natural selection, geographic isolation, and genetic drift divided them into subpopulations.

Scientists have been using SSR-based molecular markers to differentiate *M. phaseolina* isolated from strawberry and soybean plants (Reznikov et al., 2019; Sánchez et al., 2017). *M. phaseolina* isolates from different regions of Mexico, for example, incommon bean roots and others showed a high genetic diversity in between different isolated isolates when study through an AFLP (Babu et al., 2010; Jana et al., 2003). RAPD studies also showed substantial genetic distances across isolates from various geographic origins in different crops like soybean, sesame, bean, chickpea, jute, and cotton (Jana et al., 2003). Genetic associations between Indian and American *M. phaseolina* isolates, particularly from soybean, were categorized by host and country of origin using SSRs markers as studied by Jana, Sharma, and Singh (2005). Research has shown that pathogen isolates exhibit a high degree of genetic heterogeneity despite *M. phaseolina* lacking sexual reproduction. The parasexual combination of cells from dissimilar hyphae was able to produce heterokaryons, which would increase the variety already present (Almeida et al., 2003). Brazilian isolates of this pathogen shows genetic diversity even a single root shows many haplotype (Almeida et al., 2003). *M. phaseolina* is having much diversified species. The large percentage of transposons and repetitive DNA sequences in *M. phaseolina* genome has been linked to some of the species variability because it is an anamorphic species. The genome of the pathogen have repetitive sequences (2.84%) and mobile transposable elements (3.98%), which are connected to a larger spectrum of virulence as it stops the activation of genes that enhances the immune system in plants (Islam et al., 2012). For detection and quantification of *M. phaseolina* from sick plant tissues and rhizospheric soil, real-time quantitative PCR primers were developed by Babu et al. (2011). Genetic diversity and hydrolytic enzyme production, both are correlated with the variation in pathogenicity (Khan et al., 2017). To locate and identify *M. phaseolina* at the molecular-level species-specific primers and probes are used which targets the rDNA gene cluster. Highly specific primers, that are MpKF1 (5′-CCG CCA GAG GAC TAT CAA AC-3′) and MpKR1 (5′-CGT CCG AAG CGA GGT GTA TT-3′), are unique and were developed using the conserved sequences of the ITS region having 350 bp products and used to differentiate *M. phaseolina* species (Babu et al., 2007). It is clear that variation in appearance, physiology, genetics, pathogenicity, and other aspects is necessary for the fungus to have a greater ability to respond to a variety of environmental conditions.

12.6 Genetic resistance

M. phaseolina does not have a vertical resistance (R-gene based) that prevents or restricts infection and it shows the partial resistance, which does not allow the infection to occur instead it compensates the loss at the cost on the strength of plants. In soybean, cultivars that are only moderately resistant, like DT97-4290, shows partial resistance to this pathogen has been identified. No plant species has demonstrated complete resistance to *M. phaseolina* (Mengistu et al., 2013; Pawlowski et al., 2015). There are various soybean and strawberry types that are resistant to *M. phaseolina* to varying degrees (Gomez et al., 2020; Reznikov et al., 2018).Chowdhury et al. (2014) studied the nature of fungal activities in rhizosphere and when infection has taken place in roots between resistant and susceptible sesame cultivars. They found that rhizosphere of resistant variety showed less of *M. phaseolina* growth than the susceptible varieties. Hemmati et al. (2018) reported that there is the formation of adventitious roots in the crown region of soybean plants and pathogen was not able to complete its life cycle in resistant cultivars; this behavior of pathogen in the pre-penetration steps within the roots was not linked to the resistant, also there was no hyphal and microsclerotial development. Thus, *M. phaseolina* resistance is linked with mapping of QTLs, and it identified candidate genes with the potential for additional functional genomics investigation, which may aid in breeding and molecular engineering advancements against this disease (da Silva et al., 2019; Mahmoud et al., 2018; Muchero et al., 2011; Tomar et al., 2017). Mungbean lines 11160a (Dreshka et al., 1974) in India, NCM 252-10 and 40536 (Khan & Shuaib, 2007) in Pakistan, and MNUYT-317 and NM-2011 in Pakistan showed resistance by using the paper towel method (Khan et al., 2016). Mungbean lines MSJ 118, KM 4–44, and KM 4–59 in India and Azri 2006 in Pakistan showed field resistance to dry root rot (Choudhary et al., 2011; Haseeb et al., 2013). In Pakistan, urdbean lines

45718-45719, 45711-45731, 45731-45737,VH9440034-1, and VH9440034-7 had showed resistance to dry root rot in Pakistan (Iqbal et al., 2003).

In India, none of the urdbean or mungbean lines showed resilient resistance to dry root rot, but in one of the study, field resistance to this pathogen in urdbean lines IPU-96-6, IC-16511, and NO-5131 was shown (Dambal et al., 2019). While testing with genetic markers related to resistance to *M. phaseolina* in common bean, charcoal rot resistance in a common bean variety known as BAT 477, which is having two dominant genes with double-recessive epistasis (Hernandez-Delgado et al., 2009). Sharma et al. (2014) discovered resistant in sorghum cultivars which had higher expression levels of protective enzymes like chitinase and stilbene synthase, which boosted resistance to this pathogen. In beans and cowpea, quantitative trait loci (QTLs) for *M. phaseolina* resistance (Hernández-Delgado et al., 2009; Muchero et al., 2011) have been discovered. Cowpea resistance involves up to eight QTLs, with the Mac-2 major QTL accounting (40%) of the resistance variation (Muchero et al., 2011). One gene in the linked marker, 1_0853, encodes a pectin esterase inhibitor (PEI) and the role of pectins in pathogen defense, PEI has been proposed a possible gene for resistance incowpea against *M. phaseolina* (Lionetti et al., 2007; Valette-Collet et al., 2003).

Pastor-Corrales discovered that common beans are resistant to charcoal rot. The black breeding line XAN176 was developed to provide resistance to common bacterial blight and has a moderate level of resistance to charcoal rot (Hernandez-Delgado et al., 2009; Miklas et al., 1998; Reyes-Franco et al., 2006). Other Mesoamerican beans were tested for resistance to charcoal rot, including BAT477 (Hernandez-Delgado et al., 2009; Reyes-Franco et al., 2006); SEQ 12; N.8025; TLP 19; and G4523 (Miklas et al., 1999). Olaya et al. found two dominant complementary genes Mp1 and Mp2 for resistance in the BAT 477 genotype; Miklas et al. (1998) discovered three QTLs in the DOR364 Xan176 populations. According to Hernández-Delgado et al. (2009), the key regulators of BAT477 resistance are a QTL on Pv01 and recessive epistatic genes. Yan et al. (2021) examined and found the sesame defense response to *M. phaseolina* at the transcriptional level, which involves phytohormones and disease resistance-related genes type of complex process. Their findings indicated importance of the signaling pathways JA/ET and SA in sesame resistance to *M. phaseolina*. In *M. phaseolina,* exogenous siRNAs are used to target the genes-1,3-glucan synthase and chitin synthase (Fig. 12.1). Fungal cell walls are formed with the help of these targeting genes. It is worth noting that the fungal growth was reduced by siRNA treatment as demonstrated by a smaller growing area and less dense mycelia because microsclerotia developed and melanized more slowly under different treatment conditions. siRNA therapies were found to postpone the fungus maturity. Furthermore,

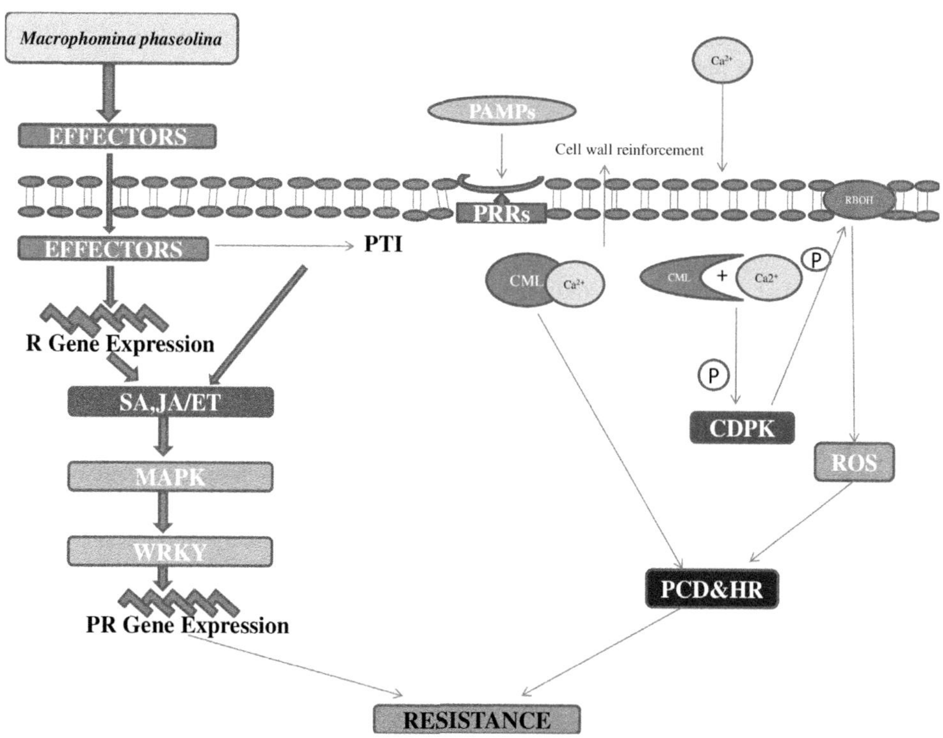

FIGURE 12.1 Major molecular pathways of defensive response in sesame resistance against *M. phaseolina*. *Modified after Yan, W., Ni, Y., Liu, X., Zhao, H., Chen, Y., et al. (2021). The mechanism of sesame resistance against Macrophomina phaseolina was revealed via a comparison of transcriptomes of resistant and susceptible sesame genotypes.* Plant Biology, 21, *159.*

when *M. phaseolina* growth was inhibited, transcript abundances of target genes were considerably reduced (Forster & Shuai, 2020a, 2020b).

Genetic resistance is a practical, cost-effective, environmentally beneficial, and economical method of managing plant diseases. As a result, research efforts to examine the genotype-genotype specific interactions between plant cultivars and *M. phaseolina* isolates will support the development of resistant cultivars as stated by Reznikov et al. (2019). The resistance in dry root rot is required due to the relatively low level of resistance existing in marketable varieties and superior reproductive resources of various crops that are commonly not affected by *M. phaseolina*. Additionally, the transcript abundances of the target genes were significantly reduced when *M. phaseolina* growth was suppressed (Forster & Shuai, 2020a, 2020b). Utilizing genetic resistance is a practical, cost-effective, environmentally friendly, and money friendly way to manage the plant diseases. The quest for developing the sources of resistance against dry root rot is necessary due to the relatively low degree of resistance present in marketable varieties and superior breeding resources of different crops, which usually get infected by the *M. phaseolina*.

12.7 Management strategies

12.7.1 Cultural practices

Soil-borne diseases are managed by using one of two major strategies: preventing the pathogen from entering new field areas; or when a field is already infected, lessening the intensity of the disease by using resistant cultivars, chemicals, and biofungicides (Lodha & Mawar, 2020). Farm hygiene and biovigilance can be used to stop the migration of overwintering microsclerotia and minimize the spread of the disease into the field. For example, in club root rot of canola, *M. phaseolina* microsclerotia gets attach to the soil particles and by this got transfer to other places (Rennie et al., 2015). In infected soil in the Sahel area, soil solarization was found effective and decreased *M. phaseolina* survival emergence of cowpea charcoal rot either alone or in combination with the millet residues (Ndiaye et al., 2007). *M. phaseolina* numbers were reduced by 80%–90% in biosolarized plots that had chicken manure added to it, and strawberry plant mortality was also decreased (Chamorro et al., 2015). Charcoal rot was controlled in soybean found under low tillage system; the tillage system forms less of *M. phaseolina* sclerotial growth than under a conventional tillage system (Mengistu et al., 2009; Perez-Brandán et al., 2012). Colder soil temperature having high volume of crop residues on its soil surface may be the cause of the formation of fewer sclerotia in the no-tillage system. Strawberry root rot severity significantly decreased after soil was fumigated with chemicals like methyl bromide: chloropicrin (50:50 at 35.9 g/m^2) and 1,3-dichloropropene: chloropicrin (63:35 at 326.7 L/ha), which decreased colonization of this pathogen (Chamorro et al., 2015). Fertilization with phosphorus showed less of diseases, while use of nitrogen fertilization increased the disease severity (Spagnoletti et al., 2018, 2020). Other cultural methods, such as rotating wheat and rice crops, can economically reduce the occurrence of *M. phaseolina* disease in South Asia. Zinc sulfate application to the soil decreased the occurrence of dry root rot and pathogen survival in beans (Ansari, 2010; Siddiqui et al., 2001). The incidence of strawberry charcoal rot has been shown to be reduced by the use of the combined technique of biosolarization and biofumigation, which reduces or stabilizes the *M. phaseolina* microsclerotia population in soil (Chamorro et al., 2015). These studies, however, were site-specific and must be verified in various ecoclimatic locations. The incidence of strawberry charcoal rot has been shown to be reduced by the use of the combined technique of biosolarization and biofumigation, which reduces or stabilizes the *M. phaseolina* microsclerotia population in soil (Chamorro et al., 2015). These studies, however, were site-specific and must be verified in various eco-climatic locations.

12.7.2 Fungicides

With the conditions favorable to the development of a disease, treatment-using fungicides may be necessary in the absence of resistance. In intensive production systems, *M. phaseolina* incidence is greatly reduced by the use of synthetic fungicides for disease management. Mungbean and urdbean seeds treated with carbendazim at a rate of 2 g active ingredient/kg seeds had less dry root rot (Athira, 2017; Kumari et al., 2012; Singh & Kumar, 2002). Additionally, thiophanate-methyl, at 1 g a.i./kg seeds, was discovered to be beneficial against dry root rot because it prevented *M. phaseolina* from growth (Athira, 2017; Singh & Kumar, 2002). Many scientists reported different fungicides, Bavistin, Saaf, and captan effective against *M. phaseolina* (Chaudhari & Chaudhari, 2012; Kar & Sahu, 2009). Various fungicides were used against root rot in cotton (Lambhate et al., 2002), guar (Jaiman & Jain, 2010), charcoal rot in groundnut (Moradia & Khandar, 2011), melon wilt (Cohen et al., 2012), castor (Parmar et al., 2017), and crown and root rot in strawberries (El-Marzoky et al., 2018) But it was discovered that thiophanate methyl and carbendazim were more efficient. Different systemic and nonsystemic fungicides were evaluated under in vitro and in vivo conditions like difenoconazole, carbendazim, azoxystrobin, benomyl, and dazome (Chamorro et al., 2015; Cohen et al., 2012; Lokesh et al., 2020; Parmar et al., 2017; Tonin et al., 2013) against *M. phaseolina*. Tebuconazole 50% + trifloxystrobin 25% WG was found good against the pathogen infection when tested under in vitro conditions (Kumari et al., 2022).

12.7.3 Biological control

Mungbean and urdbean dry root rot incidence was reduced by using bacterial formulations of *Pseudomonas aeruginosa*, *Bacillus subtilis,* and *Burkholderia* species as seed dressing and soil soaking (Ashwini et al., 2014; Latha et al., 2017; Satya et al., 2011; Siddiqui et al., 2001). Isolates of *Trichoderma* reduced mungbean dry root rot when grown on organic substrates like coir pith, groundnut shell, and press mud (Raghuchander et al., 1993). Under glasshouse and field circumstances, mungbean and urdbean dry root rot incidence was reduced by combining the *Pseudomonas fluorescens*, *B. subtilis,* and bavistin with the biocontrol agents like *Trichoderma viride* or *T. harzianum* (Kumari et al., 2012; Sajeena et al., 2004; Shahid & Khan, 2016; Thilagavathi et al., 2007). The dry root rot of mungbeans and urdbeans was also minimized by the alone putting of *T. harzianum* or *T. viride* (Ashwini et al., 2014; Athira, 2017; Jamwal et al., 2016; Kaithikeyan et al., 2015; Tetali et al., 2015). *T. harzianum* conidia applied to melon seeds decreased the occurrence of charcoal rot stems (Etabarian, 2006). To suppress *M. phaseolina* population in the crops like insesame, maize, strawberry, and soybean, PGPRs and biocontrol agents, i.e., *Bacillus* spp. and *T. harzianum* used in the form of seed coating or soil treatment, in combination with biofumigants like *Brassica* sp. and alone was found to be effective against the pathogen (El-Fik et al., 2004; Israel et al., 2005; Khalili et al., 2016; Pastrana et al., 2016). The antifungal activity of *T. harzianum* was due to different volatile substances, including isocyanide derivatives, terpenes, acetaldehyde, derivatives of alpha-pyrone, piperazine, and hydrazone derivatives (Shoaib et al., 2018), as well as polyketides and alcohols (Pitson et al., 1993), and they are also in charge of the breakdown of fungal cell walls. This could be a factor in the suppression of *M. phaseolina* (Chang et al., 2006). According to Martnez-Salgado et al. (2021), the strains of *T. koningiopsis* (T-K11) and *T. harzianum* (TH-3) reduced the disease charcoal rot in peanut crop in Puebla and Mexico. According to Kumari et al. (2012), *T. harzianum* inhibited *M. phaseolina* growth to the greatest extent (73.33%).

In field tests, both the arbuscular mycorrhizal species *Glomus claroideum* and *Glomus aggregatum* lessened dry root rot in mungbeans and urdbeans, respectively (Chandra et al., 2007). According to Jaber and Fayyadh (2019), different *Streptomyces* spp. demonstrated high antagonistic activity against *M. phaseolina*, in cowpea and mung bean under in vitro and *in vivo* conditions against the *M. phaseolina*.

12.7.4 Botanicals

Leaf extracts of different botanicals like *Eucalyptus camaldulensis, Allium* species, *Azadirachta indica,* and *Adenocalymma alliaceum* as soil amendments (Javaid & Siddique, 2011) and oils of*Launaea nudicaulis*, *Palma rosa* have been reported to reduce *M. phaseolina* in legumes (Haseeb et al., 2013; Kumari et al., 2012; Mansoor et al., 2007; Rana et al., 2014; Tandel et al., 2010). Murugapriya et al. (2011) found that *Allium* spp. (20%) by agar well diffusion method completely inhibited the mycelial growth of fungus. Due to application of neem cake, there was reduction in mortality caused by *M. phaseolina* infection in brinjal (eggplant) (Dapkekar et al., 2018). Neem cake and farmyard manure were found effective against dry root rot (Csondes et al., 2008; Latha et al., 2017). Kumari et al. (2022) reported *Allium sativum* followed by *Azadirachta indica* showed maximum mycelial growth inhibition of *M. phaseolina* in vitro.

12.8 Conclusion

It may be concluded that *M. phaseolina* is an imperative soil- and seed-borne pathogen. It consists of a wide range of hosts and a widespread geographic distribution. The goal of current chapter is to identify the genetic diversity among a large group of *M. phaseolina*. Almost all the diseases caused by this pathogen are quantitatively inherited and regulated with several genes, breeding for resistance is challenging; therefore, methods that lend a hand to clarify various resistance mechanisms also categorize resistant genotypes add to or helps to achieve greater breeding program achievement. There will be several opportunities to comprehend population genetics and fitness characteristics through morphological and genomic characterization of the pathogen. These investigations further may help producers to better understand and control the disease known as charcoal rot. Population structure of this pathogen makes us known that there have been evaluation of genetic diversity among different isolates of *M. phaseolina* from different areas of the world. The identification of heterogeneity across *M. phaseolina* isolates is necessary for directing the progress of suitable techniques and will help breeders to optimize their breeding programs to develop long-term resistance across wider geographical zones. These findings will be helpful in creating integrated charcoal rot management strategies as well as breeding programs for pulses and other crops.

References

Aboshosha, S. S., Atta Alla, S. I., El-Korany, A. E., & El-Argawy, E. (2007). Characterization of *Macrophomina phaseolina* isolates affecting sunflower growth in El-Behera governorate, Egypt. *International Journal of Agriculture and Biology, 9*, 807–815.

Almeida, A. M. R., Abdelnoor, R. V., Arias, C. A. A., et al. (2003). Genotypic diversity among Brazilian isolates of *Macrophomina phaseolina* revealed by RAPD. *Fitopatologia Brasileira, 28*(3), 279–285.

Aly, A. A., Abdel-Sattar, M. A., Omar, M. R., & Abd-Elsalam, K. A. (2007). Differential antagonism of *Trichoderma* sp. against *Macrophomina phaseolina*. *Journal of Plant Protection Research, 47*, 91–102.

Ansari, M. M. (2010). Integrated management of charcoal rot of soybean causes *M. phaseolina*. *Soybean Research, 8*, 39–47.

Ashwini, C., Giri, G. K., & Halgekar, N. Y. (2014). Efficacy of bioagents against seed borne fungi of black gram. *International Journal of Applied Biology and Pharmaceutical Technology, 5*, 56–57.

Athira, K. (2017). Efficacy of fungicide and bio-control agent against root rot of black gram (*Vigna mungo* L.) caused by *Macrophomina phaseolina* (Tassi) Goid. *International Journal of Current Microbiology and Applied Science, 6*, 2601–2607.

Babu, B. K., Mesapogu, S., Sharma, A., Somasani, S. R., & Arora, D. K. (2011). Quantitative real-time PCR assay for rapid detection of plant and human pathogenic *Macrophomina phaseolina* from field and environmental samples. *Mycologia, 103*, 466–473.

Babu, B. K., Reddy, S. S., Yadav, M. K., Sukumar, M., Mishra, V., & Saxena, A. K. (2010). Genetic diversity of *Macrophomina phaseolina* isolates from certain agro-climatic regions of India by using RAPD markers. *Indian Journal of Microbiology, 50*, 199–204.

Babu, B. K., Saxena, A. K., Srivastava, A. K., & Arora, D.]K. (2007). Identification and detection of *Macrophomina phaseolina* by using speciesspecific oligonucleotide primers and probe. *Mycologia, 99*, 797–803.

Baird, R. E., Wadl, P. A., Allen, T., McNeill, D., Wang, X., Moulton, J. K., et al. (2010). Variability of United Statesisolates of *Macrophomina phaseolina* based on simple sequence repeats and cross genus transferability to related genera with in Botryosphaeriaceae. *Mycopathologia, 170*, 169–180.

Burkhardt, A., Ramon, M. L., Smith, B., Koike, S. T., & Martin, F. (2018). Development of molecular methods to detect *Macrophomina phaseolina* from strawberry plants and soil. *Phytopathology, 108*, 1386–1394.

Chamorro, M., Miranda, L., Domínguez, P., Medina, J. J., Soria, C., Romero, F., et al. (2015). Evaluation of biosolarization for the control of charcoal rot disease (*Macrophomina phaseolina*) in strawberry. *Crop Protection, 67*, 279–286.

Chandra, S., Khare, V., & Kehri, H. K. (2007). Evaluation of arbuscular mycorrhizal fungi against *Macrophomina phaseolina* causing dry root rot of urdbean and mungbean. *Indian Phytopathology, 60*, 42–47.

Chang, K. F., Hwang, S. F., Wang, H., Turnbull, G., & Howard, R. (2006). Etiology and biological control of Sclerotinia blight of coneflower using *Trichoderma* species. *The Plant Pathology Journal, 5*, 15–19.

Chaudhari, N. D., & Chaudhari, S.,M. (2012). Evaluation of fungicides against *Macrophomina phaseolina* (Tassi.) goid causing leaf blight root rot of green gram. In *Soveniour of third global conference (ISMPP) on "Plant Pathology for food security" held at MPUAT, Udaipur January 10–13* (p. 164).

Choudhary, S., Choudhary, A. K., & Sharma, O. P. (2011). Screening of mungbean (*Vigna radiata*) genotypes to identify source of resistance to dry root rot. *Journal of Food Legumes, 24*(2), 117–119.

Chowdhury, S., Basu, A., Ray Chaudhuri, T., & Kundu, S. (2014). *In-vitro* characterization of the behaviour of *Macrophomina phaseolina* (Tassi) Goid. at the rhizosphere and during early infection of roots of resistant and susceptible varieties of sesame. *European Journal of Plant Pathology, 138*, 361–375.

Cloud, G. L., & Rupe, J. C. (1988). Preferential host selection of isolates of *Macrophomina phaseolina*. *Phytopathology, 78*(12), 1563.

Cohen, R., Omari, N., Porat, A., & Edelstein, M. (2012). Management of *Macrophomina* wilt in melons using grafting or fungicide soil application: Pathological, horticultural and economical aspects. *Crop Protection, 35*, 58–63.

Csondes, I., Baliko, K., & Degenhardt, A. (2008). Effect of different nutrient levels on the resistance of soybean to *Macrophomina phaseolina* infection in the field experiments. *Acta Agronomica Hungarica, 56*, 357–362.

Da Silva, M. P., Klepadlo, M., Gbur, E. E., Pereira, A., Mason, R. E., & Rupe, J. C. (2019). QTL mapping of charcoal rot resistance in PI 567562A soybean accession. *Crop Science, 59*, 474–479.

Dambal, G., Rashmi, D., Revanappa, S. B., Suma, M., & Saabale, P. R. (2019). Identification for resistant sources against dry root rot in black gram germplasm (*Vigna mungo* L.). *International Journal of Chemical Studies, 7*, 990–992.

Dapkekar, A. G., Mondal, B., Kadam, R. V., & Suryawanshi, A. P. (2018). Effect of organic amendments against *Macrophomina phaseolina* (Tassi) Goid. of brinjal in pot culture. *International Journal of Current Microbiological Applied Science, 7*, 95–100.

Das, I. K., Fakrudin, B., & Arora, D. K. (2008). RAPD cluster analysis and chlorate sensitivity of some Indian isolates of *Macrophomina phaseolina* from sorghum and their relationships with pathogenicity. *Microbial Research, 163*, 215–224.

Dhingra, O. D., & Sinclair, J. B. (1978). *Biology and pathology of* Macrophomina phaseolina. Viçosa/MG, Brasil: UFV, Imprensa Universitária.

Dreshka, M. V., Khare, M. N., & Singh, L. (1974). Evaluation of varieties of mungbean (*Physeolus aureus* Roxb.) for their resistance to *Rhizoctonina bataticola* (Taub.)Butler by paper towel technique. *JNKVV Research Journal, 8*, 60–62.

El-Fiki, A. I. I., Mohamed, F. G., El-Deeb, A. A., & Khalifa, M. M. A. (2004). Some applicable methods for controlling sesame charcoal rot disease (*Macrophomina phaseolina*) under greenhouse conditions. *Egypt Journal of Phytopathology, 32*, 87–101.

El-Marzoky, H. A., Abdalla, M. E., Abdel-Sattar, M. A., & Abid, M. A. (2018). Management of crown and root rot diseases in strawberry commercial fields in Egypt. *Journal of Plant Protection and Plant Pathology, 9*, 399–404.

Etabarian, H. R. (2006). Evaluation of *Trichoderma* isolates for biological control of charcoal stem rot in melon caused by *Macrophomina phaseolina. Journal of Agriculture, Science and Technology, 8*, 243–250.

Forster, H., & Shuai, B. (2020a). Exogenous siRNAs against chitin synthase gene suppress the growth of the pathogenic fungus *Macrophomina phaseolina. Mycologia, 112*, 699–710.

Forster, H., & Shuai, B. (2020b). RNAi-mediated knockdown of β-1,3-glucan synthase suppresses growth of the phytopathogenic fungus *Macrophomina phaseolina. Physiology and Molecular Plant Pathology, 110*, 101486.

Gahlot, N. (2018). *Physiological and management studies of* Macrophomina phaseolina *(Tassi) Goid. causing root rot of mungbean [*Vigna radiate *(L.) Wilczek] (Nitisha Gahlot, MSc. Thesis)*. Sri Karan NarendraAgriculture University, Jobner.

Gautam, R., Singh, S. K., & Sharma, V. (2014). RAPD and Nuclear rDNA ITS polymorphism within *Macrophomina phaseolina* isolated from arid legumes of western Rajasthan. *Proceedings of the National Academy of Science Sec B Biological Science, 84*, 171–181.

Ghosh, T., Biswas, M. K., Guin, C., & Roy, P. (2018). A review on characterization, therapeutic approaches and pathogenesis of *Macrophomina phaseolina. Plant Cell Biotechnology and Molecular Biology, 19*, 72–84.

Ghosh, T., Mukherji, N., & Basak, M. (1964). On the occurrence of a new species of *Orbilaria* Fr. *Jute Bulletin, 27*, 134–141.

Ghosh, R., Tarafdar, A., & Sharma, M. (2017). Rapid and sensitive diagnoses of dry root rot pathogen of chickpea (*Rhizoctonia bataticola* (Taub.) Butler) using loop-mediated isothermal amplification assay. *Scientific Reports, 7*, 42737.

Goidanich, G. (1947). A revision of the genus *Macrophomina* Petrak. type species: M.P. (Tassi) G. Goid. n. comb. M. P. (Maubl.) Ashby. *Ann Sperim Agrar rome, 3*, 449–461.

Gomez, A. O., De Faveri, J., Neal, J. M., Aitken, E. A. B., & Herrington, M. E. (2020). Response of strawberry cultivars inoculated with *Macrophomina phaseolina* in Australia. *International Journal of Fruit Science, 20*, 164–177.

Gupta, G. K., & Chauhan, G. S. (2005). *Symptoms, identification and management of soybean diseases. Technical Bulletin 10*. Indore, India: National Research Centre for Soybean.

Gupta, G. K., Sharma, S. K., & Ramteke, R. (2012). Biology, epidemiology and management of the pathogenic fungus *Macrophomina phaseolina* (Tassi) goid with special reference to charcoal rot of soybean (Glycine max (L.) Merrill). *Journal of Phytopathology, 160*, 167–180.

Haseeb, A. F., Sahi, S. T., Ali, S., & Fiaz, M. (2013). Response of different mungbean varieties against *Macrophomina phaseolina* (Tassi) Goid. and *in vitro* studies of plant extracts against pathogen. *Pakistani Journal of Phytopathology, 25*, 78–83.

Hemmati, P., Zafari, D., Mahmoodi, S. B., Hashemi, M., Gholamhoseini, M., & Dolatabadian, A. (2018). Histopathology of charcoal rot disease (*Macrophomina phaseolina*) in resistant and susceptible cultivars of soybean. *Rhizosphere, 7*, 27–34.

Hernández-Delgado, S., Reyes-Valdés, M., Rosales-Serna, R., & Mayek-Pérez, N. (2009). Molecular markers associated with resistance to *Macrophomina phaseolina* (Tassi) Goid. in common bean. *Journal of Plant Pathology, 91*, 163–170.

Ilyas, M. B., & Sinclair, J. B. (1974). Effects of plant age upon development of necrosis and occurrence of intraxylem sclerotia in soybean infected with *Macrophomina phaseolina. Phytopathology, 64*, 156–157.

Iqbal, S. M., Ghafoor, A., Arshad, M., & Bashir, M. (2003). Screening of urdbean (*Vigna mungo* L.) germplasm for resistance to charcoal rot disease. *Pakistan Journal of Plant Pathology, 2*, 107–110.

Iqbal, U., & Mukhtar, T. (2014). Morphological and pathogenic variability among *Macrophomina phaseolina* isolates associated with mungbean (*Vigna radiata* L.) Wilczek from Pakistan. *The Scientific World Journal Journal, 3*, 950175.

Islam, M., Haque, M., Islam, M., Emdad, E., Halim, A., & Hossen, Q. M. (2012). Tools to kill: Genome of one of the most destructive plant pathogenic fungi *Macrophomina phaseolina. BMC Genomics, 13*, 493.

Israel, S., Mawar, R., & Lodha, S. (2005). Soil solarisation, amendments and bio-control agents for the control of *Macrophomina phaseolina* and *Fusarium oxysporum* f. sp. cumini in aridisols. *Annals of Applied Biology, 146*, 481–491.

Jaber, M. A., & Fayyadh, M. A. (2019). Biological control of charcoal rot disease caused by *Macrophomina phaseolina* (Tassi) Goid. on cowpea and mung bean by some isolates of *Streptomyces* spp. *Basrah Journal of Agricultural Sciences, 32*(Spec. Issue 2), 207–219. https://doi.org/10.37077/25200860.2019.269

Jaiman, R. K., & Jain, S. C. (2010). Effect of fungicides on root rot of clusterbean caused by *Macrophomina phaseolina. Environment and Ecology, 28*, 1138–1140.

Jamwal, S., Jamwal, A., Reena, K. A., Isher, A., & Dutta, U. (2016). Management of root rot of urdbean (Phaseolus mungo) with Trichoderma spp in rainfed areas of Jammu and Kathua districts. *The Bioscan, 11*, 2947–2951.

Jana, T., Sharma, T. R., Prasad, R. D., & Arora, D. K. (2003). Molecular characterization of *Macrophomina phaseolina* and Fusarium species by a single primer RAPD technique. *Microbiological Research, 158*(3), 249–257.

Jana, T., Sharma, T. R., & Singh, N. K. (2005). SSR-based detection of genetic variability in the charcoal root rot pathogen *Macrophomina phaseolina. Microbiological Research, 109*, 81–86.

Jana, T. K., Singh, N. K., Koundal, K. R., & Sharma, T. R. (2005). Genetic differentiation of charcoal rot pathogen, *Macrophomina phaseolina*, into specific groups using URP-PCR. *Canadian Journal of Microbiology, 51*, 159–164.

Javaid, A., & Saddique, A. (2011). Management of *Macrophomina* root rot of mungbean using dry leaves manure of *Datura metel* as soil amendment. *Spanish Journal of Agricultural Science, 9*, 901–905.

Kaithikeyan, V., Brindha, S., Annadurai, B., & Gangwar, S. K. (2015). Biological control of *Macrophomina phaseolina* (Tassi) Goid. root rot in *Vigna mungo* (blackgram) with *Trichoderma* spp. *International Journal of Advanced Biological Research, 5*, 118–127.

Kar, A. K., & Sahu, K. C. (2009). Deteroition of seed quality in mung bean by *Macrophomina phaseolina* (Tassi.) Goid. and its management. *Seed Research, 37*(1/2), 139–142.

Kaur, S., Dhillon, G. S., Brar, S. K., Vallad, G. E., Chand, R., & Chauhan, V. B. (2012). Emerging phytopathogen *Macrophomina phaseolina*: Biology, economic importance and current diagnostic trends. *Critical Reviews in Microbiology, 38*, 136–151.

Khalili, E., Javed, M. A., Huyop, F., Rayatpanah, S., Jamshidi, S., & Abdul Wahab, R. (2016). Evaluation of Trichoderma isolates as potential biological control agent against soybean charcoal rot disease caused by *Macrophomina phaseolina. Biotechnology & Biotechnological Equipment, 30*, 479–488.

Khan, S. K. (2007). *Macrophomina phaseolina* as causal agent for charcoal rot of sunflower. *Mycopathology, 5*, 111–118.

Khan, A. N., Shair, F., Malik, K., Hayat, Z., Khan, M. A., & Hafeez, F. Y. (2017). Molecular identification and genetic characterization of *Macrophomina phaseolina* strains causing pathogenicity on sunflower and chickpea. *Frontiers in Microbiology, 8*, 1309.

Khan, K. A., Shoaib, A., & Akhtar, S. (2016). Response of *Vigna radiata* (L.) Wilczek genotypes to charcoal rot disease. *Mycopathologia, 14*, 1–7.

Khan, S. H., & Shuaib, M. (2007). Identification of sources of resistance in mung bean (*Vigna radiata* L.) against charcoal rot *Macrophomina phaseolina* (Tassi) Goid. In K. Z. Ahmed (Ed.), *8th African Crop Science Society Conference, El-Minia, Egypt, 27–31 October 2007.*

Kumar, P., Gaur, V. K., & Meena, A. K. (2017). Screening of different *Macrophomina phaseolina* on susceptible (RMG-62) variety of mungbean. *International Journal of Pure and Applied Bioscience, 5*, 698–702.

Kumari, N., Meena, A. K., Godara, S. L., Gahlot, V. S., & Reddy, D. K. (2022). Evaluation of bio-agents, botanicals and fungicides under *in vitro* conditions against *Macrophomina phaseolina* (Tassi) Goid causing charcoal rot of cowpea. *Biological Forum - An International Journal, 14*(2), 84–90.

Kumari, R., Shekhawati, K. S., Gupta, R., & Khokhar, M. K. (2012). Integrated management against root rot of mungbean (*Vigna radiate* L. Wilczek) incited by *Macrophomina phaseolina. Journal of Plant Pathology & Microbiology, 3*, 136.

Lakhran, L., Ahir, R. R., Choudhary, M., & Choudhary, S. (2018). Isolation, purification, identification and pathogenicity of *Macrophomina phaseolina* (Tassi) Goid. caused dry root rot of chickpea. Journal of Pharmacognosy *and* Phytochemistry, *7*, 3314–3317.

Lambhate, S. S., Chaudhari, G. K., Mehetre, S. S., & Zanjare, S. R. (2002). *In vitro* evaluation of chemicals against root rot of cotton caused by *M. phaseolina. Journal of Maharashtra Agriculture University, 27*, 9–100.

Latha, P., Karthikeyan, M., & Rajeswari, E. (2017). Development of bioformulations for the management of blackgram dry root rot caused by *Rhizoctonia bataticola* (Taub Butler). *Advances in Research, 9*, 1–12.

Lionetti, V., Raiola, A., Camardella, L., Giovane, A., Obel, N., Pauly, M., et al. (2007). Overexpression of pectin methylesterase inhibitors in *Arabidopsis* restricts fungal infection by *Botrytis cinerea. Plant Physiology, 143*, 1871–1880.

Lodha, S., & Mawar, R. (2020). Population dynamics of *Macrophomina phaseolina* in relation to disease management: A review. *Journal of Phytopathology, 168*, 1–17.

Lokesh, R., Rakholiya, K. B., & Thesiya, M. R. (2020). Evaluation of different fungicides against *Macrophomina phaseolina* (Tassi) goid. causing dry root rot of chickpea (*Cicer arietinum* L.*) in vitro. International Journal of Current Microbiology and Applied Science, 9*, 1–11.

Mahdizadeh, V., Safaie, N., & Goltapeh, E. M. (2011). Diversity of *Macrophomina phaseolina* based onmorphological and genotypic characteristics in Iran. *Plant Pathology Journal, 27*, 128–137.

Mahmoud, A. F., Abou-Elwafa, S. F., & Shehzad, T. (2018). Identification of charcoal rot resistance QTLs in sorghum using association and in silico analyses. *Journal of Applied Genetics, 59*, 243–251.

Mansoor, F., Sultana, V., & Ehteshamul-Haque, S. (2007). Enhancement of biocontrol potential of *Pseudomonas aeruginosa* and *Paecilomyceslilacinus* against root rot of mungbean by a medicinal plant*Launaeanudicaulis* L. *Pakistan Journal of Botany, 39*, 2113–2119.

Martínez-Salgado, S. J., Andrade-Hoyos, P., Parraguirre Lezama, C., Rivera-Tapia, A., Luna-Cruz, A., & Romero-Arenas, O. (2021). Biological control of charcoal rot in peanut crop through strains of *Trichoderma* spp., in Puebla, Mexico. *Plants (Basel), 10*(12), 2630.

Maublanc, A. (1905). Especes nouvelles de champignons infgrieurs. *Bulletin de la Societe Mycologique de France, 21*, 90–91.

Mayek-Pérez, N., García-Espinosa, R., López-Castaneda, C., AcostaGellegos, J. A., & Simpson, J. (2002). Water relations histopathology and growth of common bean (*Phaseolus vulgaris* L.) during pathogenesis of *Macrophomina phaseolina* under drought stress. *Physiology and Molecular Plant Pathology, 60*, 185–195.

Mayek-Perez, N., López-Castañeda, C., González-Chavira, M., et al. (2001). Variability of Mexican isolates of *Macrophomina phaseolina* on basis of pathogenesis and AFLP genotype. *Physiological and Molecular Plant Pathology, 59*(5), 257–264.

Mayek-Perez, N., Castañeda, C. L., & Gallegos, J. A. (1997). Gallegos, Variación en características culturales in vitro de aislamientos de *Macrophomina phaseolina* y su virulencia en frijol. *Agrociencia, 31*(2), 187–195.

Mengistu, A., Bond, J., Nelson, R., Rupe, J., Shannon, G., Arelli, P., et al. (2013). Identification of soybean accessions resistant to *Macrophomina phaseolina* by field screening and laboratory validation. *Plant Health Progress, 14*, 25.

Mengistu, A., Reddy, K. N., Zablotowicz, R. M., & Wrather, A. J. (2009). Propagule densities of *Macrophomina phaseolina* in soybean tissue and soil as affected by tillage, cover crop, and herbicide. *Plant Health Progress, 10*, 28.

Mihail, J. D. (1992). Macrophomina. In L. L. Singleton, J. D. Mihail, & C. M. Rush (Eds.), *Methods for research on soilborne phytopathogenic fungi* (pp. 134–136). St. Paul, MN: American Phytopathological Society.

Mihail, J. D., & Taylor, S. J. (1995). Interpreting variability among isolates of *Macrophomina phaseolina in* pathogenicity, pycnidium production, and chlorate utilization. *Canadian Journal of Botany, 73*(10), 1596–1603.

Miklas, P., Stone, V., Urrea, C. A., Johnson, E., & Beaver, J. S. (1998). Inheritance and QTL analysis of field resistance to ashy stem blight in Common Bean. *Crop Science, 38*, 916–921.

Miklas, P. N., Zapata, M., Beaver, J. S., & Grafton, K. F. (1999). Registration of four dry bean germplasms resistant to common bacterial blight: ICB-3, ICB-6, ICB-8, and ICB-10. *Crop Science, 39*, 594.

Moradia, A. M., & Khandar, R. R. (2011). Loss of yield of groundnut (*Arachis hypogaea* L.) due to dry root rot (*Macrophomina phaseolina*) and their management under in vivo condition. *International Journal of Agriculture Science, 7*, 282–285.

Muchero, W., Ehlers, J. D., Close, T. J., & Roberts, P. A. (2011). Genic SNP markers and legume synteny reveal candidate genes underlying QTL for *Macrophomina phaseolina* resistance and maturity in cowpea [*Vigna unguiculata* (L) Walp.]. *BMC Genomics, 12*, 8.

Murugapriya, E., Alice, D., & Jayamani, P. (2011). Antifungal activity of botanicals and micro-nutrients against *Macrophomina* leaf blight in mungbean. *Journal of Food Legumes, 24*(2), 113–116.

Ndiaye, M., Termorshuizen, A. J., & VanBruggen, A. H. C. (2007). Combined effects of solarization and organic amendment on charcoal rot caused by *Macrophomina phaseolina* in the Sahel. *Phytoparasitica, 35*, 392–400.

Pandey, A. K., Burlakoti, R. R., Rathore, A., & Nair, R. M. (2020). Morphological and molecular characterization of Macrophomina phaseolina isolated from three legume crops and evaluation of mungbean genotypes for resistance to dry root rot. *Crop Protection, 127*, 104962. https://doi.org/10.1016/j.cropro.2019.104962

Parmar, H. V., Kapadiya, H. J., & Bhaliya, C. M. (2017). Efficacy of different fungicides against *Macrophomina phaseolina* (Tassi) Goid. causing castor root rot. *International Journal of chemical Studies, 5*, 1807–1809.

Pastrana, A. M., Basallote-Ureba, M. J., Aguado, A., Akdi, K., & Capote, N. (2016). Biological control of strawberry soil-borne pathogens *Macrophomina phaseolina* and *Fusarium solani*, using *Trichoderma asperellum* and *Bacillus* spp. *Phytopathologia Mediterranea, 55*, 109–120.

Pawlowski, M. L., Hill, C. B., & Hartman, G. L. (2015). Resistance to charcoal rot identified in ancestral soybean germplasm. *Crop Science, 55*, 1230–1235.

Perez-Brandán, C., Arzeno, J. L., Huidobro, J., Grümberg, B., Conforto, C., Hilton, S., et al. (2012). Long-term effect of tillage systems on soil microbiological, chemical and physical parameters and the incidence of charcoal rot by *Macrophomina phaseolina* (Tassi) Goid. in soybean. *Crop Protection, 40*, 73–82.

Pitson, S. M., Seviour, R. J., & McDougall, B. M. (1993). B-glucanasafúúngica no celulolíítica: Sufisiologíía y regulacióón. *Enzyme and Microbial Technology, 15*, 178–192.

Purkayastha, S., Kaur, B., Dilbaghi, N., & Chaudhary, A. (2006). Characterization of *Macrophomina phaseolina,* the charcoal rot pathogen of cluster bean, using conventional techniques and PCR-based molecular markers. *Plant Pathology, 55*, 106–116.

Purkayastha, S., Kaur, B., Dilbaghi, N., Power, J. B., Davey, M. R., & Chaudhury, A. (2003). Studies on cultural characteristics and nitrogen assimilation in *Macrophomina phaseolina* infecting cluster bean (*Cyamopsis tetragonoloba*). Plant pathogen genomics from sequence to application. In *Proc of the British Society for Plant Pathology, Presidential Meeting, Nottingham, UK.*

Raghuchander, T., Samiyappan, R., & Arjunan, G. (1993). Biocontrol of *Macrophomina* root rot of mungbean. *Indian Phytopathology, 46*, 379–382.

Rajkumar, Bashasab, F., & Kuruvinashetti, M. S. (2007). Genetic variability of sorghum charcoal rot pathogen (*Macrophomina phaseolina*) assessed by random DNA markers. *Plant Pathology, 23*, 45–50.

Rana, M., Ashiq, R., & Virender, A. (2014). Systemic opposition of green gram against leaf blight caused by *Macrophomina phaseolina. International Journal of Mycology and Plant Pathology, 1*, 72–77.

Rennie, D. C., Holtz, M. D., Turkington, T. K., Leboldus, J. M., Hwang, S. F., Howard, R. J., et al. (2015). Movement of *Plasmodiophora brassicae* resting spores in windblown dust. *Canadian Journal of Plant Pathology, 37*, 188–196.

Reyes-Franco, M. C., Hernandez-Delgado, S., Beas- Fernandez, R., Medina-Fernández, M., Simpson, J., & Mayek-Pérez, N. (2006). Pathogenic and genetic variability within *Macrophomina phaseolina* from Mexico and other countries. *Journal of Phytopathology, 154*(7–8), 447–453.

Reznikov, S., Chiesa, M. A., Pardo, E. M., Lisi, V. D., Bogado, N., & Gonzalez, V. (2019). Soybean–*Macrophomina phaseolina*-specificinteractions and identification of a novel source of resistance. *Phytopathology, 109*, 63–73.

Reznikov, S., Vellicce, G. R., Mengistu, A., Arias, R. S., Gonzalez, V., De Lisi, V., et al. (2018). Disease incidence of charcoal rot (*Macrophomina phaseolina*) on soybean in north-western Argentina and genetic characteristics of the pathogen. *Canadian Journal of Plant Pathology, 40*, 423–433.

Sajeena, A., Salal, R. F., Seetharaman, K., & Mohan Babu, R. (2004). Evaluation of biocontrol agents against dry root rot of blackgram (*Vigna mungo*). *Journal of Mycology and Plant Pathology, 34*, 341–344.

Sánchez, S., Chamorro, M., Henríquez, J. L., Grez, J., Díaz, I., & Santos, B.d. L. (2017). Genetic and biological characterization of *Macrophomina phaseolina* (Tassi) Goid. causing crown and root rot of strawberry. *Chilean Journal of Agriculture Research, 77*, 325–331.

Sarr, M. P., Ndiaye, M. B., Groenewald, J. Z., & Crous, P. W. (2014). Genetic diversity in *Macrophomina phaseolina*, the causal agent of charcoal rot. *Phytopathologia Mediterranea, 53*(2), 250–268. https://doi.org/10.14601/Phytopathol_Mediterr-13736

Satya, V. K., Vijayasa mundeeswari, A., Paranidharan, V., & Velazhahan, R. (2011). *Burkholderia* sp. strain TNAU-1 for biological control of root rot in mungbean (*Vigna radiata* L.) caused by *Macrophomina phaseolina. Journal of Plant Protection Research, 51*, 273–278.

Shahid, S., & Khan, M. R. (2016). Biological control of root rot of mungbean plants incited by *M. phaseolina* through microbial antagonists. *Plant Pathology Journal, 15*, 27–39.

Sharma, R. C., & Bhowmik, T. P. (1986). Estimation of yield losses in groundnut due to *Macrophomina phaseolina* (Tassi) Goid. *Indian Journal of Plant Pathology, 4*, 108–112.

Sharma, I., Kumari, N., & Sharma, V. (2014). Defense gene expression in *Sorghum bicolor* against *Macrophomina phaseolina* in leaves and roots of susceptible and resistant cultivars. *Journal of Plant Interactions, 9*, 315–323.

Shoaib, A., Munir, M., Javaid, A., Awan, Z. A., & Rafiq, M. (2018). Anti-mycotic potential of *Trichoderma* spp. and leaf biomass of*Azadirachta indica* against the charcoal rot pathogen, *Macrophomina phaseolina* (Tassi) Goid. in cowpea. *Egyptian Journal of Biological Pest Control, 28*, 26.

Siddiqui, A., Ehetshamul-Haque, E., & Shaukat, S. S. (2001). Use of rhizobacteria in the control of root rot—root knot disease complex of mungbean. *Journal of Phytopathology, 149*, 337—346.

Singh, J., & Kumar, K. (2002). Location, survival, transmission and control of seed borne, *Macrophomina phaseolina* causing dry root rot and leaf blight in urdbean. *Annals of Plant Protection Science, 10*, 111—113.

Sobti, A. K., & Sharma, L. C. (1992). Cultural and pathogenic variations in isolates of *Rhizoctonia bataticola* from groundnut in Rajasthan. *Indian Phytopathology, 45*, 117—119.

Sowmya, P., Naik, V. G., & Naik, V. N. (2022). Genetic variability among South Indian isolates of *Macrophomina phaseolina*, the fungus involved in charcoal rot of mulberry and association of pathogenicity. *The Pharma Innovation Journal, SP-11*(12), 39—52.

Spagnoletti, F. N., Cornero, M., Chiocchio, V., Lavado, R. S., & Roberts, I. N. (2020). Arbuscular mycorrhiza protects soybean plants against *Macrophomina phaseolina* even under nitrogen fertilization. *European Journal of Plant Pathology, 156*, 839—849.

Spagnoletti, F. N., Leiva, M., Chiocchio, V., & Lavado, R. S. (2018). Phosphorus fertilization reduces the severity of charcoal rot (*Macrophomina phaseolina*) and the arbuscular mycorrhizal protection in soybean. *Journal of Plant Nutrition and Soil Science, 181*, 855—860.

Su, G., Suh, S. O., Schneider, R. W., & Russin, J. S. (2001). Host specialization in the charcoal rot fungus, *Macrophomina phaseolina*. *Phytopathology, 91*, 120—126.

Tandel, D. H., Sabalpara, A. N., & Pandya, J.,R. (2010). Efficacy of phytoextracts on *Macrophomina phaseolina* (Tassi.) Goid. causing leaf blight of green gram. *International Journal of Pharma Bio Sciences, 2*, 1—5.

Tančić Živanov, S., Dedić, B., Dimitrijević, A., Dušanić, N., Jocić, S., Miklič, V., et al. (2019). Analysis of genetic diversity among *Macrophomina phaseolina* (Tassi) Goid. isolates from Euro-Asian countries. *Journal of Plant Disease Protection, 126*, 565—573.

Tassi, A. (1901). Novae Micromycetum species. *Ser. III. Bull. Lab. Orto Bot. Univ. Siena, 4*, 7—12.

Tetali, S., Karpagavalli, S., & Pavani, L. S. (2015). Management of dry root rot of blackgram caused by *Macrophomina phaseolina* (Tassi) Goid. using bioagent. *Plant Archives, 15*, 647—650.

Thilagavathi, R., Saravanakumar, D., Ragupathi, N., & Samiyappan, R. (2007). A combination of biocontrol agents improves the management of dry root rot (*Macrophomina phaseolina*) in greengram. *Phytopathologia Mediterranea, 46*, 157—167.

Tok, F. M. (2019). Relationship between morphologic, phenotypic and pathogenic characteristics in *Macrophomina phaselina* isolates from cucumber plants. *International Journal of Innovative Approaches in Agricultural Research, 3*, 651—660.

Tomar, R. S., Parakhia, M. V., Rathod, V. M., Thakkar, J., Padhiyar, S. M., & Thummar, V. D. (2017). Molecular mapping and identification of QTLs responsible for charcoal rot resistance in Castor (*Ricinus communis* L.). *Indian Crops Production, 95*, 184—190.

Tonin, R. F. B., Avozani, A., Durante Danelli, A. L., Reis, E. M., Zoldan, S. M., & Garcés-Fiallos, F. R. (2013). *In vitro* mycelial sensitivity of *Macrophomina phaseolina* to fungicides Sensibilidade micelial *in vitro* de *Macrophomina phaseolina* a fungicidas. *Pesquisa Agropecuária Tropical, 43*, 460—466.

Valette-Collet, O., Cimerman, A., Reignault, P., Levis, C., & Boccara, M. (2003). Disruption of *Botrytis cinerea* pectin methylesterase gene *Bcpme1* reduces virulence on several host plants. *Molecular Plant-Microbe Interactions, 16*, 360—367.

Vandemark, G., Martnez, O. P. V., & de Alvardo, M. J. (2000). Assessment of genetic relationships among isolates of *Macrophomina phaseolina* using simplified AFLP technique and two different methods of analysis. *Mycologia, 92*, 656—664.

Yan, W., N.I., Y., Liu, X., Zhao, H., Chen, Y., et al. (2021). The mechanism of sesame resistance against *Macrophomina phaseolina* was revealed via a comparison of transcriptomes of resistant and susceptible sesame genotypes. *Plant Biology, 21*, 159.

Section III

Ecology and epidemiology

Chapter 13

Ecological niche modeling of *Macrophomina phaseolina* (Tassi) Goid. using bio-climatic and non-bio-climatic variables: a machine learning assessment

Manish Mathur[1] and Preet Mathur[2]

[1]*ICAR- Central Arid Zone Research Institute, Jodhpur, Rajasthan, India;* [2]*Computer Science Department, Jodhpur Institute of Engineering and Technology, Jodhpur, Rajasthan, India*

13.1 Introduction

The ecological niche model (ENM) is a statistical approximation for a species distribution because it integrates location data to environmental variables using statistical approaches to describe, explain, and/or predict species distribution (Sillero & Barbosa, 2021). ENMs can produce an equation that links an expected species distribution related to a group of environmental indicators or a response curve that describes how the predictors regulate species distribution. It is possible to refine the mathematical model into a cartographic model, which is a map showing habitat suitability, species occurrence probability, or species occurrence favorability. Consequently, ENMs are predicted in the environmental space and projected in the geographic space. The ecological niche of a species is the range of conditions, both biotic and abiotic, within which it can thrive. Instead, a niche is a collection of factors that help a particular species thrive (Putman & Wratten, 1984). Initially, the niche was conceived of as a species habitat requirement as well as a biotic community feature. Later, the concepts of niche variables and axes are introduced through niche formalization and hypervolume. Each dimension of hypervolume is a variable, and variables are linked so that species can be matched to environmental gradients. Many distinct regions exist in niche space, which can be summarized in a variety of ways, including habitat, food chain, and many-sided consideration (Dash, 2007). The core premise is that there are multiple relevant axes to consider, all of which have a positive correlation with species survival. Soil moisture and nitrate ion content, for example, are effective predictors of the occurrence of AM fungi (Fitzsimons et al., 2008).

The concept of hyperspace made the niche more practical and measurable. This hyperspace can be subdivided into subspaces, each of which has a collection of conditions or variables to be examined. The internal structure of a niche can be modeled by monitoring environmental variables along axes and then extrapolating species performance (population fitness) (Petersen, 2007). Response variables are niche characteristics that indicate how populations/species react. The niche preemption model includes species abundance as a response variable. Most of a given resource is taken in by the species that are very numerous (He & Tang, 2007).

The most noticeable difference between parasite/pathogen ecology and that of free-living species is that the pathogen's environment is a living organism (Begon et al., 2006). The host is both a place for the pathogen to live and a source of food for it. This means that the host includes axes of the basic niche related to the pathogen's home and food sources, as well as axes that stand in for the pathogen's consumption or physiological response. The pathogen's realized niche contains host reactions to infection as well as biotic limiting interactions, which are explored in greater depth later in the text. The result of the parasitic relationship is determined by the individual host physiology and the degree of individual trait diversity at the population level. When describing a niche, one can take one of two main approaches: (i) structural approach that takes into account the host—pathogen trophic relationship, population structure with specific pathogen attributes

Macrophomina Phaseolina. https://doi.org/10.1016/B978-0-443-15443-0.00006-1

(virulence, pathogenicity, infectivity), and host attributes (resistance, anatomical, physiological barriers), habitat description including the environmental envelope and relationships with other organisms, and (ii) functional approach that has three hierarchical levels: molecular, cellular, and ecological (gene-to-host).

Within the constraints of its genotypic and phenotypic plasticity, a pathogen can adapt to its environment by a variety of strategies, such as host alternation or saprophytic phases. The interaction between the host and the pathogen makes the pathogen's properties one of the axes of the realized niche.

An alternate strategy is to determine a pathogen's or set of pathogens' ecological niche. This suggests that a pathogen's partial niche hypervolume is a host (Gudelj et al., 2004). The interaction between the host and the guest has two outcomes: Both events redimension niche dimensions. One often discussed example is the hybridization that resulted in the creation of *Phytophthora alni*, which caused the fall of alders throughout Europe (Brasier et al., 1999).

Molecular biologists are collecting enormous amounts of biological data in the age of high-throughput and smart technologies. Machine learning (ML), a subset of artificial intelligence (AI), is one of the tried-and-true strategies for building momentum in microbiology. It has previously been used to solve complex problems like identifying therapeutic targets and vaccine candidates, detecting microorganisms that cause infectious diseases, classifying drug resistance against antimicrobial treatments, predicting disease outbreaks, and researching microbial relationships. ML has been used in all areas of microbiology, including virology, parasitology, mycology, and bacteriology (Goodswen et al., 2021). Recent studies also show that machine learning algorithms may be able to estimate accurately where a species is now and where it will be in the future in response to possible climate change scenarios (Phillips & Dudik, 2008; Sarikaya et al., 2018; Tittensor et al., 2009; Wei et al., 2018).

Macrophomina phaseolina (Tassi) Goid. is a fungal species of anamorphic Ascomycetes (Ndiaye, 2007). *Rhizoctonia bataticola* was the name given to the fungus in its micro-sclerotial condition (Taub.). *M. phaseolina* causes economic losses in high-value crops, such as *Arachis hypogea, Cicer arietinum, Cyamopsis tetragonoloba, Glycine max, Gossypium* spp., *Phaseolus vulgaris, Sorghum bicolor, Sesamum indicum, Vigna radiata, Vigna unguiculata,* and *Zea mays*. The fungus survives in soil and host crop detritus, usually via microsclerotia (Short et al., 1980). Microsclerotia are generally sheltered in crop residues after harvest and released into the soil as the crop wastes decompose. This can also persist in asymptomatic seeds as mycelium and in infected seeds as microsclerotia (Hartman et al., 1999). Clusters of microsclerotia can be found anywhere from the soil's surface to a depth of 20 cm (Mihail, 1989). These organisms can endure extreme conditions, such as low soil nutrient levels and temperatures exceeding 30°C, for up to 3 years (Dhingra & Sinclair, 1977). Sclerotia is harmed by soil C:N ratios, bulk density, and moisture content (Gupta et al., 2012). Microsclerotia can live for 2–15 years depending on environmental circumstances and their relationship with host material (Baird et al., 2003). The fungus can also live as mycelium in asymptomatic seeds or as microsclerotia in symptomatic seeds for up to 3 years (Hartman et al., 1999). As a result, recognizing the importance of host waste, seeds, and soil as inoculum sources is critical to disease management. Disease can be prevented to some extent through cultural techniques, organic amendments, seed treatment, and genetic host resistance.

However, for this species, predictive models are absent that its spatial distribution with reference to present and future climatic variables as well as nonclimatic variables like soil, land cover and habitat heterogeneity (i.e., diversity). In this study, we examined bioclimatic and NBC variables in order to foresee the existing and future distribution of *M. phaseolina* within the Indian region.

13.2 Material and methods

13.2.1 Data collections

Distributional records of this species were obtained from data repository like the Global Biodiversity Information Facility (www.gbif.org) and published literature (Dadwal & Bhartiya, 2012; Kohire et al., 2012; Kumari & Govindaiah, 2011; Lokesh et al., 2020; Mahadevakumar & Janardhana, 2016; Mallaiah & Rao, 2016; Meena et al., 2018; Nishad et al., 2018; Prakash et al., 2007; Swamy et al., 2017).

These point's coordinates were calculated employing high-resolution Google Earth satellite image data using QGIS 3.10.0 (Coban et al., 2020) software on a WGS84 coordinate system. We used Google Earth (http://ditu.google.cn/) to calculate the latitude and longitude values of occurrence reports that lacked precise geo-coordinates (Xu et al., 2021). Fig. 13.1 depicts a graphical representation of the places analyzed. The distributional locales were assembled into a CSV database using the sources mentioned above (.csv). Using the Spatial Thin window of R-based Graphical User Interface Wallace Software, duplicate records were geographically filtered and eliminated to maintain only one occurrence (Kass et al., 2018). Finally, 40 records of the species occurrence were acquired in order to construct the ENM.

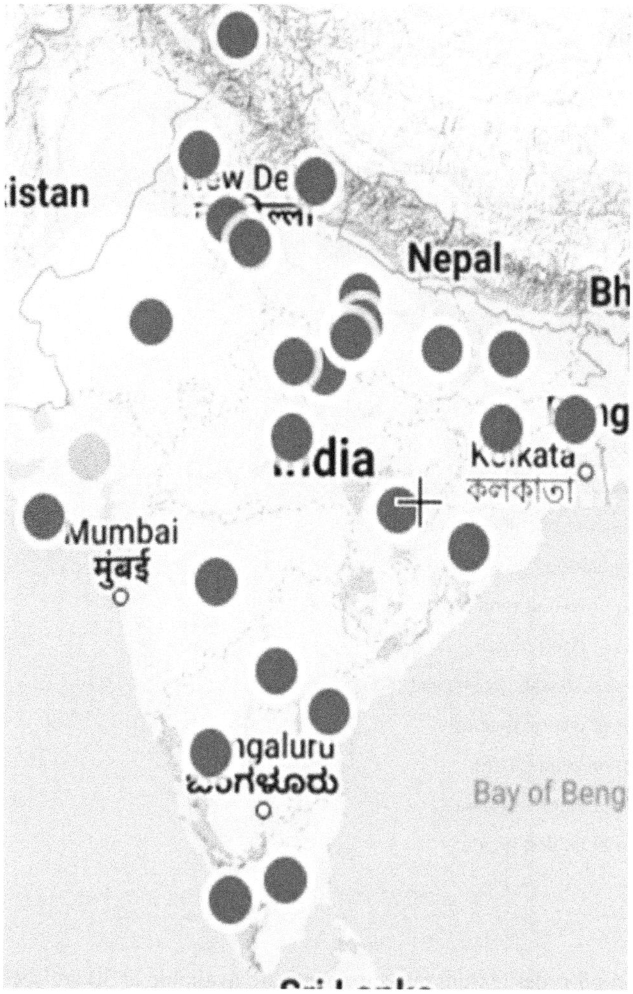

FIGURE 13.1 Existing field locations of *M. phaseolina* in India.

13.2.2 Bio-climatic and non-bio-climatic variables

The SSP126, SSP245, SSP370, and SSP585 climate change scenarios can be found in the sixth IPCC assessment report. Human action causes a decrease in global average temperature, thus we choose the SSP245 scenario in which greenhouse gas emissions are nearly the same as they are now (1970–2000). Current (Mishra et al., 2021; Rong et al., 2019) and two future scenarios (2050-time frame that represents mean values from 2041 to 2060 and 2070-time frame that represents mean values from 2061 to 2080) (Coban et al., 2020; Ye et al., 2020). Bio-climatic (BC) variables were downloaded at a spatial resolution of 30 s (~ 1 km^2) from WorldClim ver. 1.4, which is available online at www.worldclim.org (accessed on November 21st, 2021). The 19 BC variables (Hijmans et al., 2005) are one of the outputs of WorldClim Version 1.4, which was downloaded and clipped from global data for India at a spatial resolution of 30 arc sec ($\sim 1 \times 1$ km) and translated to ASCII (or ESRI ASCII) in DIVA-GIS version 7.5 (Hijmans et al., 2001; Zhang et al., 2021). Table 13.1 contains information on each BC parameter, including units and mathematical formulae.

13.2.3 Habitat heterogeneity index

Niche theory predicts a positive heterogeneity-diversity link because a more diversified site may supply more niche space and permit more species to coexist through niche partitioning. Following the textural characteristics of the Enhanced Vegetation Index (EVI, i.e., frequency distribution of pixel values) data from the Moderate Resolution Imaging Spectroradiometer (MODIS); Tuanmu and Jetz (2015) developed 14 metrics to measure global habitat heterogeneity at 1-km resolution.

TABLE 13.1 Predictive variables (current and future) bioclimatic data variables.

Code	Environmental variables	Scaling factor	Unit
BC-1	Annual mean temperature	10	°C
BC-2	Mean diurnal range (mean of monthly (max temp – min temp))	10	°C
BC-3	Isothermality (BC2/BC7) (×100)	100	–
BC-4	Temperature seasonality (standard deviation ×100)	100	–
BC-5	Max temperature of warmest month	10	°C
BC-6	Min temperature of coldest month	10	°C
BC-7	Temperature annual range (BC 5–BC 6)	10	°C
BC-8	Mean temperature of wettest quarter	10	°C
BC-9	Mean temperature of driest quarter	10	°C
BC-10	Mean temperature of warmest quarter	10	°C
BC-11	Mean temperature of coldest quarter	10	°C
BC-12	Annual precipitation	1	mm
BC-13	Precipitation of wettest month	1	mm
BC-14	Precipitation of driest month	100	mm
BC-15	Precipitation seasonality (coefficient of variation)	1	Fraction
BC-16	Precipitation of wettest quarter	1	mm
BC-17	Precipitation of driest quarter	1	mm
BC-18	Precipitation of warmest quarter	1	mm
BC-19	Precipitation of coldest quarter	1	mm

http://www.worldclim.org/bioclim

Six first-order and eight second-order texture measurements are available at 30 arc-second (~1 km at the equator), 2.5 arc-minute (~5 km), and 12.5 arc-minute (~25 km) resolutions (http://www.earthenv.org/texture). We used a 30 arc second data set for First-Order Texture Measures (Coefficient of variation = Normalized dispersion of EVI; Evenness = Evenness of EVI; Range = Range of EVI; Shannon and Simpson Indices = Diversity of EVI; Standard deviation = Dispersion of EVI) and Second Order Texture Measures (Uniformity = Orderliness of EVI; Maximum = Dominance of EVI combinations).

13.2.4 Land cover

Land cover parameters like Forest land, GRS (i.e., percent share of grass/scrub/woodland), and NVG (i.e., barren/very sparsely vegetated land) were downloaded from https://www.fao.org/soils-portal/soil-survey/soil-maps-and-databases/harmonized-world-soil-database-v12/en/ and utilized as recommended by Fischer et al. (2008).

13.2.5 Soil qualities

Rooting conditions (soil textures, bulk density, coarse fragments, vertic soil properties, and soil phases that affect root penetration and soil depth and soil volume), nutrient retention capacity (soil organic carbon, soil texture, bases saturation, cation exchange capacity, and clay fraction), oxygen availability to roots (soil drainage and soil phasing), excess salts (soil salinity, sodicity, and soil phases that affect salt conditions), and toxicity ($CaCo_3$ and $CaSO_4 \cdot 2H_2O$) were downloaded and utilized (Fischer et al., 2008) https://www.fao.org/soils-portal/data-hub/soil-mapdatabases/harmonized-world-soil-database-v12/en/ for current and future prediction of this species.

We followed the proposal of Zhao et al. (2021), who said that land cover, soil and HHI related factors are unlikely to change dramatically in the next few decades (~50 years). As a result, we used the same NBC settings for BC 2050 and 2070.

13.3 Data processing

13.3.1 Issue of multicollinearity

A multicollinearity test was performed in this study to reduce the problem of overfitting by examining the cross-correlation with the Pearson Correlation Coefficient (*r*). Furthermore, variables having a cross correlation coefficient greater than 0.85 were eliminated progressively (Pradhan, 2016). This analysis was carried out by using Niche Tool Box (Osorio-Olvera et al., 2020 https://github.com/luismurao/ntbox).

Following the approaches proposed by Kumar et al. (2006), multicollinearity among predictor variables was decreased. One of two strongly cross-correlated variables was chosen based on it is biologically relevance to the species and provides comfort in model interpretation (Kumar & Stohlgren, 2009; Padalia et al., 2014). For further analysis, only one variable from each set of significantly cross-correlated variables ($r^2 > 0.85$) was preserved (Ma & Sun, 2018). In this study, 70% and 30% of the data were given to model training and validation, respectively (Mousazade et al., 2019).

13.3.2 Species distribution modeling

Zhu et al. (2018) suggested that species distribution models can be split into two types based on the response variables they use: (i) those using presence-absence data (GLM, GAM, and CART) and (ii) those using presence-only data (ENFA, DOMAIN, GARP, MaxEnt, Bioclim, and Support Vector Machine). We used presence-only data and evaluated comparative performance of GARP, Bioclim, SVM, and MaxEnt tools with BC variables (current, 2050 and 2070) with help of OpenModeller (Munoz et al., 2009 WWW.openmodeller.sourceforge.net). Before proceeding further, we comparatively evaluated the performance of these tools by judging their visual outputs for habitat suitability-unsuitability classes. Based on graphical outputs of and classification accuracy, we found that among assessed tools, MaxEnt gave the better classified output as compared to the others. Therefore, in this work, all other ENM modeling was carried out with MaxEnt outputs only.

In this study, MaxEnt 3.4.1 software (sourced from http://www.cs.princeton.edu/%schapire/Maxent/) was used to simulate and predict the current and the future potential geographical distribution probability of this species.

MaxEnt's profile has expanded since its debut (Renner & Warton, 2013). MaxEnt is well-suited for modeling species distributions since it estimates a target probability distribution by computing the maximum entropy probability distribution (Phillips & Dudik, 2008; Qin et al., 2017). Topography, soil characteristics, land use, and biological interactions have been recognized as the primary determinants of species distributions at different geographical scales, and MaxEnt can estimate shifts in species distributions under different climate change scenarios (Abolmaali et al., 2018). Meanwhile, MaxEnt compares the presence distribution along environmental gradients to that of randomly generated points from the research area to serve as a background sample (Vitor et al., 2018).

The number of random background points in MaxEnt was set to 10,000 (Zhang et al., 2021). We set the regularization factor to 0.1 to avoid overfitting of the test data. Linear, quadratic, and hinge characteristics were employed. For model building, a total of 100 runs were planned (Flory et al., 2012). We employed the Jackknife approach in the environment parameter settings, and the other parameter settings were set to the program default values. We used a reiterative method of model generation and stepwise removal of the least contributing variables to run Jack knife analyses to assess the relative impact of each environmental element. Threshold-independent receiver-operating characteristic (ROC) analyses were used to calibrate and validate the robustness of Maxent model evaluation, and an area under the receiver operating curve (AUC) was used to estimate the accuracy of the model predictions (Elith et al., 2006; Phillips & Dudik, 2008). Model performance was evaluated by Araujo and Pearson (2005) as failing (0.5–0.6), poor (0.6–0.7), fair (0.7–0.8), good (0.8–0.9), and exceptional (0.9–1).

13.3.3 Changes in core distribution centers and area loss and gain

For graphical presentation of the Maxent output, we overlay ASCII files of Maxent on Indian natural shape file and reclass the habitats by applying "reclass" tool of spatial analysis tool window of ARC-GIS. Based on the class values, projected areas were classified into four categories: (i) high suitability (1.0–0.81), (ii) moderate suitability (0.80–0.51), (iii) low suitability (0.50–0.31), and (iv) unsuitability (0.30–0.01). With help of DIVA-GIS, these image files were converted into KML file for area (km sq.) calculation under each category (Ahmad et al., 2019).

The SDM toolkit was used to examine the centroid shift between different BC timeframes (Brown & Anderson, 2014). The projected contraction, expansion, and areas of no change in the distribution of a single species are graphically displayed when the computer calculates the distributional changes between two binary SDMs (current vs. future SDMs) (Brown et al., 2017). The area loss and gain between two-time frames, as well as between present BC and three NBC variables (soil, HHI, and land cover), were measured using QGIS's raster calculator.

13.3.4 Raster similarity analysis

A Fuzzy Kappa map comparison was carried out to ascertain the degree of similarity between adjacent cells in the two maps. The fuzzy kappa statistic is calculated by comparing individual cells on a map, with the neighborhood of each cell being assigned a numerical value between 0 (totally dissimilar) and 1 (entirely similar) to signify its degree of similarity (fully identical). This was accomplished using the Map Comparison Kit (MCK) (http://mck.riks.nl/downloads) for various time periods of particular BC.

13.4 Post maxent analysis

13.4.1 Niche overlap, range, and breadth

In the present day, niche overlap contrasted the inferred and true distributions of suitability ratings across geographic space. The Maxent output in ASCII format from each analysis was used to assess niche overlap between two analyzed parameters (related to BC and NBC). ENMTools was used to accomplish this (Warren et al., 2010). The purpose of this investigation was to depict the amount of territory retained by this species under various predictions. To depict the ecological niche overlap, Schoener's D (which evaluates the consistency of niche overlap per pair) and Hellinger's-based I (which represents the degree of overlap of the geographical distribution) values were used. The values of D and I ranged from 0 to 1. The ecological niche overlap would improve as the value increased (Ahmad et al., 2019). Different species niche variables, such as overlap and breadth, were investigated to understand better the behavior of this species' hyper-volumes in relation to various parameters. As a measure of niche width, Levin's B1: inverse concentration and B2: uncertainties (Levin, 1968; Marchan et al., 2016) were calculated for MaxEnt output of bioclimatic time frames: larger values of these indexes imply a broader niche.

With varying BC and NBC parameters, MaxEnt modeling reveals a substantial variable importance for the projected distribution of this species. For this species' ecological niche analysis, we used the first three most important variables or contributors from each Maxent analysis. NicheToolBox carried out the ENM (Osorio-Olvera et al., 2020). Ellipsoidal models were built using the centroid and covariance matrix of each species' environmental data. From the centroid to each environmental combination in the study region, Mahalanobis distances (MD) were determined. Suitability values (S) were calculated by constructing a multivariate normal distribution from MD; $S \sim \exp(-0.5 \times MD)$ and assuming that the higher suitability values were closer to the centroid (Nunez-Penichet et al., 2021). Raster maps were created using Arc-GIS and classified into optimum (1.0–0.81), moderate (0.80–0.51), marginal (0.50–0.31), and low (0.31–0.10) classes based on habitat suitability ratings.

13.4.2 Random forest clustering (RF-C)

This machine learning technique was carried out for two specific objectives: (i) variables clustering under a specific threshold and (ii) to know the variable importance in niche traits of studied species under different categories. RF-C was done as per the niche class types, i.e., optimum, moderate, marginal, and low belongs to current BC, BC-2050, BC-2070, HHI, soil, and slopes. Use of this algorithm in SDM with specific data set was advocated by Fukuda and De Baets (2016) as RF-C model accuracy is less pronounced with data prevalence.

With the default settings, the random Forest package (Liaw & Wiener, 2002) of the R software (R Development Core Team, 2011) was utilized. The number of trees to grow (ntree) was set to 500, while the number of variables randomly selected as candidates at each split (mtry) was set to the square root of the number of input variables. The elbow method was utilized to determine the number of suitable cluster numbers.

R^2, Akaike Information Criterion (AIC), Bayesian Information Criterion (BIC), and Silhouette Coefficient Indices were utilized to evaluate model precision, while variable significance was determined by the mean decline in the Gini Index.

13.4.3 Uni-dimensional reliability analysis

A URA analysis was performed to assess the internal consistency of the niche-related classes (i.e., optimum, moderate, marginal and low). This tool determines how closely connected a bunch of things is. It is regarded as scale reliability metric. Cronbach's alpha grows as the average interitem correlation increases. This tool determines whether the elements are measuring the same construct. Cronbach's alpha must be at least 0.70 to be considered acceptable. The internal consistency of the common range is low below this value. Meanwhile, the maximum predicted value is 0.90; anything above this amount is considered redundancy or duplication. In Microsoft Excel, the following formula was used to calculate it.

$$\alpha = \frac{K}{K-1}\left[1 - \frac{\sum S_Y^2}{S_X^2}\right]$$

Where: K is the number of test item; $\sum S_Y^2$ is the sum of item variance and S_X^2 is the variance of total score.

13.5 Result and discussion

13.5.1 Dataset and issue of multi-collinearity

Graphical display of spatial distributions of *M. phaseolina* within India is presented in Fig. 13.1. Forty different geographical locations were collected from different sources, which were randomly bifurcated in to 70% and 30% as training and test data set, respectively.

Results of correlation analysis among different variables of BC and NBC variables are presented in Figs. 13.2A and B, 13.3A and B, 13.4A and B. To address the issue of multicollinearity in species distribution modeling, we follow the procedures as described by Kumar et al. (2006) and Pradhan (2016). Within current bioclimatic variables, we found a significant correlation (>0.85) between temperature seasonality (BC- 4) and minimum temperature of coldest month (B-6 $R^2 = -0.89$) and between BC-4 and temperature annual range (BC-7 $R^2 = 0.94$). Similarly, precipitation of wettest quarter (BC-16) significantly correlated with BC-12 (annual precipitation $R^2 = 0.95$) and with BC-13 (precipitation of wettest month $R^2 = 0.95$). The maximum temperature of warmest month and mean temperature of warmest quarter was also significantly related with each other ($R^2 = 0.91$ Fig. 13.2A).

In step wise exclusion, we removed BC - 4, 5, 6 and 16 from SDM analysis. Similarly, highly correlated variables related to BC-2050 (Fig. 13.2B) were also removed (temperature seasonality BC-4; maximum temperature of warmest month BC-5, minimum temperature of coldest month BC-6; temperature annual range BC-7; and precipitation of wettest quarter BC- 16). Among BC variables of 2070, BC-5, BC-7, BC-11 and BC-17 were excluded (Fig. 13.3A). No such relationships were observed among variables related to Habitat Heterogeneity Indices (Fig. 13.3B) and variables related to land cover (Fig. 13.4A) and soil (Fig. 13.4B).

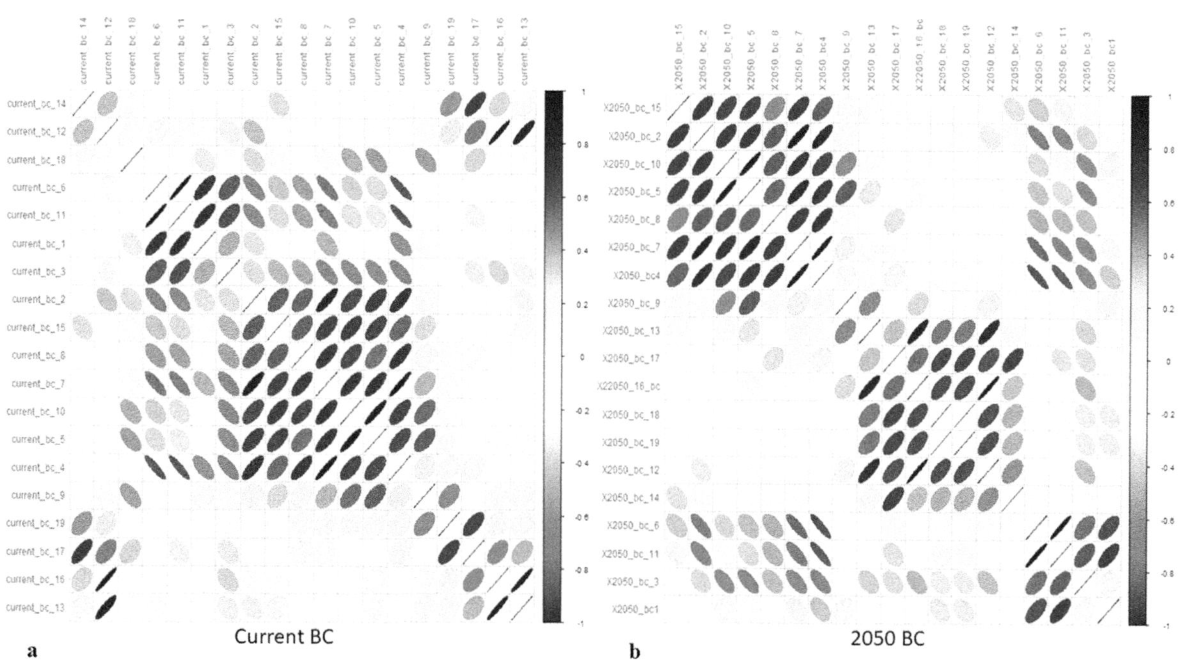

FIGURE 13.2 Correlation analysis among variables related to current (A) and future (2050, B) bioclimatic variables.

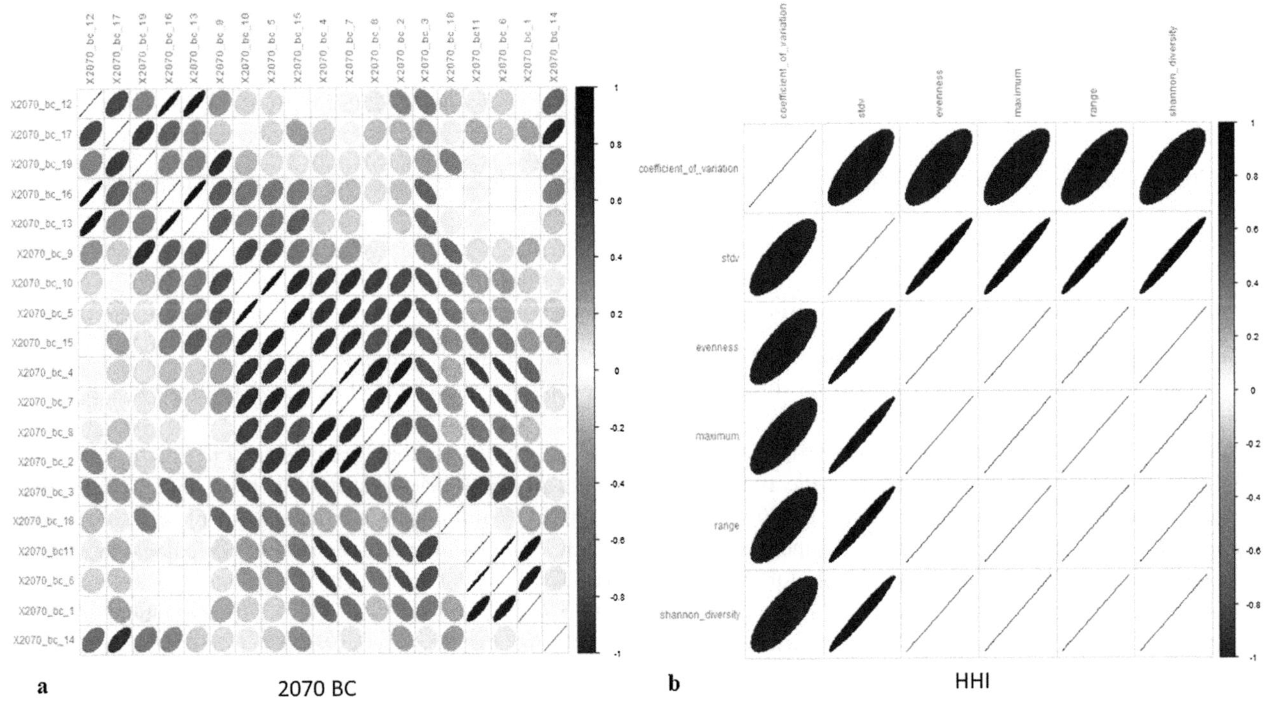

FIGURE 13.3 Correlation analysis among variables related to future (2070; A) and habitat heterogeneity indices (B).

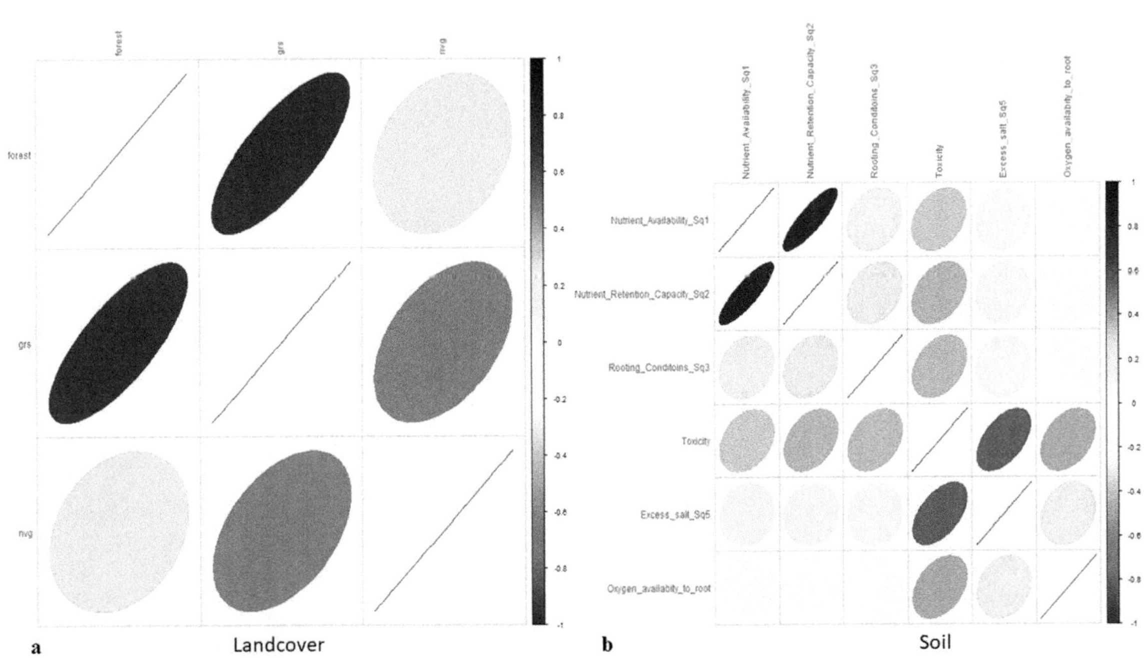

FIGURE 13.4 Correlation analysis among variables related to landcover (A) and soil (B).

13.5.2 Maxent analysis

With our presence only data, we evaluated comparative performance of GARP, Bioclim, SVM, and MaxEnt tools. Based on graphical outputs of and classification accuracy, we found that among above four tools, MaxEnt gave the better classified output as compared to others. Therefore, in this work, all other ENM modeling was carried out with MaxEnt outputs only.

13.5.2.1 Current bio-climatic BC scenario

Graphical outputs of *M. phaseolina* with current bioclimatic conditions are presented in Fig. 13.5A–D. Maxent gave the better classification of the SDM (Fig. 13.5A) as compared to bioclim (Fig. 13.5B), GARP (Fig. 13.5C), and SVM (Fig. 13.5D). We can classify with Maxent the habitat suitability into different classes like most suitable, suitable, less suitable, and not suitable. Further, model performance was judged based on area under curve (AUC) and value of this closer to one indicates better-performing models. Based on this, the model performance can be classified as poor (0.5–0.6), fair (0.6–0.7), good (0.7–0.8), and excellent (0.9–1) (Araujo & Pearson, 2005). AUC curve with current BC is presented in Fig. 13.6A with AUC values of 0.817 for training data and 0.826 with test data set (Table 13.2) suggested excellent

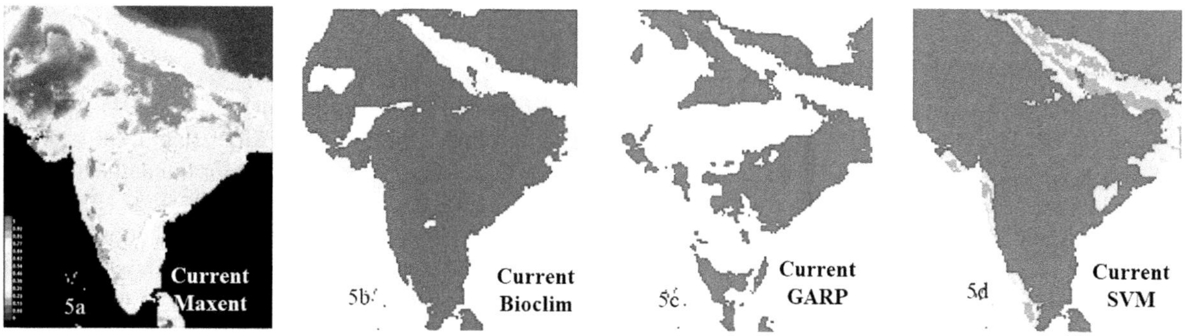

FIGURE 13.5 Graphical output of various machine learning tools for SDM of *M. phaseolina* (A–D).

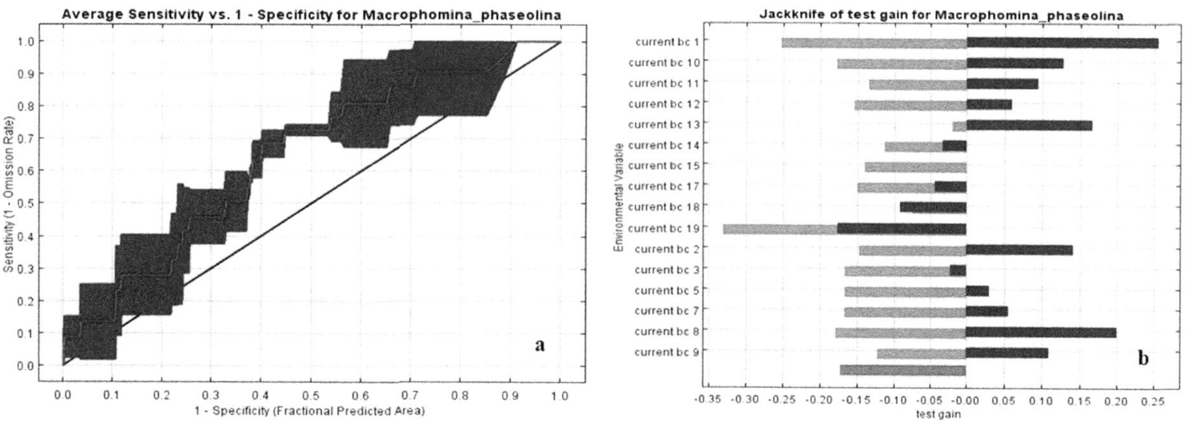

FIGURE 13.6 Area under the curve (AUC) and Jacknife test gain of current BC for *M. phaseolina* (A and B).

TABLE 13.2 Area under the curve (AUC) values of training and test data's belongs to different parameters.

Parameters	Training data	Test data
Climatic parameters		
Current BC	0.817	0.826
2050 BC	0.822	0.845
2070 BC	0.794	0.774
Nonclimatic parameters (current)		
Habitat heterogeneity index	0.806	0.788
Soil qualities	0.811	0.824
Land cover	0.806	0.832

model quality. Jacknife test gain is displayed in Fig. 13.6B that stipulated the percent contribution of each BC variable in distribution of this species. Higher percent contribution was recorded with variables like precipitation of wettest month (BC-13; 19.8%), annual precipitation (BC-12, 16.7%), precipitation of coldest quarter (BC-19, 13%), and annual mean temperature (BC-1, 12.5%).

Presence of this species significantly increases with precipitation of wettest month which is up to 250–500 mm above which the density of this specie significantly decreases (Fig. 13.7A). Similarly annual precipitation (BC-12) up to 1000 mm supports this species and above which species population decreases (Fig. 13.7B). Annual mean temperature (BC-1) from 5 to 25°C support the growth of this fungus (Fig. 13.7C), which reaches the maximum between 25 and 27°C, beyond which population start to decline. The environmental variable that decreases the gain the most when it is omitted is current BC 19 (precipitation of coldest quarter), which therefore appears to have the most information that is not present in the other variables (Fig. 13.7D).

13.5.2.2 Bio-climatic—2050 scenario

Graphical outputs of *M. phaseolina* with future (2050) bioclimatic conditions are presented in Fig. 13.8A–D. Maxent gave the better classification of the SDM (Fig. 13.8A) compared to bioclim (Fig. 13.8B), GARP (Fig. 13.8C), and SVM (Fig. 13.8D). AUC curve with 2050 BC is presented in Fig. 13.9A with AUC values of 0.822 for training data and 0.845 with test data set (Table 13.2) suggested excellent model quality. Jacknife test gain is displayed in Fig. 13.9B that stipulated the percent contribution of each BC variable in distribution of this species. Higher percent contribution was recorded with variables like annual precipitation (BC-12, 30%), precipitation of wettest month (BC-13; 27.8%), and annual mean temperature (BC-1, 15.7%). Population of this specie linearly increases with annual precipitation up to 1000 mm and remains constant between 1000 and 1800 mm (Fig. 13.10A) and beyond which it started to decline. Similarly, with precipitation of wettest month (BC-13) up to 500 mm, population increases and suddenly decline beyond 500 mm threshold (Fig. 13.10B). With annual mean temperature, peak of this species recorded between 25 and 28°C (Fig. 13.10C).

13.5.2.3 Bio-climatic—2070 scenario

Graphical outputs of *M. phaseolina* with future (2070) bioclimatic conditions are presented in Fig. 13.11A–D. Again, better classification was recorded with Maxent tool (Fig. 13.11A) compared to bioclim (Fig. 13.11B), GARP (Fig. 13.11C), and SVM (Fig. 13.11D). AUC curve with 2070 BC is presented in Fig. 13.12A with AUC values of 0.794 for training data and 0.774 with test data set (Table 13.2) suggested good model quality. Jacknife test gain is displayed in

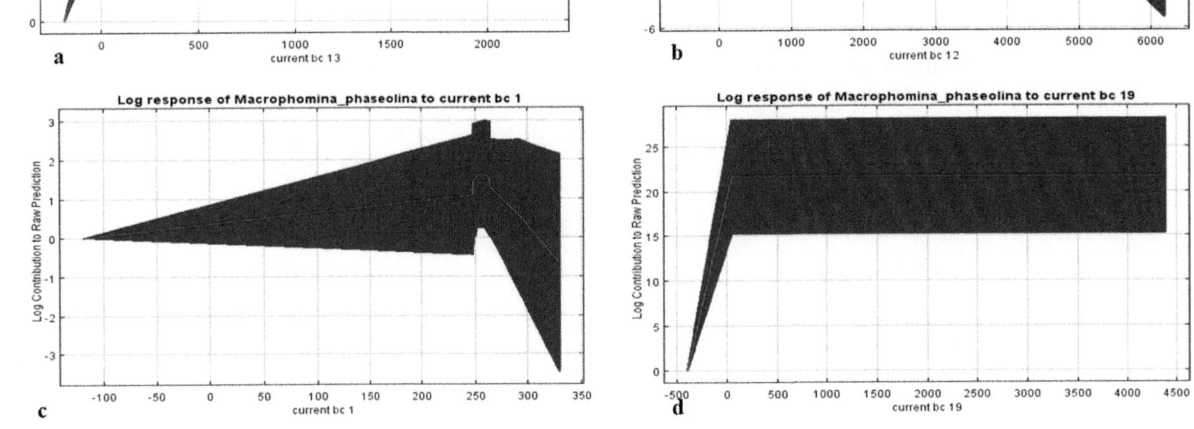

FIGURE 13.7 Impact of various current BC variables on *M. phaseolina* population (A–D).

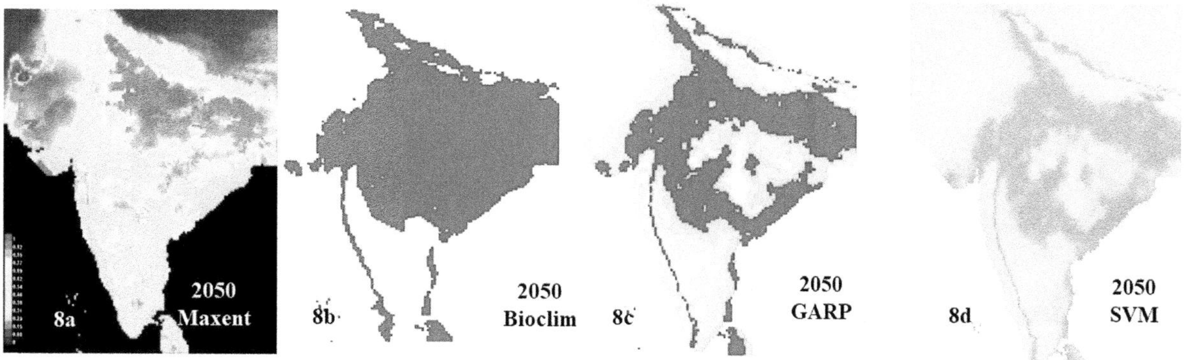

FIGURE 13.8 Graphical output of various machine learning tools for SDM of *M. phaseolina* with BC-2050 (A–D).

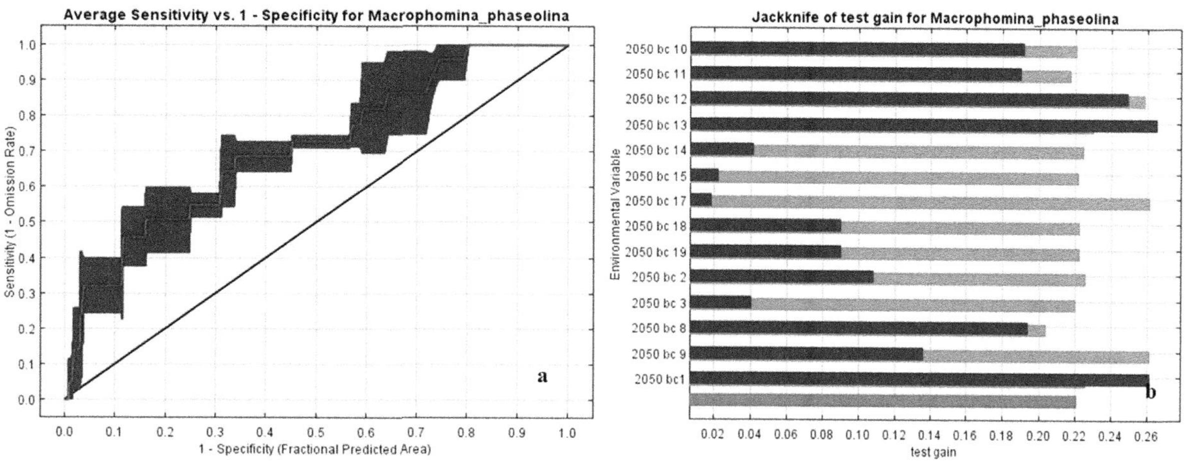

FIGURE 13.9 Area under the curve (AUC) and Jacknife test gain of 2050 BC for *M. phaseolina* (A and B).

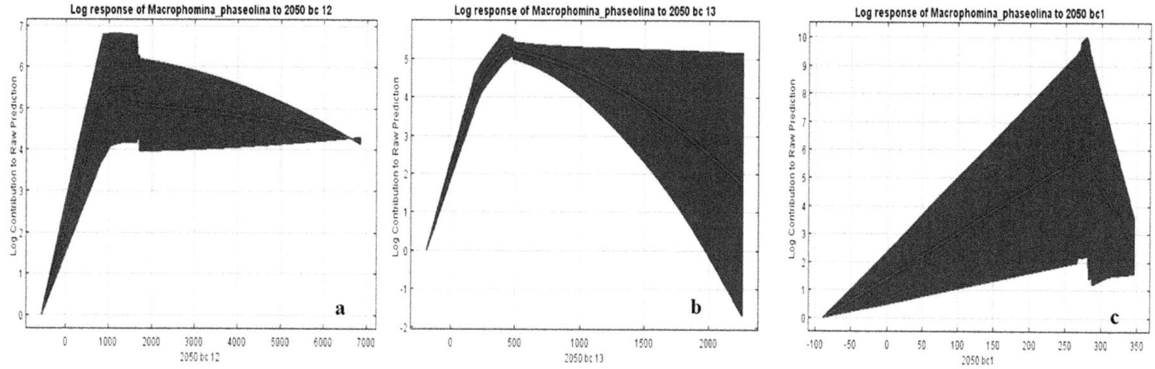

FIGURE 13.10 Impact of various 2050 BC variables on *M. phaseolina* population (A–C).

Fig. 13.12B that stipulated the percent contribution of each BC variable in distribution of this species. Higher percent contribution was recorded with variables like precipitation of wettest quarter (BC-16, 26.1%), annual mean temperature (BC-1; 15.3%), and annual precipitation (BC-12; 14.8%). With BC-16 population increases and reaches the maximum on 1000 mm after which gradual decrease was observed (Fig. 13.13A). Steady increase of population was noticed with annual mean temperature (Fig. 13.13B) which reached maximum at 30°C and showed sharp decline between 31 and 35°C. Annual precipitation between 1000 and 1500 mm was recorded as most supportive of population of this fungus (Fig. 13.13C).

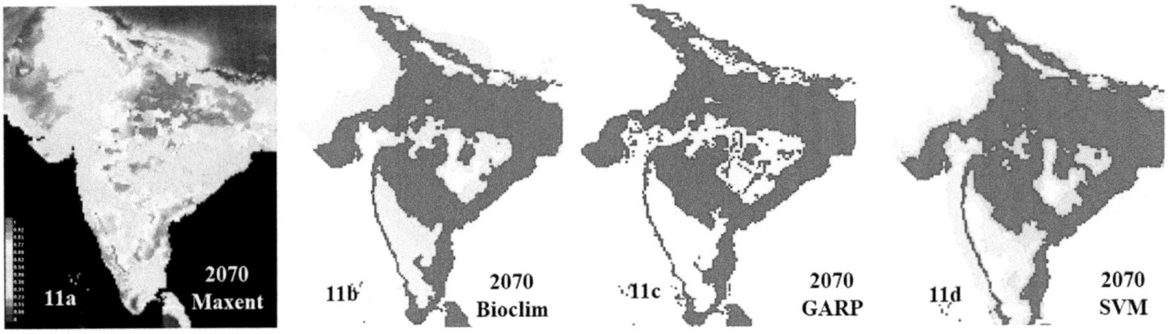

FIGURE 13.11 Graphical output of various machine learning tools for SDM of. *M. phaseolina* with BC-2070 (A–D).

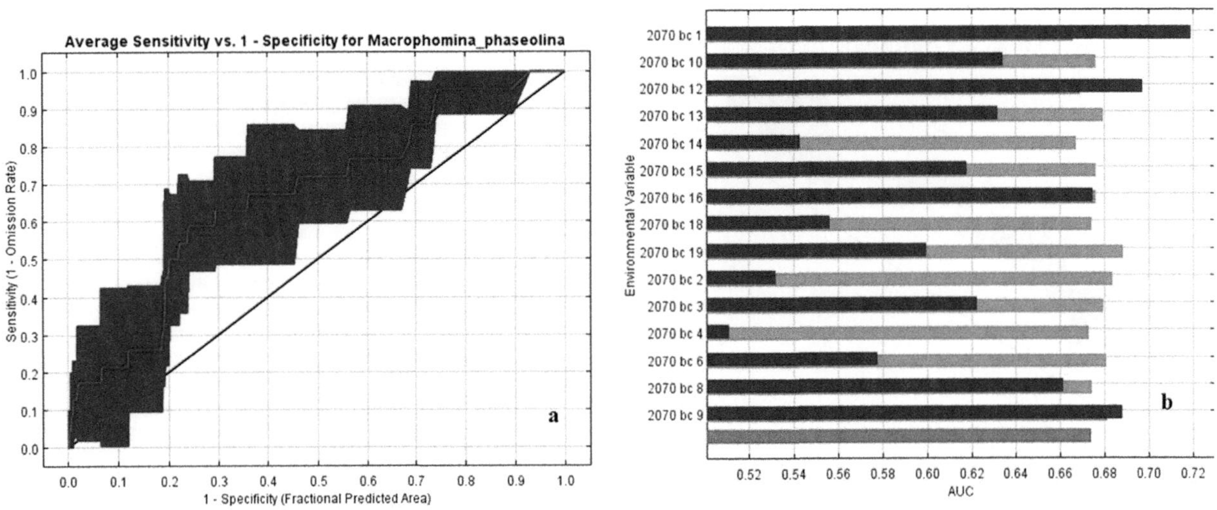

FIGURE 13.12 Area under the curveAUC) and Jacknife test gain of 2070 BC for *M. phaseolina* (A and B).

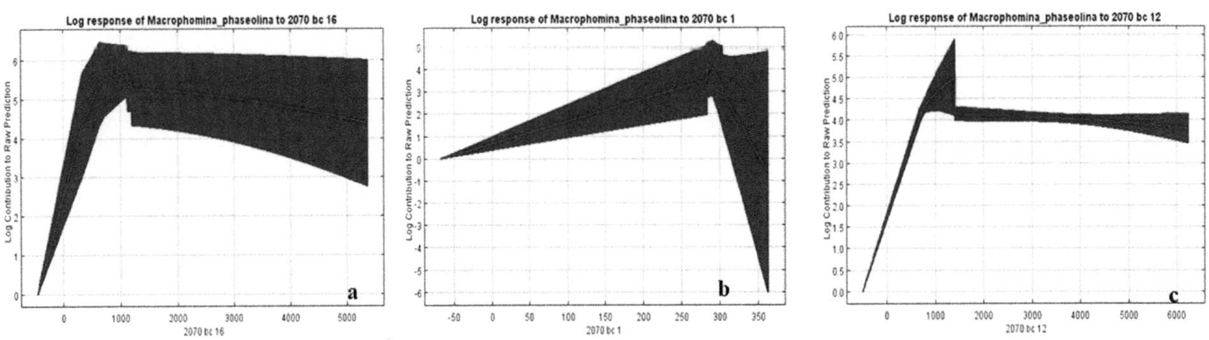

FIGURE 13.13 Impact of various 2070 BC variables on *M. phaseolina* population (A–C).

Karl and Trenberth (2003) claim that current climate change is responsible for rising global temperatures, shifting patterns of precipitation, and shrinking ice and snow cover both seasonally and year-round. These shifts have the potential to influence diseases by influencing seasonal phenology (including the stages and rates of development of the pathogen's life cycle) and population dynamics (e.g., overwintering, oversummering, survival, and changes in abundance, Eastburn et al., 2011; Juroszek & Von Tiedemann, 2013; Pautasso et al., 2012). In a study by Manici et al. (2012), the three soil-borne plant diseases *Fusarium nivale, Athelia rolfsii*, and *Macrophomina phaseolina* were classified into low, medium-high, and high temperature requirements, respectively, based on their reactions to temperature.

Future warming will significantly modify the patterns of richness and community dynamics of the total, plant pathogenic and saprotrophic fungi in wheat residues as found in the study by Wahdan et al. (2020). Impacts of the maximum and minimum air and soil temperature, rainfall, and relative humidity on population dynamics of this species under the agrarian system of soybean were examined by Teja et al. (2020). Their study suggested that air temperature from 28 to 35°C, relative humidity 65%−80%, and soil temperature of 35.6°C at 5 cm depth and 32.2°C at 10 cm depth were most suitable for growth of this species.

The relative abundance of soil-borne fungal plant diseases was studied by Delgado-Baquerizo et al. (2020) in 235 natural settings across six continents. They found that there was a positive and statistically significant association between the mean annual temperature (MAT) and the abundance of soil pathogens around the world. In line with the IPCC (2000), they found that the relative abundance of plant pathogens was substantially correlated with ecosystem type (such as forests and grasslands) and plant cover. In addition, some pathogen families are particularly sensitive to variations in precipitation and soil pH.

Our population-based approach with different time frame BC variables revealed the following trends.

- Annual precipitation is the most significant contributor for prediction of this species. Under current climatic condition, threshold for this factor was up to 1000 mm, while during 2050 BC, the population of this species would remain constant between 1000 and 1800 mm. This range will be 1000−1500 mm during BC-2070.
- Annual mean temperature is also a constant influencing factor during all the time frame. 25−27°C, 25−28°C and 28−30°C are the threshold limits for this species. Such trends suggesting gradual temporal synchronization of pathogen with prevailing temperature.
- Additionally, precipitation of wettest month (current and BC-2050) and wettest quarter (Bc-2070) will also work as common factor.

13.5.2.4 Habitat heterogeneity index

MaxEnt outputs of *M. phaseolina* with HHI variables and their AUC and Jacknife test are presented in Fig. 13.14A−C. MaxEnt output suggested probable distribution of this species all over the India in small patches and complete absence from western part specifically from Rajasthan state (Fig. 13.14A). AUC curve with HHI is presented in Fig. 13.14B with AUC values of 0.806 for training data and 0.788 with test data set (Table 13.2) which suggest good model quality. Jacknife test gain displayed in Fig. 13.14C stipulated the percent contribution of each variable of HHI for distribution of this species. Higher percent contribution was recorded with coefficient of variation (44%), followed by evenness (27.9%) and standard deviation (22.6%). Coefficient of variation (normalized dispersion of EVI) produces negative impacts on population of this species (Fig. 13.15A). In contrast, evenness is supportive for population (Fig. 13.15B), while standard deviation (dispersion of EVI) above 500 works as threshold for this species (Fig. 13.15C). Negative relationships of *M. phaseolina* with coefficient of variation (*i.e.*, greater the level of host dispersion around the mean), but positive relationship with evenness indicates that host density to a particular level is essential for survival of this species. Our HHI result suggested lesser population of this fungus at western part of the country which can be explained by the scattered availability of host or plant population (Mathur & Sundaramoorthy, 2016).

13.5.2.5 Soil factors

MaxEnt outputs of *M. phaseolina* with soil variables and their AUC and Jacknife test are presented in Fig. 13.16A−C. MaxEnt output suggested probable distribution of this species predominantly at western part specifically around Rajasthan

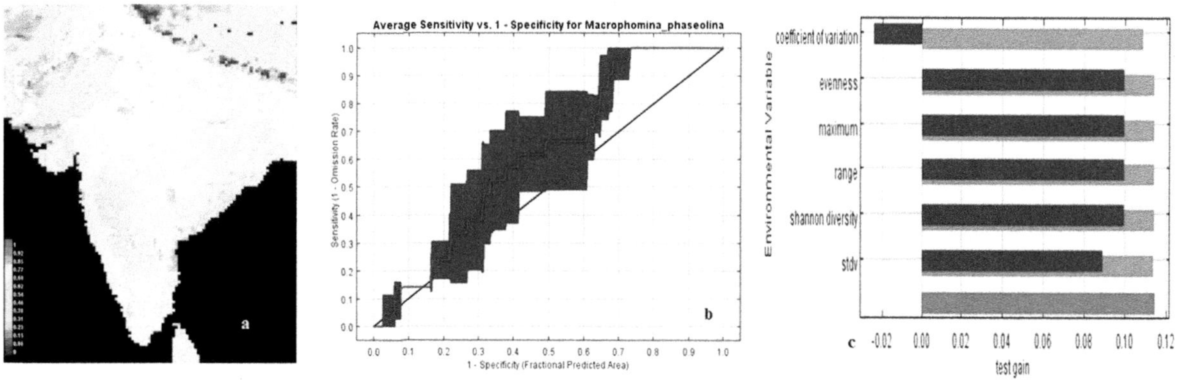

FIGURE 13.14 Maxent output (A); AUC curve (B); and Jacknife test gain (C) of HHI for *M. phaseolina* (A−C).

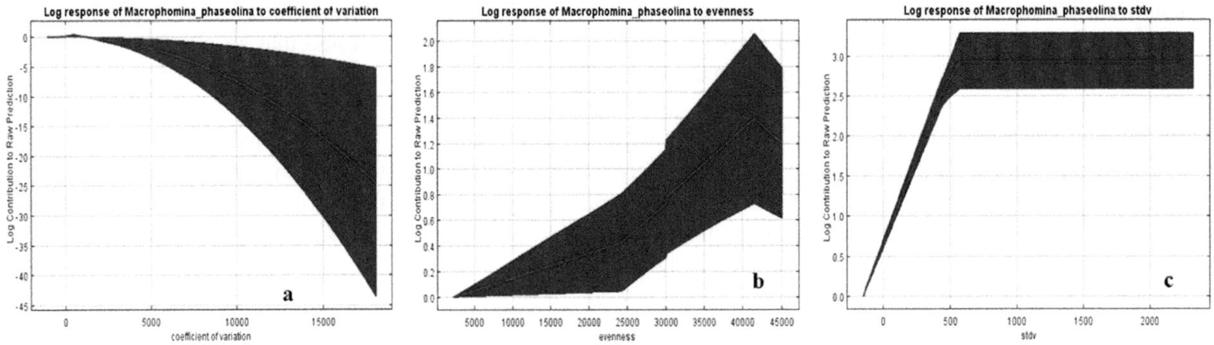

FIGURE 13.15 Impact of various HHI variables on *M. phaseolina* population (A–C).

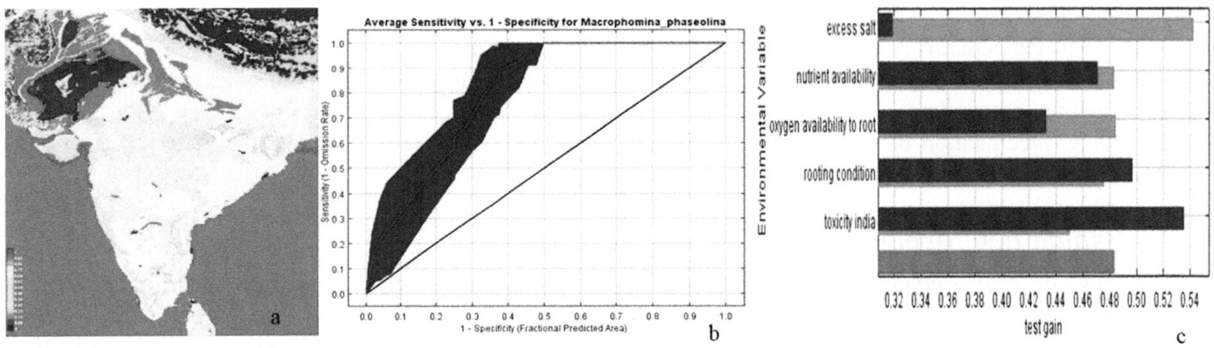

FIGURE 13.16 Maxent output (A); area under the curve (AUC) curve (B); and Jacknife test gain (C) of soil for *M. phaseolina.*

state (Fig. 13.16A). AUC curve with soil is presented in Fig. 13.16B. AUC values of 0.811 for training data and 0.824 with test data set (Table 13.2) suggest the excellent model quality. Jacknife test gain displayed in Fig. 13.16C stipulated the percent contribution of each variable of soil for distribution of this species. Higher percent contribution was recorded with excess salt (58.9%), followed by toxicity (19.8%) and rooting condition (16%). Threshold limit of these variables for this species were recorded, beyond which the population drastically decline (Fig. 13.17A–C).

Low soil moisture levels favored the features of *M. phaseolina*, which thrived in the humid tropical climate of southwestern Nigeria, while high soil moisture levels stunted the growth and pathogenicity of this fungus (Wokocha, 2000). The number of *M. phaseolina* in cowpea soil showed a similar negative relationship with soil moisture and a positive relationship with the highest soil temperature (Gupta & Gupta, 1986). According to Dhingra and Sinclair (1975), maintaining soil moisture over 60% of the soil's capacity to retain water for 3–4 weeks at a temperature of 30°C or above can minimize the amount of live microsclerotia in the field. In Northern Iran, Taliei et al. (2013) found no statistically significant relationships between the quantity of *M. phaseolina* microsclerotia and the characteristics of the soil (soil particle size, organic matter, pH, and EC). Organic matter content and *M. phaseolina* population density in the 0–7.5 cm soil layer were found to be significantly and positively correlated (Wrather et al., 1998).

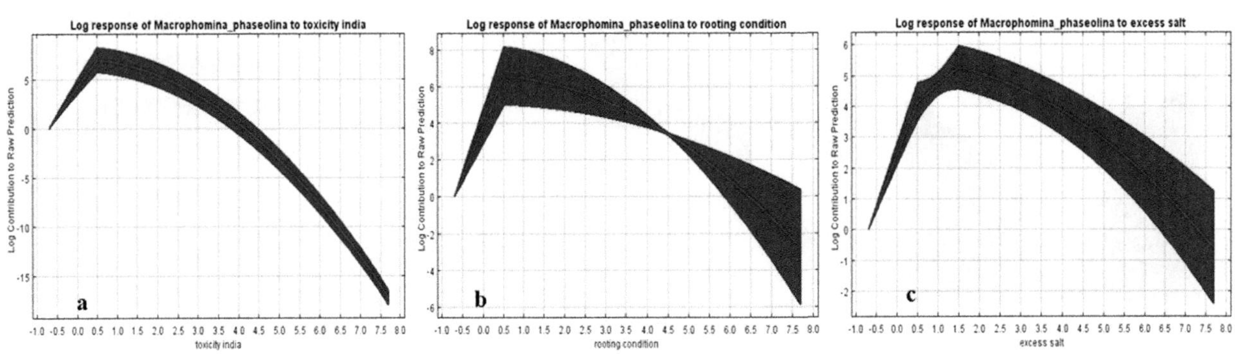

FIGURE 13.17 Impact of various soil variables on *M. phaseolina* population (A–C).

The main elements that have a noticeable impact on the survival of *M. phaseolina* microsclerotia include repeated freezing and thawing of the soil, low carbon to nitrogen ratios in the soil, and soil moisture content (Kaur et al., 2012). Drought-related increased microsclerotia production under low water potential has been shown by Dhingra and Sinclair (1977) and Olaya and Abawi (1993). High soil moisture has been negative and harmed *M. phaseolina* sclerotia's ability to survive in the soil (Dhingra & Sinclair, 1975). Masafumi et al. (2002) predicted that the population density of this species in Nagano soil after 6 years of storage was one per gram of soil, and that $t_{1/2}$ was 1.88 and P0 was 9.1 based on the half-life idea ($t_{1/2}$ = T × log2/log P0 − logP1).

You et al. (2011) came to the conclusion that the *M. phaseolina* – *Phaseolus vulgaris* pathogen-host complex will provide a greater threat to disease where the host is grown in locations with salinized soil. In a study with *Sorghum bicolor*, Goudarzi et al. (2011) discovered that *M. phaseolina* colonized the shoots and roots substantially more when the soil salinity was up to 1400 mg NaCl $kg^{-1.}$

In soybean fields, dry soil, high soil C:N ratios of additives, low bulk density, and oxygen content favor microsclerotia germination (Gupta et al., 2012). These variables affect the soil's oxygen diffusion rate and nutritional balance, which promote the germination of latent spores (Craig, 1992). The number of years that maize and soybeans were planted was also related to the density, size of the microsclerotia, and overall its population. Suryawanshi et al. (2015) reported that the application of coarse gypsum reduced the mortality caused by charcoal rot in *Helianthus annus*. Al-Hazmi et al. (2017) investigated the effects of soil moisture and texture on the interaction of the *Meloidogyne javanica—M. phaseolonia* (root-knot – charcoal root rot) disease complex in *P. vulgaris*. In terms of soil texture, a higher sand concentration correlates with more severe disease and fungal growth. Furthermore, the severity of the illness steadily declined as the soil moisture level rose, with the most plant damage occurring at a moisture level of 30% of the field capacity.

Contrary to our result of HHI, harmonized soil factors suggested predominance of this species at western part of the country (arid region with high sand proportion, Mathur et al., 2022), and such result can be explained with the sand loving properties of this fungus as experimentally proven by Al-Hazmi et al. (2017).

13.5.2.6 Land cover

MaxEnt outputs of *M. phaseolina* with land cover variables and their AUC and Jacknife test are presented in Fig. 13.18A–C, respectively. MaxEnt output suggested probable distribution of this species predominantly at eastern part specifically (Fig. 13.18A) and in small patches at southern part of the country. AUC curve with soil is presented in Fig. 13.18B. AUC values of 0.806 for training data and 0.832 with test data set (Table 13.2) suggest the excellent model quality. Jacknife test gain is displayed in Fig. 13.18C that stipulated the percent contribution of each variable of land cover for distribution of this species. Higher percent contribution was recorded with forest (60.1%), followed by GRS (31.7%) and NVG (8.2%). Threshold limit of these variables for this species were recorded, beyond which the population remains constant (Fig. 13.19A–C).

Modeling with harmonized land cover variable suggested predominance of this species at Eastern part of the country. Such results are in accordance with finding of Nazir et al. (2017) competition and facilitation studies of tree species in temperate forest of the eastern Indian Himalaya.

Thus, with results of our BC and NBC variables, we can conclude that prediction modeling of this species should be gauged through use multiple unrelated variables.

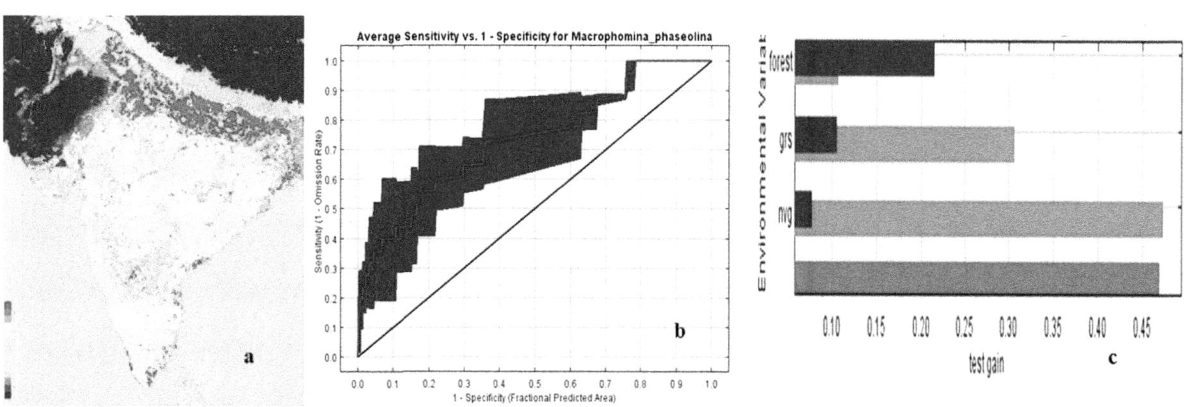

FIGURE 13.18 Maxent output (A); area under thr curve (AUC) curve (B); and Jacknife test gain (C) of land cover for *M. phaseolina*.

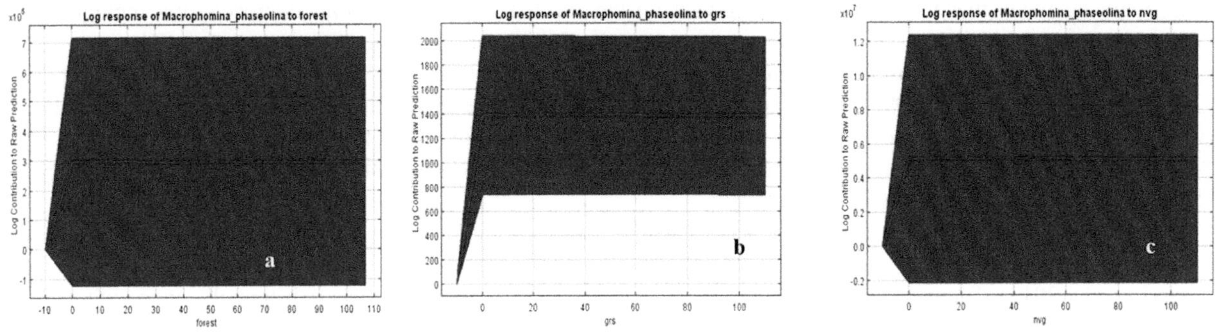

FIGURE 13.19 Impact of various land cover variables on *M. phaseolina* population (A–C).

13.5.3 Changes in core distribution centers and area loss and gain

With "reclass" tool of spatial analysis tool window of ARC-GIS, we categorized the Indian region into four classes pertains to area suitability for *M phaseolina*: (i) area of high suitability (1.0–0.81), (ii) moderate suitability (0.80–0.51), (iii) low suitability (0.50–0.31), and (iv) unsuitability (0.30–0.01). Further, with the help of DIVA-GIS, these image files were converted into KML file for area (km sq.) calculation under each category (Ahmad et al., 2019). Here we presented area with high suitability for all the studied BC and NBC variables. Among the BC time frames, the highest area under this class was recorded with BC-2070 (65.34×10^4 km^2), followed by BC-2050 (38.28×10^4 km^2) and current (30.29×10^4 km^2). While among the non-BC, highest area under this class was recorded with HHI (153.83×10^4 km^2), followed by soil (51.27×10^4 km^2) and land cover (29.77×10^4 km^2) variables.

With current BC the centroid for this species was located at N25° 15′ and E80° at Manuriya (Madhya Pradash). This centroid shifted to N26° 30′ and E79° 30′ located at Jamrehisani (Uttar Pradesh) during period of 2050 BC, which is 151.02 km away from current centroid (Fig. 13.20 Red [light gray in print version] Line). We noticed same centroid during

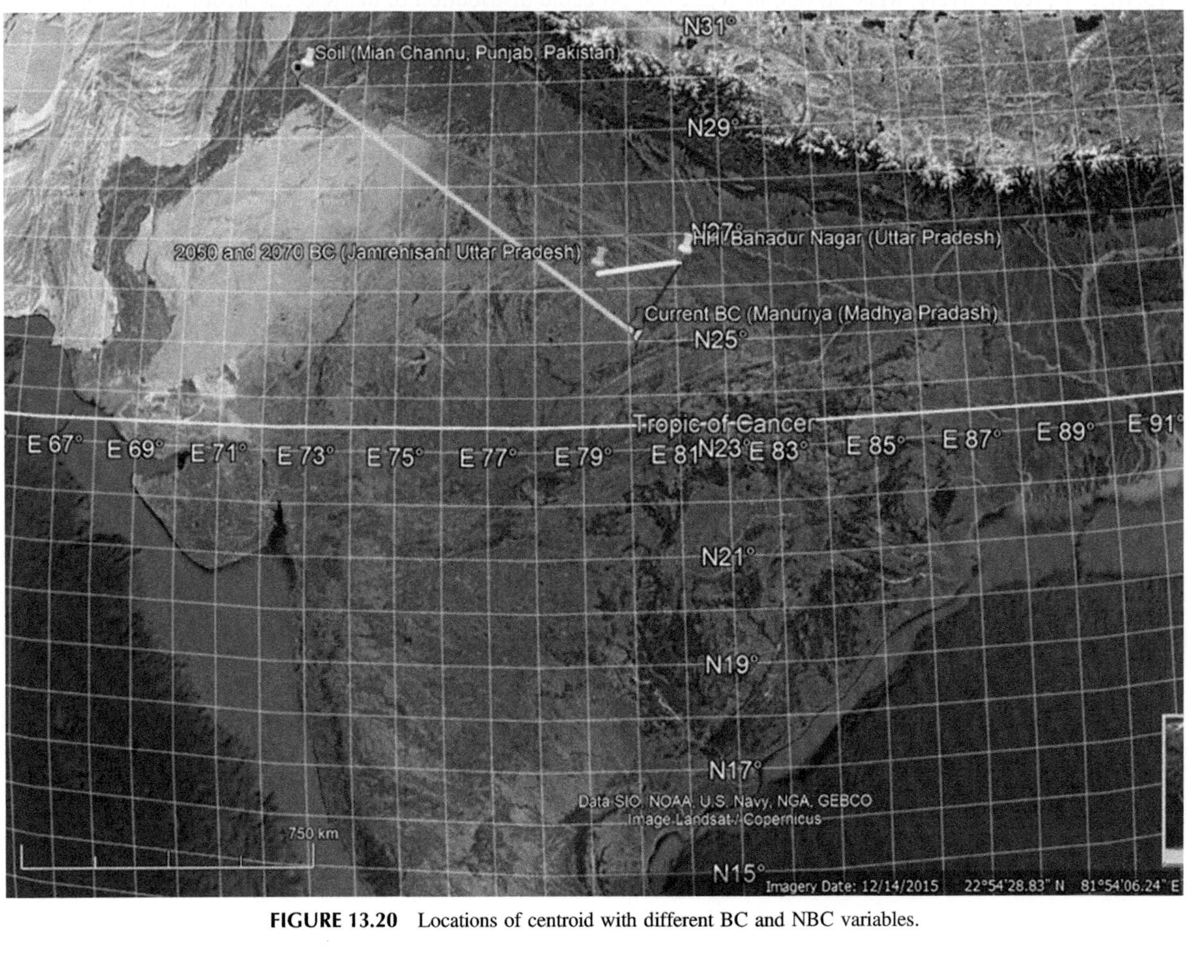

FIGURE 13.20 Locations of centroid with different BC and NBC variables.

2070 BC also. With land cover variables, the centroid is located at N26° 51′ and E81° 15′ which is located at Bahadur Nagar (Uttar Pradesh) having distance of 176.17 km from Manuriya site (Current -BC; Blue [dark gray in print version] line) and 179.20 km away from 2050 to 2070 BC (Fig. 13.20 Yellow [very light gray in print version] line). With soil variables, the centroid was located at N30° 30′ and E72° 30′ which is located outside Indian, having and aerial distance of 939.54 km (Fig. 13.20 Green [light gray in print version] line). This has the almost similar distance 951.25 km (purple [gray in print version] line) from centroid of HHI.

Area loss and gain between two-time frames and between current BC and three NBC (soil, HHI, and land cover) are presented in Figs. 13.21A–C and 13.22A–C. In raster images, white patches represent the areas gain and black patches area lost between two-time frames. Between 2050 –current BC 37.38 × 10^4 sq km^2 area is gained, while 7.97 × 10^4 sq km^2 area is lost, suggesting 78.67% of gain (Fig. 13.21A). A very higher area gain (206.79 × 10^4 sq km^2) was noticed from 2070–50 with 27.62 × 10^4 sq km^2 area lost. During these time frames, around 86% area gain was recorded (Fig. 13.21B). We also compared area gain-lost between 2070 andcurrent and noticed that around 58.39 × 10^4 sq km^2 and 10.17 × 10^4 sq km^2 area are gained and lost, respectively suggesting 81.52% area gain (Fig. 13.21C). Similarly, with current BC-land cover, variables 23.87 × 10^4 sq km^2 and 21.27 × 10^4 sq km^2 areas were recorded as gain and lost, respectively (10.89% gain Fig. 13.22A).

However, with current BC-HHI, we observed only 21.10 × 10^4 sq km^{2+} area as gained and 110.26 × 10^4 sq km^2 as lost that indicates 422.51% decrease in area (Fig. 13.22B). Such trends are also recorded between current BC and soil variables, where 39.05 × 10^4 sq km^2 and 63.58 × 10^4 sq km^2 were gain and lost, respectively, representing 62.81% lost (Fig. 13.22C).

13.5.4 Raster similarity analysis

To measure the levels of similarity between pairs of cells in two maps, a Fuzzy Kappa map comparison was done. This tool uses a cell-by-cell map comparison to determine how similar 2 cells are by using a scale from 0 (completely separate) to one to reflect how similar they are to each other (fully identical). This exercise was carried out with ASCII file of two different BC time frame generated by their respective MaxEnt models. Results are displayed in Fig. 13.23A (current BC/2050 BC), Fig. 13.23B (current BC/2070 BC) and Fig. 13.23C (2050 BC/2070 BC). We recorded low similarity between all time frame maps. Kappa values between current/2050 were 0.432; 0.315 between current/2070 and little bit higher (0.53) similarity between 2050/2070.

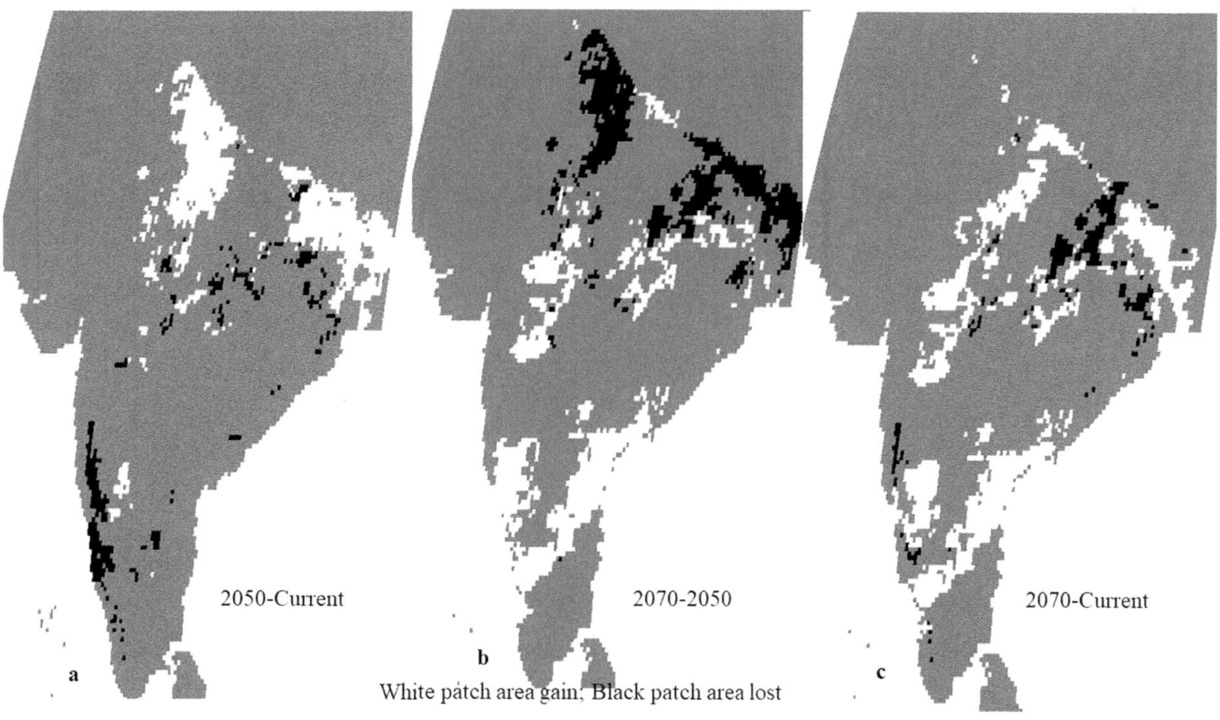

FIGURE 13.21 Area loss and gain between two-time frames (A–C).

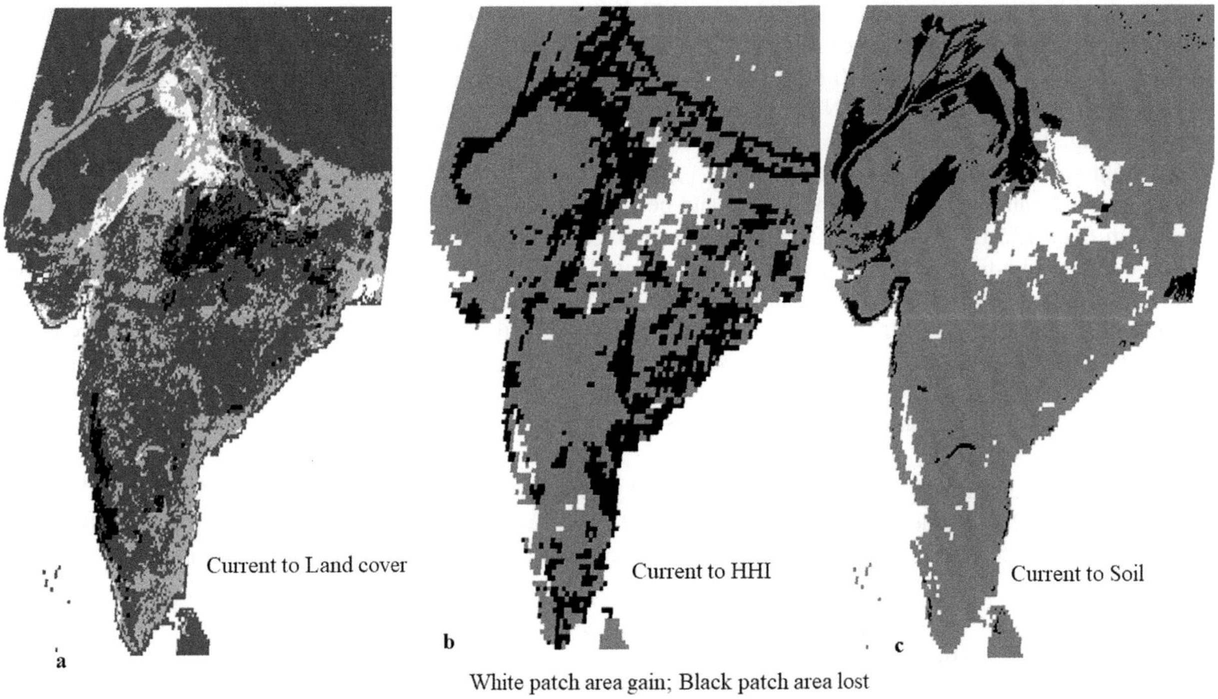

FIGURE 13.22 Area loss and gain between current BC and three NBC (land cover HHI and soil) (A–C).

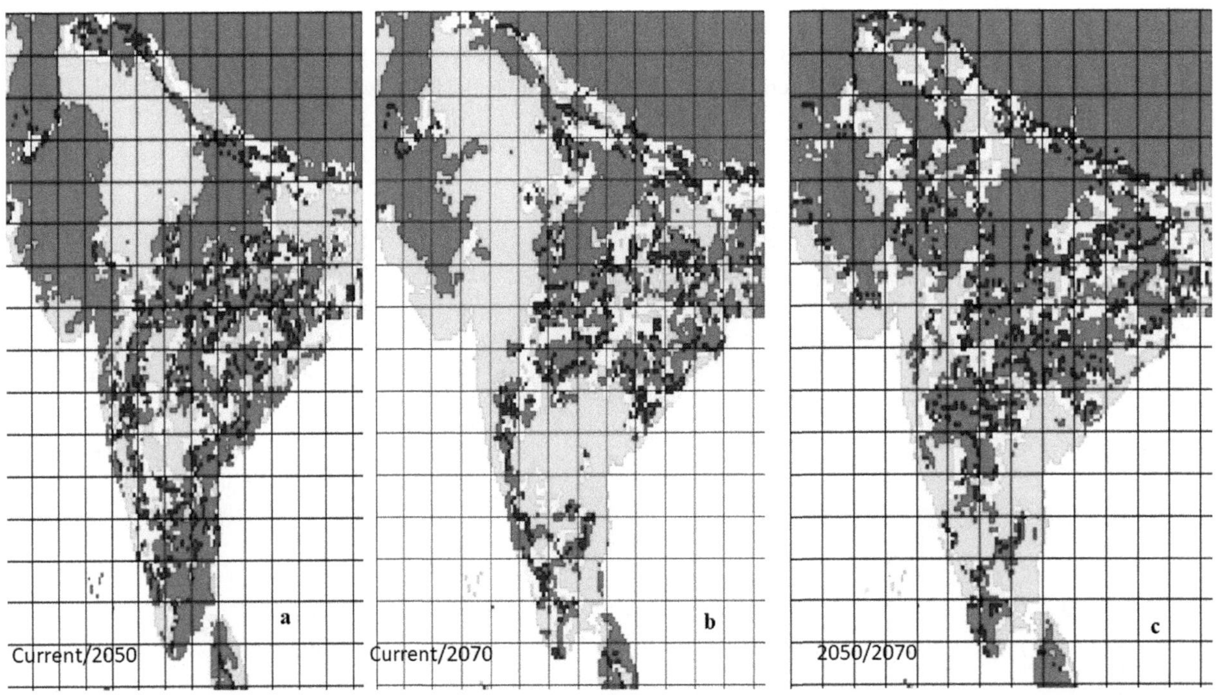

FIGURE 13.23 Kappa comparison of map similarity (A–C).

13.5.5 Niche overlap, range, and breadth

The inferred and genuine distributions of suitability ratings throughout geographic space in the present were compared using niche overlap. This analysis was done to show how much space this species would keep based on various scenarios. Schoener's D (which measures consistency of niche overlap per pair) and Hellinger's-based I were used to indicate the ecological niche overlap (which measures overlap degree of the geographical distribution). This index assumes that the MaxEnt score is proportional to abundance, in contrast to I, which "deals with the two environmental niche models as probability distributions" (Karami et al., 2020; Vanderwal et al., 2009). Values closer to 0 imply limited niche overlap, while values closer to one suggest extensive niche overlap, with Schoener's D and Hellinger's I values falling somewhere in between. Schoener's D and Hellinger's I values are listed in Tables 13.3 and 13.4, respectively. Between 2050 and 2070 B.C., the highest recorded concentrations of D (0.83 ppm) and I (0.97 ppm) were found. These indices were found to have their lowest values in the soil and HHI variables (0.33 and 0.48, respectively).

Fig. 13.24 displays the niche breadth results. The distribution of suitable habitat over the testing area is shown in this matrix. Here, the niche breadth analysis is utilized to describe how this species relationship to environmental factors is

TABLE 13.3 Schoener' *D* a measure consistency of niche overlap per pair.

Variables	2050 BC	2070 BC	Current BC	HHI	Land cover
Soil	0.37	0.39	0.39	0.33	0.67
2050 BC	–	0.83	0.77	0.70	0.55
2070 BC	–	–	0.76	0.74	0.58
Current BC	–	–	–	0.65	0.60
HHI	–	–	–	–	0.55

TABLE 13.4 Hellinger's Index *I* a measure for overlap degree of the geographical distribution.

Variables	2050 BC	2070 BC	Current BC	HHI	Land cover
Soil	0.55	0.57	0.57	0.48	0.87
2050 BC	–	0.97	0.95	0.85	0.72
2070 BC	–	–	0.95	0.89	0.75
Current BC	–	–	–	0.85	0.77
HHI	–	–	–	–	0.73

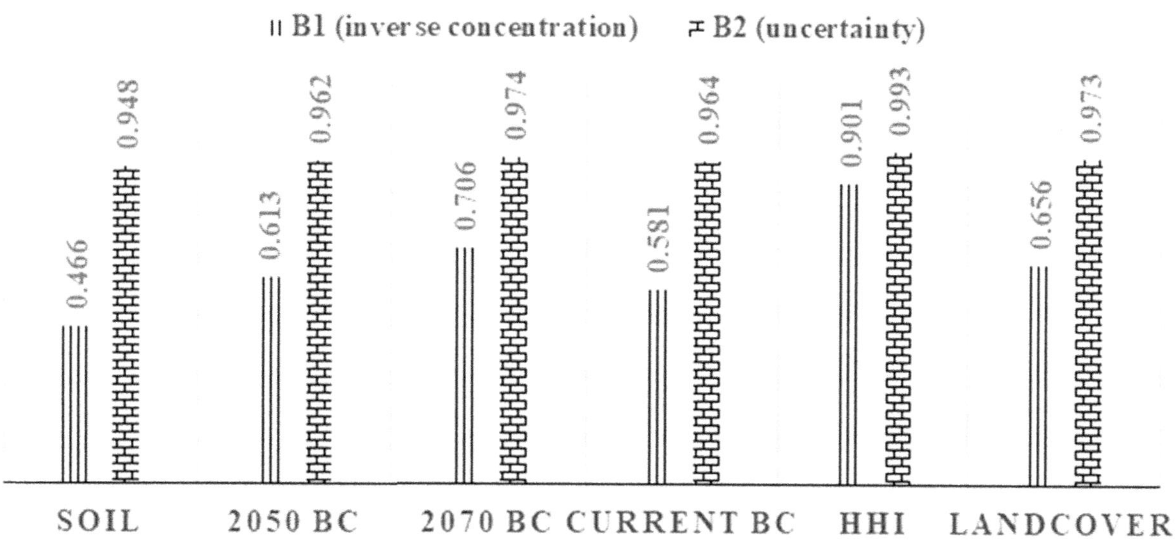

FIGURE 13.24 Niche breadth of *M. phaseolina* with various studied BC and NBC variables.

evolving through time. Similar to the D and I index, the values of the niche test vary from 0 (where only one grid cell is deemed acceptable) to 1, when every grid cell is suitable (Vorsino et al., 2013).

For this species, the highest niche breadth was recorded with HHI variables, while the lowest was recorded with soil. Among different BC time frames, this species would have the highest niche breadth during 2070 period.

Further, based on values of habitat suitability, raster maps prepared with help of Arc-GIS and classified the map under optimum (1.0–0.81), moderate (0.80–0.51), marginal (0.50–0.31), and low (0.31–0.10) classes. This exercise facilitated us to identify the shape and size of different types of niches, i.e., fundamental niche, realized niche, potential, and tolerance niche of this species.

13.5.6 Random forest clustering

Through this machine learning technique, we assessed our classification accuracy for the niche type, i.e., optimum, moderate, marginal, and low. Results were interpreted based on criteria of R^2, AIC, BIC, and Silhouette. Further, appropriate cluster number was determined by Elbow method, and the importance of each class during their respective time frame was judge based on Gini Index. Results are displayed in Tables 13.5 and 13.6. Appropriate cluster numbers during current, 2050 BC and 2070 BC were 7, 6, and 8, respectively (Fig. 13.25).

A model is said to fit well if its R^2 is close to the upper bound of one, while a model's R^2 that is close to the bottom bound of zero is thought to fit badly. The R^2 values were near to or over 0.90 for each of the climate time periods examined, indicating that the model was successful (Table 13.5).

TABLE 13.5 Random forest clustering of various categories during different timeframes.

Time frame	Parameters	Values
Current BC	R^2	0.972
	AIC	95.42
	BIC	203.76
	Silhouette	0.480
2050 BC	R^2	0.97
	AIC	79.93
	BIC	168.82
	Silhouette	0.600
2090 BC	R^2	0.982
	AIC	81.15
	BIC	192.26
	Silhouette	0.62

TABLE 13.6 Variable importance (Gini index).

	Time frames		
Classes	Current BC	2050 BC	2070 BC
Optimum	89.84	80.277	66.41
Moderate	88.69	78.27	60.97
Marginal	87.60	73.54	58.76
Low	87.30	62.77	46.04

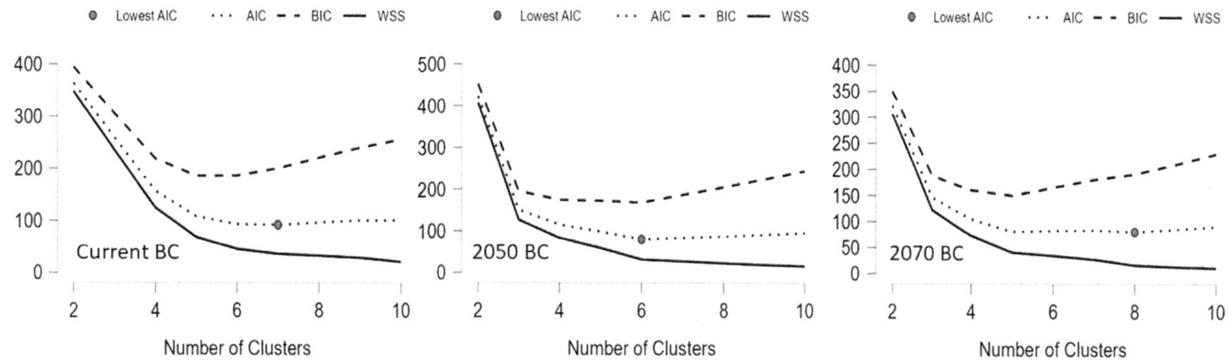

FIGURE 13.25 Elbow method for random forest clustering.

In addition to measuring a model's goodness-of-fit, the AIC and the BIC additionally penalize a model for having too many free parameters. These information requirements penalize the number of parameters in an effort to prevent overfitting. Models with lower information criteria are thought to generalize more effectively. The Silhouette score, which compares how similar each case is to its own cluster to other clusters, shows the average internal consistency of the grouping. Better model quality was indicated by low AIC values for the current time (95.42), 2050 BC (79.93), and 2070 BC (81.15); the similar tendencies were also seen with BIC (203.76, 168.82, and 192.26). The basic rule for silhouette scores is that the closer they are to the upper bound of 1, the more consistent the grouping is, and the closer they are to the lower bound of 1, the worse the match. The score for the Silhouette coefficient is restricted to values of −1 for improper clustering and +1 for dense clustering. Scores near zero suggest clusters that overlap. This measure was close to 0.480 (during current BC) and 0.60 in this investigation, indicating that cluster consistency was on par with average (Table 13.5).

The average decline in the Gini coefficient is a measure of how much each variable improved the uniformity of the resulting random forest's nodes and leaves. Greater variable relevance is indicated by a higher mean Gini drop. Table 13.6 lists the outcomes of this index. The variable relevance is in the following order for each of the analyzed climatic time frames: optimal > moderate > marginal > low.

At least three different levels can be used to measure a pathogen's habitat: the host, or sometimes a specific tissue or organ, is the restricted habitat; the host's immediate microenvironment is the extended habitat; and at least three different levels can be used to measure the habitat of the general population. Pathogen−host contact is an example of a biotic relationship. Molecular interactions determine the confined habitat, which includes the axes of both the fundamental niche and the realized niche. Diseases live in their habitat, which is made up of the target tissue, target plant species, target population, target community, target ecosystem, and target landscape. The host is also a nested entity that represents the disease's habitat. Habitat use is also nested, with biochemical control of pathogen−host interaction, susceptible individuals within a population, and the pathogen's probabilistic/stochastic interception process all affecting successful colonization and dispersal in terms of community and ecosystem, as well as some landscape features like connectivity, topography, and the spatial distribution of important environmental factors to the hosts.

13.5.7 Unidimensional reliability analysis

The findings of the classes' internal consistency are shown in Table 13.7 (optimum, moderate, marginal, and low). This tool evaluates the reliability of a scale by determining how closely connected a group of items are to one another. This was done by utilizing the Cronbach's alpha index. Cronbach's alpha has a minimum acceptable value of 0.70. Low internal consistency is considered to be anything below this value. The highest predicted value is 0.90, and anything above this

TABLE 13.7 Unidimensional reliability analysis (Cronbach's α) among classes of different timeframes.

	Optimum	Moderate	Marginal	Low
Current BC	0.904	0.911	0.968	0.995
2050 BC	0.879	0.770	0.756	0.871
2070 BC	0.938	0.694	0.788	0.865

point is seen as redundant or redundant. Internal consistency for all classes in the current BC is >0.90, indicating scale reliability. Values of this index were >0.70 for 2050 BC and 2070 BC, with the exception of moderate class for 2070 BC (Table 13.7).

Based on our RF-C and URA data, we were able to determine that different types of *M. phaseolina* niches exist within the studied regions (Fig. 13.26). Fundamental niche is the set of physical requirements and resources that a species needs to maintain self-sustaining populations, which may not be fully occupied due to antagonistic species interactions, a lack of necessary positive species interactions, or a dispersal restriction; Realized niche is the intersection of a species' fundamental niche, accessible climate or territory; Species-specific niches are the set of physical requirements and resources that a species needs to maintain self-sustaining populations. Potential distribution is the area having a climate that allows self-sustaining populations to exist, and it is a subset of the fundamental niche. The tolerance niche consists of the conditions and resources that allow for human development but prohibit a species from establishing a self-sustaining population.

Consider the ecological niche as a three-dimensional volume with each dimension representing a different biotic or abiotic element necessary for the survival of a certain species. The volume's shape, size, and evolution profoundly affect interactions between species and, by extension, the present and future distribution of the species around the globe. In spite of this, niche geometry still needs extensive scientific inquiry. Competition, trait evolution, ecological specialization, community assembly principles, and worldwide spatial distributions of species are only some of the topics that Hutchinson's model has shed light on. In essence, it is a multidimensional space volume describing the conditions under which a given species might thrive (Blonder et al., 2014; Chaloner et al., 2020; Colwell & Rangel, 2009).

Species having broad tolerance ranges are referred to as "*specialists*," whereas species with narrow tolerance ranges are generally referred to as "*generalists*" (Davison et al., 2021). These two concepts cannot be used to describe the ecology of oomycete or fungal species in its entirety. Speculating on the selective factors that might result in differential specialization in abiotic and biotic habitats is challenging. Niche specialization is uncorrelated on abiotic (temperature sensitivity) and biotic (host range) axes, and both abiotic and biotic niches exhibit low phylogenetic constraint as reported by Chaloner et al. (2020) for many species of the genus *Phytophthora*. These fungi are thought to defy the classification as either "generalists" or "specialists" because they have independently evolved specialization and occupied many niche axes. Because of their versatility, plant diseases constitute a significant threat to the forestry and agricultural industries. Contrary to what Chaloner et al. (2020) found, Davison et al. (2021) study showed that temperature and pH play a major role in determining the actual niches of AM fungal virtual taxa (VT; roughly species-level phylogroups). Significant subclades of the Glomeraceae showed distinctive niche optima (the most favorable environmental conditions for an organism). While Gigasporaceae had niche optimum conditions in high precipitation settings, Acaulosporaceae typically had niche optimum conditions in low pH and low temperature settings.

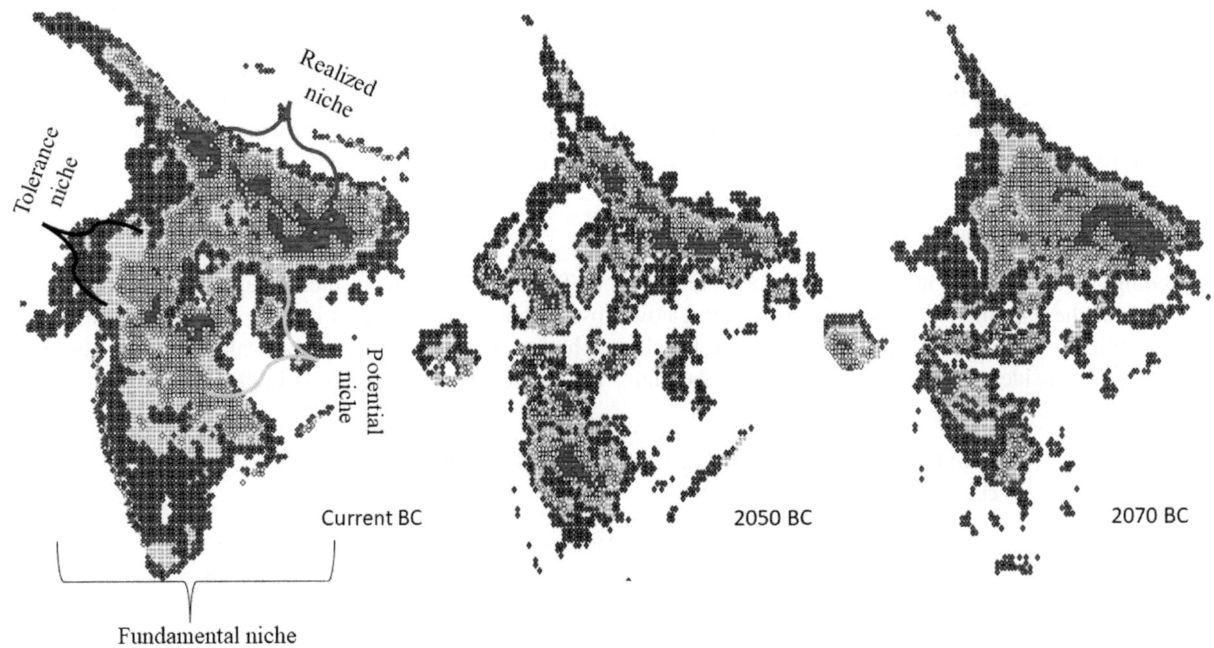

FIGURE 13.26 Fundamental, realized, potential and tolerance climatic-temporal niches of *M. phaseolina*.

According to Fodor (2011), the disease intersection concept refers to the hypervolume of space that encompasses the pathogen's effective colonization of the habitat, the pathogen's successful acquisition of host resources, and favorable regions along environmental gradients. The same axes incorporate unfavorable host areas as well as environmental ingredients that ultimately aid the pathogen for the sake of symmetry. For instance, prolonged soil flooding, which is both beneficial for soil-borne *Phytophthora* species and a protective condition for trees.

13.6 Conclusion

This study can be regarded as first compressive ecological niche modeling or species distribution modeling for a soil-borne plant pathogen *M. phaseolina*. This cosmopolitan species caused sever disease in many important crops that accelerated yield reduction and impose food constraints for growing world population. Prediction of probable suitable habitats for this species with using both BC and non-BC variables of different time frames can enhance our knowledge for its possible outbreak that can facilitate to timely place their control measure. We comparatively evaluate the impacts of different bioclimatic variables of different time frame as well as some other non-BC factors that can be served as a predisposing factor for this species by using some machine learning techniques. Such algorithms-based studies are the need of time through which pathogen adaptability and its temporal and spatial adaptability can timely be forecast and thus, control measures (cultural, chemical, biological) can be applied at appropriate schedule.

Contribution: Senior author conceptualized the chapter theme and interpretation of output of various machine learning techniques. Co-Author prepared various types of language codes in python, Java and in R scripts and convert the various file format from ASCII to KML, Raster, dbf, CSV, etc. for software like QGIS 3.10.0, Wallace, DIVA-GIS version 7.5, OpenModeller, MaxEnt 3.4.1 software, SDM toolbox, Map Comparison Kit, ENMTools, Ntbox, and randomForest R packages.

References

Abolmaali, S. M. R., Tarkesh, M., & Hossein, B. (2018). MaxEnt modelling for predicting suitable habitats and identifying the effects of climate change on a threatened species, *Daphne mucronata*, in central Iran. *Ecological Informatics, 43*, 116–123.

Ahmad, R., Khuroo, A. A., Hamid, M., Charles, B., & Rashid, I. (2019). Predicting invasion potential and niche dynamics of *Parthenium hysterophorus* (Congress grass) in India under projected climate change. *Biodiversity & Conservation, 28*(5), 2319–2344.

Al-Hazmi, A. S., Dawabah, A. A. M., Al-Nadhary, S. N., Al-Yahya, F. A., & Lafi, H. A. (2017). Influence of soil texture and moisture on the interaction of *Meloidogyne javanica* and *Macrophomina phaseolina* on green beans. *Journal of Experimental Biology and Agricultural Sciences, 5*, 148–154.

Araujo, M. B., & Pearson, R. G. (2005). Equilibrium of species' distributions with climate. *Ecography, 28*, 693–695. https://doi.org/10.1111/j.2005.0906-7590.04253.x

Baird, R. E., Watson, C. E., & Scruggs, M. (2003). Relative longevity of *Macrophomina phaseolina* and associated mycobiota on residual soybean roots in soil. *Plant Diseases, 87*(5), 563–566.

Begon, M., Townsend, C. R., & Harper, J. L. (2006). *Ecology: From individuals to ecosystems* (4th ed.). Blackwell Publishing House.

Blonder, B., Lamanna, C., Violle, C., & Enquist, B. J. (2014). The n-dimensional hypervolume. *Global Ecology and Biogeography, 23*, 595–609.

Brasier, C. M., Cooke, D. E. L., & Duncan, J. M. (1999). Origin of a new *Phytophthora* pathogen through interspecific hybridization. *Population Biology, 96*(10), 5878–5883.

Brown, J. L., & Anderson, B. (2014). Sdmtoolbox: A python–based GIS toolkit for landscape genetic, biogeographic and species distribution model analyses. *Methods in Ecology and Evaluation, 5*, 694–700.

Brown, J. L., Bennett, J. R., & French, C. M. (2017). Sdmtoolbox 2.0: The next generation python-based GIS toolkit for landscape genetic, biogeographic and species distribution model analyses. *PeerJ, 5*, e4095.

Chaloner, T. M., Gurr, S. J., & Bebber, D. P. (2020). Geometry and evolution of the ecological niche in plant-associated microbes. *Nature Communications, 11*, 1–9.

Coban, H. O., Orucu, O. K., & Arslan, E. S. (2020). MaxEnt modelling for predicting the current and future potential geographical distribution of *Quercus libani* Olivier. *Sustainability, 12*(7), 2671. https://doi.org/10.3390/su12072671

Colwell, R. K., & Rangel, T. F. (2009). Hutchinson's duality: The once and future niche. *Proceedings of the National Academy of Sciences, 106*, 19651–19658.

Craig, S. R. (1992). Tillage system and plant disease. *Soil Science, 154*, 308–315.

Dadwal, V. S., & Bhartiya, S. (2012). New report of a leaf spot disease of *Chlorophytum borivillianum* caused by *Macrophomina phaseolina* from India. *Journal of Mycology and Plant Pathology, 42*(3), 397–398.

Dash, M. C. (2007). *Fundamentals of ecology*. Tate McGraw Hill.

Davison, J., Moora, M., Semchenko, M., Adenan, S. B., Ahmed, T., Akhmetzhanova, A. A., Alatalo, J. M., et al. (2021). Temperature and pH define the realized niche space of arbuscular mycorrhizal fungi. *New Phytologist, 231*, 763–776.

Delgado-Baquerizo, M., Guerra, C. A., Cano-Diaz, C., Egidi, E., Wang, J. T., Eisenhauer, N., Singh, B. K., & Maestre, F. T. (2020). The proportion of soil-borne pathogen increases with warming at the global scale. *Nature Climate Change, 10*, 550–554.

Dhingra, O. D., & Sinclair, J. B. (1975). Survival of *Macrophomina phaseolina* sclerotia in soil: Effect of soil moisture, carbon: Nitrogen ratio, carbon sources, and nitrogen concentrations. *Phytopathology, 65*, 236–240.

Dhingra, O. D., & Sinclair, J. B. (1977). *An annotated bibliography of* Macrophomina phaseoli, *1905–1975* (p. 277). Viçosa, Brazil: Universidade Federal Viçosa.

Eastburn, D. M., McElrone, A. J., & Bilgin, D. D. (2011). Influence of atmospheric and climatic change on plant-pathogen interactions. *Plant Pathology, 60*, 54–69.

Elith, J., Geaham, H. C., & Anderson, P. R. (2006). Novel methods improve prediction of species' distributions from occurrence data. *Ecography, 29*, 129–151.

Fischer, G., Nachtergaele, F., Prieler, S., van Velthuizen, H. T., Verelst, L., & Wiberg, D. (2008). *Global agro-ecological zones assessment for agriculture (GAEZ 2008)*. Rome, Italy: IIASA, Laxenburg, Austria and FAO.

Fitzsimons, M. S., Miller, M. R., & Jastrow, J. D. (2008). Scale dependent niche axes in arbuscular mycorrhizal fungi. *Community Ecology, 158*(1), 117–127.

Flory, A. R., Kumar, S., Stohlgren, T. J., & Cryan, P. M. (2012). Environmental conditions associated with bat white nose syndrome mortality in the north-eastern United States. *Journal of Applied Ecology, 49*, 680–689.

Fodor, E. (2011). Ecological niche of plant pathogens. *Annals of Forest Research, 54*(1), 3–21.

Fukuda, S., & De Baets, B. (2016). Data prevalence matters when assessing species responses using data-driven species distribution models. *Ecological Informatics, 32*, 69–78.

Goodswen, S., Barratt, J. L. N., Kennedy, P. J., Kaufer, A., Calarco, L., & Ellis, J. T. (2021). Machine learning and application in microbiology. *FEMS Microbiology Reviews, 45*, 1–19.

Goudarzi, A., Banihashemi, Z., & Maftoun, M. (2011). Effect of salt and water stress on root infection by *Macrophomina phaseolina* and ion composition in shoot in sorghum. *Iranian Journal of Plant Pathology, 47*(3), 69–83.

Gudelj, I., Fitt, B. D. L., & van der Bosch, F. (2004). Evolution of sibling fungal plant pathogens in relation to host specialization. *Analytical and Theoretical Plant Pathology, 94*(7), 789–795.

Gupta, J. P., & Gupta, G. K. (1986). Effect of tillage and mulching on soil environment and cowpea seedling growth under arid conditions. *Soil and Tillage Research, 7*, 233–240.

Gupta, G. K., Sharma, S. K., & Ramteke, R. (2012). Biology, epidemiology and management of the pathogenic fungus *Macrophomina phaseolina* (Tassi) Goid with species reference to charcoal rot of soybean (*Glycine max*). *Journal of Phytopathology, 60*, 167–180.

Hartman, G. L., Sinclair, J. B., & Rupe, J. C. (1999). *Compendium of soybean diseases* (IV ed.). St. Paul, MN, USA: The American Phytopatholgical Society. APS Press.

He, F., & Tang, D. (2007). Estimating the niche pre-emption parameter of the geometric series. *Acta Oecologia, 33*(1), 105–107. https://doi.org/10.1016/j.actao.2007.10.001:1-3

Hijmans, R. J., Cameron, S. E., Parra, J. L., Jones, P. G., & Jarvis, A. (2005). Very high-resolution in-terpolated climate surfaces for global land. *International Journal of Climatology, 25*, 1965–1978.

Hijmans, R. J., Guarino, L., Cruz, M., & Rojas, E. (2001). Computer tools for spatial analysis of plant genetic resources data: 1. DIVA-GIS. *Plant Genetic Resources Newsletter, 127*, 15 19.

IPCC. (2000). *Special report on land use, land-use change and forestry*. Cambridge Univ. Press.

Juroszek, P., & Von Tiedemann, A. (2013). Climate change and potential future risks through wheat diseases: A review. *European Journal of Plant Pathology, 136*, 21–33.

Karami, P., Rezaei, S., Shadloo, S., & Naderi, M. (2020). An evaluation of central Iran's protected areas under different climate change scenarios (A case on Markazi and Hamedan provinces). *Journal of Mountain Science, 17*(1), 68–82.

Karl, T. R., & Trenberth, K. E. (2003). Modern global climate change. *Science, 302*, 1719–1723.

Kass, J. M., Vilela, B., Aiello-Lammen, M. E., Muscarella, R., Merow, C., & Anderson, R. P. (2018). *Wallace*: A flexible platform for reproducible modelling of species niches and distributions built for community expansion. *Methods in Ecology and Evolution, 9*, 1151–1156.

Kaur, S., Dhillon, G. S., Brar, S. K., Vallad, G. E., Chand, R., & Chauhan, V. B. (2012). Emerging phyo-pathogen *Macrophomina phaseolina*: Biology, economic importance and current diagnostic trends. *Critical Review in Microbiology, 38*(2), 136–156.

Kohire, O. D., Muluk, M. B., Paul, V. O. K., Thombere, B. B., & More, S. S. (2012). Survey of root disease of chickpea in Jalana district of Marathwada region. *International Journal of Plant Protection, 5*(20), 381–385.

Kumari, G. B. G., & Govindaiah, S. J. (2011). Incidence of stem canker (*Lasiodiplodia therbromae*) in mulberry nurseries of Kolar district. *Indian Journal of Sericulture, 50*(1), 78–81.

Kumar, S., & Stohlgren, T. J. (2009). MaxEnt modelling for predicting suitable habitat for threatened and endangered tree *Canacomyrica monticola* in New Caledonia. *Journal of Ecology and the Natural Environment, 1*, 94–98.

Kumar, S., Stohlgren, T. J., & Chong, G. W. (2006). Spatial heterogeneity influences native and non-native plant species richness. *Ecology, 87*, 3186–3199.

Levin, R. (1968). *Evolution in changing environments: Some theoretical explorations*. Princeton University Press.

Liaw, A., & Wiener, M. (2002). Classification and regression by random forest. *R News, 2*(3), 18–22.

Lokesh, B. R., Shashidhara, N., & Kantharaju, V. (2020). Survey for the incidence of wilt disease and management on wilt in chickpea. *International Journal of Agricultural Sciences, 16*(1), 91–94.

Mahadevakumar, S., & Janardhana, G. R. (2016). First report of leaf blight caused by *Macrophomina phaseolina* on *Jasminium multiflorum* in India. *Journal of Plant Pathology, 98*(1), 177. http://www.sipav.org/main/jpp/index.php/jpp/article/view/3484.

Mallaiah, B., & Rao, V. K. (2016). Survey and virulence studies on dry root of green gram [*Vigna radiata* (L.) Wilczek] incited by *Macrophomina phaseolina* (Tassi.) Goid. *International Journal of Tropical Agriculture, 34*(3), 521–527.

Manici, L. M., Donatelli, M., Gumagalli, D., Lazzari, A., & Bregaglio, S. (2012). Potential response of soil borne fungal pathogen affecting crops to a scenario of climate change in Europe. In R. Seppelt, A. A. Voinov, S. Lange, & D. Bankamp (Eds.), *International Environmental Modelling and Software Society (iEMSs) 2012 International congress on environmental modelling and software managing resources of a limited planet, sixth biennial meeting, Leipzig, Germany*. http://www.iemss.org/society/index.php/iemss-2012-proceedings.

Marchan, D. F., Refoyo, P., Fernandez, R., Novo, M., Soda-de, I., & Cosin, D. J. D. (2016). Macroecological inferences on soil fauna through comparative niche modelling: The case of Hormogastride (Annelida, Oligochaeta). *European Journal of Soil Biology, 75*, 115–122.

Masafumi, F., Tokiya, S., Hidenori, O., Hidetake, S., Masanobu, K., & Toyozo, S. (2002). Charcoal rot of watermelon newly found in Nagano and Kanagawa prefectures and pathogenicity of causal fungus *Macrophomina phaseolina*. *Annals of the Phytopathological Society of Japan, 68*(2), 148–152.

Ma, B., & Sun, J. (2018). Predicting the distribution of *Stipa purpurea* across the Tibetan plateau via MaxEnt model. *BMC Ecology, 18*, 10. https://doi.org/10.1186/s12898-018-0165-0

Mathur, M., Kumar, M., & Pandey, C. B. (2022). Impacts of soil qualities and *Prosopis juliflora* on density, canopy volume and community position of *Leptadenia pyrotechnia* in Arid region of India. *Tropical Ecology, 63*, 252–262. https://doi.org/10.1007/s42965-021-00204-y

Mathur, M., & Sundaramoorthy, S. (2016). Pattern of herbaceous species richness and productivity along gradients of soil moisture and nutrients in the Indian Thar Desert. *Journal of Arid Environments, 125*, 80–87.

Meena, B. R., Nagendran, K., Tripathi, A. N., Kumar, S., Sagar, V., Gupta, N., Rai, A. B., & Singh, B. (2018). First report of charcoal rot caused by *Macrophomina phaseolina* in *Basella alba* in India. *Plant Disease, 102*(8), 1669–1670.

Mihail, J. D. (1989). *Macrophomina phaseolina*: Spatio-temporal dynamics of inoculum and of disease in a highly susceptible crop. *Phytopathology, 79*, 848–855.

Mishra, S. N., Kumar, D., Kumar, B., & Tiwari, S. (2021). Assessing impact of varying climatic conditions on distribution of *Buchanania cochinchinensis* in Jharkhand using species distribution modelling approach. *Current Research in Environmental Sustainability*. https://doi.org/10.1016/j.crsust.2021.100025

Mousazade, M., Ghanbarian, G., Pourghasemi, H. R., Safaeian, R., & Cerda, A. (2019). Maxent data mining technique and its comparison with a bivariate statistical model for predicting the potential distribution of *Astragalus fasciculifolius* Boiss. In Fars, Iran. *Sustainability, 11*(12), 3452. https://doi.org/10.3390/su11123452

Munoz, M. E. S., Giovanni, R., Siqueira, M. F., Sutton, T., Brewer, P., Pereira, R. S., Canhos, D. A. L., & Canhos, V. P. (2009). Openmodeller: A generic approach to species potential distribution-modelling. *GeoInformatica, 15*, 111–135.

Nazir, A. P., Mathur, M., Iqbal, K., Negi, A. K., Shukla, G., & Chakravarty, S. (2017). Community dynamics: Competition and facilitation studies of tree species in temperate forest of the India Himalaya. *Indian Journal of Ecology, 44*(2), 279–288.

Ndiaye, M. (2007). *Ecology and management of charcoal rot (*Macrophomina phaseolina*) on cowpea in the Sahel (PhD Thesis)*. Wageningen, The Netherlands: Wageningen University.

Nishad, I., Srivastava, A. K., Saroj, A., Babu, B. K., & Samad, A. (2018). First report of root rot of *Nepeta cataria* caused by *Macrophomina phaseolina* in India. *Plant Disease, 102*(11), 2380–2381.

Nunez-Penichet, C., Cobos, M. E., & Soberon, J. (2021). Non-overlapping climatic niches and biogeographic barriers explain disjunct distributions of continental *Urania* moths. *Frontiers of Biogeography, 13*, 1–12.

Olaya, G., & Abawi, G. S. (1993). Effect of water potential on germination, growth, and microsclerotial production of. Macrophomina. phaseolina. *(Abstr.)*. Phytopathology, *83*, 1394.

Osorio-Olivera, L., Lira- Noriega, A., Soberon, J., Townsend, P. A., Facon, M., Contreas Diaz, R. G., et al. (2020). Ntbox: an R package with graphical user interface for modelling and evaluating multidimensional ecological niches. *Methods in Ecology and Evolution, 11*, 1199–1206.

Padalia, H., Srivastava, V., & Kushwaha, S. P. S. (2014). Modelling potential invasion range of alien invasion species, *Hyptis suaveolens* (L) in India: Comparison of MaxEnt and GARP. *Ecological Informatics, 22*, 36–43.

Pautasso, M., Döring, T. F., Garbelotto, M., Pellis, L., & Jeger, M. J. (2012). Impacts of climate change on plant diseases— opinions and trends. *European Journal of Plant Pathology, 133*, 295–313.

Petersen, T. (2007). Ecological niche modelling and understanding the geography of disease transmission. *Veterinaria Italiana, 43*(2), 393–400.

Phillips, S. J., & Dudik, M. (2008). Modelling of species distributions with Maxent: New extensions and a comprehensive evaluation. *Ecography, 31*, 161–175.

Pradhan, P. (2016). Strengthening Maxent modelling through screening of redundant explanatory bioclimatic variables with variance inflation factor analysis. *Researcher, 8*(5), 29–34.

Prakash, V. R., Venugopal, S., & Thomas, S. (2007). First report of *Macrophomina* leaf spot (*Macrophomina phaseolina*) on *Artocarpus hirsutus* in India. *Plant Archives, 7*(2), 929–930.

Putman, R., & Wratten, S. D. (1984). *Principles of ecology*. University of Chicago Press.

Qin, A., Liu, B., Guo, Q., Bussmann, R. W., Ma, F., Jian, Z., Xu, G., & Pei, S. (2017). Maxent modeling for predicting impacts of climate change on the potential distribution of *Thuja sutchuenensis* Franch., an extremely endangered conifer from southwestern China. *Global Ecology and Conservation, 10*, 139–146.

R Development Core Team. (2011). *R: A language and environment for statistical computing.* Vienna: R Foundation for Statistical Computing. http://www.R-project.org.

Renner, I. W., & Warton, D. I. (2013). Equivalence of MAXENT and Poisson point process models for species distribution modelling in ecology. *Biometrics, 69*, 274–281.

Rong, Z., Zhao, C., Liu, J., Gao, Y., Zang, F., Guo, Z., Mao, Y., & Wang, L. (2019). Modelling the effect of climate change on the potential distribution of Qinghai Spruce (*Picea crassifolia* Kom.) in Qilian mountains. *Forests, 10*(1), 62. https://doi.org/10.3390/f10010062

Sarikaya, O., Karaceylan, I. B., & Sen, I. (2018). Maximum entropy modelling (maxent) of current and future distributions of *Ips Mannsfeldi* (Wachtl, 1879) (Curculionidae: Scolytinae) in Turkey. *Applied Ecology and Environmental Research, 16*, 2527–2535.

Short, G. E., Wyllie, T. D., & Bristow, P. R. (1980). Survival of *Macrophomina phaseolina* in soil and in residue of soybean. *Phytopathology, 70*, 13–17.

Sillero, N., & Barbosa, A. M. (2021). Common mistakes in ecological niche models, International. *Journal of Geographical Information Science, 35*(2), 213–226.

Suryawanshi, A. P., Mule, V. P., Apet, K. T., Dey, U., & Kuldhar, D. P. (2015). Managing *Macrophomina phaseolina* causing charcoal rot of sunflower (*Helianthus annus*) by soil amendment. *Indian Phytopathology, 68*(2), 196–200.

Swamy, C., Naik, M. K., Amaresh, Y. S., & Jayalakshmi, S. K. (2017). Prevalence of emerging disease of *Macrophomina* stem canker of pigeon pea in north eastern Karnataka, India. *Environment and Ecology, 35*(2), 1323–1326.

Taliei, F., Safair, N., & Aghajani, M. A. (2013). Spatial distribution of *Macrophomina phaseolina* and soybean charcoal rot incidence using Geographic Information System 9A case study in Northern Iran). *Journal of Agricultural Sciences and Technology, 15*, 1523–1536.

Teja, T. S., Kelayia, D. S., & Asha, R. (2020). Impacts of environmental factors on *Macrophomina phaseolina* causing charcoal rot of soybean. *International Journal of Current Microbiology and Applied Sciences, 9*(10), 3784–3790.

Tittensor, D. P., Baco, A. R., Brewin, P. E., Clark, M. R., Consalvey, M., Hall-Spencer, J., Rowden, A. A., et al. (2009). Predicting global habitat suitability for stony corals on seamounts. *Journal of Biogeography, 36*, 1111–1128.

Tuanmu, M. N., & Jetz, W. (2015). A global, remote sensing-based characterization of terrestrial habitat heterogeneity for biodiversity and ecosystem modelling. *Global Ecology and Biogeography, 24*(11), 1329–1339.

VanDerWal, J., Shoo, J. P., Johnson, C. N., & Williams, S. E. (2009). Abundance and the environmental niche: Environmental suitability estimated from niche models predicts the upper limits of local abundance. *The American Naturalist, 174*(2), 1–11.

Vitor, H. F. G., Stephanie, D. I., Niels, R., Amaral, I. L., Salomão, R. P., de Souza Coelho, L., et al. (2018). Species distribution modelling: Contrasting presence-only models withplot abundance data. *Scientific Report, 8*, 1003. https://doi.org/10.1038/s41598-017-18927-1

Vorsino, A. E., King, C. B., Haines, W. P., & Rubinoff, D. (2013). Modelling the habitat retreat of the rediscovered endemic Hawaiian Moth *Omiodes continuatalis* Wallengren (Lepidoptera: Crambidae). *PLoS One, 8*(1), e51885. https://doi.org/10.1371/journal.pone.0051885

Wahdan, S. F. M., Hossen, S., Tanunchai, B., Schadler, M., Buscot, F., & Purahong, W. (2020). Future climate significantly alters fungal plant pathogen dynamics during the early phase of wheat litter decomposition. *Microorganisms, 8*(6), 1–17.

Warren, D. L., Glor, R. E., & Turelli, M. (2010). ENMTools: A toolbox for comparative studies of environmental niche models. *Ecography, 33*, 607–611.

Wei, B., Wang, R. L., Hou, K., Wang, X. Y., & Wu, W. (2018). Predicting the current and future cultivation regions of Carthamus tinctorius L. using MaxEnt model under climate change in China. *Global Ecology and Conservation, 18*, 1–12.

Wokocha, R. C. (2000). Effect of different soil moisture regimes on the development of the charcoal rot diseases of soybean by *Macrophomina phaseolina*. Global Journal of Pure and Applied Science*s, 6*, 599–602.

Wrather, J. A., Kendig, S. R., & Tyler, D. D. (1998). Tillage effect on *Macrophomina phaseolina* population density and soybean yield. *Plant Diseases, 82*, 247–250.

Xu, W., Jin, J., & Cheng, J. (2021). Predicting the potential geographic distribution and habitat suitability of two economic forest trees on the Loess Plateau, China. *Forests, 12*(6), 747. https://doi.org/10.3390/f12060747

Ye, X. Z., Zhao, G. H., Zhang, M. Z., Vui, X.,Y., Fan, H. H., & Liu, B. (2020). Distribution pattern of endangered plants *Semiliquidambar catayensis* (Hamamelidaceae) in response to climate change after the last interglacial period. *Forest, 11*, 434. https://doi.org/10.3390/f11040434

You, M. P., Colmer, T. D., & Barbetti, M. J. (2011). Salinity drives host reaction in *Phaseolus vulgaris* (common bean) to *Macrophomina phaseolina*. *Functional Plant Biology, 38*, 984–992.

Zhang, Y., Tang, J., Ren, G., Zhao, K., & Wang, X. (2021). Global potential distribution prediction of *Xanthium italicum* based on Maxent Model. *Scientific Reports, 11*, 16545. https://doi.org/10.1038/s41598-021-96041-z

Zhao, G., Cui, X., Sun, J., Li, T., Wang, Q., Ye, X., & Fan, B. (2021). Analysis of the distribution pattern of Chinese *Ziziphus jujuba* under climate change based on optimized biomod2 and MaxEnt models. *Ecological Indicators, 132*, 108256. https://doi.org/10.1016/j.ecolind.2021.108256

Zhu, Y., Wei, W., Li, H., Wang, B., & Yang, X. (2018). Modelling the potential distribution and shifts of three varieties of *Stipa tianschanica* in the eastern Eurasian steppe under multiple climate change scenarios. *Global Ecology and Conservation, 16*, e00501. https://doi.org/10.1016/j.gecco.2018.e00501

Chapter 14

Survival of *Macrophomina phaseolina* in plant tissues and soil

Sandeep Kumar[1], Vineet Kumar Vishnoi[1], Pankaj Kumar[2] and Ramesh Chandra Dubey[1]
[1]*Department of Botany and Microbiology, Gurukul Kangri (Deemed to be University), Haridwar, Uttarakhand, India;* [2]*Department of Microbiology, Dolphin (PG) Institute of Biomedical and Natural Sciences, Dehradun, Uttarakhand, India*

14.1 Introduction

Macrophomina phaseolina (Tassi) Goid. is a saprophytic fungus, belonging to the family Botryosphaeriaceae. It is a serious soil- as well as seed-borne pathogen of prime importance reducing crops yield all over the world. It has broad host range causing several diseases, such as charcoal rot, collar rot, basal stem rot, stem rot, stem canker, root rot, and seedling blight resulting in heavy crop losses worldwide (Gupta et al., 2020). It is found globally and infecting over 850 host plants (Saleh et al., 2010; Zhao et al., 2020). The most interesting and unusual fact about *M. phaseolina* is that it has a wide range of survival on both monocots and dicots (Pratt, 2006). *M. phaseolina* has been reported as an emerging pathogen from several decades ago in South Asia that was responsible for plant mortality by 30% early in the season and 60% by the end of the season (Peres et al., 2018). Disease incidence by *M. phaseolina* has been reported up to 90% infield and reduced production by 80% (Gupta et al., 2012; Pandey et al., 2020; Yang & Navi, 2005; Zhao et al., 2020). The plurivorous nature of *M. phaseolina* enables it to survive on several alternative lives or dead host plants and also in the absence of host crops.

However, crop management practices, such as manipulation of cropping system and crop rotation methods, are still promising and effective in limiting the pathogenicity and to reduce the inoculum density (Songa & Hillocks, 1996). Environmental and physiological factors, such as moisture, temperature, pH, and heat, affect the survival and prevalence of pathogens (Ashraf et al., 2017; Dhingra & Sinclair, 1978). The pathogen survives in diverse climatic conditions and has a broad host range, infecting a large number of plant species including legumes, nonlegumes, weeds, grasses, agricultural crops, and tree in temperate, sub-tropical, and tropical areas (Gupta et al., 2020; Wagan et al., 2018). This pathogen reduces productivity up to 90% under favorable environmental conditions (Khan, 2007). *M. phaseolina* invades and survives in seeds also while they are in the stage of development on plants in the field until maturity or invade seeds during storage. Depending on their innate resistance, different crop species have varying degrees of tolerance to *M. phaseolina*. The more vulnerable crop species succumb to the disease more quickly, allowing it to spread throughout them and raise the inoculum levels in the soil (Singh et al., 1990). *M. phaseolina* survives in soil through production of small sized prapagules called microsclerotia. There have been no effects of oxygen and carbon dioxide concentration on germination and production of microsclerotia of *M. phaseolina* that seems to be independent from natural concentration (Wyllie et al., 1984). The pathogens can spillover where it is adapted to agriculture and move to native systems (Saleh et al., 2010).

The diseases caused by *M. phaseolina* ranks as one of the most poorly understood and probably under estimated diseases. The control of disease caused by *M. phaseolina* is effectively and economically attained by utilizing the integrated control strategies. Another, important problem concerning to control *Macrophomina* is unavailability of suitable disease resistant cultivars for use in breeding (Jana et al., 2003).

14.1.1 What is survival?

Survival is a feature of an organism to withstand the unfavorable environmental conditions that prevail for short or long period. Specialized resistant bodies are produced as a means of survival, for example reproductive spores, hyphae

Macrophomina Phaseolina. https://doi.org/10.1016/B978-0-443-15443-0.00015-2

possessing special characteristics or complex vegetative structures, such as sclerotia and rhizomorphs formed by the aggregation of hyphae and mycelia. Some resting bodies are formed only when the environmental conditions become extremely adverse. The survival potential enhances longevity of an organism by retaining the ability for prolong viability in the unsuitable habitat. However, the preformed structures help in survival by allowing for mycelial growth, quick and plentiful spore production, and colonization of a wide range of substrates. The survival characteristics can also be described as genetic variability, disseminability, competitive saprophytic ability, and responsiveness (Gupta et al., 2020).

14.1.2 Survival in extreme environmental conditions

Survival of any organism in nature is its inherent potential to adverse environmental conditions. If the organisms have no skills of survival, they cannot be alive because they are not fit for living. Therefore, all organisms possess ability to withstand unfavorable conditions through different modes and gets equipped with different ways of survival. *M. phaseolina* possesses two ecologically distinct phases of life—first the active phase limited to mycelia only, and the second inactive phase represented by microsclerotial one. In the first case of the survival, it parasitizes the host or nonhost plants and continues growing actively. *M. phaseolina* often infects roots first, but pathogenesis and sclerotium production can also occur above ground favored by hot and dry circumstances (Pratt, 2006).

14.2 Survival structures

M. phaseolina produces several structures that are involved in survival, either for short or long duration. These survival structures have been described below:

14.2.1 Mycelia

M. phaseolina survives either by mycelia or by producing tiny propagules called microsclerotia in the active phase of life in the host tissues during parasitic phase. The mycelia are light brown to dark brown, thick-walled and melanized structure which survive in soil only for a short duration depending upon the environmental and biological factors (Fig. 14.1A). Huge numbers of mycelia are produced on dead and decaying organic material of host or nonhost plants until host materials are not exhausted. Secondly, mycelia parasitize the living hosts by infecting the tissues and ramify internally. However, special modification in mycelia may confer resistance and survivability (Forster & Shuai, 2020). The hyphae of *M. phaseolina* initially develop intercellularly and move through the plant's vascular tissue. The biotrophic phase of an infection frequently precedes the necrotrophic phase, which can be triggered by environmental changes, stress on the host plant or plant maturation (Twizeyimana et al., 2012).

Recently, Meena and Kadam (2021) reported that *M. phaseolina* infecting musli (*Chlorophytum borivilianum*) and causing leaf blight diseases showed dark olive black colonies that ultimately turns black; they are found embedded with abundant microsclerotia. Mycelia plugs of *M. phaseolina* forming tiny swellings and brownish necrotic lesions or cankers

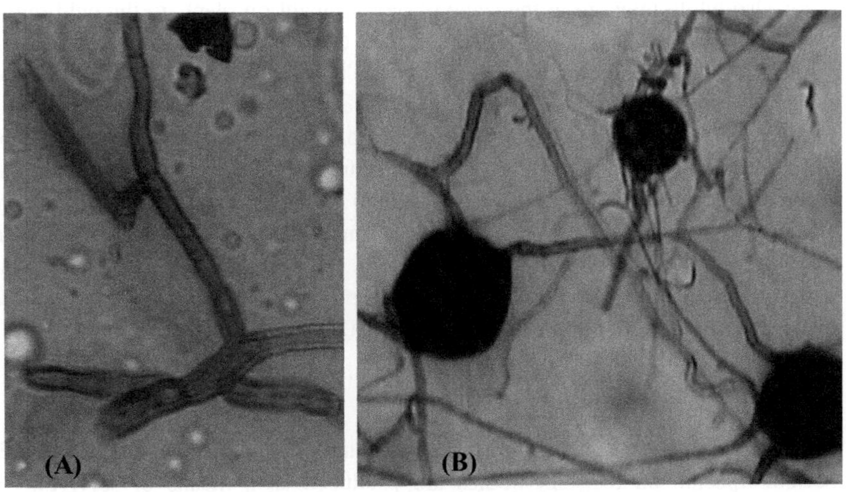

FIGURE 14.1 *Macrophomina phaseolina*; (A) dark brown [dark gray in print version] colored mycelia; (B) black colored microsclerotia bearing few mycelia.

at the site of inoculation on the stems of dogwood, which spread along the vertical and horizontal axis of stem (Mmbaga & Mackasmiel, 2018).

14.2.2 Pycniospores

Millions of clonal, asexual conidia (spores) are produced by fungal pathogen and remain dormant until the return of favorable conditions. Conidia synthesize and store transcripts in accordance with the environmental conditions prevalent prior to dormancy as a way of preparing for the future (Wang et al., 2021). *M. phaseolina* produces pycniospores within pycnidia which are formed naturally on some of the hosts during perfect stage of life cycle. Pycnidiospores are single celled, hyaline, aseptate, obovoid and measuring 13.83−17.28 × 5.39−7.28 μm (Meena & Kadam, 2021). *M. phaseolina* forms pycnidiospores on certain hosts which may enable potential aerial and seed-borne transmission (Pratt et al., 1998). Pycniospores produced in pycnidia also play an important role in dispersal and survival of the pathogen (Avilés et al., 2008). Dispersal of pycniospores is important for infection of new hosts and survival of pathogen through secondary infection. Hyperosmotic stress is one of the environmental stimuli that often trigger the condition favorable for *M. phaseolina* infection (Ghosh et al., 2018).

14.2.3 Pycnidia

Pycnidia of *M. phaseolina* are dark to greyish, globose, membranous, and 83.62−123.88 μm in diameter. Commonly, pycnidial stage is found in nature on host plant (Ma et al., 2010). Pycnidia immersed in the infected host tissues may provide conidial inoculum for secondary spread among host plants in the field. Ma et al. (2010) reported that the number of pycnidia produced by eight isolates of *M. phaseolina* were dependent on induction medium; however, peanut butter extract-saturated filter paper placed over soyanut butter extract agar induced the production of greater pycnidia and pycniospores than the other media. Further, pycnidia have also been produced on potato dextrose agar medium (Yang & Navi, 2005) as well as soybean meal agar medium (Machado, 1987). Moreover, large number of pycnidia have also been produced *invitro* by using sterilized pieces of host plant or oil seed extract, filter paper treated with peanut meal, and incubation under UV and long wave UV light (Dhingra & Sinclair, 1995).

Zhao et al. (2020) produced pycnidia of *M. phaseolina* on oatmeal agar medium under 365 nm UV light at 25°C for 120 h of incubation. The UV light affects the growth of *M. phaseolina* and induces formation of pycnidia and pycniospores *invitro*. The number of pycnidia and pycniospores produced varies with the isolates of *M. phaseolina.*

14.2.4 Sclerotia

The sclerotia are multicellular structure, each cell contains 1−3 nuclei, mitochondria, lipid droplets, etc (Wyllie & Brown, 1970). Sclerotia are dark brown to black, smooth, oval or round, made up of barrel-shaped cells (Fig. 14.1B). Size of sclerotia differs in different isolates from different regions of infected plants, that is, 85.4 × 125.7 μmat the surface and below the cortex and throughout the stalk in severe case (Amann, 1956) and 80−165 μm diameter on root and stem bases (Meena & Kadam, 2021; Sangawan et al., 2005). Melanin pigment is present on the cell wall of the sclerotia, which provides resistance against chemicals and biological disintegration. Biosynthesis of melanin is genetically controlled and is formed through polyketide pathway (Cordero & Casadevall, 2017).

Infection of *M. phaseolina* usually originates from microsclerotia present in soil or from pycniospores, produced on infected plant tissues or residue, that are splashed onto host plant parts during rain or irrigation (Twizeyimana et al., 2012). Microsclerotia in soil, infected seeds, or hosts tissues serve as primary source of inoculum. Root exudates induce sclerotia germination and infection of host root. The infective hyphae enter the plant through epidermal cells of host root or injuries caused mechanically or by living organisms. After establishment of pathogen, the mycelium penetrates the root epidermis restricting primarily in the intercellular spaces of the cortex of the primary roots. As a result, adjacent cells collapse and heavily infected plant die. In later phase, hyphae grow intracellularly through the xylem and form microsclerotia that plug the vessels of vascular bundles (Ghosh et al., 2018; Short et al., 1978). *M. phaseolina* populates roots and stems of young plants producing many sclerotia; finally, the infected plants collapse, and sclerotia persist with high survivability (Elad et al., 1986). Microsclerotia within the xylem reduce water transport leading to wilting of the leaves and reduced plant vitality (Laluk & Mengiste, 2011). Sclerotia produced in the parasitized host tissues function as units of long-term survival in soil and primary inoculum for root infection (Pratt et al., 1998).

14.2.4.1 Factors affecting sclerotia formation

There are many factors that affect sclerotia formation. Light and dark regimes slightly affect sclerotia formation in *M. phaseolina*. A slight difference in sclerotia formation was noted in the cultures incubated in light and dark regime. Further, intense sclerotia were formed at pH 3.3–7.8 and 8; however, sclerotia formation declines with production of larger sized ones. Moreover, intense sclerotia formation has been reported at higher concentration of sugar. The optimum temperature of sclerotia formation has been recorded as 31°C (Bhargava, 1965). Further, production and replacement sclerotia in soil depend on microbial activity, which in turn, is governed by soil temperature, moisture, pH, and organic matter (Lodha & Mawar, 2020).

14.3 Phases of survival

There are two phases of survival of *M. phaseolina*, active phase and passive (inactive or dormant) phase (Fig. 14.2).

14.3.1 Survival during active phase (through disease development)

Any microorganism grows and reproduces fast during favorable environmental conditions, such as availability of optimum moisture, temperature, organic materials, pH, and presence of the other organisms. This results in an increase in the number of fungal propagules, called inoculum. The inoculum density is the number of potentially infective propagules available per gram soils, and the effective inoculum density characterizes the competitive saprophytic ability (CSA) of a parasite. Therefore, for prediction of disease, ecological conditions favorable for saprophytic growth of such fungi are more important than their initial inoculum density. This is much less evident for root infecting fungi. However, disease severity is more closely correlated with CSA than the inoculum density at a time (Bouhot, 1979). The degradative capacity of a saprophyte or enzymatic potential, which involves the ability of an organism to degrade organic materials in its environment, is the key features for survival as saprophytes or parasites.

M. phaseolina produces cellulolytic and pectolytic enzymes (Dubey & Dwivedi, 1988b) and toxins (Dhingra & Sinclair, 1974b), which play a major role in organic matter decomposition. However, the saprophytic activity of *M. phaseolina* does not effectively increase its inoculum density in soil. Bohra and Lodha (1998) studied the interactive effects between *M. phaseolina* and *Fusarium solani* at different inoculum levels on seedling mortality of Jojoba. The pathogens and seedling growth were drastically affected with increase in inocula. Moreover, there are different factors that affect the survival of *M. phaseolina* during active phase. Presence of crop residues may alter the survivability of *M. phaseolina* under varied environmental conditions, such as soil temperature, moisture, and microflora.

14.3.1.1 Soil moisture

Saprophytic survival of *M. phaseolina* was found the maximum at low soil moisture content, i.e., below 40% moisture holding capacity (MHC) upto 10% MHC, which decreased with increasing moisture content (60%–80% MHC) (Sekhar et al., 1987) with considerable reduction in half-life at 80% MHC (Satischandra et al., 1980). Low saprophytic activity at

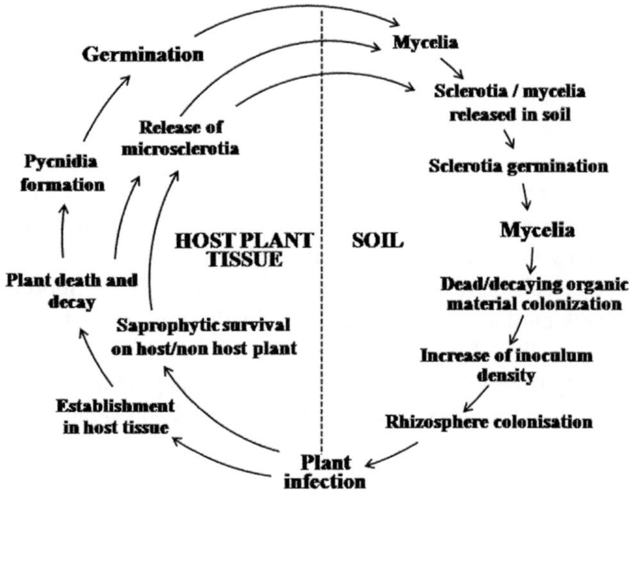

FIGURE 14.2 Life cycle of *Macrophomina phaseolina*.

high moisture level has been attributed to anaerobic condition and accumulation of carbon dioxide (Dhingra & Sinclair, 1975). Lodha (1996) established an inverse correlation between soil moisture and *M. phaseolina* population in soil.

14.3.1.2 Soil temperature

Maximum saprophytic activity and survival of the fungal pathogen have been found between 30 and 35°C (Byadgi & Hedge, 1988). The fungus cannot grow at its optimal temperature in culture or under sterile conditions without significantly increasing the density of its inoculum (Dhingra & Sinclair, 1973). *M.phaseolina* grows actively and colonizes substrates when conditions are cool (15–20°C) and moist (15%–25% MHC) or dry and warm (Dhingra & Chagas, 1981). Hopefully, these circumstances would apply during some seasons in tropical locations and in places with little rainfall. Soil temperature and moisture content seem to be the main determinants of saprophytic activity of *M. phaseolina* in soil.

14.3.1.3 Organic amendments

Presence of crop residues may alter the survivability of *M. phaseolina* under varied environmental conditions, such as soil temperature, moisture, and soil microflora. Several types of weeds have been identified as host of *M. phaseolina* and may help the survival between successive crops (Fuhlbohm et al., 2012). It has been found that oil cakes of margosa (neem), cotton, castor, groundnut, and mustard suppressed the saprophytic activities and survival of *M. phaseolina* because organic amendments in soil cause an increase in carbon and nitrogen ratios based on substrates in soil resulting in enhancement of soil microbial activities. Consequently, active growth of *M. phaseolina* is declined due to production of inhibitory microbial products in soil. Secondly, presence of inhibitory chemicals in organic materials also poses toxic effect on *M. phaseolina*. Dubey and Kumar (2003) reported the presence of good fungicidal properties in a neem-based product "azadirachtin," which was as effective as the fungicides mancozeb and bavistin. Therefore, addition of bark, cake powders and oil of neem drastically reduced the germination of sclerotia of *M. phaseolina* (Dwivedi & Chaube, 1985).

14.3.1.4 Inorganic chemicals

Colonization of dead host material in soil by *M. phaseolina* is inhibited by boron and zinc elements upon incubation for long period in soil. However, manganese, zinc, and boron favored the growth of soil microflora (Radha, 1956). Addition of salts, such as Na_2So_4, NaCl, and $CaCl_2$, in soil on variation in pH did not affect the saprophytic activity of the fungus (Satischandra et al., 1980). Lodha and Burman (2000) suggested the beneficial effect of nitrogen fertilizers in reducing the saprophytic activity and decreasing inoculum density of *M. phaseolina* in the soil.

14.3.2 Passive phase (through microsclerotia production)

The gradual shift from a parasitic (active mycelial) to a nonparasitic (inactive sclerotial) existence of *M. phaseolina* in soybean may be an indicative of moribundicity of the host (Short et al., 1978). Sclerotia are formed at certain conditions by mycelial aggregation either on dead substrates saprophytically or in the infected host tissue after pathogenesis. The mycelia penetrate the root epidermis of the susceptible hosts but are restricted to the intercellular spaces of the cortex; later on, it gets established inside the xylem vessels, parenchyma, and pith to produce intraxylem sclerotia (Dubey, 1986). Fully developed sclerotia are metabolically quiescent propagules, which no longer live at the expense of the host (Chet & Henis, 1975). Therefore, upon death and decay of infected host tissues by microorganisms, the sclerotia are gradually released into the soil (Cook et al., 1973). Infected host tissue serves as reservoir of inoculum which gives energy and protection; hence, the sclerotia are the major source of primary inoculum.

14.3.2.1 Structure and type of sclerotia

The outer layer of cells of sclerotia is surrounded by a cortex of hyphae with thicker walls. In plate cultures, sclerotia are formed by intertwining of primary branches originating from major hyphal filament. Adjacent cells enlarge to become globose and slightly compact into a single unitized structure of which the cells are cemented together by the gelatinous matrix (Wyllie & Brown, 1970). Sclerotia formation has been grouped into five different types on the basis of their mode of formation, occurrence, and morphology (Reicher & Hellinger, 1947): type 1, barrel-shaped cells along which a single hypha arise; type 2, sclerotia formed from two adjacent hyphae; type 3, sclerotia arising from the normal cells of a single hypha, type 4, sclerotia formed from two adjacent hyphae; and type 5, sclerotia formed from barrel-shaped cells of one hypha fusing with normal cells of a second one. The number of cells per sclerotium is directly related to its size, which in turn depends on nutrient availability of the substrate where the propagules are produced. Large sclerotia produce considerably more germ tubes than the small ones when germinate on culture media (Short & Wyllie, 1978).

14.4 Survival of sclerotia

Germination of *M. phaseolina* sclerotia is also affected by root exudates (Smith, 1969). After establishment, sclerotia are produced in plant tissues during parasitic phase of life. They remain entangled with mycelia integrated with host cells. After plant death, the host parts are left in the field that are subjected to microbial decomposition. Consequently, sclerotia are released in soils that survive for long period, or germinate during favorable conditions and produces secondary sclerotia utilizing nourishment from the dead organic materials (Table 14.1).

14.4.1 Survival inplant tissue

M. phaseolina has unique features to infect the host tissues and get established in the vascular bundles producing intra-xylem sclerotia. Thus sclerotia are embedded beneath the epidermis and appearing upon it also. The survival of intraxylem sclerotia depends on the duration how long they remain inside host tissues as well as biotic and abiotic factors.

14.4.2 Survival in soil

Sclerotia are produced in host tissues during the parasitic phase and discharged into the soil after the tissue decomposition in order to survive in soil (Dhingra & Chagas, 1981). The two most crucial parameters that impact the viability of *M. phaseolina* sclerotia are temperature and soil moisture. *M. phaseolina* quickly grows in the infected plants and produces a large amount of microsclerotia that plug the vessels causing the plant to wilt and die. The negative effects of high soil moisture at low temperatures, and low soil moisture at high temperatures may be applicable in the field (Islam et al., 2012). Sclerotia survive in soil and in host tissues for about 3 and 16 years, respectively. Soil fumigation that reduces inoculum levels of sclerotia in soil has been widely practiced all over the world (Chamorro et al., 2016). Sclerotia are released into soil upon death and decay of the host tissues, which may involve a passive inactivity imposed by the environmental factors or a dormancy governed by the physiology of the resting structures.

TABLE 14.1 Record of survival of *M. phaseolina* propagules in soil and host tissues.

Surviving structures	Sites	Duration of survival	References
Mycelia	Normal field/soil	7 days	Dhingra and Sinclair (1978)
	Seeds	20 weeks	Singh and Kumar (2002)
	Laboratory dry conditions	Several months	Wyllie et al. (1984)
	Host tissue	2 years	Baird et al. (2003)
	Normal Soil	More than 5 years	Short et al. (1980)
	Host tissue	Several years	Dhingra and Sinclair (1977)
Sclerotia	Normal soil	8 weeks	Baggio et al. (2019)
	Normal soil	8 months	Holliday and Punithalingam (1970)
	Normal soil	1 year	Papavizas (1977a)
	Sorghum residues	16 months	Cook et al. (1973)
	Potato tubers	18 months	Bhargava (1965)
	Field soil	21 months	Songa and Hillocks (1996)
	Host tissues	3 years	Watanabe (1973)
	Host tissue and soil-formed sclerotia	4 years	Watanabe (1973)
	Normal soil	15 years	Ghosh et al. (2018)
	Normal soil	15 years	Kaur et al. (2012), Gupta et al. (2012)
	Host tissues	16 years	Young et al. (1982)

The most striking feature of survival of the fungus is the rapid production of secondary sclerotia on dead organic matter which could presumably extend its longevity in the absence of suitable substrates. A multiplicity of factors, such as dead or living structure, function of the surviving propagules and the environmental conditions, influence the survival of the fungus in nature. Some climatic, edaphic and host factors govern repetition and infectivity, multiplication, and the survival of the fungus. Therefore, knowledge about these factors may be helpful for control of *M. phaseolina* (Dhingra & Sinclair, 1975).

14.4.3 Duration of survival

The incidence of diseases caused by *M. phaseolina* has increased in recent years because of low rainfall and increasing temperature (Deshmukh & Tiwari, 2021). Longevity of sclerotia survival also depends on the amount of melanin pigment produced by the fungal cells. However, in vitro studies with melanin substrates and inhibitors indicate that the sclerotia-producing fungi carry out typical pentaketide reaction and synthesize melanin from 1,8 dihydroxynaphalene or related metabolites (Wheeler, 1983). Furthermore, soil amended with poultry manure and sawdust probably suppress the growth of *M. phaseolina* due to increase in C:N ratio (Sheikh & Ghaffar, 2004). But in the infected plants tissues, mycelia survive for 2 years (Baird et al., 2003) and sclerotia survive for more than 16 years (Young et al., 1982).

14.4.4 Ability of survival

During saprophytic phase when the living hosts die, survivability of *M. phaseolina* depends upon the inherent potential of the fungus to compete with other microorganisms and derive nutrition for its growth and multiplication. It depends on the following two factors:

14.4.4.1 Competitive saprophytic colonization

Competitive saprophytic colonization (CSC) refers to the colonization of substrates (leaves and root/stem pieces) that has been already colonized by other microorganisms. CSC is estimated by "Cambridge method." Dhingra et al. (1976) reported that *M. phaseolina* had a high CSC at specific inoculum density. CSC of *M. phaseolina* declined after adding the organic matter in soil. However, the presence of crop residues may alter the survivability of *M. phaseolina* under varied environmental conditions, such as soil temperature, moisture, and soil microflora. Wyllie and Calvert (1969) have found that sclerotia formation in soybean tissues was conditioned solely by flowering and pod set; the other variables, such as soil temperature, moisture, etc. were subordinate in their effect. The mycelia survive in soil only for a short duration depending upon the environmental and biological factors.

14.4.4.2 Competitive saprophytic ability

Competitive saprophytic ability (CSA) is the summation of all physiological characteristics that make a saprophyte successful in competition for saprophytic colonization of substrates. *M. phaseolina* is a soil-borne facultative parasite of low saprophytic ability and a poor to moderate competitor (Dhingraet al., 1976). The competitive antagonistic saprophytic hyperparasites or other soil microorganisms can easily inhibit its growth (Dhingra & Sinclair, 1978). Further, Dhingra and Chagas (1981) found that *M. phaseolina* possesses a high degree of strong CSA and the maximum substrate colonization at 25% MHC. But substrate colonization decreases with increasing soil moisture. Increase of nitrogen in soil completely inhibits saprophytic colonization of pathogen. Reis et al. (2014) reported that *M. phaseolina* did not colonize saprophytically the dead pieces of *Avena sativa, A. strigosa, Brassica napus, Gossypium, Hordeum vulgare, Helianthus annus*, etc.

M. phaseolina has been found to have vigorous saprophytic ability and inoculum potential of cotton root-infecting cellulose decomposing fungus, and a primary and dominant colonizer of plant debris in soil (Sabet & Khan, 1969). The difference in behavior is exhibited by soil fungi to multiply by producing the dormant propagules (secondary sclerotia) to such an extent that it attains the inoculum density at disease causing levels. It has been proposed that inoculum reduction or removal in cultivated soils is not practical for farmers who do not quit producing sensitive species because the inocula are preserved, increased and replaced in cultivated and/or weed hosts season after season (Reis et al., 2014).

14.5 Factor affecting survival

There are several abiotic and biotic factors that affect the survival of *M. phaseolina*.

14.5.1 Abiotic factors

Survivability and duration of *M. phaseolina* for an extended period of several months or years depend on abiotic and biotic factors of soil. A variable duration of survival, for example, 10 months under dry storage condition and 2 months under dry to moderately moist condition has been recorded (Dubey & Upadhyaya, 2001).

14.5.1.1 Soil moisture and drought

Soil moisture plays an important role in the survival of the fungus in soil or in host tissue buried in soil. Dhingra and Sinclair (1974a) reported the reduced number of germinable sclerotia in soybean stem pieces after 3–4 weeks at 30°C in soil at 60%–100% MHC. Incubation of sclerotia in moist sandy loam soil (50% MHC) for 20 weeks reduced survivability by 43% (Papavizas, 1977a). Sclerotial population declined by 96%–99% at 66%–100% MHC as compared to population in dry soil (Dhingra & Sinclair, 1975). High moisture content did not support viability in soil (Gangopadhyay et al., 1982). Furthermore, the recoveries of the fungus from the infected soybean stem pieces decrease with increasing in the soil moisture from 10% to 20% MHC (Dhingra & Sinclair, 1974a). Survival of the fungus was adversely affected at high soil moisture (up to 80% MHC), but it survived well below 40% MHC of soils. Half-life period was considerably reduced at 80% MHC (Satischandra et al., 1980). Sclerotia are supposed to be directly parasitized by soil bacteria at high moisture levels (Dhingra & Sinclair, 1974a). Shokes et al. (1977) also reported the effect of water potential on the growth and survival of *M. phaseolina.*

In the presence of moisture, sclerotia exude nutrients; the major constituents of the sclerotial exudates are glucose and other sugars, polyols, amino acids, amino sugars and sugar-alcohols (Bristow & Lockwood, 1975). Sclerotia of *M. phaseolina* restore sufficient energy in term of nutrients that keep them viable and in turn help in germination and extension of numerous primary and secondary mycelia. The organic substances exuded on the surface of microsclerotia attract bacteria and other antagonists. Further, the bacteria accumulated in the sporosphere exhaust the available nutrients in soil and establish mycostasis with the consequences of complete colonization and lysis of *M. phaseolina* (Marquez et al., 2021). Srivastava et al. (1996) evaluated the colonization of sclerotia of *M. phaseolina* by bacteria and fungi in natural field soil at different moisture levels and temperature. They reported *Trichoderma harzianum* and *Pseudomonas fluorescens* as the most aggressive parasite of sclerotia which declined its germination by 60%–63%. The antagonists of sclerotia parasitize it by their agglutination ability with sclerotia.

Oxygen concentration of soil also affects viability of *M. phaseolina* sclerotia. Sclerotia did not remain well at high bulk density due to reduced level of oxygen. In contrast, the propagules got increased in noncompacted soil. Hence, high bulk density, high soil moisture, and reduced oxygen levels contribute significantly to reduce the survivability of *M. phaseolina* in soil on a long term basis. Increase in population in noncompacted soil facilitates porosity in soil. Therefore, enhanced sclerotia population in field as a result of conventional tillage practices has been attributed to high porosity, low bulk density, and high oxygen levels in soil (Gangopadhyay et al., 1982).

In vitro studies with sclerotia have revealed that drying and wetting adversely affect their survivability. Dry sclerotia exude more nutrients than nondried ones which enhanced the microbial activity around the sclerotial surface. These nutrients act as energy sources for microorganisms that reside near sclerotia; in turn, it strengthens microbial potential to break the defense mechanisms of the sclerotia because of being parasitized by aggressive microorganisms (Reddy et al., 1989). More sclerotia were colonized by fungi, such as species of *Fusarium, Penicillium, Trichoderma,* and *Verticillium* in alkaline than the acidic soil. Increase in loss of viability and decline in germinability of sclerotia got increased with relative humidity. Alternate drying and wetting of soil containing sclerotia did not affect their survivability (Papavizas, 1977a, 1977b). Moreover, in dried culture sclerotia can remain viable for about 1 year.

14.5.1.2 Soil temperature

Temperature plays an important role in the longevity and survival of sclerotia of *M. phaseolina.* Sclerotia suffer from thermal inactivation; for example, sclerotia survive at 50°C for 24 h, at 55°C for 8–12 min and at 60°C for 30 s (Bega & Smith, 1962). Time-temperature relationship for the inactivation and viability of sclerotia in wet and dry soils under elevated temperature in constant cycle or short term regimes was studied by Sheikh and Ghaffar (1987). They found that all the sclerotia lost its viability at a constant temperature of 55°C for 1 day, 55°C for 3 days or 1-h cycle for 14 days at 55°C. Moreover, all the sclerotia died in dry soil within 7 days when exposed to 60°C for 2 h daily but survived for 14 days at daily exposure of 55°C. Further, in dry soil, sclerotial treatment of a constant temperature of 60°C for 7 days or 2 h/day cycle for 14 days reduced the viability by 88% and 76%, respectively. Mihail and Alcorn (1984) reported that 50°C for 48 h is required for the death of *M. phaseolina* sclerotia in laboratory. Thermal inactivation of *M. phaseolina* in precolonized stem pieces of potato and strawberry was 10 min at 55°C. Time-temperature period is apparently a critical point, but the thermal death range is more useful concept than the thermal death point (Bega & Smith, 1962). With increasing the age of cultures, development of tolerance against the temperature has also been reported by Paharia and Sahai (1970). They found that *M. phaseolina* can tolerate temperature range up to 50–52°C for about 120 h in steam sterilized soils.

14.5.1.3 Amendment with heavy metals

Heavy metals have detrimental effect on the survivability of sclerotia. About 25% sclerotia of the *M. phaseolina* have been found viable in the aqueous solution of cadmium chloride at 500 μg/mL concentration for 4 days *in vitro* condition (Dubey & Dwivedi, 1985). The viability of propagules gradually declined in soils treated with the salts of cadmium, nickel and zinc for long-term incubation. Cadmium has lethal effect at 4000 μg/kg concentration; none of the propagules were recovered from the amended soil. Zn had least effect on the survival (Dubey & Dwivedi, 1988a). Further, the toxic effect of cadmium on growth of *M. phaseolina* has been found declined after addition of an ore kaolinite. The toxic effect of heavy metals was also reported at different pH and addition of zinc and manganese in the growth medium (Dubey & Dwivedi, 1985). However, survival of the fungus was unaffected in the infected host tissues by the amendment of soil with boron, manganese, and zinc (Radha, 1956).

14.5.1.3.1 Organic amendments (oil, cake, leaves, and bark)

Water-/solvent-soluble extracts from different parts, for example, leaf, seed, bark, and fruit of some higher plants have shown fungicidal properties against the survival of the sclerotia of *M. phaseolina* (Dubey & Dwivedi, 1991). A remarkable decline in the viability of sclerotia treated with leaf distillate of *Eucalyptus globulus,* neem oil cake extracts and neem oil in *in vitro* was reported by Dwivedi and Dubey (1986). The significant effect of poultry manure and sawdust on the reduction in survivability of sclerotia of *M. phaseolina* in soil has been reported by Sheikh and Ghaffar (2004). A significant reduction in survival of sclerotia treated with extracts of *Acacia arabica, Allium cepa,* and *A. sativum* was also recorded by Dubey and Dwivedi (1991). Dubey (1991) observed the most drastic effects of essential oils of *Citrus medica, Ocimum canum,* and *Pinus roxburghii* on viability of sclerotia.

Dubey et al. (2009) found that neem (*Azadirachta indica*) extract inhibited the growth and survival of sclerotia after 2 and 4 days of incubation. Sclerotia treated with neem oil failed to germinate showing 100% inhibition. Oil cake extract resulted in the maximum inhibition in sclerotia germination after 5 days followed by leaf and bark extracts. Volatiles caused more effects than the nonvolatile chemicals inhibiting 97% sclerotia germination after 5 days than the nonvolatiles of cake extract that gave negligible effects even after 2 days of incubation.

14.5.1.3.2 Amendment with pesticides and herbicides

Pesticides have been used since long time for control of diseases caused by *M. phaseolina*. Pesticides are used in different ways, e.g., soil application, seed treatment, foliar application, etc. Soil application with benomyl, brassicol (PCNB), thiram, captan, ferbam, plantavax, dithane M-45, etc. effectively controlled the pathogen either as soil drench method (Shanmugam & Govindaswamy, 1973) or as furrow application (Fulton et al., 1956). Benomyl has been found the best fungicide for the control of charcoal rot, where sclerotia population was reduced to maximum level resulting in decline in soybean seedling infection in field (Ilyas et al., 1975). Moreover, sclerotia gradually developed tolerance against the toxicity of fungicides, but survivability of such fungicide-tolerant sclerotia declined by 25%–58% as compared to non-tolerant one in control conditions (Pan & Sen, 1983).

Dwivedi and Dubey (1987a) have found that bavistin at 250 μg/g level completely eliminated the propagules (both sclerotia and mycelia) of *M. phaseolina* in soil after 30 days, while dithane M-45, and brassicol did the same at 1000 μg/g level. These fungicides were also effective in controlling the root rot of soybean seedlings in pot experiment (Dubey, 1986). Agrawal and Nema (1989) found that carbendazim reduced seed and seedling mortality caused by *M. phaseolina* significantly resulting in increased yield and control of leaf blight in black gram and green gram.

Several researchers have also focused their work on the effect of herbicides on survival of inoculum of *M. phaseolina*. A drastic change in the inoculum density of *M. phaseolina* has been noted on the treatment with herbicides viz., EPTC, dinoseb, flurodifen, and fluometuron. The population declined by 96% in sandy loam clay soil and 61% in sandy loam soil by dinoseb in 10 days after the treatment, which showed an interaction between herbicides and soils (Filho & Dhingra, 1980). Mengistu et al. (2013) studied the effect of glyphosate on disease severity of soybean in field caused by *M. phaseolina* because glyphosate is used to control weeds in agriculture field. They also found that application of glyphosate on glyphosate-resistant plant did not increase or decrease the severity of charcoal rot diseases. Further, it has been reported that *M. phaseolina* colonizes soybean root differently with different stresses induced by different herbicides. For example, stress induced by alachlor reduced root colonization but glyphosate and vernolate induced stress has no effect on root colonization (Canaday et al., 1986.). Very recently Rastija et al. (2022) reported coumarinyl Schiff bases (synthetic pesticide) as the efficient antifungal agents against *M. phaseolina.*

14.5.1.4 Soilsolarization (plasticulture, soil mulching, soil trapping, and solar heating of soil)

There are many conditions that exert stress on the host plants viz., high soil temperature, low moisture, plant wounds and flowering, etc. Solar pasteurization of soils or soil solarization for disease control is a nonchemical and economical technique that is accomplished by covering the moist soil with a transparent and thin (30–50 μm) polyethylene sheet in tropical countries where soil temperature increases up to 40–45°C in summer (Katan, 1980). By employing soil solarization, solar heat treatment naturally or in artificially infested soil reduced the inocula of *M. phaseolina* to the maximum level but not completely controlled to undetectable level due to survival of some sclerotia in soil (Dwivedi & Dubey, 1987b). In solarized soil, general soil microflora vulnerable to high temperature were also killed; only thermostatic and/or thermotolerant actinomycetes and bacterial antagonists survive indefinitely which rapidly colonize the pasteurized soil (Dwivedi & Dubey, 1987b).

Sclerotia survival depends on duration and persistence of soil temperature. The maximum temperature increases to 46–56°C by using 45 μm thick polythene sheet at 1 cm depth of field soil (Katan, 1980). Persistence of 50°C temperature for 48 h in laboratory killed *M. phaseolina* sclerotia, which were never attained during solarization in any tarp at any depth. This indicates that reduction in population achieved by soil solarization is nothing but a function of temperature. Temperature failed to exceed thermo-tolerance capacity of the fungus and allows *M. phaseolina* to remain alive at this temperature range. Moreover, the low natural population of *M. phaseolina* did not vary consistently with respect to time, depth, or treatment, and the residual inoculum has been found sufficient to cause 15%–99% loss in *Euphorbia lathyris* plants (Mihail & Alcorn, 1984). The influence of solarization on survival of *M. phaseolina* sclerotia has also been recently reported by Lodha and Mawar (2020).

Sheikh and Ghaffar (1984) have recorded the complete control of *M. phaseolina* sclerotia in artificially infested wet soil (not in dry soil) at 5 cm depth by mulching for 1 week. At this depth, temperature of the moist soil got increased from 37°C (unmulched) to 52°C (mulched soil). Mechanical composition of soil, thickness of polyethylene, duration of solarization, brightness of light, etc. govern the success of solarization. An integrated control of *M. phaseolina* employing solarization plus fungicide amendment has also been reported. Solarization in fungicide-amended soil gave the better result than the unamended one (Dubey, 1992). Solar heating of soil in summer by using transparent polyethylene mulch operates its function by: (i) heat killing of temperature sensitive soil-borne pathogens and pest, (ii) growth suppression by restoration of volatile substances, and (iii) increasing the thermophilic and/or thermo-tolerant antagonistic microorganisms in the solarized soil (Dwivedi & Dubey, 1987a, 1987b). Similarly, Israel et al. (2005) also reported the effects of soil solarization combined with crop residue incorporation, summer irrigation and amendment of biocontrol agent singly or in combination on survival of *M. phaseolina.* They found a significant reduction in diseases in residues and biocontrol agent–amended field with or without soil solarization as compared to nonamended control plots. With one summer irrigation, the disease indices in plots with a mix of mustard residues and oil cake amendment were at par with those obtained with the application of polyethylene mulching. Therefore, soil solarization is an effective, economical, eco-friendly, and nonchemical method which can be used by marginal farmers.

14.5.2 Biotic factors

There are different biotic factors that affect the survival of *M. phaseolina.*

14.5.2.1 Antagonistic fungi

Antagonism is a balancing wheel of nature, which operates with three spokes viz., amensalism, competition, and parasitism. Antagonism occurs when the pathogen stops growing after coming in contact with the antagonist. This type of contact supports suppression of pathogen where its mycelium begins to lyse backwards or the antagonist may continue its growth over the colony of the mycelium of the pathogen. Consequently, several mycelial morphological deformities occur, such as vacuolation of cytoplasm, hyphal perforation, bulging, lysis and bursting of hyphae, etc. Earlier, vacuolation and bursting of immature hyphae have been observed Dubey and Dwivedi (1988b). Biological control of root rot disease of jute by *Aspergillus versicolor* has been reported by Bhattacharyya et al. (1985). In pot culture experiment, they reported 50% decline in infection of jute by the pathogen. However, seed dressing with *Arachniotus* sp. or *A. aculeatus* control pre-and postemergence damping-off of urad bean with a total stand of 73% and 69%, respectively (Dhingra & Khare, 1973). Baird et al. (2003) reported the involvement of *Trichoderma* spp. in direct tissue degradation and nutrient depletion of the root segments; it may also act as a mycoparasite for reducing the survival of the pathogen and lowering the population density directly or indirectly and also decreasing the relative longevity of soil-borne pathogens, including *M. phaseolina*. Similarly, Triveni et al. (2015) evaluated *Trichoderma*- and *Anabaena*-based biofilms as multifunctional agents in *M. phaseolina*—infected cotton filed. *Trichoderma* spp. were found as potential biocontrol agent against *M. phaseolina* (Khan et al., 2021; Shahid & Khan, 2016).

Reduction of *Sclerotium bataticola* (syn. *M. phaseolina*) infection has been achieved with the use of fragmented mycelium of *Aspergillus flavus* on peanut shells and kernels (Jackson, 1965). Dubey and Dwivedi (1988b) have observed coagulation of cytoplasm, hyphal discoloration, cell wall dissolution, etc. of *M. phaseolina* hyphae due to metabolites secreted by *A. flavus, A. niger, Penicillium citrinum,* and *Trichoderma viride.* Hyphal coiling of *M. phaseolina* mycelium causing root-rot*in vitro* by *A. flavus, Penicillium* spp., and *T. harzianum* has also been observed (Hyder et al., 2022). Seed treatment with biocontrol fungi was found to be improved by adding different amendments to stimulate antagonist activity (Lamichhane et al., 2022; Prasad et al., 2020) (Table 14.2).

14.5.2.2 Antagonistic bacteria

Large number antagonistic bacteria, such as *Agrobacterium radiobacter, Bacillium subtilis, B. polymyxa, Erwinia herbicola, Pseudomonas fluorescens, P. aeruginosa, P. putida,* and *Xanthomonas malvacearum* are attracted toward the exudates of *M. phaseolina* in vitro and in soil through water-filled pores and cause destruction of pathogen (Kumar et al., 2012a, b; Marquez et al., 2021). Singh et al. (2008) reported that rhizospheric *B. subtilis* BN1 not only controls *M. Phaseolina* ecofriendly but also enhances the growth of *Pinus roxburghii.* Chemotactic fluorescent *P. aeruginosa* PN1 inhibited the charcoal rot in chir pine (Singh et al., 2010). *P. aeruginosa* LES4 in combination with reduced dose of chemical fertilizers not only control *M. phaseolina* but also increase growth and yield of *Sesamum indicum* (Kumar et al., 2009). Dubey et al. (2012) found that the culture filtrates of plant growth promoting *Bradyrhizobium* sp. (*Vigna*) VR1 and VR2 inhibited the growth and sclerotia germination of *M. phaseolina.* Several abnormal structures and deformities in fungal hyphae and sclerotia viz., hyphal fragmentation, valuation, cytoplasm coagulation, loss of vigor of sclerotia along with loss of melanin pigments and failure of sclerotia formation and germination were also observed. Recently, Sarli et al. (2022) showed that *Serratia proteamaculans* 136, 137, and *Burkholderia gladioli* MB39 enhanced plant growth promotion of soybean and strongly inhibited *M. phaseolina* infection by producing different antifungal compounds. Substantial work done on this aspect is summarized in Table 14.2.

14.5.2.3 Antagonistic actinobacteria

Several antagonistic actinobacteria associated with nonrhizosphere, rhizosphere, and rhizoplane regions of plants have been tested by various workers *in vitro* and in soil for their antagonistic potential against *M. phaseolina* (Díaz-Díaz et al., 2022; Dubey & Dwivedi, 1988b; Manikandan et al., 2022; Shrivastava et al., 2017). Antagonistic properties and control of collor rot of okra by *Streptomyces* sp. studied by Goel and Mehrotra (1974). The cotton seedling infection caused by *M. phaseoli* has been significantly controlled by *Streptomyces albus, S. griseus,* and *S. noursei* (Ghaffar, 1971). Shrivastava et al. (2017) recorded the biocontrol activities of *Streptomyces aureofaciens* K20 by damaging mycelia of *M. phaseolina* due to chitinase production. The ability to produce inoculum in excess and to survive, grow and proliferate in soil/seed surface is the desirable characters for being a successful antagonist (Baker & Cook, 1974). Some of actinobacterial antagonists have been summarized in Table 14.2.

14.5.2.4 Microbial products (antibiotics, antimicrobials, and biosurfactants)

Antibiosis is one of the major mechanisms of biocontrol of phytopathogens. Antibiotics act through several mechanisms, i.e., penetration of pathogen cell membrane and interaction with specific internal targets causing mycelial deformities and perforation resulting in leakage of intracellular contents. The effect of various antibiotics including agrimycin, ampicillin, griseofulvin, and tetracycline on growth of *M. phaseolina* has been studied and found that tetracycline was most effective in inhibiting the growth (73.9%) at 1000 μg/mL concentration (Dubey, 1989). Similarly, the antibiotics produced by *Pseudomonas* spp. and *Bacillus* spp. were found effective against *M. phaseolina* that reduced the growth by 61.13—69.62% over the control (Reetha et al., 2014). Khokhar et al. (2011) exploited *Penicillium simplicissimum* extracted molecules as an effective agent against *M. phaseolina* where it decreased the colony growth by 25%. The necrotrophic effect of peptide antibiotics induced abnormal contraction of cytoplasm, granulation and fragmentation of hyphal mycelia,

TABLE 14.2 Antagonists of *Macrophomina phaseolina.*

Antagonistic microorganisms	Mode of antagonism	References
Bacteria		
Azotobacter chroococcum TRA2	Growth inhibition	Maheshwari et al. (2012)
Bacillus atrophaeus	Reduced fungal growth	Salman et al. (2021)
B. cereus LPR2	Biological control	Kumar et al. (2020)
B. subtilis BN1	Chitinolytic enzyme chitinase and β-1,3-glucanase	Singh et al. (2008)
Bacillus sp. BPR7	Growth inhibition	Kumar et al. (2012b)
Bacillus sp. BSK5 and *B.subtilis* BSK17	Chitinase and β-1,3-glucanase	Dubey et al. (2013)
Brevibacterium sp. Hvs8 strain	Reduce plant mortality	Viejobueno et al. (2021)
Klebsiella pneumoniae HR1	Growth inhibition	Dey et al. (2019)
Pseudomonas aeruginosa PN1	Growth inhibition	Singh et al. (2010)
P. fluorescens	Growth inhibition	Manjunatha et al. (2013)
P. plecoglossicida and B. altitudinis	Biological control	Gopalakrishnan et al. (2011)
P. aeruginosa LES4	Growth inhibition	Kumar et al. (2009)
Paenibacillus polymyxa and *B.brevis*	Reduced disease severity in strawberry plants	Abd-El-Kareem et al. (2022)
Pseudomonas sp. PPR8	Growth inhibition	Kumar et al. (2016b)
Rhizobium leguminosarum RPN5	Biological control	Kumar et al. (2016c)
Stenotrophomonas sp. AG3 and *Exiguobacterium* sp. S58	Producing polyamines and lytic enzymes	Santos et al. (2021)
Actinomycetes		
Streptomyces griseus	Reduced infection on Sorghum	Gopalakrishnan et al. (2021)
S. tendae RDS16	Growth inhibition	Al-Askar et al. (2013)
S. aureofaciens K20	Growth inhibition	Shrivastava et al. (2017)
Streptomyces sp. S160	Reduced infection in chickpea	Yadav et al. (2014)
Actinomycetes spp.	Biological control	Saratha et al. (2018)
Actinoalloteichus cyano griseus	Reduced fungal growth in Mulberry	Saratha et al. (2022)
Streptomyces sp.	Biological control	Ruhangiz et al. (2021)
*S. rameus*GgS 48	Reduced root rot in Mung bean	Meena et al. (2022)

Fungi		
Aspergillusversicolor	Reduced pathogen by producing secondary metabolites	Khan and Javaid (2022a)
Trichoderma harzianum, Penicillium spp. and *A. flavus*	Reduced disease severity	Hyder et al. (2022)
A. fumigatus	Granulation of cytoplasm	Dubey and Dwivedi (1988a)
T. harzianum	Growth inhibition by hyphal disintegration	Shahid and Khan (2016)
Penicilliumitalicum	DNA cleavage of the fungal pathogen	Khan and Javaid (2022b)
T.harzianum	Biological control	Khalili et al. (2016)
Cephalosporium humicola	Growth inhibition	Dhingra and Khare (1973)
Trichoderma spp.	Inhibition of fungal growth	Khan et al. (2019)
Penicillium citrium	Growth inhibition	Dubey and Dwivedi (1988a)
T. harzianum and *T. viride*	Biological control	Kumari et al. (2022)
T.viride	Growth inhibition	Manjunatha et al. (2013), Khan et al. (2021)
Trichoderma sp. (TCBG-2) T. *koningiopsis* (TCBG-8)	Growth inhibition	Larralde-Corona et al. (2008)
T. harzianum	Biological control	Swehla et al. (2020)

and hyphal cell lysis of *R. bataticola* (Senthilkumar et al., 2007). The purified pyocyanin from *P. aeruginosa* TO3 not only enhanced the nodulation along with rhizobia but also leads to inhibition of *M. phaseolina* (Khare & Arora, 2011). Lorenzetti et al. (2017) studied the antimicrobial activity of two homeopathic drugs, Sepia and Arsenicum album, against *M. phaseolina* and recorded reduction in fungal growth up to 50% and 70% respectively, compared to control treatments. Derbalah et al. (2022) first time reported the inhibitory effect of Ag_2O nanoparticles as a new strategy to control *M. phaseolina* in Strawberry plants.

Biosurfactants are low-molecular-weight surface-active compounds, which are widely produced by microorganisms. Because of their natural good biodegradability, low toxicity, better foaming capacity, wide selectivity, and effectiveness in harsh situations, biosurfactants have recently attracted the attention for environmental protection, making them "green" chemicals. Surface tension at air/water interfaces and interfacial tension at oil/water surfaces are both decreased by these amphiphilic biomolecules (Santos et al., 2016). Based on their physico-chemical characteristics, biosurfactant agents can be classified as glycolipids, lipopeptides, neutral lipids, phospholipids, fatty acids, or polymeric ones (Gayathiri et al., 2022). The harsh surfactants now used in the pesticide, herbicide, and fungicide industries can be replaced by these biosurfactants since natural surfactants are used as a carbon source by soil-dwelling bacteria, which explains the biological elimination of surfactants from agricultural soil (Lima et al., 2011).

Enterococcus faecium produces glycolipid-type biosurfactant, which interferes with the microbial adhesion and showed antifungal activity (Sharma et al., 2015). They have numerous beneficial attributes, for examples they are be nontoxic, nonhazardous, biodegradable, environmental friendly, selective, and effective under extreme conditions. Microbial biosurfactants have numerous agriculture applications and unique surface-active properties. They have distinctive structures also which provide them new properties that eminent surfactants may lack (Abou & Seoud, 2008).

Biosurfactant producing antagonistic and plant growth promoting *Enterococcus* sp. BS13 inhibited hyphal growth and sclerotia formation of *M. phaseolina* causing degradation of hyphe, formation of halo cell and several deformities in mycelia causing the reduced incidence of charcoal rot with increase in growth of *Vigna mungo* (Kumar et al., 2016a). Kumar et al. (2021) reported that biosurfactant (cyclic siloxane) producing *Bacillus cereus* BS14 inhibited the growth of *M. phaseolina* and induced the growth of *V. mungo* under pot trial studies. Many biosurfactants of microbial origin have antimicrobial activity against plant pathogens; therefore, they are considered as a promising biocontrol molecules for achieving ecofriendly sustainable agriculture model (Crouzet et al., 2020; Khare & Arora, 2021; Pérez-García et al., 2011).

14.6 Conclusion

M. phaseolina is one of the most prevalent and serious phytopathogens that cause various diseases in several plant species globally. Its microsclerotia play a significant role in survival, multiplication, and infection of plants. The inoculum density directly relates to the infectivity and disease incidence. Disease severity is much more closely related to the CSC and CSA than to inoculum density. The rapid increase in population of sclerotia in soil occurs due to frequent mono-cropping with susceptible hosts, persistence of sclerotia for a long time in soil and in host tissue. Because of the several multiplicity factors, a single approach cannot be significant that affect the survival of *M. phaseolina*. It is highly difficult to eradicate the sclerotia even with high concentration of fungicides due to development of resistance. Therefore, integrated approach by adopting crop rotation, use of fertilizers, antagonists, solarization, and manipulation of soil conditions may be more fruitful for reducing the inoculum density to the maximum level.

References

Abd-El-Kareem, F., Elshahawy, I. E., & Abd-Elgawad, M. M. M. (2022). Native bacteria for field biocontrol of black root rot in Egyptian strawberry. *Bulletin of the National Research Centre, 46*, 82. https://doi.org/10.1186/s42269-022-00775-3

Abou, E. L., & Seoud, I. I. (2008). Phosphorus efficiency of *Tagetes* plant inoculated with two arbuscular mycorrhizal fungi strains. *Australian Journal Basic and Applied Science, 2*(2), 234–242.

Agrawal, S. C., & Nema, S. (1989). Effect of carbendazim on Macrophomina leaf blight of black gram and green gram. *Indian Journal of Plant Protection, 17*(1), 147–149.

Al-Askar, A. A., Rashad, Y. M., & Abdulkhair, W. M. (2013). Antagonistic activity of an endemic isolate of *Streptomyces tendae* RDS16 against phytopathogenic fungi. *African Journal of Microbiology Research, 7*(6), 509–516.

Amann, M. (1956). Ist neben *Colletotrichum atramentarium* (B. et Br.) Taub, auch *Macrophomina phaseoli* (Maubl.) Ashby am Zustandekommen der Gummiknollenwelke der Kartoffel beteiligt. *NachrBl. dtsch. PflSchDienst, Braunschweig, 8*, 25–27.

Ashraf, W., Sahi, S. T., Amer Habib, A., Khan, R., Zeshan, M. A., Intisar, A., & Ahmad, A. (2017). Sensitivity of *Macrophomina phaseolina* (Tassi) Goid. isolates of maize (*Zea mays* L.) to different temperature and pH levels. Asian Journal Agriculture and Biology, *5*(3), 133–139.

Avilés, M., Castillo, S., Borrero, C., Castillo, M. L., Zea-Bonilla, T., & Pérez-Jiménez, R. M. (2008). Response of strawberry cultivars: 'camarosa', 'candonga' and 'ventana' to inoculation with isolates of *Macrophomina phaseolina. VI International Strawberry Symposium, 842*, 291–294.

Baggio, J. S., Cordova, L. G., & Peres, N. A. (2019). Sources of inoculum and survival of *Macrophomina phaseolina* in Florida strawberry fields. *Plant Disease, 103*(9), 2417–2424.

Baird, R. E., Watson, C. E., & Scruggs, M. (2003). Relative longevity of *Macrophomina phaseolina* and associated mycobiota on residual soybean roots in soil. *Plant Disease, 87*(5), 563–566.

Baker, K. F., & Cook, R. J. (1974). *Biological control of plant pathogens*. San Francisco, C.A: WH Freeman and Company.

Bega, R. V., & Smith, R. S. (1962). Time-temperature relationships in thermal inactivation of sclerotia of *Macrophomina phaseoli. Phytopathology, 52*(7), 632.

Bhargava, S. N. (1965). Studies on charcoal rot of potato. *Journal of Phytopathology, 53*(1), 35–44.

Bhattacharyya, D., Basu, S., Chattapadhyay, J. P., & Bose, S. K. (1985). Biocontrol of Macrophominaroot-rot disease of jute by an antagonistic organism, *Aspergillus versicolor. Plant and Soil, 87*(3), 435–438.

Bohra, M. D., & Lodha, S. (1998). Effect of inoculum levels and interaction of *Macrophomina phaseolina* and *Fusarium solani* on root rot severity of jojoba seedlings. *Indian Phytopathology, 51*, 294–296.

Bouhot, D. (1979). Estimation of inoculum density and inoculum potential: Techniques and their values for disease prediction. In B. Schippers, & W. Gams (Eds.), *Soil borne plant pathogens* (pp. 21–34). New York, London: Academic Press.

Bristow, P. R., & Lockwood, J. L. (1975). Soil fungistasis: Role of spore exudates in the inhibition of nutrient-independent propagules. *Microbiology, 90*(1), 140–146.

Byadgi, A. S., & Hedge, R. K. (1988). Factors affecting survival of *Rhizoctonia bataticola* in soil. *Indian Phytopathology, 41*(1), 122–127.

Canaday, C. H., Helsel, D. G., & Wyllie, T. D. (1986). Effects of herbicide-induced stress on root colonization of soybeans by *Macrophomina phaseolina. Plant Diseases, 70*, 863–866.

Chamorro, M., Seijo, T. E., Noling, J. C., De los Santos, B., & Peres, N. A. (2016). Efficacy of fumigant treatments and inoculum placement on control of *Macrophomina phaseolina* in strawberry beds. *Crop Protection, 90*, 163–169.

Chet, I., & Henis, Y. (1975). Sclerotial morphogenesis in fungi. *Annual Review of Phytopathology, 13*, 169–192.

Cook, G. E., Boosalis, M. G., Dunkle, L. D., & Odvody, G. N. (1973). Survival of *Macrophomina phaseoli* in corn and sorghum stalk residue. *Plant Disease Reporter, 57*(10), 873–875.

Cordero, R. J., & Casadevall, A. (2017). Functions of fungal melanin beyond virulence. *Fungal Biology Reviews, 31*(2), 99–112. https://doi.org/10.1016/j.fbr.2016.12.003

Crouzet, J., Arguelles-Arias, A., Dhondt-Cordelier, S., Cordelier, S., Pršić, J., Hoff, G., et al. (2020). Biosurfactants in plant protection against diseases: Rhamnolipids and lipopeptides case study. *Frontiers in Bioengineering and Biotechnology, 8*(8), 1014. https://doi.org/10.3389/fbioe.2020.01014

Derbalah, A., Essa, T., Kamel, S. M., Omara, R. I., Abdelfatah, M., Elshaer, A., et al. (2022). Silver oxide nanostructures as a new trend to control strawberry charcoal rot induced by *Macrophomina phaseolina. Pest Management Science*. https://doi.org/10.1002/ps.7084

Deshmukh, R., & Tiwari, S. (2021). Molecular interaction of charcoal rot pathogenesis in soybean: a complex interaction. *Plant Cell Reports, 40*(10), 1799–1812. https://doi.org/10.1007/s00299-021-02747-9

Dey, S., Dutta, P., & Majumdar, S. (2019). Biological control of *Macrophomina phaseolina* in *Vigna mungo* L. by endophytic *Klebsiella pneumoniae* HR1. *Jordan Journal of Biological Sciences, 12*(2), 219–227.

Dhingra, O. D., & Chagas, D. (1981). Effect of soil temperature, moisture, and nitrogen on competitive saprophytic ability of *Macrophomina phaseolina. Transactions of the British Mycological Society, 77*(1), 15–20.

Dhingra, O. D., & Khare, M. N. (1973). Biological control of *Rhizoctonia bataticola* on urid bean. *Phytopathology, 2*(76), 23–29.

Dhingra, O. D., & Sinclair, J. B. (1973). Variation among isolates of *Macrophomina phaseoli* (*Rhizoctonia bataticola*) from different regions. *Journal of Phytopathology, 76*(3), 200–204.

Dhingra, O. D., & Sinclair, J. B. (1974a). Effect of soil moisture and carbon: Nitrogen ratio on survival of *Macrophomina phaseolina* in soybean stems in soil. *Plant Disease Reporter, 58*(11), 1034–1037.

Dhingra, O. D., & Sinclair, J. B. (1974b). Isolation and partial purification of phytotoxin produced by *Macrophomina phaseolina. Phytopathologische Zeitschrift, 80*(1), 35–40.

Dhingra, O. D., & Sinclair, J. B. (1975). Survival of *Macrophomina phaseolina* sclerotia in soil: Effects of soil moisture, carbon: Nitrogen ratios, carbon sources and nitrogen concentrations. *Phytopathology, 65*, 236–240.

Dhingra, O. D., & Sinclair, J. B. (1977). *An annotated bibliography of* Macrophomina phaseolina *1905–1975*. Vicosa, Brazil: Imprensia Universitaria, Universidade Federal de Vicosa.

Dhingra, O. D., & Sinclair, J. B. (1978). *Biology and pathology of Macrophomina phaseolina* (p. 166). Minas Gerais, Brazil: Universidade Federal de Vicosa.

Dhingra, O. D., & Sinclair, J. B. (1995). *Basic plant pathology methods* (2nd ed.). Boca Raton, FL: Lewis Publishers.

Dhingra, O. D., Tenne, F. D., & Sinclair, J. B. (1976). Method for the determination of competitive saprophytic colonization of soil fungi. *Transactions of the British Mycological Society, 66*(3), 447–456.

Díaz-Díaz, M., Bernal-Cabrera, A., Trapero, A., Medina-Marrero, R., Sifontes-Rodríguez, S., Cupull-Santana, R. D., et al. (2022). Characterization of actinobacterial strains as potential biocontrol agents against *Macrophomina phaseolina* and *Rhizoctonia solani*, the main soil-borne pathogens of *Phaseolus vulgaris* in Cuba. *Plants, 11*(5), 645. https://doi.org/10.3390/plants11050645

Dubey, R. C. (1986). *Biology of rhizospheric microfungi of soybean with emphasis on root rot caused by Macrophomina phaseolina (Tassi) Goid* (Ph.D. Thesis). India: Banaras Hindu University.

Dubey, R. C. (1989). Effect of some antibiotics on *Macrophomina phaseolina*, a charcoal rot fungus of soybean. *Hindustan Antibiotics Bulletin, 31*(1–2), 38–39.

Dubey, R. C. (1991). Fungicidal effect of essential oils of three higher plants on sclerotia of Macrophomina phaseolina. *Indian Phytopathology, 44*(2), 241–244.

Dubey, R. C. (1992). Effect of soil solarization on survival of *Macrophomina phaseolina* in fungicide amended and unamended soils. *Tropical Science, 32*, 275–279.

Dubey, R. C., & Dwivedi, R. S. (1985). Toxicity of cadmium on the growth of *Macrophomina phaseolina* causing root rot of soybean as influenced by kaolinite, pH, zinc and manganese. *Proceedings of the Indian National Science Academy, B51*, 259–264.

Dubey, R. C., & Dwivedi, R. S. (1988a). Effect of heavy metals on growth and survival of *Macrophomina phaseolina* (Tassi) Goid. *Biology and Fertility of Soils, 6*, 311–314.

Dubey, R. C., & Dwivedi, R. S. (1988b). Antagonism between *Macrophomina phaseolina* and some rhizosphere microfungi of soybean. *International Journal of tropical Plant Diseases, 6*, 89–94.

Dubey, R. C., & Dwivedi, R. S. (1991). Fungitoxic properties of some plant extracts against vegetative growth and sclerotial viability of *Macrophomina phaseolina. Indian Phytopathology, 44*(3), 411–413.

Dubey, P., Gupta, G. P., & Dubey, R. C. (2012). Culture filtrates of plant growth promoting *Bradyrhizobium* sp. (*Vigna*) strains VR1 and VR2 inhibit growth and sclerotia germination of *Macrophomina phaseolina in vitro. New York Science Journal, 5*(8), 1–9.

Dubey, R. C., Khare, S., Kumar, P., Dubey, P., & Maheshwari, D. K. (2013). Evaluation of diversity of bacilli from chick-pea rhizosphere by 16S ARDRA and assessment of their plant growth promoting attributes. *Archives of Phytopathology and Plant Protection, 46*(19), 2323–2340.

Dubey, R. C., & Kumar, R. (2003). Efficacy of azadirachtin and fungicides on growth and survival of sclerotia of *Macrophomina phaseolina* causing charcoal rot of soybean. *Indian Phytopathology, 56*(2), 216–217.

Dubey, R. C., Kumar, H., & Pandey, R. R. (2009). Fungitoxic effect of neem extracts on growth and sclerotial survival of Macrophomina phaseolina in vitro. *Journal of American Science, 5*(5), 17–24.

Dubey, R. C., & Upadhyaya, R. S. (2001). Survival and control of *Macrophomina phaseolina* (Tassi) Goid. In D. K. Maheshwari, & R. C. Dubey (Eds.), *Innovative approaches in microbiology* (pp. 169–190). Dehradun, India: Bishan Singh Mahendrapal Singh.

Dwivedi, T. S., & Chaube, H. S. (1985). Effect of oil-cakes on survival of *Macrophomina phaseolina* in soil. *Journal of Soil Biology & Ecology, 5*(2), 86–91.

Dwivedi, R. S., & Dubey, R. C. (1986). Effect of volatile and non-volatile fractions of two medicinal plants on germination of *Macrophomina phaseolina* sclerotia. *Transactions of the British Mycological Society, 87*, 326–328.

Dwivedi, R. S., & Dubey, R. C. (1987a). Effect of fungicides on survival of *Macrophomina phaseolina* in soil and in soybean stems in soil. *International Journal Tropical Plant Disease, 5*, 67–73.

Dwivedi, R. S., & Dubey, R. C. (1987b). Effect of soil solarization on the population dynamics of *Macrophomina phaseolina* (Tassi) goid, and general soil mycoflora. *International Journal of tropical Plant Diseases, 5*(1), 67–73.

Elad, Y., Zvieli, Y., & Chet, I. (1986). Biological control of *Macrophomina phaseolina* (Tassi) Goid.by *Trichoderma harzianum. Crop Protection, 5*(4), 288–292.

Filho, E. S., & Dhingra, O. D. (1980). Effect of herbicides on survival of *Macrophomina phaseolina* in soil. *Transactions of the British Mycological Society, 74*(1), 61–64.

Forster, H., & Shuai, B. (2020). Exogenous siRNAs against chitin synthase gene suppress the growth of the pathogenic fungus *Macrophomina phaseolina. Mycologia, 112*(4), 699–710.

Fuhlbohm, M. J., Ryley, M. J., & Aitken, E. A. B. (2012). New weed hosts of *Macrophomina phaseolina* in Australia. *Australasian Plant Disease Notes, 7*(1), 193–195.

Fulton, N. D., Awaddle, B., & Thomas, J. A. (1956). Influence of planting date on fungi isolated from diseased cotton seedlings. *Plant Disease Reporter, 40*, 556–558.

Gangopadhyay, S., Wyllie, T. D., & Teague, W. R. (1982). Effect of bulk density and moisture content of soil on the survival of *Macrophomina phaseolina. Plant and Soil, 68*(2), 241–247.

Gayathiri, E., Prakash, P., Karmegam, N., Varjani, S., Awasthi, M. K., & Ravindran, B. (2022). Biosurfactants: Potential and eco-friendly material for sustainable agriculture and environmental safety—a review. *Agronomy, 12*, 662. https://doi.org/10.3390/agronomy12030662

Ghaffar, A. (1971). Interactions of actinomycetes with *Macrophomina phaseoli* (Maubl.) Ashby, the cause of root rot of cotton. *Mycopathologia et Mycologia Applicata, 44*(3), 271–276.

Ghosh, T., Biswas, M. K., Guin, C., & Roy, P. (2018). A review on characterization, therapeutic approaches and pathogenesis of *Macrophomina phaseolina. Plant Cell Biotechnology and Molecular Biology, 19*, 72–84.

Goel, S. K., & Mehrotra, R. S. (1974). Biological control of the root rot and collar rot of okra (*Abelmoschus esculentus* (L.) Moench). *Annals of Microbiology, 125A*(3), 365–370.

Gopalakrishnan, S., Humayun, P., Kiran, B. K., Kannan, I. G. K., Vidya, M. S., Deepthi, K., & Rupela, O. (2011). Evaluation of bacteria isolated from rice rhizosphere for biological control of charcoal rot of sorghum caused by *Macrophomina phaseolina* (Tassi) Goid. *World Journal of Microbiology and Biotechnology, 27*(6), 1313–1321.

Gopalakrishnan, S., Srinivas, V., Naresh, N., Pratyusha, S., Ankati, S., Madhuprakash, J., Govindaraj, M., & Sharma, R. (2021). Deciphering the antagonistic effect of *Streptomyces* spp. and host-plant resistance induction against charcoal rot of sorghum. *Planta, 253*(2), 1–12.

Gupta, R. N., Saharan, H. S., Rathi, A. S., & Avtar, R. (2020). Identification of resistance sources against charcoal rot of sesame incited by *Macrophomina phaseolina* (Tassi) Goid. *International Journal of Current Microbiologyand and Applied Science, 9*(2), 1795–1801.

Gupta, G. K., Sharma, S. K., & Ramteke, R. (2012). Biology, epidemiology and management of the pathogenic fungus *Macrophomina phaseolina* (Tassi) Goid.with special reference to charcoal rot of soybean (*Glycine max* (L.) Merrill). *Journal of Phytopathology, 160*(4), 167–180.

Holliday, P., & Punithalingam, E. (1970). *Macrophomina phaseolina* [Descriptions of fungi and bacteria]. *Commonwealth Mycological Institute Descriptions of Pathogenic Fungi and Bacteria*, (28), 275.

Hyder, S., Gondal, A. S., Rizvi, Z. F., Iqbal, R., Hannan, A., & Sahi, S. T. (2022). Antagonism of selected fungal species against *Macrophomina phaseolina* (Tassi) Goid. causing charcoal rot of mungbean. *Pakistan Journal of Botany, 54*(3), 1129–1138. https://doi.org/10.30848/PJB2022-3(8

Ilyas, M. B., Ellis, M. A., & Sinclair, J. B. (1975). Evaluation of soil fungicides for control of charcoal rot of soybeans. *Plant Disease Reporter, 59*(4), 360–364.

Islam, M. S., Haque, M. S., & Islam, M. M. (2012). Tools to kill: Genome of one of the most destructive plant pathogenic fungi Macrophomina phaseolina. *BMC Genomics, 13*(1), 1–16.

Israel, S., Mawar, R., & Lodha, S. (2005). Soil solarisation, amendments and bio-control agents for the control of *Macrophomina phaseolina* and *Fusarium oxysporum* f. sp. *cumini* in aridisols. *Annals of Applied Biology, 146*(4), 481–491.

Jackson, C. R. (1965). Reduction of *Sclerotium bataticola* infection of peanut kernels by *Aspergillus flavus*. *Phytopathology, 55*(8), 934.

Jana, T., Sharma, T. R., Prasad, R. D., & Arora, D. K. (2003). Molecular characterization of *Macrophomina phaseolina* and *Fusarium* species by a single primer RAPD technique. *Microbiological Research, 158*(3), 249–257.

Katan, J. (1980). Solar pasteurization of soils for disease control: Status and prospects. *Plant Disease, 64*(5), 450–454.

Kaur, S., Dhillon, G. S., Brar, S. K., Vallad, G. E., Chand, R., & Chauhan, V. B. (2012). Emerging phytopathogen Macrophomina phaseolina: biology, economic importance and current diagnostic trends. *Critical Reviews in Microbiology, 32*(2), 136–151. https://doi.org/10.3109/1040841X.2011.640977

Khalili, E., Javed, M. A., Huyop, F., Rayatpanah, S., Jamshidi, S., & Wahab, R. A. (2016). Evaluation of *Trichoderma* isolates as potential biological control agent against soybean charcoal rot disease caused by *Macrophomina phaseolina*. *Biotechnology & Biotechnological Equipment, 30*(3), 479–488. https://doi.org/10.1080/13102818.2016.1147334

Khan, S. K. (2007). *Macrophomina phaseolina* as causal agent for charcoal rot of sunflower. *Mycopathologia, 5*, 111–118.

Khan, I. H., & Javaid, A. (2022a). Antagonistic activity of *Aspergillus versicolor* against *Macrophomina phaseolina*. *Brazilian Journal of Microbiology*. https://doi.org/10.1007/s42770-022-00782-6

Khan, I. H., & Javaid, A. (2022b). DNA cleavage of the fungal pathogen and production of antifungal compounds are the possible mechanisms of action of biocontrol agent *Penicillium italicum* against *Macrophomina phaseolina*. *Mycologia, 114*(1), 24–34. https://doi.org/10.1080/00275514.2021.1990627

Khan, I. H., Javaid, A., & Ahmed, D. (2021). *Trichoderma viride* controls *Macrophomina phaseolina* through its DNA disintegration and production of antifungal compounds. *International Journal of Agriculture and Biology, 25*(4), 888–894.

Khan, M. R., Haque, Z., Rasool, F., Salati, K., Khan, U., Mohiddin, F. A., et al. (2019). Management of root-rot disease complex of mungbean caused by Macrophomina phaseolina and Rhizoctonia solani through soil application of Trichoderma spp. *Crop Protection, 119*, 24–29.

Khare, E., & Arora, N. K. (2011). Dual activity of pyocyanin from *Pseudomonas aeruginosa*—antibiotic against phytopathogen and signal molecule for biofilm development by rhizobia. *Canadian Journal of Microbiology, 57*(9), 708–713.

Khare, E., & Arora, N. K. (2021). Biosurfactant based formulation of *Pseudomonas guariconensis* LE3 with multifarious plant growth promoting traits controls charcoal rot disease in *Helianthus annus*. *World Journal of Microbiology and Biotechnology, 37*(4), 55. https://doi.org/10.1007/s11274-021-03015-4. PMID: 33615389

Khokhar, I., Mukhtar, I., & Mushtaq, S. (2011). Antifungal effect of *Penicillium* metabolites against some fungi. *Archives of Phytopathology and Plant Protection, 44*(14), 1347–1351.

Kumar, S., Dheeman, S., Dubey, R. C., Maheshwari, D. K., & Baliyan, N. (2021). Cyclic siloxane biosurfactant-producing *Bacillus cereus* BS14 biocontrols charcoal rot pathogen *Macrophomina phaseolina* and induces growth promotion in *Vigna mungo* L. *Archives of Microbiology*. https://doi.org/10.1007/s00203-021-02492-3

Kumar, P., Dubey, R. C., & Maheshwari, D. K. (2012b). *Bacillus* strains isolated from rhizosphere showed plant growth promoting and antagonistic activity against phytopathogens. *Microbiological Research, 167*(8), 493–499. https://doi.org/10.1016/j.micres.2012.05.002

Kumar, S., Dubey, R. C., & Maheshwari, D. K. (2016a). Biosurfactant mediated biocontrol of *Macrophomina phaseolina* causing charcoal rot in *Vigna mungo* by a plant growth promoting *Enterococcus* sp. BS13. *Journal of Plant Pathology & Microbiology, 7*, 2–8.

Kumar, P., Dubey, R. C., Maheshwari, D. K., & Bajpai, V. K. (2016c). ACC deaminase producing *Rhizobium leguminosarum* RPN5 isolated from root nodules of *Phaseolus vulgaris* L. *Bangladesh Journal of Botany, 45*(3), 477–484.

Kumar, P., Dubey, R. C., Maheshwari, D. K., Park, Y. H., & Bajpai, V. K. (2016b). Isolation of plant growth promoting *Pseudomonas* sp. PPR8 from the rhizosphere of *Phaseolus vulgaris* L. *Archives of Biological Sciences, 68*(2), 363–374.

Kumari, N., Meena, A. K., Godara, S. L., Gahlot, V. S., & Reddy, S. D. K. M. (2022). Evaluation of bio-agents, botanicals and fungicides under *in vitro* conditions against *Macrophomina phaseolina* (Tassi) Goid.causing charcoal rot of cowpea. *Biological Forum - An International Journal, 14*(2), 84–90.

Kumar, P., Pahal, V., Gupta, A., Vadhan, R., Chandra, H., & Dubey, R. C. (2020). Effect of silver nanoparticles and *Bacillus cereus* LPR2 on the growth of Zea mays. *Scientific Reports, 10*(1), 20409.

Kumar, S., Pandey, P., & Maheshwari, D. K. (2009). Reduction in dose of chemical fertilizers and growth enhancement of sesame (*Sesamum indicum* L.) with application of rhizospheric competent *Pseudomonas aeruginosa* LES4. *European Journal of Soil Biology, 45*(4), 334–340.

Kumar, A., Saini, S., Wray, V., Nimtz, M., Prakash, A., & Johri, B. N. (2012a). Characterization of an antifungal compound produced by *Bacillus* sp. strain A5F that inhibits *Sclerotinia sclerotiorum. Journal of Basic Microbiology, 52*(6), 670–678.

Laluk, K., & Mengiste, T. (2011). The *Arabidopsis* extracellular unusual serine protease inhibitor functions in resistance to necrotrophic fungi and insect herbivory. *The Plant Journal, 68*(3), 480–494.

Lamichhane, J. R., Corrales, D. C., & Soltani, E. (2022). Biological seed treatments promote crop establishment and yield: A global meta-analysis. *Agronomy for Sustainable Development, 42*, 45. https://doi.org/10.1007/s13593-022-00761-z

Larralde-Corona, C. P., Santiago-Mena, M. R., Sifuentes-Rincón, A. M., Rodríguez-Luna, I. C., Rodríguez-Pérez, M. A., Shirai, K., & Narváez-Zapata, J. A. (2008). Biocontrol potential and polyphasic characterization of novel native *Trichoderma* strains against *Macrophomina phaseolina* isolated from sorghum and common bean. *Applied Microbiology and Biotechnology, 80*, 167–177. https://doi.org/10.1007/s00253-008-1532-0

Lima, T. M., Procópio, L. C., Brandão, F. D., Leão, B. A., Tótola, M. R., & Borges, A. C. (2011). Evaluation of bacterial surfactant toxicity towards petroleum degrading microorganisms. *Bioresource Technology, 102*(3), 2957–2964.

Lodha, S. (1996). Influence of moisture conservation techniques on *Macrophomina phaseolina* population, dry root rot and yield of cluster bean. *Indian Phytopathology, 49*, 342–349.

Lodha, S., & Burman, U. (2000). Efficacy of composts on nitrogen fixation, dry root rot (*Macrophomina phaseolina*) intensity and yield of legumes. *Indian Journal of Agricultural Sciences, 70*(12), 846–849.

Lodha, S., & Mawar, R. (2020). Population dynamics of *Macrophomina phaseolina* in relation to disease management: A review. *Journal of Phytopathology, 168*(1), 1–17.

Lorenzetti, E., Stangarlin, J. R., & Kuhn, O. J. (2017). Antimicrobial activity against *Macrophomina phaseolina* and the control of charcoal rot in soybeans using the homoeopathic drugs Sepia and *Arsenicum album. Arquivos do Instituto Biológico, 84*, 1–5. https://doi.org/10.1590/1808-1657000562016

Machado, C. (1987). *Macrophomina phaseolina: Biological behavior of isolates, spatial pattern of microsclerotia in the soil, and incidence on soybeans* (Ph.D. Thesis). Urbana: University of Illinois.

Maheshwari, D. K., Dubey, R. C., Aeron, A., Kumar, B., Kumar, S., Tewari, S., & Arora, N. K. (2012). Integrated approach for disease management and growth enhancement of *Sesamum indicum* L. utilizing *Azotobacter chroococcum* TRA2 and chemical fertilizer. *World Journal of Microbiology and Biotechnology, 28*(10), 3015–3024.

Ma, J., Hill, C. B., & Hartman, G. L. (2010). Production of *Macrophomina phaseolina* conidia by multiple soybean isolates in culture. *Plant Disease, 94*(9), 1088–1092.

Manikandan, A., Iruthayasamy, J., Jaivel, N., Krishnamoorthy, R., SenthilKumar, M., et al. (2022). Gamma-induced mutants of *Bacillus* and *Streptomyces* display enhanced antagonistic activities and suppression of the root rot and wilt diseases in pulses. *Biomolecular Concepts, 13*(1), 103–118. https://doi.org/10.1515/bmc-2022-0004

Manjunatha, S. V., Naik, M. K., Khan, M. F. R., & Goswami, R. S. (2013). Evaluation of bio-control agents for management of dry root rot of chickpea caused by *Macrophomina phaseolina. Crop Protection, 45*, 147–150.

Marquez, N., Giachero, M. L., Declerck, S., & Ducasse, D. A. (2021). *Macrophomina phaseolina*: General characteristics of pathogenicity and methods of control. *Frontiers of Plant Science, 12*, 634397. https://doi.org/10.3389/fpls.2021.634397

Meena, R. P., & Kadam, V. A. (2021). Characterization of *Macrophomina phaseolina* associated with leaf blight disease on *Chlorophytum borivilianum* Santapau and RR Fern. and its fungicidal susceptibility. *Journal of Applied Research on Medicinal and Aromatic Plants, 20*, 100288.

Meena, L. I., Rajeswari, E., Ahiladevi, P., Kamalakannan, A., & Kalaiselvi, T. (2022). Antifungal potential of *Streptomyces rameus* GgS 48 against mungbean root rot [*Rhizoctonia bataticola* (Taub.) Butler]. *Journal of Biosciences, 47*(10), 5–17. https://doi.org/10.1007/s12038-021-00244-5

Mengistu, A., Reddy, K. N., Bellaloui, N., Walker, E. R., & Kelly, H. M. (2013). Effect of glyphosate on Macrophomina phaseolina in vitro and its effect on disease severity of soybean in the field. *Crop Protection, 54*, 23–28.

Mihail, J. D., & Alcorn, S. M. (1984). Effects of soil solarization on *Macrophomina phaseolina* and *Sclerotium rolfsii. Plant Disease, 68*(2), 156–159.

Mmbaga, M. T., & Mackasmiel, L. A. (2018). Flowering dogwood infections with *Macrophomina phaseolina. Horticulture Science, 53*(3), 334–336.

Paharia, K. D., & Sahai, D. (1970). Effect of temperature on the viability of *Macrophomina phaseoli* from Potato. *Indian Journal of Microbiology, 10*(4), 107–110.

Pandey, A. K., Burlakoti, R. R., Rathore, A., & Nair, R. M. (2020). Morphological and molecular characterization of *Macrophomina phaseolina* isolated from three legume crops and evaluation of mungbean genotypes for resistance to dry root rot. *Crop Protection, 127*, 104962.

Pan, S., & Sen, C. (1983). Survivability of parental and fungicide tolerant isolates of *Macrophomina phaseolina. Indian Phytopathology, 36*, 484–488.

Papavizas, G. C. (1977a). Some factors affecting survival of sclerotia of *Macrophomina phaseolina* in soil. *Soil Biology and Biochemistry, 9*(5), 337–341.

Papavizas, G. C. (1977b). Survival of sclerotia of *Macrophomina phaseolina* and *Sclerotium cepivorum* after drying and wetting treatments. *Soil Biology and Biochemistry, 9*(5), 343–348.

Peres, N. A., Baggio, J. S., & Mertely, J. C. (2018). *Charcoal rot of strawberries caused by* Macrophomina phaseolina (p. 33598). Wimauma, Florida: Plant Pathology Department, UF/IFAS Extension, Gulf Coast Research and Education Center.

Pérez-García, A., Romero, D., & De Vicente, A. (2011). Plant protection and growth stimulation by microorganisms: Biotechnological applications of bacilli in agriculture. *Current Opinion in Biotechnology, 22*(2), 187–193.

Prasad, R. D., Chandrika, K. S. V. P., & Godbole, V. (2020). A novel chitosan biopolymer based *Trichoderma* delivery system: Storage stability, persistence and bio efficacy against seed and soil borne diseases of oilseed crops. *Microbiological Research, 237*, 126487.

Pratt, R. G. (2006). A direct observation technique for evaluating sclerotium germination by *Macrophomina phaseolina* and effects of biocontrol materials on survival of sclerotia in soil. *Mycopathologia, 162*(2), 121–131.

Pratt, R. G., McLaughlin, M. R., Pederson, G. A., & Rowe, D. E. (1998). Pathogenicity of *Macrophomina phaseolina* to mature plant tissues of alfalfa and white clover. *Plant Disease, 82*(9), 1033–1038.

Radha, K. (1956). Soil conditions and root diseases XVI. Colonization and survival of *Macrophomina phaseoli* (Maubl.) Ashby in trace element amended soils. *Journal of the Indian Botanical Society, 35*, 47–52.

Rastija, V., Vrandečić, K., Ćosić, J., Šarić, G. K., Majić, I., Agić, D., et al. (2022). Effects of coumarinyl schiff bases against phytopathogenic fungi, the entomopathogenic nematodes: Deeper insight into the mechanism of action. *Molecules, 27*, 2196. https://doi.org/10.3390/molecules27072196

Reddy, M. N., Anandaraj, G. V. D., & Dev, P. S. (1989). Chemical composition of sclerotial exudates of *Macrophomina phaseolina* pathogenic to groundnut. *Soil Biology and Biochemistry, 21*(7), 931–933.

Reetha, A. K., Pavani, S. L., & Mohan, S. (2014). Hydrogen cyanide production ability by bacterial antagonist and their antibiotics inhibition potential on *Macrophomina phaseolina* (Tassi.) Goid. *International Journal of Current Microbiology and Applied Sciences, 3*(5), 172–178.

Reicher, Y., & Hellinger, E. (1947). On the occurrence, morphology and parasitism of *Sclerotia bataticola*. *Palestine Journal of Botany, 6*, 107–147.

Reis, E. M., Boaretto, C., & Danelli, A. L. D. (2014). *Macrophomina phaseolina*: Density and longevity of microsclerotia in soybean root tissues and free on the soil, and competitive saprophytic ability. *Summa Phytopathologica, 40*, 128–133.

Ruhangiz, R. M., Gholamhosein, S. B., Sonia, A., & Akram, S. (2021). Isolation and molecular identification of melon (*Cucumis melo*) rhizosphere inhabitant actinomycetes with plant growth promoting activity under biotic stress caused by *Macrophomina Phaseolina*. *Journal of Agricultural Biotechnology, 13*(2), 21–44.

Sabet, K. A., & Khan, I. D. (1969). Competitive saprophytic ability and inoculum potential of cotton root-infecting fungi in five soils. *Empire Cotton Growing Review, 46*, 119–133.

Saleh, A. A., Ahmed, H. U., Todd, T. C., Travers, S. E., Zeller, K. A., Leslie, J. F., & Garrett, K. A. (2010). Relatedness of *Macrophomina phaseolina* isolates from tallgrass prairie, maize, soybean and sorghum. *Molecular Ecology, 19*(1), 79–91.

Salman, O., Bayram, F. R., Boyraz, N., & Koçak, R. (2021). Determination of the antagonistic effects of some rhizospheric bacteria against *Macrophomina phaseolina* under *in vitro* conditions. *Turkish Journal of Agriculture- Food Science and Technology, 9*(sp), 2474–2479. https://doi.org/10.24925/turjaf.v9isp.2474-2479.4882

Sangawan, M., Metha, N., & Saharan, G. (2005). *Diseases of oil seed crops* (pp. 11–15). Pune, India: Industry Publication Co.

Santos, A. P., Muratore, L. N., Solé-Gil, A., Farías, M. E., Ferrando, A., Blázquez, M. A., et al. (2021). Extremophilic bacteria restrict the growth of *Macrophomina phaseolina* by combined secretion of polyamines and lytic enzymes. *Biotechnology Reports, 32*, e00674. https://doi.org/10.1016/j.btre.2021.e00674

Santos, D. K., Rufino, R. D., Luna, J. M., Santos, V. A., & Sarubbo, L. A. (2016). Biosurfactants: *Multifunctional biomolecules* of the 21st century. *International Journal of Molecular Sciences, 17*(3), 401. https://doi.org/10.3390/ijms17030401

Saratha, M., Angappan, K., Karthikeyan, S., Marimuthu, S., & Chozhan, K. (2022). Actinoalloteichus cyanogriseus: A broad spectrum bio-agent against mulberry root rot pathogens. *Egyptian Journal of Biological Pest Control, 32*, 33. https://doi.org/10.1186/s41938-022-00532-8

Saratha, M., Angappan, K., & Tilak, M. (2018). Isolation of actinomycetes from the rhizosphere of mulberry against charcoal rot pathogen *Macrophomina phaseolina*. *Trends in Biosciences, 11*(22), 3094–3099.

Sarli, D. A., Gomes, G. T., Reznikov, S., Acosta, F. F., Fariña, J. I., Delgado, O. D., & Sánchez, L. A. (2022). Cold-adapted strains as plant growth-promoting bacteria on soybean seeds and biocontrol agents against Macrophomina phaseolina. *Journal of Applied Microbiology, 133*(5), 2835–2850. https://doi.org/10.1111/jam.15759

Satischandra, K. M., Hiremath, R. V., & Hegde, R. K. (1980). Factors affecting the survival of *Rhizoctonia bataticola* in black soil. *Plant and Soil, 54*(2), 307–312.

Sekhar, G., Anahosur, K. H., & Hegde, R. K. (1987). Competitive saprophytic survival of *Macrophomina phasieolina* causing charcoal rot of sorghum. *Indian Phytopathology, 40*, 406–407.

Senthilkumar, M., Govindasamy, V., & Annapurna, K. (2007). Role of antibiosis in suppression of charcoal rot disease by soybean endophyte *Paenibacillus* sp. HKA-15. *Current Microbiology, 55*(1), 25–29.

Shahid, S., & Khan, M. R. (2016). Biological control of root-rot on mungbean plants incited by *Macrophomina phaseolina* through microbial antagonists. *Plant Pathology Journal, 15*(2), 27.

Shanmugam, N., & Govindaswamy, C. V. (1973). Control of *Macrophomina phaseolina* root rot of groundnut. *Madras Agricultural Journal, 60*, 500–503.

Sharma, D., Saharan, B. S., Chauhan, N., Procha, S., & Lal, S. (2015). Isolation and functional characterization of novel biosurfactant produced by *Enterococcus faecium*. *Springer Plus, 4*(1), 1–14.

Sheikh, A. H., & Ghaffar, A. (1984). Reduction in viability of sclerotia of *Macrophomina phaseolina* with polyethylene mulching of soil. *Soil Biology and Biochemistry, 16*(1), 77–79.

Sheikh, A. H., & Ghaffar, A. (1987). Time-temperature relationships for the inactivation of sclerotia of *Macrophomina phaseolina*. *Soil Biology and Biochemistry, 19*(3), 313–315.

Sheikh, A. H., & Ghaffar, A. (2004). Effect of poultry manure and sawdust on survival of sclerotia of *Macrophomina phaseolina* in soil. *Pakistan Journal of Botany, 36*(2), 425–428.

Shokes, F. M., Lyda, S. D., & Jordan, W. R. (1977). Effect of water potential on the growth and survival of *Macrophomina phaseolina*. Phytopathology, *67*(2), 239–241.

Short, G. E., & Wyllie, T. D. (1978). Inoculum potential of *Macrophomina phaseolina* [causes of root and stem rot in seed crops]. *Phytopathology, 68*, 742–746.

Short, G. E., Wyllie, T. D., & Ammon, V. D. (1978). Quantitative enumeration of *Macrophomina*. *Phytopathology, 68*, 736–741.

Short, G. E., Wyllie, T. D., & Bristow, P. R. (1980). Survival of *Macrophomina phaseolina* in soil and in residue of soybean. *Phytopathology, 70*, 13–17.

Shrivastava, P., Kumar, R., & Yandigeri, M. S. (2017). In vitro biocontrol activity of halo tolerant *Streptomyces aureofaciens* K20: A potent antagonist against *Macrophomina phaseolina* (Tassi) Goid. *Saudi Journal of Biological Sciences, 24*(1), 192–199.

Singh, J., & Kumar, K. (2002). Location, survival, transmission and control of seedborne, *Macrophomina phaseolina* causing dry root rot and leaf blight in Urdbean. *Annals of Plant Protection Sciences, 10*(1), 111–113.

Singh, N., Kumar, S., Bajpai, V. K., Dubey, R. C., Maheshwari, D. K., & Kang, S. C. (2010). Biological control of *Macrophomina phaseolina* by chemotactic fluorescent *Pseudomonas aeruginosa* PN1 and its plant growth promotory activity in chir-pine. *Crop Protection, 29*(10), 1142–1147.

Singh, S. K., Nene, Y. L., & Reddy, M. V. (1990). Influence of cropping systems on *Macrophomina phaseolina* population of Soil. *Plant Disease, 74*(10), 812–814.

Singh, N., Pandey, P., Dubey, R. C., & Maheshwari, D. K. (2008). Biological control of root rot fungus *Macrophomina phaseolina* and growth enhancement of *Pinus roxburghii* (Sarg.) by rhizosphere competent *Bacillus subtilis* BN1. *World Journal of Microbiology and Biotechnology, 24*(9), 1669–1679.

Smith, W. H. (1969). Germination of *Macrophomina phaseoli* sclerotia as effected by *Pinus lambertiana* root exudate. *Canadian Journal of Microbiology, 15*(12), 1387–1391.

Songa, W., & Hillocks, R. J. (1996). Legume hosts of *Macrophomina phaseolina* in Kenya and effect of crop species on soil inoculum levels. *Journal of Phytopathology, 144*, 387–391.

Srivastava, A. K., Arora, D. K., Gupta, G., Pandey, R. R., & Lee, M. W. (1996). Diversity of potential microbial parasites colonizing sclerotia of *Macrophomina phaseolina* in soil. *Biology and Fertility of Soils, 22*, 136–140.

Swehla, A., Pandey, A. K., & Nair, R. M. (2020). Bioactivity of *Trichoderma harzianum* isolates against the fungal root rot pathogens with special reference to *Macrophomina phaseolina* causing dry root rot of mungbean. *Indian Phytopathology, 73*, 787–792. https://doi.org/10.1007/s42360-020-00288-x

Triveni, S., Prasanna, R., Kumar, A., Bidyarani, N., Singh, R., & Saxena, A. K. (2015). Evaluating the promise of *Trichoderma* and *Anabaena* based biofilms as multifunctional agents in *Macrophomina phaseolina*-infected cotton crop. *Biocontrol Science and Technology, 25*(6), 656–670. https://doi.org/10.1080/09583157.2015.1006171

Twizeyimana, M., Hill, C. B., Pawlowski, M., Paul, C., & Hartman, G. L. (2012). A cut-stem inoculation technique to evaluate soybean for resistance to *Macrophomina phaseolina*. *Plant Disease, 96*(8), 1210–1215.

Viejobueno, J., Rodríguez-Berbel, N., Miranda, L., de los Santos, B., & Camacho, M. (2021). Potential bacterial antagonists for the control of charcoal rot (*Macrophomina phaseolina*)in strawberry. *Horticulturae, 7*(11), 457. https://doi.org/10.3390/horticulturae7110457

Wagan, K. H., Khaskheli, M. I., & Lanjar, A. G. (2018). Isolation and characterization of *Macrophomina phaseolina* isolates prevailing in Sindh, Pakistan. *Pure and Applied Biology (PAB), 7*(4), 1309–1315.

Wang, F., Sethiya, P., Hu, X., Guo, S., Chen, Y., Li, A., Tan, K., & Wong, K. H. (2021). Transcription in fungal conidia before dormancy produces phenotypically variable conidia that maximize survival in different environments. *Nature Microbiology, 6*(8), 1066–1081.

Watanabe, T. (1973). Survivability of *Macrophomina phaseoli*(Maubl.) Ashby in naturally-infested soils and longevity of the sclerotia formed *in vitro*. *Annals of Phytopathology Society Japan, 39*, 333–337.

Wheeler, M. H. (1983). Comparisons of fungal melanin biosynthesis in ascomycetous, imperfect and basidiomycetous fungi. *Transactions of the British Mycological Society, 81*(1), 29–36.

Wyllie, T. D., & Brown, M. F. (1970). Ultrastructural formation of sclerotia of *Macrophomina phaseoli*. *Phytopathology, 60*(3), 524–528.

Wyllie, T. D., & Calvert, O. H. (1969). Effect of flower removal and pod set on formation of sclerotia and infection of *Glycine max* by *Macrophomina phaseoli*. *Phytopathology*, 1243–1245.

Wyllie, T. D., Gangopadhyay, S., Teague, W. R., & Blanchar, R. W. (1984). Germination and production of *Macrophomina phaseolina* microsclerotia as affected by oxygen and carbon dioxide concentration. *Plant and Soil, 81*(2), 195–201.

Yadav, A. K., Yandigeri, M. S., Vardhan, S., Sivakumar, G., Rangeshwaran, R., & Tripathi, C. P. M. (2014). *Streptomyces* sp. S160: A potential antagonist against chickpea charcoal root rot caused by *Macrophomina phaseolina* (Tassi) Goid. *Annals of Microbiology, 64*, 1113–1122. https://doi.org/10.1007/s13213-013-0750-6

Yang, X. B., & Navi, S. S. (2005). First report of charcoal rot epidemics caused by *Macrophomina phaseolina* in soybean in Iowa. *Plant Disease, 89*(5), 526.

Young, D. J., Gilbertson, R. L., & Alcorn, S. M. (1982). A new record for longevity of *Macrophomina phaseolina* sclerotia (soil-inhabiting fungi). *Mycologia, 74*, 504–505.

Zhao, X., Ni, Y., Liu, X., Zhao, H., Wang, J., Chen, Y. C., & Liu, H. (2020). A simple and effective technique for production of pycnidia and pycnidiospores by *Macrophomina phaseolina*. *Plant Disease, 104*(4), 1183–1187.

Chapter 15

Development of tolerance by *Macrophomina phaseolina* against fungicide, insecticide, and drought

Samadhan Bhagwan Dahikar and Rachana Dattatray Nagarkar
Department of Microbiology, Sanjivani Arts, Commerce and Science College, Kopargaon, Maharashtra, India

15.1 Introduction

Plants are unable to escape from unfavorable environmental challenges, both biotic and abiotic, due to their sessile nature. The biotic agents including fungi, bacteria, and viruses are too responsible for plant diseases that significantly reduce agricultural output and plant growth. The primary pathogen group that causes considerable losses in global agriculture's economy is fungi. The three categories of fungal parasites are determined by how the fungi feed on their host: biotrophs infect various areas of the plant without killing it and consume the living tissue, in contrast to necrotrophs, which use their toxins to kill host cells and feed on dead tissue. The third category of fungal infections includes hemi-biotrophic fungi, a peculiar group that uses both biotrophic and necrotrophic infection cycles to infect and spread throughout the host plants. The main economic crop losses worldwide are caused by necrotrophic diseases (Dean et al., 2012). A generalist soil-borne, destructive necrotrophic fungal pathogen found worldwide is *Macrophomina phaseolina*. The development of both pycnidia and sclerotia in host tissues and culture medium distinguishes this member of the anamorphic Ascomycetes group. It causes diseases including stem and root rot, charcoal rot, and seedling blight (Crous et al., 2006; Dhingra & Sinclair, 1978). Its formation of sclerotia and pycnidia in host tissues and culture medium distinguishes it from other anamorphic ascomycetes. It results in diseases, such stem and root rot, charcoal rot, and seedling blight (Table 15.1). Microsclerotia are the main invasive source of *M. phaseolina* according to disease cycle. For about 15 years, this sturdy structure can last in the ground (Gupta et al., 2012). Multiple hyphae that seem germinating can infect the host plant's roots while it is still at the seedling stage. Once within the roots, the fungus has an effect on the vascular system, which disrupts the flow of water and nutrients to the plants' higher sections. The senescence and yellowing of leaves that are still linked to stems by their petioles, the loss of cortical tissues from the lower stem and taproot, and the appearance of gray in these

TABLE 15.1 Diseases caused by *M. phaseolina*on its host plants.

Major cultivated hosts	Host plants	Symptoms
Food crops	Cabbage, maize, sweet potato, sorghum,	Stem rot, root rot, charcoal rot, stalk rot, seed and seedling rot, wilt, leaf spot, rotting of developing pods
Pulse crops	Green gram, common bean, chick pea	
Fiber crops	Cotton, jute	
Oil crops	Soybean, sunflower, groundnut, peanut and sesame	
Other crops	Pepper, alfalfa	

Macrophomina Phaseolina. https://doi.org/10.1016/B978-0-443-15443-0.00024-3

tissues as a result of an abundance of microsclerotia are typical indications of this ailment. The host plant may die too soon as a result of these symptoms (Marquez et al., 2021). Soybean charcoal rot causes early maturation, chlorosis, and improper pod filling. While in corn, the fungus during hot and dry conditions causes a stalk to decay. *M. phaseolina* is highly persistent in soil as microsclerotia and has a broad host range. In order to effectively control the pathogen, one must have a solid grasp of both its interactions with host plants and the mechanisms behind pathogenicity. Additionally, it has been found that virulence and phenotype or morphological variations strongly correlate (Tok, 2019). *M. phaseolina* management strategies remain to be challenging to implement despite widespread research attempts to control the diseases. In actuality, the diseases spread by this soil pathogen are caused by interactions between the host plant, the pathogen, and the biotic and abiotic factors of the environment. The purpose of this chapter is to examine the pathogenicity mechanisms of *M. phaseolina* as well as the host's defense system, which involve the production of toxins and other enzymes. The number of different disease management strategies is expanding continuously, but the biggest problem that is already emerging is *M. phaseolina's* development of tolerance or resistance to various fungicides, insecticides, and drought tolerance. As a consequence, the research on the fungicide, insecticide, and drought tolerance in *M. phaseolina* and resistance in plants due to release of some toxins and metabolites is the main emphasis of this chapter.

15.2 General characteristics of *M. phaseolina* and pathogenicity mechanisms

Hyaline hyphae with thin walls to light or dark brown hyphae with septa are found in *M. phaseolina*. The major hyphae of parent hyphae frequently branch off at a right angle, constricting at the place of origin.Microsclerotia are spherical, oval or oblong-shaped masses of hardened fungal mycelium that begin as a light brown color before maturing to a deeper shade (brown to black). Pycnidia are larger than microsclerotia, rarely observed in the environment, dark brown to black, rough, globose or irregular, beaked, and ostiolated (Lakhran et al., 2018). On synthetic medium, the fungus can display significant variation in the colors of the mycelium, the distribution of microsclerotia, the development of pycnidia, and the phenotypes of isolates (Marquez et al., 2021).

The ability of the plants to endure biotic stresses is due to their highly evolved immune systems. Pattern-triggered immunity (PTI), which is induced by pathogen-recognition receptors (PRRs), that recognize microbe-associated molecular patterns or pathogen-associated molecular patterns, protect the host from microbe infestation (MAMPs or PAMPs). In response to the detection of MAMP (PAMP/DAMP), PRRs initiate a differentiated defense response that involves calcium influx, mitogen-activated kinase (MAPK) cascades, reactive oxygen species (ROS) homeostasis, phytohormone, and TF/gene regulation. Similar to this, the hypersensitive response (HR) is activated and the R-gene is regulated by the effector triggered immunity (ETI) host defense in response to infections. The phytohormone SA is essential for several plant activities, including innate immunity. Because of SA, pathogens that are biotrophic and hemi-biotrophic find it easier to cause local and systemic acquired resistance (LAR and SAR). The SA biosynthesis pathway is triggered upon microbial infection, resulting in the manifestation of a broad-spectrum resistance known as an SAR. Pathogenesis-related (PR) proteins are activated along with SAR, which inhibits the pathogen's development (Glazebrook, 2005).

Additionally, HR is activated by the pathogen attack as a secondary plant defense mechanism. As a result, ROS such as hydrogen peroxide (H_2O_2), hydroxyl radicals, and superoxide radicals build antioxidant enzymes, which in turn scavenge these dangerous molecules to maintain a cell balance, are frequently linked to the formation of ROS. Additionally, it is unknown whether SA acts in front of or behind ROS, despite the fact that the self-amplifying feedback loop describes the precise relationship between SA and H_2O_2 quite well (Jayakannan et al., 2015). Genetically, intricate necrotroph resistance involves the fusion of phytohormones like SA, ethylene (ET), jasmonic acid (JA), and abscisic acid (ABA) (Kazans & Lyons, 2014).

The first full genome of *M. phaseolina*, edited by Islam et al. (2012), was notable for having a large number of enzymes that were involved in the breakdown of lignocellulose and cell wall polysaccharides. This discovery paved the way for further research into the cytological and molecular aspects of infection using a wide range of enzymatic and toxin tools to infect a wide variety of plants. The genome of *M. phaseolina* encodes a wide variety of genes linked to pathogenicity that may play a role in the fungus' pathogenesis (Fig. 15.1). In fact, 12% of the genes in the genome encoded are linked to genes that regulate interactions between pathogens and hosts. *M. phaseolina* is able to adhere to the host tissue, get past the initial host defense given by salicylate-1-monooxygenase, and penetrate the plant epidermis and proliferate. When a pathogen invades a host, it releases a number of toxins and cell wall-degrading enzymes (CWDEs), which, along with a breakdown in the host's defences, result in host cell death and the development of illness (Islam et al., 2012). The genome of *M. phaseolina* contains a wide variety of pathogenicity-associated genes that allow it to (A) adhere to host tissues (e.g., transglutaminase-like proteins and CBEL-cellulose binding elicitor lectin), neutralize the initial host defense (salicylate-1-monooxygenase), and invade plant tissues (B). Once inside the host, the pathogen secretes a range of different

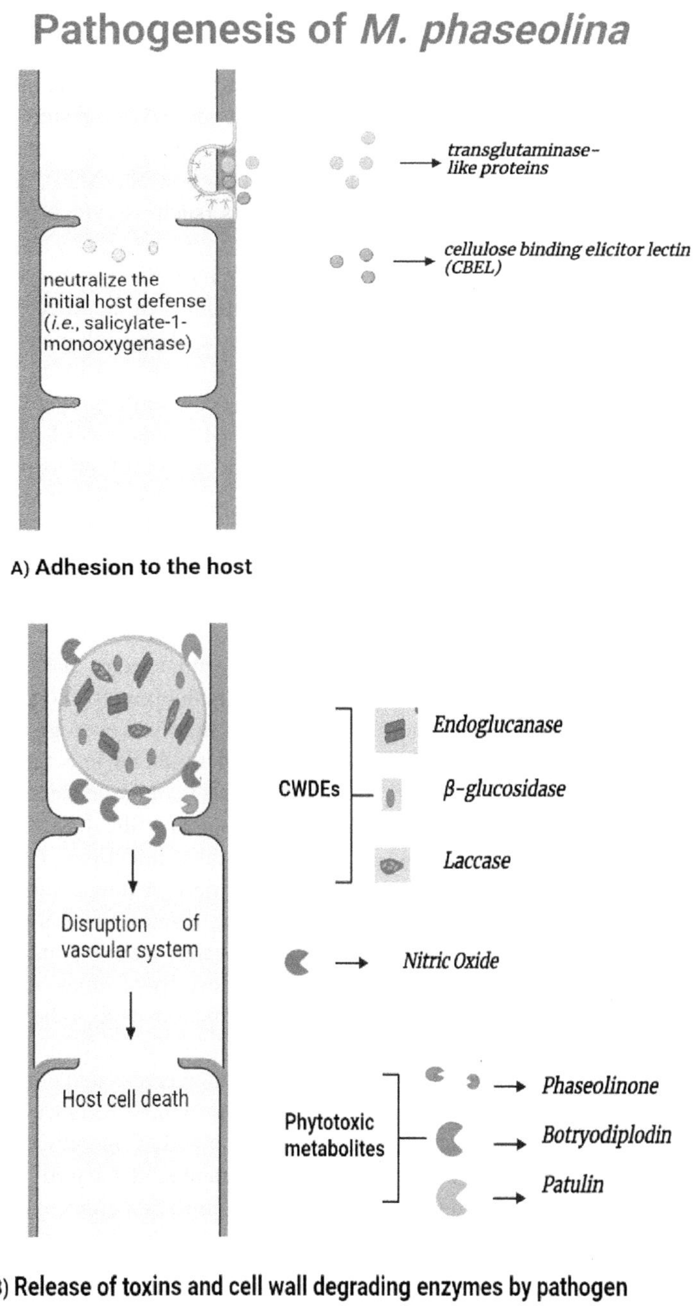

FIGURE 15.1 Pathogenesis by *M. phaseolina* (Marquez et al., 2021).

toxins and cell wall-degrading enzymes (CDWEs), which interfere with the vascular system and weaken host defense, causing host cell death, and the development of sickness (Marquez et al., 2021).

The abundance of hydrolytic enzymes that *M. phaseolina* contains allows it to infiltrate the host tissue by breaking down the polysaccharides and lignocelluloses in cell walls. Enzymes involved in the hydrolysis of cellulose include endoglucanases, exocellobio-hydrolases, and -glucosidases, whereas those involved in the breakdown of lignin include laccases, lignin peroxidases, galactose oxidases, chloroperoxidases, haloperoxidases, and heme peroxidases. It is interesting to note that *M. phaseolina's* genome had the most laccases and cellulolytic enzymes when compared to the genomes of other fungal species that have been sequenced (e.g., *Postia placenta, Ustilago maydis, Phanerochaete chrysosporium, Saccharomyces cerevisia, Aspergillus nidulans, Cryptococcus neoformans, Neurospora crassa*) (Bandara et al., 2018; Islam et al., 2012). These plant cell wall-degrading enzymes' production and activity have been shown to occur in in vitro culture conditions (Ramos et al., 2016).

Some of the phytotoxic byproducts of *M. phaseolina* that have been found include patulin, botryodiplodin, and phaseolinon, which are thought to be involved in the early stages of the infection that causes seedlings to wilt and the development of necrotic lesions on leaves and roots (Abbas et al., 2020; Bhattacharya et al., 1987; Ramezani et al., 2007; Salvatore et al., 2020). Because of this, *M. phaseolina* is more virulent and may be better able to infect a variety of hosts and tissues. The fungus' excellent tolerance to a variety of environmental circumstances also contributes to its widespread dissemination and plant infection potential (Islam et al., 2012; Salvatore et al., 2020). To combat the plant defensive response, this adaptation involves the production of detoxifying genes (such as cytochrome P450, Cof, and superoxide dismutase) (Ghosh et al., 2018).

The genome of *M. phaseolina* revealed intriguing nitric oxide synthase-like regions with conserved amino acid sequences. In vitro, nitric oxide (NO) was discovered in considerable amounts in the infected jute tissues as well as in the mycelium and surrounding medium. This shows that NO may have a crucial physiological role in the interaction between pathogens and necrotrophic hosts (Sarkar et al., 2014).

Although connections between plant cultivars (like soybean) and the competitiveness of *M. phaseolina* genotypes have been shown to occur, *M. phaseolina* is a polyphagous pathogen without indications of host specificity (Reznikov et al., 2018). In order to effectively control the pathogen, one must have a solid grasp of both its interactions with host plants and the mechanisms behind pathogenicity.

15.3 Pathogenic interactions between *M. phaseolin* and other phyllosphere microorganisms

Numerous nonpathogenic microorganisms as well as a variety of pathogenic species that coexist in complex communities pose a hazard to plants. The pathobiomes of a plant are a group of coexisting phytopathogens that have an impact on both the plant and each other (Mannaa & Seo, 2021). Pathogens that occupy the same ecological niche and either cooperate or compete for the same plant resources give rise to them. As compared to single infections, the effects of two or more phytopathogens on the same host can produce considerably diverse disease consequences. Nonpathogenic members of the communities of natural microorganisms living in the plant's phyllosphere, the habitat found above ground, or the rhizosphere, the environment around the roots, can act as a barrier against infections. For instance, recent research has revealed that the grains of the *Zea mays* (maize) plant are home to a variety of fungus and bacteria (Degani et al., 2021), some of which are known phytopathogens can be opportunists, causing disease when the conditions are favorable. The microflora of maize grains contains additional helpful bioprotective species. Before planting, microorganisms interact in the soil of commercial fields, and the results can significantly alter the severity of illness (Zhang et al., 2021). The fungus causes diseases known as maize postflowering stalk rot in conjunction with other plant pathogenic fungi like *M. phaseolina*.

Up to 45 days after sowing, *M. maydis* infection in cotton plants resulted cracks and dark red lesions on immature roots as well as enhanced growth in lateral roots. Nevertheless, no such aboveground effects were seen as cotton plants grew up to maturity. Additionally, none of these symptomatic cotton plants contained *M. maydis*. According to a recent study, *M. maydis* infection in immature cotton plants growing in a growth chamber decreased wet biomass and phenological development (Dor & Degani, 2019). The fungus had no effect on cotton growth metrics or production, however, according to 2-year's research that monitored *M. maydis* infection on plants during the whole season (Degani et al., 2020). While this outcome was attained with routine irrigation, during dry spells, *M. maydis* infection decreased the weight and height of the aboveground parts of cotton plants. When considering all of these lines of evidence, it should be assumed that this pathogen may exhibit some opportunistic behavior. In temperate, warm, and tropical locations all throughout the world, the widespread disease *M. phaseolina* affects a wide variety of domesticated and wild animals.

M. phaseolina and *M. maydis* are both largely soil- and seed-borne pathogens; however,*M. phaseolina* differs from *M. maydis* in that it possesses a highly competitive saprophytic capacity (Babu et al., 2010). Both fungi induce similar growth decrease and wilt symptoms and have a similar pathogenic activity style. These species can remain underground for extended periods of time making them difficult to regulate. The roots of the sprouts are penetrated by the fungi, which results in damping-off, root necrosis, and sprout development restriction (Tej et al., 2018). Within the plant's xylem, *M. phaseolina* and *M. maydis* may grow upward. The symptoms typically develop during dry conditions and appear later in the season, around the blossoming stage. Pathogens limit the water supply as the growing season goes on, which causes dehydration and lower yields (Cohen et al., 2016; Degani et al., 2019). The development of maize seeds is adversely impacted by *M. phaseolina* and *M. maydis* culture filtrate.

The interactions between *M. phaseolina* and *M. maydis* are particularly intriguing because both viruses are common in agricultural fields that support cotton and maize, two summer crops. Under field conditions, the co-inoculum of these two fungi has a considerable impact on the plants. When both fungi were co-inoculated into the soil, their antagonistic interactions were substantially more reflected than when each fungus was inoculated separately (Degani et al., 2020). Their coexistence within the plants encouraged growth, increased yield, and reduced disease symptoms. It is interesting to observe that stem tissue of the plants had higher levels of *M. phaseolina* DNA and lower levels of *M. maydis* DNA as a result of the dual infection. A fuller comprehension of these interactions might reveal untapped information levels about managing the infections, their virulence, and the plant's resistance and help create novel plant defense techniques. However, the symptoms of CRD were much lessened and the plants' development and yield increased when the two fungi had previously been inoculated simultaneously in the soil (Degani et al., 2020). It can be deduced that this result is the outcome of competitive interactions between the two fungi. Both *M. phaseolina* in cotton and *M. maydis* in maize plants have difficulty spreading because of their antagonistic interactions. The findings demonstrated that it is more intricate than a straightforward two-directional antagonism. There are probable variations in the CRD's severity or the virulence of the infections. *M. maydis* can act as a bioprotective endophyte in the presence of *M. phaseolina*.

The amount of CRD agent (*M. phaseolina*) infection with both infections simultaneously resulted in a considerable reduction of DNA in the plant roots, pretreating the soil with *M. maydis* 14 DBS or on the sowing day, and the growth indices and survival of the sprouts were enhanced. This outcome was obtained when just *M. maydis* inoculation on the sowing day had a negligible impact on the first repeat of the experiment. In the second experiment's replication, *M. maydis* inoculation with sowing had a significant impact on the growth and survival of the plants. In this case, the plants were safeguarded by the preinoculation of the soil with *M. phaseolina* in the 14 DBS. As a result, *M. phaseolina* and *M. maydis* exhibit mutual inhibition, which prevent one of the fungal populations from growing out of control. Since both fungi live in the same area, one theory is that hostility develops there. As a result, the fungi have less impact on plant development. An alternative hypothesis is that the inhibitory metabolites released by both fungi cause the antagonism directly (Degani et al., 2020). The cotton plants also took longer to mature under drought stress as a result of *M. maydis* infection, although crop production was unaffected (Degani et al., 2020). So, some opportunistic behavior of this pathogen in cotton could exist. The Degani's study results are in line with this conclusion. The repetition settings in the second experiment most likely favored *M. maydis* pathogenic behavior and elicited its aggressive effects. Further research will be necessary to address this assumption, the nature of these intriguing correlations, and their reasons during a whole growth span. However, it appears that the underlying mechanism in the co-infection with *M. maydis* and *M. phaseolina* is somewhat different. Previous research indicates that *M. maydis* does not actually cause illness in cotton. It lives in the roots and stems (Degani et al., 2020; Dor & Degani, 2019) that may provide some defense against other intruding soil pathogens. For example, it was demonstrated that the interaction between the cotton pathogen, *F. oxysporum* and *M. maydis* on the roots, induces a decrease in the severity of *Fusarium* cotton wilt. It is curious why *M. maydis* failed to protect the plants when it was added to the soil after *M. phaseolina*. It is probable that before the fungus had developed the inhibitory metabolites produced by *M. maydis*, it did not reach an effective dose. As a result, when added to the soil after *M. phaseolina*, *M. maydis* has a significantly reduced anti-CRD fungus effect, but when added before, it has an increased effect. Future research should support this notion, but it is also important to take into account the possibility that the other members of the soil microbiome are also involved. Given the diversity of the microorganisms that make up the soil and root microflora, the network of connections that exists between *M. maydis* and *M. phaseolina* is likely to affect how they interact! The findings of this study still offer a crucial starting point for ongoing research and have ramifications for the creation of future control techniques. The results point out to a reduced probability of yield loss in areas where *M. phaseolina* and *M. maydis* co-occur, particularly when the former occurs before the latter. *M. maydis* function as a cotton endophyte is still unknown. Aiming to understand their potential as biological control agents, researchers are now becoming more interested in asymptomatic fungal endophytes. Numerous latent pathogens were also discovered. It is known that certain latent species are antagonists of plant pathogens. Data from the research listed above and others indicate that soil-borne fungus, in particular pathogenic soil biota, are crucial in controlling plant productivity or biomass (Mommer et al., 2018). Bioassays that inoculate diverse plant species with isolates from one or two fungus may reveal host-specific pathogenic effects and offer proof-of-concept for the pathogenic potential of these underground organisms.

The research found that the pathogen's presence in the soil could have a significant impact on the ability of the agent of charcoal rot disease (CRD), *M. phaseolina,* to injure cotton sprouts. Each infection produced substantial suppression of the sprouts' appearance, development, and survival, 14 days prior to sowing. This was accompanied by elevated DNA levels of each pathogen in the roots of the sprouts. When *M. phaseolina* soil infection was carried out before *M. maydis*, similar results were found. However, *M. maydis* had the potential to behave either dangerously or as a helpful endophyte, shielding the plants from CRD. *M. maydis* shows great virulence potential in particular situations, demonstrating its true nature as an opportunist in cotton. It was interestingly shown that protecting plants from *M. maydis* required soil inoculation with *M. phaseolina,* 14 days prior to sowing. As a result, the two viruses compete with one another, and this balance may cause a mild outbreak of sickness. The findings of this work enhance our knowledge of the interaction between *M. phaseolina*, the agent that causes CRD, and *M. maydis*. They contend that the coexistence of two diseases is a crucial balancing element that may result from the rotation of the maize-cotton crop. This information should be taken into account by any agro-technical, chemical, or biological intervention control plan in this 2-year rotation farmed fields since an attempt to limit one of the partners could unintentionally help to the success of the other.

15.4 Symptoms of infection by *M. phaseolina* on different host plants

15.4.1 Soybean

The root and stem of the soybean are affected by charcoal rot. The symptoms of charcoal rot change depending on when time of year the plant is affected. Although charcoal rot often appears later in the growing season, it can affect seedlings since the plant's roots can become infected at any time of year. Infected seedlings have a reddish-brown discoloration at the soil line that moves up the stem and finally turns dark brown or even black. Infected seedlings' leaves may seem discolored or start to dry out and turn brown (Fig. 15.2). The afflicted young plants may survive in chilly, damp conditions but still harbor a latent infection that will manifest symptoms later in the season in hot and dry weather. Early maturation, non-abscission of the leaves, chlorosis and incomplete pod filling are the signs of charcoal rot in older soybeans. In the woody areas of the stems and roots, microsclerotia can be seen as small black spots. The microsclerotia are discharged into the soil as diseased tissue breaks down (Bellaloui et al., 2021).

15.4.2 Peanut

In peanuts, charcoal rot can be found in the soil or in the seeds. The fungus can spread by moving peanut seed and pods, which seem to be commonly inhabited to it. Charcoal rot on peanuts is distinguished by water-soaked lesions on the hypocotyl near to the soil line. The seedling will perish after the hypocotyl is girdled. Infected plant components can exist in any stage of development. Stem and root infections initially resemble water-soaked areas before turning a light brown color. The taproot and stem may become infected. The plant will wilt once the stem is girdled, and the fungus will quickly infiltrate the remaining tissues. The profuse production of microsclerotia causes the dead tissues to turn black. On roots, pegs, and pods, sclerotia can also be formed. The rupture could be brought on by a root-specific disease (Martínez-Salgado et al., 2021).

15.4.3 Corn

Similar to other fungal stalk rots, corn shows symptoms resembling charcoal rot. The appearance of black microsclerotia in the vascular tissue and inside the stalk rind is a telltale symptom of charcoal rot. The color of the stalk varies from gray to black. When there is a drought, corn becomes infected because the stalks begin to deteriorate. During hot and dry years, this promotes a widespread incidence and severity in corn (Cook et al., 1973).

15.4.4 Maize

Initial signs of charcoal stalk rot in maize include aberrant drying of top leaf tissues after blooming, lodging of the stalk, and premature death. The first five nodes of the lower stalk typically have a distinctive charcoal, gray-black discoloration as

FIGURE 15.2 Wilting of leaves of *Glycin max* due to charcoal rot. *Source-Agricultural farm, Rahata, Maharashtra, India.*

plants mature. Numerous black spots (microsclerotia) can be seen inside the stalk and in the shredded vascular bundles when the stalk is cut open (Kumar, 2019).

15.5 Development of tolerance by *M. phaseolina* against fungicide

While considering the crop losses due to infections caused by *M. phaseolina*, it becomes very necessary to control the fungal growth. Consequently, a variety of management techniques are in use, including biological and chemical control. Since there are no systemic fungicides that reach the root, *M. phaseolina* is challenging to chemical control. So far, no fungicides have been approved to control this infection. However, several concentrations of systemic and nonsystemic fungicides (such as carbendazim, difenoconazole, benomyl, azoxystrobin, and dazome) were assessed in vitro and in vivo against *M. phaseolina* (Chamorro, Domínguez, et al., 2015; Cohen et al., 2012; Lokesh et al., 2020; Parmar et al., 2017; Tonin et al., 2013).

15.5.1 The theory of reactive oxygen species-induced sclerotia development, based on tolerance

M. phaseolina forms the survival structure known as microsclerotia, which allows it to survive for years in soil.

15.5.1.1 True sclerotia

Ainsworth and Bisby's documentation, which describes a "sclerotium" as "a hard, frequently spherical mass of hyphae, with or without the inclusion of host tissue or soil, typically having no spores in or on it, is where the term first appeared (Ainsworth & Bisby, 1971, p. 663). Later, the real sclerotium shows a specialized fungal structure that divides the inner layer of the medulla from the middle layer of cortex and has a clear and melanized outer rind. On some plants, the actual sclerotia are layered structures that can range in size from 2 to 8 mm and 12 cm or even more. True sclerotia are simply composed of hyphal mass and have no plant tissues or soils, and they are able to create apothecia to disperse ascospores (Ainsworth & Bisby, 1950; Lokesh et al., 2020).

15.5.1.2 Pseudosclerotia

The second type of sclerotia is known as "pseudosclerotia," which describes fungal survival structures with melanized and compressed mycelia that are frequently wrapped in plant tissues or soils (Lentz & McKay, 1970). Fungal conidia may be present in pseudosclerotia, and these conidia may develop into apothecia, which serve as the main inoculum for ascospore dissemination the next season. The mummified survival structure of the *Monilia* species on stone fruits is a common example (Baard et al., 1981).

15.5.1.3 Small sclerotia

The third type of sclerotia, referred to as "little sclerotia," was first discovered by R.E. Smith (1900) in connection with an *S. sclerotiorum* infection of lettuce. The small sclerotia were envisioned as sclerotia of 0.5–2 mm, with an emphasis on its ability to infect plants via hyphae grown out from the small sclerotia instead of producing apothecia to disperse ascospores. Based on the morphological and physiological alterations, the isolate was reclassified as a new species, sclerotinia minor. The production of tiny sclerotia by this species type is already well documented.

15.5.1.4 Microsclerotia

The fourth word, "microsclerotia" refers to a tiny melanized and compacted surviving structure of *Verticillium* species and is analogous to the observation on the microscopic sclerotia. It was suggested that microsclerotia arise from a single hypha, unlike sclerotia and small sclerotia. A single hyphal lateral budding causes the creation of compact, spherical microsclerotia with melanized and hyaline cells inside. According to a theory, the inner hyaline cells undergo autolysis during differentiation and may not survive for germination. All of these concepts were said to be capable of surviving (Coley-Smith & Cooke, 1971).

Microsclerotia formation was initially observed in *M. phaseolina* secondary branching hyphae wrapping on the parent hypha, which matures into the globose structure. The initial hyphae may alternatively multiply and protrude in order to increase the cell mass for the development of microsclerotia. All of the *M. phaseolina* microsclerotia's inner cells were discovered to be melanized and capable of germination (Wyllie & Brown, 1970).

The concept of reactive oxygen species (ROS)-induced sclerotia growth was conceptualized using these fungi. A hypothesis was put forth when it was discovered that the total phospholipids' malondialdehyde content was higher in the center of a fungal colony, where more sclerotia are formed. This discovery demonstrated a relationship between lipid peroxidation, an indication of intracellular ROS stress, and the development of sclerotia (Georgiou et al., 2006). Additionally, erythroascorbate's reduced form outperformed its oxidized version by a factor of five when oxidative stress was compared in strains that formed sclerotia and those that did not. Antioxidants like ascorbic acid were routinely used in these studies to prevent the development of sclerotia. The formation of sclerotia was inhibited by the application of ROS scavengers, such ethanol and benzoate. Levels of hydrogen peroxide (H_2O_2) were also shown to be higher in association with the emergence of sclerotia, and the radical stimulator H_2O_2 increased cell proliferation and sclerotia development. It was widely acknowledged as a result of the combined results of this research that the production of ROS was substantially connected with the emergence of sclerotia. This hypothesis proposes that the induction of undifferentiated vegetative hyphae by endogenous ROS stress is necessary to initiate hyphal differentiation. Defensive mechanisms like melanin production and catalase activity would be boosted to decrease the damage caused by an ROS burst. The sclerotia growth culminates in planned cell death and autolysis to produce the O_2 in the impermeable and melanized rind, while the inner medulla stays dormant in preparation for germination under favorable conditions. In the light of the emergence of high-throughput sequencing for identifying fungal transcriptomes, numerous studies have examined whether the idea of ROS-induced sclerotia growth is applicable. The correlation between ROS burst, detoxification, and the emergence of microsclerotia was validated by the expression studies. The idea of ROS-induced microsclerotia development has been supported by the usage of many fungi, all of which belong to *Leotiomycetes* and *Sordariomycetes* classes, whereas *M. phaseolina* belongs to the *Dothideomycetes*. It is still unclear if the *Leotiomycetes, Sordariomycetes*, and *Dothideomycetes* produce microsclerotia in the same way, and whether the idea that ROS-induced sclerotia growth may be utilized to explain the microsclerotia of *M. phaseolina* is also still up for debate (Liu et al., 2022).

The RNA-sequencing (RNA-Seq) to profile gene expression during *M. phaseolina's* microsclerotia development phases because there is no much genetic research on the species. Investigations into gene ontology and differential gene expression led to the identification of ROS-related processes that took place during the development of the microsclerotia. The results of the microsclerotia inhibition and stimulation experiments supported the notion that hyphal differentiation of *M. phaseolina* to microsclerotia development is primarily stimulated by ROS. It was discovered that superoxide (O^-_2) in particular was very significant. Studies have demonstrated that O^-_2 serves as the main trigger for the synthesis of sclerotia and microsclerotia (Liu et al., 2022).

In other words, decreasing O^-_2 decreased sclerotia development. It was found that the concentration of O^-_2 in isolates that formed sclerotia was consistently higher than that in isolates that did not develop sclerotia when O^-_2 concentration and xanthine oxidase-derived O^-_2 were measured. A study repeatedly showed that *S. minor* growth on the PDA that had been modified by H_2O_2 did not result in the development of further sclerotia. One study found that the absence of H_2O_2 had no negative consequences since *S. sclerotiorum* SOD1 mutants (SsSOD1, SS1G 00699) did not show any impairment in sclerotia development (Liu et al., 2022).

Additionally, two investigations that conceptually enhanced H_2O_2 buildup by silencing *S. sclerotiorum* catalases actually decreased the production of sclerotia. SsCat1 (SS1G 02784) silence resulted in slower and smaller sclerotia production in the mutants, whereas SsCat2 (SS1G 00547) silencing resulted in fewer sclerotia, though their size and weight looked to be normal. The *S. sclerotiorum*thioredoxin1 SsTrx1 (SS1G 08534) and thioredoxin reductase SsTrr1 (SS1G 05899) mutants showed increased sensitivity and decreased capacity to manage exogenous H_2O_2, in addition to increasing H_2O_2 buildup by silencing catalases. The findings also revealed a consistent decrease in the number of sclerotia in the SsTrx1-silenced and SsTrr1-silenced mutants, suggesting that H_2O_2 buildup may not be a factor in sclerotia production. As a result, this research revealed that O_2might be able to stimulate and control the amount of sclerotia and microsclerotia production but not H_2O_2 (Liu et al., 2022).

Results demonstrate that carbendazim (50 ppm) has a significant impact on mycelial growth and the development of sclerotia, and this effect increases with the increase in the concentration of this systemic fungicide (Lokesh et al., 2020). Carbendazim inhibits the activity of tubulin, a crucial part of microtubules necessary for fungi to grow (Davidse & Flach, 1978). Additionally, in another series of tests, carbendazim uses decreased disease occurrence and raised plant survival rates (Iqbal & Mukhtar, 2020).

Although some efficacy of some chemical fumigants' has been demonstrated (Iqbal & Mukhtar, 2020; Lokesh et al., 2020); agro-environmental policies and the public's growing disapproval of agrochemicals have prompted the evaluation and comparison of chemical agents with more environmentally friendly alternatives to control plant diseases caused by *M. phaseolina* (Reznikov et al., 2016; Swamy et al., 2018; Adhikary et al., 2019). In certain research, the antifungal action of water-soluble chitosan (s-chitosan) on *M. phaseolina*, which infects jute seedlings, was evaluated, and the activity of secreted enzymes during infection was observed to change. Against this pathogen, s-chitosan displayed a fungistatic mode of action. The use of s-chitosan during *M. phaseolina* infection of jute seedlings inhibited fungal growth, and the length of the seedlings was practically identical to uninfected seedlings. A correct comprehension of the mode of action of s-chitosan as an antifungal agent was made possible by changes in the enzyme profiles of jute seedlings during the prevention of fungal infection with s-chitosan. It has been demonstrated that infected seedlings treated with s-chitosan had an increase in the activity of defense-related enzymes, such as chitosanase and peroxidase (Chatterjee et al., 2014).

According to Tonin et al. (2013), the most effective soybean seed treatments for *M. phaseolina* were: penflufen + trifloxystrobin and the active component carbendazim. The fluconazole, metalaxyl, thiram, and tolyfluanida active components are insensitive to the *M. phaseolina* isolate.

15.5.2 General mechanism of development of fungicide resistance

A fungus can develop resistance by genetic modification, which lowers its sensitivity to fungicides. Reduced sensitivity is likely to be caused by naturally occurring subpopulations of resistant people or by genetic changes that happen at low frequency (one in a million or fewer). The mycelium, sclerotia, spores, or the nucleus of a single cell with the ability to reproduce and spread are all examples of individuals in a fungal community. The mutation of one or more genes may cause the resistance characteristic (see build-up of resistance below). Changes in multiple genes are necessary to provide resistance to multisite inhibitory fungicides, which occurs less frequently than the advent of single-gene mutations that provide resistance to site-specific fungicides. Depending on the mode of action, resistance tactics may include shifting the target location, decreasing fungicide uptake, active fungicide export outside of fungus cells, and fungicide detoxification or breakdown. By administering a fungicide to a group of people from a field population and observing the toxicity response, it is possible to determine the degree of fungicide resistance in a lab setting. In circumstances when the fungus cannot be grown, toxicity reactions are often assessed as inhibition of fungus growth, spore germination, or actual plant infection. In a similar manner to how the LD50 (50% lethal dosage) is established when measuring the acute toxicity of a pesticide to rats or mice, the effective concentration (EC) that 50% suppresses growth, germination, or infection is then estimated for each sampled individual. There is typically a range of sensitivity (or resistance) to the fungicide seen when numerous members of a population are tested and examined.

Frequency distributions of individuals' sensitivity tend to be normal or bell-shaped, which is typical of many biological reactions found in nature. The population is spread out over a sensitive range when the fungicide is either recently introduced or the possibility of resistance is minimal. The presence of a small subpopulation of resistant strains and a larger subpopulation of sensitive strains, however, may potentially result in a distribution with two separate sub-populations (John & Smith, 2017).

15.5.3 Build-up of resistance

When the prevalence of resistant strains increases to the point where they dominate the population, resistance in that group becomes essential. Repeated uses of the fungicides, which place selection pressure on the population, result in the accumulation of resistant strains. The fungicide facilitates the growth of resistant strains while selectively inhibiting sensitive strains. Depending on how many genes are involved in giving resistance, this change toward resistance happens at varying rates. When a single gene mutation confers resistance, there may be an abrupt loss of control over the population due to a quick shift toward resistance (Fig. 15.3). When several genes are at play, the shift toward resistance happens

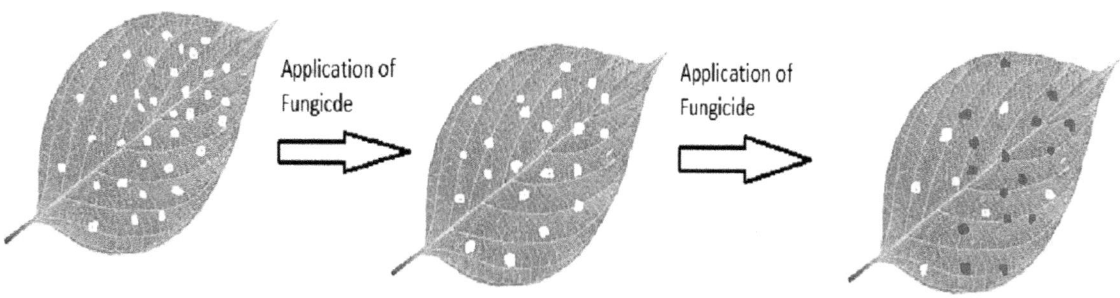

FIGURE 15.3 Selection for resistant (*red* [dark gray in print version] *spots*) fungal strains among fungicide sensitive strains (*white spots*) with repeated applications of the same fungicide active ingredient.

gradually, reducing population sensitivity as a result. The progressive change brought on by the multiple gene impact may limit the effectiveness of fungicides between applications, but there is little chance of a total loss of control. It could be challenging to discern between sensitive and resistant sub-populations with field sampling since sensitivity responses overlap during the early changes toward decreasing sensitivity. There must be widespread testing to detect the rising type of resistance (Beckerman, 2008; John & Smith, 2017).

15.6 Development of tolerance by *M. phaseolina* against insecticide

The natural soil processes that maintain soil health and fertility heavily rely on microorganisms. Pesticide use has a negative impact on bacteria and other microorganisms that are not its intended targets. Beneficial soil invertebrates called entomopathogenic nematodes (EPNs) are employed in initiatives to control insect pests. To research the selectivity of pesticides and insecticides, EPNs provide an excellent model. Acetylcholinesterase (AChE), an enzyme that catalyzes the hydrolysis of acetylcholine, a crucial neurotransmitter in insects, rodents, and humans' central nervous systems, is inhibited by some plant protection solutions' active ingredients (Rastija et al., 2022).

Additionally, research by Abbod et al. (2020) on the biological functions and mode of action of 3-butylidene phthalide (3-BPH) revealed that 3-BPH suppressed *M. phaseolina's* mycelial growth and generation of microsclerotia. 3-BPH disrupted *M. phaseolina's* ability to produce melanin, most likely by inhibiting 1,3,8 trihydroxynaphthalenereductase. It displayed inhibitory effect against phenol oxidase and acetylcholinesterase. At high doses (1000 mg/L), it also demonstrated phytotoxicity, but not at doses below 250 mg/L, which suppressed *M. phaseolina* microsclerotia. They observed that 3-BPH was more effective than the common fungicide tricyclazole at preventing *M. phaseolina* microsclerotia. At the concentrations that prevented *M. phaseolina* microsclerotia and dark pigments, it was harmless. However, it was found from their observations that there was no phytotoxicity at low concentrations and that *M. phaseolina* was not influenced.

15.7 Development of tolerance by *M. phaseolina* against drought

Many of the research findings indicate the process of resistance developed by the plants, but the there is no such evidence found against the development of tolerance in *M. phaseolina.* So, in below points, we have focused our discussion on tolerance development in plants against the infection by *M. phaseolina*. The main factor affecting yield loss globally is drought. According to reports, there is a direct association between dryness and the severity of damage. When plants are challenged by drought, charcoal rot incidence and severity are increased. Due to the effects of heat stress and drought on the severity of the disease, the overall amount of soybean yield suppression caused by charcoal rot varied across years. In certain popular soyabean genotypes, resistance to this disease was discovered to be correlated with drought tolerance. There is no known soybean genotype resistant to *M. phaseolina*, but a few have been categorized as moderately resistant due to less root colonization by this pathogen than other genotypes (Wrather et al., 2008).

15.7.1 Metabolic and transcriptomic alterations

Plants with the fungus *M. phaseolina* in their stalks can induce lodge and lose yield, especially under dry conditions. These fungi are the cause of charcoal rot in a number of the host plants. Field crops are routinely exposed to a range of biotic and abiotic challenges as well as challenging weather conditions, and this situation will only get worse as the consequences of climate change become more pronounced. Some isolates of *M. phaseolina* are viable at temperatures higher than 40°C (Sexton et al., 2016). Although they have previously been labeled as necrotrophic diseases, they may first go through a biotrophic phase (Chowdhury et al., 2017). The most detrimental diseases to production are stalk rots, which can have an incidence of up to 100% in some fields and cause lodging and large biomass losses as a result of difficult harvesting. For instance, sorghum [*Sorghum bicolor*] tolerates heat and drought and needs little nitrogen and water. Despite its resistance to many common stresses, the productivity of the plant is threatened by a number of fungal stalk infections (Smith & Frederiksen, 2000).

The plant reaction to stress involves secondary lignification of the cell walls, which is a critical step. Infection, herbivory, and abiotic stressors are all protected from by the firm, hydrophobic barrier and mechanical support that lignin offers. The biosynthesis of lignin, a complex and diversely cross-linked polymer that connects primary metabolism to secondary metabolism occurs via the phenylpropanoid pathway via phenylalanine ammonia lyase (PAL) (Scully et al., 2018; Tetreault et al., 2018). Lignin is composed primarily of phenylpropanoids. The intermediates of the phenylpropanoid cycle can enter a number of metabolic pathways, activating signal transduction systems that result in defense responses and resistance to herbivores and pathogens. The final two steps of monolignol biosynthesis, caffeic acid O-methyltransferase

(COMT) and cinnamyl alcohol dehydrogenase (CAD), are catalyzed by the sorghum genes Brown midrib (Bmr)-12 and Bmr6, respectively. The Bmr6-ref and Bmr12-ref alleles are completely functionally inactive as a result of nonsense mutations. These mutants' cell walls contain less lignin and lignin that is different in composition from the wild-type. Inhibition of any of the last stages of monolignol production caused soluble and cell wall-bound hydroxycinnamic acids to build up in Bmr6 and Bmr12 plants (Palmer et al., 2008). The study of these enzymes' functions in disease responses has been made easier by these mutations. In both well-watered and dry environments, the lignin-altered lines that were susceptible to or resistant to *M. phaseolina* were examined by Maya et al. (2021).

The monolignol biosynthesis pathway is a part of a hugely interconnected metabolic grid whose alteration can have an impact on a wide range of other processes, and the cell wall is a hub of stress surveillance and response (Scully et al., 2018). The evaluation of the associated metabolic and transcriptome variations on the response to disease and drought is made possible by changing the structure of cell walls. Enzymes in the monolignol biosynthesis pathway in plants had an impact on the co-expression of genes in multiple pathways, including plant hormone signal transduction, RNA and protein processing and turnover, transcription, and translation. It has been noted that the monolignol biosynthetic enzymes and immune system components interact. For instance, patterns of primary and secondary cell wall biosynthetic gene expression as well as hypothesized regulatory systems that react to both disease and drought imply that the location of complex links between growth, development, and defense, is the cell wall. Due to the activation of defense pathways, dry circumstances may have further prepared the bmr12 plants utilized in the study for disease resistance. Studies revealed that Monolignol mutant lines are not more prone to disease and, depending on the environment, may even be more resistant. It was shown that priming, which may also be a method to increase crop disease resistance while giving reduced lignin lines for the production of bioenergy or feed, was associated to Maya et al. (2021)'s pathway induction.

15.7.2 Role of phytohormones

It was discovered that phytohormones start defense-related signal transduction cascades. Khasin et al. (2021) tested a wide range of phytohormones in tissues that were also analyzed for hydroxycinnamic acids. Syringic acid levels in the plants were substantially lower than those of salicylic acid (SA), jasmonic acid (JA), and gibberellin A19 (GA19). They only looked at 12 plants in all, and only IAA Aspartate (IAA-Asp) was discovered. According to the study, bmr12 plants had higher levels of SA and JA and a different hydroxycinnamic acid composition than wild-type plants. As a result, lignin alteration could lead to improved stress reactions and increase resistance. If there is a water shortage, flavonoids, ROS signaling, and ethylene signaling may all be changed. These mechanisms might work together to combat disease. To go at these correlations in more detail, co-expression analysis was studied. Pathways involved in primary metabolism, protein turnover, phenylpropanoid biosynthesis, and ETI elements of plant—pathogen interaction were enriched in co-expression modules linked with fungus infection. The early defense response involved genes involved in cell wall maintenance, including laccases, peroxidases, and genes that produce phenylpropanoid. Shorter lesion durations may be caused by cell wall alteration and water restriction that overlap or activate downstream defense mechanisms. A variety of genes encoding drought response proteins, including an ortholog of the diligent protein, were also linked to water restriction. The production of tricin synthase I and acetylserotoninOmethyltransferase I was linked to water restriction. Flavonoids and phenylpropanoids are crucial early components of several plant defensive responses (Bilska et al., 2018). Several priming genes that produce chalcone synthases were also discovered to be related with water shortage indicating a potential role in the heightened immunity brought on by drought. These genes contained a rice orthologous sulfoquinovosyl transferase with flavonoid glycosylation activity (Zhan et al., 2017, 2019). The rice sulfoquinovosyl transferase was overexpressed, which increased drought resistance. Khasin et al. (2021) also investigated the relationship between bmr12 and water restriction and probable sulfoquinovosyl transferase. A coordinated response was demonstrated by the co-expression of these flavonoid biosynthesis enzymes with a number of disease-responsive PR proteins and chitinases. Fungus infection was associated with additional modules that were enriched for ribosomal proteins, protein synthesis in the ER, ubiquitin-mediated proteolysis, and proteasome. This emphasizes the critical role of protein turnover in the pathogen response as well as the increased generation of defensive enzymes and other defensive proteins that facilitate a speedier and more potent resistance response. Drought conditions may activate defense systems in bmr12 plants. Drought may be an environmental trigger of priming in bmr12 plants due to the diversity of pathways associated to defense priming in this work and in earlier studies, which raises the possibility that there may be many, interconnected methods to activate and change these pathways.

15.7.3 Synthesis of compounds—pathogenesis-related (PR) proteins and phytoalexins

When two sorghum cultivars interact with *M. phaseolina* during the first hour of infection, chitinase and stilbene synthetase are overexpressed indicating that these genes may contribute to enhancing sorghum resistance. Sharma et al. (2014) also investigated the Charcoal rot of sorghum brought on by *M. phaseolina* (Sharmaetal., 2014). *M. phaseolina* causes infections in plants at nearly all growth stages and is spread through leftover seed, soil, and plant material. The obstruction of xylem vessels caused by the colonization of the charcoal rot pathogen causes the death of the seedling. The development of dark mycelia and black microsclerotia, as well as the formation of red to brown lesions on the roots and stems, causes plant defoliation and wilting. *M. phaseolina* is challenging to eradicate because it persists as sclerotia in soil and plant detritus. However, the primary reasons for low levels of productivity in sorghum are the drought and the illness known as charcoal rot, which normally appears at the stage of grain maturity and causes severe lodging. Plants that are affected by sorghum charcoal rot lodge as they become mature. By stimulating responses like an oxidative burst in the cell, altering the composition of the cell to prevent infection, and synthesizing substances like phytoalexins and pathogenesis-related (PR) proteins from scratch, plants exhibit an innate pathogen-specific resistance. Production of PR proteins is the most crucial of all induced responses because it can boost the plant as a whole's resistance to a pathogenic onslaught. One of the PR proteins thought to be the most significant is chitinase, since it has the potential to boost the plant as a whole's resistance to a pathogenic onslaught. One of the main components of many pathogenic fungal cell walls, chitin, is hydrolyzed by chitinases. In the presence of oxygen, the chitinases degrade the chitin polymers in the cell wall, weakening the cell wall and causing fungal cells osmotically sensitive.

On the basis of gene expression, numerous investigations on plants' stress defense mechanisms have been performed. Innate defense mechanisms in sorghum against pathogens require PR proteins. In sorghum, a number of chitinases have been identified. Although it was shown that the concentration of these chitinases increased as the caryopsis developed, their antifungal action was not confirmed. The 3-deoxyanthocyanidins, a special family of flavonoid phytoalexins, are also produced by sorghum and are a crucial part of its defense system against pathogen invasion. According to studies, the enzyme stilbene synthase generates the phytoalexin resveratrol that confers plants resistance to fungi. Sorghum was the first monocot to carry the stilbene synthase (SbSTS1) gene, and it is known to help plants defend themselves from diseases (Sharma et al., 2014).

15.7.4 Tolerance due to type of soil and microsclerotia formation

Microsclerotial survival, root infection, and disease development are all influenced by a variety of environmental conditions. Microsclerotia can live for many years in dry soil but for a short time in saturated soil. Although the amount of microsclerotia (inoculum density) may be higher in soils with pH levels outside the range best for soy production, soil pH may not affect the survival of microsclerotia (Chamorro, Miranda, et al., 2015). Even though the charcoal rot fungus can infect plants early in the season and spread throughout the season, symptoms may not appear until the infected plants are under stress. The type and severity of symptoms that manifest will depend on the timing and length of these conditions, which are frequently characterized by excessive heat and drought. When it happens throughout the soybean reproductive growth phases, hot and dry conditions is when charcoal rot symptoms are most common. However, in irrigated systems and in crops with no obvious signs, illness and subsequent production loss have been seen (Liu et al., 2022).

15.7.5 Tolerance based on irrigated and nonirrigated soil types and developmental stages of plant

At the R8 stages of growth and development, *M. phaseolina* colonizes soybean roots more frequently in nonirrigated than irrigated soybeans; however, this was only true up to the R2 growth stages of soybean (Crop Protection Network, 2015).

15.7.6 Development of soybean cultivars and genotypes resistant to charcoal rot

The degree to which *M. phaseolina* colonizes the root tissue varies among soybean cultivars and lines (genotypes). For instance, Paris et al. (2006) recently released a genotype with moderate resistance to charcoal rot because *M. phaseolina* colonization of the root tissue was decreased in this genotype. Additionally, there are genotypes of soybean that are either moderately resistant (MR) or sensitive (S) to charcoal rot (Alemu et al., 2018). However, the mechanism of resistance is unknown. Under irrigated and nonirrigated circumstances, MR genotype seeds had significantly higher amounts of phenolics, seed coat lignin, isoflavones, sugars, and total boron than S genotype seeds, pointing to a potential correlation between these compounds and resistance to toxin-mediated infection (Bellaloui et al., 2012). According to the explanation above, changes in sorghum cell wall structure allow for changes in metabolism and transcriptome responses to disease and drought. Through the activation of defense mechanisms and the production of many enzymes and phytohormones, drought-related conditions in plants promote disease resistance. Therefore, more research is required to understand how the interactions between the host and the fungus in drought conditions result in the development of fungal tolerance. But based on the above research, we may infer that fungi may have evolved a tolerance to drought as a result of the production of compounds of host plants or the activation of resistance pathways.

Another study found that plants exhibit an innate pathogen-specific resistance through the de novo creation of substances like phytoalexins and pathogenesis-related (PR) proteins including chitinase and stilbene synthase enzyme, as well as an oxidative burst of cell activity. The composition of the cell wall is altered to prevent infection in addition to all other such responses. The most significant of all induced responses is the production of PR proteins since it can increase the plant's overall resistance to a pathogenic infection (Sharma et al., 2014).

15.8 Conclusion

M. phaseolina is the major problem for charcoal rot disease in many plant species all over the world. Its pathogenicity mechanisms as well as the host defensive mechanism, which include the synthesis of toxins and other enzymes, are leading to cause resistance in host plants. As a consequence, the research on the fungicide, insecticide, and drought tolerance in *M. phaseolina* and resistance in plants due to release of some toxins and metabolites are the major issues, which need to be more focused. In order to extend the active life of possibly dangerous fungicides and insecticides, growers should employ resistance mitigation strategies that crop advisers recommend.

Also, changes in cell wall structure of host plants which allow for activating the metabolism and transcriptome responses to disease and drought through the production of many enzymes and phytohormones that result in drought-related conditions in plants and promote disease resistance need to be implemented to control the conditions. Additional host resistance sources, other approaches and research are needed to determine whether the impacts of drought and *M. phaseolina* infection are additive, synergistic, or independent. The mechanism of tolerance in *M. phaseolina* requires more study.

References

Abbas, H. K., Bellaloui, N., Butler, A. M., Nelson, J. L., Abou-Karam, M., & Shier, W. T. (2020). Phytotoxic responses of soybean (*Glycine max* L.) to botryodiplodin, a toxin produced by the charcoal rot disease fungus, *Macrophominaphaseolina*. *Toxins, 12*(1), 25. https://doi.org/10.3390/toxins12010025

Abbod, M., Safaie, N., Gholivand, K., Mehrabadi, M., & Bonsaii, M. (2020). Mode of action of 3-butylidene phthalide as a competent natural pesticide. *Pesticide Biochemistry and Physiology, 164*, 228–236.

Adhikary, N. K., Chowdhury, M. R., Begum, T., & Mallick, R. (2019). Integrated management of stem and root rot of sesame (*Sesamumindicum* L.) caused by *Macrophominaphaseolina* (Tassi) Goid. *International Journal of Current Microbiology and Applied Sciences, 8*, 804–808. https://doi:10.20546/ijcmas.2019.804.089

Ainsworth, G. C., & Bisby, G. R. (1950). *A dictionary of the fungi*. Kew, England: Commonwealth Mycological Institute.

Ainsworth, G. C., & Bisby, G. R. (1971). *Dictionary of the fungi*. Commonwealth Mycological Institute.

Alemu, M., Jeffery, D. R., James, R. S., Prakash, R. A., Nacer, B., Pengyin, C., Grover, S., & Debbie, B. (2018). Effect of charcoal rot on selected putative drought tolerant soybean genotypes and yield. *Crop Protection, 105*, 90–101. https://doi.org/10.1016/j.cropro.2017.11.012

Baard, S. W., Van Wyk, P. W. J., & Pauer, G. D. C. (1981). Structure and lysis of microsclerotia of *Verticilliumdahliae* in soil. *Transactions of the British Mycological Society, 77*, 251–260. https://doi.org/10.1016/S0007-1536(81)80027-7

Babu, B. K., Saikia, R., & Arora, D. K. (2010). Molecular characterization and diagnosis of *Macrophominaphaseolina*: A charcoal rot fungus. In Y. Gherbawy, & K. Voigt K (Eds.), *Molecular identification of fungi* (pp. 179–193). New York: Springer.

Bandara, Y. M. A. Y., Weerasooriya, D. K., Liu, S., & Little, C. R. (2018). The necrotrophic fungus *Macrophominaphaseolina* promotes charcoal rot susceptibility in grain sorghum through induced host cell-wall-degrading enzymes. *Phytopathology, 108*, 948–956. https://doi:10.1094/PHYTO-12-17- 0404-R

Beckerman, J. (2008). *Understanding fungicide mobility*. Purdue Extension BP-70-W.

Bellaloui, N., Mengistu, A., Smith, J. R., Abbas, H. K., Accinelli, C., & Shier, W. T. (2021). Effects of charcoal rot on soybean seed composition in soybean genotypes that differ in charcoal rot resistance under irrigated and non-irrigated conditions. *Plants, 10*(9), 1801. https://doi.org/10.3390/plants10091801

Bellaloui, N., Mengistu, A., Zobiole, L. H. S., & Shier, W. T. (2012). Resistance to toxin-mediated fungal infection: Role of lignins, isoflavones, other seed phenolics, sugars, and boron in the mechanism of resistance to charcoal rot disease in soybean. *Toxin Reviews, 31*, 16–26. https://doi:10.3109/15569543.2012

Bhattacharya, G., Dhar, T. K., Kathleen Bhattacharya, F., & Siddiqui, K. A. (1987). Mutagenic action of phaseolinone, a Mycotoxin isolated from *Macrophominaphaseolina*. *Australian Journal of Biological Sciences, 40*, 349–353. https://doi:10.1071/BI9870349

Bilska, K., Stuper-Szablewska, K., Kulik, T., Buśko, M., Załuski, D., Jurczak, S., & Perkowski, P. (2018). Changes in phenylpropanoid and trichothecene production by *Fusariumculmorum* and *F. graminearumSensuStricto* via exposure to flavonoids. *Toxins, 10*(3), 110. https://doi.org/10.3390/toxins10030110

Chamorro, M., Domínguez, P., Medina, J. J., Miranda, L., Soria, C., Romero, F., Aranda, L., Daugovish, O., Mertely, J., & Santos, B. (2015). Assessment of chemical and biosolarization treatments for the control of *Macrophominaphaseolina* in strawberries. *Scientia Horticulturae, 192*, 361–368.

Chamorro, M., Miranda, L., Domínguez, P., Medina, J. J., Soria, C., & Romero, F. (2015). Evaluation of biosolarization for the control of charcoal rot disease (*Macrophominaphaseolina*) in strawberry. *Crop Protection, 67*, 279–286. https://doi.org/10.1016/j.cropro.2014.10.021

Chatterjee, S., Chatterjee, B. P., & Guha, A. K. (2014). International journal of biological macromolecules a study on antifungal activity of water-soluble chitosan against *Macrophominaphaseolina*. *International Journal of Biological Macromolecules, 67*, 452–457.

Chowdhury, S., Basu, A., & Kundu, S. (2017). Biotrophy-necrotrophy switch in pathogen evoke differential response in resistant and susceptible sesame involving multiple signaling pathways at different phases. *Scientific Reports, 7*(1), 17251. https://doi.org/10.1038/s41598-017-17248-7

Cohen, R., Elkabetz, M., & Edelstein, M. (2016). Variation in the responses of melon and watermelon to *Macrophominaphaseolina*. *Crop Protection, 85*, 46–51.

Cohen, R., Omari, N., Porat, A., & Edelstein, M. (2012). Management of *macrophomina* wilt in melons using grafting or fungicide soil application: Pathological, horticultural and economical aspects. *Crop Protection, 35*, 58–63. https://doi.org/10.1016/j.cropro.2011.12.015

Coley-Smith, J. R., & Cooke, R. C. (1971). Survival and germination of fungal sclerotia. *Annual Review of Phytopathology, 9*, 65–92. https://doi.org/10.1146/annurev.py .09.090171.000433

Cook, G. E., Boosalis, M. G., Dunkle, L. D., & Odvody, G. N. (1973). Survival of *Macrophominaphaseoli* in corn and sorghum stalk residue. *Plant Disease Reporter, 57*(10), 873–875.

Crop Protection Network. (2015). *A product of land grant universities*. https://doi.org/10.31274/cpn-20190621-002. Published: 02/25/2022, CPN-0002-15.

Crous, P. W., Slippers, B., Wingfiel, M. J., Rheeder, J., Marasas, W. F. O., & Philips, A. J. L. (2006). Phylogenetic lineages in the botryosphaeriaceae. *Studies in Mycology, 55*, 235–253. https://doi.org/10.3114/sim.55.1.235

Davidse, L. C., & Flach, W. (1978). Interaction of thiabendazole with fungal tubulin. *Biochimicaet Biophysica Acta (BBA)-General Subjects, 543*(1), 82–90.

Dean, R., Van Kan, J. A., Pretorius, Z. A., Hammond-Kosack, K. E., Di Pietro, A., & Spanu, P. D. (2012). The Top 10 fungal pathogens in molecular plant pathology. *Molecular Plant Pathology, 13*(4), 414–430. https://doi.org/10.1111/j.1364-3703.2011.00783.x

Degani, O., Danielle, R., & Dor, S. (2021). The microflora of maize grains as a biological barrier against the late wilt causal agent, Magnaporthiopsismaydis. *Agronomy, 11*, 965.

Degani, O., Dor, S., Abraham, D., & Cohen, R. (2020). Interactions between *Magnaporthiopsismaydis* and *Macrophominaphaseolina*, the causes of wilt diseases in maize and cotton. *Microorganisms, 8*, 249.

Degani, O., Dor, S., Movshovitz, D., & Rabinovitz, O. (2019). Methods for Studying *Magnaporthiopsismaydis*, the maize late wilt causal agent. *Agronomy, 9*, 181.

Dhingra, O. D., & Sinclair, J. B. (1978). *Biology and pathology of* Macrophominaphaseolina. Minas Gerais: Universidade Federal de Viçosa: Impresa Universitaria. https://doi:10.1016/j.ijbiomac.2014.04.008

Dor, S., & Degani, O. (2019). Uncovering the host range for maize pathogen *Magnaporthiopsismaydis*. *Plants, 8*(8), 259.

Georgiou, C. D., Patsoukis, N., Papapostolou, I., & Zervoudakis, G. (2006). Sclerotial metamorphosis in filamentous fungi is induced by oxidative stress. *Integrative and Comparative Biology, 46*, 691–712. https://doi.org/10.1093/icb/icj034

Ghosh, T., Biswas, M. K., Guin, C., & Roy, P. (2018). A review on characterization, therapeutic approaches and pathogenesis of *Macrophominaphaseolina*. *Plant Cell Biotechnology and Molecular Biology, 19*, 72–84.

Glazebrook, J. (2005). Contrasting mechanisms of defense against biotrophic and necrotrophic pathogens. *Annual Review of Phytopathology, 43*, 205–227. https://doi.org/10.1146/annurev.phyto.43.040204.135923

Gupta, G. K., Sharma, S. K., & Ramteke, R. (2012). Biology, epidemiology and management of the pathogenic fungus *Macrophominaphaseolina* (Tassi) goid with special reference to charcoal rot of soybean (*Glycine max* (L.) Merrill). *Journal of Phytopathology, 160*, 167–180. https://doi.org/10.1111/j.1439-0434.2012.01884.x

Iqbal, U., & Mukhtar, T. (2020). Inhibitory effects of some fungicides against *Macrophominaphaseolina* causing charcoal rot. *Pakistan Journal of Zoology, 52*, 709–715. https://doi.org/10.17582/journal.pjz/20181228101230

Islam, M., Haque, M., Islam, M., Emdad, E., Halim, A., & Hossen, Q. M. (2012). Tools to kill: Genome of one of the most destructive plant pathogenic fungi *Macrophominaphaseolina*. *BMC Genomics, 13*, 493. https://doi.org/10.1186/1471-2164-13-493

Jayakannan, M., Bose, J., Babourina, O., Rengel, Z., & Shabala, S. (2015). Salicylic acid in plant salinity stress signaling and tolerance. *Plant Growth Regulation, 1*, 25–40. https://doi.org/10.1007/s1072 5-015-0028-z

John, D., & Smith, D. (2017). *Fungicide resistance Management.Division of agricultural sciences and natural resources* (pp. 1–8). Oklahoma State University, 7633 http://osufacts.okstate.edu

Kazan, K., & Lyons, R. (2014). Intervention of phytohormone pathways by pathogen effectors. *The Plant Cell Online, 26*, 2285–2309. https://doi.org/10.1105/tpc.114.125419

Khasin, M., Bernhardson, L. F., O'Neill, P. M., Palmer, N. A., Scully, E. D., Sattler, S. E., & Funnell-Harris, D. L. (2021). Pathogen and drought stress affect cell wall and phytohormone signaling to shape host responses in a sorghum COMT bmr12 mutant. *BMC Plant Biology, 21*(1), 1–24.

Kumar, S. (2019). Study on charcoal rot of maize. *Journal of Advances in Scholarly Researches and Allied Education, 16*(5), 1400–1404. http://ignited.in/I/a/303810

Lakhran, L., Ahir, R. R., Choudhary, M., & Choudhary, S. (2018). Isolation, purification, identification and pathogenicity of *Macrophominaphaseolina*(Tassi) goid caused dry root rot of chickpea. *Journal of Pharmacognosy and Phytochemistry, 7*, 3314–3317.

Lentz, P. L., & McKay, H. H. (1970). Sclerotial formation in *Corticiumolivascens* and other hymenomycetes. *Mycopathologiaet Mycologia Applicata, 40*, 1–13. https://doi.org/10.1007/BF02128659

Liu, H. H., Huang, C. C., Lin, Y. H., Tseng, M. N., & Chang, H. X. (2022). Superoxide initiates the hyphal differentiation to microsclerotia formation of. *Macrophominaphaseolina.Microbiology Spectrum, 10*(1), e0208421. https://doi.org/10.1128/spectrum.02084-21

Lokesh, R., Rakholiya, K. B., & Thesiya, M. R. (2020). Evaluation of different fungicides against *Macrophominaphaseolina* (Tassi) goid.causing dry root rot of chickpea (*Cicerarietinum* L.) *in vitro*. *International Journal of Current Microbiology and Applied Sciences, 9*, 1–11. https://doi.org/10.20546/ijcmas.2020.907

Mannaa, M., & Seo, Y. S. (2021). Plants under the attack of allies: Moving towards the plant pathobiome paradigm. *Plants, 10*(1), 125.

Marquez, N., Giachero, M. L., Declerck, S., & Ducasse, D. A. (2021). *Macrophominaphaseolina*: General characteristics of pathogenicity and methods of control. *Frontiers of Plant Science, 12*, 634397. https://doi.org/10.3389/fpls.2021.634397

Martínez-Salgado, S. J., Andrade-Hoyos, P., ParraguirreLezama, C., Rivera-Tapia, A., Luna-Cruz, A., & Romero-Arenas, O. (2021). Biological control of charcoal rot in peanut crop through strains of *Trichoderma* spp. in Puebla, Mexico. *Plants, 10*(12), 2630. https://doi.org/10.3390/plants10122630

Maya, K., Lois, F. B., Patrick, M., Palmer, Nathan, A., Scully, Erin, D., Sattler, Scott, E., & Deanna, L. (2021). Pathogen and drought stress affect cell wall and phytohormone signaling to shape host responses in a sorghum COMT bmr12 mutant. *BMC Plant Biology, 21*, 391. https://doi.org/10.1186/s12870-021-03149-5

Mommer, L., Cotton, T. E. A., Raaijmakers, J. M., Termorshuizen, A. J., van Ruijven, J., Hendriks, M., Sophia, Q. R., Judith, E. M., Jan Willem, P., Elio, M. S., Annemiek, E. S., Frank, B., Hans, K., & Alex, J. D. (2018). Lost in diversity: The interactions between soil-borne fungi, biodiversity and plant productivity. *New Phytologist, 218*, 542–553.

Palmer, N. A., Sattler, S. E., Saathoff, A. J., Funnell, D., Pedersen, J. F., & Sarath, G. (2008). Genetic background impacts soluble and cell wall-bound aromatics in brown midrib mutants of sorghum. *Planta, 229*(1), 115–127.

Paris, R. L., Mengistu, A., Tyler, J. M., & Smith, J. R. (2006). Registration of soybean germplasm line DT97-4290 with moderate resistance to charcoal rot. *Crop Science, 46*, 2324–2325.

Parmar, H. V., Kapadiya, H. J., & Bhaliya, C. M. (2017). Efficacy of different fungicides against *Macrophominaphaseolina* (Tassi) Goid causing castor root rot. *International Journal of Chemical Studies, 5*, 1807–1809.

Ramezani, M., Shier, W. T., Abbas, H. K., Tonos, J. L., Baird, R. E., & Sciumbato, G. L. (2007). Soybean charcoal rot disease fungus *Macrophominaphaseolina* in Mississippi produces the phytotoxin (−)-botryodiplodin but no detectable phaseolinone. *Journal of Natural Products, 70*(1), 128–129.

Ramos, A. M., Gally, M., Szapiro, G., Itzcovich, T., Carabajal, M., & Levin, L. (2016). In vitro growth and cell wall degrading enzyme production by Argentinean isolates of *Macrophominaphaseolina*, the causative agent of charcoal rot in corn. *Revista Argentina de Microbiología, 48*(4), 267–273.

Rastija, V., Vrandečić, K., Ćosić, J., Šarić, G. K., Majić, I., Agić, D., & Molnar, M. (2022). Effects of coumarinylschiff bases against phytopathogenic fungi, the soil-beneficial bacteria and entomopathogenic nematodes: Deeper insight into the mechanism of action. *Molecules, 27*(7), 2196.

Reznikov, S., Vellicce, G. R., González, V., de Lisi, V., Castagnaro, A. P., & Ploper, L. D. (2016). Evaluation of chemical and biological seed treatments to control charcoal rot of soybean. *Journal of General Plant Pathology, 82*, 273–280. https://doi.org/10.1007/s10327- 016-0669-4

Reznikov, S., Vellicce, G. R., Mengistu, A., Arias, R. S., Gonzalez, V., & De Lisi, V. (2018). Disease incidence of charcoal rot (*Macrophominaphaseolina*) on soybean in north-western Argentina and genetic characteristics of the pathogen. *Canadian Journal of Plant Pathology, 40*, 423–433. https://doi.org/10.1080/07060661.2018.1484390

Salvatore, M. M., Félix, C., Lima, F., Ferreira, V., Naviglio, D., & Salvatore, F. (2020). Secondary metabolites produced by *Macrophominaphaseolina* isolated from eucalyptus globulus. *Agriculture, 10*, 72. https://doi.org/10.3390/agriculture10030072

Sarkar, T. S., Biswas, P., Ghosh, S. K., & Ghosh, S. (2014). Nitric oxide production by necrotrophic pathogen *Macrophominaphaseolina* and the host plant in charcoal rot disease of jute: Complexity of the interplay between necrotroph–host plant interactions. *PLoS One, 9*(9), e107348.

Scully, E. D., Gries, T., Palmer, N. A., Sarath, G., Funnell-Harris, D. L., Baird, L., Twigg, P., Seravalli, J., Clemente, T. E., & Sattler, S. E. (2018). Overexpression of SbMyb60 in Sorghum bicolor impacts both primary and secondary metabolism. *New Phytologist, 217*(1), 82–104. https://doi.org/10.1111/nph.14815

Sexton, Z. F., Hughes, T. J., & Wise, K. A. (2016). Analyzing isolate variability of *Macrophominaphaseolina* from a regional perspective. *Crop Protection, 81*, 9–13. https://doi.org/10.1016/j.cropro.2015.11.012

Sharma, I., Kumari, N., & Sharma, V. (2014). Defense gene expression in Sorghum bicolor against *Macrophominaphaseolina* in leaves and roots of susceptible and resistant cultivars. *Journal of Plant Interactions, 9*(1), 315–323.

Smith, R. E. (1900). Botrytis and sclerotinia: Their relation to certain plant diseases and to each other. *Botanical Gazette, 29*, 369–407. https://doi.org/10.1086/328001

Smith, C. W., & Frederiksen, R. A. (Eds.). (2000). *Sorghum: Origin, history, technology, and production* (Vol 2). John Wiley and Sons.

Swamy, C., Naik, M. K., Amaresh, Y. S., & Jayalakshmi, S. K. (2018). Evaluation of fungicides and bio-agents under in vitro condition against *macrophominaphaseolina*causing stem canker of pigeonpea. *International Journal of Current Microbiology and Applied Sciences, 7*, 811–819. https://doi.org/10.20546/ijcmas.2018.701.099

Tej, R., Rodríguez-Mallol, C., Rodríguez-Arcos, R., Karray-Bouraoui, N., & Molinero-Ruiz, L. (2018). Inhibitory effect of *Lyciumeuropaeum* extracts on phytopathogenic soil-borne fungi and the reduction of late wilt in maize. *European Journal of Plant Pathology, 152*, 249–265.

Tetreault, H. M., Scully, E. D., Gries, T., Palmer, N. A., Funnell-Harris, D. L., & Baird, L. (2018). Overexpression of the Sorghum bicolor SbCCoAOMT alters cell wall associated hydroxycinnamoyl groups. *PLoS One, 13*(10), e0204153. https://doi.org/10.1371/journal.pone.0204153

Tok, F. M. (2019). Relationship between morphologic, phenotypic and pathogenic characteristics in *Macrophominaphaselina* Isolates from cucumber plants. *The International Journal of Innovative Approaches in Agricultural Research, 3*, 651–660. https://doi.org/10.29329/ijiaar.2019.217.11

Tonin, R. F. B., Avozani, A., Danelli, A. L. D., Reis, E. M., Zoldan, S. M., & Garcés-Fiallos, F. R. (2013). *In vitro*mycelial sensitivity of *Macrophominaphaseolina* to fungicides. *Pesquisa Agropecuária Tropical, 43*, 460–466.

Wrather, J. A., Shannon, J. G., Carter, T. E., Bond, J. P., Rupe, J. C., & Almeida, A. M. R. (2008). Reaction of drought-tolerant soybean genotypes to *Macrophominaphaseolina. Plant Health Progress, 9*(1). https://doi.org/10.1094/PHP-2008-0618-01-RS

Wyllie, T., & Brown, M. (1970). Ultrastructural formation of sclerotia of *Macrophominaphaseolina. Phytopathology, 60*, 524–528. https://doi.org/10.1094/Phyto -60-524

Zhang, X., Xue, C., Fang, D., He, X., Wei, M., Zhuo, C., et al. (2021). Manipulating the soil microbiomes during a community recovery process with plant beneficial species for the suppression of *Fusarium* wilt of watermelon. *AMB Express, 11*, 87.

Zhan, X., Shen, Q., Chen, J., Yang, P., Wang, X., & Hong, Y. (2019). Rice sulfoquinovosyltransferase SQD2. 1 mediates flavonoid glycosylation and enhances tolerance to osmotic stress. *Plant, Cell and Environment, 42*(7), 2215–2230. https://doi.org/10.1111/pce.13554

Zhan, X., Shen, Q., Wang, X., & Hong, Y. (2017). The Sulfoquinovosyltransferase-like Enzyme SQD2.2 is involved in flavonoid glycosylation, regulating sugar metabolism and seed setting in rice. *Scientific Reports, 7*, 4685. https://doi.org/10.1038/s41598-017-04002-2

Chapter 16

Management of *Macrophomina phaseolina* by cultural practices

Shubhra Singh, Hema Prashad and Himani Gautam
Department of Entomology, Dr. Yashwant Singh Parmar University of Horticulture and Forestry, Solan, Himachal Pradesh, India

16.1 Introduction

Macrophomina phaseolina is an ascomycetous phytopathogenic fungus that has about 500 host plant species of various crops and weeds, which can be roughly summed up to approximately 100 families (Khan, 2007). It is nonspecific, globally widespread and responsible for diseases like seedling blight, rotting of stems, and roots, and damping off (Babu et al., 2007; Dhingra & Sinclair, 1978; Ghosh et al., 2018). During peak temperatures (30−35°C) and depleted soil moisture (less than 60%), the fungus causes a consequential decline in crop yields, such as sorghum and soybean, drastically affecting farmer income (Aegerter et al., 2000; Kaur et al., 2012). When the disease appears before the emergence and turns extreme, groundnut cultivars may lose up to 100% of their yield (Sharma & Bhowmik, 1986). Another vital feature of this fungus is that as per the environmental conditions, its microsclerotia can survive for a long time on infected plant material and soil, as a saprophyte for 2 years in soybean root residue and as microsclerotia for 15 years (Baird et al., 2003; Vasebi et al., 2013).

M. phaseolina is also an opportunistic intermittent human pathogen that is capable of causing infections, like cutaneous cellulitis and/or ocular keratitis in immunocompromised patients receiving antifungal prophylaxis (Arora et al., 2011; Schwartz & Kapila, 2020; Srinivasan et al., 2009; Tan et al., 2008).

16.2 Morphological characteristics

M.phaseolina (Tassi) Goid.belongs to the phylum Ascomycota, class Dothidiomycetes, order Botryosphaeriales, and family Botryosphaeriaceae. Based on morphological or genomic data, this fungus currently lacks any subspecies or physiological races (Crous et al., 2006; Dhingra & Sinclair, 1978). The fungus is famous for causing dry root or charcoal rot of pulses.

The hyphae of *M. phaseolina* are light or dark brown in color, hyaline, thin walled with septa. Its constricted branches are at a right angle from the parent hyphae and mycelium masses/microsclerotia are spherical, oval, hard, oblong when young, and darken (brown to black) with age. Pycnidia, on the other hand, are larger than microsclerotia, have a darker color, are rough, globose, or irregular in shape, are beaked, circular, central, and are surrounded by thick darkened cell walls. These structures are very rare in nature (Lakhran et al., 2018). The color of mycelia, pycnidia formation, distribution of microsclerotia, and chlorate phenotypes lead to differences in artificially grown isolates. Tok (2019) has discovered a strong link between morphological variations and virulence, suggesting that phenotypic heterogeneity could be credited to fungal responses to the environmental factors or host-specific variations (Pandey et al., 2020; Tok, 2019).

Morphological identification of *M. phaseolina* is difficult and frequently inaccurate due to isolate variation (Babu et al., 2010; Saleh et al., 2009). Even though biochemical and serological techniques are explored to identify the fungus, their specificity is limited (Srivastava & Arora, 1997). However, molecular technique advancements have provided alternative methods for identifying this fungus.

Macrophomina Phaseolina. https://doi.org/10.1016/B978-0-443-15443-0.00013-9

16.3 Disease cycle

The primary infective source of the fungus, microsclerotia invades the seedlings through their roots, causing sudden wilting and rotting of young succulent seedlings. This resistant structure can last for upto 15 years in the soil (Gupta et al., 2012). During the seedling stage, the host plant roots can get infected by various mycelial growths. Once inside the roots, the fungus starts attacking the plant vascular system, disrupting the transportation of water and nutrient to aerial parts. The petiole leaves attached to the stem begin to yellow and wither, the cortical tissues clump, and their appearance changes to gray from the stem part closer to the ground and main root due to an abundance of microsclerotia, all of which lead to the plant's premature death (Short et al., 1978; Sinclair & Backman, 1989; Smith & Carvil, 1997; Wyllie, 1988).

16.4 Host plant

The fungus is a pathogen of over 500 host plant species that comprise both crops as well as weeds. The major cultivated hosts that suffer its invasion are *Glycine* max (soybean), *Sorghum bicolor* (sorghum), *Zea mays* (corn), *Arachis hypogaea* (peanut), *Beta vulgarius, Sesamum indicum* (sesame), *Brassica oleracea* (cabbage), *Capsicum annuum* (pepper), *Cicer arietinum* (chickpea), *Citrus* spp., *Corchorus* sp., *Cucumis* spp., *Fragaria* sp. (strawberries), *Vigna unguiculata* (bean), *Vigna radiate* (mung bean), *Gossypium* sp., *Helianthus annuus* (sunflower), *Ipomoea batatas* (sweet potato), *Medicago sativa* (alfalfa), *Phaseolus* spp., *Pinus* spp., *Prunus* spp., *Psidium guajava* (guava), and *Solanumtuberosum* (potato) (Kaur et al., 2012). Recently, the white spider lilies (*Crinum asiaticum* and *Hymenocallis littoralis*) of Malaysia have also been found infected with *M. phaseolina* causing leaf blight (Huda-Shakirah et al., 2019).

16.5 Symptoms

Infected plants shrivel up and roots seem shredded and decaying (Ilyas & Sinclair, 1974). The infected roots acquire sclerotia, and distinctive dark black stripes are seen on peeling.

Macrophomina infection kills plants both before and after their emergence. After the emergence, the dark-bordered spindle-shaped lesions with light gray centers start appearing. These lesions are usually covered with microsclerotia and, in some cases, pycnidia, both as small as a pinhead. Charcoal rot infects stems as well as roots and is both seed-borne and soil-borne, differing in severity and symptoms as per the season and host plants. Though charcoal rot emerges late in the season yet it might cause early diseases of seedlings because of its soil-borne nature, infecting the plant roots at any time. Infected seedlings have a red to brown disfigurement from soil touching area that extends up the stem and may later on darken. The foliage of infected seedlings may seem off-colored, start drying out, and turn brown. If the fungus terminates the growing point, the plant may form two stems. Young infected plants may survive in cool, wet conditions, but they carry a latent infection that will apparently demonstrate overdue symptoms in hot, dry weather. Early maturation, nonabscission of the leaves, discoloration, and failure to complete pod filling are all symptoms of charcoal rot in older soybeans. Microsclerotia appear as black specks in the woody parts of stems and roots. As infected tissue decays, microsclerotia are released into the soil (Bowers & Russin, 1999).

The fungus is commonly found in peanuts and is spread by seeds and pods. The hypocotyl region closer to the soil line appears with distinguished water-soaked lesions. The seedling will die if the hypocotyl is girdled. Plant parts are susceptible to infection at all stages. The initial appearance of stem and root lesions is water-soaked that later fade to a light brown discoloration. The infection has the potential to spread to the taproot as well as the stem. The plant starts wilting once the stem is girdled, and the fungus quickly colonizes the remaining plant tissues. As microsclerotia proliferate, blackening of the dead tissues starts occurring. Sclerotia can grow on roots, pegs, and pods as well. The disease affects only the roots and can cause taproot maceration and plant yellowing and wilting (Mehan & McDonald, 1997).

In corn, the symptoms are quite indistinguishable from the other fungal rots of the stem. The presence of dark-colored microsclerotia in the vascular system and inside the stalk's rind is a tell-tale sign of charcoal rot. The stalk will range in color from gray to black. The cornis usually infected during drought conditions when corn stalks look older and begin to wither off. During hot and dry years, this causes widespread disease incidence in corn (White, 1999) (Table 16.1).

16.6 Management

Despite numerous strategic scientific efforts of controlling the diseases, the coining of an effective *M. phaseolina* management system continues to be a challenge. The interactions between the pathogen, the host plant, and biotic and abiotic components of the environment lead to the damage causing hampered yields. As per the chemical control, there are no

TABLE 16.1 Symptoms of *Macrophomina phaseolina*on various hosts.

Symptoms	Host	References
Charcoal rot	Lima bean, *Phaseolus lunatus*	Sousa et al. (2017)
Charcoal rot	Sugarcane, *Saccharum officinarum*	Leyva-Mir et al. (2015)
Charcoal rot	Soybean, *Glycine max*	Cummings and Bergstrom (2013)
Charcoal rot	Stevia. *Stevia rebaudiana*	Koehler &Shaw (2018)
Leaf and stem blight	Jasmine, *Jasminum multiflorum*	Mahadevakumar and Janardhana (2016)
Root rot	Catmint, *Nepeta cataria*	Nishad et al. (2018)
Vine decline	Musk melon, *Cucumis melo*	Walker (1994)
Crown rot	Strawberry, *Fragaria* sp.	Sanchez et al. (2013)
Crown rot	Strawberry, *Fragaria* sp.	Sharifi and Mahdavi (2011)

systemic fungicides with movement toward the root; thus, its management becomes quite difficult. Presently, no known fungicides have been approved for use against this pathogen. However, different concentrations of various systemic as well as nonsystemic fungicides, e.g., carbendazim, difenoconazole, benomyl, azoxystrobin, and dazomet, have been tested in vitro and in vivo against *M. phaseolina* (Chamorro et al., 2015; Cohen et al., 2012; Lokesh et al., 2020; Parmar et al., 2017; Tonin et al., 2013).

Early maturing varieties frequently avoid stem rot. Also, weed management depletes the overall plant stressors during a growing season. It is possible to reduce the infestation of charcoal rot by keeping the soil moist through irrigation, especially during the postflowering period. A balanced fertilizer program is advantageous since charcoal rot might result from the application of excessive nitrogen levels combined with low potassium levels. Avoiding a high plant population is advised. Management practices such as tillage, compost amendment, irrigation, the inclusion of nitrogen, and cropping system are effectively used (Fig. 16.1).

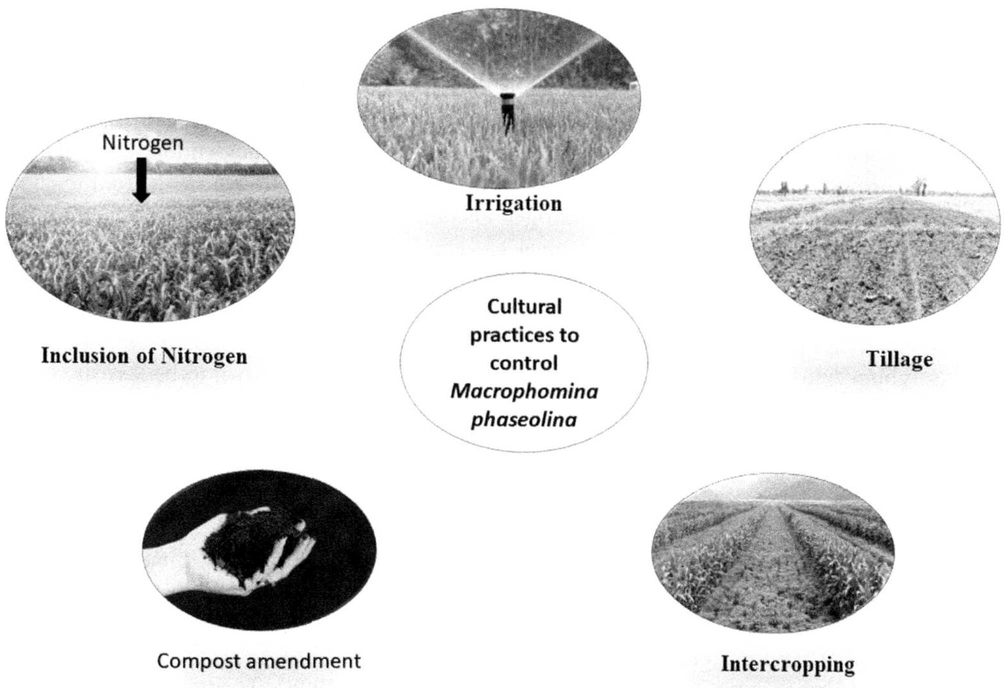

FIGURE 16.1 Cultural practices to control *Macrophomina phaseolina.*

16.7 Cultural practices

16.7.1 Compost amendment

Under a mango tree, a 3 m (long) × 1.5 m (wide) ×0.20 m (deep) compost pit was created using the approach of Ndiaye et al. (2010). The bottom was humidified first and then a 3 cm clay layer was added, followed by 5 cm of cow dung and a 20 cm layer of 90% *Pennisetum glaucum* and 10% *Digitaria exilis* straw (chopped into pieces 10 cm-long). The mixture had enough moisture before being crushed, flattened, and sprinkled with 500 g of ash. In the same manner, more layers of manure and straw were applied up to a height of 0.80 m. Millet stalks were used to cover the resulting mound. The mixture was changed over, watered, and compressed at every 15 days. Every day at 13 h GMT, a thermometer (WWT LF91 (Sartorius and Schott, the Netherlands) was used to monitor the temperature in the heap at three distinct locations that were 40–60 cm deep. The average of the four data was used to calculate the temperature of the pit. After flipping over, the temperature of the heap remained constant, the compost was deemed mature (Yacouba et al., 2001). Before composting, the quantity of sclerotia in cowpea leftovers was high (2691 sclerotia/g dry weight). From week 3 onwards, the numbers of viable sclerotia were reduced, and the pathogen became undetectable from week 9 onwards. The density of sclerotia in agricultural crop residues exposed to ambient conditions did not vary significantly. The use of compost (6 ton) single or in combination with 50 kg nitrogen, phosphorus, and potassium ha^{-1} resulted in 28–45% reduced Area-Under-the-Disease-Progress-Curves (AUDPC) and 43–66% greater cowpea yields. The microsclerotia of *M. phaseolina* were completely destroyed when the temperature in the compost pile containing the infected cowpea leftovers reached 52–60°C. Chamorro et al. (2015) in their 3 year's study found that the manuring and bio-solarization with chicken manure @ 25,000 kg/ha give similar results in reducing soil inoculum of *M. phaseolina* as compared to chemical fumigation with 1,3-dichloropropene:chloropicrin before strawberry production. They also concluded the positive effect of bio-solarization with sugar beet vinasse at 15,000 kg/ha in reducing the charcoal rot incidence.

16.7.2 Inclusion of nitrogen

Charcoal rot has been linked to nitrogen fertilizer and plant density. Charcoal rot is becoming more severe in India because of the high quantities of nitrogen fertilizer required to boost the production potential of better cultivars (Avadhani & Ramesh, 1979; Mote & Ramshe, 1980). Variations between cultivars in how plant density affects charcoal rot were documented by Patil et al. (1982). While the hybrid CSH-8R had a greater frequency of charcoal rot at 180,000 plants/ha than at 45,000 plants/ha, no changes were seen in the varieties SPV 86, SPV 265, or M 35–1. Extremely significant positive associations were found between drought stress, plant density, and nitrogen level in a factorial experiment employing line-source irrigation. High plant density tends to enhance conflict between plants for the moisture in the soil that appears to increase with dryness. The indirect influence of nitrogen on the root-to-shoot growth ratio is likely to be the cause of increased charcoal rot. Nitrogen encourages lush shoot growth at the expense of root development. We found extremely significant positive associations between drought stress, plant density, and nitrogen level in a factorial experiment employing line-source irrigation. High plant density tends to enhance plant competition for available soil moisture, and this competition appears to increase with dryness. The indirect influence of nitrogen on the root-to-shoot growth ratio is likely to be the cause of increased charcoal rot. Nitrogen encourages lush shoot growth at the expense of root development. The lack of a suitable root system limits a plant capacity to collect moisture during drought stress, while the luxuriant growth increases its water need (Ayers, 1978). Charcoal rot is less common with crop management systems that limit pathogen inoculum and promote soil water conservation. In a winter wheat-sorghum-fallow rotation, sorghum cultivated with minimal tillage (Eco fallow) suffered 11% stalk rot as compared to 39% in conventional tillage (Doupnik & Boosalis, 1975). Sorghum cultivated in a mixed crop environment has also been shown to be less susceptible to charcoal rot than sorghum grown as a single crop (Khurne et at., 1980). Increasing the nitrogen:phosphorus:potassium (NPK) application is also effective against charcoal rot in soybean (Csondes et al., 2008).

16.7.3 Irrigation

M. phaseolina has the potential to greatly lower the severity of illness in afflicted plants. Sunflower isolate of the fungus had a selective host preference for sunflower plants, which resulted in a little larger yield loss under RSM circumstances; nevertheless, microsclerotia were more concentrated in soybeans, which might lead to an increase in soil inoculum in future (Jordaan et al., 2019). The occurrence of *M. phaseolina* can be minimized by irrigation in crops like soybean (Michail et al., 1979) and French bean (Diaz-Franco & Cortinas-Escobar, 1988). Treatment of seeds with *Trichoderma viride* in combination with irrigation, during moisture stress, can help in reducing the severity of this disease by 50% (Ansari, 2010).

16.7.4 Tillage

No-tillage systems alter the availability of nutrients in the soil resulting in an increase in microbial biomass and activity as well as competing impacts. Direct planting also had better aggregate stability than traditional tillage techniques. Intensive labor in traditional tillage techniques likely disturbed soil structure and accelerating aggregate decomposition. High soil quality inhibited the establishment of *M. phaseolina*, in direct seeding systems, protecting soybean health. Suppressing the pathogen was easier in these systems because the environment favored the number and activity of soil bacteria resulting in increased nutrient exchange and the establishment of healthy plants. As a result, we discovered a substantial negative association between the occurrence of charcoal rot and various biological, chemical, and physical soil characteristics. Furthermore, there was a substantial negative association between crop productivity and illness incidence. Conservation measures, such as direct sowing, are becoming increasingly important. Direct seeding is a method of increasing microbial activity in the background, amplifying nutrient cycling, lowering the concentrations of readily accessible nutrient sources, raising microbial diversity and intensifying natural disease control (Perez-Brandan et al., 2012).

The incidence of *M. phaseolina* on soybeans was studied for four seasons in both conventional and no-till farming practices. Total precipitation between planting and harvest in the 1997/98 and 2000/01 seasons was 876.3 and 846.9 mm, respectively. During these periods, there was no substantial difference in disease incidence between the no-tillage and traditional methods. Precipitation totaled 689.9 mm in 1998/99 and 478.3 mm in 1999/00, respectively. In 1998/99, illness incidence was 43.7% in the no-tillage system and 53.1% in the traditional system. In 1999/00, the ultimate incidence rates for no-tillage and conventional systems were 68.7% and 81.2%, respectively. During these two seasons, rainfall was less than what was needed for soybean crops (840 mm), and disease incidence was substantially greater in the conventional system. In soil samples, the density of microsclerotia obtained using the traditional technique at 0–10 cm depth was greater. Although, a study of microsclerotia in roots revealed that there was no difference in years with enough rainfall. Roots from plants grown using the traditional approach, on the other hand, exhibited much more microsclerotia in dry years. Crop rotation would probably be ineffective in eliminating charcoal rot due to large host range of *M. phaseolina* and the lengthy survival duration of microsclerotia. Under these test conditions, using a no-tillage cropping method to conserve moisture of soil and slow disease progression may be a superior option for suppressing charcoal rot (Almeida et al., 2002).

16.7.5 Cropping system

Cropping history (Mueller et al., 1985), intermediate cropping (Ahmed & Ahmed, 1977), and crop rotation all have a significant impact on the amount of *M. phaseolina* inoculum in soils. Intermediate cropping involves growing the primary crop twice, and sandwiching it with a nonhost crop (Tiwari & Shroff, 1982). The population of *M. phaseolina* grew in the desert parts of India as a result of the continuous and successive cultivation of the sensitive legume guar (Lodha et al., 1990). Even though this disease has not been effectively controlled by crop rotation, lower inoculum densities were seen when soybean was utilized less frequently in rotations (Francl et al., 1988). Less *Macrophomina* stem and root rot were detected in the specific case of sesame when it was cultivated in combination with or intercropped with green gram, and greater seed production was also noted (Rajpurohit, 2002). When compared to a single cropping system, the population of sclerotia was influenced differently by rain-fed crops than it was by crop rotations (Lodha & Mawar, 2020). Sorghum or cowpea intercropping systems with pigeon peas showed a higher population of sclerotia. The population of sclerotia was high even in systems where sorghum and pigeon pea intercropped in alternate years. Therefore, it may be inferred that crop rotation, even when the other crops in the rotation are hosts of *M. phaseolina,* may be a useful strategy for minimizing charcoal rot on sensitive crops.Ndiaye et al. (2008) have concluded in their study that in an infected field, the cowpea crop should be rotated with any gramineous crop for a relatively fast decrease in *M. phaseolina* density. They also suggested continuous mono-cropping of fonio (*Digitaria exilis*) for at least 3 years in case of extreme infections. As per study conducted by Singh et al. (1990), the cropping systems that were kept fallow during the rainy season, thereafter followed by sorghum or chickpea stabilized the inoculum density of *M. phaseolina* (Singh et al., 1990). As per Naseri (2014), beans cultivated after the cultivation of the tomato or potato decrease the charcoal rot levels in comparison to when they were cropped after legumes or cereals. Interesting research was done by Almeida et al. (2008) on *M. phaseolina* isolates from soybean, corn, and sunflower for genetic diversity of microsclerotia that followed four crop rotations and uncropped soil. The results indicate that genetic differentiation can be changed through crop rotation. Tamil Nadu Agriculture University recommends the intercropping of sesamum and moth bean in the ratio of 1:1 or 2:1 for the control of stem and root rot and intercropping with moth bean in a similar ratio for controlling damping-off/root rot. In case of the groundnut cropping system, intercropping with pearl millet or sorghum (1:3) has been recommended (TNAU, 2016). In the presence of *M. phaseolina* and *Fusarium oxysporum* infection, intercropping with legume fodder, gaur (*Cyamopsis tetragonoloba*), and sesame production has been proved to give higher seed production than the crop grown alone (Ahmed & Ibrahim, 2020). As per this study, the comparatively best intercropping system was 100% sesame +100% gaur grown for two seasons.

16.7.6 Integration of various cultural practices

Cultural practices when integrated can effectively manage various diseases. Such cultural practices comprises of crop rotation and changing the sowing dates (Bristow & Wyllie, 1975), lime application, fertilizers application, deep plowing, addition of organic matters (Collins et al., 1991), applications of organic manures and micronutrients along with irrigations in various crops like sunflower and sorghum, etc., crop rotations, soil fertility, sowing density and irrigation (Ploper et al., 2001), and application of Zn in soil along with *Bradyrhizobium japonicum* and *T. viride* (Ansari, 2010).

16.8 Case study

16.8.1 Tillage practices

According to Wrather et al. (1998), no-tillage soybean fields had a higher soil population of *M. phaseolina* than fields with disc or mold broad tillage because these tillage techniques affect the size and distribution of plant remains present on the surface of the soil, which in turn affects the diversity and density of *M. phaseolina* (Baird et al., 1993). Mengistu et al. (2009) revealed that no tillage system offers a less congenial environment for the population of *M. phaseolina* and had lower colony forming units of *M. phaseolina* in soybean root and stem tissues in comparison to the conventional tillage method. Wrather et al. (1998) included population density from plant tissue and soil throughout harvest; their observations were solely on the basis of soil samples drawn during planting. Tillage may lessen the severity of *M. phaseolina* by enhancing the breakdown of microsclerotia or boosting antagonistic interactions (Baird et al., 2003). Olanya and Campbell (1988) revealed that a tillage up to 15 cm deep decreased *M. phaseolina* populations, but the efficacy of this method will differ depending on the initial amount of microsclerotia present in the soil. Wrather et al. (1998) examined the occurrence and severity of charcoal rot in relation to reduced-tillage farming strategies. Population density of *M. phaseolina* in 0–15 cm layer of soil did not vary between treatments at planting. Whereas, in both disk-tillage and no-tillage plots, the soil population density of *M. phaseolina* was considerably higher in 0–7.5 cm layer of soil than in the 7.5–15 cm layer. Similarly, no-tillage plots had considerably higher soil population densities in the 0–7.5 cm layer of soil than disk-tillage or moldboard plowed plots.

Tillage had no impact on the quantity of *M. phaseolina* infection sites in 6 week-old plant roots or the proportion of *M. phaseolina* infected segments of root in physiologically mature plant roots. The amount of *M. phaseolina* in the soil did not significantly correlate with soybean yield. Soil population density of *M. phaseolina* in the top 15 cm of soil was unaffected by tillage, and the rates of soybean root infection were comparable among tillage treatments (Almeida et al., 2001; Wrather et al., 1998).

In a study, (Gupta & Gupta, 1986) examined how tillage affected the soil environment and the growth of cowpea (*Vigna unguiculata* cv. FS-68) seedlings. The soil environment was enhanced by one disking and three disking with the help of a disc harrow up to 15 cm depth, which also raised the final seedling emergence count, but had no effect on the *M. phaseolina* population in the soil.

16.8.2 Changing the sowing dates

The effects of soybean planting date on *M. phaseolina* population density in the soil were investigated by Wrather et al. (2007). Cotton was sown on the site over the period of 10 years. Mid-April, Mid-May, and Mid-June were the planting dates from 2000 to 2003, and the maturity group three designated cultivars FFR3975 and Asgrow 3834 were very sensitive to *M. phaseolina*. Every year, the same field was used to make plots. From May 2000 to April 2004, top 15 cm of soil of each plot were sampled, and the population density of *M. phaseolina* was determined. The soil population density of *M. phaseolina* was unaffected by the date of planting, but there were notable variations between years. These findings indicate that soybean farmers should not really worry about planting dates having an impact on *M. phaseolina* soil population densities. Similarly, planting dates have no impact on *M. phaseolina* infection of muskmelon or the appearance of charcoal rot symptoms in muskmelon (Bruton et al., 1985). Smith and Carvil (1997) found that planting date had an impact on *M. phaseolina* population density in R7 stage soybean root and lower stem tissue, but they were unable to identify whether this led to variations in *M. phaseolina* population density in the soil. Bowen and Schapaugh (1989) reported that full-season maturity groups of soybeans when planted at such time, when they can avoid the hottest and driest periods during postflowering period, can effectively escape the infection of *M. phaseolina.*

16.8.3 Crop rotation

Growing nonhost crops can decrease the soil inoculum below a damage threshold level; hence, this practice reduces the prevalence of charcoal rot disease. In this regard, Ndiaye (2007, p. 114) grew fonio (*Digitaria exilis*) and millet (*Pennisetum glaucum*) continually for 3 years in plots with a natural infestation of 24–53 microsclerotia per gram of soil at the beginning of the experiment. To quantify disease incidence and severity on various soils, a susceptible variety of cowpea was also planted. From the first year, fonio and millet both reduced the microsclerotial densities in soils. Reductions under fonio were much higher after the second and third years (81% and 86%, respectively) than under millet (56% and 66%, respectively). Millet showed low densities of microsclerotia in its root systems, whereas fonio did not have *M. phaseolina* infection. After 3 years of fonio, cowpea produced much more hay and pods than millet in terms of yield. Hence, the conclusion was drawn that cowpea and gramineous crop rotation may cause a relatively quick drop in inoculum density. Fonio can be grown for 3 years to lower *M. phaseolina* densities in soil when the inoculum density is high.

Singh et al. (1990) studied the cumulative effects of 15 crop combinations and rotations on populations of sclerotia of *M. phaseolina* in the soil during the fifth year of a Vertisol cropping. Sorghum or cowpea intercropping systems with pigeon pea had higher sclerotia counts than single-cropping systems. In both the rainy and postrainy seasons, plots where sorghum and pigeon pea were intercropped constantly had the greatest sclerotia counts. Additionally, when sorghum harvested during the rainy season was followed by either safflower or chickpea, sclerotia populations increased (*Cicer arietinum*). The inoculum density of *M. phaseolina* was stabilized by cropping patterns that included fallow in the rainy season, followed by sorghum or chickpea in the postrainy season.

Todd et al. (1987) recommended 1-year crop rotation of soybean with any crops like maize, sorghum, and cotton for maintaining the populations of microsclerotia as at an acceptable level which is 15 microsclerotia per gram of soil.

16.8.4 Soil solarization and organic amendment

Ndiaye (2007) investigated the combined effects of solarization and organic amendment with millet residues or paunch contents on the survival and emergence of *M. phaseolina* in a naturally infested soil. Solarization considerably decreased the *M. phaseolina* soil inoculum by 44% in modified plots by raising the soil temperature to 50°C for at least 4 h per day in June. In solar-heated plots, paunch amendment (3 tonnes per hectare) resulted in a 66% reduction of the initial inoculum, but millet amendment had no effect on the reduction of microsclerotia in solarized plots. In the amended nonsolar heated plots, where maximum temperatures of 40–46°C were recorded, microsclerotia were slightly reduced. In plots that had been amended and/or solar-heated, charcoal rot was significantly reduced, and cowpea yields were higher than in the control. In solarized plots that have been paunched, the disease incidence was lowest, and the yield was highest. Hence, charcoal rot can be effectively managed, and cowpea yield can be improved in fields with heavy infestations of *M. phaseolina* by using combination of solar heating and organic amendment.

Rathore (2000) reported that charcoal rot disease can be managed by using organic amendments like farmyard manure, neem, and mustard cake. In another study conducted by Lodha et al. (2002), plant mortality due to *M. phaseolina* infection was reduced by 63%–72% at harvest in cluster bean by soil amendment with pearl millet compost. Muthusamy and Mariappan (1992) reported that neem or gingelly cake when applied at 3% stimulated germination and lysis thus inactivated 90% and 80% microsclerotia from soybean, respectively, when compared to the control. Maximum production of germ tube and lysis was found in coconut cake extracts.

16.9 Conclusion

M. phaeseolina is a devastating fungus that carries out disruption in the plant cycle in various forms, leading to extensive yield losses. The control methods known to date are either not very effective or are not environment friendly, leading to a new set of problems to tackle. Thus, cultural control is a revived line of treatments that may provide longer management of the fungal populations without degrading the environment. The methods discussed are designed for long time periods and thus bring out results in years but are surely effective and should be explored further as integration with other management practices.

References

Aegerter, B., Gordon, T., & Davis, R. (2000). Occurrence and pathogenicity of fungi associated with melon root rot and vine decline in California. *Plant Disease, 84*, 224–230. https://doi.org/10.1094/PDIS.2000.84.3.224

Ahmed, K. M., & Ahmed, Q. A. (1977). Intermediate cropping and their effects upon the survival of *Macrophominaphaseolina*in soil between the jute seasons. *Bangladesh Journal of Botany, 5*, 99–106.

Ahmed, M. R., & Ibrahim, M. M. A. (2020). Influence of intercropping legume fodder crop guar with sesame for reducing charcoal–rot and wilt disease complex concerning its yield and yield components. *Annals of Agricultural Science, Moshtohor, 58*, 541–552.

Almeida, A. M. R., Amorim, L., Filho, A. B., Torres, E., Farias, J. R. B., Benato, L. C., Pinto, M. C., & Valentim, N. (2002). Progress of soyabean charcoal rot under tillage and no-tillage systems in Brazil. *Fitopatologia Brasileira, 28*, 131–135.

Almeida, A. M. R., Saraiva, O. F., Farias, J. R. B., Gaudencio, C. A., & Torres, E. (2001). Survival of pathogens on soybean debris under no-tillage and conventional tillage systems. *Pesquisa Agropecuaria Brasileira, 36*, 1231–1238.

Almeida, A. M. R., Sosa-Gomez, Binneck, E., Marin, S. R. R., Zucchi, M. I., Abdelnoor, R. V., & Souto, E. R. (2008). Effect of crop rotation on specialization and genetic diversity of *Macrophominaphaseolina*. *Tropical Plant Pathology, 33*(4), 257–264.

Ansari, M. M. (2010). Integrated management of charcoal rot of soybean caused by *Macrophominaphaseolina* (Tassi) Goid. *Soybean Research, 8*, 39–47.

Arora, P., Dilbaghi, N., & Chaudhury, A. (2011). Opportunistic invasive fungal pathogen *Macrophominaphaseolina*prognosis from immunocompromised humans to potential mitogenic RBL with an exceptional and novel antitumor and cytotoxic effect. *European Journal of Clinical Microbiology & Infectious Diseases, 31*, 101–107. https://doi.org/10.1007/s10096-011-1275-1

Avadhani, K. K., & Ramesh, K. V. (1979). Charcoal rot incidence in some released and pre-released varieties and hybrids. *Sorghum Newsletter, 21*, 37–38.

Ayers, P. G. (1978). Water relations of diseased plants. In T. T. Kozlowski (Ed.), *Water deficits and plant growth* (pp. 1–60). USA: Elsevier: Academic Press.

Babu, B. K., Saikia, R., & Arora, D. K. (2010). Molecular characterization and diagnosis of *Macrophominaphaseolina*, a charcoal rot fungus. In Y. Gherbawy, & K. Voigt (Eds.), *Molecular identification of fungi* (pp. 179–193). New York, USA: Springer.

Babu, B. K., Saxena, A. K., Srivastava, A. K., & Arora, D. K. (2007). Identification and detection of *Macrophominaphaseolina* by using species-specific oligonucleotide primers and probe. *Mycologia, 99*, 797–803. https://doi.org/10.1080/15572536.2007.11832511

Baird, R. E., Bell, D. K., Sumner, D. K., Mullinix, B. G., & Culbreath, A. K. (1993). Survival of *Rhizoctoniasolani*AG-4 in residual peanut shells in soil. *Plant Disease, 77*, 973–975.

Baird, R. E., Watson, C. E., & Scruggs, M. (2003). Relative longevity of *Macrophominaphaseolina* and associated mycobiota on residual soybean roots in soil. *Plant Disease, 87*, 563–566. https://doi.org/10.1094/PDIS.2003.87.5.563

Bowen, C. R., & Schapaugh, W. T. (1989). Relationships among charcoal rot infection, yield, and stability estimates in soybean blends. *Crop Science, 29*, 42–46.

Bowers, G. R., & Russin, J. S. (1999). Soybean disease management. In L. G. Heatherly, & H. F. Hodges (Eds.), *Soybean productionin the mid-south* (pp. 231–271). Boca Raton: CRC Press.

Bristow, P. R., & Wyllie, T. D. (1975). The effect of crop rotation and date of planting on charcoal rot of soybean (Abstr). *Proceedings of Annual Meeting of Phytopathology Society, 2*, 83.

Bruton, B. D., Hartz, T. K., & Cox, E. L. (1985). Vine decline in muskmelon as influenced by cultivar and planting date. *Horticultural Science, 20*, 899–901.

Chamorro, M., Domínguez, P., Medina, J. J., Miranda, L., Soria, C., Romero, F., Aranda, J. M. L., Daugovish, O., Mertely, J., & Santos, B. D. (2015). Assessment of chemical and biosolarization treatments for the control of *Macrophominaphaseolina* in strawberries. *ScientiaHorticulturae, 192*, 361–368. https://doi.org/10.1016/j.scienta.2015.03.029

Cohen, R., Omari, N., Porat, A., & Edelstein, M. (2012). Management of *Macrophomina* wilt in melons using grafting or fungicide soil application: Pathological, horticultural and economical aspects. *Crop Protection, 35*, 58–63. https://doi.org/10.1016/j.cropro.2011.12.015

Collins, D. J., Wyllie, T. D., & Anderson, S. H. (1991). Biological activity of *M. phaseolina* in soil. *Soil Biology and Biochemistry, 23*, 495–496.

Crous, P. W., Slippers, B., Wingfield, M. J., Rheeder, J., Marasas, W. F. O., Philips, A. J. L., Artur, A., Treena, B., Paul, B., & Jojannes, G. Z. (2006). Phylogenetic lineages in the botryosphaeriaceae. *Studies in Mycology, 55*, 235–253. https://doi.org/10.3114/sim.55.1.235

Csondes, I., Baliko, K., & Degenhardt, A. (2008). Effect of different nutrient levels on the resistance of soybean to *Macrophominaphaseolina* infection in field experiments. *Acta Agronomica Hungarica, 56*, 357–362.

Cummings, J. A., & Bergstrom, G. C. (2013). First report of charcoal rot caused by *Macrophominaphaseolina* in soybean in New York. *Plant Disease, 97*, 1506-1506.

Dhingra, O. D., & Sinclair, J. B. (1978). Biology and Pathology of Macrophominaphaseolina. *Minas gerais: Universidade federal de Viçosa: ImpresaUniversitaria.*

Diaz-Franco, A., & Cortinas-Escobar, H. (1988). Effect of irrigation regimes on susceptibility of bean to *Macrophominaphaseolina*. *Journal of the Rio Grande Valley Horticultural Society, 414*, 7–50.

Doupnik, B., & Boosalis, M. G. (1975). Ecofallow reduces stalk rot in grain sorghum. *Phytopathology, 65*, 1021–1022.

Francl, L. J., Wyllie, T. D., & Rosenbrock, S. M. (1988). Influence of crop rotation on population density of *Macrophominaphaseolina* in soil infested with *Heteroderaglycines*. *Plant Disease, 72*, 760–764.

Ghosh, T., Biswas, M. K., Guin, C., & Roy, P. (2018). A review on characterization, therapeutic approaches and pathogenesis of *Macrophominaphaseolina*. *Plant Cell Biotechnology and Molecular Biology, 19*, 72–84.

Gupta, J. P., & Gupta, G. K. (1986). Effect of tillage and mulching on soil environment and cowpea seedling growth under arid conditions. *Soil and Tillage Research, 7*, 233–240.

Gupta, G. K., Sharma, S. K., & Ramteke, R. (2012). Biology, epidemiology and management of the pathogenic fungus *Macrophominaphaseolina* (Tassi) Goid.with special reference to charcoal rot of soyabean (*Glycine max* (L.) Merrill). *Journal of Phytopathology, 160*, 167–180.

Huda-Shakirah, A. R., Kee, Y. J., Hafifi, A. B. M., Mohamad Azni, N. N., Zakaria, L., & Mohd, M. H. (2019). Identification and characterization of *Macrophominaphaseolina* causing leaf blight on white spider lilies (*Crinum asiaticum* and *Hymenocallislittoralis*) in Malaysia. *Mycobiology, 47*, 408–414. https://doi.org/10.1080/12298093

Ilyas, M. B., & Sinclair, J. B. (1974). Effect of plant age upon development of necrosis and occurrenceofintraxylemsclerotia in soybean infected with *Macrophominaphaseolina. Phytopathology, 64*, 155–157.

Jordaan, E., van der Waals, J. E., & McLaren, N. W. (2019). Effect of irrigation on charcoal rot severity, yield loss and colonization of soybean and sunflower. *Crop Protection, 122*, 63–69. https://doi.org/10.1016/j.cropro.2019.04.026

Kaur, S., Dhillon, G. S., Brar, S. K., Vallad, G. E., Chand, R., & Chauhan, V. B. (2012). Biology, economic importance and current diagnostic trends. *Critical Reviews in Microbiology, 38*, 136–151. https://doi.org/10.3109/1040841X.2011.640977

Khan, S. N. (2007). *Macrophominaphaseolina* as causal agent for charcoal rot of sunflower. *Mycopath, 5*, 111–118.

Khurne, N. N., Shiwankar, S. K., & Wangikar, P. D. (1980). Effect of mixed cropping on the incidence of charcoal rot of sorghum. *Food Farming and Agriculture, 121*, 292–293.

Koehler, A. M., & Shew, H. D. (2018). First report of charcoal rot of stevia caused by *Macrophominaphaseolina* in North Carolina. *Plant Disease, 102*, 241-241.

Lakhran, L., Ahir, R. R., Choudhary, M., & Choudhary, S. (2018). Isolation, purification, identification and pathogenicity of *Macrophominaphaseolina* (Tassi) Goid caused dry root rot of chickpea. *Journal of Pharmacognosy and Phytochemistry, 7*, 3314–3317.

Leyva-Mir, S. G., Velázquez-Martínez, G. C., Tlapal-Bolaños, B., Alvarado-Gómez, O. G., Tovar-Pedraza, J. M., & Hernandez-Arenas, M. (2015). First report of charcoal rot of sugarcane caused by *Macrophominaphaseolina* in Mexico. *Plant Disease, 99*, 553-553.

Lodha, S., Mathur, B. K., & Solanki, K. R. (1990). Factors influencing population dynamics of *Macrophominaphaseolina*in arid soils. *Plant and Soil, 125*, 75–80. https://doi.org/10.1007/BF000.10746

Lodha, S., & Mawar, R. (2020). Population dynamics of *Macrophominaphaseolina* in relation to disease management: A review. *Journal of Phytopathology, 168*(1), 1–17. https://doi.org/10.1111/jph.12854

Lodha, S., Sharma, S. K., & Aggarwal, R. K. (2002). Inactivation of *Macrophominaphaseolina*propagules during composting and effect of composts on dry root rot severity and on seed yield of clusterbean. *European Journal of Plant Pathology, 108*, 253–261.

Lokesh, R., Rakholiya, K. B., & Thesiya, M. R. (2020). Evaluation of different fungicides against *Macrophominaphaseolina* (Tassi) Goid.causing dry root rot of chickpea (*Cicer arietinum* L.) *in vitro*. *International Journal of Current Microbiology and Applied Sciences, 9*, 1–11. https://doi.org/10.20546/ijcmas.2020.907

Mahadevakumar, S., & Janardhana, G. R. (2016). First report of leaf blight caused by *Macrophominaphaseolina* on *Jasminiummultiflorum* in India. *Journal of Plant Pathology, 98*, 1–6.

Mehan, V. K., & McDonald, D. (1997). Charcoal rot. In N. Kokalis-Burelle, D. M. Porter, R. Rodriguez-Kábana, D. H. Smith, & P. Subrahmanyam (Eds.), *Compendium of peanut diseases* (2nd ed., p. 94). St. Paul (MN): APS Press.

Mengistu, A., Reddy, K. N., Zablotowicz, R. M., & Wrather, A. J. (2009). Propagule densities of *Macrophominaphaseolina* in soybean tissue and soil as affected by tillage, cover crop, and herbicide. *Plant Health Progress, 10*(1). https://doi.org/10.1094/PHP-2009-0130-01-RS

Michail, S. H., Abd, E. R. M. A., & Abu, E. E. A. (1979). Seed health testing of soybean in Egypt. *Acta Phytopathologicaet Entomologica Hungarica, 14*, 371–377.

Mote, U. N., & Ramshe, D. G. (1980). Nitrogen application increases the incidence of charcoal rot in rabi sorghum cultivars. *Sorghum Newsletter, 23*, 129.

Mueller, J. D., Short, B. J., & Sinclair, J. B. (1985). Effects of cropping history, cultivar and sampling date on the internal fungi of soybean roots. *Plant Disease, 69*, 520–523. https://doi.org/10.1094/PD-69-520

Muthusamy, S., & Mariappan, V. (1992). Disintegration of microsclerotia of *Macrophominaphaseolina* (soybean isolate) by oil cake extracts. *Indian Phytopathology, 45*, 271–273.

Naseri, B. (2014). Charcoal rot of beans in diverse cropping systems and soil environments. *Journal of Plant Diseases and Protection, 121*, 20–25.

Ndiaye, M. (2007). *Ecology and management of charcoal rot (*Macrophominaphaseolina*) on cowpea in the Sahel.* PhD Thesis. Wageningen University, the Netherlands with summaries in English, French and Dutch x. ISBN 978-90-8504-664-6. https://edepot.wur.nl/121897

Ndiaye, M., Termorshuizen, A. J., & Bruggen, A. H. C. (2008). Effect of rotation of cowpea (*Vignaunguiculata*) with fonio (*Digitariaexilis*) and millet (*Pennisetumglaucum*) on *Macrophominaphaseolina*densities and cowpea yield. *African Journal of Agricultural Research, 3*, 37–43.

Ndiaye, M., Termorshuizen, A. J., & Bruggen, A. H. C. (2010). Effect of compost amendment and the biocontrol agent *Clonostachysrosea* on the development of charcoal rot (*Macrophominaphaseolina*) on cowpea. *Journal of Plant Pathology, 92*, 173–180.

Nishad, I., Srivastava, A. K., Saroj, A., Babu, B. K., & Samad, A. (2018). First report of root rot of nepetacataria caused by *Macrophominaphaseolina* in India. *Plant Disease, 102*, 23–80.

Olanya, O. M., & Campbell, C. L. (1988). Effects of tillage on the spatial pattern of microsclerotia of *Macrophominaphaseolina. Phytopathology, 78*, 217–221.

Pandey, A. K., Burlakoti, R. R., Rathore, A., & Nair, R. M. (2020). Morphological and molecular characterization of *Macrophominaphaseolina* isolated from three legume crops and evaluation of mungbean genotypes for resistance to dry root rot. *Crop Protection, 127*, 104962. https://doi.org/10.1016/j.cropro.2019.104962

Parmar, H. V., Kapadiya, H. J., & Bhaliya, C. M. (2017). Efficacy of different fungicides against *Macrophominaphaseolina* (Tassi) Goid.causing castor root rot. *International Journal of Chemical Studies, 5*, 1807–1809.

Patil, R. C., Deshamane, N. B., & Pandhare, T. M. (1982). Effect of plant density and row spacing on charcoal rot incidence in four cultivars of sorghum. *Sorghum Newsletter, 25*, 110.

Pérez-Brandan, C., Arzeno, J. L., Huidobro, J., Grümberg, B. C., Conforto, C., Hilton, S., Bending, G. D., Meriles, J. M., & Vargas-Gil, S. (2012). Long-term effect of tillage systems on soil microbiological, chemical and physical parameters and the incidence of charcoal rot by *Macrophomina phaseolina* (Tassi) Goid in soybean. *Crop Protection, 40*, 73–82. https://doi.org/10.1016/j.cropro.2012.04.018

Ploper, L. D., Gonzalez, V., de-Ramallo, N. V., Galvez, R., Devani, M., & de-Ramallo, N. V. (2001). Presence of black stem rot on soybeans in central and northeast Argentina. *Advance Agroindustrial, 22*, 30–34.

Rajpurohit, T. S. (2002). Influence of intercropping and mixed cropping with pearl millet, green gram and mothbean on the incidence of stem and root rot (*Macrophominaphaseolina*) of sesame. *Sesame Safflower Newsletter, 17*, 40–41.

Rathore, B. S. (2000). Effect of organic amendments of incidence of seed rot and seedling blight of mothbean. *Plant Disease Research, 15*, 217–219.

Saleh, A. A., Ahmed, H. U., Todd, T. C., Travers, S. E., Zeller, K. A., Leslie, J. F., & Garrett, K. A. (2009). Relatedness of *Macrophominaphaseolina* isolates from tallgrass prairie, maize, soybean and sorghum. *Molecular Ecology, 19*, 79–91. https://doi.org/10.1111/j.1365-294X.2009.04433.x

Sanchez, S., Gambardella, M., Henríquez, J. L., & Díaz, I. (2013). First report of crown rot of strawberry caused by *Macrophominaphaseolina* in Chile. *Plant Disease, 97*, 996-996.

Schwartz, R. A., & Kapila, R. (2020). *Macrophominaphaseolina:* An overlooked cutaneous infection, seed rot disease in humans. *International Journal of Dermatology, 59*(7), 796–798. https://doi.org/10.1111/ijd.14707

Sharifi, K., & Mahdavi, M. (2011). First report of strawberry crown and root rot caused by *Macrophominaphaseolina* in Iran. *Iranian Journal of Plant Pathology, 47*(4), 161.

Sharma, R. C., & Bhowmik, T. P. (1986). Estimation of yield losses in groundnut due to *Macrophominaphaseolina* (Tassi) Goid. *Indian Journal of Plant Pathology, 4*, 108–112.

Short, G. E., Wyllie, T. D., & Ammon, V. D. (1978). Quantitative enumeration of *Macrophominaphaseolina* in soybean tissues. *Phytopathology, 68*, 736–741.

Sinclair, J. B., & Backman, P. A. (1989). *Compendium of soybean diseases* (3rd ed.). St.Paul, MN: APS Press.

Singh, S. K., Nene, Y. L., & Reddy, M. V. (1990). Influence of cropping systems on *Macrophominaphaseolina*populations in soil. *Plant Disease, 74*, 812–814.

Smith, G. S., & Carvil, O. N. (1997). Field screening of commercial and experimental soybean cultivars for their reaction to *Macrophominaphaseolina*. *Plant Disease, 81*, 363–368.

Sousa, E. S., Melo, M. P., Pires, L. L., Silva, B. A., García, M. E. M., & Sousa, E. M. J. (2017). First report of *Macrophominaphaseolina* causing charcoal rot in lima bean (*Phaseolus lunatus*) in Brazil. *Plant Disease, 101*, 1551-1551.

Srinivasan, A., Wickes, B. L., Romanelli, A. M., Debelenko, L., Rubnitz, J. E., Sutton, D. A., Thompson, E. H., Fothergill, A. W., Rinaldi, M. G., Hayden, R. T., & Shenep, J. L. (2009). Cutaneous infection caused by *Macrophomina phaseolina* in a child with acute myeloid leukemia. *Journal of Clinical Microbiology, 47*, 1969–1972.

Srivastava, A., & Arora, D. (1997). Evaluation of a polyclonal antibody immunoassay for detection and quantification of *Macrophominaphaseolina*. *Plant Pathology, 46*, 785–794. https://doi.org/10.1046/j.1365-3059.1997.d01-64.x

Tan, D. H., Sigler, L., Gibas, C. F., & Fong, I. W. (2008). Disseminated fungal infection in a renal transplant recipient involving *Macrophominaphaseolina* and *scytalidiumdimidiatum*: Case report and review of taxonomic changes among medically important members of the botryosphaeriaceae. *Medical Mycobiology, 46*, 285–292.

Tiwari, A., & Shroff, V. N. (1982). Differential reaction of cotton genotypes to root rot. *Indian Phytopathology, 35*, 514–515.

TNAU. 2016. https://agritech.tnau.ac.in/org_farm/orgfarm_agridiseases.html

Todd, T. C., Pearson, C. A. S., & Schwenk, F. W. (1987). Effect of *Heteroderaglycines* on charcoal rot severity in soybean cultivars resistant and susceptible to soybean cyst nematode. *Annals of applied nematology, 1*, 35–40.

Tok, F. M. (2019). Relationship between morphologic, phenotypic and pathogenic characteristics in *Macrophominaphaseolina* isolates from cucumber plants. *International Journal of Innovatives Approaches in Agricultural Research, 3*, 651–660. https://doi.org/10.29329/ijiaar.2019.217.11

Tonin, R. F. B., Avozani, A., Durante, D. A. L., Reis, E. M., Zoldan, S. M., & Garcés-Fiallos, F. R. (2013). *In vitro* mycelialsensitivity of *Macrophominaphaseolina* to fungicides, sensibilidademicelial *in vitro* de *Macrophominaphaseolina* a fungicidas. *Pesquisa Agropecuária Tropical, 43*, 460–466. https://doi.org/10.1590/S1983-40632013000400014

Vasebi, Y., Safaie, N., & Alizadeh, A. (2013). Biological control of soybean charcoal root rot disease using bacterial and fungal antagonists *in vitro* and greenhouse condition. *Journal of Crop Protection, 2*, 139–150.

Walker, G. E. (1994). First report of *Macrophominaphaseolina* associated with vine decline in muskmelon in south Australia. *Plant Disease, 78*(6), 640.

White, D. G. (1999). In D. G. White (Ed.), Fungal stalk rots.*Compendium of corn diseases* (3rd ed.). St. Paul, MN: APS Press.

Wrather, J. A., Kendig, S. R., & Tuler, D. D. (1998). Tillage effects on *Macrophominaphaseolina* population density and soybean yield. *Plant Disease, 82*, 247–250.

Wrather, J. A., Shannon, J. G., & Mengistu, A. (2007). Impact of soybean planting date on soil population density of *Macrophominaphaseolina*. *Plant Health Progress, 8*(1). https://doi.org/10.1094/PHP-2007-0917-03-RS

Wyllie, T. D. (1988). Charcoal rot of soybeans-current status. In T. D. Wyllie, & D. H. Scott (Eds.), *Soybean diseases of the north central region* (pp. 106–113). St. Paul, MN: APS Press.

Yacouba, H., Morel, M., & Hounto, T. (2001). Valorisation par compostage de la biomasse de *Pistia stratiotes*issue de la station de lagunage de l'E.I.E.R. *Sud Technologie, 7*, 40–47.

Section IV

Biological control

Chapter 17

Biocontrol of charcoal rot by soil microorganisms

Sarita Bhutada[1] **and Mukta Shinde**[2]

[1]Department of Microbiology, Sanjivani Arts, Commerce and Science College Kopargaon, Maharashtra, India; [2]Department of Zoology, Sanjivani Arts, Commerce and Science College Kopargaon, Maharashtra, India

17.1 Introduction

Charcoal rotis the soil-borne fungal disease caused by *M. phaseolina*, which is one of the most pervasive and severe diseases of crop plants. Charcoal rot is the name given to the tiny, dark fungal growths or microsclerotia that appear in and on the lower stem and roots of afflicted plants. Yield loss from charcoal rot may vary significantly depending on when the illness first appears. Once hot and dry circumstances start, the yield loss will rise. Fields under drought stress without charcoal rot typically withstand it for additional two or more weeks as opposed to those with *M. phaseolina*. Therefore, the disease can be directly held accountable for any yield reductions brought on by this earlier plant death. Finding new methods to keep the soil moist enough is often part of management when there is insufficient rainfall.

The widespread fungus *M. phaseolina* known to infect over 500 different plant species from over 100 different families including soybeans, edible beans, corn, and sorghum (Nagaraja et al., 2021). It causes illnesses like seedling blight, charcoal rot, and stem and root rot (Ghosh et al., 2018). The disease is exacerbated by hot, dry conditions. If the infection is severe enough, plants may never develop producing small light kernels altogether.

Charcoal rot might be considered a condition caused by stress. A silver-gray lesion at the stem's base and black, pepper-like structures (microsclerotia) scattered throughout the pith are symptoms of charcoal rot, as well as abrupt wilting and death of plants. After flowering during dry conditions, charcoal rot is more prevalent. Occasionally, developing seeds on seed heads might become infected with charcoal rot. *M. phaseolina* can adapt to a variety of ecological situations due to its wide range of physiological, morphological, and pathogenic diversity (Ambrosio et al., 2015). A few of the reasons that lead to the eventual death of infected plants include exposure to vascular blockages that obstruct the flow of nutrients or the production of phytotoxic chemicals by *M. phaseolina*.

M. phaseolina can infect the plant host at any point in the growth cycle from the seedling to the postflowering stage. The first symptom that can be seen is a stain at the base of the stalk; this is an early sign of the disease. Two indicators of a charcoal rot infection include crop lodging and insufficient grain filling. The early lodging, though, is the most visible symptom of charcoal rot. Water-soaked sores are visible on the infected root and stem, and they start turning brown or black. These stalks frequently lodge and soften at the root due to light wind or plant bending. The vascular bundles break from one another and the cortical and pith tissues fall apart in an infected stalk. Due to presence of numerous minutes, black, charcoal-colored sclerotia that are found in the vascular tubes, the condition is known as "charcoal rot." The disease frequently shows symptoms after flowering; however, seedlings might occasionally catch the infection and suffer seedling blight.

Hyaline hyphae with thin walls to light or dark brown hyphae with septa are what set *M. phaseolina* apart. Parent hyphae generally branch off from the main hyphae at a right angle, characterized by constriction at the initiation. Round, oval, or oblong-shaped microsclerotia are compact masses of hardened fungal mycelium that are initially light brown in color before maturing to a deeper shade (brown to black). Pycnidia are rough, globose, uneven, beaked, ostiolate, and larger than microsclerotia. They do not frequently show up in the wild. On synthetic conditions, isolates of the fungus can have vastly different mycelium colors, microsclerotia distributions, pycnidia production, and chlorate phenotypes.

Macrophomina Phaseolina. https://doi.org/10.1016/B978-0-443-15443-0.00008-5

The isolates, however, belonged to a single species, as determined by amplification of the internal transcribed regions (Almomani et al., 2013). According to a study, some ideas suggest that morphological variation may be connected to changes in the host species or how fungus respond to their environment (Pandey et al., 2020; Tok, 2019).

17.2 Economic significance

In several countries including India, Sudan, the US, Mexico, Brazil, Australia, Ethiopia and Mali, the disease results in significant economic losses. After the wet season, it is among the major obstacles to India's sorghum output. The most severe of the various diseases that affect cowpeas is charcoal rot. Lodha & Singh (1984) found 42%–71% plant loss with negligible grain yield in plots due to charcoal rot. Surveys conducted by the investigator in 1998–99 demonstrated that it is a problem anywhere cowpea is planted, despite the fact that there are no publicly available figures of its occurrence or losses in Haryana. Information on integrated sickness management is scarce, though a number of crops, including sorghum, corn, sesame, soybeans, chickpea, cowpea, sunflower, cotton, peanut, castor, potato, strawberry, and sunflower, are susceptible to the disease charcoal rot (Table 17.1).

17.3 Disease cycle

The main source of infection for *M. phaseolina* is microsclerotia (Gupta et al., 2012). This robust construction has a maximum 15-year soil life. Many hyphae can infect the roots of the host plant while it is still in the seedling stage. Once inside the roots, fungus has an impact on the vascular system, which influences how nutrients and water reach the higher sections of the plant. The aging and yellowing of leaves that are still linked to stems by their petioles, the loss of cortical tissues from the lower stem and taproot, and the appearance of gray in these tissues as a result of an abundance of microsclerotia are the distinctive signs of this disease (Fig. 17.1). The host plant may die too soon as a result of these symptoms (Marquez et al., 2021).

17.4 Pathogenesis of *M. Phaseolina*

M. phaseolina genome contains many genes that have been linked to pathogenicity and may be essential for the pathogenesis of the fungus. In reality, 13% of the genome's genes have properties that are comparable to those of the pathogen–host interactions. *M. phaseolina* may attach to the host tissue (using cellulose binding elicitor lectin and transglutaminase-like proteins), neutralize the early host defense, enter the plant epidermis, and disperse throughout the

TABLE 17.1 Yield loss in different crops due to charcoal rot disease.

S. N.	Name of crop	Yield loss (%)
1	Sunflower	25%–90%
2	Cowpea	42%–71%
3	Sorghum	50%
4	Maize	70%
5	Soybean	50%
6	Potato	88%
7	Groundnut	100%
8	Strawberry	30%–60%
9	Sugarbeet	50%
10	Mungbean	10.8%–12.3%

Sources: Mathur, A. C. (2001). Integrated disease management of charcoal rot of cowpea.CCS Haryana Agricultural University, Das, I.K., Indira, S., Annapurna, A., Prabhakar, S., & Seetharama, N. (2008).Biocontrol of charcoal-rot in sorghum by fluorescent Psedomonads associated with the rhizosphere. Crop Protection 27,1407–1414 and Hashem, A., Fathi, E., Abdulaziz, A., Radhakrishnan, R., & Kumar, A. (2017). Plant defense approach of Bacillus subtilis (BERA 71) against Macrophomina phaseolina (Tassi) Goid. inmungbean. Journal of Plant Intercations, 12(1), 390-401. https://doi/full/10.1080/17429145.2017.1373871.

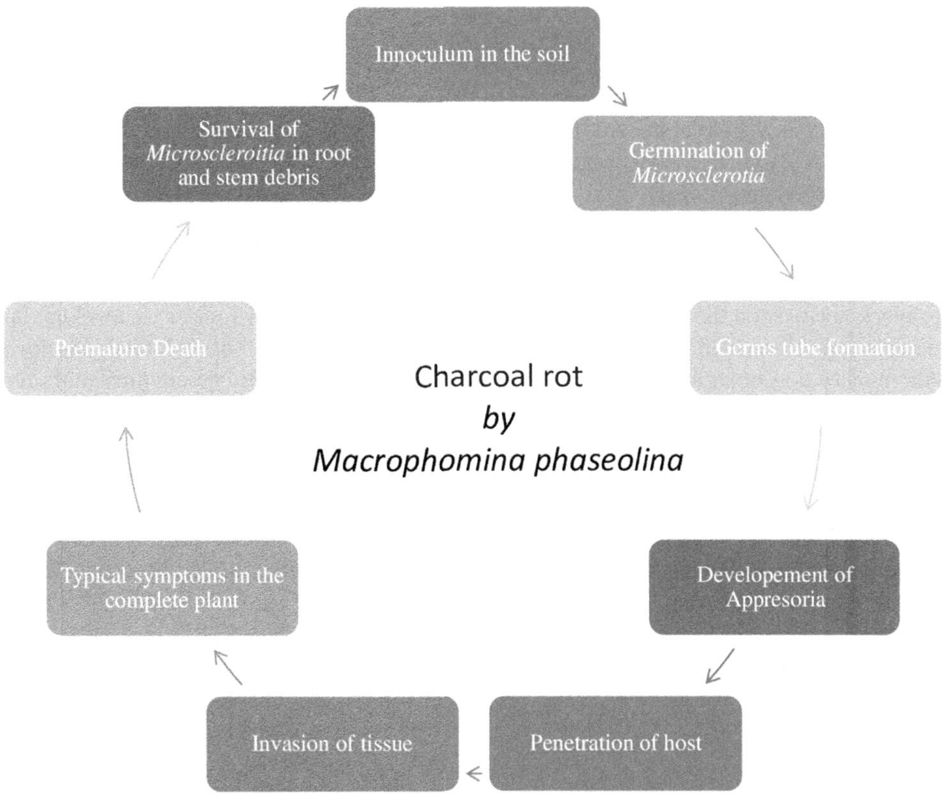

FIGURE 17.1 Charcoal rot disease cycle caused by *M. phaseolina.*

entire plant with the aid of a wide range of genes. Once inside the host, the pathogen secretes a number of toxins and cell wall—degrading enzymes, which after combined with the host defense system's disruption causes host cell death and disease. By dissolving lignocellulose and polysaccharides in cell membranes, the hydrolytic enzymes of *M. phaseolina* allow to enter the host tissue. For the hydrolysis of cellulose, it encompasses lignin peroxidases, laccases, endoglucanases, exocellobiohydrolases, and glucosidases. Galactose oxidases, chloroperoxidases, halo peroxidases, and heme peroxidases that break down lignin. It is interesting to note that as compared to the genomes of other sequenced fungal species, *M. phaseolina* has the most laccases and cellulolytic enzymes (e.g., *Phanerochaete chrysosporium*, *Cryptococcus neoformans*, *Ustilago maydis*, *Saccharomyces cerevisiae*, *Aspergillus nidulans*, *Postia placenta,* and *Neurospora crassa).* In vitro culture conditions have been shown to develop and activate these plant cell wall—degrading enzymes. *Phaseolinon, botryodiplodin,* and patulin are a few of the phytotoxic metabolites of *M. phaseolina* that have been found. A study showed the development of necrotic lesions on leaves and roots as well as the early stages of infection that cause seedlings to wilt are thought to be influenced by these substances. As a result, *M. phaseolina* is more virulent and may be better able to infect a wide variety of hosts and tissues (Islam et al., 2012; Salvatore et al., 2020). The fungal exceptional environmental adaptability contributes to both its global distribution and plant pathogenicity. Ghosh et al. (2018) showed that to counter the plant defensive reaction this adaptation entails the creation of detoxifying genes, such as cytochrome P450, Cof, and superoxide dismutase. It is interesting to note that the *M. phaseolina* genome contains areas that resemble nitric oxide synthase and have conserved amino acid sequences. In vitro, infected jute tissues, mycelium, and the surrounding medium all contained high concentrations of nitric oxide (NO). This demonstrates that NO may play a substantial physiological part in the interaction between pathogens and necrotrophic hosts. Even though *M. phaseolina* is a polyphagous pathogen without any evidence of host specialism, it interacts with a variety of plant species, including soybeans.

17.5 Symptoms

The entire plant may develop the symptoms of charcoal rot. An essential source of the initial inoculum for new areas is the infected seeds. Phase I (a) of the seedling symptoms and Phase II (b) of the charcoal rot are the two categories of symptoms.

17.5.1 Seedling phase

The pathogen invades the seed coat, cotyledons, plumule, and radicle and is known to cause pre-emergence seed deterioration. Reddish-brown staining at the hypocotyl's developing section can be seen in infected seedlings. If the infection develops through the root, the discoloration is visible at the soil line and above. In hot and dry conditions, the seedlings could perish. Infected seedlings survive in damp and cool weather, but they still harbor the infection; when hot, dry weather comes, the symptoms could recur (Mathur, 2001).

17.5.2 Charcoal rot phase

An infection leads to a brown to black necrosis on the roots of older plants. The charcoal rot phase starts to develop after mid-season. At this period, the bottom portions of the sub-epidermal tissues of stem and tap root, exhibit light-blown discoloration. The infected plants initially do not show any symptoms above ground, but in extreme circumstances, the leaves fall off. In the later phases, the leaves start to yellow and droop. Surface stem lesions may develop from the soil lineup. The fungal black pigment may be so numerous that it gives tissues a grayish-black color when the epidermis is removed. When the stem is pulled apart, the base and tap root both exhibit black streaks in the woody tissues, and the pith of the stem frequently develops sclerotia. When the disease is present in the field, it shows up as spots. The fungus colonizes dead plant materials in the soil while being a weak rival. So, the usual name for this disease stage is dry root rot. Apparently, legume stems with disease exhibit pycnidial phases. Numerous crops are susceptible to the disease including sorghum, maize, groundnuts, cotton, potatoes, soybeans, strawberries, sugar beet, jute, and sesame (Fig. 17.2A and B) (Mathur, 2001).

FIGURE 17.2 Charcoal rot in soybean (a and b).

17.6 Charcoal rot management

Few methods including crop rotation, tillage operation, irrigation, fertility and nutrition, weed control, nematode management, fungicide, and biocontrol are currently available to farmers to lessen the damage caused by charcoal rot.

Disease intensity and yield loss, as well as disease intensity and pathogen inoculum density in soil, are found to correlate. In order to reduce inoculum density, various agricultural techniques are used. Kendig et al. (2000) studied the process known as "bio-solarization," which combines bio-fumigation with solarization, has been demonstrated to be successful in reducing or stabilizing the population of *M. phaseolina* microsclerotia in soil, which lowers the incidence of charcoal rot. However, higher soil moisture levels (over 60%) lessened the severity of the disease. Spagnoletti et al. (2018, 2020) described the phosphorus-based fertilization showed a reduction, whereas nitrogen-based fertilization increased pathogenicity.

17.7 Biological control

There is a biological remedy that is both economical and environment friendly that stops phytopathogen activity by using hostile bacteria (Yang et al., 2014). *Aspergillus niger*, *Trichoderma* species, *Bacillus subtilis*, and other biocontrol agents have all been studied for the treatment of root-rot infections (Khan & Anwar, 2007; Kumar et al., 2012; Kumar et al., 2020; Pan et al., 2013). Because they can withstand a wider range of temperatures, *Trichoderma* species are among the most effective mycoparasites for suppressing soil-borne fungus (Pan & Bhagat, 2008; Siddiqui et al., 2001; Zhao et al., 2014). The adaptability of the *Bacillus* species allows them to protect plants from fungi that cause plant diseases and their ability to successfully colonize new settings. *A. niger* may also work as a microbial antagonist to combat a variety of plant diseases (Mondal et al., 2000).

Over the past few generations, there has been a rise in interest in biological control agents, plant metabolites, and elicitors of plant defenses. The most prevalent fungal and microbial plant-associated species, such as *Trichoderma* and the plant growth promoting Rhizobacteria (*Bacillus*, *Pseudomonas* and *Streptomyces*) rhizobacteria, have been discovered act as biological control agents (BCAs). Some BCAs directly attack the pathogens by stopping their growth, while the others unintentionally harm the pathogen by activating the defense mechanisms of the host plant. A more efficient and financially viable alternative to synthetic poisons in the fight against disease is biologically based pest management (BBPM), which is integrated into the current integrated pest management (IPM) plans.

Biological management is a crucial element of sustainable agriculture. The use of living things other than humans to inhibit the activity of microorganisms that cause disease or damage is a potential substitute for pesticides. A living organism always has a natural opponent striving to keep it in check, according to the theory of biocontrol. It can be characterized as the management of one or more naturally occurring organisms by changing the environment in order to reduce the number of pathogens or pests that are actively producing disease.

Earlier studies have looked into the effect of antagonistic bacteria on charcoal rot. *Rhizobacterium* strain FPT721, a rhizobacterium that encourages plant growth, was applied to seeds under controlled conditions and reduced the incidence of disease (Choudhary, 2011). When tested against the pathogen in vitro and in vivo, Senthilkumar et al. (2009) found that two soybean bacterial endophytes, *Bacillus* (strain: HKA-121) and *Paenibacillus* (strain: HKA-15), suppressed and reduced *M. phaseolina* illness, respectively. These biocontrol strains, however, are not readily available for use in soybean farming.

Numerous rhizosphere and root-associated bacteria have been identified having detrimental effects on *M. phaseolina* (Hussain & Khan, 2020; Kumar et al., 2007; Torres et al., 2016; Vasebi et al., 2013). A small number of isolates had a significant impact on disease incidence reduction and host plant growth features. Species from the genera *Bacillus*, *Pantoea*, *Pseudomonas, Stenotrophomonas*, and *Serratia* are included in this collection.

17.7.1 Fungal BCAs

Arbuscular mycorrhizal fungi (AMF) are thought to establish mutualistic relationships with 72% of plants, making them possibly the most common and oldest symbiosis on earth (Brundrett & Tedersoo, 2018; Marquezet al., 2021). AMFs dramatically modify the environment and the host plant, and it has frequently been noticed that they decrease the frequency of a variety of diseases and insects (Marquez et al., 2019; Pozo et al., 2010). Transforming plant metabolism and gene expression, especially those involved in defense mechanisms, as well as competition for nutrients with soilborne pathogens and space, are among the techniques listed (Liu et al., 2003, 2007; Marquez et al., 2018, 2019). Additionally, these fungi could prepare the host tissues for a competitive attack (Pozo & Azcón-Aguilar, 2007). Mycorrhizal symbiosis has been demonstrated to minimize the severity of disease symptoms (Doley & Jite, 2013; Marquez et al., 2018; Oyewole et al., 2017; Spagnoletti et al., 2017, 2020). Even if root infection by *M. phaseolina* or any other diseases cannot be prevented. These results were associated with a decrease in oxidative injury and an improvement in protective mechanisms, including a noticeable increase in total phenol and proline levels and superoxidase activity (Doley & Jite, 2013; Spagnoletti et al., 2017). Additionally, transcribing research reveals a connection between protection and the regulation of pathogen infection. Serine carboxipeptidase-like (SCPL) enzyme and lectin gene activation are thought to improve roots' ability to identify pathogens, allowing for an early defensive response, a decrease in infection rates, and an improvement in cell homeostasis. However, it is significant to note that 40% of the differently expressed genes in *M. phaseolina*-infected mycorrhizal soybean plants were either uncharacterized or had unknown function. Marquez et al. (2018) showed these genes could have contributed to the plant defense mechanism. To have a better knowledge of the mechanisms behind the biological regulation mediated by AMF, more research on these interactions is required. We are unaware of any studies on the impact of AMF on the pathogenicity mechanisms of *M. phaseolina*.

Phosphorus fertilization had reduced the severity of *M. phaseolina* disease, but when both combination treatments were used, a little decrease in the AMF protection was seen (Spagnoletti et al., 2018). Conversely, AMF shields soybean plants from *M. phaseolina* even when nitrogen fertilizer is present (Spagnoletti et al., 2020). A study showed the impact of integrated management on agricultural activities (Bastakoti et al., 2017; Hyder et al., 2017). *Trichoderma* species are the efficient BCAs for *M. phaseolina* and other soil-borne fungal plant diseases. The development of these saprotrophic fungus uses variety of strategies, including mycoparasitism and competition for nutrients. Additionally, a number of species have a reputation for having positive impact on health of the plant that is the ability to encourage systemic resistance or plant development (Martinez-Medina et al., 2016). The ability of *Trichoderma* species to overgrow and eradicate the pathogen mycelia during antagonistic interactions was positively correlated with *M. phaseolina* growth inhibitions. It has been observed that the induction of chitinase, β-1, three glucanase, and an increase in the total phenol level, may have assisted in limiting the pathogen's proliferation during antagonism (Gajeraet al., 2012). *M. phaseolina* may be controlled by antagonistic and growth-promoting yeast (*Brettanomyces naardensis*) (Nafady et al., 2019). In the instance of *Trichoderma longibrachiatum*, antibiosis via microbial volatile organic compounds (mVOCs) was also found, and the pathogen's growth was reduced during direct interaction (Sridharan et al., 2020). These mVOCs prevented the growth of the *M. phaseolina* mycelia by altering the mycelial structure. Longifolene, caryophyllene, and cuprenene were a few of the terpenoids whose concentrations increased as a result of interactions. The expression of novel compounds not previously produced by any of the species, such as limonene, azulene, 3-methyl-1-butanol, styrene, salicylaldehyde, undecane, and 3-methylphenol, was also a result of interactions. These substances have strong antibacterial agents and may also function as signaling molecules in microbial interactions.

17.7.1.1 Trichoderma *spp*

It has frequently been proposed that *Trichoderma* species, also referred to as green spore fungi, can function as biological control agents against diseases that are frequently caused by soil-borne infections. The occurrence of charcoal rot was reduced more successfully by *T. koningiopsis* (T-K11) and *T. harzianum* (native), *M. phaseolina* was biologically managed in field settings to lower the prevalence of charcoal rot for the peanut crop by utilizing the different isolates of *Trichoderma* (Martínez-Salgadoet al., 2021). Khan et al. (2019); it was found that *T. harzianum* successfully controls *M. phaseolina* for mung beans. *T. harzianum* enhanced the frequency of healthy plants. According to Dubey et al. (2007), *T. koningiopsis* (T-K11) was able to reduce the prevalence of charcoal rot by 76%. Using biocontrol agents to manage plant diseases is an alternative, sustainable strategy. Singh & Majumdar (1995) utilized this to cure cowpea charcoal rot. They examined the effects of *T.harzianum* and *T. viride* on *M. phaseolina* cowpea stem and root fungus, which was cultivated in vitro. The results showed that *T. viride* induced the most inhibition. In a different study, Mathur (2001) showed that *T. harzianum* reduced the development to the highest extent feasible in lab experiments. Agar disc technique was used by Mathur(2001) to observe soil colonization by *T. harzianum* and two strains of *T. viride*. The strain that faced the most opposition was successful in colonizing each of the three soils separately. The other strains fared less well. Ramann brown forest soil and Chernozem meadow soil were more successful. *T. viride* performed well on cheniozem but did poorly on soils from dark forests or meadows. On a medium with peptone and glucose, this strain was extremely competitive, but not as much on a medium containing cellulose. The rate of colonization significantly enhanced when streptomycin was added in the media. *T. herzianum* (isolates MTR-35 and local), *T. logibranchiatum* and *T. viride*, *Epicoccum nigrum*, and *Verticillium biguttatum* (isolates D-25 and M-73) were evaluated in relation to growth rate, growth intensity, and sporulation on a variety of surfaces (Singh et al., 1996). With the exception of *V. biguttatum*, all the antagonists experienced plentiful growth intensity in wheat bran + malt medium (isolates M-73). On the majority of the substrates, *T. longibrachiatum* and

T. viride produced yellow to green pigments, while *E. nigrum* produced yellow to pink pigments that were conspicuous. On wheat bran combinations, the majority of the isolates produced an abundant or moderate quantity of conidia but few on sawdust combinations.

It was discovered that factors, such as pH, carbon: nitrogen (C:N) ratio, carbon content, and harvesting time had an impact on the characteristics of *T.harzianum* spores (Agosin et al., 1997). The highest spore lifetime was found in a medium with a C:N ratio of 14 and a pH of 7.0, where the majority of the produced spores were still alive after 45 days of storage. Furthermore, they found that the spores maintained viability when kept at various relative humidity levels, indicating that they might be more adaptable in the field.

Ushamalini et al. (1997) investigated how antagonists affected *M. phaseolina* and *Fusarium oxysporum* sp. Tracheihilum's *in vitro* growth. The cowpea plants that were affected by charcoal rot and wilt served as the source of both illnesses. The effectiveness of the antagonists such as *B.subtilis*, *Pseudomonas fluorescens*, *T.viride*, *T. harziamtm*, *T. hamatum*, *T. koningi*, and *T. pseudokoningi* to inhibit the pathogens was examined by using dual-culture method. Although *T. viride* and *T. harzianum* appeared to be the most effective antagonists, all of them severely inhibited *M. phaseolina* development.

Trichoderma spp. mutants that were benomyl tolerable had their rhizosphere competency assessed by Ahmad & Baker (1988). The fungicide benomyl was tolerated by two strains of *T. harzianum* and one each of *koningii*, *polysporum*, and *viride*. There was evidence that *Trichoderma* spp. experienced a mutation that increased rhizosphere competence and linear growth rate.

According to Dawar et al. (1993), cowpea and mungbean seed pelleting with biological antagonists was effective in controlling fungi that infect the roots of the plants. *Stachybotris* and *Rhizobium meliloti* were found to be more effective to prevent *M. phaseolina* and *Fusarium* spp. from colonizing *Vigna radiata* and *R. solani* from colonizing cowpea roots when used together than when used alone.

T. viride antagonistic relationship with *M. phaseolina* which is used in the management of *V. radiata* dry root rot was assessed by Kehri & Chandra (1991). In unsterilized soil under greenhouse circumstances, *T. viride* decreased *M. phaseolina* mortality in *Vigna radiate* Cv. T-44 and Cv. PusaBaisakhi from 19 to 18% and 19 to 10%, respectively. They also suggested that sterilized soil boosted the biological control effectiveness of the antagonist. Shoots, grains, and nodules all experienced increases in dry weight between T-44 and PusaBaisakhi of 31.7, 16.6, and 00%, respectively.

Trichoderma species have been mass-multiplicated by Gangadharan & Jeyarajan (1990). *T. viride* and *T. harzianum* were mass-produced on 18 distinct substrates made from agricultural waste and by-products. The largest number of colony-forming units (CFU) was produced by both biocontrol agents on tapioca rind, followed by tapioca waste and well-decomposed farmyard manure. *T. viride* and *T. harzianum* were able to grow and sporulate well on other substrates as well.

Shahzadet al. (1991) evaluated the biological control of *Macrophomina*-infected urdbeans (*Vigna mungo*). Seeds were treated with a microbial antagonist prior to planting in two locations in Pakistan with naturally infected soil (Karachi and Islamabad). Following treatment with *Paecilomyces lilacinus*, *T. hamatum*, *T. harzianum*, *T. viride*, *Gliocladium virens*, *Aspergillus candidus*, and *Streptomyces* spp., no infection was seen in Islamabad plots; *Rhizobium meliloti* had no effect on the infection or frequency of *M. phaseolina* root colonization. With the exception of *T. harzianum*, *A. candidus*, and *R. meliloti*, all therapies were found to considerably diminish the infection in Karachi. The two antagonists, *T. harzianum*, and *A. candidus*, dramatically decreased root colonization as compared to the untreated control.

In an acidic soil environment, Raghuchander et al. (1993) reported the use of *T. viride* as row treatments 2 days before to sowing to biocontrol *Macrophamina* root rot of mungbean. They also discovered that groundnut shell medium, among the organic substrates utilized for *T. viride* proliferation supported the best chlamydospore formation, best native *Rhizobium* nodulation, and maximum yield. In comparison to treatments utilizing substrates made of pressed mud or coir pith, the number of *M. phaseolina* sclerotia and the incidence of root rot were significantly decreased in the groundnut shell treatment.

T. harzianum, *T. virid*, and *G. virens* have antagonistic effects in vitro against a number of soil-borne pathogens, including *Rhizoclonia solani* and *Pythium aphanidermatum*. Previous researchers have also reported *S. rolfsii*, *Sclerotinia sclerotiorum*, etc (Dohroo et al., 1990; Lewis & Papavizas, 1987; Mathur, 2001; Mukhopadhyay, 1987; Muller et al., 1985; Xu et al., 1993; Zheng et al., 1990).

17.7.1.2 Actinomycetes *BCAs*

A class of Gram-positive bacteria known as Actinomycetes is found in the order Actinomycetales. They are commonly referred to as imperfect fungus since they produce branching mycelium. The breakdown of organic materials and the production of secondary metabolites with commercial value depend on them, which most typically are found in soil, compost, freshwater and marine settings. Actinomycetes were found to potentially be harmful to *M. phaseolina*. The hydrolytic enzymes or antibiotics produced were the reason for this inhibition. Therefore, it may be assumed that the production of antifungal compounds may be involved in the inhibition of fungal pathogens' capacity to proliferate as hyphae given that there was no direct interaction between actinomycete colonies and fungal pathogens in the dual-culture assay. Actinomycetes (*Streptomyces* spp.) isolated from rhizosphere soil produce siderophores that inhibit the development of phytopathogens (Tokala et al., 2002). Another potential tactic for managing the phytopathogens is competition for iron. In order to get iron, rhizosphere actinomycetes must outcompete the other rhizosphere plant pathogens. Additional evidence suggests that HCN production contributes to illness prevention (Gopalakrishnan et al., 2011). Actinomycetes, especially *Streptomyces* species, are the best choice of microorganisms for biological control because they are well known for producing antifungal compounds.

17.7.2 Bacterial BCAs

The antagonistic effects of a number of rhizospheric and root-associated bacteria against *M. phaseolina* have been discovered and characterized. *Pseudomonas putida* and *Bacillus clausii* have been investigated to know the various substances that might either encourage or impede plant development for their potential to lower the severity of the charcoal rot disease in soybean crops (Yasmin et al., 2020). *Bacillus amyloliquefaciens* and *B. siamensis* have demonstrated antifungal activity; however, more research is needed to determine the precise content and molecular structure of the filtrates. As an illustration, it was found that the microsclerotia sustained deadly injuries (Hussain & Khan, 2020; Torres et al., 2016). *B. subtilis*, on the other hand, created a fungistatic impact as opposed to a fungistatic effect, which was most likely brought on by rivalry for resources or space (Torres et al., 2016). Additionally, *M. phaseolina* was inhibited by two plant growth promoting rhizobacteria (PGPR) isolates, *Pseudomonas fluorescens* and *B.subtilis,* both of which were grown in vitro (on soybean seeds) and in vivo (in a greenhouse experiment) (Simonetti et al., 2015). Fascinatingly, the full proteome of *M. phaseolina* revealed significant down regulation *in oxidative* stress defense pathways, growth, and cell wall integrity and virulence under bacterial contamination-challenged conditions. Instead, it was shown that proteins involved in energy synthesis, defense, and stress response were up-regulated. *M. phaseolina* remained dormant after the bacteria were destroyed before resuming its active life with less virulence. In this sense, it appears that bacterial contaminants suppress *M. phaseolina* growth and pathogenicity (Zaman et al., 2020).

17.7.2.1 Bacillus *species*

The most prevalent bacterial genus in agro-systems is *Bacillus*, and one of its main uses is to suppress phytopathogenic fungi in crops (Villarrealet al., 2018). In Argentina, common beans have been found to be resistant to *M. phaseolina* because of *B. amyloliquefaciens*, which also reduces the frequency of charcoal root rot on seeds (Sabateet al., 2017).

According to Hashem et al. (2017), mungbean species grown more quickly in both healthy and *M. phaseolina*-infected conditions, demonstrating the ability of the endophytic bacterium *B. subtilis* to promote plant growth and disease resistance. Due to the activation of the antioxidant system and the buildup of osmolytes, mung bean plants treated with *B. subtilis* has shown the ability to prevent oxidative membrane damage. According to this study, *B. subtilis* can be used to sustainably boost growth and productivity while preventing charcoal rot in mung beans. Pre-emergence, postemergence, seedling survival after infection with *M. phaseolina*, and disease severity were all given as percentages of 33.33, 20.33, 33.00, and 3.88, respectively. The activity of *B. subtilis*, however, significantly increased the survival rate of healthy plants, reaching 82.14%, as compared to plants infected by *M. phaseolina*.

Bojorquez-Armentaet al. (2021) claim that the *M. phaseolina*, which also harms seedling roots, stems, and leaves, is responsible for common bean charcoal rot. Four isolates, later identified as *Bacillus* spp. (BA97, BN17, BN20, and BR20), exhibited antagonistic behavior versus *M. phaseolina in vitro*, inhibiting its growth by 62.5%–85%. In a plant antagonistic experiment, isolate BN20 had the highest effect on lowering disease level. As a defense mechanism, the isolates BA97, BN17, BN20, and BR20 generated volatile chemicals. These findings imply that decreasing *M. phaseolina* agricultural harm using rhizobacterial seed bacterization for field planting may be successful.

17.7.3 Other BCAs

17.7.3.1 Elicitors of plant defense

Elicitors, which are sprayed on plants and can be either natural or synthetic, have been shown to support systemic acquired resistance (SAR) and shield them from bacterial, fungal, and viral infections. In order to inhibit *M. phaseolina* and two additional soybean diseases (*Phytophtorasojae* and *Sclerotiniasclerotiorum*), the soybean foliage has been treated with the elicitors-benzothiadiazole (BTH), chitosan (CHT), phenylalanine (PHE), and salicylic acid (SA). The results showed that the efficacy of the elicitor varied according to the genotypes, pathogens, and environmental aspects of the soybeans (Pawlowski et al., 2016). Chitosan has shown a potential dual role by inducing a defense response in jute seedlings and directly reducing *M. phaseolina* during infection. Changes in the enzyme profiles of jute after treatment with water-soluble chitosan made it simpler to comprehend the mechanism of action of this antifungal agent (s-chitosan). This is because, during *M. phaseolina* infection of jute seedlings, treatment with s-chitosan increased the activity of defense-related enzymes such chitosanase and peroxidase in infected seedlings (Chatterjee et al., 2014). Understanding the immunological responses that elicitor's trigger is crucial, the use of elicitors in plant resistance may have a negative impact on other physiological functions, which may have a negative impact on biomass and seed production, among other plant characteristics. It is crucial to distinguish between elicitors that operate as priming molecules and those that directly activate plant defenses. It is believed that plants would reply more rapidly or more forcefully to a second attack when they have previously been exposed to a stimulus in a priming setting (Conrath, 2011; Denance et al., 2013; Pare et al., 2005).

17.7.4 Innovative genetic tools

Numerous plant diseases, such as *Fusarium*, *Aspergillus*, *Verticillium*, and *Sclerotinia*, have been managed using small interfering RNA (siRNA) molecules (Mcloughlin et al., 2018). The RNAi-mediated suppression of specific target genes, based on their role in growth and/or pathogenicity, may have a significant impact on the pathogen's ability to infect the host or the severity of host symptoms. Exogenous siRNAs were employed to target the genes for 1,3-glucan synthase and chitin synthase in *M. phaseolina.* The synthesis of the fungal cell wall depends on these targeting genes. It is interesting to note that siRNA treatment of the fungi reduced growth as seen by a smaller growth area and less thick mycelium. Since microsclerotia developed and melanized more slowly under various treatment scenarios, the siRNA therapies have also been observed to postpone the fungal maturation. Additionally, the transcript abundances of target genes were decreased when *M. phaseolina* growth was suppressed (Forster & Shuai, 2020). Due to the fact that siRNAs can be employed in other cutting-edge technologies, it is imperative to select siRNAs that lessen the negative effects brought on by off-target binding in a host plant or other species. For instance, in host-delivered RNA interference (HD-RNAi), plants have genes that create siRNA that is intended to interfere with infections (Hu et al., 2015).

17.8 Conclusion

It is essential to fully understand the interactions with *M. phaseolina* that may be useful in disease control. This is due to the complexity of the interactions that occur underground between *M. phaseolina* and micro- or macro-organisms as well as with the physicochemical environmental conditions.The polyphagous pathogen *M. phaseolina*, which causes charcoal rot in many crops and persists in crop waste and soil, can significantly be suppressed by implementing fewer practices that are well integrated as opposed to those procedures alone. Therefore, the most practical strategy for integrated disease management could include seed treatment, soil application of native bio-control agents and usage of resistant varieties.The incidence of disease was found to be reduced by all *Trichoderma* species and several *Bacillus* species, which results in improving the plant growth. In addition to fungal and bacterial biocontrol agents, plant metabolites, ecofriendly plant defense elicitors are also promising antifungal agents that can easily is produced and purchased. Also new innovative techniques such as siRNA can also be used effectively.

References

Agosin, E., Voipe, D., Muiioz, C., Mailin, R., & Crawford, A. (1997). Effect of culture conditions on spore shelf life of the bio control agent *Trichoderma harzianiim. World Journal of Microbiology' and Biotechnology, 13*(2), 225–232.

Ahmad, J. S., & Baker, R. (1988). Rhizosphere competence of benomyl-tolerant mutants of *Trichoderma* spp. *Canadian Journal of Microbiology, 34*(5), 694–696.

Almomani, F., Alhawatema, M., & Hameed, K. (2013). Detection, identification and morphological characteristic of *Macrophomina phaseolina*: The charcoal rot disease pathogens isolated from infected plants in northern Jordan. *Archieves of Phytopathology and Plant Protection, 46*, 1005–1014. https://doi.org/10.1080/03235408.2012.756174

Ambrosio, M. M., Dantas, A. C., Martinez-Perez, E., Medeiros, A. C., Nunes, G. H., & Pico, M. B. (2015). Screening a variable germplasm collection of *Cucumismelo* L. for seedling resistance to *Macrophomina phaseolina. Euphytica, 206*, 287–300. https://doi.org/10.1007/s10681-015-1452-x

Bastakoti, S., Belbase, S., Manandhar, S., & Arjyal, C. (2017). *Trichoderma* species as biocontrol agent against soil borne fungal pathogens. *Nepal Journal of Biotechnology, 5*, 39–45. https://doi.org/10.3126/njb.v5i1.18492

Bojorquez-Armenta, Y. D. J., Mora-Romero, G. A., López-Meyer, M., Mendoza, I. E., Martinez, C. C., Urias, C. A., Cordero-Ramirez, J. D., & Martinez-Alvarez, J. C. (2021). Evaluation of *Bacillus* spp. isolates as potential biocontrol agents against charcoal rot caused by *Macrophomina phaseolina* on common bean. *Journal of General Plant Pathology, 87*, 377–386. https://doi.org/10.1007/s10327-021-01019-4

Brundrett, M. C., & Tedersoo, L. (2018). Evolutionary history of mycorrhizal symbioses and global host plant diversity. *New Phytologist, 220*, 1108–1115. https://doi.org/10.1111/nph.14976

Chatterjee, S., Chatterjee, B. P., & Guha, A. K. (2014). A study on antifungal activity of water-soluble chitosan against *Macrophomina phaseolina*. *International Journal of Biological Macromolecules, 67*, 452–457. https://doi.org/10.1016/j.ijbiomac.2014.04.008

Choudhary, D. K. (2011). Plant growth-promotion (PGP) activities and molecular characterization of rhizobacterial strains isolated from soybean (*Glycine max* L. Merr plants against charcoal rot pathogen, *Macrophomina phaseolina. Biotechnology Letters, 33*, 2287–2295.

Conrath, U. (2011). Molecular aspects of defence priming. *Trends in Plant Science, 16*, 524–531. https://doi.org/10.1016/j.tplants.2011.06.004

Das, I. K., Indira, S., Annapurna, A., Prabhakar, S., & Seetharama, N. (2008). Biocontrol of charcoal-rot in sorghum by fluorescent *Psedomonads* associated with the rhizosphere. *Crop Protection, 27*, 1407–1414.

Dawar, S., Shahzad, S., Tqbal, S., & Ghaffar, A. (1993). Effect of seed pelleting with biological antagonists in the control of root infecting fungi of cowpea and mungbean. *Pakistan Journal of Botany, 25*(2), 219–224.

Denance, N., Sánchez-Vallet, A., Goffner, D., & Molina, A. (2013). Disease resistance or growth: The role of plant hormones in balancing immune responses and fitness costs. *Frontiers of Plant Science, 4*, 155. https://doi.org/10.3389/fpls.2013.00155

Dohroo, N. P., Gupta, S. K., Shyam, K. R., & Sharma, B. K. (1990). Antagonistic studies on causal fungi of wirestem and stalk rot of cauliflower. *Indian Journal of Plant Pathology, 8*(1and2), 77–78.

Doley, K., & Jite, P. K. (2013). Effect of arbuscular mycorrhizal fungi on growth of groundnut and disease caused by *Macrophomina phaseolina. Journal of Experimental Sciences, 3*(9), 46–50.

Dubey, S. C., Suresh, M., & Singh, B. (2007). Evaluation of Trichoderma species against *Fusariumoxysporum* f. sp. ciceris for integrated management of chickpea wilt. *Biological Control, 40*, 118–127. https://doi.org/10.1016/j.biocontrol.2006.06

Forster, H., & Shuai, B. (2020). Exogenous siRNAs against chitin synthase gene suppress the growth of the pathogenic fungus *Macrophomina phaseolina*. *Mycologia, 112*, 699–710. https://doi.org/10.1080/00275514.2020.1753467

Gajera, H., Bambharolia, R., Patel, S., Khatrani, T., & Goalkiya, B. (2012). Antagonism of *Trichoderma* spp. against *Macrophomina phaseolina*: Evaluation of coiling and cell wall degrading enzymatic activities. *Journal of Plant Pathology and Microbiology, 3*, 149. https://doi.org/10.4172/2157-7471.1000149

Gangadhar, & Jeyarajan, R. (1990). Mass multiplication of *Trichoderma* spp. *Journal of Biological Control, 4*(1), 70–71.

Ghosh, T., Biswas, M. K., Guin, C., & Roy, P. (2018). A review on characterization, therapeutic approaches, and pathogenesis of *Macrophomina phaseolina. Plant Cell Biotechnology and Molecular Biology, 19*, 72–84.

Gopalakrishnan, S., Kiran, B., Humayun, P., Vidya, M., Deepthi, K., Jacob, S., Vadlamudi, S., Alekhya, G., & Rupela, O. (2011). Biocontrol of charcoal-rot of sorghum by actinomycetes isolated from herbal vermicompost. *African Journal of Biotechnology, 10*(79), 18142–18152. https://doi.org/10.5897/AJB11.2710

Gupta, G. K., Sharma, S. K., & Ramteke, R. (2012). Biology, epidemiology and management of the pathogenic fungus *Macrophomina phaseolina* (Tassi) Goid. with special reference to charcoal rot of soybean (*Glycine max* (L.) Merr.). *Journal of Phytopathology, 160*, 167–180.

Hashem, A., Fathi, E., Abdulaziz, A., Radhakrishnan, R., & Kumar, A. (2017). Plant defense approach of Bacillus subtilis (BERA 71) against *Macrophomina phaseolina* (Tassi) Goid. inmungbean. *Journal of Plant Intercations, 12*(1), 390–401. https://doi/full/10.1080/17429145.2017.1373871

Hu, Z., Parekh, U., Maruta, N., Trusov, Y., & Botella, J. R. (2015). Downregulation of *Fusariumoxysporum* endogenous genes by host-delivered RNA interference enhances disease resistance. *Frontiers of Chemistry, 3*, 1. https://doi.org/10.3389/fchem.2015.00001

Hussain, T., & Khan, A. A. (2020). Determining the antifungal activity and characterization of *Bacillus siamensis AMU03* against *Macrophomina phaseolina* (tassi) goid. *Indian Phytopathology, 73*, 507–516. https://doi.org/10.1007/s42360-020-00239-6

Hyder, S., Inam-ul-haq, M., Bibi, S., & Humayun, A. (2017). Novel potential of *Trichoderma* Spp. as biocontrol agent. *Journal of Entomology and Zoology Studies, 5*, 214–222.

Islam, M., Haque, M., Islam, M., Emdad, E., Halim, A., Hossen, Q. M., Hossain, M., Ahmed, B., Rahim, S., Rahman, M., Alam, M., Hou, S., Wan, X., Saito, J., & Alam, M. (2012). Tools to kill: Genome of one of the most destructive plant pathogenic fungi *Macrophomina phaseolina. BMC Genomics, 13*, 493. https://doi.org/10.1186/1471-2164-13-493

Kehri, H. K., & Chandra, S. (1991). Antagonism of *Trichodermaviride* to *Macrophomina phaseolina* and its application in the control of dry root of mung. *Indian Phytopathology, 44*(1), 60–63.

Kendig, S. R., Rupe, J. C., & Scott, H. D. (2000). Effect of irrigation and soil water stress on densities of *Macrophomina phaseolina* in soil and roots of two soybean cultivars. *Plant Disease, 84*, 895–900. https://doi.org/10.1094/PDIS.2000.84.8.895

Khan, M. R., & Anwer, A. M. (2007). Molecular and biochemical characterization of soil isolates of *Aspergillusniger* aggregate and an assessment of their antagonism against *Rhizoctoniasolani. Phytopathologia Mediterranea, 46*, 304–315.

Khan, M. R., Haque, Z., Rasool, F., Salati, K., Khan, U., Mohiddin, F. A., & Zuhaib, M. (2019). Management of root-rot disease complex of mungbean caused by *Macrophomina phaseolina* and *Rhizoctoniasolani* through soil application of Trichoderma spp. *Crop Protion, 119*, 24–29. https://doi.org/10.1016/j.cropro.2019.01.014

Kumar, P., Dubey, R. C., & Maheshwari, D. K. (2012). *Bacillus* strains isolated from rhizosphere showed plant growth promoting and antagonistic activity against phytopathogens. *Microbiological Research, 167*(8), 493–499. https://doi.org/10.1016/j.micres.2012.05.002

Kumar, V., Kumar, A., & Kharwar, R. N. (2007). Antagonistic potential of fluorescent *pseudomonads* and control of charcoal rot of chickpea caused by *Macrophomina phaseolina. Journal of Environment Biology, 28*, 15–20.

Kumar, P., Pahal, V., Gupta, A., Vadhan, R., Chandra, H., & Dubey, R. C. (2020). Effect of silver nanoparticles and *Bacillus cereus* LPR2 on the growth of Zea mays. *Scientific Reports, 10*(1), 20409.

Lewis, J. A., & Papavizas, G. C. (1987). Application of *trichoderma* and *gliocladium* in alginate pellets for control of *rhizoclonia* damping off. *Plant Pathology, 36*, 438–446.

Liu, J., Blaylock, L. A., Endre, G., Cho, J., Town, C. D., VandenBosch, K. A., & Harrison, M. J. (2003). Transcript profiling coupled with spatial expression analyses reveals genes involved in distinct developmental stages of an arbuscular mycorrhizal symbiosis. *The Plant Cell Online, 15*, 2106–2123. https://doi.org/10.1105/tpc.014183

Liu, J., Maldonado-Mendoza, I., Lopez-Meyer, M., Cheung, F., Town, C. D., & Harrison, M. J. (2007). Arbuscular mycorrhizal symbiosis is accompanied by local and systemic alterations in gene expression and an increase in disease resistance in the shoots. *The Plant Journal, 50*, 529–544. https://doi.org/10.1111/j.1365-313X.2007.03069.x

Lodha, S., & Singh, M. (1984). Quantitative determination of *Macrphomina phaseolina* (Tassi) Goid in grass-legume intercropping systems. *Annals of Arid Zone, 23*(3), 259–261.

Marquez, N., Giachero, M., Declerck, S., & Ducasse, D. (2021). *Macrophomina phaseolina*: General characteristics of pathogenicity and methods of control. *Frontiers of Plant Science, 12*. https://doi.org/10.3389/fpls.2021.634397

Marquez, N., Giachero, M. L., Gallou, A., Debat, H. J., Cranenbrouck, S., Di Rienzo, J. A., Pozo, M. J., Ducasse, D. A., & Declerck, S. (2018). Transcriptional changes in mycorrhizal and nonmycorrhizal soybean plants upon infection with the fungal pathogen *Macrophomina phaseolina. Molecular Plant-Microbe Interactions, 31*, 842–855. https://doi.org/10.1094/MPMI-11-17-0282-R

Marquez, N., Giachero, M. L., Gallou, A., Debat, H. J., Declerck, S., & Ducasse, D. A. (2019). Transcriptome analysis of mycorrhizal and nonmycorrhizal soybean plantlets upon infection with *Fusarium virguliforme*, one causal agent of sudden death syndrome. *Plant Pathology, 68*, 470–480. https://doi.org/10.1111/ppa.12964

Martinez-Medina, A., Pozo, M. J., Cammue, B. P. A., & Vos, C. M. F. (2016). Belowground defence strategies in plants: The plant–*trichoderma* dialogue. In C. M. F. Vos, & K. Kazan (Eds.), *Belowground defense strategies in plants* (pp. 301–327). Cham: Springer. https://doi.org/10.1007/978-3-319-42319-7_13

Martínez-Salgado, S. J., Andrade-Hoyos, P., Parraguirre-Lezama, C., Rivera-Tapia, A., Luna-Cruz, A., & Romero-Arenas, O. (2021). Biological control of charcoal rot in peanut crop through strains of *Trichoderma* spp., in Puebla, Mexico. *Plants, 10*(12), 2630. https://doi.org/10.3390/plants10122630

Mathur, A. C. (2001). *Integrated disease management of charcoal rot of cowpea.* CCS Haryana Agricultural University.

Mcloughlin, A. G., Walker, P. L., Wytinck, N., Sullivan, D. S., Whyard, S., & Belmonte, M. F. (2018). Developing new RNA interference technologies to control fungal pathogens. *Canadian Journal of Plant Pathology, 40*, 325–335. https://doi.org/10.1080/07060661.2018.1495268

Mondal, G., Dureja, P., & Sen, B. (2000). Fungal metabolites from *Aspergillus niger* AN27 related to plant growth promotion. *Indian Journal of Experimental Biology, 38*, 84–87.

Mukhopadhyay, A. N. (1987). Biological control of soil borne plant pathogens by *Trichoderma* spp. *Indian Journal of Mycology & Plant Pathology, 17*, 16.

Muller, J. D., Cline, M. N., Sinclair, J. B., & Jacobson, P. J. (1985). An in vitro test for evaluating efficacy of mycoparasites on sclerotia for *Sclerofinia sclerotioruin. Plant Disease, 69*(7), 584–587.

Nafady, N. A., Hashem, M., Hassan, E. A., Ahmed, H. A. M., & Alamri, S. A. (2019). The combined effect of arbuscular mycorrhizae and plant-growth-promoting yeast improves sunflower défenseagainst *Macrophomina phaseolina* diseases. *Biological Control, 138*, 104049. https://doi.org/10.1016/j.biocontrol.2019.104049

Nagaraja, A., Chethana, B. S., & Jain, A. K. (2021). Biotic stresses and their management. In M. Singh, & S. Sood (Eds.), *Millets and pseudo cereals, genetic resources and breeding advancements* (pp. 119–142). Woodhead Publishing Series in Food Science, Technology and Nutrition,Woodhead Publishing. https://doi.org/10.1016/B978-0-12-820089-6.00007-0

Oyewole, B. O., Olawuyi, O. J., Odebode, A. C., & Abiala, M. A. (2017). Influence of Arbuscularmycorrhiza fungi (AMF) on drought tolerance and charcoal rot disease of cowpea. *Biotechnology Reports, 14*, 8–15. https://doi.org/10.1016/j.btre.2017.02.004

Pan, S., & Bhagat, S. (2008). Characterization of antagonistic potential of *Trichoderma* spp. against some soil borne plant pathogens. *Journal of Biological Control, 22*, 43–49.

Pandey, A. K., Burlakoti, R. R., Rathore, A., & Nair, R. M. (2020). Morphological and molecular characterization of *Macrophomina phaseolina* isolated from three legume crops and evaluation of mungbean genotypes for resistance to dry root rot. *Crop Protection, 127*, 104962. https://doi.org/10.1016/j.cropro.2019.104962

Pan, S. R., Mukherji, & Bhagat, S. (2013). Evaluation of *Trichoderma* spp. against soil borne plant pathogens. *Annals of Plant Protection Sciences, 21*, 197–198.

Pare, P. W., Farag, M. A., Krishnamachari, V., Zhang, H., Ryu, C.-M., & Kloepper, J. W. (2005). Elicitors and priming agents initiate plant defense responses. *Photosynthesis Research, 85*, 149–159. https://doi.org/10.1007/s11120-005-1001-x

Pawlowski, M. L., Bowen, C. R., Hill, C. B., & Hartman, G. L. (2016). Responses of soybean genotypes to pathogen infection after the application of elicitors. *Crop Protection, 87*, 78–84. https://doi.org/10.1016/j.cropro.2016.04.022

Pozo, M. J., & Azcón-Aguilar, C. (2007). Unraveling mycorrhiza-induced resistance. *Current Opinion in Plant Biology, 10*, 393–398. https://doi.org/10.1016/j.pbi.2007.05.004

Pozo, M. J., Jung, S. C., & López-ráez, J. A. (2010). Impact of Arbuscular Mycorrhizal symbiosis on plant response to biotic stress: The role of plant defense mechanisms. In H. Koltai, & Y. Kapulnik (Eds.), *Arbuscular Mycorrhizas: Physiology and function* (pp. 193−207). Cham: Springer.

Raghuchander, T., Somiappan, R., & Arjunan, G. (1993). Biocontrol of *Macrophomina* root rot of mungbean. *Indian Phytopathology, 46*(4), 379−382.

Sabate, D. C., Brandan, C. P., Petroselli, G., Erra-Balsells, R., & Audisio, M. C. (2017). Decrease in the incidence of charcoal root rot in common bean (*Phaseolus vulgaris* L.) by *Bacillus amyloliquefaciens* B14, a strain with PGPR properties. *Biological Control, 113*, 1−8.

Salvatore, M. M., Félix, C., Lima, F., Ferreira, V., Naviglio, D., & Salvatore, F. (2020). Secondary metabolites produced by *Macrophomina phaseolina* isolated from eucalyptus globulus. *Agriculture, 10*, 72. https://doi.org/10.3390/agriculture10030072

Senthilkumar, M., Swarnalakshmi, K., Govindasamy, V., Lee, Y. K., & Annapurna, K. (2009). Biocontrol potential of soybean bacterial endophytes against charcoal rot fungus, *Rhizoctonia bataticola*. *Current Microbiology, 58*, 288−293.

Shahzad, S., Hiissain, S., & Dawar, S. (1991). Biological control of *macrophomina phaseolina* infection on mashbean(*vignamungo*(L.) hepper). *Pakisthan Journal of Botany, 23*(1), 131−134.

Siddiqui, I. A., Ehetshamul-Haque, S., & Shaukat, S. S. (2001). Use of *rhizobacteria* in the control of root rot-root knot disease complex of mungbean. *Journal of Phytopathology, 149*, 337−346.

Simonetti, E., Viso, N. P., Montecchia, M., Zilli, C., Balestrasse, K., & Carmona, M. (2015). Evaluation of native bacteria and manganese phosphite for alternative control of charcoal root rot of soybean. *Microbiological Research, 180*, 40−48. https://doi.org/10.1016/j.micres.2015.07.004

Singh, M., & Majumdar, V. L. (1995). Antagonistic activity of *Trichoderma* spp. to *Macrophomina phaseolina* (tassi) Goid.*in vitro*. *Environment and Ecology, 13*(2), 481−482.

Singh, J., Singh, R. S., & Singh, H. V. (1996). Selection of substrates for growth and sporulation of different antagonists *of Rhizoctonia solani* in mass culture. *Plant Disease Research, 11*(2), 132−136.

Spagnoletti, F., Carmona, M., Gómez, N. E. T., Chiocchio, V., & Lavado, R. S. (2017). *Arbuscular mycorrhiza* reduces the negative effects of *Macrophomina phaseolina* on soybean plants in arsenic-contaminated soils. *Applied Soil Ecology, 121*, 41−47. https://doi.org/10.1016/j.apsoil.2017.09.019

Spagnoletti, F. N., Cornero, M., Chiocchio, V., Lavado, R. S., & Roberts, I. N. (2020). Arbuscular mycorrhiza protects soybean plants against *Macrophomina phaseolina* even under nitrogen fertilization. *European Journal of Plant Pathology, 156*, 839−849. https://doi.org/10.1007/s10658-020-01934-w

Spagnoletti, F. N., Leiva, M., Chiocchio, V., & Lavado, R. S. (2018). Phosphorus fertilization reduces the severity of charcoal rot (*Macrophomina phaseolina*) and the arbuscular mycorrhizal protection in soybean. *Journal of Plant Nutrition and Soil Science, 181*, 855−860. https://doi.org/10.1002/jpln.201700569

Sridharan, A., Thankappan, S., Karthikeyan, G., & Uthandi, S. (2020). Comprehensive profiling of the VOCs of *Trichoderma longibrachiatum* EF5 while interacting with *sclerotiumrolfsii* and *Macrophomina phaseolina*. *Microbiological Research, 236*, 126436. https://doi.org/10.1016/j.micres.2020.126436

Tok, F. M. (2019). Relationship between morphologic, phenotypic and pathogenic characteristics in *Macrophomina phaselina* Isolates from cucumber plants. *International Journal of Innovative Approaches in Agricultural Research, 3*, 651−660. https://doi.org/10.29329/ijiaar.2019.217.11

Tokala, R. K., Strap, J. L., Jung, C. M., Crawford, D. L., Salove, M. H., Deobald, L. A., Bailey, J. F., & Morra, M. J. (2002). Novel plant-microbe rhizosphere interaction involving *Streptomyces lydicus* WYEC108 and the pea plant (*Pisumsativum*). *Applied and Environmental Microbiology, 68*, 2161−2171.

Torres, M. J., Brandan, C. P., Petroselli, G., Erra Balsells, R., & Audisio, M. C. (2016). Antagonistic effects of Bacillus subtilis subsp. subtilis and B. Amyloliquefaciens against *Macrophomina phaseolina*: SEM study of fungal changes and UV-MALDI-TOF MS analysis of their bioactive compounds. *Microbiological Research, 182*, 31−39. https://doi.org/10.1016/j.micres.2015.09.005

Ushamalini, C., Rajappan, K., & Gangadharan, K. (1997). Inhibition of *Macrophomina phaseolina* and *Fusariumoxyporium* f. sp. *Tracheiphilim* by antagonists under *in vitro* conditions. *Plant Disease Research, 12*(2), 168−170.

Vasebi, Y., Safaie, N., & Alizadeh, A. (2013). Biological control of soybean charcoal root rot disease using bacterial and fungal antagonists *in vitro* and green conditions. *Journal of Crop Protection, 2*, 139−150.

Villarreal-Delgado, M. F., Villa-Rodríguez, E. D., Cira-Chávez, L. A., Estrada-Alvarado, M. I., Parra-Cota, F. I., & de losSantosVillalobos, S. (2018). The genus *Bacillus* as a biological control agent and its implications in the agricultural biosecurity. *Mexican Journal of Phytopathology, 36*, 95−130.

Xu, T., hong, J. P., & Li, D. B. (1993). Antagonism of *Trichoderma harzianum* T82 and *Trichoderma* spp. NF9 soil-borne borne fungus pathogens. *Acta Phytopathologica Sinica, 23*(I), 63−67.

Yang, Q. Y., Jia, K., Geng, R. J., & Li, S. D. (2014). Management of cucumber wilt disease by *Bacillus subtilis* B006 through suppression of *Fusarium oxysporum* f. sp. *cucumerinum* in rhizosphere. *Plant Pathology Journal, 13*, 160−166.

Yasmin, H., Naz, R., Nosheen, A., Hassan, M. N., Ilyas, N., Sajjad, M., Anjum, S., Gao, X., & Geng, Z. (2020). Identification of new biocontrol agent against charcoal rot disease caused by *Macrophomina phaseolina* in Soybean (*Glycine max* L.). *Sustainability, 12*, 6856. https://doi.org/10.3390/su12176856

Zaman, N. R., Kumar, B., Nasrin, Z., Islam, M. R., Maiti, T. K., & Khan, H. (2020). Proteome analyses reveal *Macrophomina phaseolina*'s survival toolswhen challenged by *Burkholderia* contaminans NZ. *ACS Omega, 5*, 1352−1362. https://doi.org/10.1021/acsomega.9b01870

Zhao, P. C., Quan, Y., Wang, J. W., & Fan, S. (2014). *Bacillus amyloliquefaciens* Q-426 as a potential biocontrolagent against *Fusarium oxysporum* f. sp. *spinaciae*. *Journal of Basic Microbiology, 54*, 448−456.

Zheng, J. P., Chen, J. H., Xu, T., & Li, D. B. (1990). Hyper-parasitism of *Trichodenva* spp. on *rhizoctoniasoluni*. *Acta Aghculiurae Universitilis Zhajiangensis, 16*(Suppl. 2), 73−77.

Chapter 18

Trichoderma spp.: A bio-agent for sustainable management of *Macrophomina phaseolina*

Md. Tanbir Rubayet and Md. Khurshed Alam Bhuiyan
Department of Plant Pathology, Bangabandhu Sheikh Mujibur Rahman Agricultural University, Gazipur, Bangladesh

18.1 Introduction

Macrophomina phaseolina is an important seed- and soil-borne noxious fungus present across the globe. It can affect around 500 plant species from >100 plant families. This pathogen results in several types of diseases, such as damping off, charcoal rot, stem and root rot, and seedling blight in the major field crops (Adekunle et al., 2001; Ghosh et al., 2018). It can cause significantly yield losses in soybean, lentil, blackgram, chickpea, mungbean, jute, melon, peanut, cucumber, etc. Optimum weather condition accelerates the disease severity and yield losses of these crops. As for example, 30−35°C with <60% soil moisture can significantly reduce soybean and sorghum yields (Kaur et al., 2012). It can propagate and disseminate very rapidly under the congenial environmental conditions. *M. phaseolina* MSP-4 isolate caused 38.15% seedling mortality during the soybean cultivation under open field condition and 32.73% disease incidence at vegetative stage (Rahman et al., 2021). According to Sharma and Bhowmik (1986), *M. phaseolina* infects the groundnut at pre-emergence stage and causes 100% yield loss. *M. phaseolina* can produce a wide range of secondary metabolites, including asperlin, botryodiplodin, isoasperlin, phaseolinone, and mullein, which could hasten the pathogenesis (Khambhati et al., 2020).

M. phaseolina persists as microsclerotia, as viable structure in the infected plant parts or soil for up to 3 years (Short et al., 1980; Vasebi et al., 2013). According to Gupta et al. (2012), the primary source of inoculum such as microsclerotia of *M. phaseolina* can survive in the soil up to 15 years. When they get susceptible host along with the congenial environment swiftly they invade the plant cortical tissues and produce gray black sclerotia in the infected portion (Chowdhury et al., 2014). The infected plants at the beginning stage become yellowish, wilt, block the vascular buddle, and develop greyish color due to production of abundance microsclerotia with premature death of the seedling (Farnaz et al., 2018; Short et al., 1978). As the fungal pathogen is seed-and soil-borne in nature and remains a wide host range, it is really very tough to manage by the application of traditional methods. Despite synthetic chemical compounds are toxic for the whole ecosystem, growers randomly use chemical fungicides for minimizing the crop losses. Therefore, drawbacks of fungicides have enhanced the awareness to reject the application of synthetic products and encouraged the growers to surge dependency on bioagents for the suppression of pathogens population density and diseases control (Zhang et al., 2018).

Biological method is the potential alternative instead of chemical fungicides. Biocontrol strategy is cost effective, reduced possibility of inducing pathogen resistance, nontoxic, selective target organism, self-sustaining, easy acclimatization, and diversification of mode of action (Muller-Scharer et al., 2020). In the present time, several types of bio-agents, such as species of *Trichoderma, Talaromyces*, *Cladosporium*, *Phomopsis*, *Aspergillus*, *Penicillium, Bacillus*, *Pseudomonas*, etc. have been discovered by researchers (Bayoumy et al., 2017; Deng et al., 2018; Hossain et al., 2017; Zhao et al., 2018). Among these beneficial microorganism, *Trichoderma* spp. (*T. harzianum, T. viride, T. pseudokoningii, T. hamatum,* and *T. koningii*) are the filamentous potent fungal biocontrol agent that can be used for controlling *Sclerotium rolfsii, Rhizoctonia solani, Fusarium oxysporum,* and *M. phaseolina* (Bastakoti et al., 2017; Rahman et al., 2018, 2021; Rubayet & Bhuiyan, 2016). Enormous reports have shown that *Trichoderma* has antagonistic activity against pathogens and control

Macrophomina Phaseolina. https://doi.org/10.1016/B978-0-443-15443-0.00020-6

seed- and soil-borne diseases (Ahmed et al., 2019; Arefin et al., 2019; Das et al., 2019; Liton et al., 2019; Simi et al., 2019). It is a fast growing saprophyte that produces chitin-degrading enzymes, for instance, chitinases, glucanases, etc. (Hewavitharana et al., 2018). It also produces plethora of secondary metabolites, such as tricholin, harzianic acid, viridian, gliosoprenins, heptelidic acid, 6-pentyl—pyrone, massoilactone, peptaibols, terpenes, polyketides, gliotoxin, gliovirin, koninginins, pyrones, and nonribo (Contreras-Cornejoet al., 2016; Mukherjee et al., 2012; Vinale et al., 2008). Actually, these special characteristics of this fungi help retard the growth and development as well as drastically dwindle the pathogen likely *M. phaseolina* population density at open field condition.

A hefty number of strains of *Trichoderma* species have been tested for their ability to control *M. phaseolina* and provide insight into the mode of action throughout the last few decades. A few of them were formulated and tested in the field condition (Adekunle et al., 2006; Mishra & Dantre, 2017; Taylor et al., 1991). This chapter intends to summarize the current understanding regarding the critical function and underlying mechanisms of *Trichoderma* spp. for sustainable management of *M. phaseolina* in field soil.

18.2 General characteristics of *Macrophomina phaseolina*

M. phaseolina belongs to the family Botryosphaeriaceae. On the basis of its characterization followed by morphological as well as molecular, still there has not been recognized subspecies or physiological races of this pathogen (Crous et al., 2006; Dhingra & Sinclair, 1978). On the contrary, *M. pseudophaseolina* and *M. euphorbiicola* these two new species of *Macrophomina* have been isolated in Senegal and Brazil (Machado et al., 2019; Sarr et al., 2014). *M. pseudophaseolina* was separated from lady's finger (*Abelmoschus esculentus*), groundnut (*Arachis hypogea*), roselle (*Hibiscus sabdariffa*), and cowpea (*Vigna unguiculata*) in Senegal, and *M. euphorbiicola* was isolated from groundnut (*A. hypogea*), cotton (*Gossypium hirsutum*), castor bean (*Ricinus communis*), and physic nut (*Jatropha curcas*) in Brazil, respectively.

18.3 Morphological characteristics of *M. phaseolina*

M. phaseolina produces hyaline mycelium at the beginning stage which becomes dark or pale brown at mature stage. Usually, mycelia are profusely branched and septate. Hypha normally branches from the main branch with right angle. A prominent constriction is present at the branching base point. This fungus generates microsclerotia, which are compacted clump of rigid mycelium. Microsclerotia are spherical, oval, or oblong in shape and light brown in the juvenile stage and growing darker (brown to black) with time. They also formed fruiting body like pycnidia. Pycnidia are spherical to obpyriform (inversed pear-shape), dark brown to black in color, round or globose, ostiolated, and embedded in the host tissue with 3—4 layers of heavily pigmented thick-walled cells on the outside. It contains short conidiophores and conidia. Conidiophores are also known as phialid. Phialides are hyaline, simple, short at a pyriform (pear-shape) of cylindrical and development from the inner wall of the pycnidia. However, conidia are also known as pycnidiospores, which are small, single celled, smooth, hyaline, ellipsoid to obvoid, oblong, and thin walled (Fig. 18.1) (Lakhran et al., 2018).

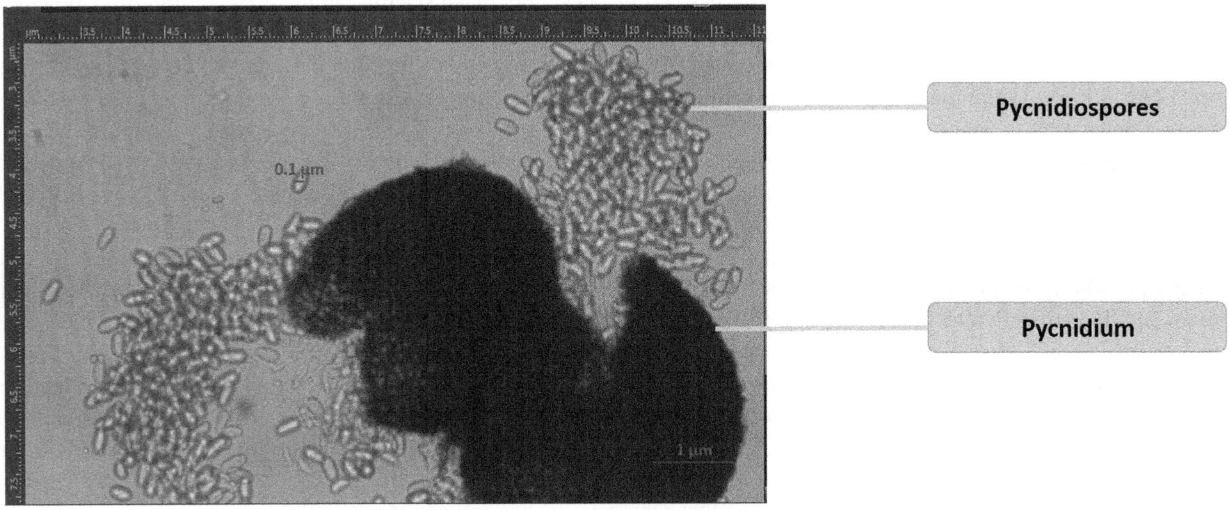

FIGURE 18.1 Different structures of *M. phaseolina* observed under compound microscope.

18.4 Life cycle of *M. phaseolina*

The resting microsclerotia remain in seed, soil, as well as plant debris for almost 15 years as a source of primary inoculum (Gupta et al., 2012). It can germinate under the availability of susceptible host along with the congenial environment (30–35°C). Further, the germ tube and appressoria are developed. The appressorium helps to take the nutrient from host. Then, pathogen penetrates into the host plant through the degradation of host tissues or natural opening pores such as stromata. After invasion inside the plant tissues, the fungus colonizes the host tissue. This situation is also responsible for development of the prominent symptoms including wilting (plugging of vascular buddle). If the seeds become infected with pathogen earlier to emergence, it will be pre-emergence mortality, and if it is after infection, it is known as postemergence mortality. When the lifespan of host completes, *M. phaseolina* recycles its life events (Marquez et al., 2021) (Fig. 18.2).

18.5 Distribution of *M. phaseolina*

M. phaseolina is a cosmopolitan notorious fungus found all over the world. It causes disease in diversified crops in the different geographical regions in the planate. According to first report of *M. phaseolina* disease in common bean

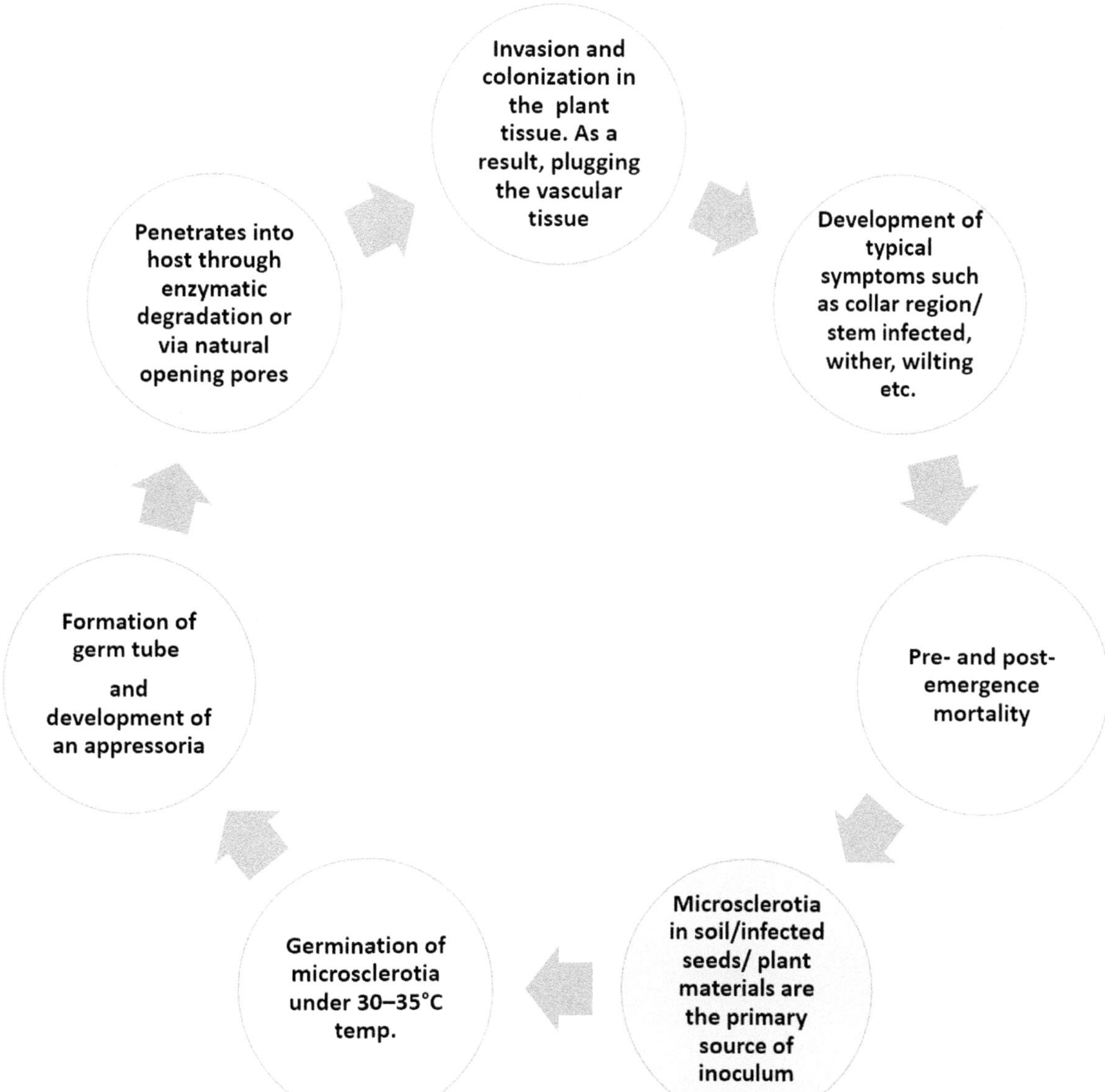

FIGURE 18.2 Disease cycle of *M. phaseolina*.

(Erper et al., 2021), mungbean (Zhang et al., 2011), peanut (Martínez-Salgado et al., 2021), melon (Jacob et al., 2013), tomato (Hyder et al., 2018), canola (Khangura & Aberra, 2009), sugar beet (Karadimos et al., 2002), strawberry (Abied & Duzan, 2020; Gerin et al., 2018), cactus (Abdel-Kader et al., 2010), immortelle (Pavlovic et al., 2015), kiwifruit (Türkkan et al., 2020), soybean (Yang & Navi, 2005), malabar spinach (Meena et al., 2018), lady's finger (Fakir & Mridha, 1985), lentil (Kouadri et al., 2021), watermelon (Fujinaga et al., 2002), lima Bean (Sousa et al., 2017), canola (Gaetán et al., 2006), bottle gourd (Kim et al., 2021), cassava (Msikita et al., 1998), and sunflower (Bokor, 2007) are presented in world map (Fig. 18.3).

18.6 Pathogenesis of *M. phaseolina*

M. phaseolina has the unique living characteristics in the active as well as dead plant tissues (Wang & Jones, 1995). The penetration and invasion into the plat tissue are facilitated by the plethora of cell wall–degrading enzymes such as endoglucanases, exocellobiohydrolases, and β-glucosidases (Marquez et al., 2021). Apart from these, this pathogen produces various types of phytotoxic metabolites including phaseolinon, mullein, botryodiplodin, azelaic acid, and patulin (Fig. 18.4). These biochemical substances play a substantial role for initiation of infection and finally develop the prominent symptoms for instance damping off, wilting, and necrotic lesions formation on foliar and ground parts of various crops (Table 18.1) (Bhattacharya et al., 1987; Khambhati et al., 2020; Ramezani et al., 2007; Salvatore et al., 2020). It addition, these also also enhance the virulence of the *M. phaseolina.* As a result, this fungus becomes more efficient about infection, increases host rage and adaptability, and is ubiquitously distributed in the nature (Islam et al., 2012; Salvatore et al., 2020). Thus, understanding the basis of the pathogenicity of *M. phaseolina* interaction with host plants is critical for management of the pathogen during profitable farming.

18.7 Management of *M. phaseolina*

Management of *M. phaseolina* is sometimes really difficult due to its long persistence in seed and soil as well as wide host range in nature. There are several management strategies that have been evaluated to curb *M. phaseolina* under the field

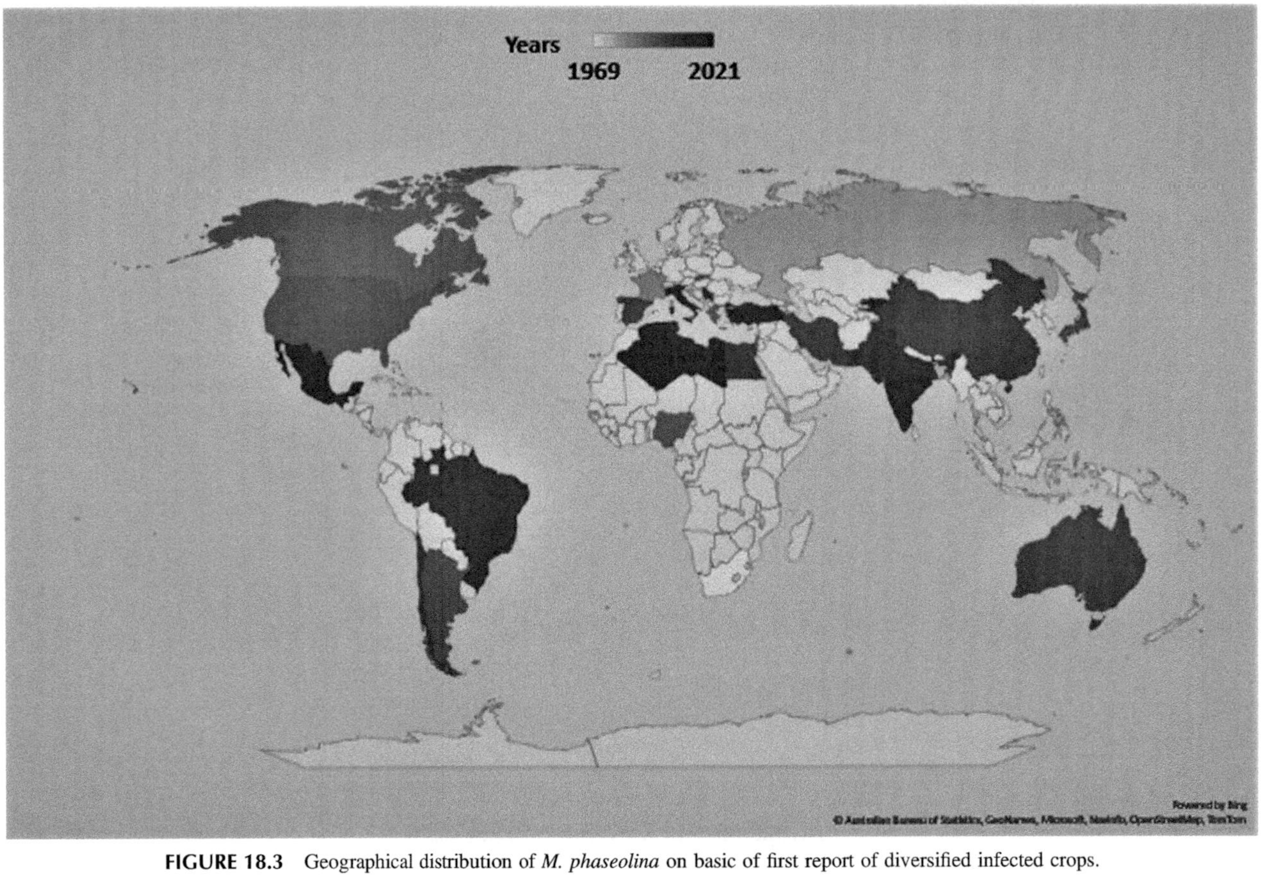

FIGURE 18.3 Geographical distribution of *M. phaseolina* on basic of first report of diversified infected crops.

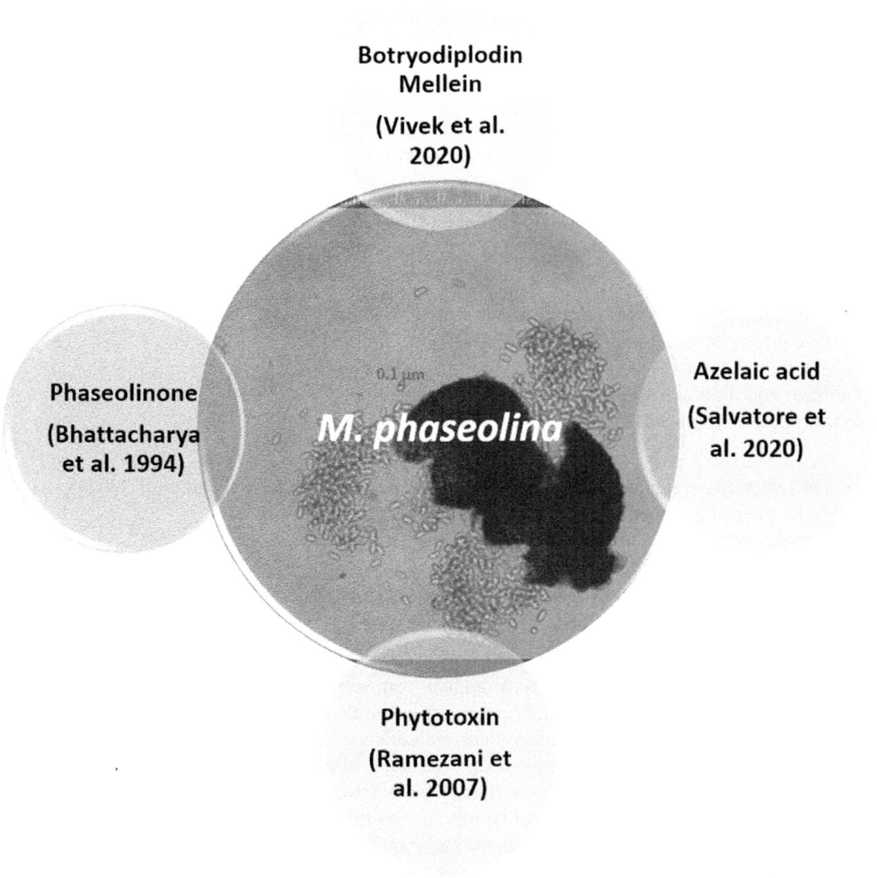

FIGURE 18.4 *M. phaseolina* secretes different metabolites during pathogenesis.

conditions. In case of cultural management techniques, eradication of alternate and collateral hosts, crop rotation, organic manure and fertilizer management, mixed cropping, filed sanitation, soil solarization and biofumigation, soil amendments such soil pH, time of sowing and harvesting, seed rate and plant population density, irrigation, and drainage are the major considering issues for effective management of the pathogen population density as well as disease suppression during the crop cultivation. On the other hand, hot water treatment (50−54°C for 10 min) is a very common physical method which is used for seed treatment before sowing in the field. Host-plant resistant is the very sophisticated and highly effective disease management strategies all over the world. But it has some drawbacks for instance time consuming, costly and poor sustainability for the longer period of times. Therefore, biological method is becoming very popular nowadays. Biocontrol is an environment friendly, cheap, and sustainable for control of plant pathogen and disease management. There are several kinds of biocontrol agents, such as arbuscular mycorrhizal fungi (Zhang et al., 2018), *Trichoderma* spp. (Hyder et al., 2018; Rahman et al., 2021; Rubayet & Bhuiyan, 2016; Simi et al., 2019), *Aspergillus* and *Penicillium* (Rojas et al., 2020), *Bacillus*, *Pseudomonas*, and *Stenotrophomonas* (Kumar et al., 2007; Mmbaga et al., 2018; Torres et al., 2016) can play the substantial role for directly inhibiting pathogen growth and dwindle their population density. Among these microorganisms, *Trichoderma* spp. are the highly antagonistic filamentous fungus against *M. phaseolina* (Bastakoti et al., 2017; Hyder et al., 2018; Rahman et al., 2021). Many researchers have confirmed that *Trichoderma* spp. reduce disease incidence and severity of *M. phaseolina* charcoal rot, dry root rot, damping off, ashy stem, collar rot of pules and oil-seed, cereals, cash crop, and vegetables (Table 18.2).

18.8 General and morphological characteristics of *Trichoderma* spp.

Trichoderma belongs to the family of Hypocreaceae. It is a filamentous beneficial fungus and survives ubiquitously in nature. It promotes plant growth, production of extracellular lytic enzymes, nonvolatile and volatile toxic metabolites, highly competitive saprophytic ability, proliferation rate, bioremediation, plant protection, and other beneficial impacts in

TABLE 18.1 Characteristics of disease symptoms caused by *M. phaseolina* on field crops.

Diseases	Symptoms	References
Damping off	The symptoms appear during the seedling stage of common bean (*Phaseolus vulgaris*) and mungbean (*Vigna radiata*) at the collar region of host-plant. Further, yellowish brown to black color thin stripes, necrotic lesions develop on the infected area and at the severe condition seedlings ultimately die.	Chandra et al. (1995)
Seedling blight	The infected seedlings of eucalyptus (*Eucalyptus globulus*) become yellowish brown to black leaves; necrotic lesions form on the top part of the stem; and the stem breaks at the collar area.	Soni et al. (1985)
Wilting	Wilt of chickpea (*Cicer arietinhum*) disease is most common in hot, dry climates during post-flowering; the tap root dries out and is devoid of fine roots; and many black pinhead microsclerotia grow beneath the bark. The vascular bundle plugging due to presence of pathogen and their activities which is hampered the photosynthesis as well as respiration cycle. As a result, plant becomes wither and finally die.	Leach and Garber (1970)
Dry root rot	The dry root rot prominent symptoms are clearly visible on the cotyledons as brown to dark spots from the time of emergence of cowpea (*Vigna unguiculata*). Cotyledons margins turn bright red to brown and finally black in color. Infected spots gradually expand and form into large necrotic lesions. However, *M. phaseolina* can also attack the roots and finally develop the severe necrotic lesions. The root tissues become dry and dead. These are the major causes of seedlings mortality.	Singh et al. (2021)
Charcoal rot	Charcoal rot disease very common in soybean (*Glycine max*) appear from seedling to blooming stage. The main symptom is the formation of brown lesions on the hypocotyl of developing seedlings. During the growth season, infected leaflets become tiny, wither, and brown in color. The taproot and ground line stem of the plant may turn striped with a pale gray tint. A smattering of minute black specks occurs under the epidermis and within the lower stem and taproot, giving the appearance of being charcoal-sprinkled. Finally, reddish-brown discoloration appears in the pith and vascular tissues of the root which is the major factor for the wilt and plant dead.	Rahman et al. (2021)
Crown rot	The visible symptoms of infected strawberry (*Fragaria ananassa*) plant consist of wilting, drying, defoliation and death of older leaves, plant stunting, and ultimately plant die. After the longitudinal dissection of the infected crown, dark brown to orange brown color vascular and cortex tissues clearly noticeable.	Koike (2008)

ecosystems. They are free-living fungi that inhabit typical soil and root environments. Their rate of multiplication is very high and comprised of about 10,000 species (Waghunde et al., 2016). Temperature range of 25–35°C and varying pH are the cozy conditions for reproduction of *Trichoderma.* It can reproduce asexually by forming chlamydospores and ascospores (Shah & Afiya, 2019). It can be easily identified by their morphological characteristics; but species-level identification of the isolates is more challenging. *Trichoderma* spp. produces white cotton growth and yellowish-green to deep green compact tufts, particularly in the center of a developing patch or on the agar surface in concentric ring-like zones. Conidiophores are irregularly formed whorls that look flask-shaped or cylindrical to nearly globose phialides. Conidia are ellipsoidal to globose, usually green, but can be hyaline, and congregate near the apex of phialides (Zhu & Zhuang, 2015) (Fig. 18.5). The conidiophore bears green conidia. However, *Trichoderma* spp. can be characterized as a most prevalent culturable fungus and plant symbiont. They can also make the mutualistic endophytic relationships with different plant species around the rhizosphere region (Bae et al., 2011). Furthermore, these microbes create a number of secondary chemical compounds that cause plant resistance. These enzymes act by altering cell wall integrity and disintegrating the polysaccharides, chitin, and glucans which are essential for fungal cell wall stiffness. Another method that has gained traction in recent years is competition via rhizosphere competence (Howell, 2003). Apart from this, it may be quite useful in plant nutrition control. According to Altomare et al. (1999), plant nutrients change from soluble to insoluble forms in soil affecting their accessibility and absorption by roots. These alterations could be influenced by microorganisms. Iron and manganese have been investigated in terms of microbial solubilization of oxidized forms of these metals as well as their effect on plant disease. At in vitro conditions, a strain of *T. harzianum* generates a significant variety of compounds that

TABLE 18.2 Management of *Macrophomina phaseolina* causing crop diseases by using *Trichoderma* spp.

Crops	Diseases	*Trichoderma* spp.	References
Vigna radiata *Vigna mungo*	Charcoal rot	*T. harzianum* *T. hamatum* *T. pseudokoningii*	Iqbal and Mukhtar (2020)
Cucumis melo	Charcoal rot	*T. harzianum* (M) *T. harzianum* (T39) *T. virens* (DAR 74290)	Etebarian (2006)
Glycine max	Charcoal rot	*T. harzianum* *T. harzianum* ISR-26	Khalili et al. (2016) Rahman et al. (2021)
Vigna unguiculata	Charcoal rot	*T. harzianum* *T. viride* *T. hamatum*	Shoaib et al. (2018)
Arachis hypogaea	Charcoal rot	*Trichoderma* spp.	Martínez-Salgado et al. (2021)
Helianthus annuus *V. radiata*	Charcoal rot	*T. harzianum*	Hussain et al. (1990)
Lens culinaris *Sesamum indicum* *V. radiata*	Charcoal rot	*T. viride*	Al-Juboory and Al-Jarah (2020)
S. indicum	Charcoal rot	*T. aureoviride*	Soufi et al. (2021)
G. max	Charcoal rot	*T. harzianum*	Khaledi and Taheri (2016)
Citrullus lanatus *G. max*	Charcoal rot	*T. viride*	Al-Juboory and Al-Jarah (2020)
Fragaria ananassa	Charcoal rot	*T. harzianum* *T. virens*	Korkom and Yildiz (2022)
Zea mays	Charcoal rot	*T. viride*	Sankar and Sharma (2001)
H. annuus	Charcoal rot	*T. harzianum* *T. viride*	Ullah et al. (2011)
H. annuus	Charcoal rot	*T. harzianum* *T. resei*	Anis et al. (2010)
Sorghum bicolor	Charcoal rot	*T. viride*	Arora and Dhurwe (2013)
S. indicum	Charcoal rot	*T. pseudokoningii*	Ibrahim and Abdel-Azeem (2014)
C. arietinum	Dry root rot	*T. viride*	Manjunatha et al. (2013)
Solanum melongena	Dry root rot	*T. harzianum* *T. hamatum* *T. viride*	Khan and Gupta (1998)
V. radiata	Dry root rot	*T. harzianum*	Basandrai et al. (2021)
V. radiata	Dry root rot	*T. harzianum*	Swehla et al. (2020)
Cajanus cajan	Dry root rot	*T. harzianum*	Lokesha and Benagi (2007)
Cicer arietinum	Dry root rot	*T. harzianum*	Khan et al. (2014)
V. unguiculata	Damping-off	*Trichoderma* spp.	Adekunle et al. (2001)
S. melongena	Damping-off	*T. harzianum*	Almammory and Matloob (2019)
Gossypium herbaceum	Ashy stem	*Trichoderma* spp.	Abd-Elsalam (2010)
Corchorus olitorius	Seedling blight Collor rot, stem rot and root rot	*T. viride*	Srivastava et al. (2010)
Capsicum frutescens	Root rot	*T. harzianum*	Hussain et al. (2013)

Continued

TABLE 18.2 Management of *Macrophomina phaseolina* causing crop diseases by using *Trichoderma* spp.—cont'd

Crops	Diseases	*Trichoderma* spp.	References
Viciafaba	Root rot	*T. harzianum, T. hamatum* *T. viride, T. atroviride* *T. virens,* and *T. koningii*	Kumari et al. (2017)
V. radiata	Root rot	*Trichoderma* spp.	Khan et al. (2019)
V. mungo	Root rot	*T. viride*	Rettinassababady et al. (2000)
A. hypogea	Root rot	*T. harzianum* (T3) *T. viride* (T1)	Sreedevi et al. (2011)
S. melongena	Root rot	*T. harzianum* *T. polysporum* *T. viride*	Ramezani (2008)

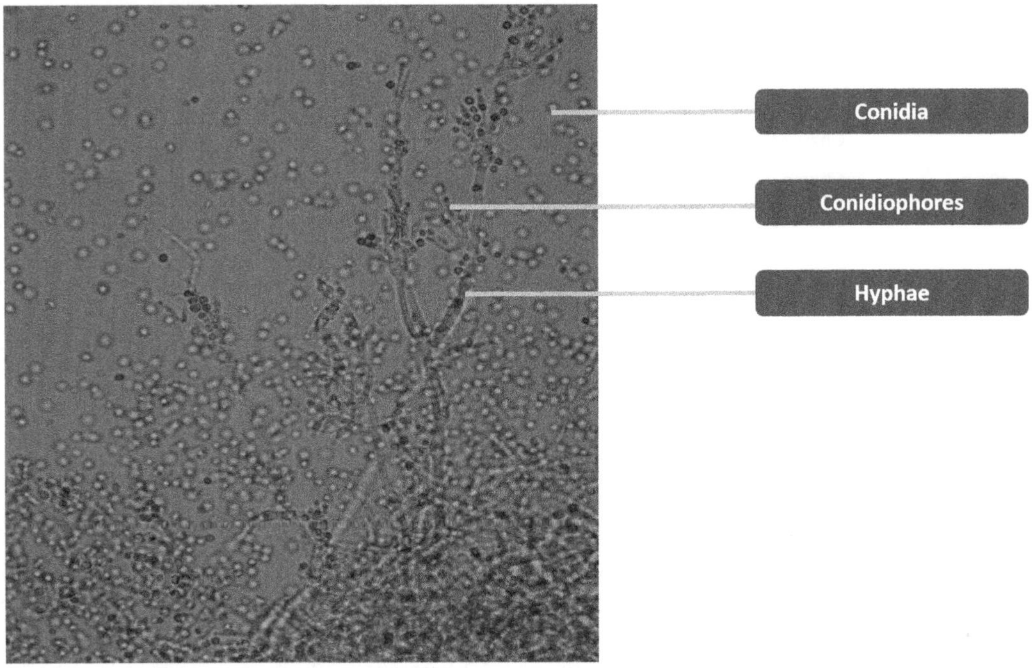

FIGURE 18.5 Different structures of *Trichoderma* spp. observed under compound microscope.

help to solubilize rock phosphate, Zn, Mn^{4+}, Fe^{3+}, and Cu^{2+} as well as boost nutrient availability and absorption. Ultimately, the genus *Trichoderma* has grown in economic relevance owing to its excellent antagonistic power against a broad variety of plant diseases and growth promotion in agricultural plants.

18.9 *Trichoderma* spp. as an effective bio-agent

Trichoderma holds the top position of beneficial microorganism due to it potentiality for minimizing plant diseases and boost up the growth of the plant and their yield. According to Monaco et al. (1991), over 20 *Trichoderma* species, including *T. harzianum, T. hamatum, T. viride, T. koningii, T. longibrachiatum, T. atroviride, T. polysporum,* and *T. Pseudokoningii* act as bio-agents against a broad variety of soil-borne and foliar plant pathogens. Literally, the efficacy and variation of *Trichoderma* spp. could be easily distinguishable by growing pattern on culture media (Fig. 18.6). The antagonism of *Trichoderma* occurs against many vegetable, cereals, and woody plant pathogens, such as *Alternaria brassicae* (Arefin et al., 2019), *Colletotrichum acutatum* and *Botrytis cinerea* (Freeman et al., 2004), *Sclerotinia sclerotiorum* (Li et al., 2005), *Phytophthora capsici* (Ezziyyani et al., 2007), *Ustilago segetum* var. *tritici* and *Cercospora*

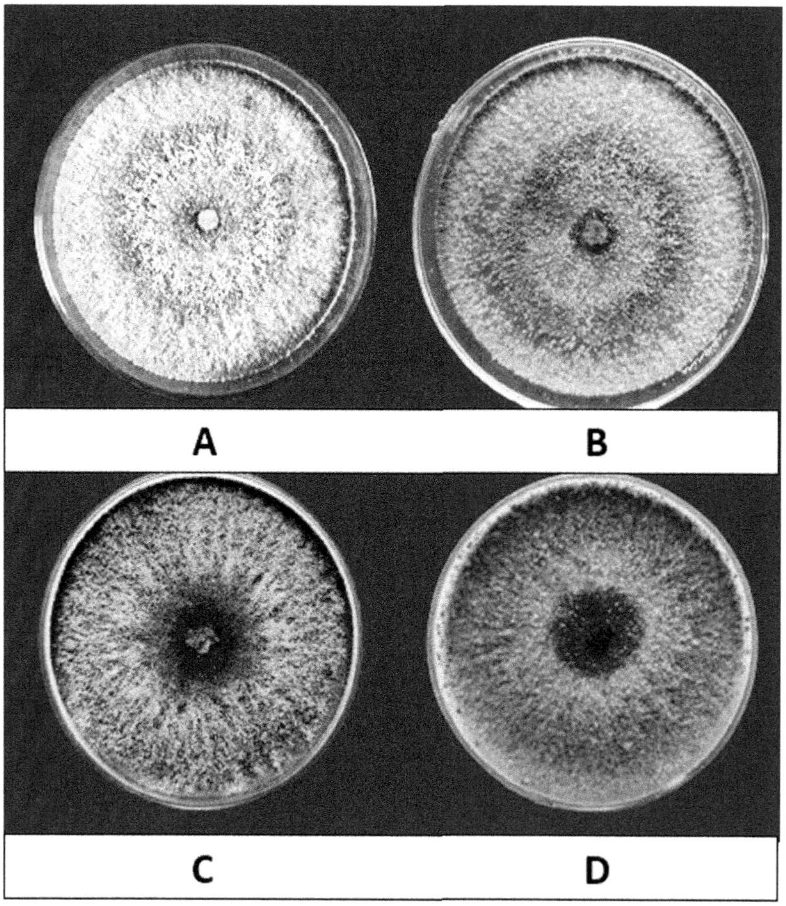

FIGURE 18.6 *Trichoderma* isolate with distinct morphological and cultural traits on PDA medium. (A) Isolate ISR-26, (B) Isolate Chan-6, (C) Isolate Pb-7, and (D) Isolate T-10.

moricola (Srivastava et al., 2016), *Meloidogyne incognita* (Khan & Mohiddin, 2018), *Fusarium ciceris*, *R. solani*, and *M. phaseolina* (Mukhopadhyay & Pan, 2012; Rahman et al., 2021). It also has significant potential as a bio-agent against root-knot and cyst-forming nematodes through direct parasitism of both eggs and larvae (Zhang et al., 2014). *T. asperellum* M2RT4 has also been shown to reduce knot and eggs formation in pineapple roots in Kenya (Kiriga et al., 2018). Several *Trichoderma* species, including *T. asperelloides, T. hamatum, T. harzianum,* and *T. viride,* showed substantial antagonistic activity against *M. incognita* infected tomato. *T. asperelloides*, for example, successfully reduced root galling up to 92.81% and up to 97.81% nematode population (Sayed et al., 2019). *Trichoderma* spp. can effectively work against many plant pathogens.

18.10 Mechanism of *Trichoderma* spp. against *M. phaseolina*

Trichoderma spp. uses a variety of antagonistic strategies to combat plant diseases, including antibiosis, mycoparasitism, competition for resources and space, plant growth stimulation, induced plant defense mechanisms, and environmental condition adjustment (Druzhinina et al., 2011) (Fig. 18.7). These mechanisms may act directly or indirectly against the causal organism, increasing yield both qualitatively and quantitatively. Biocontrol modes of action can be divided into two categories. One mechanism is direct lysis or killing of pathogen by biocontrol agent, followed by antibiosis and mycoparasitism. Another type of indirect mechanism is the exclusion of plant pathogens as a result of the presence, activity, or products of a biocontrol agent, such as competition and induced systemic resistance (Harman et al., 2004; Vinale et al., 2008).

18.10.1 Antibiosis

Antibiosis is one of the most fundamental antagonistic interactions that occur as a result of the development of diffusible antimicrobial secondary metabolites (SMs) (Singh et al., 2018). It has been observed that interactions between the host

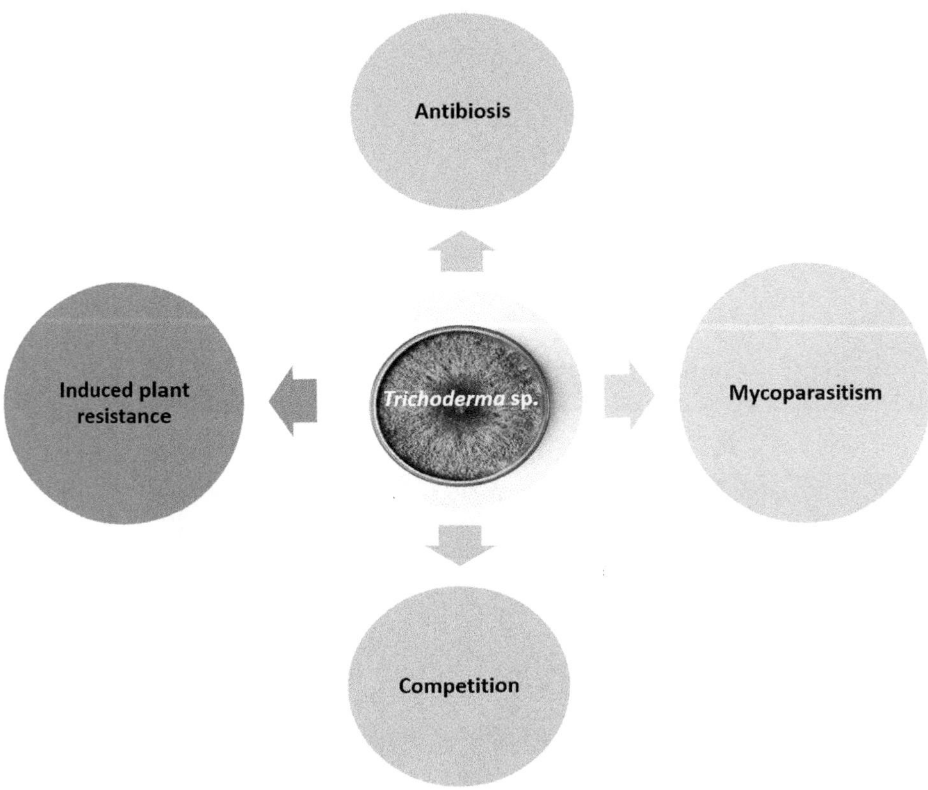

FIGURE 18.7 Major part of the mechanisms of *Trichoderma* spp. against plant pathogens.

plant, pathogens, and *Trichoderma* spp. result in *Trichoderma* producing antibiotics and low-molecular-weight chemicals that inhibit the development of phytopathogenic fungi (Gajera et al., 2013). *Trichoderma* spp. has a diverse spectrum of SMs, notably tricholin, harzianic acid, viridian, gliosoprenins, heptelidic acid, 6-pentyl—pyrone, massoilactone, peptaibols, terpenes, polyketides, gliotoxin, gliovirin, koninginins, pyrones, nonrib (Contreras-Cornejo et al., 2016; Mukherjee et al., 2012; Vinale et al., 2008). These SMs provide the biocontrol agent mycoparasitic properties, participate in plant—microbe interactions, promote host development and provoke plant defensive responses. *Trichoderma* also creates elicitors that activate plant defense systems, which may be classified as proteins or peptides (Salas-Marina et al., 2011), enzymes (chitinase, cellulase, glucanase, protease, or xylanases) (Morán-Diez et al., 2009; Vos et al., 2015), fatty acids, and lipids (Mukherjee et al., 2012).

The antagonism of *Trichoderma* is clearly distinct in confrontation assay in vitro condition (Fig. 18.8). The bio-agent does not allow growing the pathogens in the dual culture. The growth and development as well as reproduction system totally hinder due to antagonistic behavior of the *Trichoderma.* A plethora of reports regarding this content have been published in several international journals. *Trichoderma* spp., such as *T. harzianum* Pb-7 inhibited mycelial growth of *F. oxysporum* f. sp. *lentis* (89.81%), *R. solani* (86.30%), *S. rolfsii* (88.33%) (Das et al., 2019) and *Colletotrichum capsici* (82.96%) (Simi et al., 2019). According to Khan et al. (2021), *T. koningii, T. hamatum,* and *T. longipile* arrested 46—47% growth of *M. phaseolina*. But *T. viride* demonstrated 63% inhibitory efficacy against *M. phaseolina*. Harman (2006) and Hoitink et al. (2006) observed competition, antibiosis, and hyperparasitism between *M. phaseolina* and *Trichoderma* isolates in the combine culture medium under laboratory condition. *T. viride* can suppress 55% mycelial growth of *M. phaseolina* in dual culture plate (Al-Juboory & Al-Jarah, 2020). Khaledi and Taheri (2016) reported that *Trichoderma* spp. isolate produced various types of volatile and nonvolatile compounds which inhibited up to 58.67% mycelial growth of *M. phaseolina.* Additionally, this bio-agent significantly reduced the charcoal rot disease severity of soybean in greenhouse condition. According to the Khalili et al. (2016), the mycelial growth of *M. phaseolina* was 72.31% retarded by *T. harzianum. T. harzianum, T. koningiopsis,* and *T. virens* showed the antagonistic activities against *M. phaseolina* and blocked the growth and development (Mendoza et al., 2015). Cubilla-Ríos et al. (2019) found the antagonism of *Trichoderma* spp. (*T. arundinaceum, T. brevicompactum,* and *T. harzianum*) on *M. phaseolina* isolates in sesame and soybean cultivation. *T. viride* and *T. harzianum* reduced *M. phaseolina* mycelial growth by 61.1% and 64.4%, respectively, in dual culture technique (Sreedevi et al., 2011). *T. harzianum* isolates have antagonistic mode of action, which could inhibit

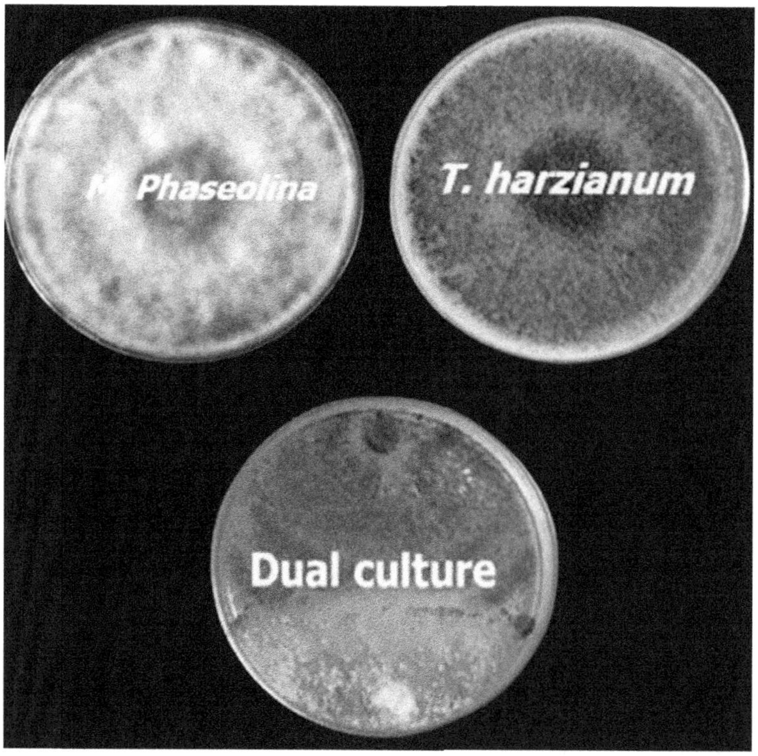

FIGURE 18.8 Antagonism effect of *T. harzianum* isolate T-10 on *M. phaseolina* in dual culture.

76.96% mycelial growth of *M. phaseolina* in dual culture plate. Additionally, it can substantially dwindle *M. phaseolina* root rot disease of mungbean in field conditions (Swehla et al., 2020). Similarly, Rahman et al. (2021) found that mycelial growth (81.85%) of *M. phaseolina* was inhibited by the presence of *T. harzianum* isolate ISR-26 in laboratory test. The isolate showed the highest efficacy against the *M. phaseolina* charcoal rot of soybean disease in the field trial.

18.10.2 Mycoparasitism

Mycoparasitism or hyperparasitism is another complicated mechanism of *Trichoderma* spp. that might have a significant impact in reducing the pathogen population density. This is the phenomenon where one microorganism directly attacks on another microorganism. *Trichoderma* spp. hyperparasitize the pathogens by secreting several types of hydrolytic and cell wall breaking enzymes including chitinases, glucanases, xylanases, cellulases, and proteinases, chitin 1,4-β-chitobiosidase, protease, N-acetyl-β-D-glucosaminidase, and glucan 1,3-β-glucosidase (Roberti et al., 2002). Lorito et al. (1994) observed that *Trichoderma* species secrete huge number of chitinolytic and glucanolytic enzymes which play a key role in mycoparasitism. Similarly, Kubicek et al. (2011) reported that exochitinases and β-1,3-glucanases that play a significant role in mycoparasitism. *T. koningiopsis* (TCBG-8) and *Trichoderma* sp. (TCBG-2) showed the significant antagonism activities against *M. phaseolina* due to produce the highest amount of exochitinase (0.1 μmol/mL/min) and glucanase (0.33 μmol/mL/min) (Larralde-Corona et al., 2008). *T. harzianum*-1 first and foremost produced zone of inhibition, then overgrow on *M. phaseolina* and finally showed strong hyperparasitism (Singh et al., 2012). *Trichoderma* hyphae hyperparasitized the pathogen hyphae and coagulated its cell protoplasm leading to shrinkage, granulated, and vacuolated protoplasm of pathogen (Pandey & Upadhyay, 2000). These hydrolytic biochemical substances are often responsible for cell wall breakdown of pathogen. The cell wall disintegration is one of the vital aspects of mycoparasitism as well biocontrol of plant disease suppression. The interaction between the bio-agent, for instance, *Trichoderma* spp. and pathogen results in coiling, formation of appressoria, production of hydrolytic enzymes and antimicrobial metabolites, killing of the host, and finally uptake of nutrients from dead organic matter (Holzlechner et al., 2016; Omann et al., 2012) (Fig. 18.9). Bio-agent gradually kills the pathogen's mycelium and decreases their population density. As a result, amount of potential inoculum disappears from the crop field. *Trichoderma* species are the most common mycoparasitic and saprophytic fungi which have mycoparasitism potential to curb the wide range of important diseases causing

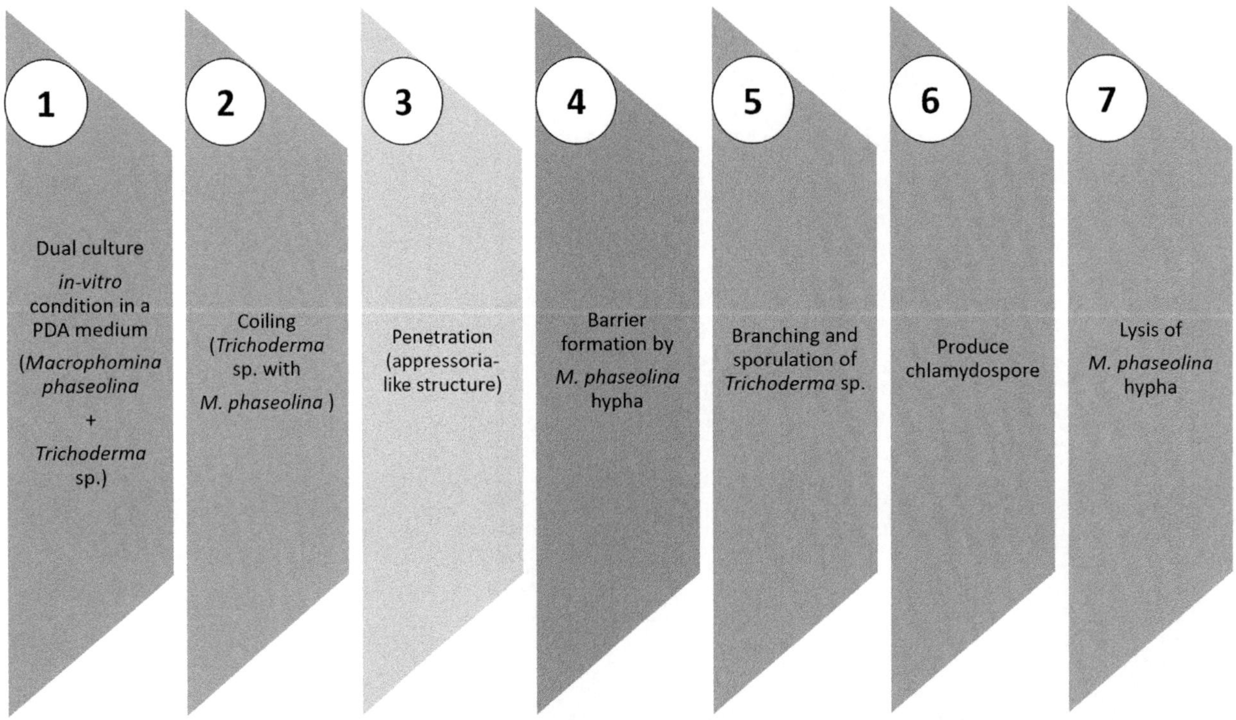

FIGURE 18.9 Different steps of *Trichoderma* spp. mycoparasitism.

phytopathogenic fungi likely *Alternaria brassicae, R. solani, S. rolfsii, F. oxysporum, C. capsici, M. phaseolina,* etc. (Arefin et al., 2019; Das et al., 2019; El-Hassan et al., 2013; Rahman et al., 2020, 2021; Rubayet & Bhuiyan, 2016).

18.10.3 Competition

Competition is a phenomenon that occurs when *Trichoderma* species and pathogen compete for limited nutrients and space because, the fugus is saprophytic and nonpathogenic in nature. They can grow faster and derive nutrients from decomposed materials (Halifu et al., 2019). It can survive an extreme environmental condition in the soil. Because, *Trichoderma* has the antagonistic mechanism which can overcome the fungistatis, such as herbicides, fungicides and pesticides or other toxic compounds in soil (Chet et al., 1997). Thus, they occupy the space quickly, which hampers the normal growth and reproduction of the plant pathogens in the same area. Pathogen suffers from nutrients starvation. This is one of the most important ways of biocontrol method for decreasing phytopathogen population density as well as plant disease management in field soil (Chet et al., 1997). However, iron is the chief nutrient for viability of filamentous fungi. Most of the fungi excrete ferric iron-specific chelators (low-molecular-weight) under iron starvation situation (Eisendle et al., 2004). *Trichoderma* produces extremely competent siderophore compounds that chelate iron and deprive pathogens from iron nutrient (Chet & Inbar, 1994). Subsequently, these substances also solubilize the iron for easily absorption by plant (Mukherjee et al., 2012). Additionally, saprophytic microorganism repeatedly breaks down the organic matter due to getting the nutrients from there. *Trichoderma* is more efficient to mobilize soil nutrient and utilization than many other beneficial fungi or bacteria. This progression might be connected to the generation of inorganic acids (fumaric, citric, and gluconic acids), which assist to down soil pH level and gear up the phosphate and micronutrients (Fe and Mn) solubilization (Vinale et al., 2008). Moreover, siderophores have various impacts and functions, such as rising pathogenicity of pathogen, intracellular iron storage, and growth inhibition of microbe during rivalry with *Trichoderma* spp. (Miethke, 2013).

Numerous researchers have already demonstrated the competitive efficacy of *Trichoderma* spp. against pathogens during crop cultivation under open field condition. *T. harzianum* isolates ISR-26, T-10, and Pb-7 decreased pre-and postemergence seedling mortality in soybean (*M. phaseolina*), potato (*S. rolfsii*), and carrot (*S. rolfsii*), respectively, in the pot and field soil (Figs. 18.10 and 18.11). The minimization of seedling mortality ultimately increases the yield and yield contributing characters of the plant. Ahmed et al. (2019) reported that application of *T. harzianum* isolate PB-7 with

FIGURE 18.10 Effect of *T. harzianum* isolates ISR-26 and T-10 on seedling mortality of soybean and potato caused *M. phaseolina* (A and B), *S. rolfsii* in earthen pot culture (C and D), (A) soil inoculated with *M. phaseolina,* (B) soil inoculated with *M. phaseolina* and *T. harzianum* isolate ISR-26, (C) soil inoculated with *S. rolfsii,* (D) soil inoculated with *S. rolfsii* and *T. harzianum* isolate T-10.

chitosan 600 ppm reduced 90.91% postemergence seedling mortality, 81.41% disease incidence, and 81.97% disease index of southern blight of carrot. Seedling mortality of bush bean caused by *F. oxysporum* f. sp. *phaseoli, R. solani,* and *S. rolfsii* had been decreased to 75.70% by the application of *T. harzianum* isolate Chan-6 with poultry manure (Liton et al., 2019). Similarly, *M. phaseolina* seedling mortality of soybean 80.47%, disease incidence 80.14% and disease severity 88.55% were minimized by the *T. harzianum* isolate ISR-26. Not only the disease suppression but also it augmented the yield by 104.42% (Rahman et al., 2021).

18.10.4 Induced plant resistance

Trichoderma has the ability to attack plant pathogens and stymie their growth by producing systemic and local resistance to a vast range of plant diseases (Harman et al., 2004; Shoresh et al., 2010). Among the many biocontrol strategies of *Trichoderma,* induced systemic resistant (ISR) is one of the most important for evading plant pathogens by inducing a plant defensive response. ISR-mediated *Trichoderma* species in agricultural plants involve the activation of a link in the defense system in various pathways. These are involved in the expression and synthesis of defensive biochemicals, either directly or indirectly. Furthermore, *Trichoderma* spp. stimulate the production of signal molecules, such as jasmonic acid (JA) and salicylic acid (SA), as well as the expression of pathogenesis-related (PR) proteins such as chitinases and glucanases. The production of phenolic antimicrobial compound is regulated by some plant defense enzymes likely phenylalanine-ammonialyase (PA), polyphenol oxidase (PPO) and peroxidase (PO) (Patel & Saraf, 2017). *Trichoderma* spp. produce huge number of molecules for instances *T. harzianum* (harzianic acid) (Vinale et al., 2014), *T. asperellum* (swollenin) (Brotmanet al., 2008), *T. virens* (hydrophobin-like elicitor) (Djonovic et al., 2007), *T. viride* (xylanas) (Rotblat et al., 2003), *T. longibrachiatum* (hydrophobin) (Ruocco et al., 2015), *T. atroviride* (phytostimulation, defense responses, JA, SA, camalexin) (Contreras-Cornejo et al., 2011), and *T. koningii* (ISR based induction of phenylalanine ammonia lyase, peroxidase, and surge in total phenolic compound) (Thangavelu & Mustaffa, 2010), which exaggerate plant defense activities during the host–pathogen interaction (Table 18.3). For example, *T. asperellum* produced 6-pentyl-α-pyrone, which triggered the activity of defense responses and reduced the development of disease symptoms in *Arabidopsis* against *Botrytis cinerea* and *Alternaria brassicicola* (Kottb et al., 2015). Similarly, 1-octen-3-ol compound was produced by *T. virens* in *Arabidopsis*. This molecule was augmented in plant resistance against pathogens by activating jasmonic acid/ethylene dependent defense pathways (Contreras-Cornejo et al., 2014; Kishimoto et al., 2007). Seed and soil amendment with *T. viride* enhanced the total phenolic compound in *M. phaseolina*–inoculated plants (Singh et al., 1998).

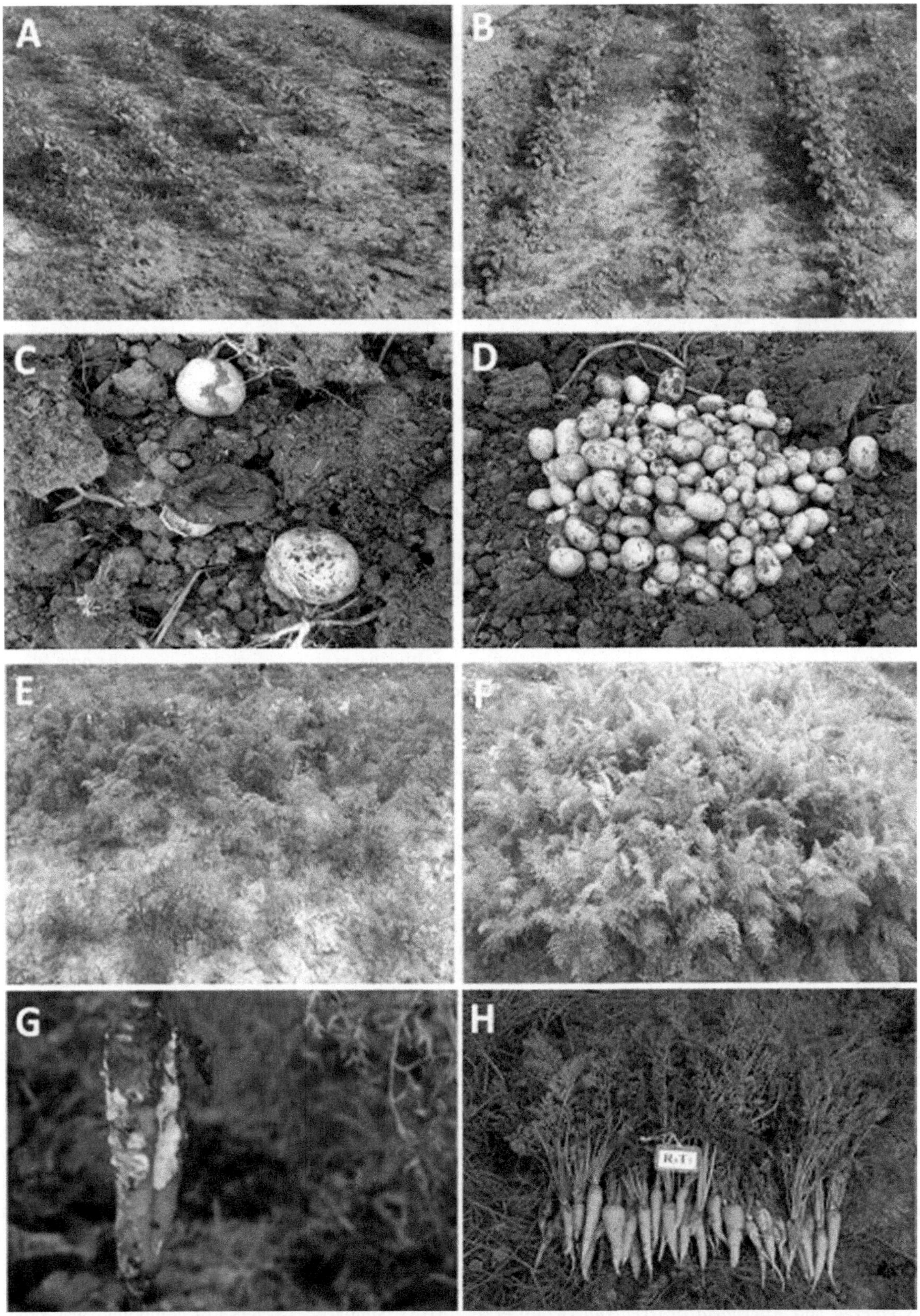

FIGURE 18.11 Effect of *T. harzianum* isolates T-10 and PB-7 on disease suppression of potato tuber and carrot, respectively caused by *S. rolfsii*. (A and E) Soil inoculated with *S. rolfsii* (B and F); soil inoculated with *S. rolfsii* and *T. harzianum* isolates T-10 and PB-7, respectively; (C and G) yield of "A and E;" (D and H) yield of "B and F." [Application of inoculum rate was same (colonized wheat grains inoculum 90 gm^2) for both pathogen (*S. rolfsii*) and bio-agents (*T. harzianum* isolate isolates T-10 and PB-7)].

TABLE 18.3 *Trichoderma* **spp. produce induced systemic resistance biochemical compounds.**

Trichoderma spp.	Bio-chemical compounds	References
T. harzianum	Phytoalexins capsidiol Phenolic compounds Harzianic acid Isoharzianic acid (iso-HA) Harzianolide Trichodiene Cellulases Hhydrophobin Ortho-dihydric phenol Tyrosinase	Ahamed and Vermette (2009) Singh et al. (2013) Vinale et al. (2014) Cai et al. (2013) Malmierca et al. (2015) Saravanakumar et al. (2016) Yu et al. (2020) Sreedevi et al. (2011) Score et al. (1997)
T. asperellum	Salicylic acid (SA) Jasmonic acid (JA) Ethylene and jasmonic acid 6-Pentyl-α-pyrone Swollenin	Elsharkawy et al. (2013) Shoresh et al. (2006) Guo et al. (2021) Kottb et al. (2015) Brotman et al. (2008)
T. virens	Terpenoidphytoalexins 1-Octen-3-ol Hydrophobin-like elicitor	Kumar and Palakshappa (2009) Kishimoto et al. (2007) Contreras-Cornejo et al. (2014) Djonovic et al. (2007)
T. viride	Xylanase	Rotblat et al. (2003)
T. longibrachiatum	Hydrophobin	Ruocco et al. (2015)
T. atroviride	Phytostimulation, defense responses, JA, SA, camalexin	Contreras-Cornejo et al. (2011)
T. viride	ISR based induction of phenylalanine-ammonialyase (PA), peroxidase (PO), and surge in total phenolic compound	Thangavelu et al. (2004)
T. harzianum		Sreedevi et al. (2011)
T. koningii		Thangavelu and Mustafa (2010)
T. peudokoningii *T. hamatum*		Thangavelu and Gopi (2015)
T. cerinum		Khare et al. (2018)
Trichoderma spp.		Bubici et al. (2019)

18.11 Bulk production of *Trichoderma* spp. for field application

The effective plant disease management technique with biological control approach relies on the creation of well-functioning material, composition, and distribution systems. *Trichoderma* mass production method should be suitable with commercial and industrial production procedures as well as field implementation. Actually, there are two main techniques for manufacturing *Trichoderma* inoculum—solid and liquid state fermentation. Solid fermentation is the most effective methodology for bulk *Trichoderma* production on a wide scale. Because a range of low-cost cereal grains are employed as substrates, including bran of rice, wheat, grass pea, maize, chick pea, millets, sorghum, tea waste, and mustard oilcake (Jeyarajan, 2006; Rubayet & Bhuiyan, 2012) (Fig. 18.12). The method of preparation of the substrate depends on the type of raw product. As for example, when inoculum of the *T. harzianum* is prepared, wheat grains are soaked in water for 12 h. After soaking, the excess water is drained out and water-soaked grains are kept into transparent polythene bags. Then, the packet material is autoclaved (121°C, 15 psi) for 30 min. Autoclaved moist wheat grain is taken up to 2/3 volume in 500 mL Erlenmeyer flask. Five mycelial discs of 5 mm diameter cut from 3 days old PDA (Potato Dextrose Agar) culture of *T. harzianum* are added in a jar. Finally, the jar is kept for incubation at room temperature for 3 weeks. It is shaken by hand at 48–72 h interval for proper colonization. The colonized dark green wheat grains are air dehydrated for 48 h and stored at 4°C for further use (Rubayet & Bhuiyan, 2016). According to the solid fermentation produces, products usually use for soil application. Sometimes, different types of well decomposed organic manures (farmyard manure, poultry refuse, and cow dung) are used as a carrier material. As a result, it not only diminishes plant disease but also

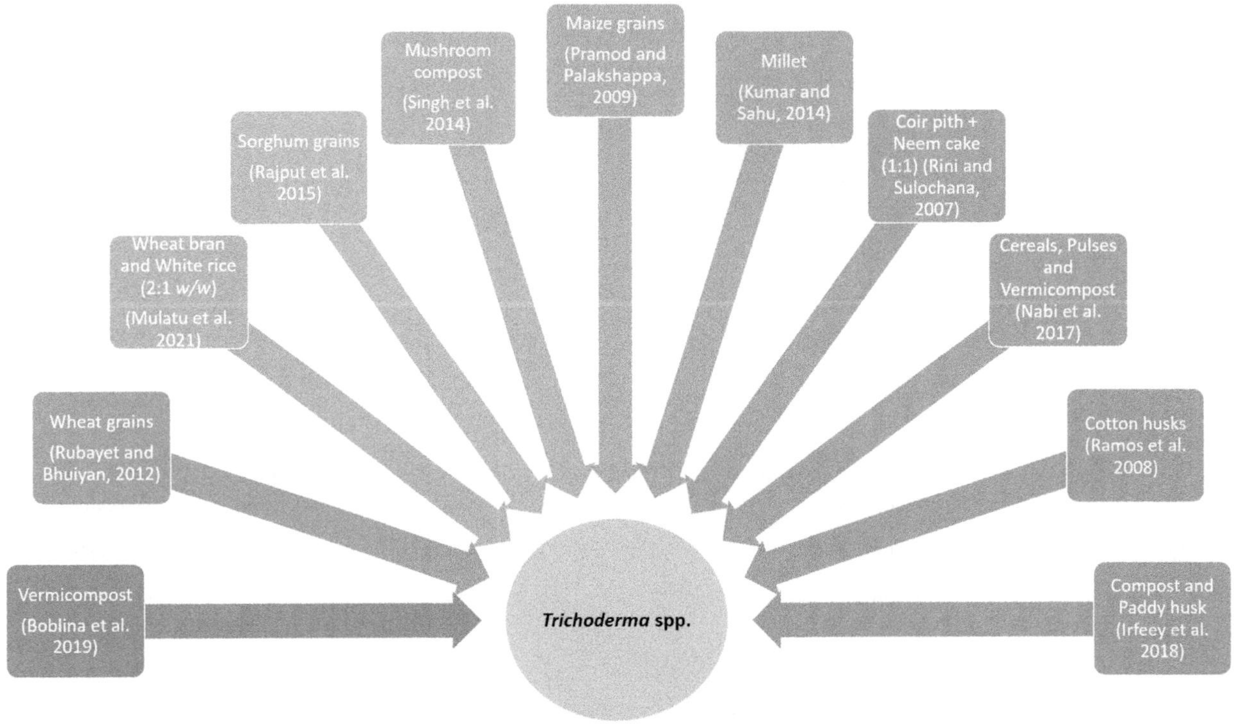

FIGURE 18.12 Suitable substrates use for bulk production of *Trichoderma* spp.

increases soil fertility concurrently. However, it has significant impact on commercial point of view but with some drawbacks. Time consumption is one of the important issues for growers during inoculum and land preparation.

Secondly, the liquid state fermentation is performed for mass production of *Trichoderma* inoculum in liquid media before applying in the open field (Waghunde et al., 2016). It is a simple and time-saving method for generating the bulk volume of *Trichoderma* spp. inoculum. According to Hassan (2014), potato dextrose broth, molasses yeast medium, and V8 juice are commonly employed as liquid substrates for mass multiplication of *T. harzianum*.

Trichoderma spp. uses to promote crop health or manage plant diseases relies on the production of commercial compositions with suitable carriers that permit *Trichoderma* spp. to persist indefinitely. There are several types of formulation, such as alginate prills (Fravel et al., 1995), agro-product (Doni et al., 2014), coffee husk (Sawant & Sawant, 1996), oil (Herrera et al., 2020), press-mud (Jeyarajan, 2006), talc (Jeyarajan et al., 1994), vermiculite-wheat bran (Lewis & Papavizas, 1991), wheat flour kaolin (Prasad & Rangeshwaran, 1998), etc., which are used for intensification the *Trichoderma* shelf life (Fig. 18.13). Shelf life of *Trichoderma* spp. indicates the viability of the bio-agent. It is very important for getting commercial successful outcome from *Trichoderma* biopesticide. Different types of additives or ingredients are used for prolonging the shelf life of *Trichoderma* (Fig. 18.13). Sriram et al. (2011) observed that the addition of 3% and 6% glycerol enhanced the shelf life of talc-based *T. harzianum* formulation (viability: $>2 \times 10^6$ CFU/g) for 7–12 months. Generally, *Trichoderma* in talc, peat, lignite, and kaolin remain active for 90–120 days but its shelf life is higher in coffee husk (>18 months) (Sankar & Jeyarajan, 1996). Environment (temperature and moisture) plays a decisive role for the activeness of *Trichoderma* formulation. Bhat et al. (2009) found that *Trichoderma* in talc-based formulations had a shelf life of up to 180 days when stored under room temperature. By including an energy source and compounds that slow down the metabolic activities of mycelium maintained as a wet paste with access to oxygen, the shelf life and competitiveness of the formulation might be significantly extended almost 12 months under room temperature (Kolombet et al., 2008).

18.12 *Trichoderma* spp. produce biopesticide compound

Application of bio-agent and their dozens of Secondary Metabolites (SMs) is one of the potential novel approaches for mitigation plant pathogen causing diseases under field condition. SMs from *Trichoderma* spp. in plant protection has been employed as an ecologically benign and efficient management technique against a number of phytopathogens (Khan et al.,

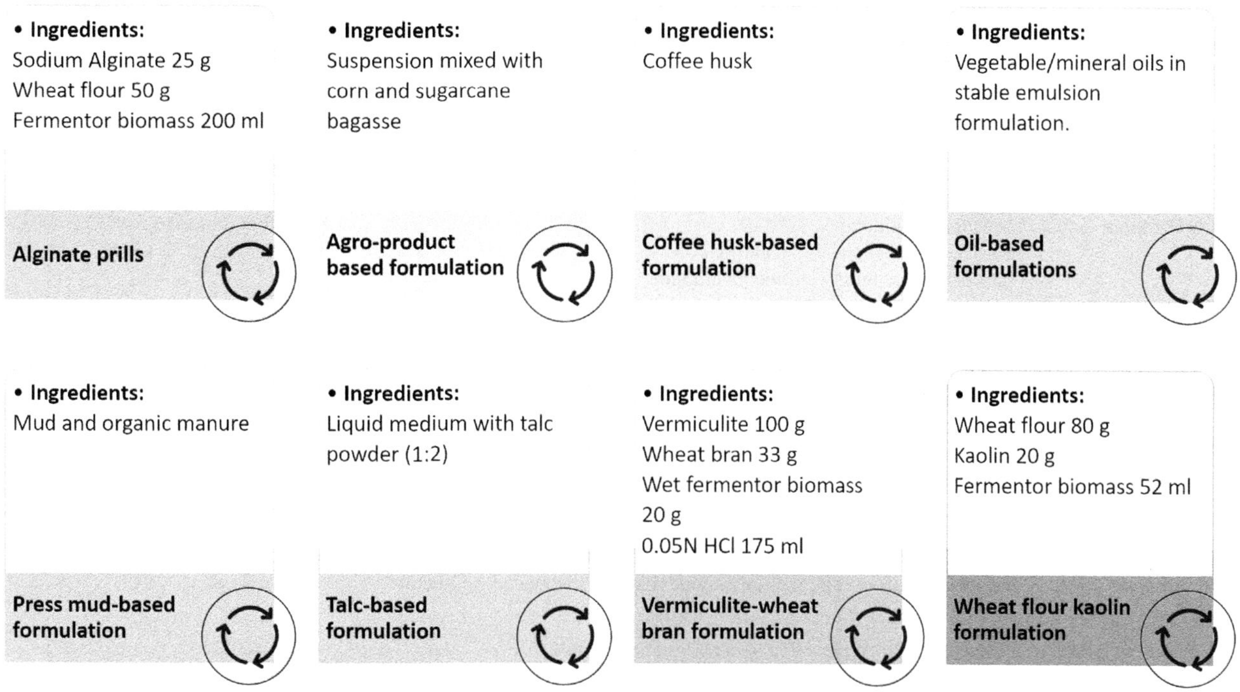

FIGURE 18.13 Different formulations and their composition for viable *Trichoderma* spp.

2020). It can alter plant metabolism and may play a momentous role in the intricate interactions of this biocontrol agent with plants and diseases. Various types of low molecular weight (3 kDa) SMs are biosynthesized by microbes and plants biochemical compounds. These accelerated the biological action during the host—-pathogen interaction (Vinale et al., 2009). These molecules have been demonstrated to not only diminish parasite proliferation and destructive actions directly but also to induce disease resistance by activating the plant host's defensive system. Literally, bioactive microbial molecule includes phytotoxins, mycotoxins, antibiotics, and pigments (colored chemicals). *Trichoderma* is capable of producing not only toxic chemicals with straight antimicrobial action against pathogens but also fungal constituents capable of stimulating the plant to manufacture its own defensive metabolites. In cotton plant, *T. virens* could produce phytoalexin and surge localized resistance (Yedidia et al., 1999). Similarly, in cucumber leaves increased phenolic glucoside levels by *T. asperellum* T-203 root colonization (Yedidia et al., 2003). The synthesized fungal SMs usually involve unusual biochemical processes that lead to many natural substances derived from a few fundamental metabolic precursors including acetyl-CoA, amino acids, and mevalonate (Zeilinger et al., 2016). A plethora of antibiotics of *Trichoderma* spp., such as harzianolide, viridin and its derivatives, T39 butenolide and deydro-harzianolide, gliotoxin, cerinolactone, gliovirin, and various peptaibols have been identified and their action has been connected with the antagonistic ability of these microorganisms (Vinale et al., 2008) (Table 18.4). In a word, these compounds could be sustainable biocidal material against pathogen population reduction as well as their activities.

18.13 Conclusions

Trichoderma spp. has the highest potential to curb the seed-and soil-borne as well as air-borne pathogens in nature. The long-lasting management of *M. phaseolina* could be possible by the application of *Trichoderma* spp. because it has complex mechanism with *M. phaseolina,* such as antagonism, hyperparasitism, competition, and induce systemic resistant molecules. Additionally, *Trichoderma* spp. contains a plethora of known and unknow secondary metabolites, which play as biopesticidal compounds. Needless to say, the efficacy of *Trichoderma* spp. has been proved against *M. phaseolina* by confrontation assay in vitro. However, *Trichoderma* spp. might be the appropriate sustainable alternative instead of chemical fungicide against *M. phaseolina. Trichoderma* spp. can help to maintain the ecological harmony in the ecosystem. But it has also some drawbacks, such as sluggish action, inability to entirely remove plant diseases, variable performance, rhizosphere incompetence and varied environment. Nevertheless, the application of *Trichoderma* spp. will be effective, ecofriendly, and economically viable for sustainable management of lethal phytopathogen *M. phaseolina.*

TABLE 18.4 *Trichoderma* spp. metabolites as a potent biopesticides against *M. phaseolina.*

Trichoderma spp.	Secondary metabolites	References
T. viride	Octadecadienoic acid, hexadecanoic acid, ethyl ester, tricosene, methyl ester	Khan et al. (2021)
T. harzianum	Volatile metabolites Peroxidase, phenylalanine ammonia-lyase, chitinase, polyphenol oxidase and β-1,3-glucanase cell free metabolites	Khalili et al. (2016) Govindappa et al. (2010)
T. viride *T. virens*	Chitinase, β-1,3-glucanase	Dubey et al. (2011)
T. atroviride	Volatile metabolites such as 5-hydroxymethylfurfural, 4H-pyran-4-one, 2,3-dihydro-3,5-dihydroxy-6-methyl-etc.	El-Benawy et al. (2020)
T. pseudokoningii	Undecane, octadecadienoic acid, decane, methyl ester and epoxylanostan	Khan et al. (2020)
T. longibrachiatum	Pentanol, hexanol, myristoyl pantothenate, bisabolol, D-alanine, and diethyl trisulphide	Sridharan et al. (2021)
T. koningii	δ-Decanolactone, 6-pentyl-α-pyranone and 6-(4-oxopentyl)-2H-pyran-2-one	Ahluwalia et al. (2014)
Trichoderma spp.	Ethylene, hydrogen cyanide, aldehydes and ketones Volatile and nonvolatile	Athira (2017) Reino et al. (2008)
T. viride	Fungistatics metabolites and emzymes	Vinothini et al. (2020)
Trichoderma spp.	Trichodermin, giotoxin, viridin Ethylene, hydrogen cyanide, aldehydes, and ketones Gliotoxin, gliovirin, viridin and its derivatives Volatile organic compound (VOC)	Harman (2006) Vey et al. (2001) Vinale et al. (2008) Knowles et al. (2019) Rajani et al. (2021)
T. virens	Gliotoxin Alanine-betanaphthylamide, pyrazine, propanoic acid, octadecatrienoic acid, glutamine, hexadecenoic acid and decane	Mukherjee et al. (2012) Emayavarman et al. (2019)
T. harzianum *T. virens*	Volatile metabolite	Barari and Foroutan (2016)

Last but not least, further molecular study should be needed to identify the more effective *Trichoderma* species against the *M. phaseolina.* Moreover, screening of the formulation and additive, congenial environment should be considered for getting viable bio-agent longer period of time. Finally, a national and international policy framework should be established to inspire the development, marketing, distribution, and use of *Trichoderma* spp. for saving the yield losses against the *M. phaseolina* during commercial crop cultivation under the open condition.

Acknowledgments

The authors express their gratefulness to the Research Management Wing, Bangabandhu Sheikh Mujibur Rahman Agricultural University, Gazipur-1706 (Bangladesh), Ministry of Science and Technology, Dhaka-1000 (Bangladesh) and National Agricultural Technology Program—Phase II Project (NATP-2), BARC, Dhaka-1215 (Bangladesh) for providing financial support to accomplish relevant research works earlier.

References

Abd-Elsalam, K. A. (2010). Genetical and biological control of cotton ashy stem caused by *Macrophomina phaseolina* in outdoor pot experiment. *Saudi Journal of Biological Sciences, 17*(2), 147—152. https://doi.org/10.1016/j.sjbs.2010.02.008

Abdel-Kader, M. M., El-Mougy, N. S., Aly, M. D. E. H., & Lashin, S. M. (2010). First report of ashy stemblight caused by *Macrophomina phaseolina* on *Aeonium canariense* in Egypt. *Journal of Plant Pathology & Microbiology, 1*, 101. https://doi.org/10.4172/2157-7471.1000101

Abied, M., & Duzan, M. A. E. A. (2020). First report of *Macrophomina phaseolina* causing crown and root rot of strawberry in Tripoli, Libya. *Journal of Applied Plant Protection, 9*(1). https://doi.org/10.21608/japp.2020.178431

Adekunle, A. T., Cardwell, K. F., Florini, D. A., & Ikotun, T. (2001). Seed treatment with *Trichoderma* species for control of damping-off of cowpea caused by *Macrophomina phaseolina. Biocontrol Science and Technology, 11*(4). https://doi.org/10.1080/09583150120067481

Adekunle, A. T., Ikotun, T., Florini, D. A., & Cardwell, K. F. (2006). Field evaluation of selected formulations of *Trichoderma* species as seed treatment to control damping-off of cowpea caused by *Macrophomina phaseolina. African Journal of Biotechnology, 5*(5), 419–424.

Ahamed, A., & Vermette, P. (2009). Effect of culture medium composition on *Trichodermareesei's* morphology and cellulase production. *Bioresource Technology, 100*(23), 5979–5987. https://doi.org/10.1016/J.BIORTECH.2009.02.070

Ahluwalia, V., Walia, S., Sati, O. P., Kumar, J., Kundu, A., Shankar, J., & Paul, Y. S. (2014). Isolation, characterization of major secondary metabolites of the Himalayan *Trichoderma koningii* and their antifungal activity. *Archives of Phytopathology and Plant Protection, 47*(9), 1063–1071. https://doi.org/10.1080/03235408.2013.829715

Ahmed, M. U., Bhuiyan, M. K. A., Hossain, M. M., Rubayet, M. T., & Khaliq, Q. A. (2019). Efficacy of chitosan and bio-agent in controlling southern blight disease of carrot caused by *Sclerotium rolfsii* and improvement the crop production. *Research in Agriculture and Veterinary Science, 3*(3), 113–125.

Al-Juboory, H. H., & Al-Jarah, N. S. (2020). Control of *Macrophomina phaseolina* by different isolates of *Trichoderma* spp. *Indian Journal of Ecology, 47*(10), 191–196.

Almammory, M. K. N., & Matloob, A. A. H. (2019). Efficiency of *Trichoderma harzianum* and bio-fertilizer bokashi and salicylic acid to control of fungi causing eggplant damping off disease. *Plant Archives, 19*(1), 73–82.

Altomare, C., Norvell, W. A., Bjorkman, T., & Harman, G. E. (1999). Solubilization of phosphates and micronutrients by the plant-growth-promoting and biocontrol fungus *Trichoderma harzianum* Rifai 1295-22. *Applied and Environmental Microbiology, 65*(7), 2926–2933. https://doi.org/10.1128/AEM.65.7.2926-2933.1999

Anis, M., JavedZaki, M., & Dawar, S. (2010). Effect of oilseed cakes alone or in combination with *Trichoderma* species for the control of charcoal rot of sunflower (*Helianthus annus* L.). *Pakistan Journal of Botany, 42*(6), 4329–4333.

Arefin, M. N., Bhuiyan, M. K. A., & Rubayet, M. T. (2019). Integrated use of fungicide, plant extract and bio-agent for management of alternaria blight disease of radish (*Raphanus sativus* L.) and quality seed production. *Research in Agriculture and Veterinary Science, 3*(1), 10–21.

Arora, M., & Dhurwe, U. (2013). Biochemical control of charcoal rot of *Sorghum bicolor* (L.) Moench. *International Journal of Current Microbiology and Applied Sciences, 2*(11), 19–23.

Athira, K. (2017). Efficacy of fungicide and bio-control agents against root rot of black Gram (*Vigna mungo* L.) caused by *Macrophomina phaseolina* (Tassi) Goid. *International Journal of Current Microbiology and Applied Sciences, 6*(10), 2601–2607.

Bae, H., Roberts, D. P., Lim, H. S., Strem, M. D., Park, S. C., Ryu, C. M., Melnick, R. L., & Bailey, B. A. (2011). Endophytic *Trichoderma* isolates from tropical environments delay disease onset and induce resistance against *Phytophthora capsici* in hot pepper using multiple mechanisms. *Molecular Plant-Microbe Interactions, 24*(3), 336–351. https://doi.org/10.1094/MPMI-09-10-0221

Barari, H., & Foroutan, A. (2016). Biocontrol of soybean charcoal root rot disease by using *Trichoderma* spp. *Agronomic Research in Moldavia, 2*(166), 41–51.

Basandrai, A. K., Pandey, A. K., Somta, P., & Basandrai, D. (2021). *Macrophomina phaseolina*–host interface: Insights into an emerging dry root rot pathogen of mungbean and urdbean, and its mitigation strategies. *Plant Pathology, 70*, 1263–1275. https://doi.org/10.1111/ppa.13378

Bastakoti, S., Belbase, S., Manandhar, S., & Arjyal, C. (2017). *Trichoderma* species as biocontrol agent against soil borne fungal pathogens. *Nepal Journal of Biotechnology, 5*, 39–45.

Bayoumy, S., Afify, A., El-Sayed, A., & Elshal, S. (2017). Antagonistic effect of *Bacillus* spp. against sugar beet pathogens fusarium wilt. *Journal of Agricultural Chemistry and Biotechnology, 8*, 177–181.

Bhat, K. A., Anwar, A., Lone, G. M., Hussain, K., & Nazir, G. (2009). Shelf life of liquid fermented product of *Trichoderma harzianum* in talc. *Journal of Mycology and Plant Pathology, 39*(2), 263–265.

Bhattacharya, G., Dhar, T. K., Kathleen Bhattacharya, F., & Siddiqui, K. A. (1987). Mutagenic action of phaseolinone, a mycotoxin isolated from *Macrophomina phaseolina. Australian Journal of Biological Sciences, 40*, 349–353. https://doi.org/10.1071/BI9870349

Bokor, P. (2007). *Macrophomina phaseolina* causing a charcoal rot of sunflower through Slovakia. *Biologia, 62*, 136–138. https://doi.org/10.2478/s11756-007-0020-9

Brotman, Y., Briff, E., Viterbo, A., & Chet, I. (2008). Role of swollenin, an expansin-like protein from *Trichoderma*, in plant root colonization. *Plant Physiology, 147*(2), 779–789. https://doi.org/10.1104/pp.108.116293

Bubici, G., Kaushal, M., Prigigallo, M. I., Cabanás, C. G. L., & Mercado-Blanco, J. (2019). Biological control agents against *Fusarium* wilt of banana. *Frontiers in Microbiology, 10*, 616. https://doi.org/10.3389/fmicb.2019.00616

Cai, F., Yu, G., Wang, P., Wei, Z., Fu, L., Shen, Q., & Chen, W. (2013). Harzianolide, a novel plant growth regulator and systemic resistance elicitor from *Trichoderma harzianum. Plant Physiology and Biochemistry, 73*, 106–113. https://doi.org/10.1016/j.plaphy.2013.08.011

Chandra, S., Kheri, H. S., & Maheshwari, S. (1995). *Macrophomina* dry root rot of leguminous crops and its management through VAM pesticides. In J. P. Verma, A. Verma, & D. Kumar (Eds.), *Detection of plant pathogens and their management* (pp. 357–364). New Delhi, India: Angkor Publishers.

Chet, I., & Inbar, J. (1994). Biological control of fungal pathogens. *Applied Biochemistry and Biotechnology, 48*(1), 37–43. https://doi.org/10.1007/BF02825358

Chet, I., Inbar, J., & Hadar, I. (1997). Fungal antagonists and mycoparasit. In D. T. Wicklow, & B. Söderström (Eds.), *The mycota IV: Environmentaland microbial relationships* (pp. 165–184). Berlin: Springer-Verlag.

Chowdhury, S., Basu, A., Ray Chaudhuri, T., & Kundu, S. (2014). *In vitro* characterization of the behaviour of *Macrophomina phaseolina* (Tassi) Goid at the rhizosphere and during early infection of roots of resistant and susceptible varieties of sesame. *European Journal of Plant Pathology, 138*, 361–375. https://doi.org/10.1007/s10658-013- 0335-z

Contreras-Cornejo, H. A., Macías-Rodríguez, L., Beltrán-Peña, E., Herrera-Estrella, A., & López-Bucio, J. (2011). *Trichoderma*-induced plant immunity likely involves both hormonal- and camalexin-dependent mechanisms in *Arabidopsis thaliana* and confers resistance against necrotrophic fungi *Botrytis cinerea*. *Plant Signaling & Behavior, 6*(10), 1554–1563. https://doi.org/10.4161/psb.6.10.17443

Contreras-Cornejo, H. A., Macías-Rodríguez, L., del-Val, E., & Larsen, J. (2016). Ecological functions of *Trichoderma* spp. and their secondary metabolites in the rhizosphere: Interactions with plants. *FEMS Microbiology Ecology, 92*(4), fiw036. https://doi.org/10.1093/femsec/fiw036

Contreras-Cornejo, H. A., Macías-Rodríguez, L., Herrera-Estrella, A., & López-Bucio, J. (2014). The 4-phosphopantetheinyl transferase of *Trichoderma virens* plays a role in plant protection against botrytis cinerea through volatile organic compound emission. *Plant and Soil, 379*, 261–274. https://doi.org/10.1007/s11104-014-2069-x

Crous, P. W., Slippers, B., Wingfield, M. J., Rheeder, J., Marasas, W. F., Philips, A. J., Alves, A., Burgess, T., Barber, P., & Groenewald, J. Z. (2006). Phylogenetic lineages in the Botryosphaeriaceae. *Studies in Mycology, 55*, 235–253. https://doi.org/10.3114/sim.55.1.235

Cubilla-Ríos, A. A., Ruíz-Díaz-Mendoza, D. D., Romero-Rodríguez, M. C., Flores-Giubi, M. E., & Barúa-Chamorro, J. E. (2019). Antibiosis of proteins and metabolites of three species of *Trichoderma* against Paraguayan isolates of *Macrophomina phaseolina*. *Agronomy Mesoamerican, 30*, 63–77. https://doi.org/10.15517/am.v30i1.34423

Das, I. R., Bhuiyan, M. K. A., Jannat, R., Kayesh, E., Rubayet, M. T., & Arefin, M. N. (2019). Effect of bio-fortified compost in controlling soil-borne diseases of lentil (*Lens culinaris* L.) and enhance the crop growth and yield. *Advances in Biology & Earth Sciences, 4*(2), 93–106. https://www.researchgate.net/publication/335107214.

Deng, J. J., Huang, W. Q., Li, Z. W., Lu, D. L., Zhang, Y., & Luo, X. C. (2018). Biocontrol activity of recombinant aspartic protease from *Trichoderma harzianum* against pathogenic fungi. *Enzyme and Microbial Technology, 112*, 35–42. https://doi.org/10.1016/j.enzmictec.2018.02.002

Dhingra, O. D., & Sinclair, J. B. (1978). *Biology and pathology of macrophomina phaseolina*. Minas Gerais: Universidade Federal de Viçosa: Impresa Universitaria.

Djonovic, S., Vargas, W. A., Kolomiets, M. V., Horndeski, M., Wiest, A., & Kenerley, C. M. (2007). A proteinaceous elicitor Sm1 from the beneficial fungus Trichoderma virens is required for induced systemic resistance in maize. *Plant Physiology, 145*(3), 875–889. https://doi.org/10.1104/pp.107.103689

Doni, F., Isahak, A., Che Mohd Zain, C. R., MohdAriffin, S., Wan Mohamad, W. N., & Wan Yusoff, W. M. (2014). Formulation of *Trichoderma* sp. SL2 inoculants using different carriers for soil treatment in rice seedling growth. *SpringerPlus, 3*, 532. https://doi.org/10.1186/2193-1801-3-532

Druzhinina, I. S., Seidl-Seiboth, V., Herrera-Estrella, A., Horwitz, B. A., Kenerley, C. M., Monte, E., Mukherjee, P. K., Zeilinger, S., Grigoriev, I. V., & Kubicek, C. P. (2011). *Trichoderma*: The genomics of opportunistic success. *Nature Reviews Microbiology, 9*(10), 749–759. https://doi.org/10.1038/nrmicro2637

Dubey, S. C., Bhavani, R., & Singh, B. (2011). Integration of soil application and seed treatment formulations of *Trichoderma* species for management of wet root rot of mungbean caused by *Rhizoctonia solani*. *Pest Management Science, 67*, 1163–1168.

Eisendle, M., Oberegger, H., Buttinger, R., Illmer, P., & Haas, H. (2004). Biosynthesis and uptake of siderophores is controlled by the PacC-mediated ambient-pH regulatory system in *Aspergillus nidulans*. *Eukaryotic Cell, 3*(2), 561–563. https://doi.org/10.1128/EC.3.2.561-563.2004

El-Benawy, N. M., Abdel-Fattah, G. M., Ghoneem, K. M., & Shabana, Y. M. (2020). Antimicrobial activities of *Trichoderma atroviride* against common bean seedborne *Macrophomina phaseolina* and *Rhizoctonia solani*. *Egyptian Journal of Basic and Applied Sciences, 7*(1), 267–280. https://doi.org/10.1080/2314808X.2020.1809849

El-Hassan, S. A., Gowen, S. R., & Pembroke, B. (2013). Use of *Trichoderma hamatum* for biocontrol of lentil vascular wilt disease: Efficacy, mechanisms of interaction and future prospects. *Journal of Plant Protection Research, 53*(1), 12–26.

Elsharkawy, M. M., Shimizu, M., Takahashi, H., Ozaki, K., & Hyakumachi, M. (2013). Induction of systemic resistance against *Cucumber mosaic virus* in *Arabidopsis thaliana* by *Trichoderma asperellum* SKT-1. *Plant Pathology Journal, 29*(2), 193–200. https://doi.org/10.5423/PPJ.SI.07.2012.01

Emayavarman, P., Renukadevi, P., Ravikesavan, R., & Nakkeera, S. (2019). Antifungal efficacy and growth promotion by *Trichoderma virens* TRI 37 and *Bacillus amyloloiquefaciens* (VB7) against *Macrophomina phaseolina*—the maize charcoal rot pathogen. *International Journal of Current Microbiology and Applied Sciences, 8*(11), 2682–2693.

Erper, I., Turkkan, M., Bobushova, S., & Ozer, G. (2021). First report of charcoal rot caused by *Macrophomina phaseolina* on common bean in Kyrgyzstan. *Journal of Plant Pathology, 103*, 1025–1026. https://doi.org/10.1007/s42161-021-00841-1

Etebarian, H. R. (2006). Evaluation of *Trichoderma* isolates for biological control of charcoal stem rot in melon caused by *Macrophomina phaseolina*. *Journal of Agricultural Science and Technology A, 8*, 243–250.

Ezziyyani, M., Requena, M. E., Egea-Gilabert, C., & Candela, M. E. (2007). Biological control of *Phytophthora* root rot of pepper using *Trichoderma harzianum* and *Streptomyces rochei* in combination. *Journal of Phytopathology, 155*(6). https://doi.org/10.1111/j.1439-0434.2007.01237.x

Fakir, G. A., & Mridha, U. (1985). Die-back, a new disease of lady's finger (*Hibiscus esculentus* L.) in Bangladesh. *Bangladesh Journal of Plant Pathology, 11*, 25–28.

Farnaz, A. A., Narmani, A., & Arzanlou, M. (2018). *Macrophomina phaseolina* associated with grapevine decline in Iran. *Phytopathologia Mediterranea, 57*, 107–111.

Fravel, D. R., Lewis, J. A., & Chittams, J. L. (1995). Alginate prill formulations of *Talaromyces flavus* with organic carriersfor biocontrol of *Verticillium dahlia*. *Phytopathology, 85*, 165–168.

Freeman, S., Minz, D., Kolesnik, I., Barbul, O., Zveibil, A., Maymon, M., Nitzani, Y., Kirshner, B., Rav-David, D., Bilu, A., Dag, A., Shafir, S., & Elad, Y. (2004). *Trichoderma* biocontrol of *Colletotrichum acutatum* and *Botrytis cinerea* and survival in strawberry. *European Journal of Plant Pathology, 110*(4), 361–370. https://doi.org/10.1023/B:EJPP.0000021057.93305.d9

Fujinaga, M., Shimizu, T., Ogiso, H., Saito, H., Kobayashi, M., & Sato, T. (2002). Charcoal rot of watermelon newly found in Nagano and Kanagawa prefectures and pathogenicity of causal fungus *Macrophomina phaseolina*. *Japanese Journal of Phytopathology, 68*(2), 148–152.

Gaetán, S. A., Fernandez, L., & Madia, M. (2006). Occurrence of charcoal rot caused by *Macrophomina phaseolina* on Canola in Argentina. *Plant Disease, 90*(4). https://doi.org/10.1094/pd-90-0524a

Gajera, H., Domadiya, R., Patel, S., Kapopara, M., & Golakiya, B. (2013). Molecular mechanism of *Trichoderma* as bio-control agents against phytopathogen system. *Current Research in Microbiology and Biotechnology, 1*(4), 133–142.

Gerin, D., Dongiovanni, C., De Miccolis Angelini, R. M., Pollastro, S., & Faretra, F. (2018). First report of *Macrophomina phaseolina* causing crown and root rot on strawberry in Italy. *Plant Disease, 102*(9). https://doi.org/10.1094/PDIS-01-18-0191-PDN

Ghosh, T., Biswas, M. K., Guin, C., & Roy, P. (2018). A review on characterization, therapeutic approaches and pathogenesis of *Macrophomina phaseolina*. *Plant Cell Biotechnology and Molecular Biology, 19*, 72–84.

Govindappa, M., Lokesh, S., Rai, R. V., Naik, V. R., & Raju, S. G. (2010). Induction of systemic resistance and management of safflower *Macrophomina phaseolina* root-rot disease by biocontrol agents. *Archives of Phytopathology and Plant Protection, 43*(1), 26–40. https://doi.org/10.1080/03235400701652227

Guo, R., Ji, S., Wang, Z., Zhang, H., Wang, Y., & Liu, Z. (2021). *Trichoderma asperellum* xylanases promote growth and induce resistance in poplar. *Microbiological Research, 248*, 126767. https://doi.org/10.1016/j.micres.2021.126767

Gupta, G. K., Sharma, S. K., & Ramteke, R. (2012). Biology, epidemiology and management of the pathogenic fungus *Macrophomina phaseolina* (Tassi) Goid. with special reference to charcoal rot of soybean (*Glycine max* (L.) Merrill). *Journal of Phytopathology, 160*, 167–180. https://doi.org/10.1111/j.1439-0434.2012.01884.x

Halifu, S., Deng, X., Song, X., & Song, R. (2019). Effects of two *Trichoderma* strains on plant growth, rhizosphere soil nutrients, and fungal community of *Pinus sylvestris* var. mongolica annual seedlings. *Forests, 10*(9). https://doi.org/10.3390/f10090758

Harman, G. E. (2006). Overview of mechanisms and uses of *Trichoderma* spp. *Phytopathology, 96*(2), 190–194. https://doi.org/10.1094/PHYTO-96-0190

Harman, G. E., Howell, C. R., Viterbo, A., Chet, I., & Lorito, M. (2004). *Trichoderma* species—opportunistic, avirulent plant symbionts. *Nature Reviews Microbiology, 2*(1), 43–56. https://doi.org/10.1038/nrmicro797

Hassan, M. (2014). Enhance suppressive effect of compost on soybean rhizoctonia root rot by soil treatment with *Trichoderma harzianum*. *Journal of Plant Physiology and Pathology, 2*(2), 1–6.

Herrera, W., Valbuena, O., & Pavone-Maniscalco, D. (2020). Formulation of *Trichoderma asperellum* TV190 for biological control of *Rhizoctonia solani* on corn seedlings. *Egyptian Journal of Biological Pest Control, 30*, 44. https://doi.org/10.1186/s41938-020-00246-9

Hewavitharana, N., Kannangara, S. D. P., & Senanayake, S. P. (2018). Isolation, identification and mass production of five *Trichoderma* spp. on solid and liquid carrier media for commercialization. *International Journal of Applied Sciences and Biotechnology, 6*, 285–293.

Hoitink, H. A., Madden, L. V., & Dorrance, A. E. (2006). Systemic resistance induced by *Trichoderma* spp.: Interactions between the host, the pathogen, the biocontrol agent, and soil organic matter quality. *Phytopathology, 96*(2), 186–189. https://doi.org/10.1094/PHYTO-96-0186

Holzlechner, M., Reitschmidt, S., Gruber, S., Zeilinger, S., & Marchetti-Deschmann, M. (2016). Visualizing fungal metabolites during mycoparasitic interaction by MALDI mass spectrometry imaging. *Journal of Proteomics, 16*(11), 1742e–1746e.

Hossain, M. M., Sultana, F., & Islam, S. (2017). Plant growth-promoting fungi (PGPF): Phytostimulation and induced systemic resistance. In D. Singh, H. Singh, & R. Prabha (Eds.), *Plant-microbe interactions in agro-ecological perspectives* (pp. 135–191). https://doi.org/10.1007/978-981-10-6593-4_6

Howell, C. R. (2003). Mechanisms employed by *Trichoderma* species in the biological control of plant diseases: The history and evolution of current concepts. *Plant Disease, 87*(1). https://doi.org/10.1094/PDIS.2003.87.1.4

Hussain, S., Ghaffar, A., & Aslam, M. (1990). Biological control of *Macrophomina phaseolina* charcoal rot of sunflower and mung bean. *Journal of Phytopathology, 130*, 157–160. https://doi.org/10.1111/j.1439-0434.1990.tb01163.x

Hussain, F., Shaukat, S. S., Abid, M., Usman, F., & Akbar, M. (2013). Pathogenicity of some important root rot fungi to the chilli crop and their biological control. *International Journal of Biology and Biotechnology, 10*(1), 101–108.

Hyder, S., Gondal, A. S., Ahmed, R., Sahi, S. T., Rehman, A., & Hannan, A. (2018). First report of charcoal rot in tomato caused by *Macrophomina phaseolina* (Tassi) Goid. from Pakistan. *Plant Disease, 102*(7), 1459. https://doi.org/10.1094/PDIS-10-17-1663-PDN

Ibrahim, M. E., & Abdel-Azeem, A. M. (2014). Management of sesame (*Sesamum indicum* L.) charcoal rot caused by *Macrophomina phaseolina* (Tassi) Goid. through the application of different control measure. *Journal of Pure and Applied Microbiology, 9*(1), 141–149.

Iqbal, U., & Mukhtar, T. (2020). Inhibitory effects of some fungicides against *Macrophomina phaseolina* causing charcoal rot. *Pakistan Journal of Zoology, 52*, 709–715. https://doi.org/10.17582/journal.pjz/20181228101230

Islam, M., Haque, M., Islam, M., Emdad, E., Halim, A., Hossen, Q. M. M., Hossain, M. Z., Ahmed, B., Rahim, S., Rahman, M. S., Alam, M. M., Hou, S., Wan, X., Saito, J. A., & Alam, M. (2012). Tools to kill: Genome of one of the most destructive plant pathogenic fungi *Macrophomina phaseolina*. *BMC Genomics, 13*, 493. https://doi.org/10.1186/1471-2164-13-493

Jacob, C. J., Krarup, C., Díaz, G. A., & Latorre, B. A. (2013). A severe outbreak of charcoal rot in cantaloupe melon caused by *Macrophomina phaseolina* in Chile. *Plant Disease, 97*(1), 141. https://doi.org/10.1094/PDIS-06-12-0588-PDN. PMID: 30722280.

Jeyarajan, R. (2006). Prospects of indigenous mass production and formulation of Trichoderma. In R. J. Rabindra, & B. Ramanujam (Eds.), *Current status of biological control of plant diseases using antagonistic organisms in India* (pp. 74–80). Bangalore: Project Directorate of Biological Control.

Jeyarajan, R., Ramakrishnan, G., Dinakaran, D., & Sridar, R. (1994). Development of products of *Trichoderma viride* and *Bacillus subtilis* for biocontrol of root rot diseases. In B. K. Dwivedi (Ed.), *Biotechnology in India* (pp. 25–36). Allahabad: Bioved Research Society.

Karadimos, D. A., Karaoglanidis, G. S., & Klonari, K. (2002). First report of charcoal rot of sugar beet caused by *Macrophomina phaseolina* in Greece. *Plant Disease, 86*(9). https://doi.org/10.1094/pdis.2002.86.9.1051d

Kaur, S., Dhillon, G. S., Brar, S. K., Vallad, G. E., Chand, R., & Chauhan, V. B. (2012). Emerging phytopathogen *Macrophomina phaseolina*: Biology, economic importance and current diagnostic trends. *Critical Reviews in Microbiology, 38*(2), 136–151.

Khaledi, N., & Taheri, P. (2016). Biocontrol mechanisms of *Trichoderma harzianum* against soybean charcoal rot caused by *Macrophomina phaseolina*. *Journal of Plant Protection Research, 56*(1), 21–31.

Khalili, E., Javed, M. A., Huyop, F., Rayatpanah, S., Jamshidi, S., & Wahab, R. A. (2016). Evaluation of *Trichoderma* isolates as potential biological control agent against soybean charcoal rot disease caused by *Macrophomina phaseolina*. *Biotechnology & Biotechnological Equipment, 30*(3), 479–488. https://doi.org/10.1080/13102818.2016.1147334

Khambhati, V. H., Abbas, H. K., Sulyok, M., Tomaso-Peterson, M., & Shier, W. T. (2020). First report of the production of mycotoxins and other secondary metabolites by *Macrophomina phaseolina* (Tassi) Goid. isolates from soybeans (*Glycine max* L.) symptomatic with charcoal rot disease. *Journal of Fungi, 6*(4), 332. https://doi.org/10.3390/jof6040332

Khan, M. R., & Gupta, J. (1998). Antagonistic efficacy of *Trichoderma* species against *Macrophomina phaseolina* on eggplant/Antagonistische Wirksamkeit von Trichoderma-Arten gegen *Macrophomina phaseolina* an Auberginen. *Zeitschrift für Pflanzenkrankheiten und Pflanzenschutz/Journal of Plant Diseases and Protection, 105*(4), 387–393. http://www.jstor.org/stable/43215256. (Accessed 10 July 2022).

Khan, M. R., & Mohiddin, F. A. (2018). *Trichoderma*: Its multifarious utility in crop improvement. In R. Prasad, S. S. Gill, & N. Tuteje (Eds.), *New and future developments in microbial biotechnology and bioengineering* (pp. 263–291). USA: Elsevier.

Khan, M. R., Ashraf, S., Rasool, F., Salati, K., Mohiddin, F. A., & Haque, Z. (2014). Field performance of *Trichoderma* species against wilt disease complex of chickpea. *Turkish Journal of Agriculture, 38*, 447–454.

Khan, M. R., Haque, Z., Rasool, F., Salati, K., Khan, U., Mohiddin, F. A., & Zuhaib, M. (2019). Management of root-rot disease complex of mungbean caused by *Macrophominaphaseolina* and *Rhizoctonia solani* through soil application of *Trichoderma* spp. *Crop Protection, 119*, 24–29. https://doi.org/10.1016/j.cropro.2019.01.014

Khan, I. H., Javaid, A., & Ahmed, D. (2021). *Trichoderma viride* controls *Macrophomina phaseolina* through its DNA disintegration and production of antifungal compounds. *International Journal of Agriculture and Biology, 25*, 888–894.

Khan, R., Najeeb, S., Hussain, S., Xie, B., & Li, Y. (2020). Bioactive secondary metabolites from *Trichoderma* spp. against phytopathogenic fungi. *Microorganisms, 8*(6), 817. https://doi.org/10.3390/microorganisms8060817

Khangura, R., & Aberra, M. (2009). First report of charcoal rot on canola caused by *Macrophomina phaseolina* in Western Australia. *Plant Disease, 93*(6), 666. https://doi.org/10.1094/PDIS-93-6-0666C

Khare, E., Kumar, S., & Kim, K. (2018). Role of peptaibols and lytic enzymes of *Trichoderma cerinum* Gur1 in biocontrol of *Fusarium oxysporum* and chickpea wilt. *Environmental Sustainability, 1*, 39e47.

Kim, S. G., Kim, T. B., & Lee, O. J. (2021). First report of *Macrophomina phaseolina* causing charcoal rot in bottle gourd in Korea. *Korean Journal of Mycology, 49*(3), 399–403. https://doi.org/10.4489/KJM.20210037

Kıriga, A. W., Haukeland, S., Kariuki, G. M., Coyne, D. L., & Beek, N. V. (2018). Effect of *Trichoderma* spp. and *Purpureocillium lilacinum* on *Meloidogyne javanica* in commercial pineapple production in Kenya. *Biological Control, 119*. https://doi.org/10.1016/j.biocontrol.2018.01.005

Kishimoto, K., Matsui, K., Ozawa, R., & Takabayashi, J. (2007). Volatile 1-octen-3-ol induces a defensive response in *Arabidopsis thaliana*. *Journal of General Plant Pathology, 73*, 35–37. https://doi.org/10.1007/s10327-006-0314-8

Knowles, S. L., Raja, H. A., Wright, A. J., Lee, A., Caesar, L. K., Cech, N. B., Mead, M. E., Steenwyk, J. L., Ries, L. N., Goldman, G. H., Rokas, A., & Oberlies, N. H. (2019). Mapping the fungal battlefield: Using *in situ* chemistry and deletion mutants to monitor interspecific chemical interactions between fungi. *Frontiers in Microbiology, 10*, 285. https://doi.org/10.3389/fmicb.2019.00285

Koike, S. T. (2008). Crown rot of strawberry caused by *Macrophomina phaseolina* in California. *Plant Disease, 92*(8), 1253.

Kolombet, L. V., Zhigletsova, S. K., Kosareva, N. I., Bystrova, E. V., Derbyshev, V. V., Krasnova, S. P., & Schisler, D. (2008). Development of an extended shelf-life, liquid formulation of the biofungicide *Trichoderma asperellum*. *World Journal of Microbiology and Biotechnology, 24*, 123–131. https://doi.org/10.1007/s11274-007-9449-9

Korkom, Y., & Yildiz, A. (2022). Evaluation of biocontrol potential of native *Trichoderma* isolates against charcoal rot of strawberry. *Journal of Plant Pathology, 104*, 671–682. https://doi.org/10.1007/s42161-022-01063-9

Kottb, M., Gigolashvili, T., Großkinsky, D. K., & Piechulla, B. (2015). *Trichoderma* volatiles effecting *Arabidopsis*: From inhibition to protection against phytopathogenic fungi. *Frontiers in Microbiology, 6*, 1–14. https://doi.org/10.3389/fmicb.2015.00995

Kouadri, M. E. A., Zaim, S., & Bekkar, A. A. (2021). First report of *Macrophomina pseudophaseolina* infecting *Lens culinaris*. *Australasian Plant Disease Notes, 16*, 26. https://doi.org/10.1007/s13314-021-00440-0

Kubicek, C. P., Herrera-Estrella, A., Seidl-Seiboth, V., Martinez, D. A., Druzhinina, I. S., Thon, M., Zeilinger, S., Casas-Flores, S., Horwitz, B. A., Mukherjee, P. K., Mukherjee, M., Kredics, L., Alcaraz, L. D., Aerts, A., Antal, Z., Atanasova, L., Cervantes-Badillo, M. G., Challacombe, J., Chertkov, O., & "..."Grigoriev, I. V. (2011). Comparative genome sequence analysis underscores mycoparasitism as the ancestral life style of *Trichoderma*. *Genome Biology, 12*, 1–15. https://doi.org/10.1186/gb-2011-12-4-r40

Kumar, T. P., & Palakshappa, M. G. (2009). Evaluation of suitable substrates for on farm production of antagonist *Trichoderma harzianum*. *Karnataka Journal of Agricultural Sciences, 22*, 115–117.

Kumar, G. A., Maharshi, J., Patel, A., Mukherjee, H. B., Singh, B. K., & Sarma. (2017). *Trichoderma*: A potential fungal antagonist to control plant diseases. *SATSA Mukhapatra Annual Technical Issue, 21*, 206–218.

Kumari, R., Khan, M. R., Bagdi, D. L., Bagri, G. K., & Bagri, D. K. (2017). Soil application of different species of *Trichoderma* for the management of charcoal rot of faba bean caused by *Macrophomina phaseolina*. *Journal of Pharmacognosy and Phytochemistry, 6*(6), 1483–1489.

Lakhran, L., Ahir, R. R., Choudhary, M., & Choudhary, S. (2018). Isolation, purification, identification and pathogenicity of *Macrophomina phaseolina* (Tassi) Goid. caused dry root rot of chickpea. *Journal of Pharmacognosy and Phytochemistry, 7*, 3314–3317.

Larralde-Corona, C. P., Santiago-Mena, M. R., Sifuentes-Rincón, A. M., Rodríguez-Luna, I. C., Rodríguez-Pérez, M. A., Shirai, K., & Narváez-Zapata, J. A. (2008). Biocontrol potential and polyphasic characterization of novel native *Trichoderma* strains against *Macrophomina phaseolina* isolated from sorghum and common bean. *Applied Microbiology and Biotechnology, 80*, 167–177. https://doi.org/10.1007/s00253-008-1532-0

Leach, L. D., & Garber, R. H. (1970). Control of rhizoctonia. In R. Parmeter, Jr. (Ed.), *Rhizoctonia solani, biology and pathology* (pp. 189–198). San Diego: University of California Press.

Lewis, J. A., & Papavizas, G. C. (1991). Biocontrol of cotton damping-off caused by *Rhizoctonia solani* in the field with formulations of *Trichoderma* spp. and *Gliocladium virens*. *Crop Protection, 10*, 396–402.

Li, G. Q., Huang, H. C., Acharya, S. N., & Erickson, R. S. (2005). Effectiveness of *Coniothyrium minitans* and *Trichoderma atroviride* in suppression of *sclerotinia* blossom blight of alfalfa. *Plant Pathology, 54*(2), 204–211. https://doi.org/10.1111/j.1365-3059.2005.01119.x

Liton, M. J. A., Bhuiyan, M. K. A., Jannat, R., Ahmed, J. U., Rahman, M. T., & Rubayet, M. T. (2019). Efficacy of Trichoderma-fortified compost in controlling soil-borne diseases of bush bean (*Phaseolus vulgaris* L.) and sustainable crop production. *Advances in Agricultural Science, 7*(2), 123–136.

Lokesha, N. M., & Benagi, V. I. (2007). Biological management of pigeonpea dry root rot caused by *Macrophomina phaseolina*. *Karnataka Journal of Agricultural Sciences, 20*, 54–56.

Lorito, M., Hayes, C. K., Di Pietro, A., Woo, S. L., & Harman, G. E. (1994). Purification, characterization, and synergistic activity of a glucan 1,3-b-glucosidase and an N-acetyl-b-glucosaminidase from *Trichoderma harzianum*. *Phytopathology, 84*, 398–405.

Machado, A. R., Pinho, D. B., Soares, D. J., Gomes, A. A. M., & Pereira, O. L. (2019). Bayesian analyses of five gene regions reveal a new phylogenetic species of *Macrophomina* associated with charcoal rot on oilseed crops in Brazil. *European Journal of Plant Pathology, 153*, 89–100. https://doi.org/10.1007/s10658-018-1545-1

Malmierca, M., Mccormick, S., Cardoza, R., Monte, E., Alexander, N., & Gutiérrez, S. (2015). Trichodiene production in a *Trichoderma harzianum* Erg1-silenced strain provides evidence of the importance of the sterol biosynthetic pathway in inducing plant defense-related gene expression. *Molecular Plant-Microbe Interactions, 28*(11), 1181–1197. https://doi.org/10.1094/MPMI-06-15-0127-R

Manjunatha, S. V., Naik, M. K., Khan, M. F. R., & Goswami, R. S. (2013). Evaluation of bio-control agents for management of dry root rots of chickpea caused by *Macrophomina phaseolina*. *Crop Protection, 45*, 147–150. https://doi.org/10.1016/j.cropro.2012.09.003

Marquez, N., Giachero, M. L., Declerck, S., & Ducasse, D. A. (2021). *Macrophomina phaseolina*: General characteristics of pathogenicity and methods of control. *Frontiers of Plant Science, 12*, 634397. https://doi.org/10.3389/fpls.2021.634397

Martínez-Salgado, S. J., Andrade-Hoyos, P., Parraguirre Lezama, C., Rivera-Tapia, A., Luna-Cruz, A., & Romero-Arenas, O. (2021). Biological control of charcoal rot in peanut crop through strains of *Trichoderma* spp., in Puebla, Mexico. *Plants, 10*(12), 2630. https://doi.org/10.3390/plants10122630

Meena, B. R., Nagendran, K., Tripathi, A. N., Kumari, S., Sagar, V., Gupta, N., Rai, A. B., & Singh, B. (2018). First report of charcoal rot caused by *Macrophomina phaseolina* in *Basella alba* in India. *Plant Disease, 102*(8). https://doi.org/10.1094/PDIS-01-18-0152-PDN

Mendoza, J. L., Pérez, M. I., Prieto, J. M., Velásquez, J. D., Olivares, J. G., & Langarica, H. R. (2015). Antibiosis of *Trichoderma* spp. strains native to northeastern Mexico against the pathogenic fungus *Macrophomina phaseolina*. *Brazilian Journal of Microbiology, 46*(4), 1093–1101. https://doi.org/10.1590/S1517-838246420120177

Miethke, M. (2013). Molecular strategies of microbial iron assimilation: From high-affinity complexes to cofactor assembly systems. *The Royal Society of Chemistry, 5*, 15. https://doi.org/10.1039/c2mt20193c

Mishra, P. K., & Dantre, R. K. (2017). Evaluation of different biological agents used as seed treatment to manage charcoal rot of soybean. *BioScience Trends, 10*, 2784–2788.

Mmbaga, M. T., Mackasmiel, L. M., & Mrema, F. A. (2018). Evaluation of biological agents for control of macrophomina root rot and powdery mildew in flowering dogwood (*Cornus florida* L.). *HortScience, 53*, 1461–1466. https://doi.org/10.21273/HORTSCI13071-18

Morán-Diez, E., Hermosa, R., Ambrosino, P., Cardoza, R. E., Gutiérrez, S., Lorito, M., & Monte, E. (2009). The ThPG1 endopolygalacturonase is required for the trichoderma harzianum-plant beneficial interaction. *Molecular Plant-Microbe Interactions, 22*(8), 1021–1031. https://doi.org/10.1094/MPMI-22-8-1021

Monaco, C., Perelló, A., Alippi, H. E., & Pasquar, A. O. (1991). *Trichoderma* spp.: A biocontrol agent of *Fusarium* spp. and *Sclerotium rolfsii* by seed treatment. *Advances in Horticultural Science, 5*, 92–95.

Msikita, W., James, B., Wilkinson, H. T., & Juba, J. H. (1998). First report of *Macrophomina phaseolina* causing pre-harvest cassava root rot in Benin and Nigeria. *Plant Disease, 82*(12). https://doi.org/10.1094/pdis.1998.82.12.1402c

Mukherjee, P. K., Buensanteai, N., Moran-Diez, M. E., Druzhinina, I. S., & Kenerley, C. M. (2012). Functional analysis of non-ribosomal peptide synthetases (NRPSs) in *Trichoderma virens* reveals a polyketide synthase (PKS)/NRPS hybrid enzyme involved in the induced systemic resistance response in maize. *Microbiology, 158*(Pt 1), 155–165. https://doi.org/10.1099/mic.0.052159-0

Mukhopadhyay, R., & Pan, S. K. (2012). Isolation and selection of some antagonistic *Trichoderma* species from different new alluvial zones of Nadia district, West Bengal. *Journal of the Botanical Society of Bengal, 66*(2), 149–152.

Muller-Schärer, H., Bouchemousse, S., Litto, M., McEvoy, P. B., Roderick, G. K., & Sun, Y. (2020). How to better predict long-term benefits and risks in weed biocontrol: An evolutionary perspective. *Current Opinion in Insect Science, 38*, 84–91. https://doi.org/10.1016/j.cois.2020.02.006

Omann, M. R., Lehner, S., Rodríguez, C. E., Brunner, K., & Zeilinger, S. (2012). The seven-transmembrane receptor Gpr1 governs processes relevant for the antagonistic interaction of *Trichoderma atroviride* with its host. *Microbiology, 158*, 107. https://doi.org/10.1099/mic.0.052035-0

Pandey, K. K., & Upadhyay, J. P. (2000). Microbial population from rhizosphere and non-rhizosphere soil of pigeon pea. Screening for antagonist and mode of mycoparasitism. *Journal of Mycology and Plant Pathology, 30*, 7–10.

Patel, S., & Saraf, M. (2017). Biocontrol efficacy of *Trichoderma asperellum* MSST against tomato wilting by *Fusarium oxysporum* f. sp. *lycopersici*. *Archive of Phytopathology and Plant Protection, 50*(5e6), 228e–238e. https://doi.org/10.1080/03235408.2017.1287236

Pavlovic, S., Ristic, D., Aleksic, G., Milosevic, D., Stevic, T., & Starovic, M. (2015). The first report of *Macrophomina phaseolina* of immortelle (*Helichrysum italicum*) in Serbia. *Plant Disease, 99*(9). https://doi.org/10.1094/PDIS-12-14-1322-PDN

Prasad, R. D., & Rangeshwaran, R. (1998). A modified liquid medium for mass production of *Trichoderma*by fermentation process. In *Abstracts of national symposium and eco-friendly approaches in the management of pant diseases* (p. 26). Karnataka, India: Shimoga.

Rahman, M. T., Bhuiyan, M. K. A., Karim, M. A., & Rubayet, M. T. (2018). Screening of soybean resistance genotypes against *Fusarium oxysporum, Macrophomina phaseolina, Rhizoctonia solani* and *Sclerotium rolfsii*. *Research in Agriculture and Veterinary Science, 2*(3), 139–156.

Rahman, M. T., Rubayet, M. T., & Bhuiyan, M. K. A. (2020). Integrated management of rhizoctonia root rot disease of soybean caused by *Rhizoctonia solani*. *Nippon Journal of Environmental Science, 1*(7), 1018. https://doi.org/10.46266/njes.1018

Rahman, M. T., Rubayet, M. T., Khan, A. A., & Bhuiyan, M. K. A. (2021). Integrated management of charcoal rot disease of soybean caused by *Macrophomina phaseolina*. *Egyptian Journal of Agricultural Research, 99*(1), 10–19. https://doi.org/10.21608/ejar.2021.37644.1019

Rajani, P., Rajasekaran, C., Vasanthakumari, M. M., Olsson, S. B., Ravikanth, G., & Uma Shaanker, R. (2021). Inhibition of plant pathogenic fungi by endophytic *Trichoderma* spp. through mycoparasitism and volatile organic compounds. *Microbiological Research, 242*, 126595. https://doi.org/10.1016/j.micres.2020.126595

Ramezani, H. (2008). Biological control of root-rot of eggplant caused by *Macrophominaphaseolina*. *American-Eurasian Journal of Agricultural & Environmental Sciences, 4*(2), 218–220.

Ramezani, M., Shier, W. T., Abbas, H. K., Tonos, J. L., Baird, R. E., & Sciumbato, G. L. (2007). Soybean charcoal rot disease fungus *Macrophomina phaseolina* in Mississippi produces the phytotoxin (–)-botryodiplodin but no detectable phaseolinone. *Journal of Natural Products, 70*, 128–129. https://doi.org/10.1021/np060480t

Reino, J. L., Guerriero, R. F., Hernàndez-Galà, R., & Collado, I. G. (2008). Secondary metabolites from species of the biocontrol agent *Trichoderma*. *Phytochemistry Reviews, 7*(1), 89–123. https://doi.org/10.1007/s11101-006-9032-2

Rettinassababady, C., Ramadoss, N., & Ramanadane, T. (2000). Effect of culture filtrates of *Trichoderma viride* isolates on germination of black gram and sclerotia of *Macrophomina phaseolina*. *Seed Research, 28*, 193–195.

Roberti, R., Zakrisson, E., Flamigni, F., De Vero, L., & Cesari, A. (2002). Antagonistic fungi producing hydrolytic enzymes, active in degrading the cell wall of some foot rot pathogens (*Fusarium* spp.) of wheat. *Journal of Plant Diseases and Protection, 109*(1), 101–108. http://www.jstor.org/stable/43215425.

Rojas, E. C., Jensen, B., Jørgensen, H. J. L., Latz, M. A. C., Esteban, P., Ding, Y., & Collinge, D. B. (2020). Selection of fungal endophytes with biocontrol potential against *Fusarium* head blight in wheat. *Biological Control, 144*, 104222.

Rotblat, B., Enshell-seijffers, D., Gershoni, J. M., Schuster, S., & Avni, A. (2003). Identification of an essential component of the elicitation active site of EIX protein elicitor. *The Plant Journal, 32*, 1049–1055. https://doi.org/10.1046/j.1365-313X.2002.01490.x

Rubayet, T., & Bhuiyan, M. K. A. (2012). Compatibility of an isolate of *Trichoderma harzianum* with fungicides and organic amendments. *Bangladesh Journal of Plant Pathology, 28*(1 &2), 63–66.

Rubayet, M. T., & Bhuiyan, M. K. A. (2016). Integrated management of stem rot of potato caused by *Sclerotium rolfsii*. *Bangladesh Journal of Plant Pathology, 32*(1&2), 7–14.

Ruocco, M., Lanzuise, S., Lombardi, N., Woo, S. L., Vinale, F., Marra, R., Varlese, R., Manganiello, G., Pascale, A., Scala, V., Turrà, D., Scala, F., & Lorito, M. (2015). Multiple roles and effects of a novel *Trichoderma hydrophobin*. *Molecular Plant-Microbe Interactions, 28*(2), 167–179. https://doi.org/10.1094/MPMI-07-14-0194-R

Salas-Marina, M. A., Silva-Flores, M. A., Uresti-Rivera, E. E., Castro-Longoria, E., Herrera-Estrella, A., & Casas-Flores, S. (2011). Colonization of *Arabidopsis* roots by *Trichoderma atroviride* promotes growth and enhances systemic disease resistance through jasmonic acid/ethylene and salicylic acid pathways. *European Journal of Plant Pathology, 131*(1), 15–26. https://doi.org/10.1007/s10658-011-9782-6

Salvatore, M. M., Félix, C., Lima, F., Ferreira, V., Naviglio, D., Salvatore, F., Duarte, A. S., Alves, A., Andolfi, A., & Esteves, A. C. (2020). Secondary metabolites produced by *Macrophomina phaseolina* isolated from eucalyptus globulus. *Agriculture, 10*(3), 72. https://doi.org/10.3390/agriculture10030072

Sankar, P., & Jeyarajan, R. (1996). Biological control of sesamum root rot by seed treatment with *Trichoderma* spp. and *Bacillus subtilis*. *Indian Journal of Mycology & Plant Pathology, 26*, 147–153.

Sankar, P., & Sharma, R. C. (2001). Management of charcoal rot of maize with *Trichodermaviride*. *Indian Phytopathology, 54*(3), 390–391.

Saravanakumar, K., Fan, L., Fu, K., Yu, C., & Wang, M. (2016). Cellulase from *Trichoderma harzianum* interacts with roots and triggers induced systemic resistance to foliar disease in maize. *Scientific Reports, 6*, 35543. https://doi.org/10.1038/srep35543

Sarr, M. P., Ndiaye, M., Groenewald, J. Z., & Crous, P. W. (2014). Genetic diversity in *Macrophomina phaseolina* the causal agent of charcoal rot. *Phytopathologia Mediterranea, 53*, 250–268. https://doi.org/10.14601/Phytopathol_Mediterr13736

Sawant, I. S., & Sawant, S. D. (1996). A simple method for achieving high cfu of *Trichoderma harzianum* on organic wastes for field applications. *Indian Phytopathology, 9*, 185–187.

Sayed, M., Abdel-rahman, T., Ragab, A., & Abdellatif, A. (2019). Biocontrol of root-knot nematode *Meloidogyne incognita* by chitinolytic *Trichoderma* spp. *Egyptian Journal of Agronematology, 18*(1). https://doi.org/10.21608/ejaj.2019.52842

Score, A. J., Palfreyman, J. W., & White, N. A. (1997). Extracellular phenoloxidase and peroxidase enzyme production during interspecific fungal interactions. *International Biodeterioration & Biodegradation, 39*(2–3), 225–233. https://doi.org/10.1016/S0964-8305(97)00012-7

Shah, M. M., & Afiya, H. (2019). Introductory chapter: Identification and isolation of *Trichoderma* spp.—their significance in agriculture, human health, industrial and environmental application. In M. M. Shah, U. Sharif, & T. R. Buhari (Eds.), Trichoderma-*the most widely used fungicide*. https://doi.org/10.5772/INTECHOPEN.83528

Sharma, R. C., & Bhowmik, T. P. (1986). Estimation of yield losses in groundnut due to *Macrophomina phaseolina* (Tassi) Goid. *Indian Journal of Plant Pathology, 4*, 108–112.

Shoaib, A., Munir, M., Javaid, A., Awan, Z. A., & Rafiq, M. (2018). Anti-mycotic potential of *Trichoderma* spp. and leaf biomass of *Azadirachta indica* against the charcoal rot pathogen, *Macrophomina phaseolina* (Tassi) Goid in cowpea. *Egyptian Journal of Biological Pest Control, 28*, 26. https://doi.org/10.1186/s41938-018-0031-6

Shoresh, M., Gal-On, A., Leibman, D., & Chet, I. (2006). Characterization of a mitogen-activated protein kinase gene from cucumber required for trichoderma-conferred plant resistance. *Plant Physiology, 142*(3), 1169–1179. https://doi.org/10.1104/pp.106.082107

Shoresh, M., Harman, G. E., & Mastouri, F. (2010). Induced systemic resistance and plant responses to fungal biocontrol agents. *Annual Review of Phytopathology, 48*, 21–43. https://doi.org/10.1146/annurev-phyto-073009-114450

Short, G. E., Wyllie, T. D., & Ammon, V. D. (1978). Quantitative enumeration of *Macrophomina phaseolina* in soybean tissues. *Phytopathology, 68*, 736–741.

Short, G. E., Wyllie, T. D., & Bristow, P. R. (1980). Survival of *Macrophomina phaseolinain* soil and in residue of soybean. *Phytopathology, 70*, 13–17.

Simi, S. A., Jannat, R., Rubayet, M. T., & Bhuiyan, M. K. A. (2019). Efficacy of bio-fortified compost in controlling anthracnose disease of chilli caused by *Colletotrichum capsici* and improvement the crop production. *Scholars Academic Journal of Bioscience, 7*(12), 482–489.

Singh, R., Maurya, S., & Upadhyay, R. S. (2012). Antifungal potential of *Trichoderma* species against *Macrophomina phaseolina*. *Journal of Agricultural Technology, 8*(6), 1925–1933.

Singh, R., Ong-Abdullah, M., Low, E. T., Manaf, M. A., Rosli, R., Nookiah, R., Ooi, L. C. L., Ooi, S., Chan, K. L., Halim, M. A., Azizi, N., Nagappan, J., Bacher, B., Lakey, N., Smith, S. W., He, D., Hogan, M., Budiman, M. A., Lee, E. K., & "..."Sambanthamurthi, R. (2013). Oil palm genome sequence reveals divergence of interfertile species in old and new worlds. *Nature, 500*, 335–339.

Singh, A., Shukla, N., Kabadwal, B. C., Tewari, A. K., & Kumar, J. (2018). Review on plant Trichoderma-pathogen interaction. *International Journal of Current Microbiology and Applied Sciences, 7*(02), 2382e–2397e. https://doi.org/10.20546/ijcmas.2018.702.291

Singh, R., Sindhan, G., Parashar, R., & Hooda, I. (1998). Application of antagonists in relation to dry root rot and biochemical status of chickpea plants. *Plant Disease Research, 13*, 35–37.

Singh, M., Singh, J., Maurya, S., Kumar, S., Meena, A. K., Sharma, P., & Lakhran, L. (2021). Root rot disease incited by *Macrophomina phaseolina* in arid legumes and their management: A review. *Legume Research, LR, 4714*, 1–7.

Soni, K. K., Dodwal, V. S., & Jamaluddin. (1985). Charcoal rot and stem rot of Eucalyptus. *European Journal of Forest Pathology, 15*, 397–401.

Soufi, E., Safaie, N., Shahbazi, S., & Mojerlou, S. (2021). Biological control of sesame charcoal rot by gamma-irradiated mutants of *Trichoderma aureoviride* and their RAPD-PCR fingerprinting. *Journal of Crop Protection, 10*(4), 723–744.

Sousa, E. S., Melo, M. P., Pires, L. L., Silva, B. A., García, M. E. M., Sousa, E. M. J., Mota, J. M., & Beserra, J. E. A. (2017). First report of *Macrophomina phaseolina* causing charcoal rot in lima bean (*Phaseolus lunatus*) in Brazil. *Plant Disease, 101*(8). https://doi.org/10.1094/PDIS-11-16-1661-PDN

Sreedevi, B., Charitha, D. M., & Saigopal, D. V. R. (2011). Induction of defense enzymes in *Trichoderma harzianum* treated groundnut plants against *Macrophomina phaseolina*. *Journal of Biological Control, 25*(1), 67–73.

Sridharan, A. P., Sugitha, T., Karthikeyan, G., Nakkeeran, S., & Sivakumar, U. (2021). Metabolites of *Trichoderma longibrachiatum* EF5 inhibits soil borne pathogen, *Macrophomina phaseolina* by triggering amino sugar metabolism. *Microbial Pathogenesis, 150*, 104714. https://doi.org/10.1016/J.MICPATH.2020.104714

Sriram, S., Panchakshari, K., Maheshwarappa, R., & Savitha, J. (2011). Extended shelf-life of liquid fermentation derived talc formulations of *Trichoderma harzianum* with the addition of glycerol in the production medium. *Crop Protection, 30*(10), 1334–1339.

Srivastava, M., Kumar, V., Shahid, M., Pandey, S., & Singh, A. (2016). *Trichoderma*—a potential and effective bio fungicide and alternative source against notable phytopathogens: A review. *African Journal of Agricultural Research, 11*(5), 310–316.

Srivastava, R. K., Singh, R. K., Kumar, N., & Singh, S. P. (2010). Management of *Macrophomina* disease complex in jute (*Corchorus olitorius*) by *Trichoderma viride*. *Journal of Biological Control, 24*, 77–79.

Swehla, A., Pandey, A. K., & Nair, R. M. (2020). Bioactivity of *Trichoderma harzianum* isolates against the fungal root rot pathogens with special reference to *Macrophomina phaseolina* causing dry root rot of mungbean. *Indian Phytopathology, 73*, 787–792. https://doi.org/10.1007/s42360-020-00288-x

Taylor, A. G., Min, T. G., Harman, G. E., & Jin, X. (1991). Liquid coating formulations for the application of biological seed treatments of *Trichoderma harzianum*. *Biological Control, 1*, 16–22.

Thangavelu, R., & Gopi, M. (2015). Field suppression of Fusarium wilt disease in banana by the combined application of native endophytic and rhizospheric bacterial isolates possessing multiple functions. *Phytopathologia Mediterranea, 54*(2), 241e–252e.

Thangavelu, R., & Mustaffa, M. (2010). A potential isolate of *Trichoderma viride* NRCB1 and its mass production for the effective management of fusarium wilt disease in banana. *Tree and Forestry Science and Biotechnology, 4*(2), 76–84.

Thangavelu, R., Palaniswami, A., & Velazhahan, R. (2004). Mass production of *Trichodermaharzianum* for managing fusarium wilt of banana. *Agriculture, Ecosystems & Environment, 103*(1). https://doi.org/10.1016/j.agee.2003.09.026

Torres, M. J., Brandan, C. P., Petroselli, G., Erra-Balsells, R., & Audisio, M. C. (2016). Antagonistic effects of *Bacillus subtilis* subsp. *subtilis* and *B. amyloliquefaciens* against *Macrophomina phaseolina*: SEM study of fungal changes and UV-MALDI-TOF MS analysis of their bioactive compounds. *Microbiology Research, 182*, 31–39. https://doi.org/10.1016/j.micres.2015.09.005

Türkkan, M., Benli, H.İ., Yõlmaz, Ö., Özer, G., Yaman, M., Şahin, N., & Erper, I. (2020). First report of charcoal rot caused by *Macrophomina phaseolina* on kiwifruit in Turkey. *Journal of Plant Pathology, 102*, 535. https://doi.org/10.1007/s42161-019-00428-x

Ullah, M. H., Khan, M. A., Sahi, S. T., & Habib, A. (2011). Evaluation of antagonistic fungi against charcoal rot of sunflower caused by *Macrophomima phaseolina* (Tassi) Goid. *Journal of Applied Sciences Research, 2*(12), 1175–1184.

Vasebi, Y., Safaie, N., & Alizadeh, A. (2013). Biological control of soybean charcoal root rot disease using bacterial and fungal antagonists in vitro and greenhouse condition. *Journal of Crop Protection, 2*, 139–150.

Vey, A., Hoagland, R. E., & Butt, T. M. (2001). Toxic metabolites of fungal biocontrol agents. In T. M. Butt, C. Jackson, & N. Magan (Eds.), *Fungi as biocontrol agents: Progress, problems and potential* (pp. 311–346). Bristol: UK: CAB International.

Vinale, F., Flematti, G., Sivasithamparam, K., Lorito, M., Marra, R., Skelton, B. W., & Ghisalberti, E. L. (2009). Harzianic acid, an antifungal and plant growth promoting metabolite from *Trichoderma harzianum. Journal of Natural Products, 72*(11), 2032–2035. https://doi.org/10.1021/np900548p

Vinale, F., Manganiello, G., Nigro, M., Mazzei, P., Piccolo, A., Pascale, A., Ruocco, M., Marra, R., Lombardi, N., Lanzuise, S., Varlese, R., Cavallo, P., Lorito, M., & Woo, S. (2014). A novel fungal metabolite with beneficial properties for agricultural applications. *Molecules, 19*, 9760–9772. https://doi.org/10.3390/molecules19079760

Vinale, F., Sivasithamparam, K., Ghisalberti, E. L., Marra, R., Barbetti, M. J., Li, H., Woo, S. L., & Lorito, M. (2008). A novel role for *Trichoderma* secondary metabolites in the interactions with plants. *Physiological and Molecular Plant Pathology, 72*(1–3). https://doi.org/10.1016/j.pmpp.2008.05.005

Vinothini, K., Renganathan, P., Balabaskar, P., & Sivakumar, T. (2020). *In vitro* efficacy various isolates of *Trichoderma Viride* and *Pseudomonas Fluorescens* against *Macrophomina phaseolina* causing sesame root rot. *Plant Archives, 20*(1), 1163–1168.

Vos, C. M., De Cremer, K., Cammue, B. P., & De Coninck, B. (2015). The toolbox of *Trichoderma* spp. in the biocontrol of *Botrytis cinerea* disease. *Molecular Plant Pathology, 16*(4), 400–412. https://doi.org/10.1111/mpp.12189

Waghunde, R. R., Shelake, M. R., Rahul, & Ambalal, N. S. (2016). *Trichoderma*: A significant fungus for agriculture and environment. *African Journal of Agricultural Research, 11*, 1952–1965. https://doi.org/10.5897/AJAR2015.10584

Wang, H., & Jones, R. W. (1995). A unique endoglucanase-encoding gene cloned from the phytopathogenic fungus *Macrophomina phaseolina. Applied and Environmental Microbiology, 61*, 2004–2006.

Yang, X. B., & Navi, S. S. (2005). First Report of charcoal rot epidemics caused by *Macrophominaphaseolina* in soybean in Iowa. *Plant Disease, 89*(5). https://doi.org/10.1094/pd-89-0526b

Yedidia, I., Benhamou, N., & Chet, I. (1999). Induction of defense responses in cucumber plants (*Cucumis sativus* L.) by the Biocontrol agent *Trichoderma harzianum. Applied and Environmental Microbiology, 65*(3), 1061–1070. https://doi.org/10.1128/aem.65.3.1061-1070.1999

Yedidia, I., Shoresh, M., Kerem, Z., Benhamou, N., Kapulnik, Y., & Chet, I. (2003). Concomitant induction of systemic resistance to *Pseudomonas syringae* pv. *lachrymans* in cucumber by *Trichoderma asperellum* (T203) and accumulation of phytoalexins. *Applied and Environmental Microbiology, 69*, 7343–7353.

Yu, C., Dou, K., Wang, S., Wu, Q., Ni, M., Zhang, T., Lu, Z., Tang, J., & Chen, J. (2020). Elicitor hydrophobin hyd1 interacts with ubiquilin1-like to induce maize systemic resistance. *Journal of Integrative Plant Biology, 62*, 509–526. https://doi.org/10.1111/jipb.12796

Zeilinger, S., Gruber, S., Bansal, R., & Mukherjee, P. K. (2016). Secondary metabolism in *Trichoderma*—chemistry meets genomics. *Fungal Biology Reviews, 30*(2), 74–90. https://doi.org/10.1016/J.FBR.2016.05.001

Zhang, S., Gan, Y., Xu, B., & Xue, Y. (2014). The parasitic and lethal effects of *Trichoderma longibrachiatum* against *Heterodera avenae. Biological Control, 72*, 1–8. https://doi.org/10.1016/j.biocontrol.2014.01.009

Zhang, S., Xu, B., Zhang, J., & Gan, Y. (2018). Identification of the antifungal activity of *Trichoderma longibrachiatum* T6 and assessment of bioactive substances in controlling phytopathogens. *Pesticide Biochemistry and Physiology, 147*, 59–66. https://doi.org/10.1016/j.pestbp.2018.02.006

Zhang, J. Q., Zhu, Z. D., Duan, C. X., Wang, X. M., & Li, H. J. (2011). First report of charcoal rot caused by *Macrophomina phaseolina* on mungbean in China. *Plant Disease, 95*(7), 872. https://doi.org/10.1094/PDIS-01-11-0010

Zhao, L., Xu, Y., & Lai, X. (2018). Antagonistic endophytic bacteria associated with nodules of soybean (*Glycine max* L.) and plant growth-promoting properties. *Brazilian Journal of Microbiology, 49*(2), 269–278. https://doi.org/10.1016/j.bjm.2017.06.007

Zhu, Z. X., & Zhuang, W. Y. (2015). Three new species of *Trichoderma* with hyaline ascospores from China. *Mycologia, 107*(2), 328–345. https://doi.org/10.3852/14-141

Chapter 19

Integrated pest management of the disease caused by *Macrophomina phaseolina*

Ankita Chatterjee
Department of Biotechnology, School of Applied Sciences, REVA University, Bangalore, Karnataka, India

19.1 Introduction

Macrophomina phaseolina, a potent fungal pathogenic fungus, dwells in soil and widely present across the world. The fungus has been infecting almost 500 species of plants belonging to more than 100 families. Most often the pathogenic fungus is observed to causes diseases like charcoal rot, stem rot, root rot, seedling blight, etc. (de Sousa Linhares et al., 2020; Ghosh et al., 2018). The effect of *M. phaseolina* has been observed to be devastating on the jute plants. The fungus targets any parts of the plants during any growth phase ranging from seed germination to harvesting (Das et al., 2015; Marquez et al., 2021). The stem canker disease caused by the pathogen was found to be one of the major soyabean plants effecting diseases which cause a huge economic loss in countries like India, Argentina, China, the United States, Bolivia, and many more. In recent studies, the *Macrophomina* stem canker has also been reported to occur in pigeon pea plants in India (Swamy et al., 2018). The fungus is heterogenous in nature when host specificity is concerned, i.e., they are able to cause infection within both monocots and dicots. It can be transmitted from one host to another via infected seeds and seed to seedling transmission. *M. phaseolina* appears to brown or black or gray on different growth medium and with time the colony color becomes darker. Aerial mycelium formation is one of prominent culture characteristic feature of the fungus. The hyphae are observed to septate. The vegetative hyphae are featured with the presence of barrel-shaped cells separated by septum. Branching of the mycelium occurs at right angle or acute angles of the parent hyphae. The hyphal appearance is observed to be translucent in the initial growth phase, but later with time, it turns to dark coloration. Presence of sclerotia, usually with black pigmentation, is observed within the hyphae or embedded on the agar surface or within the agar. On observing the reverse side of the agar plate, dark brown and black sclerotia are noticed (Abed-Ashtiani et al., 2018). The fungus is reported to produce spherical or flattened pycnidia within diameter range of 100–200 μm. The pycnidia basically is the asexual fruiting body of *M. phaseolina*. In the initial phase, pycnidia are embedded into the tissues of the host plant but later as it grows and protrudes from the tissues. Pycnidia contains circular ostiole which is located in the central surrounded by darkish brown thick-walled cells (Gaetán et al., 2006). The innermost layer of pycnidia has a translucent appearance, whereas the outer layer is darker in color. The conidiophores appear to be branched and septate with characteristic translucent and ovoid features. The tip of the conidiophores is round and surrounded by a thin membranous layer. *M. phaseolina* secretes different hydrolytic enzymes, most importantly, the carbohydrate active enzymes (CAZymes), which help in degradation of the cell wall of plants further invading in the plant tissues. The genome of *M. phaseolina* encodes 362 CAZymes, which comprises of carbohydrate esterases (65), carbohydrate binding molecules (6), glycosyl transferases (56), glycoside hydrolases (219), and polysaccharide lyases (16). The enzymes remain dispersed throughout the genome, rather than clustering to each other. The carbohydrate degradation characteristic of the pathogen is due to presence of high number of glycoside hydrolases. Cellulose breakdown is carried out due to the presence of endoglucanases, exocellobiohydrolases, and beta-glucosidases (Bandara et al., 2018).

This fungal pathogen consists of a strong group of enzymes, such as chloroperoxidases, haloperoxidases, hemeperoxidases, galactose oxidases, laccase, and lignin peroxidases for lignin degradation. Laccase concentration in

Macrophomina Phaseolina. https://doi.org/10.1016/B978-0-443-15443-0.00002-4

M. phaseolina is comparatively higher than other fungal phytopathogens. *M. phaseolina* is reported to the second fungal strain to show lignin peroxidase activity (Ghosh et al., 2018; Martinez et al., 2004). Oxidoreductases, like, alcohol oxidases, and cellobiose dehydrogenases, play important roles in degradation of lignocellulose. Involvement of distinguishable genes responsible for pathogen host interaction enhances the pathogenicity of the fungus. The genes function during a range of events occurring throughout the process of pathogenesis, beginning from attachment and adhesion of the pathogen to the host followed by signal transduction, breakdown of cell wall, and biosynthesis of purine as well as mycotoxin. The pathogen is protected from the host's metabolites by the help of ATP-binding cassette. Detoxification genes responsible for encoding cytochrome p450, Cofprotein, and superoxide dismutase are present in the pathogen. The fungus is also observed to contain enzymes for synthesis of polyketide antibiotic as well as to show resistance toward tetracycline. All the genes, encoding for the mentioned proteins and enzymes together, act for the pathogenicity of the fungus (Kaur et al., 2012; Morschhäuser, 2010).

The control of *M. phaseolina* using chemical methods has been explored greatly. However, the imbalance caused in the environment by the application of chemicals is a concern for maintaining the soil fertility. Thus, to maintain ecological balance and inhibit or control the pathogens, integrated pest management (IPM) is implemented in several cases. IPM is an aggregation of chemical, biological, and physical techniques in enhancing the crop quality while targeting only the harmful pests (Deguine et al., 2021).

19.2 Mechanism of action of IPM

Over the years, chemical pesticides are over exposed to the crops and agricultural land for protection of crops. The extensive application of pesticides is linked to intolerable environmental and health impacts as well as imbalance in the crop management. The exposure of chemicals for controlling insects and pests has now resulted in a wide range of herbicide and pesticide resistant species. With every passing year, the active ingredients for management of crops are decreasing because of increased rate of resistance. Thus, to curb the situation, integrated pest management is suggested as an alternative to the application of chemicals. In the process of integrated pest management, the chemical and biological system works in combination to each other rather than individual application. The idea of integrated pest management was proposed by entomologists due to application of insecticides and pesticides that impact negatively the environment and crop protection. Fig. 19.1 represents the general principles of integrated pest management which are followed in its implementation.

The application of biological control in combination with host plant resistance might be effective in controlling pest management and rot of the crops. The induction of partial or complete resistance in the crops against the pathogen need to be examined and explored thoroughly for implementation of integrated pest management in large scale (Table 19.1).

19.3 Success and failure of IPM

The biodiverse ecological conditions aid in creating the stability in the nature. In agriculture, the theory of diversity-stability has been proven to be of much importance. It has been scientifically proven that pest management has not only enhanced the multispecies cropping arrangement but also increased the habitat of noncrop farm resulting in growth of beneficial crops and species. Several ways are followed to utilize crops and noncrops for contribution in management of pest. Trap cropping has been reported to be an advantageous tool for suppressing pest activity. In this case, the main crop is protected by attracting the pests toward a more acceptable host plant. Once the pests concentrate on the trap crop, appropriate control mechanism is applied on the pests for further reducing their activity on the main crop. Thus, implementation of this technique might require around 10%–15% of the crop area for plantation of the trap crop as well (Holden et al., 2012). IPM involves regular protection measures for pests and diseases. In case of small-scale farmers, the identification of the pest is successfully carried out, but the problem lies in recognition of the diseases and their diagnosis. Thus, in certain cases, it is advised for timely intervention of the farmers, which might be helpful in diagnosis of the disease at earlier stage. Conventionally, integrated pest management is designed to be fit for application and implementation of any scale of agriculture or crop production; however, in practice, it has been observed to be much suitable for large-scale production (Chouinard et al., 2016). Though there are various opportunities favoring the implementation of integrated pest management, the greatest challenge faced is the lack of knowledge as well as suitable technical assistance among the farmers. Though, agricultural workshops and trainings are often conducted by various organizations, yet it has been noticed that the workshops include only the preliminary information on insecticides and pesticides focusing on pest and weed management. However, lack of practice and experience among the farmers might be one of the reasons for misuse of fertilizers and pesticides. Limitation of IPM-related research reflects to be one of the serious barriers in successful

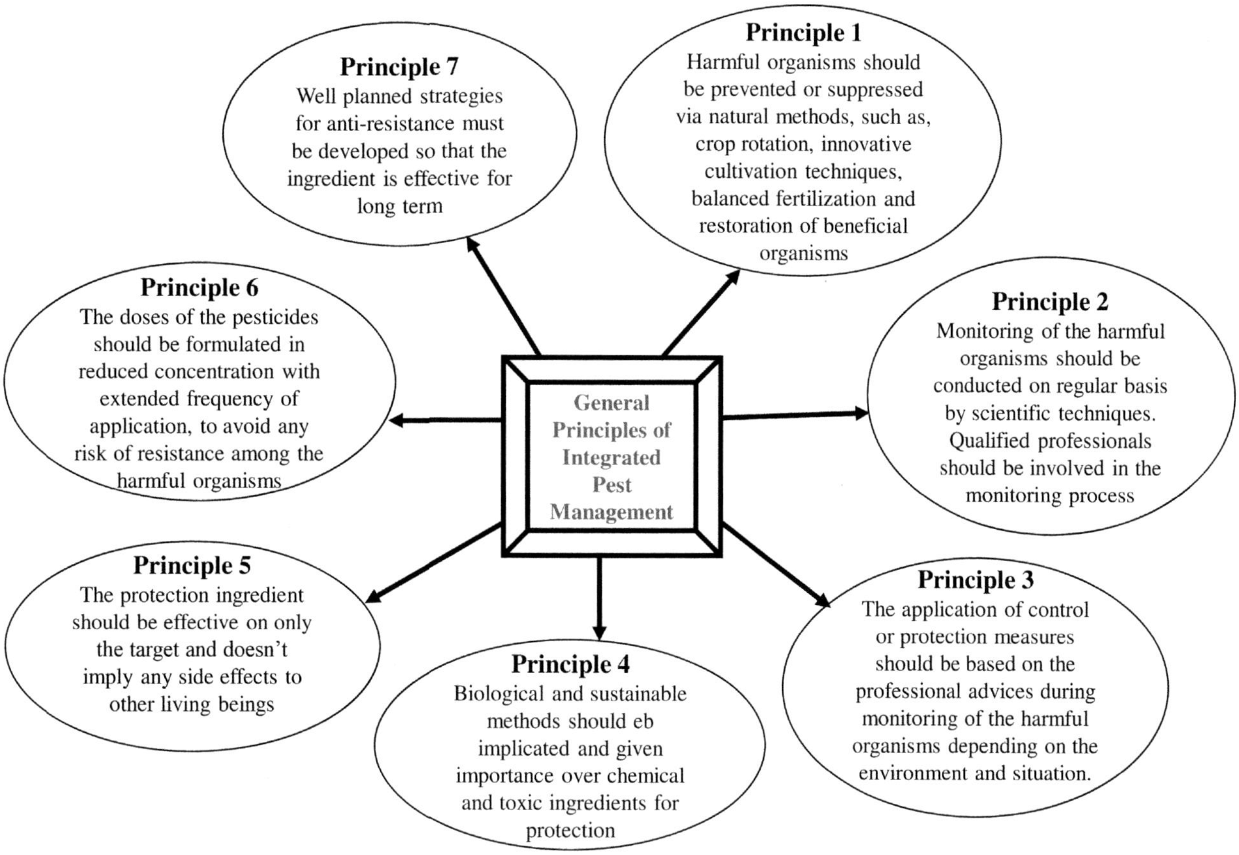

FIGURE 19.1 General principle of integrated pest management (Barzman et al., 2015).

TABLE 19.1 Components of integrated pest management (IPM).

Components	Description
Pest management	Pest management strategies should be planned efficiently so that the process of inhibiting or elimination of the pests does not cause any economic loss. Categories of pest management strategies include: • host plant resistance • biological control • behavioral regulations • cultural control • physical control • microbial control
Resources and applicable knowledge	Before initiation of an integrated pest management process, proper knowledge regarding the pests and their specific characteristics should be known. Available resources for damages done to the crops by the pest and their management must be studied in detail before deciding on implementation of an integrated pest management method.
Planning, organization and implementation	Manual collection of available information and segregating them followed by planning for the implementation is important. Maintaining a record about the damage caused by pest and their treatment depending on specific seasons and other environmental factors are necessary.
Communication and outreach	Communication, research, and outreach of the integrated pest management model is one of the most important components of IMP. This helps in conveying the information accurately followed by the anticipation of the damage so that proper strategies can be framed.

implementation of IPM. Small-scale farmers limit the implementation of pest management strategies due to the cost as well as availability of pack sizes according to the scale. In case of small-scale farmers, they are unable to negotiate with the agricultural suppliers like the large-scale producers. This can be dealt if associations of individual farmers are created to expand the bargaining limit (Grasswitz, 2019). Though, the present scenario shows IPM not to be suitable for small scale farmers, it might be improved if easier implementation of the process and proper knowledge are provided to the farmers.

19.4 Difficulties arising in IPM

For successful implementation of IPM, development of new strategies is of much importance. Although, concept of IPM has been circulated across the globe, and farmers are also aware about it yet the compulsory implementation of IPM still has certain drawbacks. Pesticide resistance is a major concern observed in agricultural practices these days. Though in IPM principles, it is clearly advised that doses of the pesticides can be altered depending on the development of resistance, but in practical field, it is not easy to carry out with every incidence. Reduction of doses is encouraged economically. Farmers are encouraged to use combination of active ingredients in reduced doses to fight the resistance development. According to IPM strategies, professional users are the ones to decide whether to apply measures for protecting the plants and the right time to do so is based on the monitoring. In some cases, this rule is followed, particularly when cereals and seeds are affected by the infectious agents. The challenge is faced during seed dressing because the prediction of seed germination time is not always appropriate, which leads to the drug administration at a wrong time. Apart from these, implementation of IPM requires establishment of a threshold level, which is extremely time consuming as well as expensive. In many cases, threshold levels are unavailable due to unavailability of enough experimental data from agricultural processes. IPM strategies focus on implementation of biological ingredients than chemicals for protection of crops. Application of biological control agents is not observed in many areas. This might be due to the reason that biological originated ingredients require more knowledge and efforts rather than application of chemical fertilizers and pesticides (Matyjaszczyk, 2019).

19.5 Control measures

19.5.1 Chemicals used for control of *M. phaseolina* infection

Chemical ways of treating *M. phaseolina* have been practiced for ages along with the fertilizers. Chemicals are applied for fumigation of soil before and after crop plantation. In addition, seed treatment with chemicals is also effective in the management of the diseases. Earlier, it has been reported that captan, benomyl, thiram, thiobendazole, thiophanatemethyl, and triforine showed effective decrease in viable microsclerotia formation when mixed with soil before plantation of crops (Ilyas et al., 1976). Application of benomyl acts in reducing the numbers of microsclerotia to a great extent. However, phytotoxicity exhibited by benomyl and thiobendazole is a major drawback as observed in regular usage for controlling the pathogen (Ilyas et al., 1975). According to Kittle and Gray (1982), sodium methyldithiocarbamate is a proven chemical agent for soil fumigation of jute crop plantation. Methyl bromide also enhances the crop yield and acts in reduction of microsclerotia formation (Watanabe et al., 1970). Application of chemical fungicides in treatment during seedling phase is effective in reduction of *M. phaseolina* spread across the plantation. Thiophanate methyl and furcarbanil at 1000 ppm concentration are effective in controlling the growth of *M. phaseolina* (Vir et al., 1972). It has also been observed that captafol and mancozeb efficiently inhibit the infection of the pathogen in root and seedling of jute (Singh et al., 1990). Though application of chemicals has been effective in many cases for controlling the widespread of the pathogen, yet their application is not advisable since they are nonprofitable and difficult to execute in large scale (Chamorro et al., 2016; Reznikov et al., 2016).

19.5.2 Biological control for controlling *Macrophomina phaseolina*

Biocontrol agents were also considered as an effective measure to reduce or inhibit the growth of the pathogenic *M. phaseolina* (Zaki & Mahmood, 1993). *Bacillus subtilis, Pseudomonas fluorescens, Klebsiella pneumoniae, Trichoderma harzanium,* and *T. viride* are few examples of bacteria and fungi which are capable of reducing the sclerotia germination (Dey et al., 2019). *Trichoderma* species were found effective in seed treatment of moong, sunflowers, and egg plants in inhibiting the growth of the pathogen (Ramezani, 2008). Plant growth promoting nitrogen fixing bacteria, e.g., *Bradyrhizobium* sp. and *Rhizobium meliloti,* act against the pathogen and aid in growth of groundnut plants (Deshwal et al., 2003). *Aspergillus* sp. has been reported to be effective against charcoal rot disease in black gram seeds (Eswaran & Mishra, 2004). Arbuscular mycorrhizal (AM) fungi are known to establish one of the primitive symbiotic relationships in the environment. AM fungi increases crop quality by inhibiting growth of a number of pests and insects by about 70% (Brundrett & Tendersoo, 2018). It has been observed that AM fungi are able to reduce root infection and the severity of disease caused by *M. phaseolina* (Marquez et al., 2018). Along with the reduction of spreading of disease, it has been reported that the response mechanism of the plants is enhanced due to the activity of the AM fungi (Spagnoletti et al., 2017). According to recent studies, AM fungi when combined with plant growth promoting yeast improve disease response ability of the sunflower plants (Nafady et al., 2019).

Certain microbial species, such as *Aspergillus*, *Trichoderma,* and *Gliocladium,* which are able to produce various volatile and nonvolatile metabolites effectively, induce plant growth by reducing the microsclerotia production. *Trichoderma* metabolites extracted using chloroform contain gliotoxin, which is responsible for antiphytopathogenic activity. Apart from *Trichoderma*, *Gliocladium virens* also produces certain antifungal metabolites that inhibit the microsclerotium. These microbes and their metabolites could be used for formulation of biopesticides for application in field trials. Apart from the biopesticides, application of agrochemicals and conventional breeding techniques and involvement of molecular biology, transgenic technology, and genetics are also effective mechanisms. Gene isolation and transformation of genetic materials in the development of transgenic crops resistant to fungal diseases is one of the most promising mechanisms of controlling the rots caused by the pathogen.

In recent days, major primary crops like wheat, jute, rice, and legumes are transformed genetically through genetic engineering. Genetic engineering helps in exploring the molecular characteristics of the protection mechanism of the plants, which further helps in developing new strategies for protection of plants against the diseases (Punja et al., 2007). The introduction of serine carboxypeptidase like genes and lectin genes are experimented to induce pathogen recognition ability in the plants which help them in initial defense mechanism and enhance cell homeostasis. AM fungi, when combined with phosphate fertilizers and treated for diseased plants, has shown the reduced severity of disease in the plants (Spagnoletti et al., 2020).

19.6 Integrated pest management for control of *M. phaseolina*

Effective control of *M. phaseolina* can be conducted by using the integrated pest management system. The process has been reported to be efficient and safe for the environmental system. The basic concept of IPM is application of the complete information regarding the life cycle and mode of interaction of the pests in the ecosystem, thereby searching for financially feasible means to reduce the hazards caused. The IMP is focused on managing the crops fixing the target as agricultural ecosystem sustainability and maintaining the biological equilibrium. IPM was introduced for disease control depending on the economic, ecological, and social principles. *T. viride, T. harzanium* and *P. fluorescens* have the ability to enhance the growth of chickpea plant and control the dry root rot caused by *M. phaseolina* (Manjunatha et al., 2013). *T. harzanium* integrated with mustard oil cake has been proven to have treatment ability against *M. phaseolina* invasion in soybean causing charcoal rot in the plants (Rahman et al., 2021). Methyl jasmonate along with ethylene has been reported to be an effective mechanism to induce partial resistance against the pathogen in the leguminous plant *Medicago truncatula.* Molecular studies of the interaction between host and pathogen showed that upon treating with the mentioned combination, the genes responsible for synthesis of isoflavonoids and flavonoids were up-regulated in shoots of the host and thus could help the host in partially resisting the charcoal rot disease (Gaige et al., 2010).

19.7 Economic aspects of IPM

Integrated approach of pest management enhances the cost effectiveness of the agricultural products. Initiation of IPM practices would effectively play a role in the economic growth of the farmers. A study revealed that introduction of IPM in four districts of Bangladesh has reduced the production cost of the agricultural process. Implementation of IPM significantly decreased the rate of applying pesticide during the process indicating that the farmers are economically gaining profit as well as producing safe quality products (Rahman et al., 2018). Apart from the economic excellences of implementing IPM, it also significantly reduces the risk of soil and ground water contamination.

19.8 Conclusion

M. phaseolina causes a wide range of seed-borne diseases. The organism can be found over a wide geographical distribution and can infect a large variety of plants. In this chapter, complete focus was provided on the disease cycle and the present control measures against the growth of *M. phaseolina.* Application of environment friendly biopesticide would be an effective way of controlling the spread of the diseases. However, for applying genetic engineering techniques in the control measure of the pathogen, in depth studies are required regarding the bio-safety issues of utilizing the transgenic crops.

References

Abed-Ashtiani, F., Narmani, A., & Arzanlou, M. (2018). *Macrophomina phaseolina* associated with grapevine decline in Iran. *Phytopathologia Mediterranea, 57*(1), 107–111.

Bandara, Y. M. A. Y., Weerasooriya, D. K., Liu, S., & Little, C. R. (2018). The necrotrophic fungus *Macrophomina phaseolina* promotes charcoal rot susceptibility in grain sorghum through induced host cell-wall-degrading enzymes. *Phytopathology, 108*(8), 948–956.

Barzman, M., Bàrberi, P., Birch, A. N. E., Boonekamp, P., Dachbrodt-Saaydeh, S., Graf, B., & Sattin, M. (2015). Eight principles of integrated pest management. *Agronomy for Sustainable Development, 35*(4), 1199–1215.

Brundrett, M. C., & Tedersoo, L. (2018). Evolutionary history of mycorrhizal symbioses and global host plant diversity. *New Phytologist, 220*(4), 1108–1115.

Chamorro, M., Seijo, T. E., Noling, J. C., De los Santos, B., & Peres, N. A. (2016). Efficacy of fumigant treatments and inoculum placement on control of *Macrophomina phaseolina* in strawberry beds. *Crop Protection, 90*, 163–169.

Chouinard, G., Firlej, A., & Cormier, D. (2016). Going beyond sprays and killing agents: Exclusion, sterilization and disruption for insect pest control in pome and stone fruit orchards. *Scientia Horticulturae, 208*, 13–27.

Das, I. K., Kishore, B. B., Srivastava, A. K., Kumar, S., & Arora, D. K. (2015). Ecology, biology and management of *Macrophomina phaseolina*: An overview. *Agriculturally Important Microorganisms, 1*, 193–206.

de Sousa Linhares, C. M., Ambrósio, M. M. Q., Castro, G., Torres, S. B., Esteras, C., de Sousa Nunes, G. H., & Picó, B. (2020). Effect of temperature on disease severity of charcoal rot of melons caused by *Macrophomina phaseolina*: Implications for selection of resistance sources. *European Journal of Plant Pathology, 158*(2), 431–441.

Deguine, J. P., Aubertot, J. N., Flor, R. J., Lescourret, F., Wyckhuys, K. A., & Ratnadass, A. (2021). Integrated pest management: Good intentions, hard realities. A review. *Agronomy for Sustainable Development, 41*(3), 1–35.

Deshwal, V. K., Dubey, R. C., & Maheshwari, D. K. (2003). Isolation of plant growth-promoting strains of *Bradyrhizobium* (*Arachis*) sp. with biocontrol potential against *Macrophomina phaseolina* causing charcoal rot of peanut. *Current Science, 84*(3), 443–448.

Dey, S., Dutta, P., & Majumdar, S. (2019). Biological control of *Macrophomina phaseolina* in *Vigna mungo* L. by endophytic *Klebsiella pneumoniae* HR1. *Jordan Journal of Biological Sciences, 12*(2).

Eswaran, A., & Mishra, R. (2004). Mycoparasitic potential and variability of *Aspergillus* spp. to manage dry root rot of blackgram (*Vigna mungo* L.). *Indian Journal of Plant Protection, 32*(2), 89–92.

Gaetán, S. A., Fernandez, L., & Madia, M. (2006). Occurrence of charcoal rot caused by *Macrophomina phaseolina* on canola in Argentina. *Plant Disease, 90*(4), 524–524.

Gaige, A. R., Ayella, A., & Shuai, B. (2010). Methyl jasmonate and ethylene induce partial resistance in *Medicago truncatula* against the charcoal rot pathogen *Macrophomina phaseolina*. *Physiological and Molecular Plant Pathology, 74*(5–6), 412–418.

Ghosh, T., Biswas, M. K., Guin, C., & Roy, P. (2018). A review on characterization, therapeutic approaches and pathogenesis of *Macrophomina phaseolina*. *Plant Cell Biotechnology and Molecular Biology, 19*, 72–84.

Grasswitz, T. R. (2019). Integrated pest management (IPM) for small-scale farms in developed economies: Challenges and opportunities. *Insects, 10*(6), 179.

Holden, M. H., Ellner, S. P., Lee, D. H., Nyrop, J. P., & Sanderson, J. P. (2012). Designing an effective trap cropping strategy: The effects of attraction, retention and plant spatial distribution. *Journal of Applied Ecology, 49*(3), 715–722.

Ilyas, M. B., Ellis, M. A., & Sinclair, J. B. (1975). Evaluation of soil fungicides for control of charcoal rot of soybeans. *Plant Disease Reporter, 59*(4), 360–364.

Ilyas, M. B., Ellis, M. A., & Sinclair, J. B. (1976). Effect of soil fungicides on *Macrophomina phaseolina* sclerotium viability in soil and in soybean stem pieces. *Phytopathology, 66*(3), 355–359.

Kaur, S., Dhillon, G. S., Brar, S. K., Vallad, G. E., Chand, R., & Chauhan, V. B. (2012). Emerging phytopathogen Macrophomina phaseolina: biology, economic importance and current diagnostic trends. *Critical Reviews in Microbiology, 38*(2), 136–151.

Kittle, D. R., & Gray, L. E. (1982). Response of soybeans and soybean pathogens to soil fumigation and foliar fungicide sprays. *Plant Diseases (USA), 66*, 213–215.

Manjunatha, S. V., Naik, M. K., Khan, M. F. R., & Goswami, R. S. (2013). Evaluation of bio-control agents for management of dry root rot of chickpea caused by *Macrophomina phaseolina*. *Crop Protection, 45*, 147–150.

Marquez, N., Giachero, M. L., Declerck, S., & Ducasse, D. A. (2021). *Macrophomina phaseolina*: General characteristics of pathogenicity and methods of control. *Frontiers of Plant Science, 12*, 634397.

Marquez, N., Giachero, M. L., Gallou, A., Debat, H. J., Cranenbrouck, S., Di Rienzo, J. A., … Declerck, S. (2018). Transcriptional changes in mycorrhizal and nonmycorrhizal soybean plants upon infection with the fungal pathogen *Macrophomina phaseolina*. *Molecular Plant-Microbe Interactions, 31*(8), 842–855.

Martinez, D., Larrondo, L. F., Putnam, N., Gelpke, M. D. S., Huang, K., Chapman, J., & Rokhsar, D. (2004). Genome sequence of the lignocellulose degrading fungus *Phanerochaete chrysosporium* strain RP78. *Nature Biotechnology, 22*(6), 695–700.

Matyjaszczyk, E. (2019). Problems of implementing compulsory integrated pest management. *Pest Management Science, 75*(8), 2063–2067.

Morschhäuser, J. (2010). Regulation of multidrug resistance in pathogenic fungi. *Fungal Genetics and Biology, 47*(2), 94–106.

Nafady, N. A., Hashem, M., Hassan, E. A., Ahmed, H. A., & Alamri, S. A. (2019). The combined effect of arbuscular mycorrhizae and plant-growth-promoting yeast improves sunflower defense against *Macrophomina phaseolina* diseases. *Biological Control, 138*, 104049.

Punja, Z. K., De Boer, S. H., & Sanfaçon, H. (Eds.). (2007). *Biotechnology and plant disease management*. U.K: CAB International.

Rahman, M. S., Norton, G. W., & Rashid, M. H. A. (2018). Economic impacts of integrated pest management on vegetables production in Bangladesh. *Crop Protection, 113*, 6–14.

Rahman, M. T., Rubayet, M., Khan, A. A., & Bhuiyan, M. (2021). Integrated management of charcoal rot disease of soybean caused by *Macrophomina phaseolina*. *Egyptian Journal of Agricultural Research, 99*(1), 10–19.

Ramezani, H. (2008). Biological control of root-rot of eggplant caused by *Macrophomina phaseolina*. *American-Eurasian Journal of Agricultural and Environmental Sciences, Dubai, 4*, 218–220.

Reznikov, S., Vellicce, G. R., González, V., de Lisi, V., Castagnaro, A. P., & Ploper, L. D. (2016). Evaluation of chemical and biological seed treatments to control charcoal rot of soybean. *Journal of General Plant Pathology, 82*(5), 273–280.

Singh, S. N., Srivastava, S. K., Bhargava, P. K., & Khare, M. N. (1990). Chemical control of seedling mortality in cv. JS 72-44 and bacterial pustule in cv. Punjab-1 of soybean (*Glycine max* (L.) Merrill). *Legume Research, 13*(1), 17–20.

Spagnoletti, F., Carmona, M., Gómez, N. E. T., Chiocchio, V., & Lavado, R. S. (2017). Arbuscular mycorrhiza reduces the negative effects of *M. phaseolina* on soybean plants in arsenic-contaminated soils. *Applied Soil Ecology, 121*, 41–47.

Spagnoletti, F. N., Cornero, M., Chiocchio, V., Lavado, R. S., & Roberts, I. N. (2020). Arbuscular mycorrhiza protects soybean plants against *Macrophomina phaseolina* even under nitrogen fertilization. *European Journal of Plant Pathology, 156*(3), 839–849.

Swamy, C., Naik, M. K., Amaresh, Y. S., Chennappa, G., & Jayalakshmi, S. K. (2018). Morphological diversity and molecular characterization of *Macrophomina phaseolina* causing stem canker of pigeonpea in north eastern Karnataka. *Journal of Plant Disease Sciences, 13*(2), 91–98.

Vir, D. S., Gangopadhyay, S., & Gaur, A. (1972). Evaluation of some systemic fungicides and antibiotics against *Macrophomina phaseolina*. *Pesticides, 6*, 25–26.

Watanabe, T., Smith, R. S., Jr., & Snyder, W. C. (1970). Populations of *Macrophomina phaseoli* in soil as affected by fumigation and cropping. *Phytopathology, 60*(12), 1717–1719.

Zaki, A., & Mahmood, I. (1993). Biological control of *Meloidogyne incognita* race 3 and *Macrophomina phaseolina* by *Paecilomyces lilacinus* and *Bacillus subtilis* alone and in combination on chickpea. *Fundamental and Applied Nematology, 16*, 215–218.

Section V

Nonchemical control

Chapter 20

Management of charcoal rot by soil solarization and biosolarization

Siva Raagini Polakala, K. Shiva Raghava Rayalu, Mohano Behara, A. Sasi Kala, N. Hari Prasad and K. Jaya Chandra

Department of Microbiology, PVKN Government Degree College (A) Chittoor, Andhra Pradesh, India

20.1 Introduction

The soil-borne fungus *Macrophomina phaseolina* is found throughout the globe. It affects around 500 plant species from roughly 100 families and has a broad host range and great soil persistence. It causes charcoal rot in the stems and roots, as well as seedling blight (Dhingra & Sinclair, 1978; Ghosh et al., 2018). This fungus may cause large-production crop losses in many plants including soybean, strawberry, and sorghum whether temperature are high (30−350°C) and soil moisture levels are low (less than 60%), reducing farmers' profitability (Khan et al., 2012).

M. phaseolina a compact mass of hardened fungal mycelium is a spherically oval or long extended filamentous fungus, light brownish in the early stages and becomes darkens (brown-blackish) with ageing pycnidia are infrequently observed in the natural condition which is greater than microsclerotia (dark brownish-black, beaked oscillated, and rough irregular) (Lakhran et al., 2018).

20.2 Principle of solarization

The basic premise of soil solarization is that it elevates temperatures in damp soil to fatal levels, affecting the life of particular organisms. Other environmental parameters are altered by the heating process, as are biological differences in the soil, which indirectly influence soil-borne pests and beneficial organism viability (Katan, 1981). The importance of high soil temperatures and the quantity of heat retained (temperature duration) impacts the likelihood of thermal death effects on soil-borne pests (Katan, 1987) and weed seed (Stapleton, 2000).

The very general way of soil solarization is contingent on mulching moisturized soil with transparent polyethene sheets. The soil mulching is incumbent for at least four to 6 weeks of duration to get a successful effect depending on the pest, climatic conditions, soil characteristic, polyethenehene properties (Katan, 1981, 1987; Rubin & Benjamin, 1983). The pest population and environmental conditions are wild variables, while soil moisture and polyethylene properties property can be altered as required. Soil per treatment and suitable polyethylene technology can overcome the unfavourable environmental conditions induced in some regions or in some seasons. This leads to increased weed sensitivity and decreased soil mulched duration (Stevens et al., 1991).

Temperature conductance increases in wet soil and the sensitivity of microorganisms toward the toxic agent. Hence, "wet heating" is more advisable than "dry heating" to control the pest. This applies in weed control because moist seed's metabolic activities are high (Iddumu et al., 2020; Shlevin et al., 2004). In the solarization process, pretreatment or drip irrigation improves water capacity during mulching; thereby improving the efficacy (Umair et al., 2019). In the solarization process, the soil surface becomes wet through drip irrigation, which is essential for maintaining and permitting the heat to penetrate deeper layers. Reliable soil preparations result in a smoothie soil surface and better mulching and prevent plastic tearing (Fig. 20.1).

Macrophomina Phaseolina. https://doi.org/10.1016/B978-0-443-15443-0.00012-7

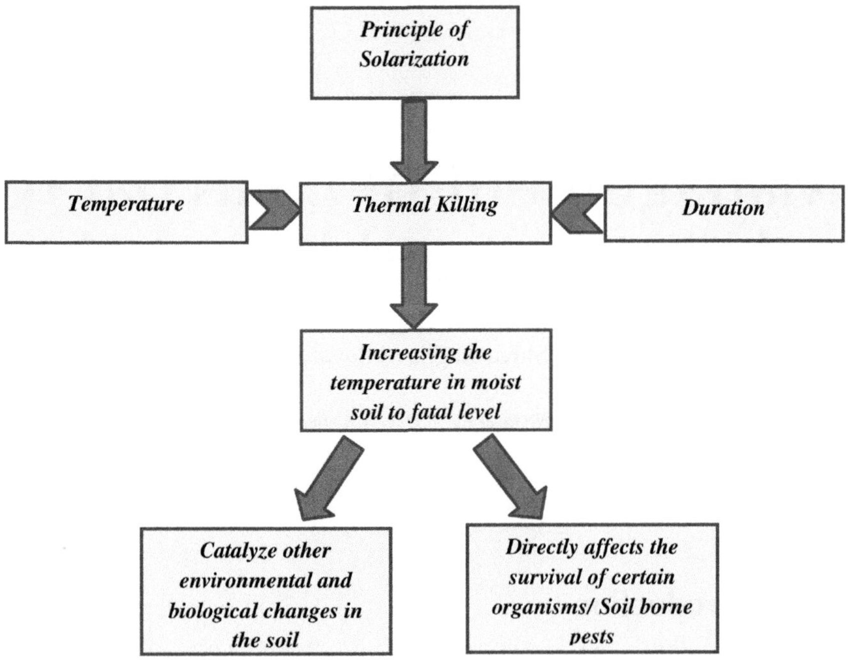

FIGURE 20.1 Principle of solarization.

20.3 Solarization and bio solarization

To give successful and efficient substitute to the agricultural soil fumigation, the ecological pest control technique is called soil solarization or soil sterilization. This is indispensable to advanced field technology. Soil solarization is a nonchemical, nonhazardous, and nonpesticidal method that is implemented by farmers in various parts of the globe. It presents several pest control based on trapping solar radiations by firmly covering the soil by placing transparent sheets on moisture soil in times of elevated temperature. The plastic sheets permit the solar energy to be caught in the soil, thereby, heating the upper levels (Simmons et al., 2013). It is an eco-friendly method by using the sun's power, and hence it is also called "solar heating."

Instead of employing hazardous substances like insecticides and fumigants, biosolarization uses organic waste to help the eliminate pests during soil solarization. Both solarization and bio-solarization on a plastic film encase damp soil that is passively heated by solar radiations in warm weather. Many ground studies have demonstrated that under the right conditions, these methods can minimize soil-borne pests (Butler et al., 2012; García-Raya et al., 2019). Solarization and bio-solarization are two strongly associated techniques that have a major impact on the anaerobic soil disinfection process (ASD). Many organic farmers have worked in high-quality environments and have embraced these technologies, which have been utilized effectively as soil disinfection treatments (Stapleton et al., 2008).

Biosolarization is the purposeful addition of organic soil amendments to achieve literal solarization. For the past 30 years, mostly organic farmers have employed a broad range of organic additives for their biosolarization treatments (Gamliel & Stapleton, 1993; Shea et al., 2021).

Solarization in the hot summer months may raise soil temperatures to dangerously high levels and effectively controlling a broad ranges of soil-borne diseases, such as bacteria, weeds, and insects. Soil solarization is based on the use of solar energy to heat the moistened soil in the upper layer up to 40–55°C when mulched with transparent polyethylene sheets resulting in thermal death of the pathogen (Dwiwedi & Dubey, 1987). The fundamental focus in the pest management technique is thermal killing. However, in a few conditions, biological and chemical processes are also included. Thermal killing, on the other hand, is effective on heavy soils that comprise clay and loam or a blend of the two. It can store more water than light soils for longer periods of time, allowing it to create steam every day it was needed to kill bacteria in the soil. Solarization is less effective on sandy soil, as it drains out water faster and produces less steam. Drip irrigations, followed by a transparent plastic cover and frequent water, is used to improve the benefits of solarization in sandy soils. During the hottest part of the year, the solarization period should be kept to a minimum of 4–6 weeks; the longer the soil is subjected to the heat created by the process of solarization, the more undesired organisms will be killed (Kapoor, 2013).

20.4 Disease cycle

Over 500 cultivated commercially significant wild crop species, such as sorghum, bean, soybean, cotton, peanut, cowpea, maize, and others, are affected by *M. phaseolina*, which causes root rot, seedling blight, and root and stem rot. Exudates from roots cause microsclerotia to germinate and infect hosts' roots. Infectious hyphae penetrate the plant through wounds or root epidermal cells (Adam, 1986; Bouhot, 1967, 1968; Dhingra & Sinclair, 1978; Diourte et al., 1995; Gray et al., 1990; Hall, 1991).

The mycelium enters the root of the epidemic in the pathogenesis early phases and is mostly limited to the intercellular spaces of the cortex of the major roots. As a result, nearby cells are harmed, and extensively infected plantlets can perish. Intracellular fungal hyphae proliferate. Necrotic lesions were observed on the branches, stems, and peduncles of the infected plants'. The development of fungal toxins, such as phaseolinone, and the plugging of fungal tissue into host arteries cause highly infected plants to die prematurely. Mycelia colonization and sclerotia formation in host tissues continue after the plant death until the tissues dry. Mycelia and microsclerotia generated in contaminated plant materials, including plant remnants, are used to spread the infection. The main surviving propagules in soil are Microsclerotia's host stems and roots. After root and plant debris degrade, microsclerotia are produced into the soil (Fig. 20.2). They are mainly

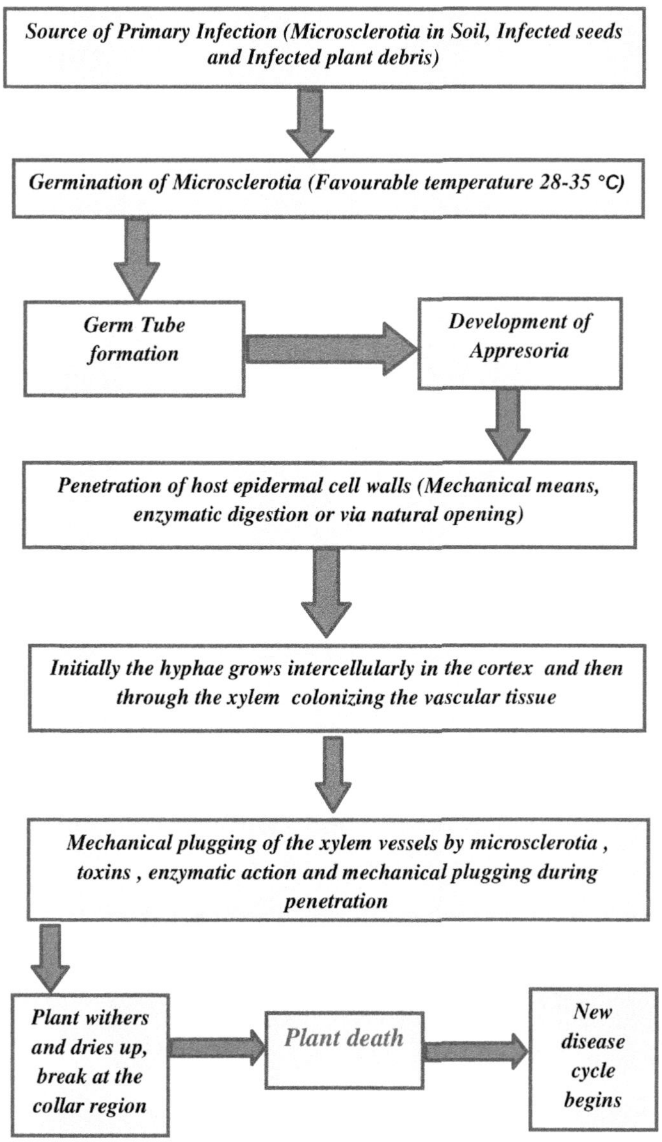

FIGURE 20.2 Disease cycle of *M. phaseolina* (flow chart).

found on the soil surface at a depth of 0–20 cm and in clusters (Baird et al., 2003; Cook et al., 1973; Papavizas, 1977; Short et al., 1978).

20.5 Mechanism of soil solarization

All combinations of magnitude radiation and heat movement in soil result in soil temperature. In the soil depth, the red, black, and clear polyethylene (PE) films were able to produce higher soil temperature than the nonsolarized films. The PE sheets reduced convective heat loss, resulting in an increase in soil temperature, and, therefore, the thermal energy balance may be affected by the soil surface. In plots covered with PE films, soil temperature might reach its highest point. Incoming solar radiation is directed to the soil surface to penetrate the PE films, where it is turned into heat energy. Heat energy is transported deeper into the soil by passing through the soil surface. Heat loss is greater at the soil surface than in the deeper subsurface (Nwankwo & Ogagarue, 2012). Furthermore, solar radiation is emitted to the ground in the form of short wave energy, which travels through transparent PE films in the soil surface until the soil surface is modified causing heat energy to be absorbed and the soil temperature to rise. Conduction is the process of heat transfer in the soil. The heat displacement in the soil has happened from layer to layer as the depth of the soil has increased. Heat is transmitted from the soil surfaces to the deeper layers. Temperature fluctuations in the soil depth will be greatest on the soil surfaces and decrease as the soil depth is increased. The highest temperature on the soil surface will be reached at the peak of solar radiation intensity (Fig. 20.3) (Paiman et al., 2020).

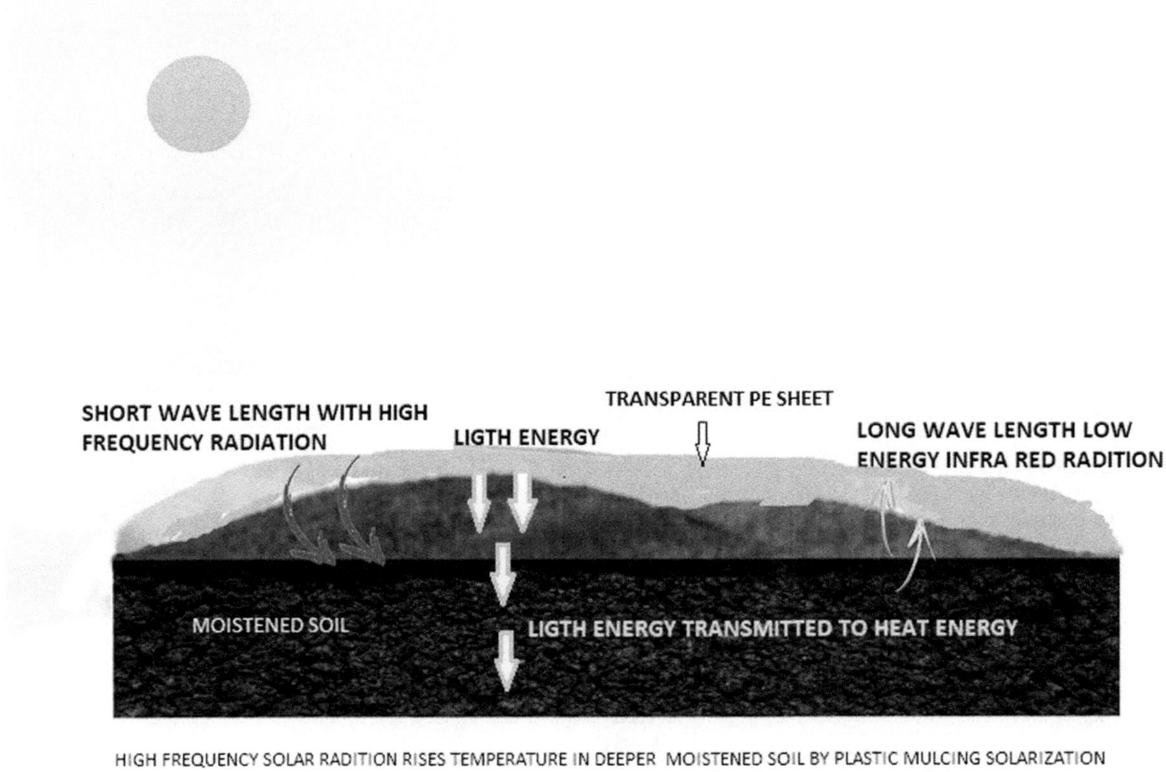

FIGURE 20.3 Mechanism of soil solarization.

20.6 The process of thermal death

The process involves the following:

a. Solar energy to the soil raises soil depth temperatures, which reach a maximum later on in the day and are sustained for lengthier durations,
b. The best time for soil mulching is when the weather is suitable, which may be discovered by experimentation by tarping the soil and recording the temperatures.
c. To allow biological activity during solarization while improving heat conduction in the soil. To improve the temperature sensitivity of the target species, sufficient soil moisture is necessary during solarization. Immediately tarping, a single irrigation can hydrate the soil. Irrigation is normally not required unless in the case of soft soils, where it may lower soil temperature if done at night.
d. As soon as the plastic is removed, it is critical to properly prepare the soil for planting.
e. If the climatic circumstances are favourable and the soil is maintained damp, the soil is mulched with thin, translucent polyethylene sheets. Sprayable plastics can be used to replace plastic mulching in the soil.
f. Pathogen control is achieved within 20–60 days after solarization in diverse parts of the world. Extending the solarization time allows for the treatment of deeper soil layers and heat-resistant bacteria.
g. Solarization in the soil causes biological, physical, and chemical changes that impact insect management, plant development, and productivity (Fig. 20.4) (Katan & Gamliel, 2014).

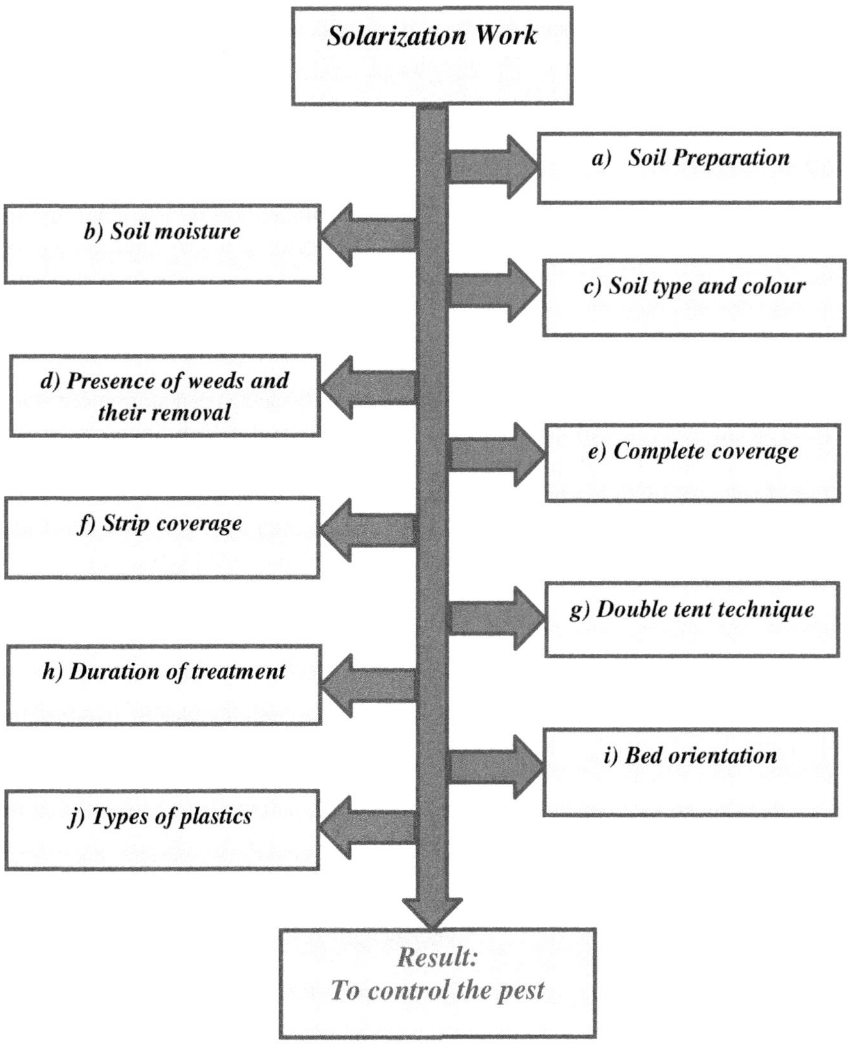

FIGURE 20.4 Working of solarization.

Heavy soils, such as clay, loam, or mixes of these, perform best for soil solarization. They can store far more water than light soils and emit steam on a daily basis. In order to eliminate diseases, weedy seeds, and insect eggs in the soil, steam is required. On sandy soil, solarization may be less successful in maximizing the advantages. Therefore, set up drip irrigation lines beneath the transparent plastic covers and fill it with water on a regular basis.

Moisture in the soil promotes temperature conductivity. As a result, "wet heating" is preferable to "dry heating" for insect management. This is also true for weed management, owing to the fact that damp seeds have highly advanced metabolic activities (Shlevin et al., 2004). A portion of the sun's energy passes through the translucent PE, is absorbed by the soil surfaces, and is then changed to save heat. Physical and chemical qualities, such as colour, thickness, UV protection, wavelength transmission, and stability vary across PE sheets. Depending on the pest, soil features, weather circumstances, and PE qualities, soil mulching took four to 6 weeks to complete. While pest populations and ecological conditions are difficult to control, soil moisture, and PE characteristics may be altered (Katan, 1981, 1987; Rubin & Benjamin, 1983).

20.6.1 Site preparations

The area that will be solarized should first be free from weeds and trash. It is beneficial to till the location to maximize heat penetration into the upper 6.0 inch of soil. Old roots, sticks, and other debris must be eliminated to prevent the plastic from tearing.

20.6.2 Soil preparation

When the plastic sheeting is applied to create a flat soil surface, solarization is most effective. Disking, rototillering, or rotating the soil to break up clods and then smoothing the surfaces are the first steps in soil preparation. Check for weeds, large rocks, or any other objects or debris that may lift or puncture the plastic and remove them before proceeding.

20.6.3 Soil moisture content on soil solarization

Weed seeds and pathogens cannot absorb water sufficiently to make them accessible to the higher heat if the soil is very dry (<70.0% of field capacity). Soil moisture content had a substantial impact on the link between hourly increase in soil temperature and heat flux, and soil solarization efficiency declined as moisture content increased (Greenberger et al. 1976).

20.6.4 Soil type and color

Solarization acts as fine heavy soil that comprises of clay loam or a mixture of clay loam and sand. It can store more water than light soils and last long enough to create steam every day to kill soil bacteria. In sandy soil, which drains faster and creates less steam, solarization may be less successful. Lay drip irrigation, followed by a transparent plastic cover and frequent water use, to optimize the effect of solarization in sandy soils. Dark-colored soils absorb more sun energy and achieve greater temperatures during solarization than lighter-colored soils. However, organic material, such as manure and adding black materials such as charcoal and to a light loam soil with only 1° increase in maximum temperatures, may have the same limited impact.

20.6.5 Presence of weeds and their removal

For weed control, the strategy is effective both in the short and long term. If weeds grow through the plastic, it shows the technique was done wrong. When the temperature is hot enough to kill particular species, the soil solarization strategy for weed management is effective. Due to the increased tolerance of summer plants, it will be more effective in suppressing winter weeds than summer weeds. Due to the deep subsurface penetration of perennial seeds and roots, soil solarization for weed control of perennials is less successful than for annuals (Khan et al., 2003).

Weed control by solarization also aids in the reduction of disease or pest populations that may be present on surrounding wild plants. Because, it favors traditional knowledge to intensive agricultural methods; soil solarization is in line with permaculture, regenerative agriculture, and organic farming ideals. Another benefit is that it encourages permanent plants, whereas solarization is more effective with annuals (Batlla & Arnold, 2007).

20.6.6 Control of pest and charcoal root rot

After heating on a daily basis, solarization alters the soil chemistry killing numerous plant pathogens, nematodes, and weed-seeds and seedlings. Many species that can endure solarization are weakened by the heat making them more susceptible to heat-resistant bacteria and fungi that operate as biological enemies. *M. phaseolina* lived for 19 h at 50°C before being completely dead at 20°C. It began to lose viability after 60 min at 55°C and 17 h at 50°C. Solarization did not diminish the possibility of *M. phaseolina* in the field at soil depths of 10.0 or 20.0 cm, but it did result in a considerable (66%) reduction in survival at a 5.0 cm of soil depth.

20.6.7 Orientation of beds

If raised beds are directed toward north-south rather than east-west, soil heating will be more constant, and pest management will be better. Solarization is most successful when there is no slope and the slope faces south or southwest. Insect control on north-facing slopes is inefficient due to the lower temperatures. Cultivation deeper than 3.0 inches following soil solarization must be avoided since it may bring weed seed and pathogens into the upper soil layer resulting in a major weed and disease problem.

20.6.8 Uses of plastic mulch for soil solarization for the control of charcoal root rot

Plastic sheets can be placed by hand or by machine. By burying the sides of the plastic sheeting in a small trench around the treated area, the edges should be attached to the earth. Plastic is spread in two ways—comprehensive coverage, which treats the whole area or fields to be planted or strip coverage, which treats only the beds or chosen areas of the area (Ben-Yaphet et al., 1987).

20.6.9 Complete coverage

If the soil is heavily infected with nematodes, perennial weeds or other pathogens like bacteria, fungi, protozoa, actinomycetes, etc., total covering is recommended to reduce the risk of reinfection from dirt carried to the plants by cultivation or irrigation water applied in furrows. Plastic sheeting is laid down in a continuous pattern throughout the whole planting field to ensure that it is completely covered. The edges of the sheets are either bonded together with ultraviolet-resistant adhesive or secured in the soil by burying both edges by laying adjacent strips of plastic. Anchoring the edges instead of gluing them together is initially less costly. The sheets' ends should be buried in the ground. After complete covering, thorough tillage should be avoided to prevent taking untreated soil to the surface (Gamliel & Becker, 1996).

20.6.10 Strip coverage

In the strip coverage, various crop rows can be planted. Strips of plastic are put to prepared beds in the shape of strips. Strip must be a min. of 30 inches broad and a maximum of 5 feet wide.

20.6.11 The double-tent techniques

Two layers of PE (with a distance of 3−7 cm between the sheets) are used to mulch the soil, which has been proven to be more effective than one single PE sheet increasing soil temperature by an additional 10°C over single-layer solarization. It increases both the economic expense and the environmental risk of PE contamination, and should only be utilized in exceptional circumstances (McGovern et al., 2004).

20.6.12 Duration of treatment

Plastic sheets must be left in place for 4−6.0 weeks to allow the soil to heat to its maximum depth. Leave the plastic in place for 6 weeks to manage the moisture resistance. Taking the temperature of the soil is either unnecessary or ineffective, according to experience. Solarization with minimal wind or cloud cover is the greatest risk in a time of intense solar radiation. The ideal months for coastal locations are May−June or August−September, with transitional times when fog or wind are at their lowest.

In terms of solarization, transparent or clear plastic is more effective than black plastic because black plastic does not heat the soil as much as clear plastic. Clear plastic is frequently used for mulching. Its major action is to inhibit the development of weeds. Black plastic may combine some solarization advantage with residual weed management in some regions where solarization is inefficient due to limited sun radiation or a large weed infestation; the treatment duration must be increased for the maximum effects. Other hues of plastics, such as brown or green, enable some soil heating, but not to the same extent as clear plastic, which necessitates longer treatment durations (Merfield, 2019).

20.6.13 Types of plastic

As the temperature rises, the thin plastic will thicken. Although 1 mil (0.001 inch [0.025 mm]) polyethylene plastic is efficient and cost-effective; it is not extremely resistant to tearing or piercing by animals. If holes or tears in the plastic need to be repaired using clear patching tape in windy areas, users should use 1.5–2 mil (0.038–0.050 mm) thick plastic sheets. To protect sheets from becoming brittle and difficult to remove from the field, use an ITV inhibitor, as well as to extend the life of the plastic. Plastic is used in solarization. Hand-laid sheets can be reused several times; however, reuse is less efficient if the plastic is unclean or dusty. Plastic sheets are modified and enabled by an extra ingredient. Thermodynamic death is important in pest management, but chemical and biological processes are also used. The highest soil temperature and the quantity of heat accelerated are used to calculate the efficacy of thermal death (Stapleton 2000; Simmons et al., 2013).

Organic additions, which are less expensive than most chemicals used as soil fumigants, along with soil solarization (bio-fumigation), raise soil temperature by 1–30°C and increase insect control by generating and accumulating harmful volatile chemicals. Before widening the spectrum of pests managed by solarization, it was necessary to use less expensive and more ecologically friendly mulching technologies (Youssef, 2015).

20.7 Encouragement of beneficial soil organisms

Soil pests are destroyed with the soil solarization approach. Many beneficial soil organisms, fortunately, may survive or recolonize the soil rapidly afterward. Mycorrhizal fungi, as well as fungi and bacteria that parasitize plant diseases and stimulate plant development, are important among these beneficial microorganisms. Solarized soils may be more disease resistant than nonsolarized or fumigated soils as a result of the population shift in favor of these advantages.

20.8 Beneficial effects of solarization

The greatest heat is influenced by soil solarization near the soil's top, and it diminishes as depth increases. At a depth of 2 inches (5 cm), the greatest temperature of solarized soil in the field varies from 108° to 131°F (42–55°C) and at 18 inches (45 cm), it ranges from 90° to 99°F (32–37°C) (45 cm). In the top 4–12 inches of soil, soil pests are most effective (10–30 cm). In greenhouses, solarized soil may reach 140°F (60°C) at 4 inches (10 cm) deep and 127°F (53°C) at 8 inches (20 cm). Dwiwedi and Dubey (1987) recorded the maximum temperature of 54°C at 2 cm depth in solarized soil and recorded the maximum reduction in survival of *M. phaseolina* in tarped soil of sun-lighted plot.

20.9 Bio-solarization

Bio-solarization, a "near relative" of solarization, is an alternative to soil disinfection treatment that involves the use of organic amendments (bio-fumigation), such as animal manure or integrated cover crop residue in combination with soil solarization (Simmons et al., 2016).

20.10 Principle

Bio-solarization is based on the addition of organic material to the soil prior to the passive solar heating processes. The organic material heats the damp soil, causing it to break down and release Allelo-chemicals and other bio-toxin decomposition products, which then appear temporarily in the soil. The pseudo-fumigation effect is enhanced by the combinations of organic material breakdown, soil heating and enough soil moisture, which can result in increased soil disinfection efficacy (Fig. 20.5) (Gamliel & Stapleton, 1997).

20.11 Applications of bio-solarization

The benefit of utilizing bio-solarization is that it combines two separate forms of pesticidal activity: soil heating and organic material breakdown, allowing undesirable organisms to be removed with little or no impact on the beneficial biotic

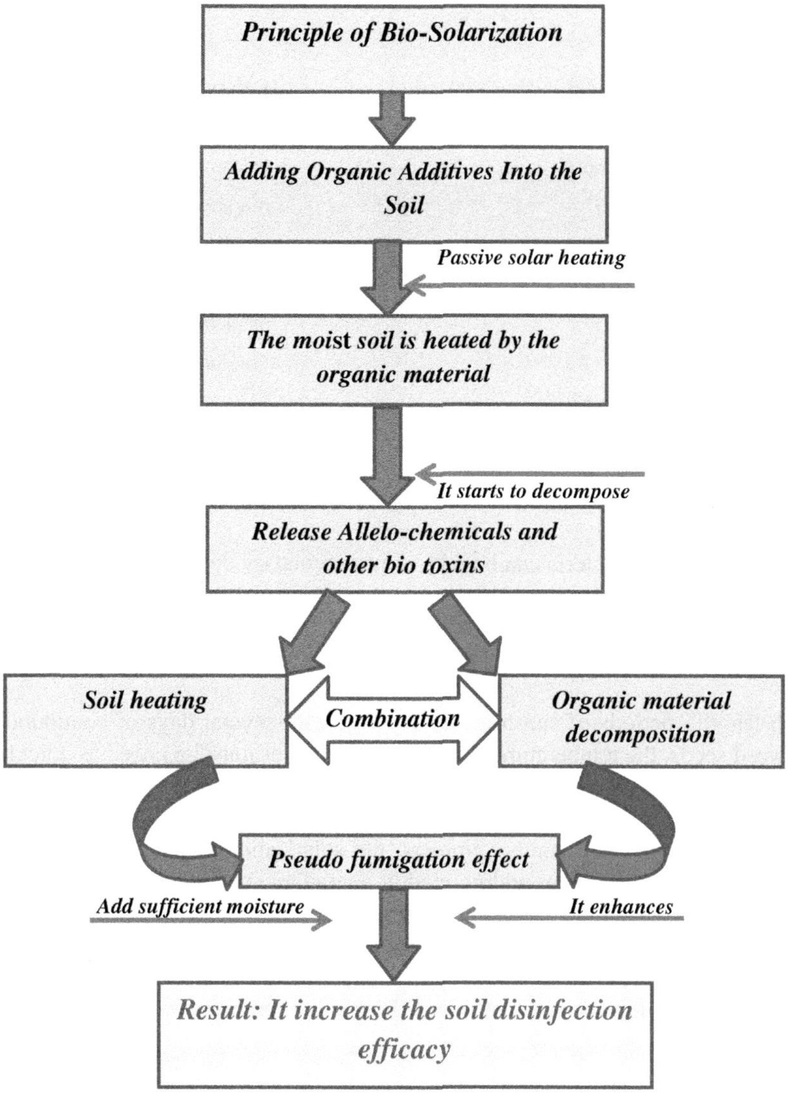

FIGURE 20.5 Principle of bio-solarization.

population. Maintaining a facultative aerobic state in the bio-solarization process, with a component of the soil microbial community respiring and going through regular metabolism can supply extra biological warmth to the soil, which is accomplished through solar energy accumulation (Simmons et al., 2013). Field soil is not a homogeneous material; there are several examples of micro-aerobic and/or anaerobic conditions dominating in deeper soil under saturated circumstances in conjunction with stiff soil layers constraining the water boundary layer, and so on.

Organic producers of high-value crops (especially strawberries) in coastal parts of California, which are less amenable to soil heating than inland areas, are presently using the soil-borne pest management strategy. Plastic mulch is used to keep the soil holding organic additions wet (Butler et al., 2012). The primary goal is to establish and sustain extractable and fermentative soil conditions in order to render pest organisms inactive. The generation of bio toxic organic acids in soil has been identified as one of ASD's mechanisms of action. The ASD approach can be employed when the temperature is low, or the weather conditions are not suitable for a predictable bio-solarization treatment. Organic additives, such as plant-based composts and crop fertilizers, are preferred (Table 20.1).

The usage of animal-based products comes with a lot of restrictions; therefore, sticking with plant-based modifications has a lot of advantages (Gamliel & Stapleton, 1993). In a recent study, solid waste materials from the food processing sector were combined with green waste compost, as well as pomaces from treating red wine grapes and/or white wine grapes, tomatoes. The chemical and microbiological makeup of soil is changed when organic materials are introduced;

TABLE 20.1 Applications of bio-solarization.

Crops	Solarization period	Yield increase (%)	References
Tomato	2 months	195.5	Mauromicale et al. (2005)
Potato	1 month	86.3	Ngakou et al. (2006)
Carrot	9 weeks	157.4	Marenco and Lustosa (2000)
Cabbage	2 weeks	130.2	Haidar and Sidahmed (2000)
Broccoli	–	64.0	Wang et al. (2008)
Strawberry	7 weeks	37.4	Stapleton et al. (2005)
Lettuce	2 months	23.9	Candido et al. (2011)
Soybean	2 months	80.0	Megueni et al. (2006)
Wheat	5 weeks	27.0	Singh et al. (2003)

plant seed and survival structures of fungus, nematodes, and bacteria can be forced into secondary dormancy that improve biocidal effects on pest species living in soil (Fernández-Bayo et al., 2018).

20.12 Limitations

Solarization is a traditional approach in which lengthy periods of sunshine are present, with several days of continuous sunshine being especially crucial. To destroy weed seeds, the temperature in the initial soil layer must be raised as quickly as possible at the start of solarization process. Weeds will grow and push the plastic sheet upwards, lowering the warming impact of solar radiation on the soil significantly. As a result, solarization is an inefficient process in some areas.

However, with rising temperatures, particularly very hot and sunny summers, the solarization process may become feasible in some areas. The bio-solarization approach, which involves adding readily degradable organic matter to the soil before covering it with plastic film, can also improve the efficiency of the process. Another choice is organic versus conventional farming, and the expense of soil solarization in practice is comparatively long period of the procedure and the climatic reliance that prevent widespread implementation. In high-income crops, soil solarization as a nonchemical weed control method has been shown to be much more profitable and lucrative than MB or other treatment.

20.13 Conclusion

Soil solarization's usefulness has been proved in a variety of agro-ecosystems. The significance of soil solarization as a feasible environmental integrated pest management tool in agricultural production systems is highlighted by worldwide changes and the constant rise in natural ecosystem erosion, particularly in areas with high solar radiations, but also in cloudy climate.

Enhanced soil temperature transfers in the vertical soil profiles using environmentally acceptable large-scale application and cheaper mulching technologies with novel plastic compositions. This advancement in technology should allow soil solarization to be used outside of the growing season, making it ideal for marginal climatic locations as well as less lucrative crops. These enhancements will increase the range of controlled pests and shorten the time of the procedure.

Soil solarization fits into the foundations of sustainable agriculture methods by enabling effective and ecologically friendly pest management for numerous soil borne pests, as well as more competitive market positions and better pricing for pesticide-free products. Solarization is a more convenient and cost-effective strategy than alternative control strategies as evidenced by a cost study. Another major benefit of soil solarization is that it may be integrated into more sophisticated pest management schemes, as it is technically compatible with most other existing control approaches.

References

Adam, T. (1986). *Contribution à la connaissance des maladies du niébé (Vigna unguiculata (L.) Walp.) au Niger avec mention spéciale au Macrophomina phaseolina (Tassi) Goïd* (p. 117). Université de Renne I. Thèse de doctorat.

Baird, R. E., Watson, C. E., & Scruggs, M. (2003). Relative longevity of *Macrophominaphaseolina* and associated mycobiota on residual soybean roots in soil. *Plant Disease, 87*(5), 563–566. https://doi.org/10.1094/PDIS.2003.87.5.563. PMID: 30812959.

Batlla, D., & Arnold, R. L. B. (2007). Predicting changes in dormancy level in weed seed soil banks: Implications for weed management. *Crop Protection, 26*(3), 189–197.

Ben-Yaphet, Y., Stapleton, J. J., Wakeman, R. J., & DeVay, J. A. (1987). Comparative effects of soil solarization with single and double layers of polyethylene film on survival of *Fusarium oxysporum* F. SP. *vasinfectum. Phytoparasitica, 15*, 181–185.

Bouhot, D. (1967). Etude du Macrophomina phaseoli sur arachide. *Tropical Agriculture, 22*, 1165–1171.

Bouhot, D. (1968). Le Macrophomina phaseoli sur les plantes cultivées au Sénégal. Agr. Tropic, *23*, 1172–1181.

Butler, D. M., Rosskopf, E. N., Kokalis-Burelle, N., Albano, J. P., Muramoto, J., & Shennan, C. (2012). Exploring warm-season cover crops as carbon sources for anaerobic soil disinfestation (ASD). *Plant and Soil, 355*, 149–165.

Cook, G. E., Boosalis, M. G., Dunkle, L. D., & Odvody, G. N. (1973). Survival of *Macrophominaphaseoli* in corn and sorghum stalk residue. *Plant Disease Reporter, 57*, 873–875.

Dhingra, O. D., & Sinclair, J. B. (1978). *Biology and pathology of macrophomina phaseolina* (p. 166). Viçosa, Minas Gerais, Brazil: Universidade Federal de Viçosa: Impresa Universitaria.

Diourte, M., Starr, J. L., Jeger, M. J., Stack, J. P., & Rosenow, D. T. (1995). Charcoal rot (Macrophomina phaseolina) resistance and the effects of stress on disease development in sorghum. *Plant Pathology, 44*, 196–202.

Dwiwedi, R. S., & Dubey, R. C. (1987). Effect of soil solarization on the population dynamics of *Macrophomina phaseolina* (Tassi) Goid. and general soil mycoflora. *International Journal of Tropical Plant Diseases, 5*, 67–73.

Fernández-Bayo, J. D., Randall, T. E., Harrold, D. R., Achmon, Y., Hestmark, K. V., Su, J., et al. (2018). Effect of management of organic wastes on inactivation of *Brassica nigra* and *Fusariumoxysporum* f.sp.lactucae using soil biosolarization. *Pest Management Science, 74*(8), 1892–1902. https://doi.org/10.1002/ps.4891

Gamliel, A., & Becker, E. (1996). *A method for applying plastic mulch.* Israel Patent No. 118787; USA Patent No. 6, 270,291 B2.

Gamliel, A., & Stapleton, J. J. (1993). Effect of chicken compost or ammonium phosphate and solarization on pathogen control, rhizosphere microorganisms, and lettuce growth. *Plant Disease, 77*, 886–891.

Gamliel, A., & Stapleton, J. J. (1997). Improvement of soil solarization with volatile compounds generated from organic amendments. *Phytoparasitica, 25*, S31–S38.

García-Raya, P., Ruiz-Olmos, C., Marín-Guirao, J. I., Asensio-Grima, C., Tello-Marquina, J. C., & de Cara-García, M. (2019). Greenhouse soil biosolarization with tomato plant debris as a unique fertilizer for tomato crops. *International Journal of Environmental Research and Public Health, 16*(2), 279. https://doi.org/10.3390/ijerph16020279

Ghosh, T., Biswas, M. K., Guin, C., & Roy, P. (2018). A review on characterization, therapeutic approaches, and pathogenesis of *Macrophomina phaseolina. Plant Cell Biotechnology and Molecular Biology, 19*, 72–84.

Gray, F. A., Kolp, B. J., & Mohamed, M. A. (1990). A disease survey of crops grown in the Bay Region of Somalia, East Africa. *FAO Plant Protection Bulletin, 38*, 39–47.

Greenberger, A., Alon, H., & Grinstein, A. (1976). Solar heating by polyethylene mulching for control of diseases caused by soil-borne pathogens. *Phytopathology, 66*(5), 683–688.

Hall, R. (Ed.). (1991). *Compendium of bean diseases* (p. 73). St. Paul, MN: American Phytopathological Society.

Iddumu, V., Vijayan, V., & Mehta, S. (2020). Soil solarization an eco-friendly approach to control harmful soil inhabiting organisms. *Agriculture Food & E-News Letter, 2*(6), 141–142.

Kapoor, R. T. (2013). Soil soalraization eco-friendly technology for farmers in agriculture for pest management. In *Proceedings of 2nd international conference on advances in biological and pharmaceuticals sciences (ICABPS 2013).*

Katan, J. (1981). Solarization heating (solarization) of soil for control of soil borne pests. *Annual Review of Phytopathology, 19*, 211–236. https://doi.org/10.1146/annurev.py.19.090181.001235

Katan, J. (1987). Soil solarization. In I. Chet (Ed.), *Innovative approaches to plant disease control* (pp. 77–105). New York, USA: John Wiley and Sons, Inc.

Katan, J., & Gamliel, A. (2014). Plant health management: Soil solarization. In *Encyclopedia of agriculture and food systems* (pp. 460–471).

Khan, M. A., Marwat, K. B., Amin, A., Nawaz, A., Khan, R., Khan, H., et al. (2012). Soil solarization: An organic weed-management approach in cauliflower. *Communications in Soil Science and Plant Analysis, 43*(13), 1847–1860. https://doi.org/10.1080/00103624.2012.684822

Khan, A.,R., Srivastva, R. C., Ghorai, A. K., & Singh, S. R. (2003). Efficient soil solarization for weed control in rainfed up land rice ecosystem. *International Agrophysics, 17*, 99–103.

Lakhran, L., Ahir, R. R., Choudhary, M., & Choudhary, S. (2018). Isolation, purification, identification and pathogenicity of *Macrophominaphaseolina* (Tassi) cause dry root rot of chickpea. *Journal of Pharmacognosy and Phytochemistry, 7*, 3314–3317.

McGovern, R. J., McSorley, R., & Wang, K. J. (2004). Optimizing bed orientation and number of plastic layers for solarization in Florida. *Annual Proceedings of the Soil Science Society of Florida, 63*, 92–95.

Merfield, C. N. (2019). *An analysis and overview of regenerative agriculture.* Report number 2-2019 MRSNZ (pp. 1–20). The BHU Future Farming Centre.

Nwankwo, C., & Ogagarue, D. (2012). An investigation of tem-An investigation of temperature variation at soil depths in parts of southern Nigeria. *American Journal of Environmental Engineering, 2*(5), 142–147. https://doi.org/10.5923/j.ajee.20120205.05

Paiman, Yudono, P., Sunarminto, B. H., & Indradewa, D. (2020). Soil solarisation for control of weed propagules. *Journal of Engineering Science & Technology, 15*(1), 139–151.

Papavizas, G. C. (1977). Some factors affecting survival of sclerotia of *Macrophominaphaseolina* in soil. *Soil Biology and Biochemistry, 9*, 337–341.

Rubin, B., & Benjamin, A. (1983). Solar heating of the soil: Effect on weed control and on soil- incorporated herbicides. *Weed Science, 31*, 819–825.

Shea, E., Fernandez-Bayo, J. D., Pastrana, A. M., & Simmons, C. W. (2021). Identification and evaluation of volatile organic compounds evolved during solarization with almond hull and shell amendments. *Journal of the Air & Waste Management Association, 71*(3), 400–412. https://doi.org/10.1080/10962247.2020.1846637

Shlevin, E., Mahrer, Y., & Katan, J. (2004). Effect of moisture on thermal inactivation of soilborne pathogens under structural solarization. *Phytopathology, 94*, 132–137.

Short, G. E., Wyllie, T. D., & Ammon, V. D. (1978). Quantitative enumeration of Macrophomina phaseolina in soybean tissues. *Phytopathology, 68*, 736–741.

Simmons, C. W., Guo, H., Claypool, J. T., Marshall, M. N., Perano, K. M., Stapleton, J. J., et al. (2013). Managing compost stability and amendment o soil to enhance soil heating during soil solarization. *Waste Management, 33*(5), 1090–1096.

Simmons, C. W., Higgins, B., Staley, S., Joh, L. D., Simmons, B. A., Singer, S. W., et al. (2016). The role of organic matter amendment level on soil heating, organic acid accumulation, and development of bacterial communities in solarized soil. *Applied Soil Ecology, 106*, 37–46.

Stapleton, J. J. (2000). Soil solarization in various agricultural production systems. *Crop Protection, 19*, 837–841.

Stapleton, J. J., Molinar, R. H., Lynn-Patterson, K., McFeeters, S. K., & Shrestha, A. (2005). Soil solarization provides weed control for limited-resource and organic growers in warmer climates. *California Agriculture, 59*(2), 84–89.

Stapleton, J. J., Wilen, C. A., & Molinar, R. H. (2008). *Soil solarization for gardens and landscapes*. Pest Notes, Publication 74145. University of California Agriculture and Natural Resources.

Stevens, C., Khan, V. A., Brown, J. E., Hochmuth, G., Splittstoesser, W. E., & Granberry, D. M. (1991). Plastic chemistry and technology as related to strawberry. *Asian Journal of Plant Sciences, 3*, 731–735.

Umair, M., Hussain, T., Jiang, H., Ahmad, A., Yao, J., Qi, Y., et al. (2019). Water-saving potential of subsurface drip irrigation for winter wheat. *Sustainability, 11*(10), 2978. https://doi.org/10.3390/su11102978

Youssef, M. M. A. (2015). Biofumigation as a promising tool for managing plant parasitic nematodes. A review. *Scientia Agriculturae, 10*, 115–118.

Chapter 21

Biosurfactant producing plant growth–promoting bacteria: eco-friendly approaches for charcoal rot management

Mohd. Mogees Ansari[1], Tanya Singh[1], Basudev Majhi[1], Sankalp Misra[2] and Puneet Singh Chauhan[1]

[1]*Microbial Technology Division, Council of Scientific and Industrial Research-National Botanical Research Institute (CSIR-NBRI), Lucknow, Uttar Pradesh, India;* [2]*Faculty of Biosciences, Institue of Bioscience and Technology, Shri Ramswaroop Memorial University, Lucknow-Deva Road, Barabanki, Uttar Pradesh, India*

21.1 Introduction

Biosurfactant (BS) has been referred to as a surface-acting substance that can enhance surface–surface interactions by generating micelles in their natural habitats, such as those generated by plants, microorganisms, and animals Rahman and Gakpe (2008). Surface-active molecules produced by microbes having low molecular weight are termed as BSs. BSs are divided into six categories based on their physical–chemical properties—fatty acids, polymeric, neutral lipids, phospholipids, glycolipids, and lipopeptides (Pacwa-Plociniczak et al., 2011). BSs are made from renewable resources as compared to synthetic surfactants, and they are safe or harmless, recyclable, have excellent surface activity, efficient in difficult conditions, and can be recycled through regeneration. The pharmaceutical, cosmetics, petroleum, and food sectors might all benefit from the use of BSs. Surfactin, a powerful lipopeptide BS, is considered a multifunctional bioactive molecule with the capacity to be used in wound healing and is used for improved oil recovery. Furthermore, it has been proven to have antimycoplasma, antiviral, anticancer, and antifungal properties. Surfactin is used as a bioprocessing agent for soil and water remediation. These characteristics of surfactin indicate its possible industrial uses (Sen, 2010). Recently, BSs and microorganisms that produce BSs have been employed to make nanoparticles. The production of nanomaterials through the use of BSs is evolving as a part of green chemistry (Kiran et al., 2011). Microbubble technology, which has numerous uses in cellular imaging, disease diagnosis, drug delivery systems, water filtration, and sewage treatment, is made more stable by the use of BSs. BSs produced by microorganisms are used for motility, biofilm formation, nutrient transport facilitation, competitive colonization, quorum sensing, and microbe–host interaction (Stringlis et al., 2018). Through modification of bioavailability and surface area of hydrophobic water-insoluble compounds and by attaching contaminants like toxic metals, BSs can enhance soil quality, resulting in higher productivity and the rehabilitation of polluted soils.

Many of these BSs are beneficial to humanity and are important tools for biotechnology (Xu et al., 2011). Numerous patents on BS-producing bacteria, particularly *Pseudomonas* spp., *Acinetobacter* spp., *Bacillus* spp., and *Candida* spp., as well as on BS kinds, manufacturing methods, and commercial applications, have been filed, demonstrating the potential of these biomolecules (Shete et al., 2006).

Macrophomina phaseolina (Tassi) Goid., a soil-borne fungal pathogen, is reported to affect ~500 plant species by causing charcoal rot disease. The significance of BSs in the control of fungal infections in some plant species has been investigated (Rodrigues et al., 2021) (Fig. 21.1).

Macrophomina Phaseolina. https://doi.org/10.1016/B978-0-443-15443-0.00019-X

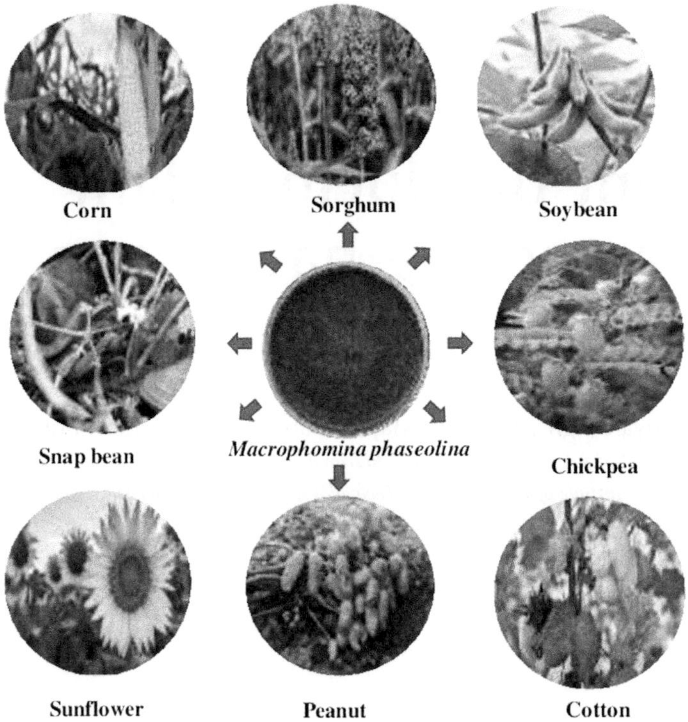

FIGURE 21.1 Schematic representation of the charcoal rot disease in major crop plants.

21.2 Classification

Classification of BSs is mainly on the basis of source of generation and its chemical structure. BSs with low molecular weight include lipopeptides and glycolipids, while amphipathic polysaccharides, lipoproteins, and lipopolysaccharides are high-molecular-weight BS. The BS with low-molecular-weight is good in lowering the surface and interfacial tensions, while the high-molecular-weight BS is good in stabilizing emulsions (Gayathiri et al., 2022). Particulate surfactant, polymeric surfactant, and phospholipids are some other BSs (Desai & Banat, 1997).

21.2.1 Glycolipid

21.2.1.1 Rhamnolipid

The rhamnolipid is made up of molecules of rhamnose attached to b-hydroxydecanoic acid. So far, seven rhamnolipid homologs are discovered (Abalos et al., 2001). The following compounds are produced by *P. aeruginosa* viz., rhamnolipid one and two are L-rhamnosyl-L-rhamnosyl-b-hydroxydecanoyl (Desai & Banat, 1997). *Pseudomonas aeruginosa* L2-1 rhamnolipids exhibited 100% emulsification in case of soybean oil and 69% in case of crude oil (Costa et al., 2010). Rhamnolipid also has antibacterial properties against a variety of microorganisms, making it important in the pharmaceutical industry (Magalhães & Nitschke, 2013). *Pseudomonas* sp. produces heterogeneous combinations of rhamnolipids in a few circumstances (Pornsunthorntawee et al., 2010). *P. aeruginosa* may synthesize rhamnolipids from sugars as carbon sources (Takenaka et al., 2007). Rhamnolipids have been reported to improve the desorption efficiency of phenanthrene demonstrating its utility in the cleanup of PAH-contaminated soils in cold climates (Wu et al., 2017).

21.2.1.2 Sophorolipids

A glycosidic bond attaches a dimeric carbohydrate called sophorose to a long-chain hydroxyl fatty acid to form glycolipids. They can be open (acidic sophorolipids) or closed (lactonicsophorolipids) in nature. Gorin et al. (1961) discovered new glucosides, sophorolipids, and sophorosides (SSs) with interesting metal coordination characteristics, including novel bolaformsophorolipids and sophorosides (Castelein et al., 2021). The physicochemical actions of sophorolipid compounds are influenced by systemic and structural differences. For example, acidic forms have better water solubility and foaming ability (Van Renterghem et al., 2018). The sophorolipids generated by *Candida bombicola* ATCC 22,214 totally suppressed the growth of *Propionibacterium acne, Staphylococcus xylosus*, *Bacillus subtilis,* and *Streptococcus mutants* (Kim et al., 2002). Despite the fact that the temperature and pH are varied, there is noticeable stability. Surface-active characteristics have been seen at pH values of 6–9 and temperatures of 20–90°C. The antibacterial capabilities of sophorolipids (SLs) were used against antibiotic resistance. The combination of SL and tetracycline entirely stopped *Staphylococcus aureus* from growing (Joshi-Navare & Prabhune, 2013). These SLs also interfered with biofilm formation (Raza et al., 2007). Bolaformsophorolipids could be a fascinating biological possibility with a variety of properties.

21.2.2 Lipoproteins and lipopeptides

Lipopeptides are cyclic lipopeptides having a lipid component and 8–17 amino acids (Desai & Banat, 1997). *Streptomyces* sp. DPUA1566 has potential applications in bioremediation procedures as well as the pharmaceutical and cosmetic sectors (Khanna & Pattnaik, 2019). Ramani et al. (2012) found that a lipoprotein produced by *Pseudomonas gessardii* has strong metal ion removing characteristics; a lipopeptide made by *Bacillus atrophaeus* 5-2a has good crude oil removal efficiency (Zeraik & Nitschke, 2010).

21.2.3 Fatty acids, phospholipids, and neutral lipids

Fatty acid and phospholipid are produced by a variety of microbes, including bacteria and yeast, and these are in high demand in the current scenario due to their diverse and effective qualities as BSs (Gayathiri et al., 2022). *Thiobacillus thiooxidans*, a sulfur-reducing bacterium, produces this BS (Desai & Banat, 1997; Gayathiri et al., 2022).

21.2.4 Polymeric biosurfactants

The components of the polymeric surfactants derived from various bacterial species are a complex mixture of biopolymers, such as proteins, polysaccharides, and lipopolysaccharides (Gayathiri et al., 2022). Polysaccharides are covalently bonded to fatty acids via o-ester linkages (Desai & Banat, 1997). Some examples of polymeric BSs are mannoprotein, emulsan, and liposan (Desai & Banat, 1997).

21.2.4.1 Emulsan

Under low concentrations, the pure form of emulsan has emulsifying properties. *Acinetobacter calcoaceticus* PTCC1318 produces emulsan with crude oil–degrading characteristics (Amani & Kariminezhad, 2016). They are made readily available for microbial access by covering the hydrophobic substrate. Except in their pure forms, aliphatic and aromatic hydrocarbon mixtures can be emulsified effectively in balanced form. The emulsifying activities of emulsan are determined by the fatty acid components (Rodrguez-López et al., 2016). When polysaccharides and proteins were mixed, emulsan of *Acinetobacter calcoaceticus* BD4 exhibited optimum emulsification activity; however, when separated, no activity was observed (Kaplan et al., 1987).

21.2.4.2 Liposan

C. lipolytica produces this emulsifier, which is made up of carbohydrates and proteins. It is widely used in the food and cosmetics industries. It creates stable emulsions with edible oils (Gayathiri et al., 2022).

21.2.4.3 Mannoprotein

Mannoprotein, a significant component of *Saccharomyces cerevisiae* cell wall, is an excellent bioemulsifier (Cameron et al., 1988). They were able to produce stable emulsions with a variety of hydrocarbons and other chemicals, implying that they could be used as cleaning agents. Mannoproteins are classified structurally and enzymatically based on their chemical makeup and function. The structural mannoproteins, formed by mannopyranosyl linked to a tiny protein, are the most common, while the enzymatic mannoproteins, which include more protein moieties, are the most effective emulsifiers.

21.2.5 Particulate biosurfactant

This surfactant is made up of the vesicles and fimbriae produced by *Acinetobacter* spp. (Desai & Banat, 1997). BS's hydrophilic and hydrophobic moieties allow them to accumulate in distinct phases of surfaces. Spherical BS vesicles are produced at concentrations higher than the critical micelle concentration (CMC) (Pornsunthorntawee et al., 2009). BSs have efficiently reduced the adhesion of biofilm formers. RHL prevents *L. monocytogenes* from adhering to surfaces.

21.3 Properties

21.3.1 Surface and interface activity

A good surfactant must be efficient and effective in order to be useful. The CMC measures efficiency, while surface and interfacial tensions determine effectiveness (Barros et al., 2007). BSs have a CMC of 1−2000 mg/L, while surface tensions and interfacial (oil/water) are 30 and 1 mN/m, respectively. Surfactants can reduce water surface tension from 72 to 35 mN/m and n-hexadecane interfacial tension from 40 to 1 mN/m, respectively.

21.3.2 Biodegradability and low toxicity

Biosurfactants proved to be an effective technique for bioremediation that did not harm the environment. They are much more environment-friendly and safer than synthetic surfactants. In a comparison investigation using Marlon A-350, a synthetic surfactant, and a BS generated by *P. aeruginosa*, in contrast to the BS, the synthetic surfactant was highly hazardous in all assays and characteristics (Muthusamy et al., 2008). When applied to the seeds of *Solanum gilo, Brassica oleracea* and *Cichorium intybus,* the BS synthesized by *Candida sphaerica* lowered the surface tension (Santos et al., 2019).

21.3.3 Emulsion forming and emulsion breaking

Emulsifiers and de-emulsifiers are two types of BSs. Emulsion is a heterogeneous system made of an immiscible liquid dispersed in another liquid as droplets with a diameter greater than 0.1 mm. Emulsions have limited stability; however, adding BSs can result in a stable emulsion that lasts months or even years (Velikonja & Kosaric, 1993). Few surfactants may be good surface tension reducers but not good emulsifiers, whereas others may not be good surface tension reducers but excellent emulsifiers. Few emulsifiers with a greater emulsifying ratio have been described from *Saccharomyces cerevisiae* and *Acinetobacter* sp. (Alcantara et al., 2014).

21.3.4 Temperature, pH, ionic strength, and tolerance

Environmental factors, such as pH and temperature, have no major effect on many BSs and their surface activity. BSs can withstand sodium salt concentration up to 10%, whereas manufactured surfactants are inactivated by just 2% NaCl. A lipopeptide produced by *B. subtilis* LB5a was able to survive for 6 months at 121°C and high-salt conditions without decreasing its surface activity. The BS maintains emulsification activity at pH 4−10, high temperatures up to 120°C, and a NaCl content of up to 10% (w/v) (Zahir & Arshad, 2004). The BS generated by *C. lipolytica* UCP0988 had similar effects but only under extreme pH circumstances (Luna et al., 2013). Even after treatment at temperatures up to 50°C, the BSs generated by *B. licheniformis* STK01, *B. subtilis* STK02, and *P. aeruginosa* STK03 displayed emulsifying capabilities (Amodu et al., 2014). *P. aeruginosa* RS29 produced a BS with stable temperature, pH, and saline stability. Even despite these harsh conditions, it was credited with high foaming and emulsifying activities (Rufino et al., 2008).

21.4 Role of microbes

Biosurfactants are a class of surfactants that are derived from living organisms, primarily bacteria and fungi. These substances are either by-products of metabolic processes or products of the surface chemistry of the cells. The majority of BSs are synthesized by aerophilic microbes in aquatic environments using hydrocarbons, polysaccharides, lipids, and oil as carbon sources that come from yeast (*Candida* and *Pseudozyma*), fungi (*Aspergillus* and *Fusarium*), and bacteria (*Pseudomonas, Bacillus,* and *Acinetobacter*).

21.4.1 Elements influencing the production of biosurfactants

Both the BS's composition and its ability to emulsify are influenced by the culture environment and BS producing strains. In addition to determining the amount of BS produced, chemical, and physical factors, such as aeration, divalent cations, temperature, and pH also affect the kind of polymer that forms (Salihu et al., 2009). These variables include dietary restrictions, nutritional needs, the kinds of carbon and nitrogen sources used, the C:N ratio, and other variables (Fig. 21.2).

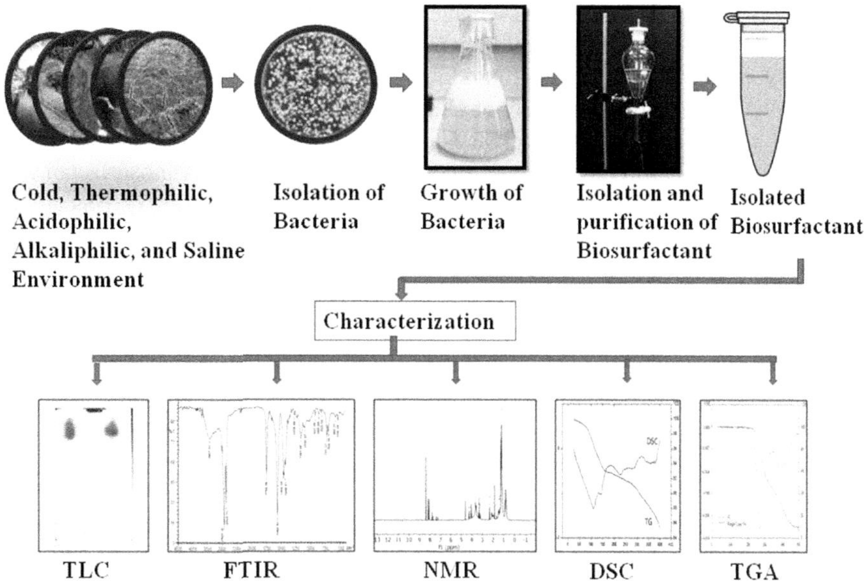

FIGURE 21.2 Isolation, screening and characterization of biosurfactants.

21.4.1.1 Carbon sources

The characteristics of the carbon substrate have an impact on and affect the quantity and quality of BS synthesis (Rahman & Gakpe, 2008). According to reports glucose, sucrose, glycerol, diesel, and crude oil are good sources of carbon substrate for the production of BSs.

21.4.1.2 Nitrogen sources

Due to its importance for microbial growth and the production of enzymes and protein synthesis, nitrogen plays a significant role in the medium used to produce BSs. The production of BSs has made use of a variety of nitrogen molecules, such as urea, peptone, yeast extract, ammonium sulfate, ammonium nitrate, sodium nitrate, meat extract, malt extracts, etc.

21.4.1.3 Environmental factors

Environmental factors have a huge impact on the yield and properties of the resulting BS. It is always essential to optimize the bioprocess in order to produce significant amounts of BSs because the product may be impacted by variations in temperature, pH, aeration, or agitation temperatures between 25 and 30°C with a pH of 8.0 (Desai et al., 1997).

21.4.1.4 Aeration and agitation

The transfer of oxygen from the gaseous form to the aqueous form is aided by both aeration and agitation, which are two important factors that affect the activity of BSs. The maximum surfactant production value (45.5 g/L) occurred when the dissolved oxygen content was maintained at 50% of saturation and the air flow rate was one vvm (Adamczak & Bednarski, 2000).

21.4.1.5 Role in charcoal rot disease management

BSs have antibacterial and antifungal activity against a variety of infectious bacteria and fungi (Kumar & Johri, 2012). *Bacillus* species producing BSs show a superior replacement for biocontrol agents against *M. phaseolina*, bioinoculants for promoting plant growth and an increase in the yield of legume crops using eco-friendly techniques. *B. cereus* BS14 was shown to have cytopathic effects on the fungus and to be effective at limiting the development of *M. phaseolina*. It is fascinating to observe how the pure BS of *B. cereus* BS14 affects the pathogen *M. phaseolina* (Kumar et al., 2021). The research on antimicrobial activity and characterization of *Bacillus* over *M. phaseolina* demonstrated the biocontrol of pathogens by such metabolites (Hussain & Khan., 2020; Kumar & Dubey, 2016).

A *Bacillus sonorensis* MBCU2 isolated from soil produced BSs, which have been reported to have antagonistic properties against *M. phaseolina;* further, it was modified with vermin compost (Pandya & Saraf, 2015). *M. phaseolina* was inhibited by a BS produced by *B. cereus* BS14, which synthesized a lipopeptidebiosurfactant and had antifungal efficacy against *M. phaseolina.* The BSs produced by *B. cereus* BS14 are of the cyclic siloxanes type and are crucial in the reduction of the charcoal rot disease as well as the promotion of the growth of the *V. mungo* plant. *B. cereus* BS14 will be developed as a commercially successful biological control agent (Kumar et al., 2021).

Enterococcus sp. BS13 was shown to have a greater ability to release BSs and enhance plant growth and reduce charcoal rot disease. *Enterococcus* sp.BS13 produced glycolipid type BS and this biosurfactan at displayed antifungal activity against the pathogen *M. phaseolina* (Kumar et al., 2016; Teichmann et al., 2007). BSs are a better biodegradable and environmentally friendly alternative to chemical surfactants. Enterococci generate various substances that restrict the development of *M. phaseolina. Enterococcus faecium* has biocontrol capability against *M. phaseolina.*

The species of fluorescent *Pseudomonas* are widely recognized for their capacity to produce rhamnolipids, initially discovered by Jarvis and Johnson (1949). Rhamnolipids permeate the fungal membrane, increasing fluidity and thereby significantly affecting membrane stability (Monnier et al., 2019). In addition to rhamnolipid synthesis, *P. guariconensis* strain LE3 also produced lytic enzymes and antifungal metabolites, which together contribute to its antagonistic action. The BS produced by *P. guariconensis* strain LE3 was examined for multidimensional biocontrol and promoting plant growth qualities in order to create a bioformulation based on BS that can increase crop output and soil condition, resulting in sustainable agriculture (Toro et al., 2013). Effective biocontrol agents have numerous qualities that help them in competition with local soil microorganisms. The occurrence of genes antibiotic-coding phenazine 1-carboxylic acid (PCA), 2,4-diacetylphloroglucinol (DAPG) and pyocyanin, as well as the lytic enzymes chitinase and endoglucanase, demonstrated *P. guariconensis* strain LE3 as a potent antagonist against *M. phaseolina.* Important secondary metabolites generated by *Pseudomonas* species include PCA and its derivative, pyocyanin, which has broad-spectrum antifungal and antibacterial properties (Zhang et al., 2017).

M. phaseolina causes charcoal rot of sunflower, which diminishes head weight and diameter with catastrophically affecting the other plant development indices, which is a serious economic loss (Ijaz et al., 2013; Khan et al., 2017). With a variety of PGP qualities, *P. guariconensis* strain LE3 makes an excellent bioinoculant and is the best choice for the development of BS-based bioformulations to increase agricultural productivity and sustainability (Table 21.1).

TABLE 21.1 **List of microorganisms producing biosurfactants against *Macrophomina phaseolina*.**

PGPR	Crop	References
Azospirillum brasilense	*Fragariaananassa*	Viejobueno et al. (2021)
Bacillus sonorensis	*Vignamungo*	Kumar et al. (2021)
Bacillus spp. MRF and MR-11(2)	Zea mays	Pal et al. (2001)
Bacillus amyloliquefaciens BS14	*Phaseolus vulgaris*	Sabate et al. (2017)
Bacillus subtilis	–	Castaldi et al. (2021)
Bacillus subtilis CRN-16	*Vignaradiata*	Sharma et al. (2018)
Bacillus vallismortis	*Oryzasativa*	Gopalakrishnan et al. (2011)
Enterococcus sp. BS13	*Vignamungo*	Kumar et al. (2016)
Pseudomonas sp. EM85	*Zea mays*	Pal et al. (2001)
Psedomonas aeruginosa	–	Verma et al. (2011)
Pseudomonas fluorescens 9	*Glycine max*	Zilli et al. (2018)
Pseudomonas fluorescens	*Glycine max*	Castaldi et al. (2021)
Pseudomonas putida CRN-09	*Vigna radiate*	Sharma et al. (2018)
Pseudomonas stutzeri (KX574858)	*Glycine max*	Mufti and Bano (2019)
Pseudomonas putida (KX574857)	*Glycine max*	Mufti and Bano (2019)
Pseudomonas EKi	–	Khare and Arora (2011)

21.5 Future perspectives and conclusions

BSs are regarded as the multifunctional biomolecules of 21stcentury, owing to their wide range of applications, which include everything from daily life to industrial applications. Several microbial strains have been identified and analyzed for BS capacity, and new BS compounds are discovered and recovered every day. The BS's delicate character, low stability, and expensive production cost appear to be a significant impediment to its widespread adoption in industries. Various alternate sources of carbon or nitrogen, which are fundamental requirements for microbial development, have been used in the past to minimize the production costs, and researches have been effective to some extent. However, the reduced production of BSs remains a limiting factor. As a result, a thorough investigation of specific aspects, such as medium composition, biosynthesis pattern, ambient conditions and growth, is required for the large-scale production of BS molecules for industrial applications and economic reasons.

BSs outperform synthetic surfactants in terms of emulsification, foaming, and surface activity. BSs are also more stable than synthetic surfactants under a larger range of environmental variables, such as pH, temperature, moisture, and salinity. Thus, BS's prospective role and application in biocontrol seems encouraging, while more research in management of charcoal rot disease is needed.

Acknowledgments

The authors acknowledge the Director, CSIR National Botanical Research Institute for providing facilities and support during the study. This work is supported by the CSIR-Network (MLP0048 and MLP0049) and an in-house project (OLP109) funded by the Council of Scientific and Industrial Research, New Delhi, India.

References

Abalos, A., Pinazo, A., Infante, M. R., Casals, M., Garcia, F., & Manresa, A. (2001). Physicochemical and antimicrobial properties of new rhamnolipids produced by *Pseudomonas aeruginosa* AT10 from soybean oil refinery wastes. *Langmuir, 17*(5), 1367–1371.

Adamczak, M., & Bednarski, W. (2000). Influence of medium composition and aeration on the synthesis of biosurfactants produced by *Candida antartica. Biotechnology Letters, 22*, 313–316.

Alcantara, V. A., Pajares, I. G., Simbahan, J. F., & Edding, S. N. (2014). Downstream recovery and purification of a bioemulsifier from *Sacchromycescerevisiae* 2031. *Philippine Agriculture Scientist, 96*, 349–359.

Amani, H., & Kariminezhad, H. (2016). Study on emulsification of crude oil in water using emulsan biosurfactant for pipeline transportation. *Petroleum Science and Technology, 34*(3), 216–222.

Amodu, O. S., Ntwampe, S. K., & Ojumu, T. V. (2014). Emulsification of hydrocarbons by biosurfactant: Exclusive use of agro-waste. *Bioresources, 9*(2), 3508–3525.

Barros, F. F. C., Quadros, C. P. D., MarósticaJúnior, M. R., & Pastore, G. M. (2007). Surfactin: Chemical, technological and functional properties for food applications. *Química Nova, 30*, 409–414.

Cameron, D. R., Cooper, D. G., & Neufeld, R. J. (1988). The mannoprotein of *Saccharomyces cerevisiae* is an effective bioemulsifier. *Applied and Environmental Microbiology, 54*(6), 1420–1425.

Castaldi, S., Petrillo, C., Donadio, G., Piaz, F. D., Cimmino, A., Masi, M., Evidente, A., & Isticato, R. (2021). Plant growth promotion function of *Bacillus* sp. strains isolated from salt-pan rhizosphere and their biocontrol potential against *Macrophomina phaseolina. International Journal of Molecular Sciences, 22*(7), 3324.

Castelein, M., Verbruggen, F., Van Renterghem, L., Spooren, J., Yurramendi, L., Du Laing, G., Boon, N., et al. (2021). Bioleaching of metals from secondary materials using glycolipid biosurfactants. *Minerals Engineering, 163*, 106665.

Costa, S. G., Nitschke, M., Lépine, F., Déziel, E., & Contiero, J. (2010). Structure, properties and applications of rhamnolipids produced by *Pseudomonas aeruginosa* L2-1 from cassava wastewater. *Process Biochemistry, 45*(9), 1511–1516.

Desai, J. D., & Banat, I. M. (1997). Microbial production of surfactants and their commercial potential. *Microbiology and Molecular Biology Reviews, 61*(1), 47–64.

Gayathiri, E., Prakash, P., Karmegam, N., Varjani, S., Awasthi, M. K., & Ravindran, B. (2022). Biosurfactants: Potential and eco-friendly material for sustainable agriculture and environmental safety a review. *Agronomy, 12*(3), 662.

Gopalakrishnan, S., Humayun, P., Kiran, B. K., Kannan, I. G. K., Vidya, M. S., Deepthi, K., & Rupela, O. (2011). Evaluation of bacteria isolated from rice rhizosphere for biological control of charcoal rot of sorghum caused by *Macrophomina phaseolina* (Tassi) Goid. *World Journal of Microbiology and Biotechnology, 27*(6), 1313–1321.

Gorin, P. A. J., Spencer, J. F. T., & Tulloch, A. P. (1961). Hydroxy fatty acid glycosides of sophorose from *Torulopsis magnoliae. Canadian Journal of Chemistry, 39*(4), 846–855.

Hussain, T., & Khan, A. A. (2020). Determining the antifungal activity and characterization of Bacillus siamensis AMU03 against Macrophomina phaseolina (Tassi) Goid. *Indian Phytopathology, 73*(3), 507–516.

Ijaz, S., Sadaqat, H. A., & Khan, M. N. (2013). A review of the impact of charcoal rot (*Macrophomina phaseolina*) on sunflower. *The Journal of Agricultural Science, 151*(2), 222–227.

Jarvis, F. G., & Johnson, M. J. (1949). A glyco-lipide produced by *Pseudomonas aeruginosa. Journal of the American Chemical Society, 71*(12), 4124–4126.

Joshi-Navare, K., & Prabhune, A. (2013). A biosurfactant-sophorolipid acts in synergy with antibiotics to enhance their efficiency. *BioMed Research International, 2013*, 512495. https://doi.org/10.1155/2013/512495

Kaplan, N., Zosim, Z., & Rosenberg, E. (1987). Reconstitution of emulsifying activity of *Acinetobactercalcoaceticus* BD4 emulsan by using pure polysaccharide and protein. *Applied and Environmental Microbiology, 53*(2), 440–446.

Khanna, S., & Pattnaik, P. (2019). Production and functional characterization of food compatible biosurfactants. *Applied Food Science Journal, 3*, 1–4.

Khan, A. N., Shair, F., Malik, K., Hayat, Z., Khan, M. A., Hafeez, F. Y., & Hassan, M. N. (2017). Molecular identification and genetic characterization of *Macrophominaphaseolina* strains causing pathogenicity on sunflower and chickpea. *Frontiers in Microbiology, 8*, 1309.

Khare, E., & Arora, N. K. (2011). Physiochemical and structural characterization of biosurfactant from fluorescent *Pseudomonas* with biocontrol activity against *Macrophomina phaseolina*. In *Proceedings of the 2nd Asian PGPR conference* (pp. 104–109).

Kim, K., Dalsoo, Y., Youngbum, K., Baekseok, L., Doonhoon, S., & Eun-Ki, K. I. M. (2002). Characteristics of sophorolipid as an antimicrobial agent. *Journal of Microbiology and Biotechnology, 12*(2), 235–241.

Kiran, G. S., Selvin, J., Manilal, A., & Sujith, S. (2011). Biosurfactants as green stabilizers for the biological synthesis of nanoparticles. *Critical Reviews in Biotechnology, 31*(4), 354–364.

Kumar, S., Dheeman, S., Dubey, R. C., Maheshwari, D. K., & Baliyan, N. (2021). Cyclic siloxane biosurfactant-producing *Bacillus cereus* BS14 biocontrols charcoal rot pathogen *Macrophomina phaseolina* and induces growth promotion in *Vigna mungo* L. *Archives of Microbiology, 203*(8), 5043–5054.

Kumar, S., & Dubey, R. C. (2016). Application of biosurfactant producing PGPR in agriculture: A mini review. *New York Science Journal, 9*(8), 44–55.

Kumar, S., Dubey, R. C., & Maheshwari, D. K. (2016). Biosurfactant-mediated biocontrol of *Macrophomina phaseolina* causing charcoal rot in *Vigna mungo* by a plant growth promoting *Enterococcus* sp. BS13. *Journal of Plant Pathology & Microbiology, 7*, 2–8.

Kumar, A., & Johri, B. N. (2012). Antimicrobial lipopeptides of *Bacillus*: Natural weapons for biocontrol of plant pathogens. In T. Satyanarayana, & B. Johri (Eds.), *Microorganisms in sustainable agriculture and biotechnology* (pp. 91–111). Dordrecht: Springer. https://doi.org/10.1007/978-94-007-2214-9_6

Luna, J. M., Rufino, R. D., Sarubbo, L. A., & Campos-Takaki, G. M. (2013). Characterisation, surface properties and biological activity of a biosurfactant produced from industrial waste by *Candida sphaerica* UCP0995 for application in the petroleum industry. *Colloids and surfaces B: Biointerfaces, 102*, 202–209.

Magalhães, L., & Nitschke, M. (2013). Antimicrobial activity of rhamnolipids against *Listeria monocytogenes* and their synergistic interaction with nisin. *Food Control, 29*(1), 138–142.

Monnier, N., Furlan, A. L., Buchoux, S., Deleu, M., Dauchez, M., Rippa, S., & Sarazin, C. (2019). Exploring the dual interaction of natural rhamnolipids with plant and fungal biomimetic plasma membranes through biophysical studies. *International Journal of Molecular Sciences, 20*(5), 1009.

Mufti, R., & Bano, A. (2019). PGPR-induced defense responses in the soybean plant against charcoal rot disease. *European Journal of Plant Pathology, 155*(3), 983–1000.

Muthusamy, K., Gopalakrishnan, S., Ravi, T. K., & Sivachidambaram, P. (2008). Biosurfactants: Properties, commercial production and application. *Current Science, 94*(6), 736–747.

Pacwa-Płociniczak, M., Płaza, G. A., Piotrowska-Seget, Z., & Cameotra, S. S. (2011). Environmental applications of biosurfactants: Recent advances. *International Journal of Molecular Sciences, 12*(1), 633–654.

Pal, K. K., Tilak, K. V. B. R., Saxena, A. K., Dey, R., & Singh, C. S. (2001). Suppression of maize root diseases caused by *Macrophomina phaseolina, Fusarium moniliforme* and *Fusariumgraminearum* by plant growth promoting rhizobacteria. *Microbiological Research, 156*(3), 209–223.

Pandya, U., & Saraf, M. (2015). Isolation and identification of allelochemicals produced by *B. sonorensis* for suppression of charcoal rot of *Arachis hypogaea* L. *Journal of Basic Microbiology, 55*(5), 635–644.

Pornsunthorntawee, O., Chavadej, S., & Rujiravanit, R. (2009). Solution properties and vesicle formation of rhamnolipid biosurfactants produced by *Pseudomonas aeruginosa* SP4. *Colloids and Surfaces B: Biointerfaces, 72*(1), 6–15.

Pornsunthorntawee, O., Wongpanit, P., & Rujiravanit, R. (2010). Rhamnolipid biosurfactants: Production and their potential in environmental biotechnology. In R. Sen (Ed.), *Biosurfactants.Advances in experimental medicine and biology* (Vol 672, pp. 211–221). New York, NY: Springer. https://doi.org/10.1007/978-1-4419-5979-9_16

Rahman, P. K. S., & Gakpe, E. (2008). Production, characterisation and applications of biosurfactants-review. *Biotechnology, 7*(2), 360–370. https://doi.org/10.3923/biotech.2008.360.370

Ramani, K., Jain, S. C., Mandal, A. B., & Seka ran, G. (2012). Microbial induced lipoprotein biosurfactant from slaughterhouse lipid waste and its application to the removal of metal ions from aqueous solution. *Colloids and Surfaces B: Biointerfaces, 97*, 254–263.

Raza, Z. A., Khan, M. S., & Khalid, Z. M. (2007). Evaluation of distant carbon sources in biosurfactant production by a gamma ray-induced *Pseudomonas putida* mutant. *Process Biochemistry, 42*(4), 686–692.

Rodrigues, A. I., Gudiña, E. J., Teixeira, J. A., & Rodrigues, L. R. (2021). Biosurfactants as biocontrol agents against mycotoxigenic fungi. In H. Sarma, & M. N. V. Prasad (Eds.), *Biosurfactants for a sustainable future: Production and applications in the environment and biomedicine* (pp. 465–490).

Rodríguez-López, L., Vecino, X., Barbosa-Pereira, L., Moldes, A. B., & Cruz, J. M. (2016). A multifunctional extract from corn steep liquor: Antioxidant and surfactant activities. *Food & Function, 7*(9), 3724–3732.

Rufino, R. D., Sarubbo, L. A., Neto, B. B., & Campos-Takaki, G. M. (2008). Experimental design for the production of tensio-active agent by *Candida lipolytica*. *Journal of Industrial Microbiology and Biotechnology, 35*(8), 907–914.

Sabate, D. C., Brandan, C. P., Petroselli, G., Erra-Balsells, R., & Audisio, M. C. (2017). Decrease in the incidence of charcoal root rot in common bean (*Phaseolus vulgaris* L.) by *Bacillus amyloliquefaciens* B14, a strain with PGPR properties. *Biological Control, 113*, 1–8.

Salihu, A., Abdulkadir, I., & Almustapha, M. N. (2009). An investigation for potential development of biosurfactants. *Microbiology and Molecular Biology Review, 3*, 111–117.

Santos, E. F., Teixeira, M. F. S., Converti, A., Porto, A. L. F., & Sarubbo, L. A. (2019). Production of a new lipoprotein biosurfactant by *Streptomyces* sp. DPUA1566 isolated from lichens collected in the Brazilian Amazon using agroindustry wastes. *Biocatalysis and Agricultural Biotechnology, 17*, 142–150.

Sen, R. (2010). Surfactin: Biosynthesis, genetics and potential applications. *Biosurfactants, 672*, 316–323.

Sharma, C. K., Vishnoi, V. K., Dubey, R. C., & Maheshwari, D. K. (2018). A twin rhizospheric bacterial consortium induces systemic resistance to a phytopathogen *Macrophominaphaseolina* in mung bean. *Rhizosphere, 5*, 771–775.

Shete, A. M., Wadhawa, G., Banat, I. M., & Chopade, B. A. (2006). *Mapping of patents on bioemulsifier and biosurfactant: A review*.

Stringlis, I. A., Zhang, H., Pieterse, C. M., Bolton, M. D., & de Jonge, R. (2018). Microbial small molecules–weapons of plant subversion. *Natural Product Reports, 35*(5), 410–433.

Takenaka, S., Tonoki, T., Taira, K., Murakami, S., & Aoki, K. (2007). Adaptation of *Pseudomonas* sp. strain 7-6 to quaternary ammonium compounds and their degradation via dual pathways. *Applied and Environmental Microbiology, 73*(6), 1797–1802.

Teichmann, B., Linne, U., Hewald, S., Marahiel, M. A., & Bölker, M. (2007). A biosynthetic gene cluster for a secreted cellobiose lipid with antifungal activity from *Ustilago maydis*. *Molecular Microbiology, 66*(2), 525–533.

Toro, M., Ramirez-Bahena, M. H., Cuesta, M. J., Velazquez, E., & Peix, A. (2013). *Pseudomonas guariconensis* sp. nov., isolated from rhizospheric soil. *International Journal of Systematic and Evolutionary Microbiology, 63*(12), 4413–4420.

Van Renterghem, L., Roelants, S. L., Baccile, N., Uyttersprot, K., Taelman, M. C., Everaert, B., & Soetaert, W. (2018). From lab to market: An integrated bioprocess design approach for new-to-nature biosurfactants produced by *Starmerella bombicola*. *Biotechnology and Bioengineering, 115*(5), 1195–1206.

Velikonja, J. O. R. A. N., & Kosaric, N. A. I. M. (1993). Biosurfactants in food applications. In N. Kosaric, & F. V. Sukan (Eds.), *Biosurfactants: Production, properties, applications* (Vol 48, pp. 419–446). CRC Press.

Verma, S., Nain, L., Saxena, J., & Sharma, V. (2011). Screening of Pgp traits and biocontrol potential of Pseudomonas aeruginosa SL-72. *Annals of Plant Protection Sciences, 19*(2), 414–417.

Viejobueno, J., Albornoz, P. L., Camacho, M., de los Santos, B., Martínez-Zamora, M. G., & Salazar, S. M. (2021). Protection of strawberry plants against charcoal rot disease (*Macrophomina phaseolina*) induced by *AzospirillumBrasilense*. *Agronomy, 11*(2), 195.

Wu, Y. S., Ngai, S. C., Goh, B. H., Chan, K. G., Lee, L. H., & Chuah, L. H. (2017). Anticancer activities of surfactin and potential application of nanotechnology assisted surfactin delivery. *Frontiers in Pharmacology, 8*, 761.

Xu, Q., Nakajima, M., Liu, Z., & Shiina, T. (2011). Biosurfactants for microbubble preparation and application. *International Journal of Molecular Sciences, 12*(1), 462–475.

Zahir, Z. A., & Arshad, M. (2004). Perspectives in agriculture. *Advances in Agronomy, 81*, 97.

Zeraik, A. E., & Nitschke, M. (2010). Biosurfactants as agents to reduce adhesion of pathogenic bacteria to polystyrene surfaces: Effect of temperature and hydrophobicity. *Current Microbiology, 61*(6), 554–559.

Zhang, L., Tian, X., Kuang, S., Liu, G., Zhang, C., & Sun, C. (2017). Antagonistic activity and mode of action of phenazine-1-carboxylic acid, produced by marine bacterium *Pseudomonas aeruginosa* PA31x, against *Vibrio anguillarum* in vitro and in a Zebrafish in vivo model. *Frontiers in Microbiology, 8*, 289.

Zilli, C. G., Carmona, M. A., Simonetti, E., Santa-Cruz, D. M., Yannarelli, G. G., & Balestrasse, K. B. (2018). Biocontrol *of Macrophomina phaseolina* (Tassi) goid.: Differential production of H_2O_2 and in the relationship pathogen–PGPR in soybean seedling. *Biocontrol Science and Technology, 28*(4), 416–422.

Chapter 22

Management of *Macrophomina phaseolina* using nanoparticles

Anam Choudhary[1], Nasreen Musheer[2] and Sabiha Saeed[1]
[1]*Department of Plant Protection, Faculty of Agricultural Sciences, A.M.U., Aligarh, Uttar Pradesh, India;* [2]*School of Agriculture, Glocal University, Saharanpur, Uttar Pradesh, India*

22.1 Introduction

Macrophomina phaseolina has morphological (Mayek-Perez et al., 2002), genetic, pathogenic, and physiological diversity (Mahdizadeh et al., 2011) and are distributed in varied environmental conditions from arid to tropical conditions (Srinivasan et al., 2009). The occurrence and prevalence of *M. phaseolina* along with plant disease severity are expected to be increased with the global warming. It also helps in the dispersal of the pathogen from the old areas to the new areas (Pandey & Basandrai, 2021). Farmers have used various approaches to lessen the effect of the disease on the mungbean plants. The use of cultural practices such as altering planting dates could be helpful but not very effective method for the management (Bruton et al., 1987). The pathogen persists in the soil/or weeds due to the wide host range even though the crop rotation is used (Srivastava et al., 2008). Development of resistant varieties using host resistance by identifying sources for resistant germplasm is not easy (Divya Rani & Sudini, 2013). The pathogen is hard to control as pathogen has long-term survival ability as it survives in soil and spread with the seed. Several inorganic and organic chemicals are being used in the suppression of *M. phaseolina,* but the excessive use of chemical pesticides causes harmful effects on the environment and humans (Meena et al., 2020). Furthermore, these fungicides also cause the emergence of resistant strains of fungus due to their single target site mechanism (Fones et al., 2020) along with reduced nitrogen fixation and bioaccumulation of toxic chemicals in the environment (Khan et al., 2021). Hence, there is a crucial requirement to reduce the extreme usage of chemical pesticide in the mungbean fields. So, efforts are made to find alternative strategy with higher potential and eco-friendly method to control this soil-borne pathogen.

Recent effective and environment friendly alternatives that are used are plant extracts (Sales et al., 2016), essential oils (Santamarina et al., 2017) and biological control methods (Bazioli et al., 2019). But these have certain drawbacks, such as high acquisition costs, delays in the effect, and vulnerability due to their excessive use (Raveau et al., 2020). Another current development in agriculture, nanotechnology has emerged as promising tool for reducing the harmful effect of the chemical pesticides and enhancing the quality and productivity of crops with the minimal impact on the ecology (Fig. 22.1). It deals with the production and use of nanoparticles with the size ranging from 1−100 nm (Rajan, 2004, pp. 1−179) and larger surface area. The activity of nanoparticles is enhanced multiple times due to superior physical and chemical properties as compared to their precursors (Shobha et al., 2014). The application of nanotechnology in the field of plant pathology helps in the development of potential and effective method for controlling the plant pathogens. These nanoscale materials having excellent antifungal properties are used as manufacturing agents due to their easy production and handling (Abd-Elsalam, 2021). They target the plant pathogen interactions and bound to the cell wall by causing deformity, which leads to the death of the pathogen.

Among the recent advancement in agricultural sciences, metallic nanomaterials play a significant role in crop protection because of its unique physical and chemical properties, huge surface to volume ratio, structural stability, and strong affinity to their target. Nanoparticles such as Ag, CaO, Cu, MgO, MgO, SiO_2, TiO_2, and ZnO are used for the protection of plants from the microorganisms (Dwivedi et al., 2016). In many aspects, nanoparticles are considered as exceptional antimicrobial agent as they have greater surface area to volume ratio (Ingle et al., 2013) for their attachment and fast penetration

Macrophomina Phaseolina. https://doi.org/10.1016/B978-0-443-15443-0.00014-0

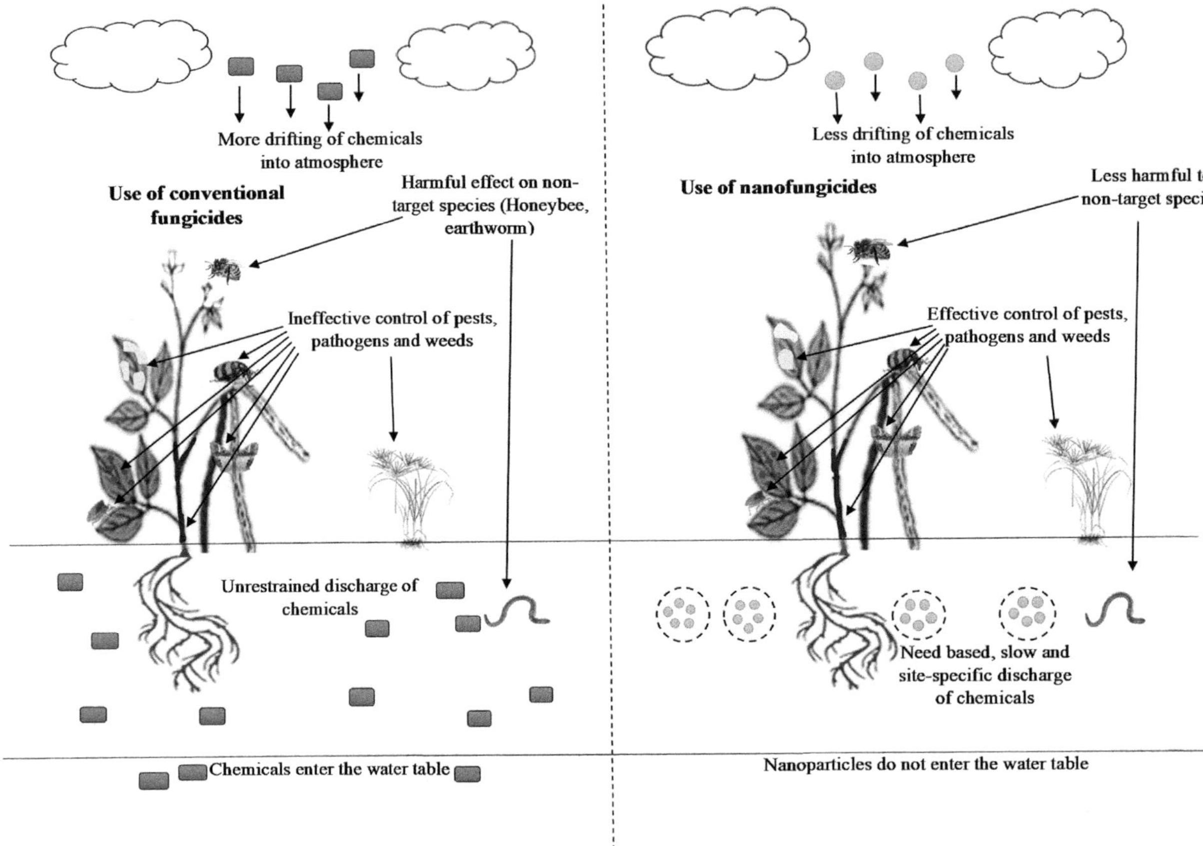

FIGURE 22.1 Schematic representation of the effect of chemical fungicides and nanobiofungicides on the environment and target and nontarget species.

inside the cell (Gurunathan et al., 2014). Various types of shapes, such as spherical, nanosized rods, tetragonals, triangles, hexagonal, pentagonal, elliptical, and irregular, were reported by Philip (2010). In this chapter, recent research on the use of nanoparticles for the sustainable management of the charcoal rot has been discussed.

22.2 Synthesis of nanoparticles

Two approaches are used for the synthesis of nanoparticles viz. top-down and bottom up. Top-down strategy involves the synthesis of nanoparticles from bulk material into smaller parts using physical methods. But bottom-up strategy involves the synthesis of nanoparticles as essential material from the smaller molecules. In this technique, chemical method and biological method (combined with metallic substance for the production of nanoscale materials) are used. Synthesis of nanoparticle is broadly grouped as physical, chemical, and biological methods using both top-down and bottom-up approaches. The strategies for the synthesis of nanoparticles are shown in Fig. 22.2.

22.2.1 Chemical method

The chief constituents for the synthesis include metallic precursors, reducing agents (inorganic and organic both), and stabilizing agents. Inorganic reducing agents used are hydrazine, hydroquinone, sodium citrate, and sodium borohydride. Organic reducing agents used are ascorbic, aniline, N,N-dimethylformaldehyde (DMF), dimethyl sulfoxide (DMSO), formaldehyde, tannic acid, and ethylene glycol (EG) (Zhang et al., 2016). The stabilizing agents are used to avoid agglomeration of nanoparticles.

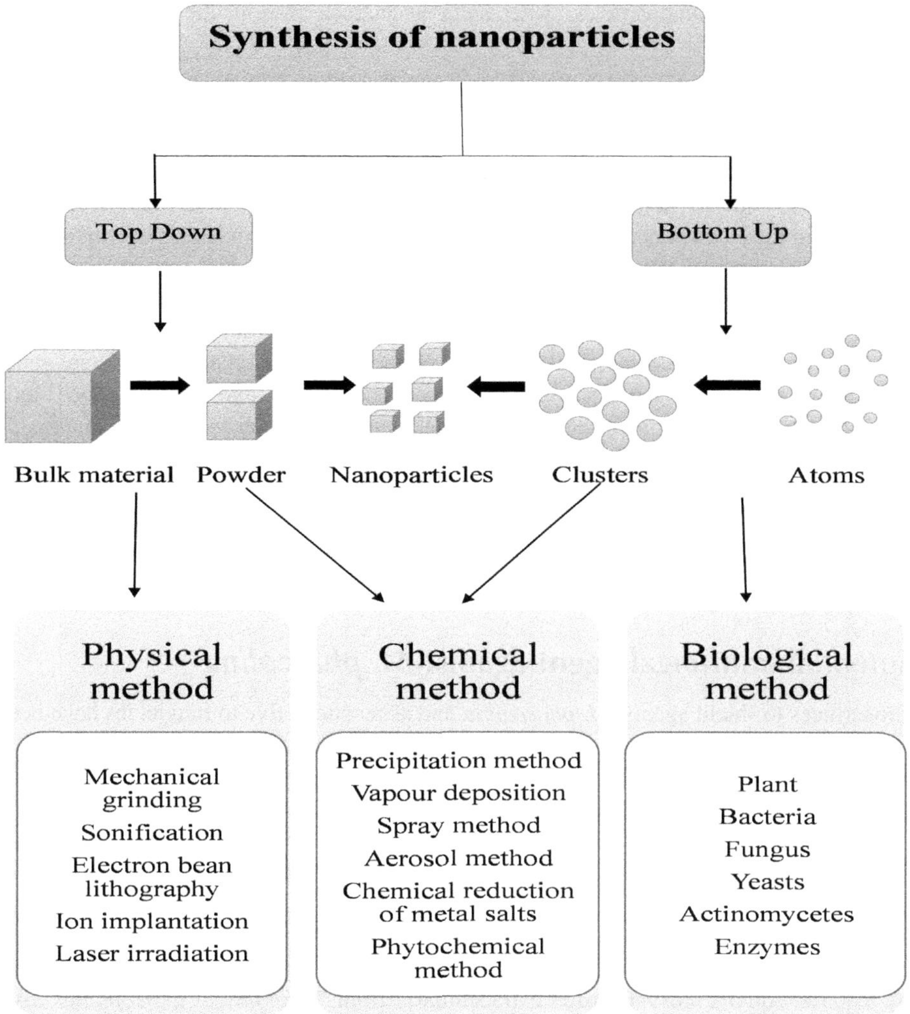

FIGURE 22.2 Methods used for the synthesis of nanoparticles using physical, chemical, and biological methods.

22.2.2 Physical method

Nanoparticle synthesis is done by top-down method, and the materials are decreased in size using high energy ball milling, arc discharge, irradiation, ultrasonication, microwave (MV), electrochemical method, CO_2 laser radiation, laser ablation, spray drying, conventional evaporation-condensation, pulsed electron evaporation, etc. Integration of nanoparticles is done by using a tube heater at barometric weight by evaporation condensation. The use of evaporation condensation and laser removal is the vital methodologies applied in physical method. The vaporization of source material into bearer gas contained in pontoon was done using heater. These build up procedures are utilized to synthesize various nanoparticles such as Ag, Au, Cd, PbS (Mathur et al., 2017). The synthesis of nanoparticles uses physical or mechanical energy using modern instruments and high energy.

22.2.3 Biosynthesis/green method

The synthesis of nanoparticles using green method is advantageous over chemical and physical method as they are cost effective, easy characterization, and less probability of failure (Abdelghany et al., 2018). The toxic metabolites are produced by physical and chemical methods causing stress to the environment. Plant extracts, fungi, bacteria, viruses, actinomycetes, algae, vitamins, and amino acids are used for the biosynthesis of nanoparticles (Baruwati et al., 2009). Glucose and sucrose are used as both stabilizers and reductants. The metallic salts are reacted with organisms/extracts followed by their biological reduction to convert into metallic nanoparticles. After suitable characterization, the nanoparticles are readily available (Mittal et al., 2013).

22.3 Application of nanoparticles

The nanoparticles are applied directly to the seeds, roots, or foliage to protect them against the disease. Metallic nanoparticles, such as Ag (Mahakham et al., 2017) and Cu (Nguyen et al., 2022), can be used for seed protection and seed coating. Nanoparticles are heteroaggregate in the environment due to their physiochemical active materials having reactive nature. This helps in the effective release of compound in the environment through seed coating (Lowry et al., 2012). Nanoparticles interact with the cells and are incorporated in different cell components. During germinationor in later stages (An et al., 2020; Mahakham et al., 2017), several biochemical pathways and gene expression profiles are initiated/altered by nano-priming of seeds (Chandrasekaran et al., 2020; Salama, 2019). Nano priming of seeds regulates the resistant genes against biotic stress and improves the innate immunity. Nanoparticles also reduces the cell damage of seeds under biotic stress conditions by reducing the ROS levels of seeds by increasing the enzymatic activities, such as catalase, superoxide dismutase, and guaiacolperoxidase (Guha et al., 2018). The expression of plant genes of peroxidase, polyphenoloxidase, and phenylalanine ammonia lyase is increased. Furthermore, pathogenesis-related (PR) genes (PR1 and PR5) associated with salicylic acid pathways are overexpressed in plants. Also, nanoparticles serve as co-factors/signals by enhancing the transcription of genes regulation associated to phytohormones and response to biotic stress (Chandrasekaran et al., 2020; Ye et al., 2020).

22.4 Nanoparticles as potential antifungal agent against *M. phaseolina*

Numerous studies on the use of nanoparticles to shield against *M. phaseolina* and as an alternative to fungicides have been described (Table 22.1). The nanoform of fungicides trifloxystrobin 25% and tebuconazole 50% (75 WG) was used by Kumar et al. (2016) to control *M. phaseolina.* These nanoformulations are more chemically reactive and enhanced the inhibitory effect than large sized particles of fungicides. Nanofungicides caused 100% mycelial growth inhibition at 25 ppm concentration as compared to the commercial fungicide, which caused only 76.3% inhibition at the same concentration. The nanoparticle affected hyphae and formation of sclerotia causing alteration and destruction of hyphal cells as well as abnormal/lack of sclerotia.

Ruiz-Romeroaet al. (2018) investigated the silver nanoparticle biosynthesized from *Yucca shilerifera* against *M. phaseolina* infecting strawberries. The AgNPs showed inhibitory action against the pathogen by damaging the fungal hyphae and reducing the growth by 67.05%. AgNPs biosynthesized extracellularly from *Streptomyces griseoplanus* SAI-25 inhibited the mycelium of pathogen infecting sorghum with 13 mm zone of inhibition at 1000 μg/mL (Vijayabharathi et al., 2018). AgNPs biosynthesized from *Justica spicigera* using 100 mg/mL concentration inhibited the mycelial growth to 79.77% (Bernardo-Mazariegos et al., 2019). Leaf extract of *Amaranthus retroflexus* was used for the biosynthesis of AgNPs by Bahrami-Teimoori et al. (2017). These NPs prevented the growth of *M. phaseolina* by 71.09% at 400 μg/mL. The application of AgNPs causes structural changes and alterations in the hyphae and cell wall and also affected the germination of the microsclerotia. AgNPs biosynthesized from *Portucaoleracea* showed antifungal effect on the pathogen by reducing the mycelial growth by 82.39%–84.09% (Al-Otibi et al., 2022). The hypha become flaccid and hyphal branches become condensed with shrunken surface. Biosynthesized AgNPs from peel of *Citrus sinensis* gave 100% inhibition of the pathogen at 100 ppm concentration. In greenhouse experiments, the incidence of charcoal rot disease in faba bean was reduced to 10.7% (Mohamed & Elshahawy, 2022). AgNPs synthesized from *T. harzianum* reduced the linear growth of the pathogen by 4.167 cm when applied at 100 μg/mL concentration. Cotton seedlings treated with 100 and 200 μg/mL of AgNPs affected their survival rate by 76.67%–93.33% (Zaki et al., 2022). *Zea mays* seeds treated with AgNPs biosynthesized from *Tegetes erectus* increased the germination percentage by 87.5% (Kumar et al., 2020). Biosynthesized AgNPs from *Moringa oleifera* exhibited 100% mycelial inhibition at 50 ppm concentration. Faba bean seeds soaked in these nanoparticles (0.05% concentration) before sowing reduced the incidence of charcoal rot disease to 10.50% (Mohamed et al., 2021).

TABLE 22.1 Nanoparticles used for the management of *Macrophominaphaseolina*.

Source for synthesis	Nano particles	Size (nm)	Shape	References
Acalphya indica	Ag	10–50	Spherical	Krishnaraj et al. (2012)
Amaranthus retroflexus	Ag	10–32	Spherical	Bahrami-Teimoori et al. (2017)
Citrus sinensis	Ag	32–47	Spherical	Mohamed and Elshahawy (2022)
Justica spicigera	Ag	86 –100	Spherical	Bernardo-Mazariegos et al. (2019)
Moringa oleifera	Ag	24	Spherical	Mohamed et al. (2021)
Ocimum sanctum	Cu	25	Rod, cylindrical and elliptical	Shende et al. (2016)
Portuca oleracea	Ag	117.4	Spherical	Al-Otibi et al. (2022)
Streptomyces griseoplanus	Ag	19.5 –20.9	Spherical	Vijayabharathi et al. (2018)
Tagetes erectus	Ag	20–60	Oval	Kumar et al. (2020)
Trichoderma harzianum	ZnO	8–25	Hexagonal, spherical, rod	Zaki et al. (2021)
Trichoderma harzianum	Ag	6–15	Spherical	Zaki et al. (2022)
Yucca schidigera	Ag	729	–	Ruiz-Romeroa et al. (2018)
Chemical method	Chitosan	50	Spherical	El-Mohamedy et al. (2019)
Chemical method	Ag	100	Spherical	Vrandecic et al. (2020)
Chemical method	Zn	8–210	Spherical to hexagonal	Kalia et al. (2021)
Physical method	Trifloxystrobin 25% + Tebuconazole 50% (75 WG)	108 –130	Irregular	Kumar et al. (2016)

Silver (20–80 nm), titanium dioxide (85–100 nm), and zinc oxide (35–45 nm) were also tested by Shyla et al. (2014) to control the pathogen. Among these silver nanoparticles showed antifungal affect as compared to titanium dioxide and zinc oxide. Saharan et al. (2013) and Divya et al. (2017) demonstrated the use of chitosan nanoparticles for the management of *M. phaseolina* causing inhibition of mycelial growth. The complete reduction of mycelial growth was observed at 0.05%–0.1% concentrations (El-Mohamedy et al., 2019). Zinc nanoparticles caused fragmentation of hyphae and cytological changes in *M. phaseolina* infecting maize in agar wall assay (Kalia et al., 2021). ZnO(zinc oxide) nanoparticles biosynthesized from *Trichoderma harzianum* caused 100% mycelial inhibition at all the three concentrations (20, 40, 100 μg/mL) using in vitro assay (Zaki et al., 2021). ZnONPs at 200 μg/mL controlled the disease incidence in cotton up to 91.11% under greenhouse conditions. Chitosan and Cu-chitosan nanoparticles with stability and uniform size lead to their higher antifungal activity against the pathogen (Saratale et al., 2018).

Shende et al. (2016) reported the use of copper nanoparticles biosynthesized from *Ocimum sanctum* exhibiting antifungal activity against *M. phaseolina* by producing highly reactive hydroxyl radicals causing damage to proteins, lipids, DNA, etc. The antimicrobial activity of CuNPs against the pathogen showed that minimum inhibitory concentration (MIC) of 0.03 mg/mL reduced the inhibition zone to 12.5 mm. Antifungal activity of GO causing inhibition of the growth of *M. phaseolina* infecting *Lens culinaris* at three different concentration (125, 250, 500 ppm) was observed by Guroo et al. (2016). The highest mycelial inhibition of 78.2% was recorded at 500 ppm as compared to the other concentrations. The nanoparticles caused the disruption of cell wall leading to leakage of electrolytes.

22.5 Mechanism of action

The exact mechanism for the inhibition of pathogen by nanoparticles is not completely collected and explained. Still, indication of metal ion release (Panicker et al., 2020), induced oxidative stress (Joe et al., 2018), and nonoxidative stress models (Anicic et al., 2018) are followed.

Nanoparticles with negative or low negative charges bind to cell membrane of the pathogen with negative charge due to the electrostatic attraction. It causes depolarization of membrane along with the disruption of cell wall of the mycelium resulting in cellular deformation. Injury causes pore formation in cell membrane and disturbs the phospholipids, proton motive force (PMF), and membrane potential leading to the leakage of electrolytes (Chen et al., 2014). The activation of membrane potential helps in the spore germination (Harris, 2005), but their perturbation and leakage of electrolytes cause inactivation of spore germination. The nanoparticles deposits on the surface of spores causing plasmolysis and also inhibits the water uptake (Wang et al., 2014). It also affects the cellular components including damage of DNA, enzymatic activity inhibition, cell cycle arrest, and generation of reactive oxygen species (ROS) leading to death (Fig. 22.3).

Various components of ROS include free radicals for example singlet oxygen ($^1O_2$), superoxide (O_2^-), carbon dioxide radical (CO_2^-), carbonate (CO_3^-), hydroxyl ($HO^\cdot$), hydroperoxyl (HO_2), hydrogen peroxide (H_2O_2), alkoxyl (RO) and peroxyl (RO_2) and non-radical for example hypochlorite (OCl^-), hypochlorous acid (HOCl), hypobromous acid (HOBr), ozone (O_3), nitric acid (NO), peroxy nitrous acid (ONOOH), peroxy nitrite ($ONOO^-$), peroxy nitrate (O_2NOO^-), peroxomonocarbonate ($HOOCO_2^-$), and organic peroxides (ROOH) (Khan et al., 2021). Furthermore, they may also form irregular pits on the cell surface of the pathogen facilitating their entry in the periplasmic space and inside the cell (Gahlawat & Choudhury, 2019). TEM analysis showed rupture of cell wall, swelling of nuclear and cytoplasmic contents (Elbeshehy et al., 2015). DNA damage includes breakage of single strands, breakage of double strands, cross-linking with proteins deletions, mutations, etc (Soenen et al., 2011). Higher concentrations of nanoparticles cause oxidative stress leading to the breakage of single and double strands, nitrogen base, and lesions in pentose sugar (Singh et al., 2019).

22.6 Conclusion and future aspects

Nanobiofungicides are considered as next generation eco-friendly and safe method for the plant disease management. Nanoparticles possess antimicrobial activity against the pathogen and can be used as an effective measure in a long term as compared to chemical fungicides. They possess properties such as requirement of a lesser quantity of dose, durability, and high effectiveness. Nanoparticles are advantageous to the plant as well as farmers and have no significant adverse effect on the soil microflora, germination of seed and root to shoot ratio. Seed treatment with nanoparticles supports the establishing of plants and helps in providing protection against the disease. This can be used as an alternative to commercial fungicide as they advance the selective toxicity and pesticide resistance in the pathogen. Use of nanoparticles would help in reducing the dependency of agriculturists, pathologists, and farmers on the environment toxic chemical fungicides.

The size, concentration, and duration of exposure of nanoparticles should be considered before seed priming as they cause side effects, such as inhibition of germination, decreased development of plants, harmful alterations of cell structure and metabolism, and alterations in the interaction between root and microbiota. Nontarget organisms, such as mineral solubilizers and symbionts, are also affected by the direct soil and foliar application of nanoparticles. Therefore, before applying these nanoparticles, critical analysis should be done on their effect on nontarget organisms under field conditions. The use of nanoparticles for the management of diseases is in initial stages. Additional studies based on the use of nanoparticles on the diagnosis of disease, physiology of host and pathogen, interaction, and process of infection will help for the development of less damaging novel nanoparticles. Hence, the apprehension of biosafety should be spoken. Collaborative and multidisciplinary research provides a platform to use nanotechnology for their application in plant protection.

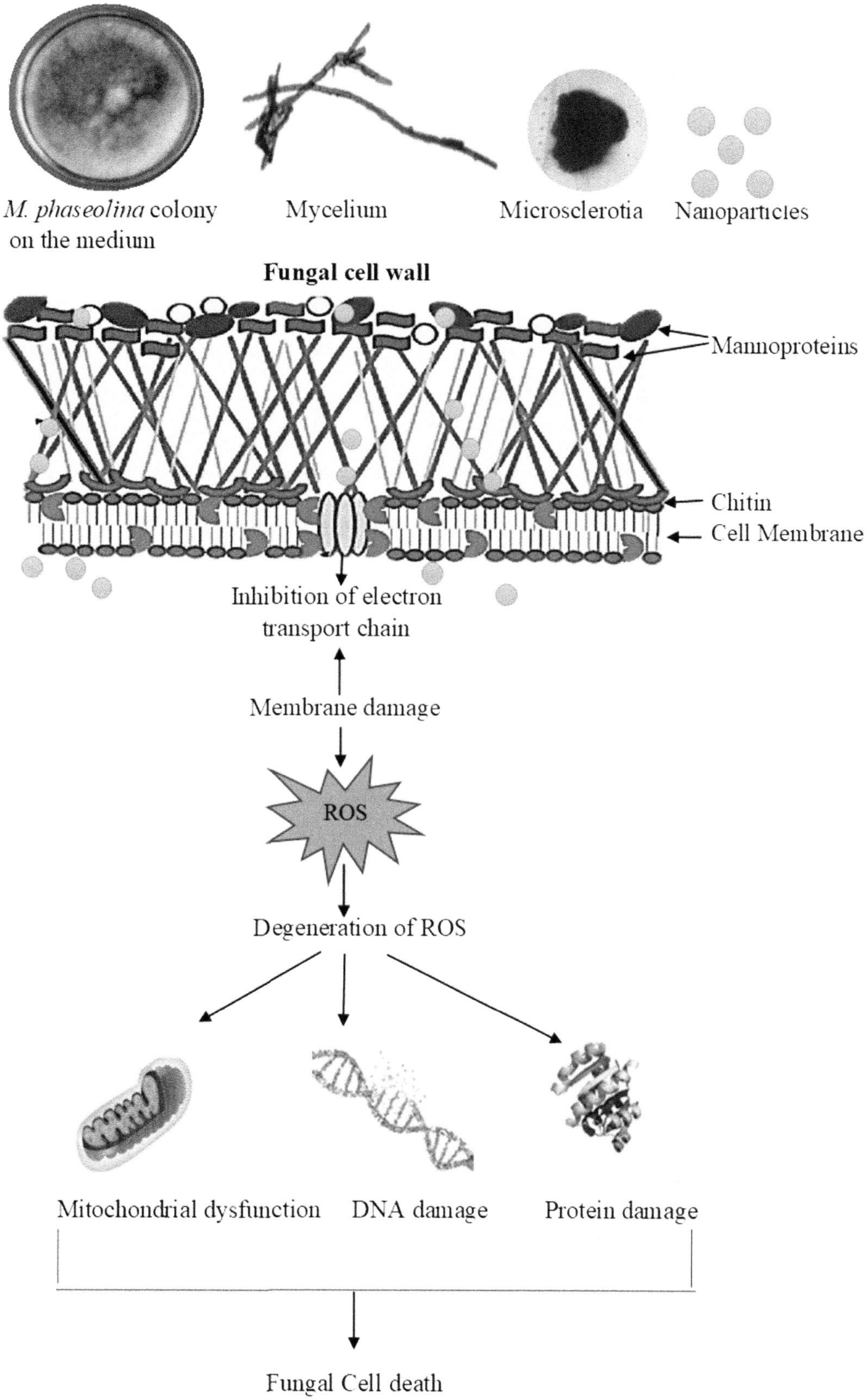

FIGURE 22.3 Diagrammatic representation of the mechanism used by the nanoparticles to inhibit the fungal pathogen.

References

Abd-Elsalam, K. (2021). Special issue: Fungal nanotechnology. *Journal of Fungi, 7*(8), 583.

Abdelghany, T. M., Al-Rajhi, A. M. H., Al Abboud, M. A., Alawlaqi, M. M., Magdah, A. G., Helmy, E. A. M., & Mabrouk, A. S. (2018). Recentadvances in green synthesis of silver nanoparticles and theirapplications: About future directions. *Bionanoscience, 8*, 5–16.

Al-Otibi, F., Alfuzan, S. A., Alharbi, R. I., Al-Askar, A. A., AL-Otaibi, R. M., Al Subaie, H. F., & Moubayed, N. M. S. (2022). Comparative study of antifungal activity of two preparations of green silver nanoparticles from *Portulacaoleracea* extract. *Saudi Journal of Biological Sciences, 29*, 2772–2781.

An, J., Hu, P., Li, F., Wu, H., Shen, Y., White, J. C., Tian, X., Li, Z., & Giraldo, J. P. (2020). Emerging investigator series: Molecular mechanisms of plant salinity stress tolerance improvement by seed priming with cerium oxide nanoparticles. *Environment Science & Nanotechnology, 7*, 2214–2228.

Anicic, N., Vukomanovic, M., Koklic, T., & Suvorov, D. (2018). Fewer defects in the surfaces lows the hydrolysis rate, decreases the ROS generation potential, and improves the non-ROS antimicrobial activity of MgO. *Small, 14*(26), 1800205.

Bahrami-Teimoori, B., Nikparast, Y., Hojatianfar, M., Akhlaghi, M., Ghorbani, R., & Pourianfar, H. R. (2017). Characterisation and antifungal activity of silver nanoparticles biologically synthesised by *Amaranthusretroflexus* leaf extract. *Journal of Experimental Nanoscience, 12*, 129–139.

Baruwati, B., Polshettiwar, V., & Varma, R. S. (2009). Glutathione promoted expeditious green synthesis of silver nanoparticles in water using microwaves. *Green Chemistry, 11*, 926–930.

Bazioli, J. M., Belinato, J. R., Costa, J. H., Akiyama, D. Y., Pontes, J. G. D. M., Kupper, K. C., Augusto, F., De Carvalho, J. E., & Fill, T. P. (2019). Biological control of citrus postharvest phytopathogens. *Toxins, 11*, 460.

Bernardo-Mazariegos, E., Valdez-Salas, B., González-Mendoza, D., Abdelmoteleb, A., Camacho, O. T., Duran, C. C., & Gutiérrez-Miceli, F. (2019). Silver nanoparticles from *Justicia spicigera* and their antimicrobial potentialities in the biocontrol of foodborne bacteria and phytopathogenic fungi. *Revista Argentina de Microbiología, 51*(2), 103–109.

Bruton, B. D., Jeger, M. J., & Reuveni, R. (1987). *Macrophominaphaseolina* infection and vine decline in cantaloupe in relation to planting date, soil environment and plant maturation. *Plant Disease, 71*, 259–263.

Chandrasekaran, U., Luo, X., Wang, Q., & Shu, K. (2020). Are there unidentified factors involved in the germination of nanoprimed seeds? *Frontiers of Plant Science, 11*, 832.

Chen, J., Peng, H., Wang, X., Shao, F., Yuan, Z., & Han, H. (2014). GO exhibits broad-spectrum antimicrobial activity against bacterial phytopathogens and fungal conidia by intertwining and membrane perturbation. *Nanoscale, 6*, 1879–1889.

Divya Rani, V., & Sudini, H. (2013). Management of soilborne diseases in crop plants: An overview. *International Journal of Plant, Animal and Environmental Sciences, 3*(4), 156–164.

Divya, K., Vijayan, S., George, T. K., & Jisha, M. (2017). Antimicrobial properties of chitosan nanoparticles: Mode of action and factors affecting activity. *Fibers and Polymers, 18*, 221–230.

Dwivedi, S., Saquib, Q., Al-Khedhairy, A. A., & Musarrat, J. (2016). Understanding the role of nanomaterials in agriculture. In D. P. Singh, H. B. Singh, & R. Prabha (Eds.), *Microbial inoculants in sustainable agricultural productivity* (pp. 271–288). New Delhi: Springer.

El-Mohamedy, R. S. R., Abd El-Aziz, M. E., & Kamel, S. (2019). Antifungal activity of chitosan nanoparticles against some plantpathogenic fungi *in vitro*. *Agriculture Engineering International: CIGR Journal, 21*(4), 201–209.

Elbeshehy, E. K. F., Elazzazy, A. M., & Aggelis, G. (2015). Silver nanoparticles synthesis mediated by new isolates of *Bacillus* spp.; nanoparticle characterization and their activity against Bean Yellow Mosaic Virus and human pathogens. *Frontiers in Microbiology, 6*, 453.

Fones, H. N., Bebber, D. P., Chaloner, T. M., Kay, W. T., Steinberg, G., & Gurr, S. J. (2020). Threats to global food security from emerging fungal and oomycete crop pathogens. *Natural Food, 1*, 332–342.

Gahlawat, G., & Choudhury, A. R. (2019). A review on the biosynthesis of metal and metal salt nanoparticles by microbes. *RSC Advances, 9*, 12944–12967.

Guha, T., Ravikumar, K. V. G., Mukherjee, A., Mukherjee, A., & Kundu, R. (2018). Nanopriming with zero valentiron (NZVI) enhances germination and growth in aromatic rice cultivar (*Oryzasativa*cv. Gobindabhog L.). *Plant Physiology and Biochemistry, 127*, 403–413.

Guroo, J. A., Khan, M., Ahmad, A., Azam, A., & Siddiqui, Z. A. (2016). Management of *Meloidogyne incognita* and *Macrophominaphaseolina* by grapheneoxide on *Lens culinaris*. *Acta Phytopathologica et Entomologica Hungarica, 51*(1), 43–56.

Gurunathan, S., Han, J. W., Kwon, D. N., & Kim, J. H. (2014). Enhanced antibacterial and anti-biofilm activities of silver nanoparticles against Gram-negative and Gram-positive bacteria. *Nanoscale Research Letters, 9*(1), 1–17.

Harris, S. D. (2005). Morphogenesis in germinating *Fusariumgraminearum* macroconidia. *Mycologia, 97*, 880–887.

Ingle, A. P., Duran, N., & Rai, M. (2013). Bioactivity, mechanism of action, and cytotoxicity of copper-based nanoparticles: A review. *Applied Microbiology and Biotechnology, 98*(3), 1001–1009.

Joe, A., Park, S. H., Kim, D. J., Lee, Y. J., Jhee, K. H., Sohn, Y., & Jang, E. S. (2018). Antimicrobial activity of ZnO nanoplates and its Ag nanocomposites: Insight into an ROS-mediated antibacterial mechanism under UV light. *Journal of Solid State Chemistry, 267*, 124–133.

Kalia, A., Kaur, J., Tondey, M., Manchanda, P., Bindra, P., Alghuthaymi, M. A., Shami, A., & Abd-Elsalam, K. A. (2021). Differential antimycotic and antioxidant potentials of chemically synthesized zinc-based nanoparticles derived from different reducing/complexing agents against pathogenic fungi of maize crop. *Journal of Fungi, 7*, 223.

Khan, M., Khan, A. U., Hasan, M. A., Yadav, K. K., Pinto, M. M. C., Malik, N., Yadav, V. K., Khan, A. H., Islam, S., & Sharma, G. K. (2021). Agronanotechnology as an emerging field: A novel sustainable approach for improving plant growth by reducing biotic stress. *Applied Sciences, 11*, 2282.

Krishnaraj, C., Ramachandran, R., Mohan, K., & Kalaichelvan, P. (2012). Optimization for rapid synthesis of silver nanoparticles and its effect on phytopathogenic fungi. *Spectrochimica Acta Part A: Molecular and Biomolecular Spectroscopy, 93*, 95–99.

Kumar, G. D., Natarajan, N., & Nakkeeran, S. (2016). Antifungal activity of nanofungicideTrifloxystrobin 25% + tebuconazole 50% against *Macrophominaphaseolina*. *African Journal of Microbiological Research, 10*, 100–105.

Kumar, P., Pahal, V., Gupta, A., Vadhan, R., Chandra, H., & Dubey, R. C. (2020). Effect of silver nanoparticles and *Bacillus cereus* LPR2 on the growth of Zea mays. *Scientific Reports, 10*, 20409.

Lowry, G. V., Gregory, K. B., Apte, S. C., & Lead, J. R. (2012). Transformations of nanomaterials in the environment. *Environmental Science & Technology, 46*, 6893–6899.

Mahakham, W., Sarmah, A. K., Maensiri, S., & Theerakulpisut, P. (2017). Nanopriming technology for enhancing germination and starch metabolism of aged rice seeds using phytosynthesized silver nanoparticles. *Scientific Reports, 7*, 8263.

Mahdizadeh, V., Safaie, N., & Aghajani, M. A. (2011). New hosts of *Macrophominaphaseolina* in Iran. *Journal of Plant Pathology, 93*, 4.

Mathur, P., Jha, S., Ramteke, S., & Jain, N. K. (2017). Pharmaceutical aspects of silver nanoparticles. *Artificial Cells, Nanomedicine, and Biotechnology, 46*, 1–12.

Mayek-Pérez, N., García-Espinosa, R., López-Castañeda, C., Acosta-Gallegos, J. A., & Simpson, J. (2002). Water relations, histopathology and growth of common bean (Phaseolus vulgaris L.) during pathogenesis of *Macrophominaphaseolina* under drought stress. *Physiological and Molecular Plant Pathology, 60*, 185–195.

Meena, R. S., Kumar, S., Datta, R., Lal, R., Vijayakumar, V., Brtnicky, M., Sharma, M. P., Yadav, G. S., Jhariya, M. K., & Jangir, C. K. (2020). Impact of agrochemicals on soil microbiota and management: A review. *Land, 9*(2), 34.

Mittal, A. K., Chisti, Y., & Banerjee, U. C. (2013). Synthesis of metallic nanoparticles using plant extracts. *Biotechnological Advances, 31*, 346–356.

Mohamed, Y. M. A., & Elshahawy, I. E. (2022). Antifungal activity of photo-biosynthesized silver nanoparticles (AgNPs) from organic constituents in orange peel extract against phytopathogenic Macrophomina phaseolina. *European Journal of Plant Pathology, 162*, 725–738.

Mohamed, Y. M. A., Osman, S. A., Elshahawy, I. E., Soliman, G. M., & AhmedA, M. A. (2021). Charcoal rot and root-knot nematode control on faba bean by photosynthesized colloidal silver nanoparticles using bioactive compounds from *Moringaoleifera* leaf extract. *Journal of Plant Protection Research, 61*(4), 414–429.

Nguyen, D. V., Nguyen, H. M., Le, N. T., Nguyen, K. H., Le, H. M., Nguyen, A. T., Dinh, N. T. T., Hoang, S. A., & Ha, C. V. (2022). Copper nanoparticle application enhances plant growth and grain yield in maize under drought stress conditions. *Journal of Plant Growth Regulations, 41*, 364–375.

Pandey, A. K., & Basandrai, A. K. (2021). Will *Macrophominaphaseolina* spread in legumes due to climate change? A critical review of current knowledge. *Journal of Plant Diseases and Protection, 128*, 9–18.

Panicker, S., Ahmady, I. M., Han, C., Chehimi, M., & Mohamed, A. A. (2020). On demand release of ionic silver from gold-silver alloy nanoparticles: Fundamental antibacterial mechanisms study. *Materials Today Chemistry, 16*, 100237.

Philip, D. (2010). Green synthesis of gold and silver nanoparticles using *Hibiscus rosasinensis*. *Physica E: Low-Dimensional Systems and Nanostructures, 42*(5), 1417–1424.

Rajan, M. S. (2004). *Nano: The next revolution*. New Delhi, India: National Book Trust, 81-237-7961-5.

Raveau, R., Fontaine, J., & Sahraoui, A. L. H. (2020). Essential oils as potential alternative biocontrol products against plant pathogens and weeds: A review. *Foods, 9*, 365.

Ruiz-Romeroa, P., Valdez-Salasb, B., Gonzalez-Mendozaa, D., & Mendez-Trujillob, V. (2018). Antifungal effects of silver phytonanoparticles from *Yucca shilerifera* against strawberry soil-borne pathogens: *Fusariumsolani* and Macrophomina phaseolina. *Mycobiology, 46*(1), 47–51.

Saharan, V., Mehrotra, A., Khatik, R., Rawal, P., Sharma, S. S., & Pal, A. (2013). Synthesis of chitosan-based nanoparticles and their in vitro evaluation against phytopathogenic fungi. *International Journal of Biological Macromolecules, 62*, 677–683.

Salama, D. M. (2019). Effect of Zinc Oxide nanoparticles on the growth, genomic DNA, production and the quality of common dry bean (*Phaseolus vulgaris*). *Biocatalysis and Agricultural Biotechnology, 11*, 101083.

Sales, M. D. C., Costa, H. B., Fernandes, P. M. B., Ventura, J. A., & Meira, D. D. (2016). Antifungal activity of plant extracts with potential to control plant pathogens in pineapple. *Asian Pacific Journal of Tropical Biomedicine, 6*, 26–31.

Santamarina, M. P., Ibáñez, M. D., Marqués, M., Roselló, J., Giménez, S., & Blázquez, M. A. (2017). Bioactivity of essential oils in phytopathogenic and post-harvest fungi control. *Natural Product Research, 31*, 2675–2679.

Saratale, R. G., Saratale, G. D., Shin, H. S., Jacob, J. M., Pugazhendhi, A., Bhaisare, M., & Kumar, G. (2018). New insights on the green synthesis of metallic nanoparticles using plant and waste biomaterials: Current knowledge, their agricultural and environmental applications. *Environmental Science and Pollution Research, 25*, 10164–10183.

Shende, S. S., Gaikwad, N. D., & Banso, S. D. (2016). Synthesis and evaluation of antimicrobial potential of copper nanoparticle against agriculturally important Phytopathogens. *International Journal of Biology Research, 1*(4), 41–47.

Shobha, G., Moses, V., & Ananda, S. (2014). Biological synthesis of copper nanoparticles and its impact. *International Journal of Pharmaceutical Science Invention, 3*(8), 6–28.

Shyla, K. K., Natarajan, N., & Nakkeeran, S. (2014). Antifungal activity of zinc oxide, silver and titanium dioxide nanoparticles against *Macrophomina phaseolina*. *Journal of Mycology and Plant Pathology, 44*, 268–273.

Singh, A., Joshi, N. C., & Ramola, M. (2019). Magnesium oxide nanoparticles (MgONPs): Green synthesis, characterizations and antimicrobial activity. *Research Journal of Pharmacy and Technology, 12*, 4644.

Soenen, S., Rivera-Gil, P., Montenegro, J. M., Parak, W. J., De Smedt, S., & Braeckmans, K. (2011). Cellular toxicity of inorganic nanoparticles: Common aspects and guidelines for improved nanotoxicity evaluation. *Nano Today, 6*, 446–465.

Srinivasan, A., Wickes, B. L., Romanelli, A. M., Debelenko, L., Rubnitz, J. E., Sutton, D. A., et al. (2009). Cutaneous infection caused by *Macrophomina phaseolina* in a child with acute myeloid leukemia. *Journal of Clinical Microbiology, 47*, 1969–1972.

Srivastava, P. C., Singh, A. P., Kumar, S., Ramachandran, V., Srivastava, M., & D'souza, S. F. (2008). Desorption and transformation of zinc in a mollisol and its uptake by plants in a rice-wheat rotation fertilized with either zinc-enriched biosludge from molasses or with inorganic zinc. *Biology and Fertility of Soils, 44*, 1035–1041.

Vijayabharathi, R., Sathya, A., & Gopalakrishnan, S. (2018). Extracellular biosynthesis of silver nanoparticles using *Streptomyces griseoplanus* SAI-25 and its antifungal activity against *Macrophomina phaseolina*, the charcoal rot pathogen of sorghum. *Biocatalyst Agriculture and Biotechnology, 14*, 166–171.

Vrandečić, K., Ćosić, J., Ilić, J., Ravnjak, B., Selmani, A., Galić, E., Pem, B., Barbir, R., Vrček, I. V., & Vinković, T. (2020). Antifungal activities of silver and selenium nanoparticles stabilized with different surface coating agents. *Pest Management Science, 76*, 2021–2029.

Wang, X., Liu, X., Chen, J., Han, H., & Yuan, Z. (2014). Evaluation and mechanism of antifungal effects of carbon nanomaterials in controlling plant fungal pathogen. *Carbon, 68*, 798–806.

Ye, Y., Cota-Ruiz, K., Hernández-Viezcas, J. A., Valdés, C., Medina-Velo, I. A., Turley, R. S., Peralta-Videa, J. R., & Gardea-Torresdey, J. L. (2020). Manganese nanoparticles control salinity-modulated molecular responses in *Capsicum annuum* L. through priming: A sustainable approach for agriculture. *ACS Sustainable Chemical Engineering, 8*, 1427–1436.

Zaki, S. A., Ouf, S. A., Abd-Elsalam, K. A., Asran, A. A., Hassan, M. M., Kalia, A., & Albarakaty, F. M. (2022). Trichogenic silver-based nanoparticles for suppression of fungi involved in damping-off of cotton seedlings. *Microorganisms, 10*, 344.

Zaki, S. A., Ouf, S. A., Albarakaty, F. M., Habeb, M. M., Aly, A. A., & Abd-Elsalam, K. A. (2021). *Trichoderma harzianum*-mediated ZnO nanoparticles: A green tool for controlling soil-borne pathogens in cotton. *Journal of Fungi, 7*, 952.

Zhang, X. F., Liu, Z. G., Shen, W., & Gurunathan, S. (2016). Silver nanoparticles: synthesis,characterization, properties, applications, and therapeuticapproaches. *International Journal of Molecular Sciences, 17*, 1534.

Section VI

Bio-management

Chapter 23

Biomanagement of *Macrophomina phaseolina* causing charcoal rot of plants

Kriti Joshi[1], Ragini Gothalwal[2] and Naveen Joshi[1]

[1]*Department of Biotechnology and Microbiology SJHS Gujarati Innovative College of Commerce and Science, Indore, Madhya Pradesh, India;*
[2]*Department of Biotechnology and Bioinformatics Centre Barkatullah Vishwavidhyalaya, Bhopal, Madhya Pradesh, India*

23.1 Introduction

Macrophomina phaseolina (Tassi) Goid. is among the most catastrophic soil-borne fungal pathogen inciting disease in more than 500 monocot and dicot plants. It belongs to the family Botryosphaeriaceae exhibiting heterogeneous specificity of hosts. It exists in the form of pycnidial stage while its sclerotial stage is known as *Rhizoctonia bataticola* (Taub.) Butler. Being a necrotrophic fungus, it causes various diseases viz., charcoal rot, root rot, damping off, seedling blight, stem rot, and collar rot in agronomical principal crops such as soybean, maize, bean, sorghum, sunflower, sesame, cotton, chickpea, and groundnut (Arenas et al., 2021; Khaledi & Taheri, 2016). Crop plants growing in subtropical and tropical countries having semi-arid climate are more prone to its infection. Forest trees viz. Pinus, Softwood, Abies, Cassia, and fruit trees-strawberry, *Cocos nucifera,* Coffea spp., Citrus spp., etc. also get infected by this polyphagous fungus. Diseases caused by it have created an alarming situation in Europe, Africa, North America, South America, and Asia due to its wide host range. It is known to infect roots, branches, stem, leaves, seeds, and petioles. Temperature and pH affect its growth. Optimum temperature of 30—35°C and optimum pH of 7.0 are required for its proper growth. This phytopathogen is saprophytic in nature which resides as microsclerotia in the soil for long duration. Survival of microsclerotia in the soil may be for 2—15 years which depends on environmental conditions and their contact with tissues of host plant (Gupta et al., 2012). Microsclerotia are minuscule, smooth, jet black, and thick-walled cells consisting of black hyphae. Nutritional composition of substrates determines microsclerotial size. They may be oval, curved, globose, oblong, or elliptical in shape. Hyphal filaments merge together leading to development of pycnidia. Pycnidia is globose or flattened asexual fruiting body varying in size from 100 to 200 μm in diameter (Bhandari, 2017). The dark brown thick-walled cells surround pycnidia bearing membrane and rod-shaped long conidiophores. Phaseolinone is the fungal toxin secreted by *M. phaseolina* known to destroy the crop plants when severity of infection is high. It has wide geographical distribution owing to its morphological, pathogenic, genetic, and physiological variability leading to its adaptability to different environmental conditions (Reznikov et al., 2018). Several techniques are being used to control this fungus like chemical fungicides, but best eco-friendly approach is the biomanagement i.e., application of beneficial microbes such as *Bacillus* spp., *Azotobacter* spp., *Pseudomonas* spp., etc. to control the charcoal rot and enhance the crop yield.

23.2 Disease cycle

The primary source of inoculum of *M. phaseolina* responsible for infection is microsclerotia (hard, small, black, or grayish-black structures), which germinate at 20—40°C. Dry and hot weather and low soil moisture favors disease development. There are four phases in the life cycle of *M. phaseolina*, germinating, penetrating, parasitic, and saprophytic phases (Dhingra & Sinclair, 1977; Pandey et al., 2021). After penetration through multiple hyphae during the course of parasitic phase, it secretes enzymes and toxins which degrade the root tissues colonizing the roots and stem (Islam et al., 2012). Toxins viz., botryodiplodin and phaseolinone produced by *M. phaseolina* are known to be involved in pathogenesis. Both of these toxins were found to be responsible for charcoal rot in soybean in southern USA (Abbas et al., 2019). In the

Macrophomina Phaseolina. https://doi.org/10.1016/B978-0-443-15443-0.00007-3

saprophytic phase, it infects plant's vascular system there by obstructing with water and nutrients transport to leaves, which results in senescence and yellowing of leaves (Dave et al., 2020), arrests plant growth affecting its yield (Lodha & Mawar, 2020). Degradation of plant cells leads to necrosis causing black discoloration of the roots and premature plant death.

23.3 Symptoms

M. phaseolina causes infection in plants at any developmental stages. However, charcoal rot symptoms may be observed during seedling to flowering stage. Hypocotyl of emerging seedlings develops brown lesions (Rubayet et al., 2021). Leaves become yellow, premature leaf drop, brown to red discoloration on roots and stem and reduced vigor is observed. In the later stage, wilting, senescence, and premature plant death occur. A large number of microsclerotia are formed on lower stem and tap roots, giving it charcoal like appearance. Development of chlorotic lesions on mature plants takes place, and dead leaves remain adhered to the petiole (Luna et al., 2017; Reznikov et al., 2018). Its symptoms remain prominent till flowering stage.

23.4 Yield loss

Charcoal rot caused by *M. phaseolina* is an economically paramount disease leading to higher yield losses worldwide. In urdbean, dry root rot caused around 29% yield loss (Kulkarni et al., 2019). The pathogen infects kidney bean and mungbean seeds, causing severe loss of seed germination and viability (Sarita & Singh, 2014). Stored mungbean seeds get deteriorated to about 2%–36%, reported from different South Asian countries including India (Ashwini & Giri, 2014). Its increasing incidence in mungbean has made it yield-limiting disease in South Asian Countries (Singh et al., 2020). Seventy percent loss in soybean yield due to charcoal rot has been reported in India. But mortality percentage of different *M. phaseolina* isolates ranged between 32% and 96% against soybean cultivar Monnetta (Pancheshwar et al., 2016). In South Africa, 40%–70% yield loss of soybean and maize has been reported (Waals et al., 2019). Similarly, in India, about 64% yield loss in sorghum has been reported under disease favoring conditions (Das et al., 2008; Hotkar et al., 2021). It mainly affects grain and fodder yield of sorghum (Bannur & Chattannavar, 2019).

23.5 Management of *M. phaseolina*

Management of charcoal rot is rather difficult due to long persistence of *M. phaseolina*, broad host range, and saprophytic ability. The control measures used for its management include practices like crop rotation, tillage, irrigation, chemical control, and biological control. Crop rotation and intercropping strategies are not considerable in controlling infection owing to its wide range of hosts. Irrigation did not show any major effect on reduction of microsclerotial densities, thus providing no prevention against its infection (Marquez et al., 2021). Similarly, inconsistent effect of tillage has been observed on charcoal rot incidence (Luna et al., 2017). Chemical control relies on the use of chemical fungicides which cause pollution, is not environmental friendly and show toxicity. The injudicious use of chemical pesticides in agriculture has resulted in inimical effects on the humans and animal's health; developed resistance in pests and created ecological imbalances. Due to agro-environmental policies and negative impact of these chemicals, emphasis is to be laid on more sustainable alternatives to control plant diseases. The best measure to combat phytopathogens is biological control which is a bio-based, eco-friendly safe strategy with practical and economical approach.

23.5.1 Biological control

Beneficial microorganisms able to antagonize plant pathogens are employed in biological control to combat plant diseases. These living agents leading to reduction in the number or disease causing ability of pathogens are known as biocontrol agents (BCAs). Since they act against plant pathogens, they are also known as antagonists. They pose an effect on metabolic activities and population of soil-borne pathogens. Bacteria, nonpathogenic fungi, viruses, and actinomycetes are mostly used as BCAs. An efficient biocontrol agent is the one which has broad spectrum activity against diseases and should be safe for all life-forms. Application of BCAs provides protection to the crops throughout the crop period is economically feasible and target specific. They exhibit dual action of controlling diseases and promoting plant growth.

Biocontrol agents suppress the effect of plant pathogens by using a bewildering array of mechanisms. The consistency and efficacy of biological control can be improved by understanding the mechanisms underlying their antagonistic activity. The different mechanisms employed by BCAs are induced systemic resistance, rivalries for nutrients and space, antibiosis, lytic enzymes production, HCN, and hyperparasitism (Fig. 23.1).

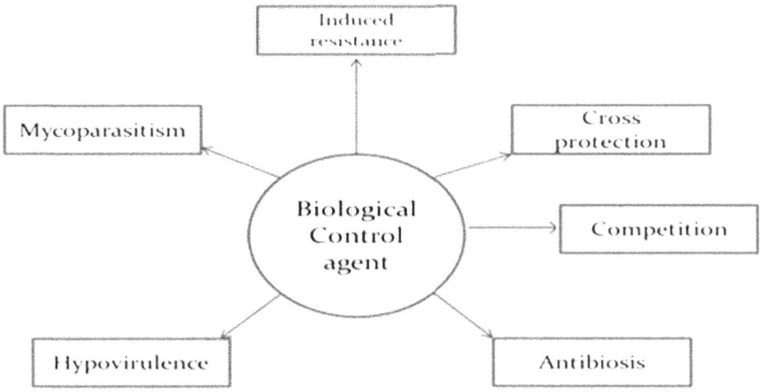

FIGURE 23.1 Mechanisms of biocontrol agent (Dave et al., 2020).

23.5.1.1 Competition

Biocontrol agents compete with pathogens for nutrients and space for successful establishment in the environment. Since soil is limiting in soluble form of nutrients required for microbial growth, competition for resources, such as nutrients, oxygen, and site on roots is a general and but obvious phenomenon among microbes. This results in starvation and thus is responsible for death of soil borne pathogens. BCAs outcompete phytopathogens for nutrients or occupation of site on plant roots (Babalola et al., 2017). For competition to occur, biocontrol agent and the pathogen should have same requirement for resources. It is an indirect interaction between pathogens and antagonists, which limits the binding of a pathogen to the plant for its proliferation. Both the BCA and plant compete for availability of carbon and trace nutrients, such as Fe, Zn, Mn, and Cu. It is thought that competition for carbon inhibits fungal spores (Dave, 2021). BCAs have more efficient uptake systems for substances as compared to pathogens leading to competition for micronutrients. Fe is essential for microbial growth and is limiting in soil. Antagonistic microbes produce high-propinquity Fe-binding chelators called siderophores, which help in the uptake of siderophore-Fe-complex making iron unavailable to pathogens. *Pseudomonas syringae* BAF.1 was found to be producer of siderophore and inhibited *Fusarium oxysporum* (Qu et al., 2017).

23.5.1.2 Antibiosis

Production of antibiotics in sufficient quantity is one of the most promising strategies used by the biocontrol agents to prevent proliferation of phytopathogens. Antibiotics being target specific exhibit systemic mode of action successfully control phytopathogens. Also they do not get easily washed off by rain, therefore, remain effective for longer duration. Some of the antibiotics produced by BCAs are pyoluteorin, 2,4-DAPG (2,4- diacetyl phloroglucinol), pyrrolnitrin, zwittermycin, iturin, bacillomycin, phenazines, mycosubtilin, etc. 2,4-DAPG produced by *Pseudomonas* species is reported to be one of the most efficient antibiotic in inhibiting plant pathogens (Babalola et al., 2012). Various isoforms of iturin produced by *Bacillus* sp. had shown inhibitory activity against *M. phaseolina* (Dave, 2021).

23.5.1.3 Production of lytic enzymes and volatile compounds

PGPR acting as biocontrol agents secrete extracellular lytic enzymes such as protease and chitinase cellulase that reduce the polymeric compounds such as proteins, chitin, cellulose, hemicelluloses, and DNA, thereby interfering in the growth of pathogens. Lytic enzymes may interrupt cell wall synthesis by altering its steps or inhibit it or even perforate cell membrane (Arora et al., 2020). All the components of fungal cell wall are hydrolyzed by various enzymes produced by BCAs. Biopesticide preparation utilizing BCA producing lytic enzymes or adding crude enzyme to it can enhance its efficacy. Chitinase produced by *B. subtilis* inhibited *Rhizoctonia solani* (Saber et al., 2015). *B. amyloliquefaciens* showed antagonistic activity against *Clavibacter michiganensis* owing to the production of cellulose, chitinase, protease, and lipase enzymes (Gautam et al., 2019). As a part of β-1,3-glucanases, chitinases, and PR proteins can actuate systemic resistance in plants on activation (Babalola et al., 2017). Volatile substances are categorized into diverse chemical groups—alcohols, aldehydes, ketones, lactones, and esters produced as a result of metabolic reactions. Their diffusion takes place to a large distances leaving inhibitory effect on pathogens. Factors such as soil properties, plant exudates, and microorganism determine their field application (Fincheira & Quiroz, 2018). Volatile compounds such as hydrogen cyanide are produced by BCAs to control phytopathogens. HCN inhibits the activity of cytochrome *c* oxidase and other metalloenzymes. It shows higher toxicity even at picomolar concentration. Since it is produced in low quantity, this mechanism needs to be coupled with other measures for effective biocontrol. HCN production is synergistic to other biocontrol strategies used by the same bacterium as evinced by the improved efficiency of antifungal compounds produced by it (Babalola et al., 2017). HCN production by the bacteria belonging to genera *Pseudomonas, Bacillus, Rhizobium, Aeromonas,* and *Alcaligenes* have been reported (Das et al., 2017).

23.5.1.4 Mycoparasitism

Mycoparasitism is a direct mechanism in which one fungus (host) acts as a source of nutrition for another fungus (parasite). Mycoparasite may establish itself on either actively growing hyphae or resting structure of the host such as sclerotia (Mukherjee et al., 2021). Mostly, it is employed by fungal BCAs as an antifungal trait reported rarely in bacteria. Production of enzymes and secondary metabolites controls this mechanism. Biocontrol agent coils around the pathogen, grows toward it and finally attacks cell wall/cell membrane of the pathogen. *Trichoderma* exhibits mycoparasitism as one of the most important biocontrol strategy to control phytopathogens. Khalili et al. (2016) observed mycelial interactions of *Trichoderma harzianum* T12 on *M. phaseolina* attacking its cell wall contributing to its antifungal activity.

23.5.1.5 Induced systemic resistance

Rhizospheric microbes alleviate the deleterious effects of phytopathogens by inducing a systemic resistance mechanism in plants. Initiation of plant defense mechanisms by external agents before infection is known as induced systemic resistance (ISR), but it is called systemic acquired resistance (SAR) when triggered by a localized infection (Miljakovic et al., 2020). It is an indirect antagonistic strategy wherein BCAs produce certain chemical stimuli, such as jasmonic acid, or ethylene leading to the biochemical changes activating resistance mechanisms in plants. The chemical stimuli are classified under two types—pathogen-associated molecular pattern (PAMP) produced by plants and microbe-associated molecular pattern (MAMP) produced by beneficial microbes. Various metabolites viz., siderophores, antibiotics like DAPG, pyocyanin, VOCs like 2R, 3R-butanediol, and biosurfactants produced by rhizobacteria aid in inducing systemic resistance in plants. Phytoalexins, phenolic compounds, and reactive oxygen species (ROS) are also produced in ISR. Endochitinase and several other compounds released by *Trichoderma* interact with plants and induce plant defense-related proteins (Dar et al., 2013). *Bacillus* species are known to synthesize antioxidant defense enzymes, such as polyphenol oxidase, peroxidase, superoxide dismutase, and phenylalanine ammonia-lyase, thereby eliciting ISR. Anthracnose disease of chili was suppressed by *Bacillus* species through the activation of defense enzymes and accumulation of phenolic compounds (Jayapala et al., 2019).

23.6 Biomanagement of *M. phaseolina*

Diseases caused by *M. phaseolina* can be efficiently managed through biological control exploiting the use of beneficial microbiota of the soil belonging to plant growth promoting bacteria and fungi (Table 23.1).

23.6.1 Bacteria as BCAs

It was reported that bacteria residing in rhizosphere and soil possess antagonistic properties against *M. phaseolina*. Bacterial isolates demonstrated their potential to reduce disease incidence and also showed plant growth promoting properties. Bacterial species belonging to *Pseudomonas, Pantoea*, and *Bacillus* have dual properties of growth promotion and reduced disease incidence.

Indigenous bacteria of *Pseudomonas* and *Bacillus* species were isolated from soybean rhizosphere and studied for antifungal activity by Leon et al. (2009). They determined the antifungal properties against *M. phaseolina, Pythium ultimum, Fusarium oxysporum, Fusarium solani,* and *Sclerotinia minor.* The potent isolates identified as *Pseudomonas* species BNM296 and *Pseudomonas* sp. BNM297, *Bacillus amyloliquefaciens* BNM340, and *Bacillus cereus* BNM343 inhibited *M. phaseolina* along with *P. ultimum, F. oxysporum, F. solani* and *S. minor. P. fluorescens* BNM296, and *B. amyloliquefaciens* BNM340 showed PGPR traits, i.e., auxin production, phosphate solubilization, and seedling emergence rate. The isolates also produced hydrolytic enzymes, siderophores, and volatile compounds. *Pseudomonas fluorescens* strains-BHUPf4, BHUPf5, BHUPf6, and BHUPsb inhibited *Rhizoctonia bataticola* and *Sclerotinia sclerotiorum* causing root rot and stem rot, respectively, in chickpea. *Pseudomonas fluorescens* BHUPf4 caused maximum inhibition of *R. bataticola* by 67% and of *S. sclerotiorum* by 66% (Patel et al., 2011).

P. fluorescence from different rhizospheric agricultural fields has been found to be a strong antagonist against *R. bataticola*. Belkar and Gade (2012) evaluated the antagonistic potential of acidic and alkaline *P. fluorescence* strains in

TABLE 23.1 Various BCAs used against *M. phaseolina*.

Host plants	Organism	BCA	Experiment	References
Soybean	Bacteria	*Bacillus* sp. and *Pantoea agglomerans*	In vitro and in vivo	Vasebi et al. (2013)
		Pseudomonas sp. *Bacillus* spp.	In vitro and in vivo	Simonetti et al. . (2015)
		P. stutzeri and *P. putida*	In vitro	Bano and Mufti (2019)
		Pseudomonas putida and *Bacillus clausii*	In vivo	Yasmin et al. (2020)
		B. amyloliquefaciens B. subtilis, B. tequilensis and *K. flava*	In vitro and in vivo	Dave (2021)
	Fungi	*Trichoderma harzianum*	In vitro and in vivo	Khaledi and Taheri (2016)
		AMF	In vitro and in vivo	Spagnoletti et al. (2017, 2020)
Chickpea	Bacteria	*Pseudomonas fluorescens*	In vitro and *in- vivo*	Manjunatha et al. (2013)
	Fungi	*Trichoderma viride*	In vitro and in vivo	Manjunatha et al. (2013)
		Trichoderma viride	In vivo	Lakhran and Ahir (2018)
Groundnut	Bacteria	*Paenibacillus illinoisensis, Bacillus subtilis* and *Pseudomonas psychrotolerans*	In vivo	Ali et al. (2015)
		Bacillus cereus	In vitro and *in- vivo*	Sanjeevkumar et al. (2020)
	Fungi	AMF	In vivo	Doley and Jite (2013)
Sunflower	Bacteria	*Azotobacter chroococcum, Klebsiella pneumonia,* and *Azospirillum brasilense*	In vitro and in vivo	Komy et al. (2020)
	Fungi	*Trichoderma* spp.	In vitro	Sridharan et al. (2020)
		AMF and yeast	In vitro and in vivo	Nafady et al. (2019)
Urdbean	Fungi	*Glomus aggregatum*	In vitro	Chandra et al. (2011)
		T. viride	In vitro	Athira (2017)

opposition to *R. bataticola,* a causative agent of root rot in soybean and *Sclerotium rolfsii*. Alkaline isolate *P. fluorescence* Pf3 was observed as the strongest antagonist showing 74.7% and 67.04% inhibition against *S. rolfsii* and *R. bataticola,* respectively after 7 days incubation. Acidic isolates were less effective as compared to alkaline isolates. Siderophore production by these strains might be the mechanism involved in inhibiting *R. bataticola.* It reveals the efficacy of *P. fluorescence* to be used as potent BCA against *M. phaseolina. Pseudomonas* strain PfCHA0 produced antibiotics 2,4-DAPG, pyoluteorin, pyrrolnitrin and phenazines, and volatile compound HCN that curbed the mycelial growth of *R.bataticola* causing dry root rot in *Vigna radiata.* Chaubey (2012) also detected and characterized the genes coding for antibiotics through RP-HPLC.

Bacillus species and *Pantoea agglomerans* were used as biocontrol agents against *M. phaseolina* causing charcoal rot disease in soybean in vitro and in greenhouse conditions by Vasebi et al. (2013). Percentage inhibition of *M. phaseolina* was observed as *Bacillus* species BIN-63.3%, *P. agglomerans*-73.8%. Both the BCAs increased the growth indices of soybean under greenhouse in the plants infested with pathogen. Two major antifungal metabolites, 2-hydroxyphenazine (2-OH-PHZ) and phenazine-1-carboxylic acid (PCA), produced by *P. fluorescence* Os25 were evaluated for their ability to inhibit *R. bataticola* and *F. oxysporum.* PCA inhibited *R. bataticola* by 40% at different concentrations (Sujatha & Ammani, 2013). Also a significant decrease in fungal biomass with increasing concentration of PCA was observed. It indicates the potential of PCA (phenazines) obtained from biocontrol agent *P. fluorescence* Os25 to be developed as an important biofungicide.

Paenibacillus illinoisensis, Bacillus subtilis, and *Pseudomonas psychrotolerans* isolated from the groundnut rhizosphere inhibited *M. phaseolina, F. oxysporum,* and *F. solani* infecting chickpea. All the three isolates efficiently reduced disease severity and increased plant biomass in contrast to control under greenhouse conditions (Ali et al., 2015). The mechanism of antibiosis in antagonism against *R. bataticola* causing DRR in chickpea under in vitro conditions was studied using a rhizospheric bacterium *Pseudomonas* strain Ps14c and *Bacillus* isolate B20b (Kumari et al., 2015). Their culture filtrates and diffusible, extracted, and thermostable antimetabolites from these strains inhibited *R. bataticola.*

Rhizospheric *Bacillus* strain CaB5 from chickpea, which showed maximum inhibition of *R. bataticola,* was studied to target the genes for antibiotic production viz., fengycin, iturin, bacillomycin D and surfactin by Kodoth et al. (2015). The isolate CaB5 increased the vigor index of chickpea seedlings as compared to control. Also crude antibiotic extracts inhibited the mycelial growth of the pathogen by 40%.

P. polymyxa and *B. subtilis* have been reported to antagonize *R. bataticola* and *F. oxysporum* due to the production of proteases, chitinases, cellulases, and β-1,3-glucanases suggesting lytic enzymes as the potential antiphytopathogenic trait along with siderophores. Both the isolates were found to be potent plant growth promoters (Shivannavar et al., 2015). Antifungal metabolites 3-isobutyl hexahydropyrrolo (1,2-a) pyrazine-1,4 dione and three benzylhexahydro -pyrrolo [1,2-a] pyrazine −1,4 dione produced by *P. protegens* inhibited *M. phaseolina* and other phytopathogens (Saini et al., 2016). Bioformulations of *Trichoderma* (TL1), *P. fluorescens* [PfUL(A)] and *Bacillus subtilis* (BsOP2) along with FYM and neem cake reduced dry root rot incidence in blackgram under field conditions (Latha et al., 2017).

Antagonistic activity of *Stenotrophomonas* and *Serratia* against *M. phaseolina* was also reported (Dave et al., 2021; Mmbaga et al., 2018). The PGPR strains, *P. stutzeri* (KX574858) and *P. putida* (KX574857) were studied for induction of systemic tolerance in Ajmeri and NARC soybean cultivars infected with charcoal rot (Bano & Mufti, 2019). Both the strains reduced disease severity in both cultivars. *P. stutzeri* also induced antioxidant and defense enzymes, phenyl ammonia lyase, superoxide dismutase, peroxidase, and polyphenol oxidase in both the cultivars. Disease suppression ability of *P. stutzeri* is significantly attributed to systemic resistance induction in infected soybean plants.

A concoction of rhizobacteria, *Azotobacter chroococcum* ZCR, *Klebsiella pneumoniae* KPR, and *Azospirillum brasilense* SBR was studied for antagonism in sunflower against root-rot complex pathogens (*M. phaseolina, R. solani* and *F. solani*) in vitro and under field conditions. Seed treatment and soil drenching significantly reduces disease incidence (44−55%) (Komy et al., 2020). These rhizobacteria also promoted plant growth via phosphate solubilization, IAA production, and nitrogen fixation resulting in improved yield and increased oil content in sunflower. *Pseudomonas putida* and *Bacillus clausii* suppressed *M. phaseolina* infection in soybean, increased germination and growth of soybean. *B. clausii* mitigated disease (97%) and enhanced content of proline and activities of peroxidase, superoxide dismutase, ascorbate peroxidase, and catalase in soybean roots and shoots. Both the strains also increased jasmonic acid and salicylic acid contents in treated plants (Yasmin et al., 2020). Their potential formulation as effective biocontrol agent can help to mitigate diseases caused by *M. phaseolina.*

Bacillus amyloliquefaciens IDK22, *B. subtilis* IDK23, *B. tequilensis* IKAK46, *B. subtilis* IKNK51, and *Kocuria flava* ISKK65 isolated from rhizosphere of soybean showed strong antagonistic activity against *M. phaseolina* causing charcoal rot in soybean (Dave, 2021). All the five isolates inhibited growth of the fungus both in vitro and in vivo. *Bacillus* strains used in the present study produced lytic enzymes, such as cellulase, chitinase and protease resulting in significant inhibition of the pathogen. Lipopeptidal antibiotic iturin and its isoforms were also produced by these *Bacillus* strains with *B. tequilensis* IKAK46 producing three different isoforms showing antagonism against *M. phaseolina*. Maximum HCN production by an actinobacterium *K. flava* ISKK65 indicates its toxicity against this pathogen. Also metabolite extracted from it showing antifungal activity against *M. phaseolina* was identified as IAA suggesting its indirect role in antagonism (Dave, 2021). In vitro plant growth promotion activities of these bacterial species were also studied wherein all of them solubilized phosphorus and produced IAA significantly.

Charcoal-based bioformulations of *B. amyloliquefaciens* IDK22, *B. subtilis* IDK23, *B. tequilensis* IKAK46, *B. subtilis* IKNK51, and *K. flava* ISKK65 reduced disease incidence of charcoal rot in soybean exhibiting their excellent biocontrol efficiency in vivo. *B. tequilensis* IKAK46 bioinoculant increased dry weight of roots maximally in plants infected with *M. phaseolina*. Results indicate that all the five bacteria are efficient biocontrol agents antagonizing this devastating pathogen (Dave, 2021). Dual benefit of biocontrol and PGPR activities of these bacteria can be commercialized to develop as effective biofungicide with biofertilizers.

23.6.2 Fungi as BCAs

Trichoderma spp. are widely used effective BCAs for several soil-borne fungal pathogens. They have developed various antagonistic mechanisms, such as nutrient competition, mycoparasitism, and production of antibiotic (Marquez et al., 2021). *Trichoderma* is a saprotrophic fungus that can establish easily in different types of soil and its population persists for longer time at detectable levels. Some species of *Trichoderma* are known for plant growth promotion or persuade systemic resistance (Martinez-Medina et al., 2016). Strains of *Trichoderma* sp. produce a variety of microbe-associated molecular patterns, which enhance plant immune response (Khaledi & Taheri, 2016). Some species lead to induction of cellular and molecular changes in plants conferring resistance to various biotic and abiotic stresses (Kumar, 2013). *Trichoderma harzianum* T7 and T14 reduced disease incidence of charcoal rot caused by *M. phaseolina* in soybean both in vitro and in vivo. Inhibition was known to be due to formation of volatile and non-volatile metabolites. Also increased phenolics and POX activity in soybean plants were observed inducing plant defense response against *M. phaseolina* (Khaledi & Taheri, 2016). *T. viride* reduced root rot incidence of urdbean and also its culture filtrate exhibited strong inhibition of *M. phaseolina* sclerotial germination (Athira, 2017). In vitro efficacy of *T. viride* against *M. phaseolina* causing root rot in sesame was evaluated; isolate *Tv*3 inhibited the pathogen maximally (Vinothini et al., 2020).

Arbuscular mycorrhizal fungi (AMF) were reported to reduce disease incidence thereby controlling phytopathogens effectively (Jain & Pundir, 2019). The antipathogenic mechanisms employed by AMF include competition for space and photosynthetic products inducing defense response and activating disease resistance system in plants. An AMF species *Glomus claroideum* inhibited dry root rot fungus *M. phaseolina* infecting mungbean, while *Glomus aggregatum* was found effective against its infection in urdbean (Chandra et al., 2011).

Reduction in infection of *M. phaseolina* on roots and severity of disease symptoms have been found upon application of AMF. A significant increase in proline and phenol content activated defense response and decreased oxidative damage was observed contributing to antifungal activity of AMF (Spagnoletti et al., 2017). AMF are effective against *M. phaseolina* in soybean under nitrogen fertilization conditions (Spagnoletti et al., 2020).

Yeast *Brettanomyces naardensis* showed antagonistic activity against *M. phaseolina* and also promoted plant growth. It colonized hyphae of the fungus leading to malformation and damage (Nafady et al., 2019).

23.7 Conclusion

M. phaseolina being a devastating pathogen causing considerable yield loss in many agronomic important crops needs to be controlled at a faster pace. Different strategies to reduce its disease severity have been used so far. The ubiquitous nature and longer survivability in soil in the form of sclerotia make its control difficult. Chemical control is unsafe for environment, cause toxicity and disturb other beneficial microbes being nontarget specific; is not a feasible measure to manage pathogens. Owing to the negative effects of chemical fungicides their use should be limited. Integrated pest management is the only long term solution for controlling soil-borne plant diseases. Combined effect of various cultural, chemical and biological measures will definitely help to combat pathogens. Amongst all, biological control is the most effective component of IPM utilizing the beneficial microflora of soil that should be well adapted to the environmental conditions of the crop plants. Biocontrol agents mainly bacterial and fungal species have a wide array of bewildering mechanisms antagonistic to *M. phaseolina* and other plant pathogens. Bacteria belonging to the species of *Bacillus*, *Pseudomonas*, *Burkholderia, Serratia, Stenotrophomonas, Kocuria,* and *Trichoderma,* and Arbuscular mycorrhiza fungi exhibit strong biocontrol efficiency against various pathogens. These BCAs also pose other beneficial effects on plant growth resulting in increased yield and crop productivity. Biological control is a failsafe and environmental friendly strategy to manage this catastrophic fungus. Dual actions of biocontrol and plant growth promotion of these BCAs increase their efficacy in reducing disease incited by *M. phaseolina* and thus improving crop production. Formulation of such beneficial BCAs as biofungicide is the step toward sustainable agriculture and will help in fulfilling the gap of demand and supply.

References

Abbas, H. K., Bellaloui, N., Accinelli, C., Smith, J. R., & Shier, W. T. (2019). Toxin production in soybean (*Glycine max* L.) plants with charcoal rot disease and by *Macrophomina phaseolina*, the fungus that causes the disease. *Toxins, 11*, 645.

Ali, Z., Inam-ul-Haq, M., Tahir, M. I., Hayat, R., Khalid, R., Ashfaq, M., Jamil, M., & Naseem, S. (2015). Bioefficacy of rhizobacterial isolates against root infecting fungal pathogens of chickpea (*Cicer arietinum* L.). *Journal of Plant Pathology & Microbiology S, 3*, 1–8. https://doi.org/10.4172/21577471.S3-011

Arenas-Romero, O., Martinez-Salgado, S. J., Andrade-Hoyos, P., Lezama, C. P., Rivera-Tapia, A., & Luna-Cruz, A. (2021). Biological control of charcoal rot in peanut crop through strains of Trichoderma spp., in Puebla, Mexico. *Plants, 10*, 2630. https://doi.org/10.3390/plants10122630

Arora, N. K., Mishra, P., Mishra, J., & Dwivedi, S. K. (2020). Microbial enzyme in biocontrol of phytopathogens. In N. K. Arora, et al. (Eds.), *Microorganisms for sustainability* (Vol. 11, pp. 259–285). Singapore: Springer Nature. https://doi.org/10.1007/978-981-15-1710-5_10

Ashwini, C., & Giri, G. (2014). Control of seed borne fungi in greengram and blackgram through bio-agents.International. *Journal of Applied Biology and Pharmaceutical Technology, 5*, 168–170.

Athira, K. (2017). Efficacy of fungicide and bio-control agent against root rot of black gram (*Vigna mungo* L.) caused by *Macrophomina phaseolina* (Tassi) Goid. *International Journal of Current Microbiology and Applied Science, 6*, 2601–2607.

Babalola, O. O., Ngoma, L., & Ahmad, F. (2012). Ecophysiology of plant growth promoting bacteria. *Scientific Research and Essays, 7*(47), 4003–4013. https://doi.org/10.5897/SRE12.646

Babalola, O. O., Olanrewaju, O. S., & Glick, B. R. (2017). Mechanisms of action of plant growth promoting bacteria. *World Journal of Microbiology and Biotechnology, 33*, 197. https://doi.org/10.1007/s11274-017-2364-9, 1-16.

Bannur, V. A., & Chattannavar, S. N. (2019). Sources of resistance for charcoal rot of sorghum caused by *Macrophomina phaseolina* (Tassi) Goid. *International Journal of Current Microbiology and Applied Sciences, 8*(12), 2539–2545.

Bano, A., & Mufti, R. (2019). PGPR-induced defense responses in the soybean plant against charcoal rot disease. *European Journal of Plant Pathology, 155*, 983–1000.

Belkar, Y. K., & Gade, R. M. (2012). Biochemical characterization and growth promotion activities of Pseudomonas fluorescence. *Journal of Plant Disease Sciences, 7*(2), 170–174.

Bhandari, P. (2017). *M.Sc Thesis: Developing QPCR assay to quantify disease severity in Medicago Truncatula due to* Macrophomina phaseolina *infection*. Kansas: Wichita State University.

Chandra, S., Khare, V., & Kehri, H. K. (2011). Evaluation of arbuscular mycorrhizal fungi against *Macrophomina phaseolina* causing dry root rot of urd and mung bean. *Indian Phytopathology, 60*(1), 553–555.

Chaubey, S. (2012). *PhD Thesis: Studies on the antifungal metabolite production and biocontrol traits of* Vigna radiata-*growth promoting strains of* Fluorescent pseudomonads. Baroda, India: MSU.

Dar, N. A., Junaid, J. M., Bhat, T. A., Bhat, A. H., & Bhat, M. A. (2013). Commercial biocontrol agents and their mechanism of action in the management of plant pathogens. *International Journal of Modern Plant and Animal Sciences, 1*(2), 39–57.

Das, I. K., Indira, S., Annapurna, A., Prabhakar, S., & Seetharama, N. (2008). Biocontrol of charcoal-rot in sorghum by fluorescent *Psedomonads* associated with the rhizosphere. *Crop Protection, 27*, 1407–1414.

Das, K., Prasanna, R., & Saxena, A. K. (2017). *Rhizobia*: A potential biocontrol agent for soilborne fungal pathogens. *Folia Microbiologica, 62*(5), 425–435. https://doi.org/10.1007/s12223-017-0513-z

Dave, K. (2021). *PhD thesis: Evaluation of antimicrobial compound involved in antagonism isolated from rhizospheric bacteria from soybean*. Bhopal, India: Barkatullah Vishwavidhyalay.

Dave, K., Gothalwal, R., Singh, M., & Joshi, N. (2020). Facets of rhizospheric microflora in biocontrol of phytopathogen *Macrophomina phaseolina* in oil crop soybean. *Archives of Microbiology, 203*(2), 405–412. https://doi.org/10.1007/s00203-020-02046-z

Dave, K., Singh, M., & Gothalwal, R. (2021). Antagonistic Efficiency of indigenous rhizospheric bacteria against *Macrophomina phaseolina* causing charcoal rot in Soybean. *Research Journal of Agricultural Sciences, 12*(2), 470–473.

Dhingra, O. D., & Sinclair, J. B. (1977). *An annotated bibliography of* Macrophomina phaseolina*, 1905-1975* (p. 244). Viscosa, Brazil: Universided Federal.

Doley, K., & Jite, P. K. (2013). Effect of arbuscular mycorrhizal fungi on growth of groundnut and disease caused by *Macrophomina phaseolina. Journal of Experimental Sciences, 2012*, 46–50.

Fincheira, P., & Quiroz, A. (2018). Microbial volatiles as plant growth inducers. *Microbiological Research, 208*, 63–75.

Gautam, S., Chauhan, A., Sharma, R., Sehgal, R., & Shirkot, C. K. (2019). Potential of *Bacillus amyloliquefaciens* for biocontrol of bacterial canker of tomato incited by *Clavibacter michiganensis* ssp. michiganensis. *Microbial Pathogenesis, 130*, 196–203.

Gupta, G. K., Sharma, S. K., & Ramteke, R. (2012). Biology, epidemiology and management of the pathogenic fungus *Macrophomina phaseolina* (Tassi) Goid with special reference to charcoal rot of Soybean (*Glycine max* (L.) Merrill). *Journal of Phytopathology, 160*, 167–180. https://doi.org/10.1111/j.1439-0434.2012.01884.x

Hotkar, S., Shubha, S., Devaranavadagi, S., Kolhar, B. C., & Rathod, S. C. (2021). Assesment of promising varieties of sorghum for charcoal rot and yield potentiality. *The Pharma Innovation Journal, 10*(12), 1259–1262.

Islam, M. S., Haque, M. S., Islam, M. M., Emdad, E. M., Halim, A., Hossen, Q. M. M., Hossain, M. Z., Ahmed, B., Rahim, S., Rahman, M. S., Alam, M. M., Hou, S., Wan, X., Saito, J. A., & Alam, M. (2012). Tools to kill: Genome of one of the most destructive plant pathogenic fungi. *Macrophomina phaseolina. BMC Genomics, 13*, 1–16.

Jain, P., & Pundir, R. K. (2019). Biocontrol of soil phytopathogens by arbuscular mycorrhiza – a review. In A. Varma, & D. Choudhary (Eds.), *Mycorrhizosphere and pedogenesis*. Singapore: Springer. https://doi.org/10.1007/978-981-13-6480-8_14

Jayapala, N., Mallikarjunaiah, N., Puttaswamy, H., Gavirangappa, H., & Ramachandrappa, N. S. (2019). Rhizobacteria *Bacillus* spp. induce resistance against anthracnose disease in chili (*Capsicum annuum* L.) through activating host defense response. *Egyptian Journal of Biological Pest Control, 29*(45), 1–9.

Khaledi, N., & Taheri, P. (2016). Biocontrol mechanisms of *Trichoderma harzianum* against soybean charcoal rot caused by *Macrophomina phaseolina. Journal of Plant Protection Research, 56*(1), 21–31. https://doi.org/10.1515/jppr-2016-0004

Khalili, E., Javed, M. A., Huyop, F., Rayatpanah, S., Jamshidi, S., & Abdul Wahab, R. (2016). Evaluation of *Trichoderma* isolates as potential biological control agent against soybean charcoal rot disease caused by *Macrophomina phaseolina. Biotechnology and Biotechnological Equipment, 30*, 479–488.

Kodoth, P. S., Mohan, R., Devadason, A., & Raguchander, T. (2015). Exploiting novel rhizosphere *Bacillus* species to suppress the root rot and wilt pathogens of chickpea. *African Journal of Microbiology Research, 9*(15), 1098–1104.

Komy, M. H. E., Hassouna, M. G., Abou-Taleb, E. M., Al- Sarar, A. S., & Abobakr, Y. (2020). A mixture of *Azotobacter, Azospirillum*, and *Klebsiella* strains improves root-rot disease complex management and promotes growth in sunflowers in calcareous soil. *European Journal of Plant Pathology, 156*, 713–726.

Kulkarni, S., Shobharani, M., & Raja. (2019). Integrated management of dry root rot of blackgram caused by *Rhizoctonia bataticola. International Journal of Current Microbiology and Applied Science, 8*, 853–858.

Kumar, S. (2013). *Trichoderma*: A biological control weapon for managing plant diseases and promoting sustainability. *International Journal of Agricultural Sciences and Veterinary Medicine, 1*(3), 106–121.

Kumari, P., Khanna, V., Sharma, P., & Kumari, S. (2015). Metabolites in control of *Rhizoctonia bataticola. Journal of Pure and Applied Microbiology, 9*(1), 749–756.

Lakhran, L., & Ahir, R. R. (2018). In-vivo evaluation of different fungicides, plant extracts, bio control agents and organics amendments for management of dry root rot of chickpea caused by *Macrophomina phaseolina. Legume Research, 43*(1), 140–145. https://doi.org/10.18805/LR-3939

Latha, P., Karthikeyan, M., & Rajeshwari, E. (2017). Development of bioformulations for the management of blackgram dry root rot caused by *Rhizoctonia bataticola* (Taub Butler). *Advances in Research, 9*(4), 1–12.

Leon, M., Yaryura, P. M., Montecchia, M. S., Hernandez, A. I., Correa, O. S., Pucheu, N. L., Kerber, N. L., & Garcia, A. F. (2009). Antifungal activity of selected indeginous *Pseudomonas* and *Bacillus* from the soybean rhizosphere. *International Journal of Microbiology, 2009*, 1–9.

Lodha, S., & Mawar, R. (2020). Population dynamics of *Macrophomina phaseolina* in relation to disease management: A review. *Journal of Phytopathology, 168*, 1–17.

Luna, M. P. R., Mueller, D., Mengistu, A., Singh, A. K., Hartman, G. L., & Wise, K. A. (2017). Advancing our understanding of charcoal rot in soybeans. *Journal of Integrated Pest Management, 8*(1), 1–8.

Manjunatha, S. V., Naik, M. K., Khan, M. F. R., & Goswami, R. S. (2013). Evaluation of bio-control agents for management of dry root rot of chickpea caused by Macrophomina phaseolina. *Crop Protection, 45*, 147–150.

Marquez, N., Giachero, M. L., Declerck, S., & Ducasse, D. A. (2021). *Macrophomina phaseolina*: General characteristics of pathogenicity and methods of control. *Frontiers of Plant Science, 12*, 634397. https://doi.org/10.3389/fpls.2021.634397

Martinez-Medina, A., Pozo, M. J., Cammue, B. P. A., & Vos, C. M. F. (2016). Belowground defence strategies in plants: The plant–*Trichoderma* dialogue. In C. M. F. Vos, & K. Kazan (Eds.), *Below ground defence strategies in plants* (pp. 301–327). Cham: Springer. https://doi.org/10.1007/978-3-319-42319-7_13

Miljakovi, D., Marinkovi, J., & Baleševic̈-Tubi, B. (2020). The significance of *Bacillus* spp. in disease suppression and growth promotion of field and vegetable crops. *Microorganisms, 8*(1037), 1–19. https://doi.org/10.3390/microorganisms8071037

Mmbaga, M. T., Mackasmiel, L. M., & Mrema, F. A. (2018). Evaluation of biological agents for control of *Macrophomina* root rot and powdery mildew in flowering dogwood (*Cornus florida* L.). *HortScience, 53*, 1461–1466. https://doi.org/10.21273/HORTSCI13071-18

Mukherjee, P. K., Mendoza, A. M., Zeilinger, S., & Horwitz, B. A. (2021). Mycoparasitism as a mechanism of *Trichoderma*-mediated suppression of plant diseases. *Fungal Biology Review, 39*(2022), 15–33.

Nafady, N. A., Hashem, M., Hassan, E. A., Ahmed, H. A. M., & Alamri, S. A. (2019). The combined effect of arbuscular mycorrhizae and plant-growth promoting yeast improves sunflower defense against *Macrophomina phaseolina* diseases. *Biological Control, 138*, 104049. https://doi.org/10.1016/j.biocontrol.2019.104049

Pancheshwar, D. K., Varma, R. K., Singh, G., & Mishra, S. (2016). Screening of *Rhizoctonia bataticola* isolates against soybean cultivars for source of resistance. *International Journal of Agricultural Sciences, 8*(2), 952–954.

Pandey, A. K., Basandrai, A. K., Somta, P., & Basandrai, D. (2021). *Macrophomina phaseolina*-host interface: Insights into an emerging dry root rot pathogen of mungbean and urdbean, and its mitigation strategies. *Plant Pathology, 70*, 1263–1275. https://doi.org/10.1111/ppa.13378

Patel, D. P., Singh, H. B., Shroff, S., & Sahu, J. (2011). Antagonistic efficiency of *Pseudomonas* strains against soil borne disease of chickpea crop under in vitro and in vivo. *Elixir Agriculture, 30*, 1774–1777.

Qu, J., Yu, S., Teng, C., Liang, J., Song, T., Dong, L., Bai, X., & Jin, Y. (2017). Characterization of siderophore produced by *Pseudomonas syringae* BAF.1 and its inhibitory effects on spore germination and mycelium morphology of *Fusarium oxysporum*. *Journal of Microbiology, 55*, 877–884.

Reznikov, S., Vellicce, G. R., Mengistu, A., Arias, R. S., Gonzalez, V., Lisi, V. D., Gabriela, G. M., Lourdes, R. C. M., Pardo, E. M., Castagnaro, A. P., & Ploper, L. D. (2018). Disease incidence of charcoal rot (*Macrophomina phaseolina*) on soybean in northwestern Argentina and genetic characteristics of the pathogen. *Canadian Journal of Plant Pathology, 40*(3), 423–433. https://doi.org/10.1080/07060661.2018.1484390

Rubayet, Md T., Rehman, M. T., Khan, A. A., & Bhuiyan, M. K. A. (2021). Integrated management of charcoal rot disease of soybean caused by *Macrophomina phaseolina*. *Egyptian Journal of Agricultural Research, 99*(1), 10–19. https://doi.org/10.21608/ejar.2021.37644.1019

Saber, W. I., Ghoneem, K. M., Al-Askar, A. A., Rashad, Y. M., Ali, A. A., & Rashad, E. M. (2015). Chitinase production by *Bacillus subtilis* ATCC 11774 and its effect on biocontrol of *Rhizoctonia* diseases of potato. *Acta Biologica Hungarica, 66*, 436–448.

Saini, S., Verma, A., Kumar, A., Prakash, A., Ramesh, A., & Johri, B. N. (2016). Identification and characterization of antifungal metabolite producing *Pseudomonas protegens* strain BNJ-SS-45 isolated from rhizosphere of wheatcrop (*Triticum aestivum* L.). *International Journal of Applied and Pure Science and Agriculture, 2*(6), 69–76.

Sanjeevkumar, K., Kumar, K.,S., Balabaskar, P., Sivakumar, T., Kannan, R., & Saravanan, K. R. (2020). Efficacy of seed plus soil application of *Bacillus cereus* on the root rot (*Macrophomina phaseolina* (tassi.) Goid.) incidence and plant growth of groundnut. *Plant Archives, 20*, 1217–1221.

Sarita, B. A. K., & Singh, R. (2014). Study of seed borne mycoflora of mungbean treated with potassium nitrate during storage. *Journal of Advanced and Applied Science Research, 5*, 11–13.

Shivannavar, C. T., Patil, S., Bheemaraddi, M. C., & Gaddad, S. M. (2015). Antiphyotpathogenic and plant growth promoting attributes of *Bacillus* strains isolated from rhizospheric soil of chickpea. *Journal of Agricultural Science and Technology A, 17*, 1365–1377.

Simonetti, E., Viso, N. P., Montecchia, M., Zilli, C., Balestrasse, K., & Carmona, M. (2015). Evaluation of native bacteria and manganese phosphite for alternative control of charcoal root rot of soybean. *Microbiological Research, 180*, 40–48. https://doi.org/10.1016/j.micres.2015.07.004

Singh, S., Kumar, V., Dhanjal, D. S., & Singh, J. (2020). Biological control agents: Diversity, ecological significances, and biotechnological applications. In J. Singh, & A. Yadav (Eds.), *Natural bioactive products in sustainable agriculture* (pp. 31–44). Singapore: Springer. https://doi.org/10.1007/978-981-15-3024-1_3

Spagnoletti, F., Carmona, M., Gómez, N. E. T., Chiocchio, V., & Lavado, R. S. (2017). Arbuscular mycorrhiza reduces the negative effects of *Macrophomina phaseolina* on soybean plants in arsenic-contaminated soils. *Applied Soil Ecology, 121*, 41–47. https://doi.org/10.1016/j.apsoil.2017.09.019

Spagnoletti, F. N., Cornero, M., Chiocchio, V., Lavado, R. S., & Roberts, I. N. (2020). Arbuscular mycorrhiza protects soybean plants against *Macrophomina phaseolina* even under nitrogen fertilization. *European Journal of Plant Pathology, 156*, 839–849.

Sridharan, A., Thankappan, S., Karthikeyan, G., & Uthandi, S. (2020). Comprehensive profiling of the VOCs of *Trichoderma longibrachiatum* EF5 while interacting with *Sclerotium rolfsii* and *Macrophomina phaseolina*. *Microbiological Research, 236*, 126436. https://doi.org/10.1016/j.micres.2020.126436

Sujatha, N., & Ammani, K. (2013). *In vitro* antibiosis of *Fluorescent Pseudomonads* against *Fusarium oxysporum* and *Rhizoctonia bataticola* and effect of the antifungal metabolite on fungal biomass. *International Journal of Biopharma Research, 2*(5), 134–140.

Vasebi, Y., Safaie, N., & Alizadeh, A. (2013). Biological control of soybean charcoal root rot disease using bacterial and fungal antagonists *in vitro* and greenhouse condition. *Journal of Crop Protection, 2*(2), 139–150.

Vinothini, K., Renganathan, P., Balabaskar, P., & Sivakumar, T. (2020). *In vitro* efficacy various isolates of *Trichoderma Viride* and *Pseudomonas fluorescens* against *Macrophomina phaseolina* causing sesame root rot. *Plant Archives, 20*(1), 1163–1168.

Waals, J. E. V., Jordaan, E., & McLaren, N. W. (2019). Effect of irrigation on charcoal rot severity, yield loss and colonization of soybean and sunflower. *Crop Protection, 122*, 63–69.

Yasmin, H., Naz, R., Nosheen, A., Hassan, M. N., Ilyas, N., Sajjad, M., Anjum, S., Gao, X., & Geng, Z. (2020). Identification of new biocontrol agent against charcoal rot disease caused by *Macrophomina phaseolina* in soybean (*Glycine max* L.). *Sustainability, 12*(6856), 1–24. https://doi.org/10.3390/su12176856

Chapter 24

Economic aspects for eco-friendly strategies to minimize disease incidences and yield loss due to *Macrophomina phaseolina*

Shiuly Bhowmick, Anshika Verma, Srishti Kar, Gauri Rai and Puneet Singh Chauhan
Microbial Technology Division, Council of Scientific and Industrial Research-National Botanical Research Institute (CSIR-NBRI), Lucknow, Uttar Pradesh, India

24.1 Introduction

Macrophomina phaseolina is widespread throughout the world and affects more than 100 families and at least 500 different plant species, including strawberry, maize, sorghum, potato, soybean, chickpea, and pigeon pea, and causes a variety of diseases worldwide (Kumari & Sharma, 2014; Reznikov et al., 2018; Upadhyay et al., 2020). The source of primary inoculum is generally microsclerotia, which play an important role in *M. phaseolina* survival and spread, as well as illness start and progression. Charcoal rot affects plants at all phases of development. The illness appears as dark lesions on the epicotyls and hypocotyls, resulting in seedling death owing to xylem vascular occlusion. *M. phaseolina* produced red to brown symptoms on the roots and stems, as well as black microsclerotia and dark mycelia. Severe infections induce defoliation and wilting, leading to significant losses in yield up to 100% (Bashir & Malik, 1988; Iqbal & Mukhtar, 2020a). *M. phaseolina* causes root rot, seedling blight, charcoal rot, ashy stem blight, wilt, collar rot, dry rot, pod rot, and seed rot in a variety of crops (Kumar et al., 2020).

Although synthetic fungicides provide significant and immediate treatment of plant infections, their application is harmful to human beings, livestock, and to the whole ecosystem (Iwuala et al., 2022). Furthermore, the efficacy of these chemical fungicides as seed treatments is short-lived, and the soil application is costly causing an imbalance in soil microbial communities that interferes with the regular activities of beneficial organisms. In addition, the pathogen develops resistance against the fungicides due to repeated and extensive usage of the same fungicides for a pathogen (Iqbal et al., 2014). The use of alternative methods, such as traditional agronomical practices, amendment of organic matters in soil, use of plant-derived compounds, plant treatment with elicitors, biological control agents, selection of resistant cultivars and other integrated approaches, besides synthetic agrochemicals for management of charcoal rot, can prove to be a holistic approach. These approaches not only help to minimize the pathogen infection but also can improve soil quality, plant health, and productivity without hampering the environment and human health (Hossain et al., 2012; Satya et al., 2011; Thakur & Sohal, 2013). Although these alternatives are well known and safer to apply yet, it is difficult to use for common peoples especially in developing countries like India due to lack of awareness and insufficient technology. Farmers always seek for a product that is easy to handle, has a good shelf life, and is economical. Nevertheless, in the present scenario intense research is going on worldwide to make better formulations and technologies for production of pesticides and fungicides having no adverse impact on plant, soil, environment, and human beings (Lahlali et al., 2022). However, in India, biopesticides account for only 2.91% of the total pesticide markets, it is predicted to increase at a rate of only around 2.4% per year in the coming years (Bharti et al., 2021). The global pesticide market was valued at about $42.8 billion in 2009. Out of which, biopesticide covers only about 3.74% (Yoon et al., 2013). North America has a lead role in

Macrophomina Phaseolina. https://doi.org/10.1016/B978-0-443-15443-0.00021-8

biopesticide production reaching USD 539 million in 2015 and predicted to achieve USD 1.67 billion by 2022 (Arthurs & Dara, 2019).

24.2 Use of chemical fungicides

The word "fungicide" originated from two latin words, viz., "fungus" and "caedo." The word "caedo" means "to kill" and therefore fungicides are such biochemical compounds, which have the capacity to eradicate fungus (Morton & Staub, 2008). Fungicides are antimycotic drugs used to kill or inhibit the growth of parasitic fungal isolates that dwell in the soil and affect the plants. The application of chemical fungicides is a common management technique for the charcoal rot pathogen, although they only have a little impact on disease control (Chamorro et al., 2015; Tonin et al., 2013). Additionally, synthetic agrochemicals harm the environment and endanger human health by having mutagenic and carcinogenic effects (Banaras et al., 2021).

Broadly, they are grouped into three categories based on mobility, namely contact, translaminar, and systemic.

(i) ***Contact fungicides:*** These fungicides are not distributed throughout the plant but eradicate the fungus as they come in their contact (Elliott et al., 2015). Upon application, these fungicides reside on the surface or sprayed and do not go deeper and hence need repeated applications as the plant grows.
(ii) ***Systemic fungicides:*** These are the most potent fungicides because they enter the plant and disperse themselves among its vessels (Mohamed et al., 2018). Typically, these disseminate throughout the tissues of the plant or travel upward inside the plant (i.e., are locally systemic). They may have long-lasting activity, in contrast to contact fungicides (Edgington et al., 1980).
(iii) ***Translaminar fungicides:*** These fungicides move from the top, sprayed area of the plants to the lower, unsprayed area of the plant (Klittich & Ray, 2013).

The major groups of fungicide used to control infection of *M. phaseolina* are Carbendazim, Captan, and Mancozeb. These fungicides have a broad-spectrum effect on plants, soil, microbes associated with plants, and ultimately on human beings (Rouabhi, 2010). Carbendazim is a broad-spectrum systemic fungicide, widely used for preventing diseases caused by pathogenic fungal isolates in agriculture, forestry, and veterinary medicine (Farag et al., 2011). Due to their extensive and repeated use, they have been categorized in the hazardous category of chemicals by the World Health Organization and have induced severe and delayed toxic effects on human beings, soil microbiome, and other terrestrial and aquatic life forms (Goodson et al., 2015). Carbendazim inhibits the microtubule formation in fungal and mammalian cells by interfering with the beta tubulin, causing impaired segregation of chromosomes during cell division leading to mutagenic toxicity or genotoxicity (Lacey & Watson, 1985; Pacheco et al., 2012). The nitrogenase activity of certain rhizobacteria such as *Rhizobium leguminosarum* was dropped by 80% due to the presence of high levels of carbendazim and thiram mixture (Wang et al., 2009, 2016). Mixture of carbendazim and mancozeb has induced deleterious effects on the nitrogen fixing and nitrifying microflora (Fawole et al., 2010). Reduction in enzymatic activities, distortion of the cellular structures, and disruption of microbial community structure were some of the adverse and toxic effects caused by fungicide on the soil microbiome.

Mancozeb is a nonsystemic, broad spectrum, contact fungicide, with preventive activity. It is an ethylene(bis) dithiocarbamate fungicide and easily degraded into ethylene thiourea which stays in the soil and affects various mammalian organ systems primarily causing toxicity through skin, leading to dermatitis and dermal sensitization (McConnell et al., 1998). The nitrification of diammonium hydrogen phosphate in soil showed that nitrate-nitrogen decreased with a greater rate with the increase in Mancozeb concentrations, and hence Mancozeb causes inhibitory effect on nitrification (Walia et al., 2014).

Captan (N-trichloromethyl-thio-tetrahydroftalimide) is a nonsystemic, broad range protectant fungicide used on fruits, vegetables, and leguminous plants (Gordon, 2010). It is a nonspecific thiol reactant, which works to inhibit respiration of various fungal and bacterial species (Tomlin, 2000). People handling these fungicides should wear protective gears. The handlers are exposed to Captan predominately by two routes, oral and dermal, causing dermatitis and conjunctivitis (US Department of Health and Human Services, 1993). Although the acute toxicity of this class of fungicides is modest, yet they can irritate biological tissues when they come into touch with them. Turkey's excessive Captan use decreased pepper plant germination rates (Tort & Turkyilmaz, 2003). Despite the use in agricultural settings, Captan showed negative impacts on aerobic diazotrophs and nitrifying microorganisms. It also poses an adverse impact on the nitrogen cycle and the microbial equilibrium of the soil microbiome (Martõnez-Toledo et al., 1998) (Table 24.1).

TABLE 24.1 Chemical fungicides used to prevent *M. phaseolina* in susceptible host.

Host plant	Disease/microbe associated	Chemicals/fungicide used	References
Soybean	Charcoal rot/ *Macrophominaphaseolina*	Carbendazim and Penflufen + Trifloxystrobin	Tonin et al. (2013); Reznikov et al. (2018)
Strawberry	Charcoal rot/ *Macrophominaphaseolina*	Carbendazim	Chamorro et al. (2015)
Green Gram and Black gram	Root rot/ *Macrophominaphaseolina*	Matalaxyl + Mancozeb, Copper + Mancozeb, Carbendazim	Iqbal and Mukhtar (2020a)
Castor	Root rot/ *Macrophominaphaseolina*	Carbendazim(95%), Propineb, Mancozeb (99.9%), Captan(85.63%) and Difenoconazole	Parmar et al. (2017)
Sesame	Root rot/ *Macrophominaphaseolina*	Tebuconazole + Trifloxystrobin, Hexaconazole 5% + Captan 70, Mancozeb 63% + Carbendazim 12%	Karibasappa et al. (2020)
Pigeon pea	Stem canker/ *Macrophominaphaseolina*	Mancozeb, Copper hydroxide, and acombi products of Carbendazim 12% + Mancozeb 63%.	Swamy et al. (2018)
Chick pea	Dry root rot/ *Macrophominaphaseolina*	Thiophanate methyl, Hexaconazole, Chlorothalonil and Mancozeb	Lokesh et al. (2020)

24.3 Eco-friendly approaches for disease management of *M. Phaseolina*

It is extremely difficult to control the infections when pathogens are soil-borne. In addition to harming the environment and people's health, synthetic chemicals used to treat plant diseases damage the environment. Fungicides have been used extensively to combat plant diseases, which has led to an increase in toxins harmful to both humans and the environment (Iwuala et al., 2022). Therefore, extensive research has been going on to find out an alternative method to reduce the deadly effect of *M. phaseolina* on valuable crops in a sustainable manner since the last 2 decades (Lahlali et al., 2022). Drenching of organic matters in the soil, application of plant-derived compounds, selection of resistant cultivars, and use of biological antagonists are some alternative control measures of charcoal rot disease, which are eco-friendly and cost-effective as well.

24.4 Agronomic practices

Some common traditional agronomic practices, such as soil irrigation, drought management, crop rotation, solarization, and fertilization intended to improve the soil environment as the *M. phaseolina* has persistent nature and survives for 10–15 years by producing microsclerotia. Microsclerotia store nutrients in it and help the pathogen to survive even in a lower nutrient level in the soil and compete well with other soil microbes (Gupta et al., 2012). It has also been observed that fertilization with P can effectively reduce the incidence of soybean charcoal rot, whereas higher N level may induce the disease severity (Spagnoletti et al., 2018). In addition, low C:N ratio and high soil bulk density adversely affect the survival of the sclerotia. The survival of the microsclerotia can also be reduced by optimizing the soil moisture and temperature as it grows faster when the temperature reaches 35°C and high soil moisture 10%–20% in the soil. Significant reduction of microsclerotia may be brought by maintaining field moisture above 60% at 30°C (Gupta et al., 2012). The survival frequency of *M. phaseolina* was greater at soil surface, whereas lowest activity was observed at 7.6 and 25.4 cm depth. The reason behind this is the occurrence of cellulose-degrading fungal isolates, such as *Trichoderma* sp. and *Basidiomycetes,* remain more active in this region because of optimum soil moisture and temperature (Baird et al., 2003).

Crop rotation, intercropping, and lay period methods are less effective due to broad host range and strong persistence of mirosclerotia (Francl et al., 1988). In the instance of sesame produced as an intercropped crop with mung bean, there was decreased incidence of *Macrophomina* stem and root rot and enhanced production of seed when compared to single sesame cultivation (Rajpurohit, 2002).

24.5 Organic amendments

The external input of organic matter in soil is one of the ancient methods to increase soil nutrient quality, soil aeration, improve plant growth, and to check pathogenic activity. It also allows the recruitment of advantageous soil microbes and rhizosphere to enhance soil properties, and plant growth and productivity. Organic matters, such as compost are rich sources of carbon, phosphorus, potassium, and nitrogen. Amendment of organic matter promotes nitrification process, soil cation exchange capacity, and retention of essential nutrients, improved water-holding capacity, and reduced nitrogen leaching. Additionally, it has been reported that composted organic matter provides resistance from different disease causing microorganisms (Vida et al., 2020). Other than direct effect on growth promotion, organic amendment also shows indirect effect on plant defense mechanisms by restructuring soil microflora, which produces a number of defense enzymes and antimicrobial proteins, thereby reducing the susceptibility of the pathogen. For this kind of defense mechanism micronutrient, such as Zn, an essential element which act as a cofactor and precursor for enzyme activation, important for redox reaction, protein synthesis, energy transfer, and assimilation of nitrogen in plants (Iwuala et al., 2022; Khan et al., 2018).

24.6 Selection of resistant varieties

M. phaseolina persists in soil as sclerotia and plant detritus making it challenging to eradicate. However, the drought and the disease known as charcoal rot, which typically manifests itself at the stage of grain maturity and causes severe lodging, are the main causes of poor levels of productivity in host plants. According to Gaige et al. (2010), a hot and dry environment could make the disease worse and result in yield losses of up to 100%. Under ideal circumstances, this disease can result in lodging and a loss of 23%–64% of grain output (Mughogho, 1984). Therefore, there is a need to search for relatively cost effective and eco-friendly alternatives. It is difficult to treat these diseases as the pathogen is a polyphagic necrotroph and can survive in the soil for long time. An integrated approach that includes resistant cultivars is one of the most effective ways to manage disease incidences and the level of severity in different valuable crop plants (Elmerich et al., 2022). There are a number of resistant crop varieties that have been reported viz.,Chickpea (BGD-98), (RSG-143), soybean (JS335, JS95-60, Munasqa RR), strawberry (Orly, Tamir, Rotmy), and mungbean (MSJ-118, BPMR-145) (Ashraf et al., 2005; Kumar & Gaur, 2020; Kumari & Sharma, 2014; Pickel et al., 2020; Reznikov et al., 2019; Thombre & Kohire, 2018; Upadhyay et al., 2020).

Chowdhury et al. (2014) has characterized the behavior of the pathogen in susceptible and resistant varieties of sesame and found delayed infection and colonization of *M. phaseolina* including delayed emergence of fungal hyphae and production of microsclerotia on the surface of roots and in the close proximity of rhizosphere. A higher amount of peroxidase activity was also reported in chickpea after infection, which is a crucial enzyme for the synthesis of lignin and terpenoids, which are necessary for the production of phytoalexins, and it may play an important part in the resistance mechanism of plants. These mechanisms of phenols and indoles are among the numerous reactions that peroxidase catalyzes (Aravind & Brahmbhatt, 2018; Ashraf et al., 2005). The stilbene synthase genes are regulating the level of production of phytoalexin. A number of abiotic and biotic stressors were identified to stimulate SbSTS1 gene expression. The constitutive expression of SbSTS1 demonstrates that stilbene synthase can be increased in heterologous plant systems to boost overall resistance to fungal infections. As a result, the signal pathway of these defense-related genes appears to be conserved throughout a wide range of plants (Sharma et al., 2013).

Host resistance may be the best option for disease treatment that is both cost-effective and environmentally friendly. Following successful field trials, the resistant genotypes could be used as resistant donors for producing resistant cultivars. Understanding the genetics of the resistance trait in the genotypes is crucial. It would also be necessary to map populations derived from resistant sources in order to identify the gene(s) that would enable marker-assisted selection (Pandey et al., 2020).

24.7 Use of botanicals

Volatile oils that have been extracted from plants are described as natural phyto-sanitary agents that attempt to replace or limit the use of conventional fungicides (Marquez et al., 2021). Due to their lower environmental impact than synthetic fungicides, use of natural agents to manage fungal diseases in plants is thought to be a useful alternative. Numerous higher plants and their constituents have proven to be safe and nonphytotoxic while successfully controlling plant diseases. The majority of plants engage in "allelopathic" biochemical interactions that are both stimulatory and inhibitory in nature with the other plants and microbes. Higher plants are particularly effective at preventing the phytopathogens from infecting crops through root exudates.

Recent research has demonstrated the potential for using natural compounds derived from plants and other organisms as fungicides (Akbar et al., 2020; Li et al., 2018). Higher plants produce a number of secondary metabolites. These include flavonoides, alkaloids, tannins, terpenoides, vitamins, organic acids, phenols, glycosides, linoleic acid, and antioxidants (Arif et al., 2022). Plant secondary metabolites may operate in several ways, including by preventing the development of fungal cell walls, rupturing cell walls, impairing the function of fungal mitochondria, and preventing the synthesis of fungal proteins (Aftab et al., 2019). Plant derivatives, such as coumarin compounds isolated from different medicinal plants, are reported to have antifungal activity against a number of plant pathogens including *M. phaseolina.* Rastija et al. (2021) stated that coumarin derivatives are most effective to fungal pathogens, such as *M. phaseolina* and *Sclerotia sclerotiorum* although it has no effect on beneficial soil nematodes and bacteria. Another experiment with coumarin derivatives showed 78% growth inhibition at a concentration of 1000 μg mL^{-1} (Sahni et al., 2020). For example, aqueous extract of *Allium sativum* exhibited 100% mycelial growth reduction in *M. phaseolina* (Iwuala et al., 2022).

Medicinally important plants are known to be a home for many biologically active compounds that possess a variety of properties viz., antifungal, antibacterial, and antiviral activities (Harish et al., 2008; Iwuala et al., 2022). Aqueous extract of *Citrus aurantifolia* was assessed in vitro for inhibitory activity that showed dose dependent growth inhibition against dry root rot of gingelly (Balamurugan, 2014). In the same way, methanolic extract of *Sorghum halepense* was tested against charcoal rot disease of cowpea in different concentrations and found that the shoot extract significantly reduced fungal biomass upto 61% (Arshad et al., 2012). Iqbal et al. (2014) tested 20 antagonistic plant samples. Out of all the test plants, *A. indica*, *C.copticum*, and *N. sativa* demonstrated the greatest increase in the emergence of mungbean plants as compared to control. In contrast, it revealed that *L. inermis*, *C. fistula,* and *M. piperita* were the least effective at lessening the pathogen's damage (Table 24.2).

24.8 Immune responses triggered by elicitors

Elicitors are natural or synthetic compounds that trigger a plant's systemic acquired resistance system. Elicitors act as signal molecules, which trigger defense activity in plants at low concentrations. Depending on the chemical utilized, different metabolic pathways are triggered in treated plants (Marquez et al., 2021). Salicylic acid, methyl salicylate,

TABLE 24.2 Antifungal properties of different plant extracts.

Plant species	Nature of extract	Reported fungal pathogens	References
Polyalthialongifolia	Methanolic	*Aspergillus* sp.	Satish et al. (2007)
Flourensiacernua	Lanolin and cocoa	*Rhizoctoniasolani*	Castillo et al. (2010)
Aloe vera leaf	Hydroalcoholic	*B. gladiolorum, F. oxysporum, H. pruneti P. gladioli*	Rosca-Casian et al. (2007)
Aloe vera gel	crude	*Aspergillusniger, A. flavus, Alternariaalternata, Drechslerahawaiiensis, Penicilliumdigitatum*	Sitara et al. (2011)
Emblicaofficinalis	Ethanolic Methanolic	*Fusariumequiseti Candida albicans*	Hossain et al. (2012)
Daturastramonium	Alcoholic	*A.flavus, A.niger, Fusariumculmorum, Rhizopusstolonifer,*	Sharma et al. (2013)
Eucalyptus citriodora Eucalyptus globulus	Methanolic Essential oil	*Trichophytonmentagrophytes, Microsporumgypseum,*	Musyimi and Ogur (2008)
Azadirachtaindica	Methanolic	*Alternariasolani*	Jabeen et al. (2013)
Lawsoniainermis	Crude	*C. albicans*	Habbal et al. (2005)
Moringaoleifera	Aqueous	*Fusariumoxysporum, F. solani, A. solani, A. alternata, R. solani, Sclerotiumrolfsii, Macrophominaphaseolina.*	El-Mohamedy and Abdalla (2014)
Bacopamonnieri	Ethyl acetate	*P. notatum A.niger C. albicans*	Ghosh et al. (2007)
Ocimum sanctum	Essential oil	*C. albicans C. tropicalis*	Khan et al. (2010)

benzothiadiazole, benzoic acid, chitosan, and other chemicals that influence the formation of phenolic compounds and the activation of numerous enzymes involved in defense mechanisms in plants are examples of chemical elicitors that are frequently investigated (Ezzat et al., 2014; Thakur & Sohal, 2013). Their use in agricultural practices may reduce the need for chemical pest control, promoting the growth of sustainable agriculture. The incorporation of plant elicitors lead to the production of AOS, biosynthesis of phytoalexins, reinforcement of the plant cell wall associated with phenylpropanoid, callose deposition, biosynthesis of defense-related enzymes, and the accumulation of PR proteins having antimicrobial properties (Van Loon & Van Strien, 1999; Zakir, 2018). Seed elicitor priming is a potent approach that modifies seed metabolism and signaling pathways, affecting all stages of plant life including germination and establishment of seedlings. A few advantages include enhanced plant growth and productivity, and improved food nutritional quality. By influencing signaling of biochemical pathways, the balance of ROS, phytohormones involved in plant growth promotion, elicitor-priming increases stress, and disease resistance while lowering the need for pesticides and fertilizers (Rayanoothala et al., 2022). For the use of elicitors in agriculture, it should be safe, biodegradable, and cost effective. Further, the selection of a specific elicitor and its dose of application made according to the plants subjected to prior stimulus. Therefore, it is crucial to distinguish between elicitors that operate as priming molecules and those that directly activate plant defences.

24.9 Use of biological antagonists

The biological control of plant infections by the use of hostile microorganisms fits well within the framework of integrated disease management and is now widely used, becoming a viable option for the successful soil-borne plant disease management (Kumar et al., 2020). These are broadly termed as "biological control agents" (BCAs). BCAs produce allele-chemicals, which include diffusible metabolites, lytic enzymes, and volatile compounds, which interfere with phyto-pathogens, and thus inhibit the growth and development of pathogen (Lahlali et al., 2022). Several fungi and bacteria are known as antagonists of root-rot fungi. Soil amendment and seed treatment with *Burkholderia* sp., *Pseudomonas aeruginosa* and *Bacillus subtilis* reduced the root rot in mungbean significantly and enhanced plant growth and seedlings vigor (Satya et al., 2011; Siddiqui et al., 2001). Some *B. cereus* strains produce antibiotics capable of suppressing rhizosphere fungal infections. Extracellular antibiotics produced by *B. subtilis* were found suppressive to some plant diseases (Podile et al., 1988). *Bacillus* sp. also produces a variety of additional metabolites, such as chitinases and other cell wall—degrading enzymes, volatiles and chemicals that induce plant resistance mechanisms (Kumar et al., 2020). Many species of *Trichodermaviz.,T. viride, T. hamatum,* and *T. virens* have been found potentially useful in the biocontrol of *M. phaseolina* (Iqbal et al., 2014). A group of *Trichoderma* spp. were reported to improve seed germination, plant height, yield, and reduce growth of fungal mycelia and sclerotia. Considering the cost of chemicals used for seed dressing, which is roughly three times higher to that of naturally produced biocontrol agents, using these antagonistic fungi as seed treatment will have no substantial impact on cost of production. In addition, the ongoing usage of these antagonistic microbes will grow their rhizosphere population, increasing the soil's disease suppressiveness (Iqbal & Mukhtar, 2020b).

24.10 Integrated approach

A combination of reduced amount of fungicide plus other biological control agents with other treatments against the infectious fungal pathogens including *M. phaseolina* is a prominent method, which can reduce infection by 100% (Muthukumar & Raj, 2020). Rahman et al. (2021) performed an integrated approach and an in vitro experiment of different organic amendments, fungicides, and bioagents to determine their role in controlling *M. phaseolina* infection and improving yield in soybean. Among these, mustard oil cakes showed the maximum growth inhibition (50.92%) at a concentration of 3% followed by sesame oil cake (35.30%) at concentration of 3%. A combination of seed treated with fungicide + *T. harzianum* + organic amendments + soil inoculated with pathogen found to decrease disease incidences (6.50%), reduced seedling mortality (2.19%), and increased disease resistance (80%) and yield (104%). Similarly, Jamal et al. (2021) used endophytic yeast along with neem oil cake as bio-controlling agent against root rotting fungi *F. oxysporum, F. solani, R. solani,* and *M.phaseolina*. Oil cakes made of Mahua and Neem resulted in a significant result in controlling spore germination and mycelial growth of *B. oryzae* (Harish et al., 2008) (Fig. 24.1).

24.11 Challenges

Basic formulation and fermentation process can make biopesticides easier to use for the common person. However, there are some major challenges for the production of biopesticides on a larger scale at commercial level. This is due to lack of sufficient technology to deal with the issue of short product shelf life as well as acceptable procedures for large-scale

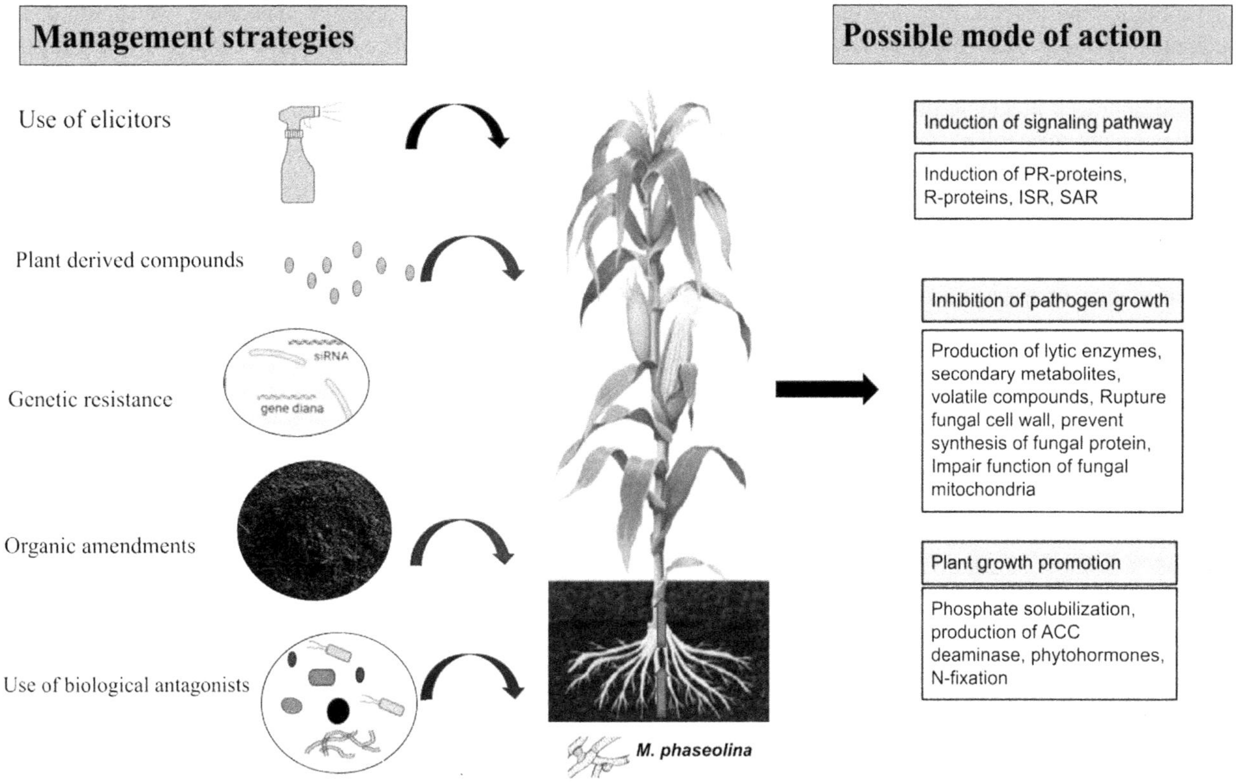

FIGURE 24.1 Eco-friendly strategies to manage *M. phaseolina* infection in host plants and their possible mode of action.

manufacture. A workable approach to this issue is to maintain a high spore density in order to achieve optimal sporulation efficiency. As a result, streamlining the entire production process is necessary to provide a productive and high-quality final product (Stamenković-Stojanović et al., 2019). Although the usefulness of biopesticide is well known, yet the market acceptance of such products is still in its infancy. The development and large-scale availability of plant sources are also crucial for the commercialization of plant-based biopesticide. Additionally, the active ingredients of plants also make the task of formulating a biopesticide very difficult (Essiedu et al., 2020).

Due to technical challenges, environmental concerns, or economic attractiveness, many farmers, especially in developing nations, are reluctant to adopt such practices. Some farmers favor formulae in granules over ones that need more expensive equipment or cold storage because they are easier to store and apply by machines. Due to a lack of technological knowledge and the scarcity of effective technologies, farmers are discouraged from employing the biological control technique (Barratt et al., 2018). In this situation, information exchange between technology developers and end users is mostly dependent on training and field demonstrations. In order to persuade farmers to utilize biological control, technology must be straightforward to evaluate and prepare while offering advantages over competing solutions in terms of cost, demand, and supply efficiency (Lahlali et al., 2022).

24.12 Conclusions

Infection of *M. phaseolina* causes huge loss of economically important food crops worldwide and the control measures we have been using, do not always give 100% results. Responsible management of disease caused by *M. phaseolina* is important because it has a direct and indirect impact on the environment and human beings. Although chemical fungicides give instant results for controlling deadly phytopathogens like *M. phaseolina,* yet it is not sustainable and economical. Its toxicity level also depends on the dose and the type of fungicide we use. Traditional agronomic practices are a sustainable way to control fungal pathogens but it is not reliable every time as it depends on the interaction between the physicochemical environment of the soil and the host. Interaction of other pathogenic microbes present in the soil with *Macrophomina* also influences the degree of virulence it causes. The incorporation of chemical or natural elicitors and plant-derived compounds can reduce the necessity of chemical fungicide in agriculture. It also depends on the susceptibility and immune response of the host itself to fight against such deadly enemies. Therefore, we should develop such plant varieties which have resistant traits against phytopathogens including *M. phaseolina*. Biological antagonists are a more prominent tool for controlling phytopathogens due to its long-term efficacy against target plant diseases. An integrated approach including physiological and genetic manipulation of the biocontrol mechanism, use of the mixture of beneficial plant metabolites, manipulation of the environment provides an additive effect to improve the efficacy of the management strategies.

Acknowledgments

The authors acknowledge the Director, CSIR National Botanical Research Institute for providing facilities and support during the study. This work is supported by the CSIR-Network (MLP0048 and MLP0049) and an in-house project (OLP109) funded by the Council of Scientific and Industrial Research, New Delhi, India.

References

Aftab, A., Yousaf, Z., Javaid, A., Riaz, N., Younas, A., Rashid, M., Bushra, H., Shamsheer, & Arif, A. (2019). Antifungal activity of vegetative methanolic extracts of Nigella sativa against Fusarium oxysporum and *Macrophomina phaseolina* and its phytochemical profiling by GC-MS analysis. *International Journal of Agriculture and Biology, 21*, 569–576.

Akbar, M., Sherazi, I. N., Khalil, T., Iqbal, M. S., Akhtar, S., & Khan, S. N. (2020). Identification of antifungal compounds from slender amaranth. *Planta Daninha, 38*. https://doi.org/10.1590/S0100-83582020380100063

Aravind, T., & Brahmbhatt, A. B. (2018). Estimation of biochemical factors responsible for resistance to root and collar rot (*Macrophomina phaseolina* (Tassi) Goid.) in okra. *Journal of Applied and Natural Science, 10*(4), 1266–1270.

Arif, Y., Bajguz, A., & Hayat, S. (2022). *Moringa oleifera* extract as a natural plant biostimulant. *Journal of Plant Growth Regulation*, 1–16. https://doi.org/10.1007/s00344-022-10630-4

Arshad, J., Syeda, F. N., & Amna, S. (2012). Antifungal activity of methanolic extracts of Sorghum halepense against *Macrophomina phaseolina. African Journal of Microbiology Research, 6*(28), 5814–5818.

Arthurs, S., & Dara, S. K. (2019). Microbial biopesticides for invertebrate pests and their markets in the United States. *Journal of Invertebrate Pathology, 165*, 13–21.

Ashraf, M. S., Khan, T. A., & Hasan, S. (2005). Reaction of chickpea varieties to *Macrophomina phaseolina* and their effect on peroxidase activity. *Pakistan Journal of Botany, 37*(3), 761.

Baird, R. E., Watson, C. E., & Scruggs, M. (2003). Relative longevity of *Macrophomina phaseolina* and associated mycobiota on residual soybean roots in soil. *Plant Disease, 87*(5), 563–566.

Balamurugan, S. (2014). *In vitro* antifungal activity of *Citrus aurantifolia* Linn plant extracts against phytopathogenic fungi *Macrophomina phaseolina*. *International Letters of Natural Sciences, 13*, 70–74.

Banaras, S., Javaid, A., & Khan, I. H. (2021). Bioassays guided fractionation of *Ageratum conyzoides* extract for the identification of natural antifungal compounds against *Macrophomina phaseolina. International Journal of Agriculture and Biology, 25*, 761–767.

Barratt, B. I. P., Moran, V. C., Bigler, F., & Van Lenteren, J. C. (2018). The status of biological control and recommendations for improving uptake for the future. *BioControl, 63*(1), 155–167.

Bashir, M., & Malik, B. A. (1988). Diseases of major pulse crops in Pakistan—a review. *International Journal of Pest Management, 34*(3), 309–314.

Bharti, D., Singh, R., Arora, D., & Arora, C. (2021). Role of phytopesticides in sustainable agriculture. In S. Singh, P. Singh, S. Rangabhashiyam, & K. K. Srivastava (Eds.), *Global climate change* (pp. 341–359). USA: Elsevier: Academic Press. https://doi.org/10.1016/B978-0-12-822928-6.00011-3

Castillo, F., Hernández, D., Gallegos, G., Mendez, M., Rodríguez, R., Reyes, A., & Aguilar, C. N. (2010). In vitro antifungal activity of plant extracts obtained with alternative organic solvents against *RhizoctoniasolaniKühn. Industrial Crops and Products, 32*(3), 324–328.

Chamorro, M., Miranda, L., Domínguez, P., Medina, J. J., Soria, C., Romero, F., & De los Santos, B. (2015). Evaluation of biosolarization for the control of charcoal rot disease (*Macrophomina phaseolina*) in strawberry. *Crop Protection, 67*, 279–286.

Chowdhury, S., Basu, A., Ray Chaudhuri, T., & Kundu, S. (2014). *In vitro* characterization of the behaviour of *Macrophomina phaseolina* (Tassi) Goid. at the rhizosphere and during early infection of roots of resistant and susceptible varieties of sesame. *European Journal of Plant Pathology, 138*(2), 361–375.

Edgington, L. V., Martin, R. A., Bruin, G. C., & Parsons, I. M. (1980). Systemic fungicides: A perspective after 10 years. *Plant Disease, 64*(1), 19–23.

El-Mohamedy, R. S., & Abdalla, A. M. (2014). Evaluation of antifungal activity of *Moringa oleifera* extracts as natural fungicide against some plant pathogenic fungi in vitro. *Journal of Agricultural Technology, 10*(4), 963–982.

Elliott, M., Shamoun, S. F., & Sumampong, G. (2015). Effects of systemic and contact fungicides on life stages and symptom expression of *Phytophthoraramorum* in vitro and in plants. *Crop Protection, 67*, 136–144.

Elmerich, C., Pandey, A. K., Vemula, A., Rathore, A., & Nair, R. M. (2022). Blackgram-*Macrophomina phaseolina* interactions and identification of novel sources of resistance. *Plant Disease*. https://doi.org/10.1094/PDIS-11-21-2588-RE

Essiedu, J. A., Adepoju, F. O., & Ivantsova, M. N. (2020). Benefits and limitations in using biopesticides: A review. In , *Vol 2313. AIP conference proceedings* (p. 080002). AIP Publishing LLC, 1.

Ezzat, A., Szabó, Z., & Nyéki, J. (2014). Induce the plant resistance to pathogen infection. *International Journal of Horticultural Science, 20*(1−2), 89−93.

Farag, A., Ebrahim, H., ElMazoudy, R., & Kadous, E. (2011). Developmental toxicity of fungicide carbendazim in female mice. *Birth Defects Research Part B: Developmental and Reproductive Toxicology, 92*(2), 122−130.

Fawole, O. B., Aluko, M., & Olowonihi, T. E. (2010). Effects of a Carbendazim-Mancozeb fungicidal mixture on soil microbial populations and some enzyme activities in soil. *Agros, 10*(1−2).

Francl, L. J., Wyllie, T. D., & Rosenbrock, S. M. (1988). Influence of crop rotation on population density of *Macrophomina phaseolina* in soil infested with *Heterodera Glycines*. *Plant Disease, 72*, 760−764.

Gaige, A. R., Ayella, A., & Shuai, B. (2010). Methyl jasmonate and ethylene induce partial resistance in *Medicago truncatula* against the charcoal rot pathogen *Macrophomina phaseolina*. *Physiological and Molecular Plant Pathology, 74*(5−6), 412−418.

Ghosh, T., Maity, T. K., Bose, A., Dash, G. K., & Das, M. (2007). Antimicrobial activity of various fractions of ethanol extract of *Bacopamonnieri* Linn. aerial parts. *Indian Journal of Pharmaceutical Sciences, 69*(2), 312.

Goodson, W. H., III, Lowe, L., Carpenter, D. O., Gilbertson, M., Manaf Ali, A., Lopez de Cerain Salsamendi, A., & Pavanello, S. (2015). Assessing the carcinogenic potential of low-dose exposures to chemical mixtures in the environment: The challenge ahead. *Carcinogenesis, 36*(Suppl. 1_1), S254−S296.

Gordon, E. B. (2010). Captan and folpet. In R. Krieger (Ed.), *Hayes' handbook of pesticide toxicology* (3rd ed., pp. 1915−1949). Academic Press. https://doi.org/10.1016/B978-0-12-374367-1.00090-2

Gupta, G. K., Sharma, S. K., & Ramteke, R. (2012). Biology, epidemiology and management of the pathogenic fungus *Macrophomina phaseolina* (Tassi) Goid with special reference to charcoal rot of soybean (Glycine max (L.) Merrill). *Journal of Phytopathology, 160*(4), 167−180.

Habbal, O. A., Al-Jabri, A. A., El-Hag, A. H., Al-Mahrooqi, Z. H., & Al-Hashmi, N. A. (2005). In-vitro antimicrobial activity of *Lawsoniainermis Linn* (henna). A pilot study on the Omani henna. *Saudi Medical Journal, 26*(1), 69−72.

Harish, S., Saravanakumar, D., Radjacommare, R., Ebenezar, E. G., & Seetharaman, K. (2008). Use of plant extracts and biocontrol agents for the management of brown spot disease in rice. *BioControl, 53*(3), 555−567.

Hossain, M. M., Mazumder, K., Hossen, S. M., Tanmy, T. T., & Rashi, M. J. (2012). In vitro studies on antibacterial and antifungal activities of *Emblicaofficinalis*. *International Journal of Pharmaceutical Sciences and Research, 3*(4), 1124.

Iqbal, U., & Mukhtar, T. (2020a). Inhibitory effects of some fungicides against *Macrophomina phaseolina* causing charcoal rot. *Pakistan Journal of Zoology, 52*(2), 709.

Iqbal, U., & Mukhtar, T. (2020b). Evaluation of biocontrol potential of seven indigenous *Trichoderma* species against charcoal rot causing fungus, *Macrophomina phaseolina*. *Gesunde Pflanzen, 72*(2), 195−202.

Iqbal, U., Mukhtar, T., & Iqbal, S. M. (2014). In vitro and in vivo evaluation of antifungal activities of some antagonistic plants against charcoal rot causing fungus, *Macrophomina phaseolina*. *Pakistan Journal of Agricultural Sciences, 51*(3), 691−696.

Iwuala, E., Unung, O., Olajide, O., Odjegba, V., & Alam, A. (2022). Ameliorating effect of zinc and manure application on the growth of *Macrophomina phaseolina* (Tassi) Goid.infected genotypes of *Arachis hypogea* L. *Archives of Phytopathology and Plant Protection, 55*(1), 28−43.

Jabeen, K., Hanif, S., Naz, S., & Iqbal, S. (2013). Antifungal activity of *Azadirachta indica* against *Alternaria solani*. *Journal of Lipid Science and Technology, 1*(1), 89−93.

Jamal, A., Farhat, H., Urooj, F., Rahman, A., Irfan, M., & Ehteshamul-Haque, S. (2021). Characterization of endophytic yeast and its suppressive effect on root rotting fungi of tomato under neem cake soil amendment. *Egyptian Journal of Biological Pest Control, 31*(1), 1−12.

Karibasappa, C. S., Bhat, B. N., & Rao, S. C. (2020). Evaluation of fungicides against *Macrophomina phaseolina* inciting root rot of sesame. *International Journal of Current Microbiology and Applied Sciences, 9*(4), 1293−1299.

Khan, A., Ahmad, A., Manzoor, N., & Khan, L. A. (2010). Antifungal activities of *Ocimum sanctum* essential oil and its lead molecules. *Natural Product Communications, 5*(2), 345−349.

Khan, K. A., Shoaib, A., Arshad Awan, Z., Basit, A., & Hussain, M. (2018). *Macrophomina phaseolina* alters the biochemical pathway in *Vigna radiata* chastened by Zn^{2+} and FYM to improve plant growth. *Journal of Plant Interactions, 13*(1), 131−140.

Klittich, C. J., & Ray, S. L. (2013). Effects of physical properties on the translaminar activity of fungicides. *Pesticide Biochemistry and Physiology, 107*(3), 351−359.

Kumar, K. S., Balabaskar, P., Sivakumar, T., Kannan, R., & Saravanan, K. R. (2020). Bio-efficacy of culture filtrate of *Bacillus cereus* against o the growth of *Macrophomina phaseolina* causing root rot of groundnut and different organic amendments on the survivability of *Bacillus cereus*. *Plant Archives, 20*(1), 1547−1550.

Kumar, P., & Gaur, V. K. (2020). Effect of *Macrophomina phaseolina* on rhizospheric soil factors of resistant and susceptible varieties of Mungbean. *Journal of Pharmacognosy and Phytochemistry, 9*(1), 352−354.

Kumari, N., & Sharma, V. (2014). Reaction of chickpea varieties to Macrophomina phaseolina and their effect on peroxidase activity. In R. Kharwar, R. Upadhyay, N. Dubey, & R. Raghuwanshi (Eds.), *Microbial diversity and biotechnology in food security* (pp. 471−477). New Delhi: Springer. https://doi.org/10.1007/978-81-322-1801-2_42

Lacey, E., & Watson, T. R. (1985). Structure-activity relationships of benzimidazole carbamates as inhibitors of mammalian tubulin, in vitro. *Biochemical Pharmacology, 34*(7), 1073−1077.

Lahlali, R., Ezrari, S., Radouane, N., Kenfaoui, J., Esmaeel, Q., El Hamss, H., & Barka, E. A. (2022). Biological control of plant pathogens: A global perspective. *Microorganisms, 10*(3), 596.

Li, D., Luong, T. T. M., Dan, W. J., Ren, Y., Nien, H. X., Zhang, A. L., & Gao, J. M. (2018). Natural products as sources of new fungicides (IV): Synthesis and biological evaluation of isobutyrophenone analogs as potential inhibitors of class-II fructose-1, 6-bisphosphate aldolase. *Bioorganic and Medicinal Chemistry, 26*(2), 386–393.

Lokesh, R., Rakholiya, K. B., & Thesiya, M. R. (2020). Evaluation of different fungicides against *Macrophomina phaseolina* (Tassi) Goid. causing dry root rot of chickpea (*Cicer arietinum L.*) *in vitro*. *International Journal of Current Microbiology and Applied, 9*, 901–911.

Marquez, N., Giachero, M. L., Declerck, S., & Ducasse, D. A. (2021). *Macrophomina phaseolina*: General characteristics of pathogenicity and methods of control. *Frontiers of Plant Science, 12*, 634397.

Martõnez-Toledo, M. V., Salmeron, V., Rodelas, B., Pozo, C., & Gonzalez-Lopez, J. (1998). Effects of the fungicide Captan on some functional groups of soil microflora. *Applied Soil Ecology, 7*(3), 245–255.

McConnell, L. L., LeNoir, J. S., Datta, S., & Seiber, J. N. (1998). Wet deposition of current-use pesticides in the Sierra Nevada mountain range, California, USA. *Environmental Toxicology and Chemistry: International Journal, 17*(10), 1908–1916.

Mohamed, H. I., El-Beltagi, H. S., Aly, A. A., & Latif, H. H. (2018). The role of systemic and non-systemic fungicides on the physiological and biochemical parameters in plant: Implications for defense responses. *Fresenius Environmental Bulletin, 27*(12), 8585–8593.

Morton, V., & Staub, T. (2008). A short history of fungicides. *APSnet Features, 308*.

Mughogho, L. K. (1984). Sorghum root and stalk rots a critical review. In *Proceedings of the consultative group discussion on research needs and strategies for control of sorghum root and stalk rot diseases, in 'Sorghum Root and Stalk Rots'* (p. 272). India: ICRISAT.

Musyimi, D. M., & Ogur, J. A. (2008). Comparative assessment of antifungal activity of extracts from *Eucalyptus globulus* and *Eucalyptus citriodora*. *Research Journal of Phytochemistry, 2*(1), 35–43.

Muthukumar, A., & Raj, T. S. (2020). Organic amendments as a disease control strategy for soil-borne and foliar diseases. *Organic Farming for Sustainable Agriculture, 2020*(1), 1–10.

Pacheco, S. E., Anderson, L. M., Sandrof, M. A., Vantangoli, M. M., Hall, S. J., & Boekelheide, K. (2012). Sperm mRNA transcripts are indicators of sub-chronic low dose testicular injury in the Fischer 344 rat. *PLoSE ONE, 7*(8), 44280.

Pandey, A. K., Burlakoti, R. R., Rathore, A., & Nair, R. M. (2020). Morphological and molecular characterization of *Macrophomina phaseolina* isolated from three legume crops and evaluation of mungbean genotypes for resistance to dry root rot. *Crop Protection, 127*, 104962.

Parmar, H. V., Kapadiya, H. J., & Bhaliya, C. M. (2017). Efficacy of different fungicides against *Macrophomina phaseolina* (Tassi) Goid causing castor root rot. *International Journal of Communication Systems, 5*, 1807–1809.

Pickel, B., Dai, N., Maymon, M., Elazar, M., Tanami, Z., Frenkel, O., & Freeman, S. (2020). Development of a reliable screening technique for determining tolerance to *Macrophomina phaseolina* in strawberry. *European Journal of Plant Pathology, 157*(4), 707–718.

Podile, A. R., Kumar, B. S. D., & Dube, R. C. (1988). Antibiosis of *Rhizoctonia* against some plant pathogens. *Indian Journal of Microbiololgy, 28*, 108–111.

Rahman, M. T., Rubayet, M., Khan, A. A., & Bhuiyan, M. (2021). Integrated management of charcoal rot disease of soybean caused by Macrophomina phaseolina. *Egyptian Journal of Agricultural Research, 99*(1), 10–19.

Rajpurohit, T. S. (2002). Influence of intercropping and mixed cropping with pearl millet, green gram and moth bean on the incidence of stem and root rot (*Macrophomina phaseolina*) of sesame. *Sesame and Safflower Newsletter, 17*, 40–41.

Rastija, V., Vrandečić, K., Ćosić, J., Majić, I., Šarić, G. K., Agić, D., & Molnar, M. (2021). Biological activities related to plant protection and environmental effects of coumarin derivatives: QSAR and molecular docking studies. *International Journal of Molecular Sciences, 22*(14), 7283.

Rayanoothala, P. S., Mahapatra, S., & Das, S. (2022). Alterations in seedling vigour in mungbean selected genotypes under pre sowing yeast and chitosan as an elicitor seed treatment against *Macrophomina phaseolina* in India. *The Pharma Innovation Journal, SP-11*(1), 464–467.

Reznikov, S., Chiesa, M. A., Pardo, E. M., De Lisi, V., Bogado, N., González, V., & Castagnaro, A. P. (2019). Soybean-*Macrophomina phaseolina*-Specific interactions and identification of a novel source of resistance. *Phytopathology, 109*(1), 63–73.

Reznikov, S., Vellicce, G. R., Mengistu, A., Arias, R. S., Gonzalez, V., Lisi, V. D., & Ploper, L. D. (2018). Disease incidence of charcoal rot (*Macrophomina phaseolina*) on soybean in north-western Argentina and genetic characteristics of the pathogen. *Canadian Journal of Plant Pathology, 40*(3), 423–433.

Rosca-Casian, O., Parvu, M., Vlase, L., & Tamas, M. (2007). Antifungal activity of *Aloe vera* leaves. *Fitoterapia, 78*(3), 219–222.

Rouabhi, R. (2010). *Introduction and toxicology of fungicides*. INTECH Open Access Publisher.

Sahni, T., Sharma, S., Arora, G., & Verma, D. (2020). Synthesis, characterization and antifungal activity of a substituted coumarin and its derivatives. *Pesticide Research Journal, 32*, 39–48.

Satish, S., Mohana, D. C., Ranhavendra, M. P., & Raveesha, K. A. (2007). Antifungal activity of some plant extracts against important seed borne pathogens of *Aspergillus* sp. *Journal of Agricultural Technology, 3*(1), 109–119.

Satya, V., Vijayasamundeeswari, A., Paranidharan, V., & Velazhahan, R. (2011). *Burkholderia sp.* strain TNAU-1 for biological control of root rot in mungbean (*Vigna radiata L.*) caused by Macrophomina phaseolina. *Journal of Plant Protection Research, 51*(3), 273–278.

Sharma, R. A., Sharma, P., & Yadav, A. (2013). Antimicrobial screening of sequential extracts of *Daturastramonium L. International Journal of Pharmacy and Pharmaceutical Sciences, 5*(2), 401–404.

Siddiqui, I. A., Ehetshamul-Haque, S., & Shaukat, S. S. (2001). Use of rhizobacteria in the control of root rot–root knot disease complex of Mungbean. *Journal of Phytopathology, 149*(6), 337–346.

Sitara, U., Hassan, N., & Naseem, J. (2011). Antifungal activity of *Aloe vera* gel against plant pathogenic fungi. *Pakistan Journal of Botany, 43*(4), 2231–2233.

Spagnoletti, F. N., Leiva, M., Chiocchio, V., & Lavado, R. S. (2018). Phosphorus fertilization reduces the severity of charcoal rot (*Macrophomina phaseolina*) and the arbuscular mycorrhizal protection in soybean. *Journal of Plant Nutrition and Soil Science, 181*(6), 855–860.

Stamenković-Stojanović, S., Karabegović, I., Beškoski, V., Nikolić, N., & Lazić, M. (2019). *Bacillus* based microbial formulations: Optimization of the production process. *Hemijska Industrija, 73*(3), 169–182.

Swamy, C., Naik, M. K., Amaresh, Y. S., & Jayalakshmi, S. K. (2018). Evaluation of fungicides and bio-agents under in vitro condition against Macrophomina phaseolina causing stem canker of pigeonpea. *International Journal of Current Microbiology and Applied Sciences, 7*(1), 811–819.

Thakur, M., & Sohal, B. S. (2013). Role of elicitors in inducing resistance in plants against pathogen infection: A review. *International Scholarly Research Notices, 2013*, 1–10.

Thombre, B. B., & Kohire, O. D. (2018). Integrated management of *Macrophomina* blight of mungbean (v*ignaradiata L.*) caused by *Macrophomina phaseolina* (tassi) goid. *Indian Phytopathology, 71*(3), 423–429.

Tomlin, C. D. S. (2000). The pesticide manual: A world compendium. In *12^th^Edition pesticide manual* (pp. 14–502). British Corp Protection.

Tonin, R. F. B., Avozani, A., Danelli, A. L. D., Reis, E. M., Zoldan, S. M., & Garcés-Fiallos, F. R. (2013). In vitro mycelial sensitivity of *Macrophomina phaseolina* to fungicides. *Pesquisa Agropecuária Tropical, 43*, 460–466.

Tort, N., & Turkyilmaz, B. (2003). Physiological effects of captan fungicide on pepper (*Capsicum annuum L.*) plant. *Pakistan Journal of Biological Sciences, 6*(24), 2026–2029.

Upadhyay, S., Singh, A. K., Tripathi, M. K., Tiwari, S., Tripathi, N., & Patel, R. P. (2020). In vitro selection for resistance against charcoal rot disease of soybean [*Glycine max (L.) Merrill*] caused by *Macrophomina phaseolina* (Tassi) Goid. *Legum. Research International Journal*, 1–7. https://doi.org/10.18805/LR-4440

US Department of Health and Human Services. (1993). *Hazardous substances data bank (HSDB, online database)*. Bethesda, MD: National Toxicology Information Program, National Library of Medicine.

Van Loon, L. C., & Van Strien, E. A. (1999). The families of pathogenesis-related proteins, their activities, and comparative analysis of PR-1 type proteins. *Physiological and Molecular Plant Pathology, 55*(2), 85–97.

Vida, C., de Vicente, A., & Cazorla, F. M. (2020). The role of organic amendments to soil for crop protection: Induction of suppression of soilborne pathogens. *Annals of Applied Biology, 176*(1), 1–15.

Walia, A., Mehta, P., Guleria, S., Chauhan, A., & Shirkot, C. K. (2014). Impact of fungicide mancozeb at different application rates on soil microbial populations, soil biological processes, and enzyme activities in soil. *The Scientific World Journal, 2014*, 1–9.

Wang, Y. S., Huang, Y. J., Chen, W. C., & Yen, J. H. (2009). Effect of carbendazim and pencycuron on soil bacterial community. *Journal of Hazardous Materials, 172*(1), 84–91.

Wang, C., Wang, F., Zhang, Q., & Liang, W. (2016). Individual and combined effects of tebuconazole and carbendazim on soil microbial activity. *European Journal of Soil Biology, 72*, 6–13.

Yoon, M. Y., Cha, B., & Kim, J. C. (2013). Recent trends in studies on botanical fungicides in agriculture. *Plant Pathology Journal, 29*(1), 1.

Zakir, M. (2018). Systemic acquired resistance (SAR) and its application in crop plants improvement to biotic stresses. *International Journal of Research Studies in Science, Engineering and Technology, 5*(1), 17–24.

Index

'*Note:* Page numbers followed by "f" indicate figures and "t" indicate tables.'

C

D

N

Q

R

S

T

9780443154430